43rd Riso International Symposium on Materials Science: Composite for Wind Energy: Manufacturing, Operation and End-of-life 2023

IOP Conference Series: Materials Science and Engineering
Volume 1293

Roskilde, Denmark
4 - 7 September 2023

ISBN: 978-1-7138-8504-7
ISSN: 1757-8981

Printed from e-media with permission by:

Curran Associates, Inc.
57 Morehouse Lane
Red Hook, NY 12571

Some format issues inherent in the e-media version may also appear in this print version.

Institute of Physics
Dirac House, Temple Back
Bristol BS1 6BE UK

Phone: 44 1 17 929 7481
Fax: 44 1 17 920 0979

techtracking@iop.org

Additional copies of this publication are available from:

Curran Associates, Inc.
57 Morehouse Lane
Red Hook, NY 12571 USA
Phone: 845-758-0400
Fax: 845-758-2633
Email: curran@proceedings.com
Web: www.proceedings.com

TABLE OF CONTENTS

IOP Conf. Series: Materials Science and Engineering 1293 (2023) 011001 doi:10.1088/1757-899X/1293/1/011001

PREFACE

Proceeding of the 43rd Risø International Symposium on Materials Science: *Composites for wind energy: Manufacturing, Operation and End-of-life*

L.P. Mikkelsen, J. Beauson

The focus of the 43rd Risø Symposium is on composites for wind energy. This symposium takes a life cycle perspective and addresses manufacturing, performance during operation , and end-of-life of composites for wind energy. Over the last 40 years, wind turbine blades have grown an order of magnitude. Today, the longest blades exceed 100 meters and weigh 50 tons. Because of this growth in blade size, the cost of wind energy can now compete with fossil-based energy sources in market terms.

As society is thriving towards a zero-emission future, the wind energy sector is foreseen to expand further. Longer wind turbine blades with improved overall lifetime, reliability, recyclability, sustainability, operability, and maintainability are some of the objectives set for this component. To address these upcoming and ambitious requirements, the contributions are dedicated to the manufacturing, operation, and performance, as well as end-of-life strategies for wind turbine blades.

The Proceedings contain 6 invited papers and 36 contributed papers.

The 43rd International Risø Symposium on Materials Science is organized by the Division of Wind Energy Materials and Components at the Department of Wind and Energy Systems, Technical University of Denmark (DTU). We want to thank all those at DTU who have assisted in the preparations for the symposium. We appreciate additionally the support from the international advisory committee consisting of Douglas Cairns, US; Kristofer Gamstedt, SE; John W. Hutchinson, US; Yasuo Kitane, JP; Kristiina Oksman, SE; Kim Pickering, NZ; Marino Quaresimin, IT; Jon Spangenberg, DK; Michael Sutcliffe, UK; Yentl Swolfs, BE; Ramesh Talreja, US; Michael Wisnom, UK; Philip Withers, UK.

We gratefully acknowledge financial support from the following sources: Danish Center for Applied Mathematics and Mechanics, Otto Mønsteds Fond and Knud Højgaards Fond.

Peer Review Statement

All papers published in this volume have been reviewed through processes administered by the Editors. Reviews were conducted by expert referees to the professional and scientific standards expected of a proceedings journal published by IOP Publishing.

- **Type of peer review:** Double Anonymous
- **Conference submission management system:** Morressier
- **Number of submissions received:** 44
- **Number of submissions sent for review:** 43
- **Number of submissions accepted:** 43
- **Acceptance Rate (Submissions Accepted / Submissions Received × 100):** 97.7
- **Average number of reviews per paper:** 2.1
- **Total number of reviewers involved:** 24
- **Contact person for queries:**
  Name: Lars P. Mikkelsen
  Email: lapm@dtu.dk
  Affiliation: Associate Professor

IOP Conf. Series: Materials Science and Engineering 1293 (2023) 012001 doi:10.1088/1757-899X/1293/1/012001

Building blocks towards progressive fatigue damage modeling of a whole wind turbine blade

W Van Paepegem and M Hajikazemi

Department of Materials, Textiles and Chemical Engineering, Faculty of Engineering and Architecture, Ghent University, Technologiepark Zwijnaarde 46, Ghent, Belgium

Wim.VanPaepegem@Ugent.be , Mohammad.Hajikazemi@Ugent.be

Abstract. This paper presents a concise summary of the recent fatigue damage modeling research in the *Mechanics of Materials and Structures* (MMS) research group at Ghent University, which utilizes physics-based models and microscopic experiments to develop components for a progressive fatigue damage model of laminated composite structures. The paper discusses the evolution of ply cracking and delamination under cyclic loading and highlights distinct modeling elements for stress analysis and fatigue damage evolution characterization. It also explores the experiments required for model calibration and validation, including the use of digital image correlation to quantify damage mechanisms. Additionally, the paper discusses the use of neural networks for efficient and robust modeling of damage effects and outlines the remaining steps required to develop a comprehensive fatigue design tool. Finally, the recent developments with the BladeMesher software of the research group provide a platform to integrate this fatigue design tool for complete wind turbine blades.

1. Introduction

The use of continuous fiber composite laminates in wind turbine design has grown in popularity due to their superior mechanical properties. These laminates, made of high-strength fibers embedded and oriented in multiple directions in a polymer matrix, offer a unique combination of strength, stiffness, and lightweight characteristics, ideal for wind turbine blades. Continuous fiber laminates are also highly resistant to fatigue and environmental degradation[1], allowing them to withstand the extreme loads and harsh weather conditions that wind turbines face. However, their use presents challenges such as the expensive design iterations, anisotropic material properties, and extensive experimental testing.
While modeling can provide insights into the material's behavior and reduce the cost and time associated with testing, it should be validated with experimental data. Computer modeling can simulate the behavior of composite materials under different loading conditions and provide insight into their mechanical properties without the need for extensive experimental testing[2]. Additionally, modeling can provide a better understanding of the anisotropic behavior of composite materials[3], allowing designers to optimize the layup to achieve the desired mechanical properties. However, while modeling has several advantages, it should be noted that it is only as accurate as the underlying material models and input parameters used. Therefore, it is crucial to ensure that the modeling results are validated with experimental data to ensure their reliability and accuracy.

Estimating the life time of a wind turbine blade or any composite structural element when subjected to fatigue or creep deformations is a difficult undertaking. While metal failure is defined as the separation of the metal into multiple pieces, composite failure follows a series of localized damage effects such as ply cracking, delamination, fiber/matrix debonding, and fiber fracture. These damage mechanisms are interactive and involve statistical uncertainty, leading to a slow deterioration of composite performance that ultimately results in catastrophic failure. To create a useful design methodology for predicting the lifespan of composite structures, it is essential to be able to predict the initiation and progression of damage leading to failure with confidence. Over the years, the Mechanics of Materials and Structures (MMS) research group at Ghent University has undertaken both theoretical and experimental activities to develop the necessary components for a progressive fatigue damage model of laminated composite structures. In parallel, the group has developed a BladeMesher software, that can generate shell, solid and hybrid shell/solid meshes of a full wind turbine blade in a few minutes, without any manual operations. Such high-quality meshes, providing converged local stress fields, are essential to be combined with fatigue damage models.

This paper primarily focuses on some of the physics-based developed models, along with the experiments performed that were supported by microscopic observation of damage mechanisms. Additionally, the paper outlines the remaining steps required to increase confidence in implementing the models for larger structural components while maintaining computational efficiency.

2. Modeling of ply cracking under multiaxial fatigue loading

The occurrence of stress concentration caused by geometrical discontinuities (such as holes or free edges) or manufacturing defects can initiate and interact with various damage mechanisms. However, in areas where these discontinuities are absent, ply level damage progression in laminated composites typically begins with the formation of ply cracks. Ply cracking degrades all the in-plane, out-of-plane and bending laminate stiffnesses and changes its coefficients of thermal expansion, moisture absorption, and the natural vibration frequency. Although ply cracking may not be a critical damage mechanism in most applications, it can significantly degrade the stiffness of laminates, such as Glass/Epoxy laminates used in wind turbine blades, and increase deformation. In extreme cases, this can lead to blade failure and impact with the tower. Ply cracking can also trigger the development of other damage mechanisms that are critical from the perspective of final fracture.

Three distinct modeling ingredients are typically required for prediction of ply cracking initiation, propagation and its effect on laminate properties under multiaxial fatigue loading conditions. It is highly challenging to predict the number of cycles required for the initiation of a measurable-sized ply crack, due to the involvement of intricate microscopic mechanisms that contribute to the formation of small ply cracks. These mechanisms operate at the interfaces of fiber and matrix, causing debonded areas that can grow in various directions, including through-thickness, width, and between plies, leading to ply cracks and delaminations. Additionally, the variable quality of the fiber/matrix interface, uneven spacing of fibers together with non-uniform distribution of residual stresses contribute to significant scatter in ply crack initiation times, requiring a statistical approach. To address this problem, a practical solution is to consider the ply crack initiation phase of fatigue damage as a known parameter (e.g. a negligibly small crack density if the laminate is not damaged), allowing predictive modeling to focus on ply crack growth and catastrophic failure onset.

2.1. Ply cracking fatigue evolution law

Expanding on the idea proposed in reference [4,5], an energy balance approach can be applied to predict unstable fatigue crack growth [6]. To begin with, an energy balance formulated based on effective laminate properties needs to be considered for the region around the crack tip during cyclic deformation. Although most of the cyclic deformation generates localized heat, some of it also contributes to the degradation of the material in the fracture process zone. As fatigue crack growth can be attributed to the dissipation of energy in the failure process zone at the crack tip, energy concepts can be used to define the range of effective (laminate level) energy release rate, ΔG, as follows:

$$\begin{aligned}
&\Delta G = \Delta G_{\mathrm{I}} + \Delta G_{\mathrm{II}} + \Delta G_{\mathrm{III}}, \text{ where} \\
&\Delta G_{\mathrm{I}} = \left(\sqrt{G_{\mathrm{I}}(\rho)} - \sqrt{G_{\mathrm{min,I}}(\rho)}\right)^2, \\
&\Delta G_{\mathrm{II}} = \left(\sqrt{G_{\mathrm{II}}(\rho)} - \sqrt{G_{\mathrm{min,II}}(\rho)}\right)^2, \\
&\Delta G_{\mathrm{III}} = \left(\sqrt{G_{\mathrm{III}}(\rho)} - \sqrt{G_{\mathrm{min,III}}(\rho)}\right)^2,
\end{aligned} \tag{1}$$

where G_{I}, G_{II} and G_{III} are the effective strain energy release rates at the maximum load of the cycle for different deformation modes and G_{min} is the effective energy release rate at the minimum load of the stress cycle. The parameter ρ is the effective ply crack density. It is essential to define an effective ply crack density for fatigue loading because, unlike quasi-static loading, ply cracks may not propagate unstably through the width or ply thickness. This definition suggests that the gradual growth of numerous small ply defects is equal to a consistent density ρ of fully developed ply cracks through the thickness and width, from a macroscopic perspective. The definition provided above for effective energy release rate range has two valuable attributes. Firstly, it considers the impact of R-ratio or the maximum and minimum applied stresses. Another desirable feature of this definition is that $\Delta G \rightarrow 0$ as $\sigma \rightarrow \sigma_{\mathrm{min}}$. Finally, a fatigue crack growth law can be assumed of the form

$$\frac{d\rho}{dN} = C_{\mathrm{I}}\left(\Delta G_{\mathrm{I}} - \Delta G_{\mathrm{th}}\right)^{n_{\mathrm{I}}} + C_{\mathrm{II}}\left(\Delta G_{\mathrm{II}} - \Delta G_{\mathrm{th}}\right)^{n_{\mathrm{II}}} + C_{\mathrm{III}}\left(\Delta G_{\mathrm{III}} - \Delta G_{\mathrm{th}}\right)^{n_{\mathrm{III}}}, \tag{2}$$

where C_{I}, C_{II}, C_{III}, n_{I}, n_{II} and n_{III} are material constants that should be calibrated using fatigue experiments on laminates with microscopic ply crack quantifications. Furthermore, the fatigue energy release rate threshold, denoted by ΔG_{th}, represents the point at which fatigue crack growth ceases. This value may also vary depending on the deformation mode (shown as a single value for simplicity) and is reached when the effective energy release rate falls below it, due to laminate stiffness reduction. Experimental evidence, such as [7], suggests that cracks in off-axis plies do not grow continuously and reach a limit in number at a certain cyclic load level. This number of cracks remains relatively constant for the rest of the specimen's lifespan. The equation for fatigue ply crack growth in Eq. (2) bears resemblance to the widely-used empirical Paris Law for fatigue crack growth in metals. However, it is important to note that there is no clear physical justification for choosing this particular form of the fatigue ply crack growth law in Eq. (2), other than the fact that it is expressed in terms of energy parameters that take into account the complex triaxial stress fields present at the tips of ply cracks and laminate anisotropy.

2.2. Stress analysis in the presence of ply cracking

A crucial modeling tool is yet required to predict the stress distribution in the existence of ply cracks, which aids in computing the effective energy release rate range in various deformation modes and estimating the influence of fatigue-induced ply cracks on the properties of the laminate. Ply cracks tend to manifest in large numbers, often interacting with other cracks in the same or different plies (Figure 1).

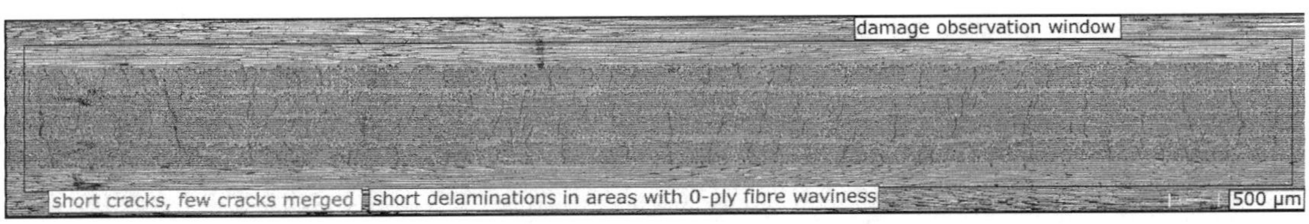

Figure 1. Post mortem microscopic edge analysis [8] of a fatigued Glass/Polypropylene $[0_2/90_2]_s$ laminate under uniaxial fatigue loading with 115 MPa amplitude after 500,000 cycles, R = 0.1, f = 2 Hz. Cracks are highlighted in red, delamination in blue.

Considering the complex patterns of ply cracks that appear during the fatigue loading, the use of Finite Element Method (FEM) is impractical. Additionally, the stress distribution must be continuously updated throughout the fatigue loading process as a result of crack propagation. Furthermore, numerous calculations are required to calibrate the parameters in Eq. (2) and during the design process, different laminate lay-ups and ply thicknesses must be assessed. As a result, a robust stress analysis modeling tool is required that not only matches the accuracy of FEM but also provides a feasible run-time for the stress field, contingent upon the crack patterns and laminate specifications. To meet this demand, the variational approach [5] has been extensively developed over the years to handle intricate crack patterns in both symmetric and non-symmetric laminates under multiaxial in-plane[6], out-of-plane[7], and bending [8] loading conditions. It considers the impact of residual stresses and non-uniformity in crack distributions, which could be caused by the presence of manufacturing defects. The variational model involves (Figure 2) incorporating the geometry of the damaged state as a perturbation to the undamaged geometry.

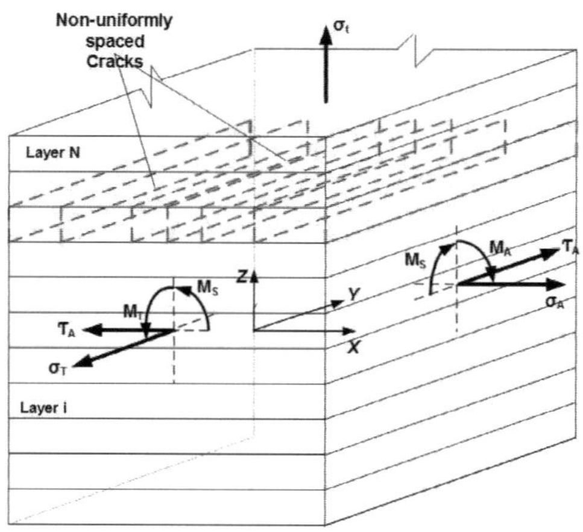

Figure 2. Variational models for laminates with arbitrary lay-ups under multiaxial in-plane (σ_A, σ_T, τ_A), out-of-plane (σ_t) and bending (M_A, M_T, M_s) loads with non-uniform ply cracking in a single orientation.

A limited set of assumptions are then made about the 3D stress field, and equilibrium equations are solved under periodic boundary conditions that correctly represent the stress states. The reader should refer to Ref. [4,5] to find detailed information about the calculation of ply cracking energy release rate in terms of laminate properties and the assumptions made. The run time of these models is a split second and extensive validation with highly refined FEM models has been done by UGent-MMS.

3. Modeling of delaminations at the tip of ply cracks under multiaxial fatigue loading

Ply cracks typically trigger delaminations due to stress concentrations at their tips. Delamination is a significant form of damage that worsens the effects of ply cracks, leading to largely reduced stiffness and stress redistribution. This damage is particularly problematic in multiaxial loading conditions where out-of-plane deformations occur, as it significantly impacts out-of-plane laminate properties. Delamination can initiate ply cracks in adjacent plies and cause fiber breakages, leading to catastrophic failure. Its detection during inspection is difficult, making it a challenging damage mechanism.

As previously mentioned, cracks in a ply reach saturation at a certain cyclic load level, but this does not imply that delamination only occurs after ply crack saturation. Recent experiments on glass/epoxy cross-ply laminates have revealed that micro-scale delamination (100 μm) occurs immediately after an off-axis crack fully penetrates the ply thickness [9]. This is attributed to high stress fields caused by the presence of ply cracks at the interface, which include a significant peeling component, near the crack tip. Therefore, considering a very small preexisting delamination length at the tips of ply cracks can be helpful to focus on fatigue delamination growth.

3.1. Delamination fatigue evolution law

After initiating at the tip of a ply crack, a delamination propagates during cyclic loading. It is reasonable to assume that the delamination spreads perpendicular to the off-axis crack faces in a self-similar manner, at least when far from free edges. In comparison to fatigue ply crack growth, the development of delaminations at the ply crack tips is more akin to the growth of a fracture mechanics crack, where the delamination increment during each loading cycle is infinitesimal. Thus, it is reasonable to consider a new delamination growth law that is analogous to the empirical Paris Law. This would involve characterizing the rate of delamination growth in terms of delamination energy release rate range ΔG_{eff} as follows

$$\frac{dD}{dN} = C_{\mathrm{D}}\left(\Delta G_{eff}\right)^{\mathrm{n}_D} \quad where \qquad (3)$$

$$\Delta G_{eff} = \left(\sqrt{G(D)} - \sqrt{G_{\min}(D)}\right)^2, \quad G = G_{\mathrm{I}} + G_{\mathrm{II}} + G_{\mathrm{III}}.$$

where C_D, and n_D are material constants that should be calibrated using fatigue experiments on laminates with microscopic quantification of delamination and where G_I, G_{II}, G_{III} concern different deformation modes. Equation (3) suggests that the total energy release rate is utilized to characterize delamination growth primarily to minimize the number of calibration parameters. However, it is important to consider a similar discussion as that for the case of ply cracking fatigue growth. As long as it is feasible to calculate the energy release rate for delamination under various deformation modes, it is possible to adopt any form of delamination fatigue evolution law, provided there are sufficient well-monitored experiments to calibrate the necessary inputs.

3.2. Stress analysis in the presence of delamination at the tips of ply cracks

It is exceedingly challenging to calculate the stress field, energy release rate range, and subsequent degradation of laminate properties in the interactive presence of both ply cracks and delaminations. The use of finite element method (FEM) for this purpose would be computationally inefficient due to the necessity of mesh refinements at the tips of both ply cracking and delamination. This problem would have been less severe if only a few ply cracks and delaminations existed in a single ply and/or interface. However, this complex damage pattern becomes increasingly intricate during fatigue loading as more cracks and delaminations appear even in adjacent plies. The variational models previously introduced for predicting the stress field in the presence of ply cracking only have recently been significantly expanded to accurately account for the impact of delaminations. This model [10] is applicable to general symmetric laminates subjected to general triaxial loading (see Figure 3).

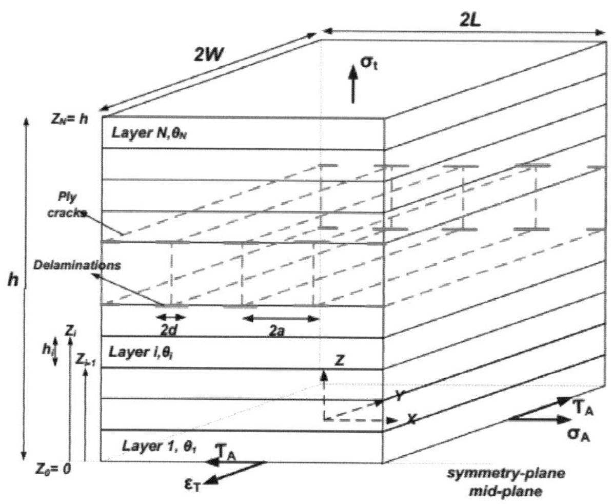

Figure 3. Developed variational models for laminates with symmetric lay-ups under multiaxial in-plane (σ_A, σ_T, τ_A), out-of-plane (σ_t) containing uniform ply cracking (ρ=1/2a) and delamination (D=2d) in a single orientation.

It is important to recognize that while Eqs. (2) and (3) display the energy release rates as a function of effective crack density and delamination length, they are interdependent. In high cycles, both damage mechanisms require stress transfer to occur simultaneously. The reader should refer to Ref. [10] for detailed information about the model and its assumptions.

4. Modeling of ply cracking and delamination in multiple off-axis plies and interfaces

As previously discussed, calculating the energy release rates through stress transfer analysis becomes increasingly challenging when dealing with complex patterns of ply cracking and delamination, particularly when they occur in multiple plies and interfaces. Construction of a unit cell with several cracks in different orientations is almost impossible due to the requirement of applying periodic boundary conditions at the same time in different orientations. However, the evolution laws presented in Eq. (2) and (3) offer a valuable advantage in that they utilize laminate effective properties to accurately capture all the details of these defects without relying on the stress field of each individual crack.

In order to handle complex damage patterns involving ply cracking and delaminations in various orientations, the variational models described earlier are utilized in tandem with a homogenization approach. Through homogenization, a cracked ply with or without its delaminated interfaces can be replaced with an uncracked ply with perfect interfaces possessing adjusted elastic properties. As a result, the actual properties of the damaged laminate and the homogenized laminate remain the same. This technique of modified effective properties of a cracked ply can expand the scope of stiffness reduction models to tackle ply cracks and delaminations occurring in multiple orientations. In reference [4], it is demonstrated that to effectively homogenize ply cracking effects in symmetric laminates, changes in four parameters are necessary: the in-plane ply transverse stiffness E_{22}, in-plane ply shear stiffness G_{12}, out-of-plane transverse shear stiffness G_{23}, and a coupling term φ_{26}, specifically for ply cracking. Analytical solutions [11] can be derived to express these parameters in terms of laminate geometry, average crack spacing considering uniform and non-uniform distributions [12]. When dealing with more intricate damage patterns that involve delamination, and when subjected to complex bending loads, a greater number of parameters are required for homogenization. In such cases, deriving simple analytical solutions is not feasible. To address these scenarios, neural networks are utilized and trained using data generated by running variational models [3,10,13] for multiple cases. The outcome of the trained neural networks is then used to calculate both the laminate properties and energy release rate range as a function of effective crack density and delamination length.

5. Fatigue experiments with in-situ microscopic observation of damage mechanisms

In-situ fatigue damage quantification in composite laminates is cumbersome specially in non-transparent thermoplastic or carbon fiber laminates, however, it provides vital information for developing reliable fatigue crack and delamination growth laws and for their validation. Recently, the advances in optical measurements, namely Digital Image Correlation (DIC), are utilized to detect automatically crack density in glass/epoxy [14] and glass/propylene [8] multidirectional laminates with different lay-ups under quasi-static [15] and fatigue loading conditions.

In order to measure full-field strains and assess the extent of damage mechanisms in specimens, two 3D-DIC (Stereo Digital Image Correlation) systems are utilized, as shown in Figure 4. The experimental setup involves capturing in-situ images synchronized with data acquired from the testing machine during quasi-static and fatigue experiments. DIC analysis using strain and displacement fields is performed to detect cracks and obtain optimal results. Subsequently, an automated program is developed to evaluate crack density and weighted crack density from the edge and top of the specimen, respectively. The details can be found in Refs. [8,14,15].

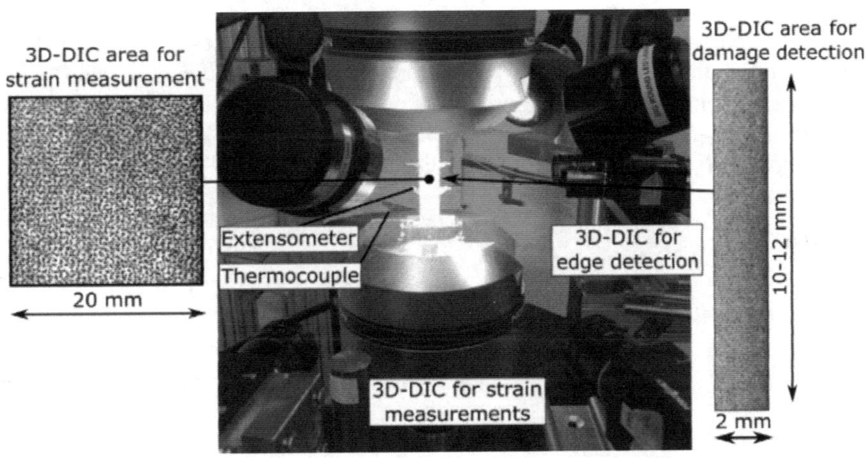

Figure 4. Fatigue setup and images taken by the cameras on the front and edge surfaces [8].

6. Results and discussions

The first step for validation of the approach is to assess its quality in predicting the laminate effective properties as a function of fatigue driven ply cracks. For comparison with experiments, a $[0_2/90_2]_s$ laminate made of Glass/Propylene, which was tested in UGent-MMS [8], is first considered. This laminate was tested under uniaxial fatigue loading with R=0.1, f = 2 Hz, at two different amplitude loads σ_A=115 MPa, 150 MPa. For each case and at specific loading cycles, the in-plane axial stiffness E_A and in-plane axial Poisson's ratio υ_A were measured together with the average value of effective crack density in the off-axis ply while delamination was reported to be negligible. We have used exactly the reported average crack densities [8] together with the provided input material properties [8] to predict elastic constants as a function of loading cycles. Figs. 5a and b, show, respectively, the variation of the in-plane axial stiffness and Poisson's ratio. It is now useful to focus on the prediction of effective ply cracking density evolution as a function of applied cycle using the fatigue crack growth law in Eq. (2).

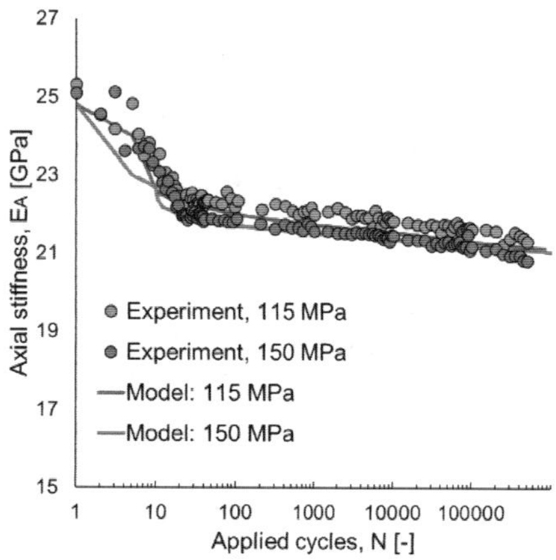

(a)

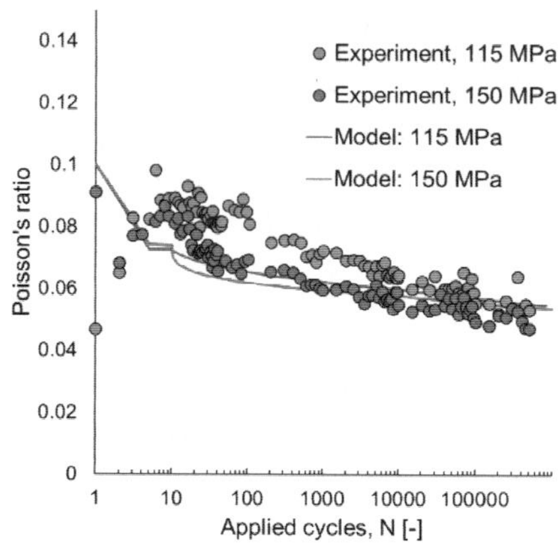

(b)

Figure 5. Fatigue behaviour of $[0_2/90_2]_s$ glass/propylene samples tested at 115 MPa and 150 MPa for 500 000 cycles, a) dynamic stiffness reduction, b) Poisson's ratio evolution.

To accomplish this, experimental data from the $[0_2/90_2]_s$ laminate tested under 115MPa amplitude loading are utilized for model calibration. It is important to note that in a cross-ply laminate under uniaxial loading, ply cracking growth at the homogenized level is exclusively in mode I. Therefore, a single experiment is adequate for determining C_I and n_I in Eq. (2). The calibrated evolution law is subsequently employed to predict the fatigue evolution of ply cracks under higher amplitude loading conditions, as depicted in Figure 6.

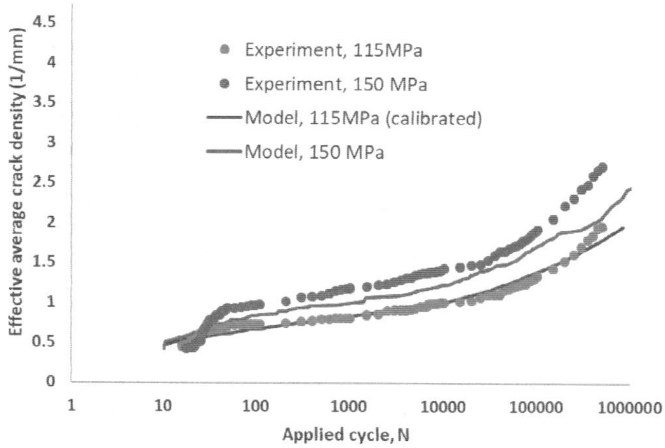

Figure 6. Fatigue ply cracking growth in 90-ply of $[0_2/90_2]_s$ glass/propylene samples tested at 115 MPa and 150 MPa for 500 000 cycles.

To further compare with experimental data, particularly when both ply cracking and delamination occur during fatigue loading, two glass/epoxy laminates - a $[0_2/90_4]_s$ and a $[0_2/45_4]_s$ - previously tested in Ref. [9,16], were examined. These laminates were subjected to uniaxial fatigue loading with various amplitude loads. Normalized in-plane axial stiffness (E_A/E_{A0}) and normalized in-plane axial Poisson's ratio (υ_A/υ_{A0}) were measured for each sample at specific loading cycles, as well as the average crack density in the off-axis ply and the length of delamination at the interface between the off-axis ply and the 0° ply. The reported average crack densities and delamination lengths [9,16], along with the provided input material properties, were used to forecast elastic constants as a function of loading cycles. Figure

7a illustrates the reduction of the normalized in-plane axial stiffness (E_A/E_{A0}) for two $[0_2/90_4]_s$ laminate samples tested under uniaxial cycling load with amplitude $\sigma_{A,\max} = 120MPa$ and load ratio $R = 0.05$. The modeling results are also demonstrated for each sample separately due to slight variations in crack densities and delamination lengths reported for each sample at different cycles. Figure 7b illustrates the variation of the normalized in-plane axial stiffness (E_A/E_{A0}) and normalized in-plane axial Poisson's ratio (υ_A/υ_{A0}) for the $[0_2/45_4]_s$ laminate tested under uniaxial cycling load with amplitude $\sigma_{A,\max} = 120MPa$ and load ratio $R = 0.05$. As seen in Figure 7a, the variational approach predictions for the in-plane axial stiffness of the $[0_2/90_4]_s$ samples are in good agreement with the experimental results, despite some scattering between the experimental results for different samples of the same type. Additionally, the predictions in Figure 7b correspond well with the experimental outcomes for the $[0_2/45_4]_s$ laminate.

In Ref. [17], experimental characterization of ply cracking propagation was conducted in the off-axis plies of different laminates such as $[0/90_2]_s$, $[0/75_2]_s$, $[0/60_2]_s$, and $[0/75_3]_s$, subjected to uniaxial cyclic loading with varying amplitudes. Consequently, the crack density in the off-axis plies was quantified with respect to the applied cycles. For unbalanced laminates, ply cracks grow under mixed mode conditions, while for cross-ply laminates, cracks form under mode I deformation during uniaxial loading. Hence, in order to fully calibrate the model, experiments on both the cross-ply laminate and an arbitrary laminate with an off-axis ply other than 90-degree are required. The experimental crack density measurements on the $[0/90_2]_s$ and $[0/75_2]_s$ samples were utilized to calibrate C_I, n_I and C_{II}, n_{II} in Eq. (2), respectively.

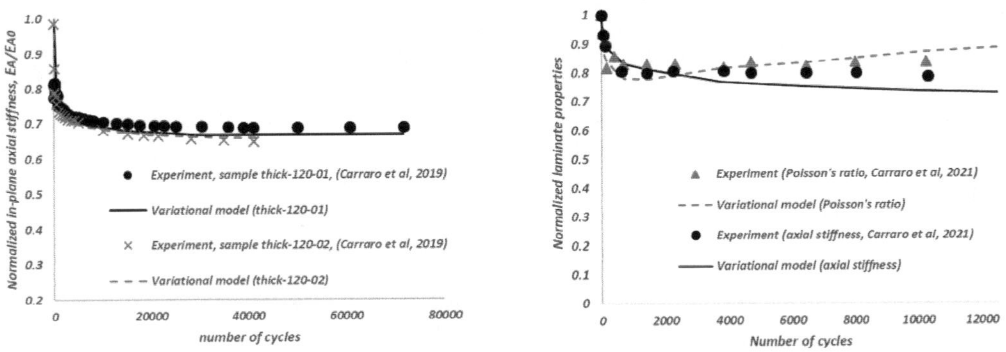

(a) (b)

Figure 7. Variation of normalized (a) in-plane axial stiffness E_A/E_{A0} in two samples with $[0_2/90_4]$s lay-up (b) In-plane axial stiffness E_A/E_{A0} and in-plane Poisson's ratio υ_A/υ_{A0} in one sample with $[0_2/45_4]_s$ lay-up, as a function of cycles which all were tested under a cyclic load $\sigma_{A,\max} = 120MPa$ *and* $R = 0.05$.

Once the model is calibrated, it can predict the crack density evolution in other laminates. Ply crack modeling in these laminates is illustrated in Figure 8, showing that the model can accurately predict the crack density evolution under cyclic loading when cracks propagate under mixed mode loading conditions.

Further validation of the model is required, particularly in predicting delamination fatigue growth and ply cracking fatigue threshold together with capability of neural networks in prediction of complex damage patterns.

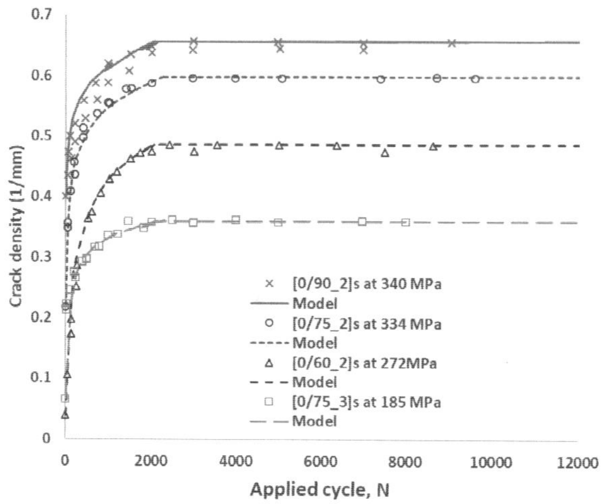

Figure 8. Variation of crack density as a function of cycles in a carbon/epoxy laminates with different lay-ups under a cyclic loading with different amplitude at R=0 .

7. Outlook

The prediction of damage mechanisms in complex geometries presents a significant challenge. One way to address this challenge is by using finite element techniques to analyze complex structures, such as wind turbine blades. This paper's developments can serve as a useful test data at the coupon level, providing inputs for anisotropic continuum damage models that can be integrated into FEM analyses. The benefits of using the methods described in this paper over actual coupon level experiments are not limited to cost and time efficiencies. Semi-analytical approaches can provide information on the damage state of laminates under various loading conditions, including those that are difficult or impossible to simulate experimentally. To develop appropriate constitutive laws that can be used as an anisotropic continuum damage model in FEM analyses at the structural level, it is also important to generate high-quality meshes, that can accurately resolve the local stress geometries.

Over the past years, UGent-MMS has developed a BladeMesher software [18] (Figure 9), that can generate in a fully automated way high fidelity FEM models of full wind turbine blades. The models can be fully meshed with (linear or quadratic) solid elements (Figure 10), shell elements or a hybrid combination. Cohesive elements can be embedded in the adhesive bondlines to simulate delamination/debonding in the bondlines.

The BladeMesher software [18] has been successfully used for (i) simulation of static certification tests, [19], (ii) fluid-structure-interaction (FSI) simulations with a CFD solver [20,21], (iii) accurate torsional stiffness prediction of large blades[22], and (iv) global-local analysis of adhesive and bolted joints. This software platform opens up the opportunity to couple the advancements of physics-based, but extremely computationally efficient fatigue models with the structural analysis of large wind turbine blades.

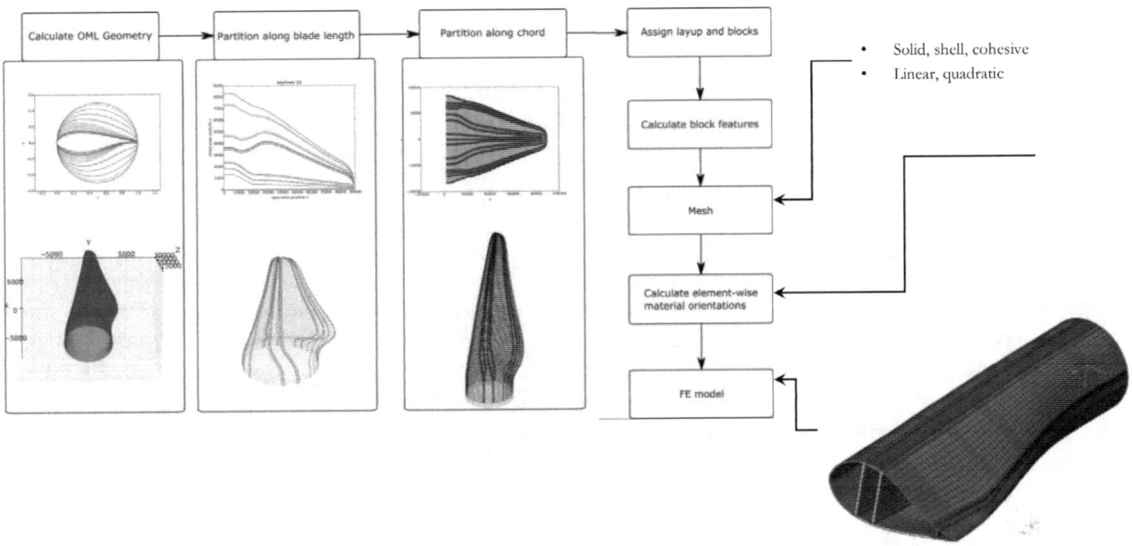

Figure 9. Workflow of the UGent BladeMesher software.

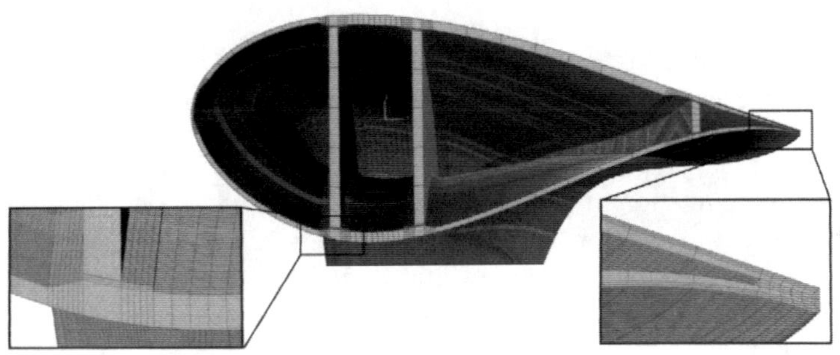

Figure 10. Quality of solid element meshes generated by the UGent BladeMesher software.

References

[1] Talreja R 2014 Assessment of the fundamentals of failure theories for composite materials *Composites Science and Technology* **105** 190–201

[2] McCartney L N 2000 Model to predict effects of triaxial loading on ply cracking in general symmetric laminates *Composites Science and Technology* 25

[3] Hajikazemi M, Garoz D and Van Paepegem W 2019 Model to accurately predict out-of-plane shear stiffness reduction in general cracked laminates *Composites Science and Technology* **179** 88–96

[4] Hajikazemi M, McCartney L N and Van Paepegem W 2020 Matrix cracking initiation, propagation and laminate failure in multiple plies of general symmetric composite laminates *Composites Part A: Applied Science and Manufacturing* **136** 105963

[5] Hajikazemi M 2021 17 - Physics-based methodology for predicting ply cracking and laminate failure in symmetric composite laminates under multiaxial loading condition *Multi-Scale Continuum Mechanics Modelling of Fibre-Reinforced Polymer Composites* ed W Van Paepegem (Woodhead Publishing) pp 509–53

[6] McCartney L N 2008 Energy methods for fatigue damage modelling of laminates *Composites Science and Technology* **68** 2601–15

[7] Tong J 2002 Characteristics of fatigue crack growth in GFRP laminates *International Journal of Fatigue* **24** 291–7
[8] Sommer J, Hajikazemi M, De Baere I and Van Paepegem W 2023 Fatigue behaviour of thermoplastic glass/polypropylene composite cross-ply laminates: An experimental study with in-situ damage observations and numerical validation *Composites Part B: Engineering* **252** 110494
[9] Carraro P A, Maragoni L and Quaresimin M 2019 Characterisation and analysis of transverse crack-induced delamination in cross-ply composite laminates under fatigue loadings *International Journal of Fatigue* **129** 105217
[10] Hajikazemi M, Ahmadi H, McCartney L N and Van Paepegem W 2022 A variational approach for accurate prediction of stress and displacement fields and thermo-elastic constants in general symmetric laminates containing ply cracking and delamination under general triaxial loading *International Journal of Solids and Structures* **254–255** 111917
[11] Ahmadi H, Hajikazemi M and Van Paepegem W 2020 Closed-form formulae for prediction of homogenized ply-properties and laminate thermo-elastic constants in symmetric laminates containing ply cracks in multiple orientations *Composite Structures* **241** 112061
[12] Hajikazemi M, McCartney L N, Van Paepegem W and Sadr M H 2018 Theory of variational stress transfer in general symmetric composite laminates containing non-uniformly spaced ply cracks *Composites Part A: Applied Science and Manufacturing* **107** 374–86
[13] Hajikazemi M, McCartney L N, Ahmadi H and Van Paepegem W 2020 Variational analysis of cracking in general composite laminates subject to triaxial and bending loads *Composite Structures* **239** 111993
[14] Miskdjian I, Hajikazemi M and Van Paepegem W 2020 Automatic edge detection of ply cracks in glass fiber composite laminates under quasi-static and fatigue loading using multi-scale Digital Image Correlation *Composites Science and Technology* **200** 108401
[15] Sommer J, Hajikazemi M, De Baere I and Van Paepegem W 2022 Experimental and numerical damage characterization of glass/polypropylene multidirectional laminates under quasi-static loading condition *Composites Science and Technology* **227** 109569
[16] Carraro P A, Maragoni L and Quaresimin M 2021 Stiffness degradation of symmetric laminates with off-axis cracks and delamination: an analytical model *International Journal of Solids and Structures* **213** 50–62
[17] Mohammadi B and Pakdel H 2018 Fatigue driven matrix crack propagation in laminated composites *Materials & Design* **146** 108–15
[18] Peeters M, Santo G, Degroote J and Van Paepegem W 2018 High-fidelity finite element models of composite wind turbine blades with shell and solid elements *Composite Structures* **200** 521–31
[19] Peeters M, Santo G, Degroote J and Van Paepegem W 2018 Comparison of Shell and Solid Finite Element Models for the Static Certification Tests of a 43 m Wind Turbine Blade *Energies* **11** 1346
[20] Santo G, Peeters M, Van Paepegem W and Degroote J 2019 Dynamic load and stress analysis of a large horizontal axis wind turbine using full scale fluid-structure interaction simulation *Renewable Energy* **140** 212–26
[21] Santo G, Peeters M, Van Paepegem W and Degroote J 2020 Fluid–Structure Interaction Simulations of a Wind Gust Impacting on the Blades of a Large Horizontal Axis Wind Turbine *Energies* **13** 509
[22] Tavares R P, Bouwman V and Van Paepegem W 2022 Finite element analysis of wind turbine blades subjected to torsional loads: Shell vs solid elements *Composite Structures* **280** 114905

Composite materials in wind energy: design, manufacturing, operation, and end-of-life

D. Todd Griffith[1], Dongyang Cao[1], Hongbing Lu[1], Dong Qian[1]

[1]Mechanical Engineering Department, University of Texas at Dallas, Richardson, Texas, USA 75080

tgriffith@utdallas.edu

Abstract. Wind energy has consistently grown over recent decades and has grown in many aspects including in terms of installed capacities, turbine size, blade length, and grid penetration. Along with this, wind energy is one of the largest producers of composite structures, and as a result is one of the largest users of composite materials. For wind energy applications, composite materials require high reliability, low-cost, and near-term and future industry goals are to reuse or recycle composite materials. These requirements are quite challenging as wind energy faces a challenging operating environment, which puts great pressure on wind blade materials over their entire life cycle. This paper aims to examine the challenges of composite materials for wind energy applications, and to highlight a few research studies that offer potential new solutions and new insights across the entire life cycle of composites for wind energy systems ranging from the design phase, to manufacturing, to operation, and finally to the end-of-life phase.

1. Introduction

The growth in both wind turbine size and volume of wind turbine deployments has driven very high usage of composite materials, and as a result the wind energy industry has been the largest industrial consumer of composite materials worldwide [1] for many years. This growth in turbine and blade size over the past decades [2] has put greater pressure on technical requirements for materials and their associated manufacturing processes for wind turbine blades, and this has translated to pressures in all life-cycle phases encountered by blade designers, manufacturers, and wind farm owners and operators.

Wind turbine blades are a true composite of multiple materials including various fiber types, resins, core materials, adhesives, and protective coatings [3]. Fiber reinforced composites in blades are based on glass fibers or carbon fibers, with various fiber orientations (uniaxial, biaxial, triaxial, etc.), and infused with different resins (epoxy, thermoplastics, etc.). Core types typically include balsa, PVC foam core, and PET foam core. All these materials are strategically placed by designers in the blade structure to meet strict and challenging structural requirements, while minimizing weight and cost [4]–[6].

As wind turbines and blade grow larger, a number of trends have emerged. With longer blades, manufacturing and logistical challenges have grown [7]. To restrict blade mass and improve structural performance, greater use of carbon fiber has been seen in blades [8]. Longer and more slender, flexible blades are also more susceptible to aero-elastic instability [9], [10]. To explore new markets and to address these technical, logistical, and manufacturing challenges, several new turbine concepts have been investigated (downwind [11], two-bladed [12], vertical axis wind turbines (VAWTs) [13]) and new material systems (pultrusion [8], among others) are being investigated. For operating machines, digitization has grown in the wind energy field including the use of digital twins [14]–[16] for operations. Further, new materials systems are also under investigation to specifically address end-of-life decisions and offer more recycling options for blade materials [17].

This paper aims to examine the challenges of composite materials for wind energy applications, and to highlight some studies that offer potential new solutions and new insights across the entire life cycle of composites for wind energy systems ranging from the design phase, to manufacturing, to operation, and finally to end-of-life (Fig. 1).

The outline of the paper is as follows. Section 2 presents an overview of materials selections for wind energy and examines material selection and design for a series of 100-meter blade designs exploring different material choices. In Section 3, manufacturing is examined with a focus on resin uptake effects on the skin/core interface of blade materials explored via experimental and numerical methods. Operational aspects are the focus of Section 4 where materials issues related to digital twins are examined. Section 5 concludes the life cycle with a look at some end-of-life studies for more sustainable materials and manufacturing of recyclable blades by fusion joining.

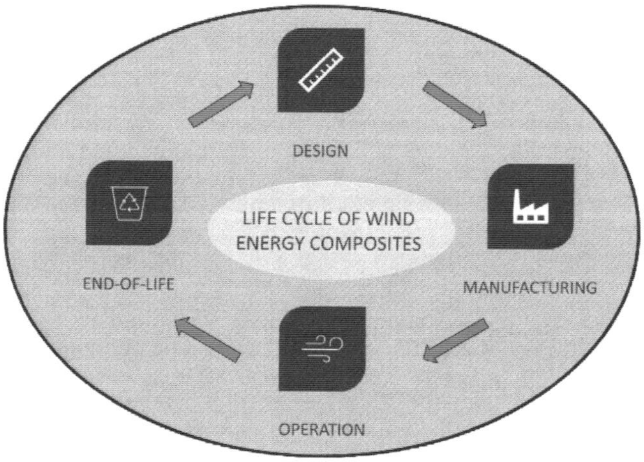

Figure 1: Life cycle Loop for Wind Blade Composite Materials: Design, Manufacturing, Operation, and End-of-Life Phases

2. Design of Blades: design studies for material selection in large rotor designs

2.1. Material Selection and Design Considerations

Composite wind turbine blades are designed to meet a wide range of structural requirements, in addition to constraints on cost, manufacturability, transportability, inspectability, etc. The structural requirements primarily include: 1) strength evaluation, 2) tower clearance (deflection evaluation), 3) fatigue evaluation, 4) buckling evaluation, and 5) assessment of blade dynamics for resonance and aero-elastic stability. These five primary structural requirements [6] are assessed in a design loop by structural designers [4], [5] in a typical blade design process until all structural requirements are met, and additional requirements on cost, manufacturability, and other factors are also met.

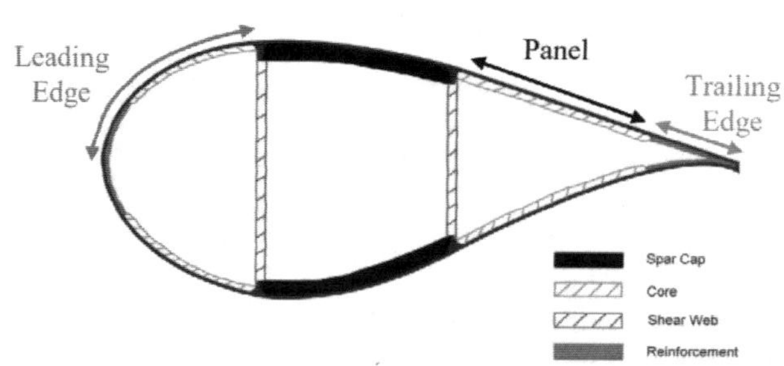

Figure 2: Representative Blade Cross-section

Meeting the structural requirements and manufacturability constraints in blade designs relies on selection of suitable candidate materials as well as strategic placement of these materials into the blade structure, which we now discuss. The inboard region of the blade, starting at the blade root and extending to roughly maximum chord (referred to as the root buildup) is mostly comprised of fiberglass laminates (e.g., triax). For blade outboard sections, from roughly maximum chord to the blade tip, a representative blade cross-section in shown in Fig. 2, where the principal areas of these outboard cross-sections are labeled: 1) spar caps (uniaxial laminates), 2) panels (sandwich composites of normally double bias and core materials), 3) shear webs (also sandwich composites), and 4) reinforcement of the leading edge and trailing edge (various laminates). A designer is thus faced with selection of materials and placement of those materials in layers into these various regions of the blade structure (root buildup, spar caps, core panels, shear webs, leading edge reinforcement, and trailing edge reinforcement, as well as the blade skins, adhesive bonds, and overlays).

The mainstream type of wind turbine in the market today is a three-bladed, upwind horizontal axis wind turbine (HAWT) configuration. However, several different, and to some extent new, turbine types are being investigated to either address the challenges of modern wind blades (e.g., technical, manufacturing, logistical, or others) or to develop machines tailored for new environments such as extreme-scale designs, shallow-water offshore, and design for floating offshore conditions. As a result, designers and developers are investigating a number of new turbine rotor types including two-bladed designs [12], downwind architectures [11], [18], and vertical axis rotors [13], [19], in contrast to the conventional three-bladed, upwind, horizontal axis rotor. Fig. 3 shows a comparison of these different rotor configurations, where from left to right are upwind, downwind, two-bladed, three-bladed, and then a vertical axis wind turbine rotor type.

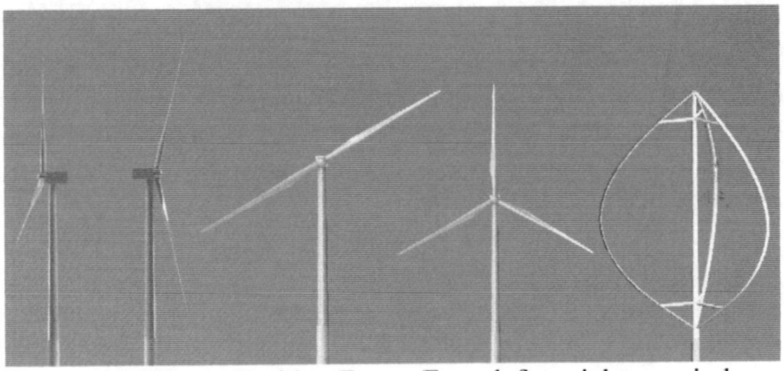
Figure 3: Different Turbine Types. From left to right: upwind, downwind, two-bladed, three-bladed, vertical axis.

2.2. SNL100 100-meter Blade Series: Materials and Manufacturing Outlook

A series of 100-meter blade designs for a 13.2 MW horizontal axis wind turbine were developed and made publicly available as reference models in the 2011-2014 timeframe [5], [20]–[22]. The SNL100 blade series provided an opportunity to examine material choices for, at the time, future large blades; however, today blades of 100-meter length have become a reality. In this series of four 100-meter blades, first an all-glass baseline was designed [20], referred to as SNL100-00 (the baseline blade), then a change from glass to carbon spar was designed [21], referred to as SNL100-01 (carbon spar blade), next a new core material strategy was studied [22], referred to as SNL100-02 (core strategy blade) and in the fourth and final design, a new design incorporating the prior new materials plus new flatback airfoils was designed [5], referred to as SNL100-03 (final, optimized design).

Through this 100-meter blade study, several key design drivers for large blades were identified and documented including gravitational fatigue loading, weight growth, panel buckling, and aero-elastic stability. The mass of the baseline blade (SNL100-00) was 114 tons, while successive blade masses were 74 tons, 59 tons, and finally 49 tons, for the SNL100-01, SNL100-02, and final SNL100-03, blade designs, respectively, which demonstrated the mass and cost reduction benefits of new materials and progressively more innovative designs.

This SNL100 series study demonstrated the feasibility of large blade structures, and the potential to design large blades that meet structural requirements while being lightweight, manufacturable, and aero-elastically stable at 100-meter scale. Design drivers for large blades were identified, as briefly summarized above, however, more details on all the findings and recommendations for the SNL100 series can be found in [5], for the interested reader.

Regarding material challenges in large blades, the findings of the SNL100 series studies provided an early look at carbon spar caps in blades (SNL100-01, SNL100-02, and SNL100-03) and showed that they offer significant weight reduction (but with a cost penalty). In addition, a new core material strategy for large blades using advanced core materials (SNL100-02 study) was found to be advantageous as utilizing balsa in critical buckling areas and PET foam in the non-critical buckling areas can provide both weight reduction as well as a sustainable core material solution in that balsa core is regrowable and PET foam is recyclable. Further, the growing significance of gravitational fatigue loading in 100-meter scale blades pointed to greater need to reinforce the blade in the edgewise direction.

3. Manufacturing: experimental mechanics of skin/core interphase for blade materials

Sandwich composite is a composite material that is fabricated by attaching two thin and stiff skins to a lightweight and thick core with high thermal, and acoustic insulation, and a high strength-to-weight ratio. Due to the many advantages sandwich composites possess compared with other structural materials, this material was intensively used in the wind blades [23], [24]. Skin/core interphase is always the first place of fracture initiated in the sandwich composites; it is necessary to fully understand the mechanism behind it. This section contains three subsections and focuses on the mechanism of skin/core interphase and guides researchers on how to design sandwich structures properly.

3.1. Nanoindentation measurement of core–skin interphase viscoelastic properties in a sandwich glass composite

3.1.1. Introduction. Low debonding toughness in the skin/core interphase region of the sandwich composites is always considered one of the shortcomings [25]–[27]. Furthermore, it is an understanding of the mechanical properties of the skin/core interphase region in the nanoscale, which can illustrate how a low debonding toughness formed in the skin/core interphase [28]–[30]. To address this issue, this work uses the viscoelastic-nanoindentation technique through the Berkovich nanoindenter tip to visualize Young's modulus spatial distribution at the interphase region. A glass skin/polyvinyl chloride (PVC) foam core sandwich composite was prepared by vacuum-assisted resin transfer molding (VARTM). Under a loading rate of 1 mN s−1, the time-averaged Young's modulus change from higher values (1.4 GPa) close to the glass skin region to the smaller values (0.8 GPa) close to the foam core was captured. This work proves the viscoelastic-nanoindentation technique is a sufficient and high-resolution way to benefit the analysis of the interphase, such as stress distribution calculations, which can eventually determine the bonding quality between skin and core and be beneficial to the design process of such sandwich composites.

3.1.2. Results and Discussion. The fabrication procedure, glass skin/PVC foam core sandwich composite, and spatial distribution of stiffness in the skin/core interphase region are shown in Fig. 4. Details of the sample preparation, viscoelastic-nanoindentation technique, and viscoelastic properties of the skin/core interphase region can be found in the previous work [31].

The spatial distribution of the time-averaged Young's modulus on the indented area against its optical micrograph is shown in Figs. 4(e) and 4(f). The viscoelastic analysis of nanoindentation data gives a better accuracy than Oliver and Pharr's approach owing to the nature of epoxy resin [32]–[34]. Compared with the mechanical data sheet, the interphase region was softened by the PVC foam core instead of stiffened by the glass skin [35]. The reason for the stiffness softening in the indented area is a relatively larger contact area between the cured epoxy and PVC cell walls than glass skin. Furthermore, the indented area is not close enough to the glass skin, and a semisphere void on the left edge of the indented area can be other reasons for stiffness softening in the interphase. Overall, a significant change in modulus across a short distance (240 μm) proves that nanoindentation is a high-resolution approach compared with other mechanical characterization approaches for sandwich composites.

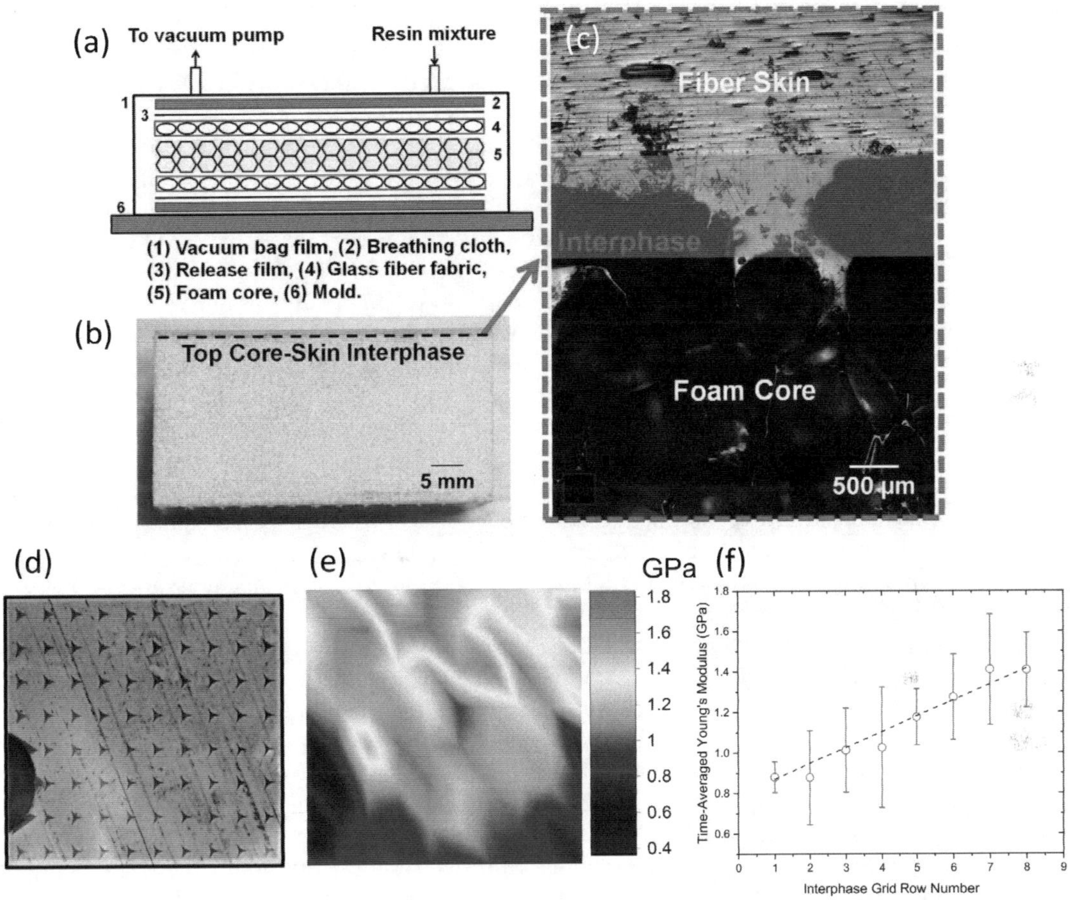

Figure 4: The fabrication procedure, glass skin/PVC foam core sandwich composite, and spatial distribution of stiffness in the skin/core interphase region: (a) schematic of the VARTM setup; (b) the cross-section of the glass skin/PVC foam sandwich composite; (c) a 5× optical micrograph of the skin/core interphase region of the sandwich composite; spatial distribution of the average Young's modulus in the interphase region: (d) a 5× optical micrograph of residual indents from 8×10 nanoindentations by Berkovich tip on the skin/core interphase region of the sandwich composite; (e) spatial distribution of time-averaged Young's modulus in the same area shown in (d); (f) time-averaged Young's modulus at different interphase grid rows from nanoindentation measurements in the same area shown in (d).

3.1.3. Conclusion. The time-averaged Young's modulus was computed through the nanoindentation-viscoelastic technique. The captured stiffness gradients through the skin/core interphase prove the effectiveness of this approach. A high correlation between the time-averaged Young's modulus mapping and the optical micrograph of the same indented area was obtained (softening due to imperfection and the larger contact area between PVC foam core and skin/core interphase). This work suggested that strengthening the debonding toughness of the sandwich composites should target the foam core side of the skin/core interphase region.

3.2. The effect of resin uptake on the flexural properties of compression molded sandwich composites

3.2.1. Introduction. A high stiffness-to-weight ratio is desired in composite structures [36]–[38] such as wind blades. Modifying the resin uptake of sandwich composites through compression molding (CM) is an easy and convenient way to maintain a high stiffness-to-weight ratio. This work investigates the flexural properties of sandwich composites made with glass skins/PVC foam cores (H60 and H80; smooth (PSC) and grooved and perforated (GPC)) at several resin uptakes. When the resin uptake of H80 GPC sandwich composites reduces by 11.0%, the specific flexural strength and modulus increase from 82.04 to 90.70 kN · m/kg and 6.03 to 7.13 MN · m/kg, respectively, which benefit most from resin uptake reduction comparing with the rest three types of core sandwich composites. Adequate ranges of resin uptakes (32% to 38% and 40% to 45%) are determined for the H80 PSC and GPC sandwich composites when reaching a high stiffness-to-weight ratio while preventing resin starvation, respectively. The findings suggested that applying the low-resin uptake within the appropriate range of sandwich composites in the wind blades leads to lighter blades and thus lower cyclic gravitational loads.

3.2.2. Results and Discussion. The schematic of the cross-section view of the wind blade; images of glass skin/PVC foam sandwich composites, digital image correlation (DIC), and failure modes; the effect of resin uptake on flexural properties of sandwich composites are shown in Fig. 5. Details of the sample preparation, experimental techniques (like three-point bending, etc.), mechanical properties of the PVC foam core, and finite element modelling (FEM) of sandwich composites under three-point bending can be found in the previous work [39].

Figure 5(a) shows a schematic diagram of sections of a wind turbine blade where sandwich composites are used (leading edge panel, trailing edge panel, and shear web) [3]. Core molding conditions are shown in Fig. 5(b). The grooves and perforations facilitated a fast flow and a proper saturation of resin in the skin/core interphase and inside of the core, which secured a good bonding and high flexibility between the skin and core. The smooth core is chosen as the benchmark for specimens with low resin uptake considerations. Figure 5(c) proves the size of the grooves of the sandwich composite is inversely correlated to mold compression. Figure 5(d) reveals resin starvation in sandwich composites due to two reasons: 1. non-uniform spatial distribution; 2. the resin precipitate to the bottom mold owing to gravity. Figures 5(e)-5(h) show the flexural and specific flexural strength and modulus as a function of resin uptake for four types of core sandwich composites. Compared with glass skins, resin contributes much less to stiffness than weight gain. As a result, lower resin uptake is desired in the adequate resin uptake range to maximize the strength-to-weight ratio of the sandwich structure. A sharp discontinuity in the flexural strength and modulus of the H60 GPC core sandwich composites as a function of resin uptake curves is noted. This phenomenon is due to the cellular structure shrinkage of cores by a high mold compression concentration during the CM process. In Figs. 5(i)-5(l), a high first principal strain concentrates on the bottom section of the core, and early structure failure (core shearing crack and bottom skin-core debonding) was observed in the H60 GPC sandwich composites. Furthermore, for the H80 PSC sandwich composites, the first principal strain is uniformly distributed on the core, leading to top skin crack and core crush failure. We can conclude that when the GPC core is used in the sandwich composite construction, a high core stiffness is needed to encounter the stress concentration issue.

3.2.3. Conclusion. The effects of resin uptake on the flexural properties of sandwich composites were investigated under bending conditions. A proper resin uptake regime has been found for four types of core sandwich composites with a high stiffness-to-weight ratio and without resin starvation exists. The results reported in this work could be beneficial to the design of large wind turbine blades to reach a higher stiffness-to-weight ratio and lower material cost than the current state-of-the-art fabrication routine by reducing resin consumption.

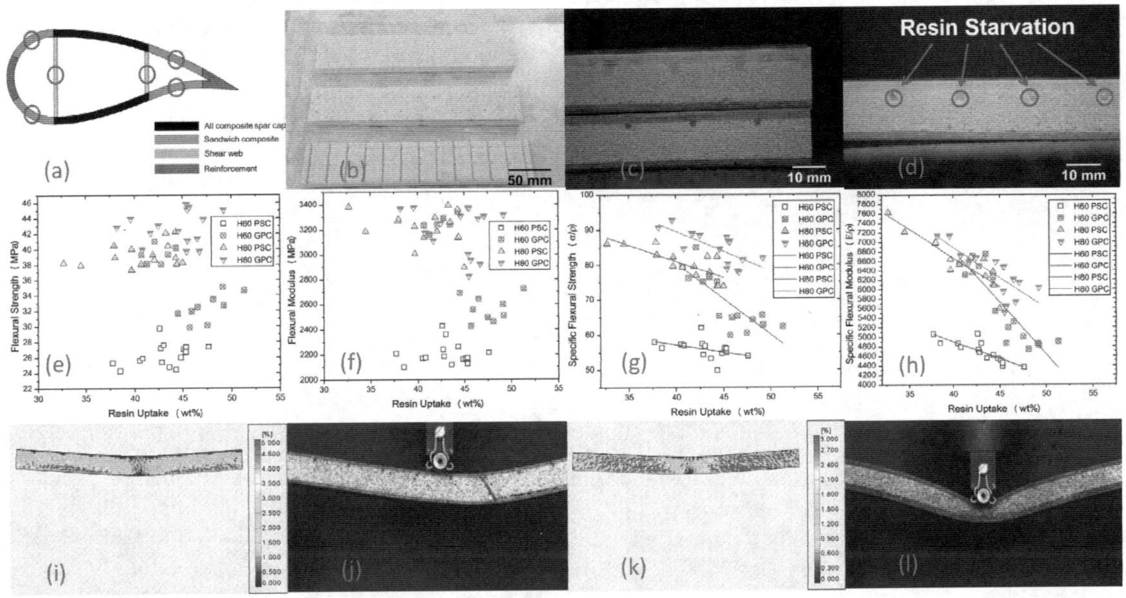

Figure 5: The schematic of the cross-section view of the wind blade; images of glass skin/PVC foam sandwich composites, DIC, and failure modes; the effect of resin uptake on flexural properties of sandwich composites: (a) cross-sectional view of the wind turbine blade (note: the red circle is indicating the sandwich composites zone.) (adapted from Griffith et al. [20]); (b) H80 PSC and GPC core with the same density ($\rho = 0.08$ g/cm^3) (from top to bottom: smooth core; grooved and perforated core top and bottom view); (c) a front view of the sandwich composites; (d) resin starved sandwich composites; (e)-(h) plots of flexural strengths and modulus; specific flexural strength and modulus as a function of resin uptake for the four types of core sandwich composites (note: units of specific flexural modulus: E/ρ and specific flexural strength: σ/ρ are kN · m/kg); (i) and (k) Three-point bending of a H60 GPC (resin uptake: 41.30%) and H80 PSC core (resin uptake: 39.66%) sandwich composite under DIC configurations at fracture initiation stage; (j) and (l) Failure modes of sandwich composites; core type: 1. H60 GPC core (core shearing crack and bottom skin-core debonding) and 2. H80 PSC core (top skin crack and core crush).

3.3. Bending and shear improvements in 3D-printed core sandwich composites through modification of resin uptake in the skin/core interphase region

3.3.1. Introduction. Fused deposition modelling (FDM) printed polymers are rarely used as a structural material. As a result, a new manufacturing routine is needed to improve the incorporation of FDM-printed polymers in composite structures. In this work, glass skin/FDM printed polylactic-acid (PLA) core sandwich composites are prepared by the CM process, which provides a good manufacturing strategy for skin/core interphase modification. A significant improvement was found compared with the optimized resin uptake (optimized resin uptake range: 20.43%–22.86 wt%) 3D-printed PLA core sandwich composite and lowest performance sandwich composite (Improvement: in-plane shear strength (~34%)/modulus (~29%), out-of-plane shear strength (~25%)/modulus (~31%), specific peak bending load (~19%)). The 3D-printed cores are suitable for replacing balsa cores in sandwich structures for many applications with a satisfactory strength-to-weight ratio.

3.3.2. Results and Discussion. Flexural properties of the 3D-printed beams, in-plane shear properties, DIC images, and failure mode of 3D-printed core sandwich composites are shown in Fig. 6. Details of the sample preparation, experimental techniques (like four-point bending, etc.), 3D-printing parameters, shear analysis of 3D-printed core sandwich composites, and comparison of published works about sandwich composites can be found in the previous work [40].

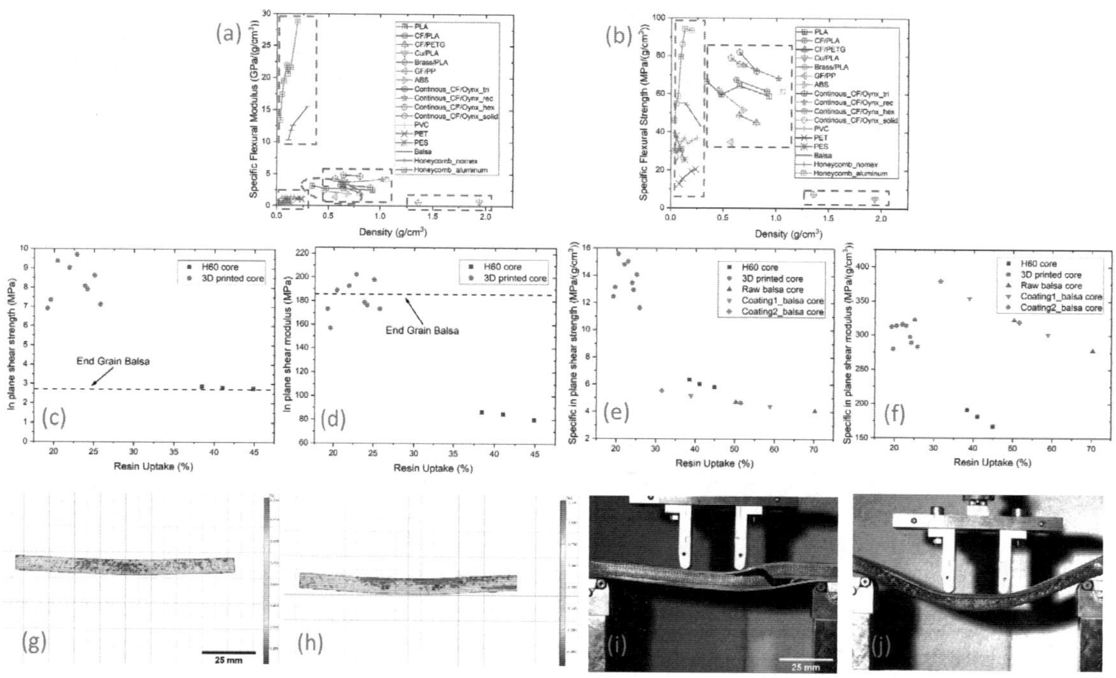

Figure 6: Flexural properties of the 3D-printed beams, in-plane shear properties, DIC images, and failure mode of 3D-printed core sandwich composites: (a) and (b) specific flexural properties as a function of the density of 3D-printed polymer beams, foam, and honeycomb structures [41]–[45], note: the red dashed line zone is indicating the conventional material, the blue dashed line zone is indicating the 3D-printed material; (c)-(f) in-plane shear and specific in-plane shear properties as a function of resin uptake of compression-molded sandwich composites; (g) and (h) first order principal strain distribution of the PLA lattice core sandwich composites ((g) 19.31% resin uptake, (h) 23.97% resin uptake, respectively) under four-point bending loads at fracture initiation stage; (i) and (j) failure modes of the PLA lattice and PVC foam core sandwich composites under four-point bending tests; note: (i) skin/core debonding and core shearing failure (PLA), (j) skin/core debonding, core shearing, and core crushing failure (PVC foam).

A proper thermoplastic polymer should be chosen for the shear properties of sandwich composites investigation. In Figs. 6(a) and 6(b), PLA materials stand out compared with the rest of the 3D-printed beams and most of the conventional core materials (except aluminum core) due to their high strength and stiffness-to-weight ratio, and cost-effectiveness. The in-plane shear strength and stiffness of balsa core sandwich composites are considered as 2.72 and 186 MPa with 0.152 g/cm^3 nominal core density [46], [47]. The 3D-printed core sandwich composites have 2–3 times higher in-plane shear strength and similar in-plane shear stiffness than balsa core sandwich composites, and 3–3.5 times higher specific in-plane shear strength and similar specific in-plane shear stiffness than balsa core sandwich composites. Furthermore, the 3D-printed core could be more cost-effective because this core could save resin during manufacture compared with the balsa core. In Figs. 6(g) and 6(h), a low resin uptake in sandwich composites could lead to strain concentration in the core due to an insufficient amount of resin in the skin/core interphase, and an adequate level of resin uptake could introduce a proper amount of resin to achieve a sufficient shear stress transfer from the skin to the core. From images of failure modes, we can conclude that the PLA core possesses a higher stiffness than the H60 PVC foam core. However, the PVC foam core can sustain a larger energy absorption due to better ductility. In general, the PVC foam core should be stiff enough to transfer the shear loads from one skin to the other skin. The resin uptake in the skin/core interphase of the 3D-printed core sandwich composites should be properly modified to prevent catastrophic structural failure.

3.3.3. Conclusion. The shear properties of two types of core sandwich composites fabricated by the FDM/CM process were investigated by three-point bending and four-point bending tests and derived through the first-order shear deformation theory. In the 3D-printed core sandwich composites, the optimized resin uptake regime of PLA core sandwich composites is determined (20.43% to 22.86%). Based on our investigation, the bending and shear properties of 3D-printed core sandwich composites are sensitive to resin uptake in the skin/core interphase due to a different chemical bonding mechanism from conventional core sandwich composites. For sustainability and recyclability of material selection, although balsa is a naturally grown material, balsa will be impregnated by resin owing to its open cell structure leads to a low recycling efficiency. Overall, a suitable strategy of skin/core interphase enhancement for 3D-printed core sandwich composites is needed to achieve a continuous shear stress flow between the skin and core.

4. Operation: digital twin models of blade and turbines

Accurate numerical tools (models or codes) are, of course, essential for wind turbine design, to simulate inflow and operating conditions, compute loads, evaluate structural/material designs, etc. However, these tools can also be quite useful for wind farm operations – namely for asset management to support decision-making throughout each turbine's lifetime [16], [48], [49]. These numerical design tools can be used to develop a model that represents the physical asset (the operating wind turbine), and such a model is referred to as a "digital twin". Digital twin models of operating wind turbines offer one very promising technology for decision support in wind farm management.

Wind turbine aero-elastic models (such as OpenFAST) can be used to support decision-making in wind farm operations, which includes a wide range of possible use cases including [49]:

- RCA (root cause analysis) of material damage identified during inspection,
- life-extension studies for aging components to assess the remaining life of a component,
- assessing the integrity of a structural repair, and
- assessment of potential component upgrades, among many other uses.

A digital twin model provides information and insight to support decision-making for "what-if scenarios" and uses cases, such as those listed above, for fleet operations and maintenance in a virtual setting.

Figure 7: Digital twin concept for a wind turbine

4.1. Turbine-level digital twin models

Figure 7 illustrates the concept of a turbine-level digital twin model, where all major components are modeled including the tower, blades, drivetrain, etc. Such a model is multi-disciplinary as aerodynamics, structures, structural dynamics, control systems, wind inflow conditions, etc. are all captured in a single model. Some recent research has focused on how to create the digital twin model based on technical specifications and wind turbine data streams [49]. For example, aerodynamics models were developed in [49] for blades based on power curve measurements and coefficient of thrust data or specifications, while structural models of blades, for example, were created based on mass properties and blade bending frequencies. These structural models were developed in such a way as to represent the mass properties and structural dynamics properties of the actual wind turbine by modifying the material properties within the model.

4.2. Multi-fidelity digital twin models

While turbine level models are quite useful for decision support tasks such as loads analysis, and life-extension studies, for tasks such as evaluation of the severity of structural damage or evaluation of structural integrity (or value of doing) a repair a high-fidelity model is needed. This need for high-fidelity models within a digital twin framework motivated development of a multi-fidelity digital twin model [14], where at the turbine-level mid- to low-fidelity models are used to model the entire system (similar to the prior section) and at the component-level high-fidelity models are used for more detailed analysis. As an example, consider a multi-fidelity digital twin model for a wind turbine blade (as illustrated in Fig. 8).

The multi-fidelity digital twin model has the capability to address components at the material level, and in the case of a wind turbine blade, can address each material in the blade composite at the layer or constituent level. For a blade, the multi-fidelity approach starts with a high-fidelity design representation of the blade geometry and internal structure (composite layup) (e.g., using a blade modeling or CAD program such as NuMAD [50]). This high-fidelity model can be analyzed in finite element programs to perform any number of high-fidelity analyses: static (strength and deflection), modal, buckling, fatigue, etc., to provide digital twin-based decision support. A low-fidelity model (such as a beam model) can be developed based on the high-fidelity specification, and in this way both fidelities of the model are consistent in the multi-fidelity digital twin model.

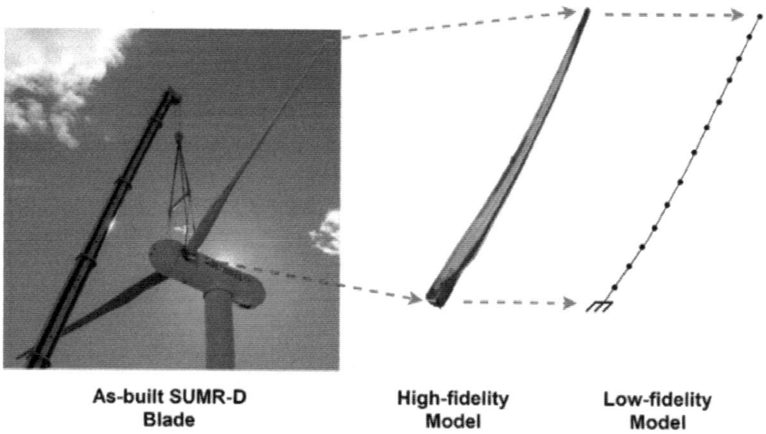

Figure 8: Multi-fidelity digital twin of a wind blade [14]

The multi-fidelity digital twin method presented in [14] is based on following a blade throughout its lifetime starting with the design definition (design basis), gathering information in the manufacturing phase, further gathering data from lab-scale or proof testing, and also collecting data in the operational phase, all to further refine and update the model. This process of data gathering and model refinement for the

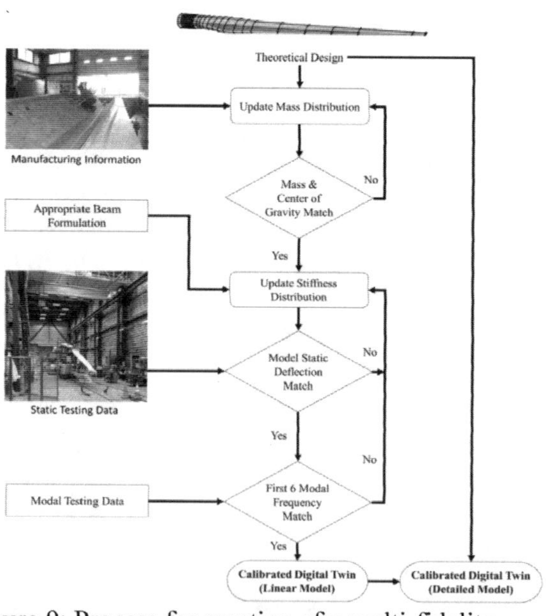

Figure 9: Process for creation of a multi-fidelity digital twin: Design basis, Manufacturing data, Lab testing data, and Operational data. [14]

various stages is detailed in the flowchart of Fig. 9.

A case study of developing a multi-fidelity digital twin for a prototype-scale wind turbine blade (called, SUMR-D [6]) is presented in [14] that starts with a design-stage blade digital twin (with low- and high-fidelity models), then these models are updated with new data at it comes available during the manufacturing stage (based on any deviations in the as-built blade), then a post-manufacturing lab-scale structural testing phase where both static load-deflection and modal testing data are collected for model refinement, and ultimately then into service in the operational phase. As noted, this multi-fidelity structural model offers all the benefits of the low-fidelity turbine-level model, with the additional capabilities of the high-fidelity model, to improve certainty regarding the state of health of the wind turbine component, in this case the blade. In the case of the SUMR-D blade [6], [14], the development and validation of the multi-fidelity digital twin for the blade pre-operation (i.e., post lab-scale testing) provided an updated, and accurate model that was then used to design more accurate control systems and to improve safety in operating this wind turbine.

5. End-of-Life: design for recycling and sustainability

5.1. Fusion-joining of additive-manufactured PLA segments for recyclable and sustainable structures application: Introduction

The demand to involve thermoplastic polymers in engineering structures (like wind blades) has been increasing due to a growing issue of environmental pollution from composite wastes and the advantages of thermoplastic polymers (like segmentation, structural repair, fusion-joining, etc.) [51]–[53]. Fusion-joining is a potential solution for three cases: 1. upscaling the 3D printed parts limited by the building volume; 2. structural repair to reduce labor cost; 3. on-site fabrication of segmented structure to reduce the transportation cost [54]–[57]. However, each fusion-joining method possesses pros and cons and researchers still have not confined it to one universally accepted process for thermoplastic structures [58]–[60].

In this work, novel displacement-controlled resistance welding and hot plate welding processes were used. Three types of metal mesh, namely 30% (open area fraction)/ 0.11 mm (open size) Ni-Cu, 34%/0.07 mm Ni-Cu, and 36%/0.25 mm Co-Ni, were used as the heating elements in the displacement-controlled resistance welding. A pair of heating molds under the shaft and socket shape were coupled with a heating pan to achieve a uniform heating region for two 3D-printed segments for the hot plate welding method. The optical micrographs and X-ray micro-computed tomography of the fusion-joined PLA slender beams show that there are no voids formed in the bonding region. After modification of the fusion-joining process, displacement-controlled resistance welding and hot plate welding can achieve 94% and 114 % welding strength efficiency (compared with the 3D-printed single continuous beams), respectively. The two processes possess a great potential to be dominant in fusion joining the thermoplastic polymers with fine surface finishing and good internal bonding quality to reduce plastic wastes, achieve easy and rapid structural repair and easy transportation, hence increasing sustainability, recyclability, and economic efficiency of a broad range of polymer and composite structures.

5.2. Results and Discussion

Manufacturing process; images of the optical micrograph, X-ray micro-computed tomography, failure modes; mechanical properties of the fusion-joining process (displacement-controlled resistance welding, hot plate welding) are shown in Fig. 10. Details of the sample preparation, experimental techniques (like three-point bending, etc.), 3D-printing parameters, analysis of the fusion-joining process, annealing, and comparison of published works about fusion-joining processes can be found in these literature [61], [62].

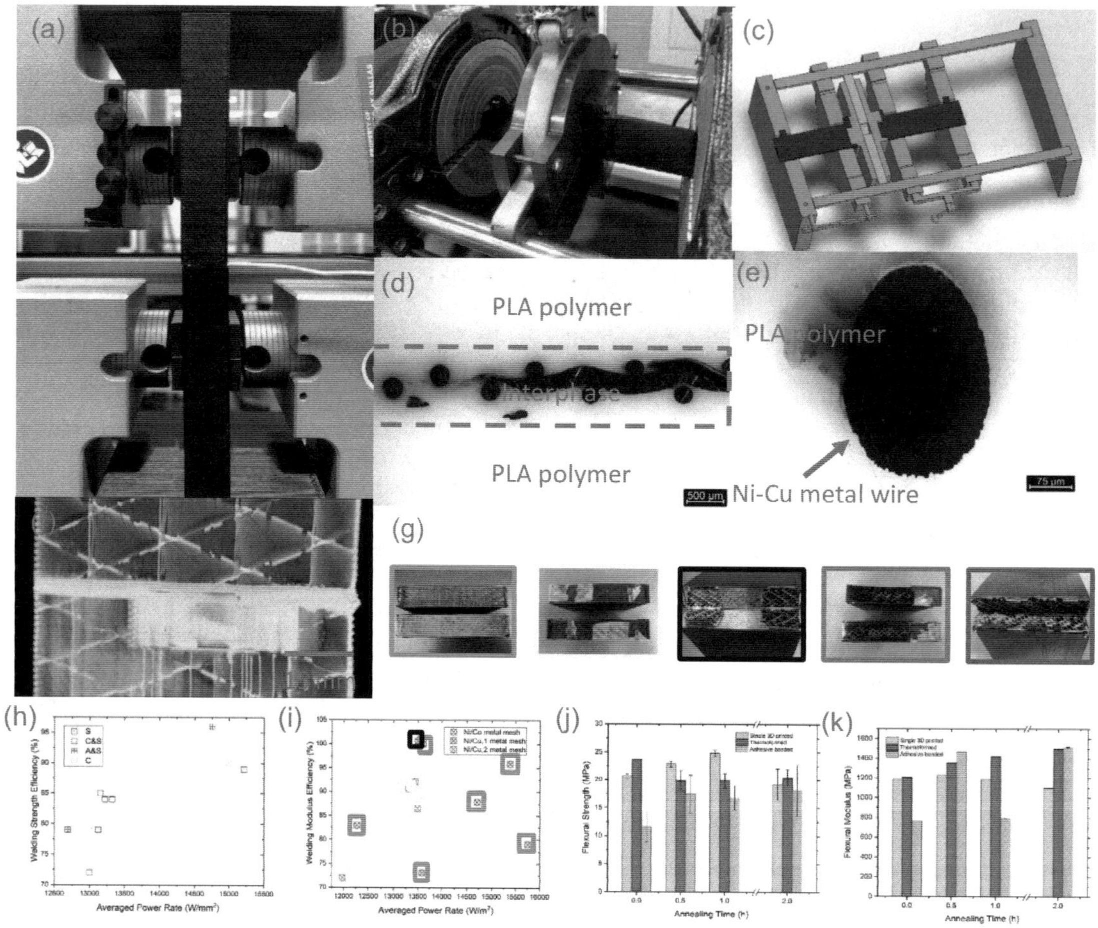

Figure 10: Manufacturing process; images of the optical micrograph, X-ray micro-computed tomography, failure modes; mechanical properties of the fusion-joining process (displacement-controlled resistance welding, hot plate welding): (a) displacement-controlled resistance welding setup; (b) and (c) hot plate welding setup; (d) and (e) micrographs of the resistance-welding region; (f) X-ray micro-computed tomography of hot plate welded beam; (g) failure modes of displacement-controlled resistance-welded 3D-printed beams under three-point bending: substrate failure (S): blue; cohesive & substrate failure (C&S): red; adhesive & substrate failure (A&S): black; cohesive failure (C): yellow; adhesive failure (A): green; (h) and (i) welding strength & modulus efficiency as a function of averaged power rate of the displacement-controlled resistance-welded PLA beams by three types of metal meshes; (j) and (k) flexural modulus and strength as a function of annealing time of three types of the hot plate welded beams.

Figures 10(a)-10(c) show the displacement-controlled resistance-welding and hot plate welding setup for the PLA samples. Micrographs and X-ray micro-computed tomography in the bonding region for displacement-controlled resistance welded and hot plate welded beams under different magnifications were investigated, respectively. No voids were detected, and a relatively good bonding was achieved. Based on the previous investigation, the failure mode of the slender beam possesses a strong relationship with the failure modes [63], [64]. The rank of welding strength efficiency will correlate to the following failure modes: substrate failure > cohesive & substrate failure > adhesive & substrate failure > cohesive failure > adhesive failure. Figure 10(g) proves that a relatively good bonding quality is achieved for the displacement-controlled resistance welded beams with a 94%

welding strength efficiency. In the case of the hot-plate-welded beams, no annealing was required to maximize the strength-to-weight ratio but to maximize the modulus-to-weight ratio of hot-plate-welded beams. A 114% welding strength efficiency was achieved for the hot plate welded beams.

5.3. Conclusion

The bonding quality of displacement-controlled resistance-welded and hot plate welded slender PLA beams was investigated by three-point bending experiments. An investigation of the parameters of the resistance welding process showed that the power output, resistance welding time, initial compressive pressure, displacement rate in the displacement-controlled process, and total travelled displacement play a critical role in resistance welding. PLA is worthy to bring into civil engineering structures due to many advantages compared with a thermoset resin, especially in recyclability and processability. A few subjects remain to be studied in the future regarding the displacement-controlled resistance-welding process, especially on the economic efficiency when applied to a broader range of composite or 3D-printed structures.

As for the hot plate welding process, we proved that hot plate welding is a fast, high dimensional accuracy, and low polymer deterioration fusion-joining process for 3D-printed polymer structures through X-Ray micro-computed tomography. The effect of the annealing time on the flexural properties of the 3D-printed short beams was investigated using three-point bending experiments. In a constant-temperature annealing, 70 °C is an adequate temperature as it causes a relatively small shrinkage below 30 mins. However, annealing under 70 °C beyond 30 mins was not beneficial to the 3D-printed PLA structure owing to the high level of linear shrinkage. Furthermore, thermal buckling and edge wrapping can easily occur owing to a low stiffness-to-length ratio of the 3D-printed beams. The hot plate welding process remains uncertain regarding the economic efficiency during the engineering structure construction process.

6. Concluding Remarks

In summary, wind energy has consistently grown over recent decades and has grown in many aspects including in terms of installed capacities, turbine size, blade length, and grid penetration. As a result, the wind energy field is today, and will continue to be, one of the largest industrial users of composite materials. Thus, composite materials for wind energy systems requires a great deal of attention, and new research is needed across the entire life cycle from the design phase, to manufacturing, to operation, and finally to end-of-life. This paper examined the challenges of composite materials for wind energy applications, and highlighted recent research studies that offer new solutions and new insights across this life cycle of composite materials from design, to manufacturing, to operation, and to end-of-life.

Design: Composite wind turbine blades are designed to meet a wide range of structural requirements, in addition to constraints on cost, manufacturability, transportability, etc. Thus, material considerations in large blade designs were the focus of Section 2. Several key design drivers for large blades were identified, based on studies of a series of four, progressively designed, 100-meter blades (the SNL100 100-meter series), which include gravitational fatigue loading, weight growth, panel buckling, and aero-elastic stability. Regarding material challenges in large blades, the findings of the SNL100 series studies provided an early look at carbon spar caps in blades offering weight reduction with a cost penalty. Further, a new core material strategy was found to be advantageous for large blades where balsa was utilized in critical buckling areas and PET foam in the non-critical buckling areas, as a result this provides both weight reduction as well as a sustainable core material solution in that balsa core is regrowable and PET foam is recyclable.

Manufacturing: In Section 3, manufacturing is examined with a focus on the skin/core interphase of sandwich composites, which are used extensively in panels and shear webs of wind turbine blades. Three studies were presented, and in the first study, the nanoindentation technique was shown to be a sufficient and high-resolution technique to measure bonding quality in the skin/core interphase. In the second study, the effect of resin uptake in various PVC core perforation and groove types was examined. Here, optimal levels of resin uptake were identified to maximize mechanical properties while minimizing extra resin mass and cost. In the third study, bending and shear improvements in 3D-printed core sandwich composites were examined with focus on modification of resin uptake in the skin/core interphase region. Here, the potential for 3D-printed core to replace traditional core materials was shown based on strong mechanical performance of the sandwich composites, high recyclability, and potential for lower resin uptake and less resin usage with 3D-printed core.

Operation: Digital twins for blades and turbines were presented in Section 4, where the digital twins offer decision support for wind farm operations and maintenance, for example, to support RCA (root cause analysis) of material damage identified during inspection, life-extension studies for aging components to assess the remaining life of a component, or integrity of a structural repair. A novel multi-fidelity digital twin framework, encompassing high- and low-fidelity component models, was presented to improve certainty in structural analysis based on a digital twin with addition of high-fidelity models. The multi-fidelity digital was demonstrated for a subscale 21-meter prototype blade following the blade from design, through manufacturing and lab-scale testing, and ultimately into operation.

End-of-Life: Section 5 concludes the life cycle with an examination of design for recycling based on fusion joining of additively-manufacturing PLA segments for wind blades. Here, two joining processes were investigated including resistance welding with metal implants and hot plate welding (no implants). The two joining processes both possess a great potential to be dominant in fusion joining thermoplastic polymers with fine surface finishing and good internal bonding quality while reducing plastic wastes and increasing sustainability, recyclability, and economic efficiency of a broad range of polymer and composite structures.

7. References

[1] “Composites end markets: Energy (2023) | CompositesWorld.” https://www.compositesworld.com/articles/composites-end-markets-energy-2023 (accessed Aug. 12, 2023).

[2] R. Wiser *et al.*, “Land-Based Wind Market Report: 2022 Edition,” Aug. 2022, doi: 10.2172/1882594.

[3] L. Mishnaevsky, K. Branner, H. N. Petersen, J. Beauson, M. McGugan, and B. F. Sørensen, “Materials for Wind Turbine Blades: An Overview,” *Materials (Basel)*, vol. 10, no. 11, Nov. 2017, doi: 10.3390/MA10111285.

[4] “IEC 61400-1 - Wind energy generation systems – Part 1: Design requirements | GlobalSpec.” https://standards.globalspec.com/std/13202600/IEC%2061400-1 (accessed Aug. 12, 2023).

[5] D. Griffith and P. W. Richards, “The SNL100-03 Blade: Design Studies with Flatback Airfoils for the Sandia 100-meter Blade.,” Albuquerque, NM, and Livermore, CA (United States), Sep. 2014. doi: 10.2172/1159116.

[6] S. Yao *et al.*, “A gravo-aeroelastically scaled wind turbine rotor at field-prototype scale with strict structural requirements,” *Renew Energy*, vol. 156, pp. 535–547, 2020, doi: 10.1016/j.renene.2020.03.157.

[7] S. Yao *et al.*, “Aero-structural design and optimization of 50 MW wind turbine with over 250-m blades,” *Wind Engineering*, vol. 46, no. 1, pp. 273–295, Feb. 2022, doi: 10.1177/0309524X211027355/ASSET/IMAGES/LARGE/10.1177_0309524X211027355-FIG19.JPEG.

[8] B. L. Ennis *et al.*, “Optimized Carbon Fiber Composites in Wind Turbine Blade Design,” Nov. 2019, doi: 10.2172/1592956.

[9] S. G. Horcas, N. N. Sørensen, F. Zahle, G. R. Pirrung, and T. Barlas, “Vibrations of wind turbine blades in standstill: Mapping the influence of the inflow angles,” *Physics of Fluids*, vol. 34, no. 5, p. 54105, May 2022, doi: 10.1063/5.0088036/2847187.

[10] M. Chetan, S. Yao, and D. Todd Griffith, "Flutter Behavior of Highly Flexible Two-and Three-bladed Wind Turbine Rotors", doi: 10.5194/wes-2021-132.

[11] P. Bortolotti, H. Ivanov, N. Johnson, G. E. Barter, P. Veers, and N. Namura, "Challenges, opportunities, and a research roadmap for downwind wind turbines," *Wind Energy*, vol. 25, no. 2, pp. 354–367, Feb. 2022, doi: 10.1002/WE.2676.

[12] L. Y. Pao *et al.*, "Control co-design of 13 MW downwind two-bladed rotors to achieve 25% reduction in levelized cost of wind energy," *Annu Rev Control*, vol. 51, pp. 331–343, Jan. 2021, doi: 10.1016/J.ARCONTROL.2021.02.001.

[13] D. T. Griffith *et al.*, "Design Studies for Deep-Water Floating Offshore Vertical Axis Wind Turbines.," Jun. 2018, doi: 10.2172/1459118.

[14] M. Chetan, S. Yao, and D. T. Griffith, "Multi-fidelity digital twin structural model for a sub-scale downwind wind turbine rotor blade," *Wind Energy*, vol. 24, no. 12, pp. 1368–1387, Dec. 2021, doi: 10.1002/WE.2636.

[15] K. Branner, E. di Lorenzo, S. Vettori, P. Berring, P. U. Haselbach, and B. Peeters, "Executable digital twin demonstrator of wind turbine blade." ISMA, 2022. Accessed: Aug. 12, 2023. [Online]. Available: https://orbit.dtu.dk/en/publications/executable-digital-twin-demonstrator-of-wind-turbine-blade

[16] H. Solman, J. K. Kirkegaard, M. Smits, B. Van Vliet, and S. Bush, "Digital twinning as an act of governance in the wind energy sector," *Environ Sci Policy*, vol. 127, pp. 272–279, Jan. 2022, doi: 10.1016/J.ENVSCI.2021.10.027.

[17] M. J. Leon, "Recycling of wind turbine blades: Recent developments," *Curr Opin Green Sustain Chem*, vol. 39, p. 100746, Feb. 2023, doi: 10.1016/J.COGSC.2022.100746.

[18] C. Qin *et al.*, "The Science of Making Torque from Wind (Active rotor coning for a 25 MW downwind offshore wind turbine," *TORQUE 2022) Journal of Physics: Conference Series*, vol. 2265, p. 32022, 2022, doi: 10.1088/1742-6596/2265/3/032022.

[19] D. T. Griffith *et al.*, "A study of rotor and platform design trade-offs for large-scale floating vertical axis wind turbines," *J Phys Conf Ser*, vol. 753, no. 10, p. 102003, Sep. 2016, doi: 10.1088/1742-6596/753/10/102003.

[20] D. T. Griffith and T. D. Ashwill, "The Sandia 100-meter All-glass Baseline Wind Turbine Blade : SNL100-00," *Baseline*, no. June, pp. 1–67, 2011.

[21] D. Todd Griffith, "SANDIA REPORT The SNL100-01 Blade: Carbon Design Studies for the Sandia 100-meter Blade," 2013. Accessed: May 03, 2021. [Online]. Available: http://www.ntis.gov/help/ordermethods.asp?loc=7-4-0#online

[22] D. T. Griffith, "The SNL100-02 Blade : Advanced Core Material Design Studies for the Sandia 100- meter Blade," *Sand2013-10162*, no. November, 2013.

[23] P. W. Richards, D. T. Griffith, and D. H. Hodges, "Aeroelastic design of large wind turbine blades considering damage tolerance," *Wind Energy*, vol. 20, no. 1, pp. 159–170, Jan. 2017, doi: 10.1002/WE.1997.

[24] S. B. Johnson, M. Chetan, D. T. Griffith, and J. A. Sherwood, "Development of high-fidelity design-driven wind blade manufacturing process models to investigate labor predictions in wind blade manufacture," *Wind Energy*, vol. 25, no. 8, pp. 1313–1331, Aug. 2022, doi: 10.1002/WE.2731.

[25] P. Bortolotti, D. Berry, W. S. Carron, T. Anderson, and M. Chann, "Toward the Advanced Manufacturing of Land-Based Wind Turbine Blades," Jan. 2023, doi: 10.2514/6.2023-2092.

[26] R. C. Østergaard, B. F. Sørensen, and P. Brøndsted, "Measurement of Interface Fracture Toughness of Sandwich Structures under Mixed Mode Loadings," *Journal of Sandwich Structures and Materials*, vol. 9, pp. 445–466, 2007, doi: 10.1177/1099636207070578.

[27] G. B. Chai and S. Zhu, "A review of low-velocity impact on sandwich structures:," *Proceedings of the Institution of Mechanical Engineers, Part L: Journal of Materials: Design and Applicationsv*, vol. 225, no. 4, pp. 207–230, Aug. 2011, doi: 10.1177/1464420711409985.

[28] A. Truxel, F. Avileés, L. A. Carlsson, J. L. Grenestedt, and K. Millay, "Influence of face/core interface on debond toughness of foam and balsa cored sandwich," *Journal of Sandwich Structures and Materials*, vol. 8, no. 3, pp. 237–258, 2006, doi: 10.1177/1099636206062570.
[29] V. Birman and G. A. Kardomateas, "Review of current trends in research and applications of sandwich structures," *Compos B Eng*, vol. 142, pp. 221–240, Jun. 2018, doi: 10.1016/J.COMPOSITESB.2018.01.027.
[30] D. Wannarong and T. Singhanart, "A Review of Sandwich Composite Structures with 3D Printed Honeycomb Cores," *Engineering Journal*, vol. 26, no. 6, pp. 27–39, Jun. 2022, doi: 10.4186/ej.2022.26.6.27.
[31] D. Cao, S. Malakooti, V. N. Kulkarni, Y. Ren, and H. Lu, "Nanoindentation measurement of core–skin interphase viscoelastic properties in a sandwich glass composite," *Mech Time Depend Mater*, 2020, doi: 10.1007/s11043-020-09448-y.
[32] W. C. Oliver and G. M. Pharr, "An improved technique for determining hardness and elastic modulus using load and displacement sensing indentation experiments," *J Mater Res*, vol. 7, no. 6, pp. 1564–1583, Jun. 1992, doi: 10.1557/JMR.1992.1564.
[33] G. Huang and H. Lu, "Measurement of Young's relaxation modulus using nanoindentation," *Mechanics of Time-Dependent Materials 2007 10:3*, vol. 10, no. 3, pp. 229–243, Jan. 2007, doi: 10.1007/S11043-006-9020-3.
[34] H. Lu, B. Wang, J. Ma, G. Huang, and H. Viswanathan, "Measurement of creep compliance of solid polymers by nanoindentation," *Mechanics Time-Dependent Materials*, vol. 7, no. 3–4, pp. 189–207, 2003, doi: 10.1023/B:MTDM.0000007217.07156.9B/METRICS.
[35] Hexion, "Datenblatt: EPIKOTE RIMR 135 and EPIKURE Curing Agent MGS RIMH 134–RIMH 137," *Technical Data Sheet*, vol. 8246, no. August 2006, pp. 1–9, 2006, [Online]. Available: http://www.metyx.com/wp-content/uploads/PDF_Files/Hexion/TDS/TDS RIMH 137.pdf
[36] L. Thomas, "Advanced materials for wind turbine blade-A Review," 2018. Accessed: May 15, 2021. [Online]. Available: www.sciencedirect.comwww.materialstoday.com/proceedings
[37] P. Brøndsted, H. Lilholt, and A. Lystrup, "Composite materials for wind power turbine blades," *Annu Rev Mater Res*, vol. 35, no. 1, pp. 505–538, 2005, doi: 10.1146/annurev.matsci.35.100303.110641.
[38] P. Freere, M. Sacher, J. Derricott, and B. Hanson, "A Low Cost Wind Turbine and Blade Performance," *Wind Engineering*, vol. 34, no. 3, p. 289, 2010.
[39] D. Cao *et al.*, "The effect of resin uptake on the flexural properties of compression molded sandwich composites," *Wind Energy*, p. we.2661, Jul. 2021, doi: 10.1002/WE.2661.
[40] D. Cao, D. Bouzolin, H. Lu, and D. T. Griffith, "Bending and shear improvements in 3D-printed core sandwich composites through modification of resin uptake in the skin/core interphase region," *Compos B Eng*, p. 110912, Jul. 2023, doi: 10.1016/J.COMPOSITESB.2023.110912.
[41] DIVINYCELL, "Datasheet: Mechanical properties of DIVINYCELL H," 2022.
[42] D. Divinycell, "Technical Characteristics DIVINYCELL ® PN Physical Characteristics DIVINYCELL ® PN," 2021.
[43] 3A Composite, "Datasheet: BALTEK ® SBC," 2022, Accessed: Sep. 23, 2022. [Online]. Available: www.3ACcorematerials.com
[44] CEL Components, "Datasheet: Nomex Aeronautical grade - CEL Components," 2022. https://www.honeycombpanels.eu/en/products/honeycomb/nomex-aeronautical-grade-en (accessed Sep. 23, 2022).
[45] Hexcel, "Datasheet: HexWeb ® Aluminum Flex-Core ® Formable Aluminum Honeycomb," 2017.
[46] BALSASUD®, "BALSASUD® Core - Product Data Sheet," 2016.
[47] BALSASUD®, "BALSASUD® Core Balsa core resin uptake - Product Data Sheet," p. 727248, 2000.

[48] A. Haghshenas, A. Hasan, O. Osen, and E. T. Mikalsen, "Predictive digital twin for offshore wind farms," *Energy Informatics*, vol. 6, no. 1, pp. 1–26, Dec. 2023, doi: 10.1186/S42162-023-00257-4/FIGURES/19.

[49] L. Haus, "New Methods for Digital Twin Modelling of Wave and Wind Energy Systems," 2020, Accessed: Aug. 12, 2023. [Online]. Available: https://utd-ir.tdl.org/handle/10735.1/9016.

[50] J. C. Berg and B. R. Resor, "SANDIA REPORT Numerical Manufacturing And Design Tool (NuMAD v2.0) for Wind Turbine Blades: User's Guide." [Online]. Available: http://www.ntis.gov/help/ordermethods.asp?loc=7-4-0#online

[51] S. Deng, L. Djukic, R. Paton, and L. Ye, "Thermoplastic-epoxy interactions and their potential applications in joining composite structures - A review," *Composites Part A: Applied Science and Manufacturing*, vol. 68. Elsevier Ltd, pp. 121–132, 2015. doi: 10.1016/j.compositesa.2014.09.027.

[52] M. Kazemi, S. Faisal Kabir, and E. H. Fini, "State of the art in recycling waste thermoplastics and thermosets and their applications in construction," *Resour Conserv Recycl*, vol. 174, p. 105776, Nov. 2021, doi: 10.1016/J.RESCONREC.2021.105776.

[53] A. Pegoretti, "Towards sustainable structural composites: A review on the recycling of continuous-fiber-reinforced thermoplastics," *Advanced Industrial and Engineering Polymer Research*, vol. 4, no. 2, pp. 105–115, Apr. 2021, doi: 10.1016/J.AIEPR.2021.03.001.

[54] C. Ageorges, L. Ye, and M. Hou, "Advances in fusion bonding techniques for joining thermoplastic matrix composites: A review," *Compos Part A Appl Sci Manuf*, vol. 32, no. 6, pp. 839–857, 2001, doi: 10.1016/S1359-835X(00)00166-4.

[55] R. E. Murray, J. Roadman, and R. Beach, "Fusion joining of thermoplastic composite wind turbine blades: Lap-shear bond characterization," *Renew Energy*, vol. 140, pp. 501–512, 2019, doi: 10.1016/j.renene.2019.03.085.

[56] M. Dubé, P. Hubert, A. Yousefpour, and J. Denault, "Resistance welding of thermoplastic composites skin/stringer joints," *Compos Part A Appl Sci Manuf*, vol. 38, no. 12, pp. 2541–2552, Dec. 2007, doi: 10.1016/J.COMPOSITESA.2007.07.014.

[57] A. S. Escalera Mendoza, S. Yao, M. Chetan, and D. T. Griffith, "Design and analysis of a segmented blade for a 50 MW wind turbine rotor:," *https://doi.org/10.1177/0309524X211069393*, vol. 2022, no. 4, pp. 1146–1172, Jan. 2022, doi: 10.1177/0309524X211069393.

[58] D. Stavrov and H. E. N. Bersee, "Resistance welding of thermoplastic composites-an overview," *Compos Part A Appl Sci Manuf*, vol. 36, no. 1, pp. 39–54, Jan. 2005, doi: 10.1016/J.COMPOSITESA.2004.06.030.

[59] X. Xiong *et al.*, "Resistance welding technology of fiber reinforced polymer composites: a review," *J Adhes Sci Technol*, vol. 35, pp. 1593–1619, 2021, doi: 10.1080/01694243.2020.1856514.

[60] S. T. Holmes and J. W. Gillespie, "Thermal Analysis for Resistance Welding of Large-Scale Thermoplastic Composite Joints," *Journal of Reinforced Plastics and Composites*, vol. 12, no. 6, pp. 723–736, 1993, doi: 10.1177/073168449301200609.

[61] D. Cao, D. Bouzolin, C. Paniagua, H. Lu, and D. , T. Griffith, "Effect of Process Parameters on the Mechanical Performance of Fusion-joined Additively-manufactured Segments," *(Under Review)*, 2023.

[62] D. Cao, D. Bouzolin, H. Lu, and D. , T. Griffith, "High Welding Strength of Additive-Manufactured Polylactic-acid (PLA) Sustainable Structures Resistance Welded by Metallic Implants ," *(Under Review)*, 2023.

[63] M. Arias and G. Ziegmann, "The impulse resistance welding: a new technique for joining advanced thermoplastic composite parts," in *41th international SAMPE symposium*, 1996. Accessed: Aug. 17, 2021. [Online]. Available: https://utdallas.illiad.oclc.org/illiad/pdf/410477.pdf

IOP Conf. Series: Materials Science and Engineering 1293 (2023) 012002 doi:10.1088/1757-899X/1293/1/012002

[64] E. Talbot, "MANUFACTURING PROCESS MODELLING OF THERMOPLASTIC COMPOSITE RESISTANCE WELDING," Montreal, Jun. 2005.

LCM of thermoset-based fiber reinforced composites for large-scale applications: are process kinetics and structural properties antagonistic?

Véronique Michaud

Laboratory for Processing of Advanced Composites (LPAC), Ecole Polytechnique Fédérale de Lausanne (EPFL), Lausanne, Switzerland,

veronique.michaud@epfl.ch

Abstract. This article reviews the main methods to manufacture large-scale composite parts, with a focus on Liquid Composite Molding techniques of thermoset-based fiber reinforced structural parts. As this process relies on the impregnation of a dry textile stack, this manufacturing step is crucial in terms of part production rate, and part quality. To increase the process kinetics, a large effort has been devoted to increase the permeability of the textile preforms, while keeping a similar fiber content. An increase of almost two orders of magnitude can be attained if the textile shows a strong separation of scales between densely packed tows and large intra-two spaces. This however leads to a potential degradation in the resulting structural properties, particularly in dynamic mode due to the presence of the resin rich pockets. Alternative solutions emerge, which may help reach a cost-effective compromise.

1. Introduction

When manufacturing large-scale thermoset-based long fiber reinforced polymer composite (FRP) structural parts such as wind turbine blades, aircraft fuselages, rocket fairings, boat hulls or masts, several additional challenges arise as compared to the manufacturing of smaller parts:(i) the risk of occurrence of a scrap part should be reduced to a minimum due to the large amount of materials involved, (ii) the molds should be as light as possible to avoid extremely heavy weights and costs, (iii) the impregnation process kinetics should be as fast as possible so as to avoid premature cure before impregnation is complete and limit personnel and equipment costs, (iv) the cure temperature and time should be well adjusted to avoid uncontrolled exotherms in thicker parts and uneven cure, and the applied pressure should ideally remain low, to avoid the use of large scale autoclaves or compression presses which also strongly increase the cost, in particular if production volumes remain low.

In early stages of large structural component production, an easy-to-access production method was the wet hand lay-up technique, where a dry fiber reinforcement is placed in the mold and impregnated with the pre-mixed resin using rollers and brushes. The resin is left to polymerize in open air, or under a vacuum bag. This technique requires minimal equipment and knowledge, but requires intensive labor, leads to generally poor-quality composites or lack of reproducibility, and presents environmental issues, due to the contact with resin and evacuation to the open air. As a result, two alternative techniques have reached maturity in the last 10-20 years: (i) prepreg technologies, whereby the fiber reinforcement is initially impregnated with resin (and generally sold as such in rolls), then positioned layer by layer in the mold, by hand or using automated deposition techniques, before curing under vacuum or moderate pressure (ii) liquid composite molding (LCM) technologies, whereby the fiber reinforcement is positioned dry into the mold (again by hand or using automated techniques, sometimes preformed beforehand), and reactive resin is forced to flow into the dry fiber preform, under a pressure difference imposed by a vacuum outlet, or a pressurized inlet, alternatively under a prescribed flow rate imposed by a volumetric pump, before curing. These two methods are currently used in most manufacturing processes for large scale structural parts, although filament winding techniques are also used for pressure vessels or cylindrical parts, and additive manufacturing of large parts is gaining interest, but still limited to pilot scale demonstration purposes [1–3].

In aeronautics and space, prepreg use is dominant due to the very high quality and reproducibility requirements, with tight control of the resin content, and due to the existing portfolio of available qualified prepreg materials. In some cases, high pressure autoclaves are still used (for example in the production of airplane wings), although intensive research has been performed over several decades in developing out-of-autoclave techniques to increase cost-efficiency, sometimes with great success, such as for the Ariane and Vega rocket fairings [4]. LCM is also gaining importance, with some breakthrough examples, for example the Airbus A380 pressure bulkhead made by resin film infusion [5], wings of the Airbus A220 by Resin Transfer Infusion (previously Bombardier C Series, infused then autoclaved, [6]) or other wing structures [7].

In boat building, both prepreg and LCM technologies are encountered, depending on the cost-performance ratio, carbon fiber prepregs being mostly used in high performance racing yachts and in masts, whereas infusion of glass fiber textiles (or combination of textiles and shorter fiber mats) is common for pleasure yachts or speed boats [3].

Wind turbine blade manufacturing is the other obvious example of a large composite part, with some blades reaching over 100 m long. Both prepreg and LCM technologies have been applied successfully to blade skins and spars production, however the trend is now towards an increased use of LCM technologies as they gain further maturity, removing the need to transport and store large amount of reactive prepreg materials. Vacuum Assisted Resin Transfer Molding (VARTM), where the fabric is held within a rigid mold, is currently the most wide-spread and cost effective method for producing wind turbine rotor blades, with room temperature curing resins (and potential higher temperature post-cure, around 120°C) [2,8].

A major difference between prepreg and LCM techniques resides in the impregnation step, which constitutes a major part of the production step. As a result, a large effort has been focused on increasing the kinetics of impregnation, with emphasis on the increase of textile permeability, while keeping the fiber content identical. However, the potential impact on the resulting structural properties needs to be considered, to reach best compromise, but both are seldom evaluated in parallel. This article will provide some insight on the main principles of LCM processes, and of the strategies to increase the permeability of the reinforcement, focusing on structural parts with about 50% fiber content. We will then address the increased risks of reduced part quality, in terms of manufacturing defects, but also in terms of structural properties. Several alternative solutions will finally be addressed.

2. Liquid composite molding principles

LCM processes are generally described following the approach developed in soil science and petroleum engineering, for the flow of a viscous fluid in a porous medium. Two important differences with flows in soils are that the fluid is generally much more viscous than water (about 500 times) and the porous medium is not homogeneous and very anisotropic, but constituted of fiber bundles, that are assembled into textiles. As a result, the pore size distribution is generally bimodal, with large inter-tow spaces (up to mm), and narrow intra-tow channels (a few microns). Nonetheless, a continuum mechanics approach is generally followed to describe the porous medium, with a representative volume element (RVE) which contains the initially present air, fibers, and infiltrated liquid, in respective volume fractions V_a, V_f, V_l, such that:

$$V_a + V_f + V_l = 1 \quad (1)$$

By similarity with soil science, the fluid phase saturation S is defined as:

$$S = \frac{V_f}{(1-V_f)} \quad (2)$$

where (1-V_f) is the initial preform open porosity; S varies from 0 to 1 between a fully dry and a fully saturated preform.

The main geometrical descriptors of the dry porous medium are its fiber volume fraction, its specific surface, and the pore size distribution (as well as pore connectivity). For a simple arrangement of aligned cylinders, packed on a square array, the maximum fiber volume fraction that can be attained is $V_{fmaxs} = \pi/4$, whereas for a hexagonal array, $V_{fmaxh} = \pi/2\sqrt{3}$.

The pore size distribution is a statistical descriptor of the porous medium that was historically measured through gradual filling of the media with non-wetting fluids [9]. As X-Ray Computed tomography became more readily available in the recent years, 3D imaging of an RVE is now becoming routinely accessible, leading to precise geometrical descriptions of the porous medium, including the presence of non-connected pores ([10–15]). As mentioned, textiles are most often formed of tows or yarns that are assembled into a preform, the porous medium is thus generally described by a bi-modal pore size distribution, as illustrated in Figure 1, and the dual scale nature of the pore size distribution can lead to exacerbated capillary/hydrodynamic effects, as compared to more uniform porous media such as encountered in soil science [16,17].

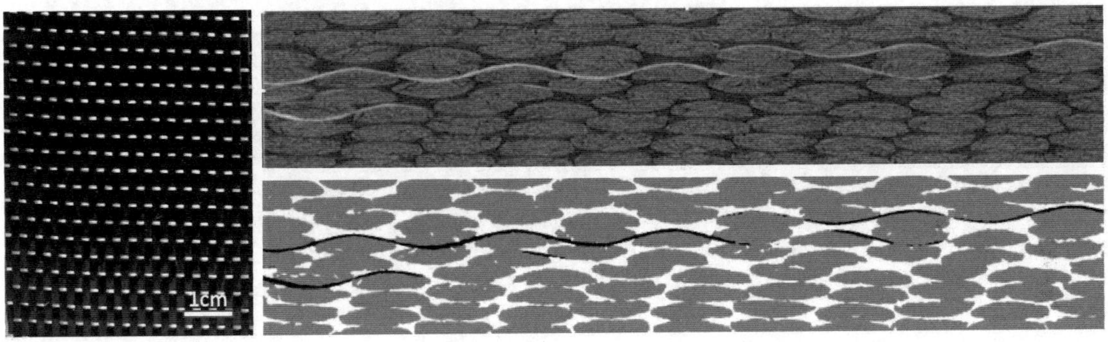

Figure 1. Example of the structure of a carbon UD textile with stabilizing glass yarns. Left, macroscopic view, right top, X Ray-CT cut of a fabric stack, right bottom, the same image after segmentation showing the tows in grey and inter-tow space in white (and glass tow in black) [18].

The underlying physical phenomena for LCM processes include capillary or surface phenomena, transport of fluid, heat, and mass, the mechanics of preform deformation before and during infiltration, matrix chemical cross-linking, and potential matrix/reinforcement chemical reaction during and after the process. For a complete description, the reader is referred to recent reviews [17,19].

In the following, for the sake of simplicity, we consider the general case of infiltration by a liquid of a rigid porous preform (constant volume fraction) in which all initial porosity is interconnected (no closed pores). We do not treat heat or mass transfer (as induced by a chemical reaction) since these often take place after impregnation. We also assume that the densities of liquid and solid phases are constant; in most cases of non-compressible liquid fluids, this is a reasonable assumption.

Mass conservation equation is written for the fluid phase as:

$$\frac{\partial(1-V_f)S}{\partial t} + \nabla((1-V_f)Su_l) = 0 \quad (3)$$

where u_l is the average local velocity of the liquid within the pores.

The momentum equation is generally written using Darcy's law:

$$(1-V_f)Su_l = -\frac{K}{\eta}\nabla P \quad (4)$$

where η is the fluid viscosity and P is the pressure in the liquid. K is the permeability of the porous medium, it is a tensor, with units: m², describing the ease for the fluid to pass through the pore space. Following usage in soil science for multiphase flow in porous media, it is generally assumed that the influence of saturation S on the value of permeability can be decoupled from the influence of the porous medium intrinsic geometry, by defining two parameters K_{sat} and k_r, such that $K = K_{sat} \cdot k_r$ where K_{sat} is the saturated permeability tensor, with units: m², which depends on the volume fraction and spatial arrangement of the fibers, and k_r is called the relative permeability, function of S only, and is a scalar varying from 0 to 1. The left-hand side of Eq. (4) is called the superficial velocity, or filtration velocity, which was initially defined by Darcy as the ratio of the volumetric flow rate Q out of a porous medium, over the cross-section of this porous medium, A. Integration of these equations for multiphase flow with the relevant boundary conditions leads to a description of the flow kinetics, and progressive saturation, provided that the hydraulic functions describing pressure and relative permeability versus saturation are identified [20].

For fully saturated flow in a rigid porous medium, the filtration velocity u_0 is simply written as:

$$\frac{Q}{A} = u_0 = (1 - V_f)u_l \tag{5}$$

As a result, for unidirectional saturated flow (k_r=1) under constant applied fluid pressure at the mold inlet, impregnation time for in plane flow along a given distance L is:

$$t = \frac{\eta(1-V_f)}{2K_{sat}\Delta P} L^2 \tag{6}$$

where ΔP is the pressure difference between outlet and inlet, which in case of a saturated flow assumption, should also include the capillary pressure drop acting at the flow front [20–22]. To minimize the impregnation time, while keeping the viscosity of the resin and the fiber volume fraction similar, strategies will include increasing the pressure gradient (but this requires heavier equipment and molds, and may lead to fiber movement), decreasing the impregnation length, for example by optimizing the presence of inlet points [23] and favoring through thickness impregnation, as performed in vacuum infusion thanks to high permeability flow grids placed on top of the textile. Another strategy is to increase the textile permeability, where a 10-fold increase leads to a 10-fold reduction of the time for impregnation. Efforts have been numerous to increase permeability, which is the subject of this review.

3. In-plane permeability measurement

Permeability measurement relies on the measure of the flow kinetics through the porous medium of interest, under well controlled conditions with a well-known Newtonian fluid [24]. Even if the method seems very simple, experimental data for permeability tend to vary by almost one order of magnitude, as reported in the first permeability round robin [25], although with a stricter set of guidelines, results fall within about 20% variation, attributed to intrinsic variability of the textile stack, as well as human error [26]. In addition, when performing 1D, in plane permeability measurement, two methods can be used, based on the measurement of the flow front progression, or on the flow rate through a fully saturated textile, as shown in figure 2.

When permeability is estimated from the position of the flow front L(t), Eq.(6) is used to evaluate its value, often neglecting progressive saturation and any capillary effects at the flow front, assuming $\Delta P = \Delta P_{app} = P_f - P_i$, where P_i is the inlet pressure, and P_f the pressure in the preform just ahead of the flow front (atmospheric pressure in general). Thus, this value, often named K_{unsat} to reflect the fact that it measured during flow (and that saturation is taking place during measurement), is computed as:

$$K_{unsat} = -\frac{\eta\,(1-V_f)}{2\,\Delta P_{app}} \frac{L^2}{t} \tag{7}$$

This value is distinguished from the saturated permeability, K_{sat}, measured from the flow rate Q_{out} at the outlet, after the fluid has filled the whole length of preform L_f:

$$K_{sat} = -\frac{\eta\, Q_{out}\, L_f}{A\,\Delta P_{app}} \tag{8}$$

where A is the cross-section area of the fabric stack [27].

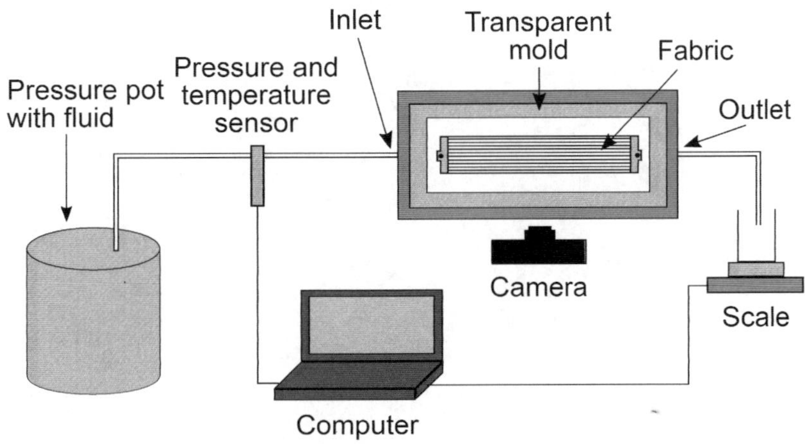

Figure 2. Schematic of in-plane permeability measurement set-up, for measurement under constant applied pressure, with unsaturated flow measurement through a transparent top with a camera, and saturated flow measurement through the flow rate from the outlet after full saturation (from [15]).

As shown by Salvatori [15], the relation between these two values depends on the magnitude of the capillary pressure drop ΔP_γ acting at the flow front, since $\Delta P = \Delta P_{app} + \Delta P_\gamma$, the ratio of unsaturated to saturated permeability is thus:

$$R_s = \frac{K_{unsat}}{K_{sat}} = 1 + \frac{\Delta P_\gamma}{\Delta P_{app}} \quad (9)$$

4. How to increase permeability while keeping the same fiber volume fraction?

The saturated permeability of textiles is a strong function of the fiber volume fraction and its orientation with respect to flow direction. In packed fiber networks, flow along the direction of fibers is not always easier than across, and early permeability models, based on the concepts of hydraulic radius, or on computing the flow past regular arrays of cylinders, illustrated rather well this relationship, as shown Figure 3, using Gebart's and Happel's models presented in Ref. [28]. As a result, for a fiber radius of 8 μm typical of glass fibers, the models predict a permeability for flow transverse to the direction of fibers of about 1.10^{-12} m^2, and for flow parallel to the fibers ranging from 7.10^{-14} to 2.10^{-12} m^2. As pointed out by Karaki [29], who also evaluated other permeability models, it is rather difficult obtain realistic values from these models, although they represent a practical and analytically calculated first estimate of permeability ranges. The larger variation in models for flow parallel to the fibers stems from different assumptions leading to pressure build-up during flow. These models are in general not very realistic, except at rather high fiber volume fractions, as the fibers are rarely arranged on a regularly spaced pattern. As mentioned earlier and shown Figure 1, the porous medium is constituted of densely packed yarns and large gaps in between, which form channels that tend to dominate the permeability value. As a result, experimental permeability values are generally higher. Rieber [30] conducted an experimental analysis of 19 woven glass textiles (plain, twill and satin weave), observing values ranging between 1.10^{-11} m^2 and 1.10^{-10} m^2 for V_f=50%. They highlighted the influence of the weave density and yarn density on the K(V_f) slope, and the role of the weft/warp ratio on the anisotropy of the in-plane permeability tensor, such that an increase in the permeability in one direction does not always directly relate to an increase in the other in-plane direction. In the permeability benchmark, experimental permeabilities were also found in the range $0.5\text{-}1 \cdot 10^{-10}$ m^2 for V_f=50% [31].

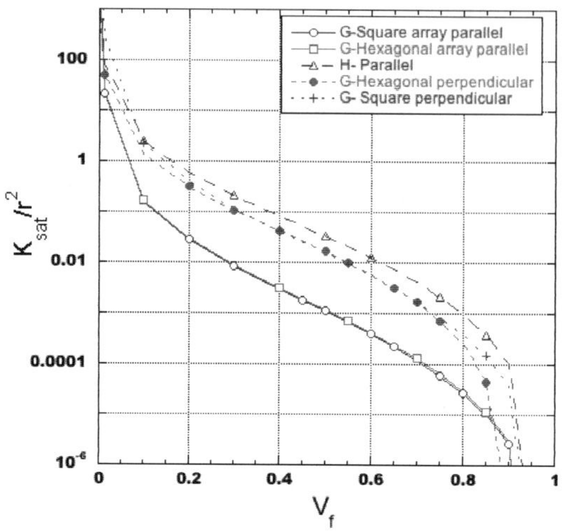

Figure 3. Ratio of permeability over the square of the fiber radius, for flow parallel or orthogonal to the fibers and for a square or a hexagonal array, as a function of overall fiber volume fraction (models given in Ref. [28]).

To model textile permeability, internal geometry of the pore space, including pore size distribution, must be considered. Analytical geometric models have been proposed, where the unit cell is greatly simplified so that a tractable solution of the equations can be written for micro-scale flow within the tow and meso-scale flow between tows, the two being then assembled into a full permeability model using circuit theories or direct flow models [32–40]. These models are useful for rapid evaluation of the influence of tow permeability, or intra-tow channel size on the permeability. A first model by Pillai and Advani in 1995 [32] highlighted the role of the dual scale in the permeability estimation, and was one of the first to indicate that the tow permeability can be neglected (porous cylinders can be treated as solid cylinders) in cases where the separation of scales was sufficiently large. This was confirmed in many following studies, for example Syerko et al. [41] performed a parametric study varying the ratio k/h^2, where k is the permeability in the tow, and h the inter-layer pore size, and Rougier developed an analysis for natural fiber composites [40], illustrated in Figure 4, where a permeability increase by almost two orders of magnitude is predicted when the tow volume fraction increases to 70%, while the overall fiber content remains identical.

As a result from the need to increase impregnation kinetics, in particular if rather high viscosity matrices are used [42] or if large parts are to be manufactured, models have been extended not only to predict permeability of a given reinforcement, but also to engineer new reinforcements with improved permeability [32,35,41]. Several modifications to non-crimp fabrics have thus been reported, with the general aim to increase the intra-tow (or intra-layer) spaces and creates flow channels to enhance permeability. Following an optimization scheme, Syerko et al. reported an increase in in-plane permeability, for a volume fraction of 50%, from $7 \cdot 10^{-11}$ m² for a regular quasi-UD NCF up to $1.1 \cdot 10^{-10}$ m² [41]. Martin et al. [43] performed an experimental study of the influence of design parameters (stich spacing, pattern and weft tow lineal weight) on the permeability tensor for the same quasi-UD NCF glass fabric, representative of materials used in wind turbine blade applications. The UD tows forming the warp layer were stitched with a PET thread onto a weft backing layer, and three variations were investigated as compared to the reference fabric: "A" with a lower stitch spacing (from 5 to 2.5 mm), "B" with an asymmetric stitch pattern (as compared to symmetric) and "C" with a different backing layer (with more spaces as compared to the reference). The results for the full permeability tensor highlighted that a reduction in stitch length (leading to tighter tows) improved the permeability, only below a threshold volume fraction of about 55%, after which the tow permeability started to play

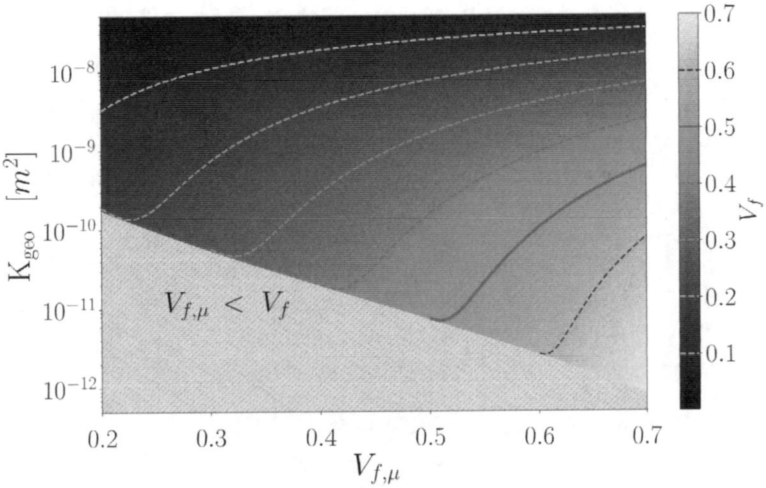

Figure 4. Estimate of the permeability K_{sat} (called here K_{geo}) of a quasi-UD flax fabric as a function of the overall V_f and the local tow volume fraction $V_{f,\mu}$. The red line shows the permeability increase for an overall V_f=50%, when $V_{f,\mu}$ increases [40].

a stronger role. All other parameters showed an unbalanced or minor influence overall, demonstrating the potential adverse effects acting in the spatial directions.

The concept of tight tows separated by large channels was pushed further with the development of highly permeable fabrics for the impregnation of thermoplastic molten matrices, such as G-Ply, as reported by Salvatori et al. [15]. As shown in Figure 5, in comparison to a more regular woven fabric, large flow channels were designed along the x direction of the fabric. As a result, permeability values increased from the usual 10^{-10}m^2 range around V_f=50%, almost to the goal of 10^{-9}m^2. However, in these cases, the flow front during impregnation was strongly affected by the presence of the flow channels, developing a large unsaturated zone with strong delay of the tow impregnation, and thus increased risks of tow porosity, if care is not taken to fully saturate the part, for example by adding a dam towards the flow outlet. As a result, the ratio R_s as defined in Eq. (9) remained close to one above a certain threshold of the flow velocity, where the tow permeability (and capillary effects) did not exert any major influence on the overall permeability value.

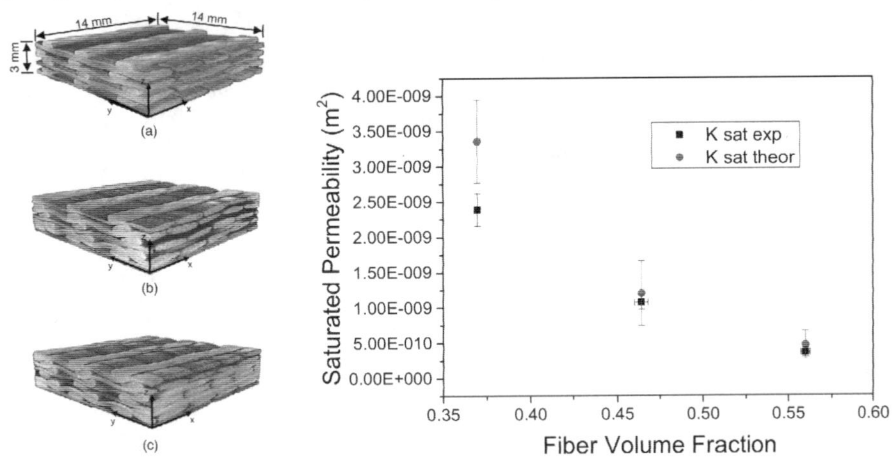

Figure 5. Left: 3D images reconstructed from computer X-ray tomography of G-PLY fabric with (a) 4, (b) 5 and (c) 6 layers stacked in a 3 mm thick mold. Right: Resulting permeability values (adapted from [15]).

5. Influence on the mechanical properties and possible way forward?

Very few reports focus on a direct comparison between the (in-plane) permeability of fabrics and the resulting composite mechanical properties, and it is difficult to establish a direct relation between these two, since the manufacturing method and the choice of resin strongly influence the part properties. However, some general trends can be highlighted.

First, as indicated above, increasing permeability by increasing the separation of scales between tightly packed tows and larger channels, leads to a larger unsaturated region during impregnation, which may increase the risk of porosity in the tow region, if the overall resin velocity is such that viscous flow dominates [20]. This can be minimized by a careful control of the processing conditions, with sufficient resin bleeding, or the presence of less permeable regions towards the resin outlet, to fully saturate the part. Another potential risk, created by the presence of large resin pockets, is to promote residual stresses or micro-cracks to take place during cure or mostly during cool-down of the part. Overall, these resin rich regions have been consistently documented to negatively affect the performance of FRP, as recently reviewed by Mahmood et al. [44], who summarized the role of resin rich regions as location for crack initiation and propagation, as well as their negative effects on static as well as dynamic properties. As a conclusion, they recommend exerting caution when modifying the fiber architecture, and to carefully assess the potential change in structural properties.

A direct comparison was performed by Ohtani et al [45], for several glass NCF with different constructions and tension in the stitches, on permeability and tensile strength, demonstrating a direct correlation between permeability increase (by a factor 5) and reduction in strength from 550 to 485 MPa. Another thorough comparison was performed by Martin [46], on the same glass NCF as presented in Ref. [43]. The fabric with higher permeability "A", due to a reduced stitching length, showed a more uniform rigidity value, but a worse fatigue performance than the reference, due to the resin rich regions which led to higher energy dissipation and damage localization. Rimmel et al. [47] also evaluated the role of stitches on transverse permeability and mechanical properties in textiles positioned by Dry Fiber Placement; results again showed an inverse correlation between increased permeability (from 2.5 10^{-14} m^2 up to 10^{-12} m^2 for 50% fiber content), and reduced bending modulus and strength (up to 30%), with the observation again that cracks initiate at the location of stitches. It is also worth noting that stitches, since they are formed of drawn polymer fibers, may also change their shape with temperature, and thus lead to a change in permeability if resin flow takes place at a temperature above the Tg of the stitching yarns (which would first contract upon heating, then easily creep to loosen the bundles). This implies that permeability should be measured at a temperature close to the processing temperature.

A potential solution to this dilemma lies in the selection of material used for the stitching yarns. These could be engineered such that the presence of large channels could be ensured during infusion, if they can tightly pack the fiber bundles, and then melt or dissolve into the resin and let the initially compressed fiber tows expand and homogenize the fiber distribution. As an example, Beier et al. [48,49] proposed an elegant solution with the use of phenoxy or polyamide stitches, which melt and phase-separate within the epoxy resin to allow partial rearrangement of the reinforcement fibers, while toughening the matrix phase. They compared biaxial NCF textiles that are stitched either with polyester as reference material or with the novel yarns, processed by RTM with RTM6 aerospace grade epoxy resin. They did not quantify the permeability but performed microstructural observations, indicating that the dissolvable yarns had indeed disappeared; however, some resin rich regions remained, highlighting the potential damage or irreversible fiber movement created by the stitching operation, which cannot be fully recovered by fiber rearrangement, even though the stitches melted into the epoxy resin pool. Fracture toughness of the resin was improved by 20% when using 10 wt% phenoxy polymer added to the RTM6 resin as compared to the neat RTM6, showing the potential benefit of the addition of the thermoplastic phase into the resin. In composites, compression, and tensile moduli, as well as tensile strength were not much affected, while compression strength was improved by about 10% as compared to polyester stitched NCF, and Tg slightly reduced, indicating the potential double effect from homogenization of the fiber structure and matrix modification. All values remained however lower than a non-stitched reference, indicating that fiber misalignment nonetheless remained. This solution seems at this time possibly too costly for wind turbine blade applications, although potentially very promising if dissolvable stitching yarns can be further developed, together with the appropriate processing temperature (to avoid premature yarn melting) and kinetics to ensure that the yarns dissolve and that the bundles have time to expand before the resin viscosity increases too much.

As an alternative, dissolvable second phases could be introduced directly within the fabric stack. Beier [49] also evaluated the combination of dissolvable stitching yarns and powdered binders, so as to limit the negative effect of the increased stitching density. Results were very promising, including fatigue resistance, showing a potential route towards preform optimization by a careful selection of binder and stitching material. Also, as expected, the energy release rate in mode I, G_{Ic}, was well improved when using polyamide (3 times higher than reference) or phenoxy stitches (2 times higher than reference) with bindered crack plane, as compared to the reference non-stitched material. A more radical approach was demonstrated by Salvatori who created large flow channels to increase the kinetics of thermoplastic RTM processes [50,51], by the use of 3D printed polymer spacers, which could collapse and melt after impregnation and allow the compressed fabric stack to expand though thickness to form a more homogenous structure. This again required the use of a dam to stop the channel flow towards the outlet, however this approach could form a suitable concept to optimize flow paths by the introduction of fusible spacer lines. Another, less intrusive approach involves the use of spherical polymer beads, which could be inserted into the fabric between layers, and again melt or dissolve into the resin phase after impregnation. It was shown that spherical inclusions can increase permeability of the fabric stack, if their size lies within an adequate range to increase bundle distortion, whereas introduction of smaller beads reduces permeability, as they nest within the spaces between tows [13].

It is worth mentioning that a very common method to increase permeability is to insert a more permeable layer, for example a fiber mat with a reduced volume fraction and thus increased permeability, within the fabric stack, but this comes at the cost of a reduced overall fiber content and mechanical performance in rigidity, hence requires a careful part optimization [52].

Finally, if the fabric is to be left unperturbed, the least perturbing approach to increase the flow kinetics without directly acting on the fiber distribution inside the fabric is to introduce flow channels within the mold [3,53,54], which can be placed in an optimal way to increase the overall impregnation kinetics, depending on the part shape and permeability tensor of the textile stacks. This is of course very similar to the use of highly permeable layers commonly used in vacuum infusion processing of large-scale composite parts, to impregnate mostly through thickness, with the drawback of a heavy use of consumables and some unused resin which remains in this sacrificial layer.

6. Conclusions

This non-exhaustive review highlighted the antagonism between process kinetics and structural properties that may arise when producing large parts with LCM. On one hand, flow models have demonstrated that a clearly suitable approach to increase permeability of textiles while preserving a similar fiber content is to increase the separation of scales between highly packed tows and large flow channels. On the other hand, the presence of large channels influences the flow patterns, thereby creating large unsaturated zones and increasing the risk of incomplete tow impregnation. In addition, resin rich regions in the final part have a detrimental effect on the strength and fatigue life of the part, by increasing the risk of crack formation, localization, and propagation. As a result, processing and structural properties are strongly linked and should be analyzed together, to reach a suitable compromise for a given application. The creation of preferential channels in the mold, or the use of more permeable layers are current effective methods to adjust the flow kinetics but may also lead to less structural parts. Alternatively, novel fabrics with dissolvable stitches, binders or second phases could allow a permeability increase during the impregnation step, followed by a homogenization of the fiber distribution during the curing phase, and thus form elegant solutions to this almost irresolvable conflict.

7. References

[1] Martinez-Marquez D, Florin N, Hall W, Majewski P, Wang H, Stewart RA. State-of-the-art review of product stewardship strategies for large composite wind turbine blades. Resour Conserv Recycl Adv 2022;15:200109. https://doi.org/10.1016/j.rcradv.2022.200109.
[2] Mishnaevsky L, Branner K, Petersen HN, Beauson J, McGugan M, Sørensen BF. Materials for wind turbine blades: An overview. Materials (Basel) 2017;10:1–24. https://doi.org/10.3390/ma10111285.
[3] Brouwer WD, Herpt ECFC Van, Labordus M. Vacuum injection moulding for large structural applications 2003;34:551–8. https://doi.org/10.1016/S1359-835X(03)00060-5.
[4] Santschi Y, Eaton N, Verheyden S, Michaud V. Structural optimisation of payload fairings. Eur. Sp. Agency, (Special Publ. ESA SP, vol. 691 SP, 2012.
[5] Black S. https://www.compositesworld.com/articles/an-elegant-solution-for-a-big-composite-part 2003.
[6] Gardiner G., Bombardier wins award for resin transfer infusion wing, (15/10/2019), CompositesWorld, https://www.compositesworld.com/news/bombardier-wins-uk-royal-academy-of-engineering-award-for-resin-transfer-infusion-wing.
[7] Sloan J., Large high volume infused composite structures on the aerispace horizon, Composites World(2019) https://www.compositesworld.com/articles/large-high-volume-infused-composite-structures-on-the-aerospace-horizon.
[8] Schubel PJ. Cost modelling in polymer composite applications: Case study - Analysis of existing and automated manufacturing processes for a large wind turbine blade. Compos Part B Eng 2012;43:953–60. https://doi.org/10.1016/j.compositesb.2011.11.036.
[9] Bear J. Dynamics of fluids in porous media. Elsevier. New York: 1972.
[10] Straumit I, Lomov S V, Wevers M. Quantification of the internal structure and automatic generation of voxel models of textile composites from X-ray computed tomography data. Compos Part A Appl Sci Manuf 2015;69:150–8. https://doi.org/https://doi.org/10.1016/j.compositesa.2014.11.016.
[11] Straumit I, Hahn C, Winterstein E, Plank B, Lomov S V., Wevers M. Computation of permeability of a non-crimp carbon textile reinforcement based on X-ray computed tomography images. Compos Part A Appl Sci Manuf 2016;81:289–95. https://doi.org/10.1016/J.COMPOSITESA.2015.11.025.
[12] Soltani P, Zarrebini M, Laghaei R, Hassanpour A. Prediction of permeability of realistic and virtual layered nonwovens using combined application of X-ray μCT and computer simulation. Chem Eng Res Des 2017;124:299–312. https://doi.org/10.1016/J.CHERD.2017.06.035.
[13] Caglar B, Orgéas L, Rolland du Roscoat S, Sozer EM, Michaud V. Permeability of textile fabrics with spherical inclusions. Compos Part A Appl Sci Manuf 2017;99:1–14. https://doi.org/10.1016/j.compositesa.2017.03.031.
[14] Ali MA, Umer R, Khan KA, Cantwell WJ. In-plane virtual permeability characterization of 3D woven fabrics using a hybrid experimental and numerical approach. Compos Sci Technol 2019;173:99–109. https://doi.org/10.1016/j.compscitech.2019.01.030.
[15] Salvatori D, Caglar B, Teixidó H, Michaud V. Permeability and capillary effects in a channel-wise non-crimp fabric. Compos Part A Appl Sci Manuf 2018;108. https://doi.org/10.1016/j.compositesa.2018.02.015.
[16] Michaud V. A Review of Non-saturated Resin Flow in Liquid Composite Moulding processes. Transp Porous Media 2016. https://doi.org/10.1007/s11242-016-0629-7.
[17] Teixidó H, Staal J, Caglar B, Michaud V. Capillary Effects in Fiber Reinforced Polymer Composite Processing: A Review. Front Mater 2022;9:1–24. https://doi.org/https://doi.org/10.3389/fmats.2022.809226.

[18] Teixido Pedarros H. Visualization and modelling of dynamic flow in fibrous preforms for liquid composite molding. PhD thesis 10112, EPFL, 2023.

[19] Verrey J, Michaud V, Månson J-AE. Dynamic capillary effects in liquid composite moulding with non-crimp fabrics. Compos Part A Appl Sci Manuf 2006;37. https://doi.org/10.1016/j.compositesa.2005.04.011.

[20] Teixidó H, Broggi G, Caglar B, Michaud V. Measurement and modelling of dynamic fluid saturation in carbon reinforcements. Compos Part A Appl Sci Manuf 2023;169:107520. https://doi.org/10.1016/j.compositesa.2023.107520.

[21] Michaud V. A Review of Non-saturated Resin Flow in Liquid Composite Moulding processes. Transp Porous Media 2016;115:581–601. https://doi.org/10.1007/s11242-016-0629-7.

[22] Teixidó H, Staal J, Caglar B, Michaud V. Capillary Effects in Fiber Reinforced Polymer Composite Processing : A Review 2022;9:1–24. https://doi.org/10.3389/fmats.2022.809226.

[23] Chai BX, Eisenbart B, Nikzad M, Fox B, Blythe A, Blanchard P, et al. Simulation-based optimisation for injection configuration design of liquid composite moulding processes: A review. Compos Part A Appl Sci Manuf 2021;149:106540. https://doi.org/10.1016/j.compositesa.2021.106540.

[24] Dei Sommi A, Lionetto F, Maffezzoli A. An Overview of the Measurement of Permeability of Composite Reinforcements. Polymers (Basel) 2023;15. https://doi.org/10.3390/polym15030728.

[25] Arbter R, Beraud JM, Binetruy C, Bizet L, Bréard J, Comas-Cardona S, et al. Experimental determination of the permeability of textiles: A benchmark exercise. Compos Part A Appl Sci Manuf 2011;42. https://doi.org/10.1016/j.compositesa.2011.04.021.

[26] Vernet N, Ruiz E, Advani S, Alms JB, Aubert M, Barburski M, et al. Experimental determination of the permeability of engineering textiles: Benchmark II. Compos Part A Appl Sci Manuf 2014;61:172–84. https://doi.org/10.1016/j.compositesa.2014.02.010.

[27] Michaud V. Permeability properties of composite reinforcements. Compos Reinf Optim Perform 2020:443–72. https://doi.org/10.1016/B978-0-12-819005-0.00014-9.

[28] Michaud V. Permeability properties of reinforcements in composites. In: Boisse P, editor. Compos. Reinf. Optim. Perform., Woodhead Publishing limited; 2011, p. 431–58.

[29] Karaki M, Younes R, Trochu F, Lafon P. Progress in Experimental and Theoretical Evaluation Methods for Textile Permeability. J Compos Sci 2019;3:73. https://doi.org/10.3390/jcs3030073.

[30] Rieber G, Jiang J, Deter C, Chen N, Mitschang P. Influence of textile parameters on the in-plane Permeability. Compos Part A Appl Sci Manuf 2013;52:89–98. https://doi.org/10.1016/J.COMPOSITESA.2013.05.009.

[31] Vernet N, Ruiz E, Advani S, Alms JB, Aubert M, Barburski M, et al. Experimental determination of the permeability of engineering textiles: Benchmark II. Compos Part A Appl Sci Manuf 2014;61. https://doi.org/10.1016/j.compositesa.2014.02.010.

[32] Pillai KM, Advani SG. Numerical and Analytical Study to Estimate the Effect of Two Length Scales upon the Permeability of a Fibrous Porous Medium. vol. 21. KluwerAcademic Publishers; 1995.

[33] Boming W, Lee LJ. A Simplified In-Plane Permeability Model for Textile Fabrics, Polymer Composites, 21(5), 2000, pp. 660-685.

[34] Endruweit A, Long AC. Composites : Part A A model for the in-plane permeability of triaxially braided reinforcements. Compos Part A 2011;42:165–72. https://doi.org/10.1016/j.compositesa.2010.11.003.

[35] Salvatori D, Caglar B, Teixidó H, Michaud V. Permeability and capillary effects in a channel-wise non-crimp fabric. Compos Part A Appl Sci Manuf 2018;108:41–52. https://doi.org/10.1016/j.compositesa.2018.02.015.

[36] Lundström T. S. The permeability of non-crimp stitched fabrics. Compos Part A Appl Sci Manuf 2000;31:1345–53.

[37] Chen ZR, Ye L, Lu M. Permeability predictions for woven fabric preforms. J Compos Mater 2010;44:1569–86. https://doi.org/10.1177/0021998309355888.
[38] Shou D, Ye L, Fan J. Longitudinal permeability determination of dual-scale fibrous materials. Compos Part A Appl Sci Manuf 2015;68:42–6. https://doi.org/10.1016/j.compositesa.2014.09.018.
[39] Zhou F, Kuentzer N, Simacek P, Advani SG, Walsh S. Analytic characterization of the permeability of dual-scale fibrous porous media. Compos Sci Technol 2006;66:2795–803. https://doi.org/https://doi.org/10.1016/j.compscitech.2006.02.025.
[40] Rougier V. Modélisation multi-échelle de l'imprégnation d'un milieu fibreux : morphologie, mouillage et perméabilité. PhD thesis, Normandie Université, 2021. https://www.theses.fr/2021NORMLH02.
[41] Syerko E, Binetruy C, Comas-Cardona S, Leygue A. A numerical approach to design dual-scale porosity composite reinforcements with enhanced permeability. Materials and Design, 2017;131:307–22. https://doi.org/10.1016/j.matdes.2017.06.035.
[42] Studer J, Dransfeld C, Jauregui Cano J, Keller A, Wink M, Masania K, et al. Effect of fabric architecture, compaction and permeability on through thickness thermoplastic melt impregnation. Compos Part A Appl Sci Manuf 2019;122:45–53. https://doi.org/10.1016/J.COMPOSITESA.2019.04.008.
[43] Martin B, Comas-Cardona S, Binetruy C, Billon N, Bouvard JL, Lucas P. Influence of fabrics' design parameters on the morphology and 3D permeability tensor of quasi-unidirectional non-crimp fabrics. Compos Part A Appl Sci Manuf 2016;90:470–9. https://doi.org/10.1016/j.compositesa.2016.08.013.
[44] Mahmood AS, Summerscales J, James MN. Resin-Rich Volumes (RRV) and the Performance of Fibre-Reinforced Composites : A Review, J. Compos.Sci. 2022, 6 (2), 53. https://doi.org/10.3390/jcs6020053.
[45] Ohtani A, Agata S, Kitamura M, Nakai A. Mechanical and permeability properties of NCF with different knitting condition. IOP Conf Ser Mater Sci Eng 2018;406. https://doi.org/10.1088/1757-899X/406/1/012061.
[46] Martin B. Etudes expérimentales de l'influence des paramètres de conception des renforts NCF unidirectionnels sur leurs propriétés de transport et leurs performances mécaniques, PhD thesis, École Nationale Supérieure des Mines, 2015. https://www.theses.fr/2015ENMP0084
[47] Rimmel O, May D, Mitschang P. Impact of stitching on permeability and mechanical properties of preforms manufactured by dry fiber placement. Polym Compos 2019;40:1631–42. https://doi.org/10.1002/pc.24911.
[48] Beier U, Wolff-Fabris F, Fischer F, Sandler JKW, Altstädt V, Hülder G, et al. Mechanical performance of carbon fibre-reinforced composites based on preforms stitched with innovative low-melting temperature and matrix soluble thermoplastic yarns. Compos Part A Appl Sci Manuf 2008;39:1572–81. https://doi.org/10.1016/j.compositesa.2008.06.003.
[49] Beier U, Sandler JKW, Altstädt V, Spanner H, Weimer C. Mechanical performance of carbon fibre-reinforced composites based on stitched and bindered preforms. Compos Part A Appl Sci Manuf 2009;40:1756–63. https://doi.org/10.1016/j.compositesa.2009.08.012.
[50] Salvatori D, Caglar B, Michaud V. 3D spacers enhance flow kinetics in resin transfer molding with woven fabrics. Compos Part A Appl Sci Manuf 2019;119. https://doi.org/10.1016/j.compositesa.2019.01.023.
[51] Gomez C, Salvatori D, Caglar B, Trigueira R, Orange G. Resin Transfer molding of High-Fluidity Polyamide-6 with modified Glass-Fabric preforms, Compos Part A Appl Sci Manuf 2021;147. https://doi.org/10.1016/j.compositesa.2021.106448.
[52] Luo G, Chen K, Hsu C. Acquisition of Key Vacuum-Assisted Resin Transfer Molding Parameters through Reverse Scanning for Application in the Manufacturing of Large Fiber-Reinforced-Plastic Products, International Journal of Polymer Science Volume 2023 | Article ID 7927196 | https://doi.org/10.1155/2023/7927196; 2023.

[53] Wang J, Simacek P, Advani SG. Use of medial axis to find optimal channel designs to reduce mold filling time in resin transfer molding. Compos Part A Appl Sci Manuf 2017;95:161–72. https://doi.org/10.1016/j.compositesa.2017.01.003.

[54] Kracke C, Nonn A, Koch C, Nebe M, Schmidt E, Bickerton S, et al. Interaction of textile variability and flow channel distribution systems on flow front progression in the RTM process. Compos Part A Appl Sci Manuf 2018;106:70–81. https://doi.org/10.1016/j.compositesa.2017.12.010.

Acknowledgments

I would like to acknowledge the current and recent LPAC colleagues involved with permeability measurement or models for their contributions and fruitful discussions: V. Rougier, H. Teixido, J. Staal, B. Caglar, D. Salvatori, C. Gomez, R. Trigueira as well as funding sources: SNF n° 200021_182669, and the Swiss Competence Center for Energy Research: Mobility.

IOP Conf. Series: Materials Science and Engineering 1293 (2023) 012004 doi:10.1088/1757-899X/1293/1/012004

Proposed holistic strategy for mechanical recycling of wind turbine blades for 3D printing and compression molding

Larry Lessard[1], Zhengshu Yan[1] and Javad Nasiry[2]

[1] McGill University, Department of Mechanical Engineering, 817 Sherbrooke St. West, Montreal, H3A 0C3 Canada
[2] McGill University, Faculty of Management, Montreal, Canada 1001 Sherbrooke St. West, Montreal, H3A 1G5 Canada

Larry.Lessard@mcgill.ca

Abstract. There are many solutions being proposed for recycling discarded wind turbine blades, but most solutions are not very ecological, involving either burning or harsh chemicals. In this research, a solution is proposed that has a very low environmental impact and has potential of being scaled to a very large industrial process. The entire wind turbine blade should be recycled and transformed into other products that are recycled and that have economic benefits in the long term. Two of the output products from the current recycling strategy are recycled 3D printing filament and recycled reinforced pellets for compression molding. The research outlines the process, the numerical simulations that help optimize the output and the outline for planning to bring the recycling solution from a small industrial scale to a large scale. This represents ongoing research in this important field of composite recycling.

1. Motivation

1.1 Turbine blade recycling

The extensive use of composites for wind turbine blades and the accelerating growth of the wind energy industry has led to significant quantities of material waste and undesirable environmental impact [1, 2]. Although most (around 85%) of wind turbines (like housing and tower parts) are made of recyclable materials like copper and steel [3], non-biodegradable materials such as fiberglass, plastic polymer, and core materials are used to make these high-performance wind turbine blades to enhance their efficiency [4]. Turbine blades have a life span of approximately 25 years, and many of the initial generations of turbines are already reaching their end-of-life. Therefore, used wind turbine blades have become a significant source of composite waste generation [2]. It is predicted that the annual amount of waste from end-of-life wind turbine blades will increase from over 50,000 tonnes in 2021 to over 200,000 tonnes worldwide by 2034 [5].

The most common routes to dispose of end-of-life turbine blades involve landfill and incineration, which pollute the soil, underground water, and air [6-8]. The gradual lower availability of landfill sites, the sharp increase in landfill cost, and imminent recycling legislation further reduce the above methods' future viability [5, 6, 9, 10]. In addition, due to the use of hazardous materials, the high cost, the inevitable substantial loss in mechanical properties, and the inability to handle large quantities of waste, chemical and thermal techniques show some promise but reveal a lot of drawbacks as solutions for recycling [11-15]. As the issue of composite waste from the rotor blades has become more pressing, a more ecological methodology with a high applicability and commercialization ability is at a premium.

1.2 Predictive models

The use of predictive models such as numerical models, finite element models, etc., are necessary to facilitate the design process and material assessment in the recycling solution of wind turbine blades. Some of the numerical models are helpful because the methods are well-studied, and there are no restrictions on the composite's geometry, material properties, and fiber angles [16]. As the current recycling solution proposed here uses discontinuous fiber recycled composites, the focus of this research is specifically on these types of materials. For discontinuous random fiber reinforcement composites, the material behaviour and performance are affected by several factors: fiber aspect ratio, fiber orientation, reinforcement shape, etc. [17-19]. Moreover, other factors, such as voids and particles, reduce the accuracy of numerical models. In this sense, using finite element analysis (FEA) can provide good potential property prediction for stiffness and strength, but only if they include all pertinent aspects of the problem.

Explicit modelling of a 3D-printed composite component with all the features mentioned above can be computationally prohibitive. Enormous computational resources and impracticalities are required to obtain a convergent solution when trying to model fibers, particles, and voids in the resulting recycled material. To overcome this hurdle, mechanical analysis methods can be beneficial to obtain the effective mechanical properties of the material by using methods that involve a Representative Volume Element (RVE) [20]. An effective RVE must be sufficiently large to include enough variety of reinforcements. In addition, the effective properties of the RVE must statistically represent the actual material properties on a macroscopic scale [21]. At the same time, increasing the size of the RVE comes with a higher computational cost. Therefore, the size of RVE should be minimized for computational efficiency but at the same time must guarantee accuracy.

1.3 Life cycle analysis

To ensure the sustainable development of wind power, it is essential to better understand the potential life-cycle environmental impact of managing the waste generated during the manufacture, operation and end-of-life of wind turbines. The optimization of potential recycling site locations for blade waste is an integral part. Although this is the least developed aspect of the current research project, it is worth mentioning even at this stage because of its essential nature in comprehensively developing this large industrial recycling operation.

1.4 Application to 3D printing and compression molding

Most commercial 3D printing used PLA, ABS, or some similar thermoplastic material as the 3D printing feedstock. Although some 3D printing is done with continuous reinforcing filament, this technique remains a niche operation. The bulk of 3D printing users would greatly benefit from a 3D printing material that has enhanced reinforced properties as well as "predictable" properties. This is the motivation for the 3D printing aspect of this project. However, this entails producing 3D printing filament with an ideal reinforcing fiber average length to maximize the benefits. Fiber lengths that fall outside of the ideal range would be used to make other reinforced pellets that would be used in compression molding of parts. Industrial applications for compression molding would be targeted. In this way, all the fibers produced by the grinding operations would find their way into one of the two categories of manufacturing methods (3D printing and/or compression molding).

2. Objective

The aim of this research includes but is not limited to three phases: I. Develop an optimum process for recycling glass fibers from scrap wind turbine blades as reinforcements in 3D printing and compression molding based on low environmental impact of the process and high consistent improvements in the properties of the recycled materials. II. Establish novel algorithms and predictive models to facilitate the design process and material assessment based on Phase I. III. Propose life cycle assessments for the above recycling solutions (Phase I) and optimize a plan for recycling site locations in Canada. The distinct project phases are defined and summarized below. Previous work and relevant results can be found in [22-27].

I. **Develop an optimum process for recycling glass fibers from end-of-life wind turbine blades as reinforcements in 3D printing and compression molding.**
   a) Raw material acquisition: Grind turbine blade pieces to obtain recyclate fibers with a cost-effective and efficient method.
   b) Sieving: Sieve the recyclate fibers to obtain fibers featuring a suitable length range for both the 3D printing process and compression molding. Optimize the sieving process to obtain the most fibers within a short sieving time.
   c) Pelletizing: Pelletize fibers for the use of 3D printing and compression molding.
   d) 3D printing: Develop and validate the optimum extrusion process parameters to produce filaments with a consistent diameter for use in 3D printing.
      - Evaluate the compatibility of the recycled glass fibers with the base polymer.
      - Eliminate air bubbles in the preform by dehydrating the polymer.
      - Optimize the extrusion process parameters to produce filaments with a tolerance of ± 0.05 mm in diameter.
      - Conduct Differential Scanning Calorimetry (DSC) and rheology tests on the pure and reinforced thermoplastic filaments to identify the processing and operation window.
      - Compare the mechanical properties of the pure and reinforced thermoplastic filaments with the available commercial filament.
      - Optimize the design and manufacturing parameters of 3D printing.
      - Print tensile test specimens in compliance with ASTM standards to determine the tensile properties of the parts made of pure and reinforced thermoplastics.
      - Measure other important properties through bending tests, impact tests, fracture toughness tests, etc.
   e) Compression molding: Develop and validate the optimum parameters of the compression molding process to produce parts with enhanced mechanical properties.

- Evaluate the compatibility of the recycled glass fibers with the base polymer.
- Eliminate air bubbles by dehydrating the polymer.
- Conduct Differential Scanning Calorimetry (DSC) and rheology tests on the pure and reinforced thermoplastic preforms to identify the processing and operation window.
- Compare the mechanical properties of the pure and reinforced thermoplastic parts with the available commercial parts.
- Optimize the design and manufacturing parameters of compression molding.
- Prepare tensile test specimens in compliance with ASTM D638 standards to determine the tensile properties of the parts made of pure and reinforced thermoplastics.
- Measure other important properties through bending tests, impact tests and fracture toughness tests.

II. Establish novel algorithms and predictive models to facilitate the design process and material assessment of the above recycling solutions.

a) Novel algorithms: Propose new modified random sequential adsorption (MRSA) methods for the efficient generation of the RVEs with hybrid and arbitrary-geometry reinforcements while satisfying geometric periodicities and no intersections among entities without tedious looping or Boolean calculations.
b) Predictive models: Establish representative volume element (RVE) models and perform stress and strength analysis of recycled fibre composites fabricated by 3D printing and compression molding, making use of PYTHON, ABAQUS, Moldflow and PAM-RTM.
c) Boundary conditions: Improve the PYTHON code from EasyPBC to enable the ease of enforcing periodic boundary conditions (PBCs) without periodic meshes.
d) Structural characteristics: Fibers, resin particles, voids, etc. will be characterized by using Microscopy, Micro-CT, ImageJ and Avizo.

III. Propose life cycle assessments for the above recycling solutions and optimizing a plan for recycling site locations in Canada (as an example of a collection of wind turbines)

a) Data acquisition: Collect and verify data related to turbine blade waste in Canada.
b) End-of-life models: Establish mathematical models to predict the amount of blade waste in Canada from now until at least 2050.
c) Impact assessment: Access the impact of incineration and landfill of blade waste.
d) Recycling sites: Use existing data and machine learning techniques to initially plan and optimize the location of recycling sites in Canada.
e) Mechanical recycling: Access the impact of the above recycling solutions.

3. Proposed methodology

3.1 Phase I

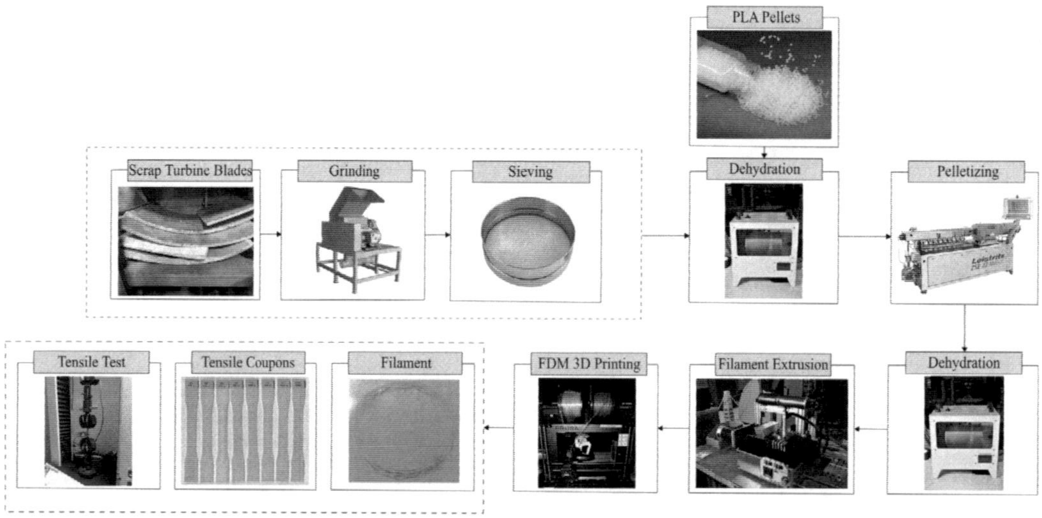

Figure 1. The proposed methodology for recycling wind turbine blades.

The proposed methodology for the recycling of wind turbine blades can be found in **Figure 1**. A three-stage grinding methodology will be applied to obtain the fibers for the filament extrusion. Of the three known methods for recovering fibers (grinding, pyrolysis, and chemical methods like solvolysis), grinding is the method that uses up the least energy by far and thus has the lowest environmental impact on the fiber-recovery process. This is according to a study by Hani et al. [28] published in 2022 and the resulting energy results are shown below in **Figure 2.**

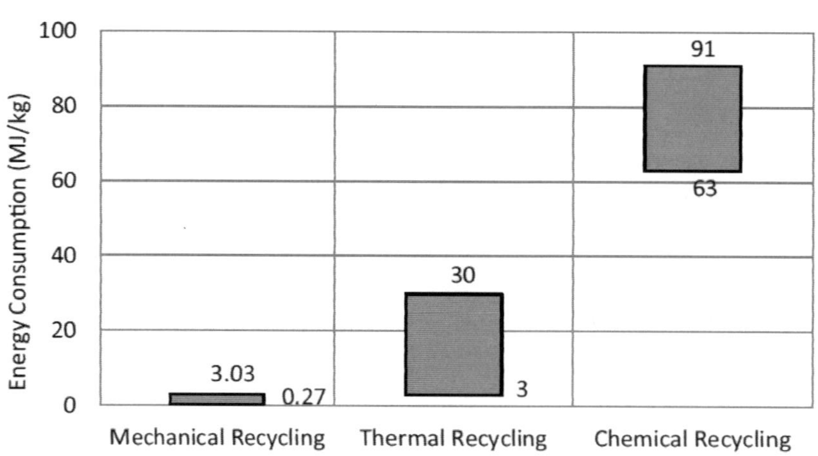

Figure 2. Energy consumption for three different methods for recovering fibers from composite materials [28].

a) Grinding: The scrap turbine blades will be first cut down into smaller pieces (20 cm × 20 cm) and placed in a grinding machine (ECOWOLF, INC.) which contains a hammer mill system and a metal classifier with a hole diameter of 3 mm.
b) Double sieving: A double-sieving mechanism will be demonstrated to classify the recycled materials (a mixture of fibers and resin powder) from the grinder, and two grades of granulated materials will be obtained. The granulated materials obtained from the initial grinding process will be sieved through a stainless-steel screen with a hole diameter of 0.1 mm. The larger-sized recycled material will be then re-fed into the sieve shaker for the second sieving operation to extract more fine fibers. Details on the exact range of ideal fiber sizes and the factors which determine the optimum results is proprietary information (patent pending) which cannot be revealed in the current research paper.
c) Visual inspection: Microscopy will be carried out for the recycled materials after sieving to guarantee the fibers are in the desired length range before the filament extrusion process.

In the end, over 60% of the ground recycled material that comes out of the sieve should be usable for the filament extrusion process with a nozzle diameter of 0.7 mm. The remaining sieved residue could be either reground and reused or processed with larger nozzle diameters.

Fibers obtained from turbine blade waste and PLA pellets (Ingeo 4043D, Natureworks LLC, Blair, Nebraska) will be fed into a dehydrator machine for a drying process of 4 hours at 60°C to reduce the moisture content of the pellets to below 250 ppm. The materials will be then placed in a twin-screw extruder (Leistritz ZSE18HP-40D, Nuremberg, Germany) with 8 subzones and a pelletizer to produce glass fiber-reinforced pellets. The pellets will be re-dried and fed into a single screw extruder (FilaFab, D3D Innovations Limited, Bristol, UK) to produce glass fiber-reinforced filaments with a diameter of 1.75 ± 0.05 mm. A laser micrometer with an accuracy of ± 2 µm will be used to monitor the filament diameter along the extrusion. The single and twin-screw extruder parameters are summarized in **Table 1**.

Table 1. The single and twin-screw extruders parameters (recommended for the specific extruder that was used in the test program)

Twin screw pelletizer	Screw speed (rpm)	90
	Subzone 1-2 Temp. (°C)	190
	Subzone 3 Temp. (°C)	185
	Subzone 4 Temp. (°C)	180
	Subzone 5 Temp. (°C)	175
	Subzone 6-8 Temp. (°C)	170
Single screw filament maker	Screw speed (rpm)	25
	Die temperature Temp. (°C)	210
	Winder speed (rpm)	1

Once filaments are prepared, glass fiber-reinforced specimens with a total thickness of 3.36 mm and a $[0]_{24}$ 3D printing direction sequence will be prepared using a 3D printer Prusa i3 Mk2S. The manufacturing process and design parameters are summarized in **Table 2**. These are typical values only, that were used to get good printing results with the specific 3D printing equipment used in this project.

Table 2. Manufacturing and design parameters for specimen 3D printing.

Manufacturing parameter	Value	Manufacturing parameter	Value
Print direction	XYZ	Nozzle diameter (mm)	0.4
Raster angle (°)	0	Nozzle temperature (°C)	215
Layer height (mm)	0.14	Cooling	No fan cooling
Bed temperature (°C)	60	Infill (%)	100
Print speed (mm/min)	2400	Filament diameter (mm)	1.75

Fibers within an undesired length range (longer fibers) will be used in the compression molding process. As stated before, it is essential to try to recycle all the fibers resulting from the grinding process. Those fibers that are not optimal for 3D printing filament will nevertheless still be useful for making compression molding pellets. Corresponding simulation work will be carried out via Moldflow and PAM-RTM. Plan details will be enriched as the project progresses.

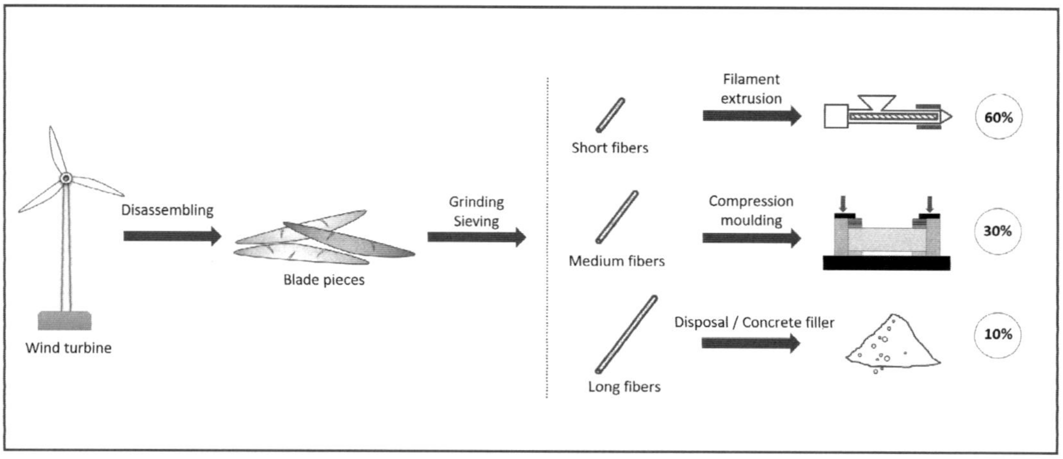

Figure 3. Recycling process showing the three resulting waste streams. These results are with the currently optimized process and are subject to changes (improvements) in the future.

In the end, after careful sieving operations, around 60% of the generated recyclates are short fibers, which will be used for filament extrusion. About 30% are medium fibers which will be used for compression molding while the remaining 10% will be either disposed or used as concrete filler (see **Figure 3**).

3.2 Phase II

The generation of an RVE cuboid, containing fibers and other entities (inner voids and resin particles) is essential to FE analysis of 3D printed filament material and compression molding pellet material.

The RVE cubic volume will be considered first, which will be assigned a size obtained through a sensitivity analysis and comparison with experimental data.

The fibers are modelled as straight cylinders with circular cross-sections of constant radius. This is due to small aspect ratios, low volume fraction [29], and a simplification because non-constant radius fibers would have an insignificant effect on the elastic properties [30]. The constant radius is set as the mean value of the actual fiber radius measured from experiments, and the length range of fibers is obtained following the classification operation. The orientation, distribution, and content of fibers from experiments are also verified. Specifically, Scanning Electron Microscopy (Hitachi UHR Cold-Emission FE-SEM SU8000) and optical microscopy (Nikon, Tokyo, Japan) is used to characterize the fiber orientation and fiber distribution in the filament. The filament is sectioned longitudinally and transversely, where the former is used to check the fiber orientation and the latter reveals the fiber distribution. The samples sectioned longitudinally are immersed in an epoxy resin, ground, and polished in preparation for the microstructural analysis. Grinding is done using 120 grit, 220 grit and 600 grit sandpaper. Polishing is performed using a 10 μm diamond slurry, followed by a 5 μm diamond slurry, and finished with a 0.3 μm alumina suspension. To verify the fiber volume content and fiber orientation in 3D space, a micro-CT scan is performed, and the sample entities are reconstructed in Avizo 3D software. The generation of the fibers is realized through different modified random sequential adsorption (MRSA) algorithms by using ABAQUS and PYTHON scripts.

Inner voids and resin particles emerge during the filament extrusion and 3D printing processes. A micro-CT scan is performed on sample materials, and the sample entities are reconstructed in Avizo 3D software to characterize the inner voids and resin particles. The diameters of the voids and resin particles in the RVE model are assumed to be the average values obtained from the software.

A homogenization procedure is usually applied to the RVE to obtain the overall elastic properties. To apply the periodic boundary conditions on the RVE planes, an identical mesh pattern needs to be created on the opposite faces of the RVE through a copy mesh scheme. However, due to the complex structure of the model, a periodic mesh cannot always be guaranteed. Instead, an interpolation technique is implanted into the plugin tool EasyPBC [31] to impose PBC with non-periodic meshes.

FE analysis of 3D-printed recycled composite parts made from PLA and recycled glass fibers will be performed and the effects of fiber content, fiber aspect ratio, fiber orientation, local fiber density, inner voids, resin particles, etc., will be discussed [27].

An example of the RVE hierarchical structure of recycled glass fiber-reinforced specimens is shown below (**Figure 4**) [27]. All entities with different geometries are generated in the RVE frame while maintaining geometric periodicity. Fibers with different orientations (UD: unidirectional, RD: randomly directional), voids and resin particles are considered in this RVE model.

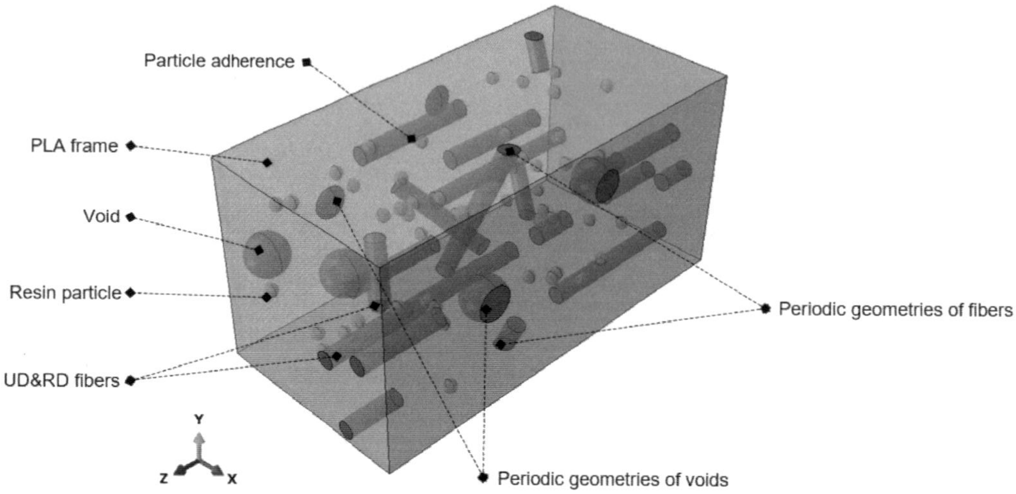

Figure 4. Example of RVE showing all relevant entities: fibers, particles and voids [27].

3.3 Phase III

The initial step before applying the above recycling solutions is the concentration and management of end-of-life blades from different wind farms within the interested area, which can be formulated as a facility location problem (FLP) of recycling sites.

By optimizing the site locations of recycling centres, life cycle assessment with aspects of the dynamic amount of waste resources, transportation among wind farms and recycling sites, relative energy cost and carbon footprint can then be studied.

Facility location plays a crucial role in the strategic planning of various public and private organizations [32]. When addressing the issue of blade waste, it is essential to take into account multiple factors, including distance and cost from the corresponding demand points. In this specific context, the demand points are represented by wind farm locations in Canada.

Among the various models for facility location, the covering problem stands out as a popular and compelling choice. Typically, in covering problems, the services provided by facilities are determined based on the distance between the customer and the facilities. Specifically, a customer can receive service from any facility located within a predetermined coverage distance or coverage radius [33]. Numerous scenarios can be considered as covering problems, such as determining the optimal number and locations of public schools, police stations, libraries, hospitals, public buildings, post offices, parks, military bases, radar installations, branch banks, shopping centers, and waste-disposal facilities [34]. Similarly, the problem of establishing recycling points for blade waste can also be viewed as a covering problem.

The dynamics of facility location problems encompass the potential modification of existing facilities or the development of new ones over time. These dynamic aspects must be taken into consideration to effectively manage changing parameters. To ensure the practical applicability and implementation readiness of relevant models, it is crucial to incorporate the uncertainty of parameters. This allows for a balanced assessment between the benefits brought about by facility location changes and the associated costs of potential modifications.

In the case of recycling sites for blade waste, a notable challenge arises from the inconsistent supplies that fluctuate over decades. Given this scenario, it is common for organizations to contemplate relocating their facilities within a specific time horizon, without causing disruptions to ongoing activities. Thus, the possibility of facility relocation should be factored into the decision-making process, allowing for careful evaluation of associated implications and trade-offs.

Clustering plays an important role in data analysis that is used as a common method in modern scientific research. Several clustering approaches include Centroid-based clustering, Density-based clustering, Distribution-based clustering, Hierarchical clustering, etc.

The proper approach for the blade waste problem could be considered as a density-based clustering (**Figure 5**) since the local density of wind farms and corresponding waste in different regions varies in Canada. In the example shown in **Figure 5**, yellow and blue circles can be considered as wind farms in different regions, while grey spots can be assumed as possible recycling sites in this given optimization problem. Note that the location of the recycling sites is subject to change (relocation) as the waste supply changes.

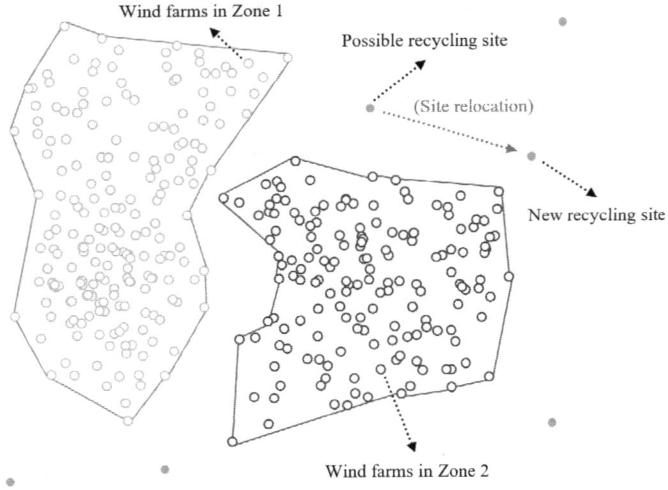

Figure 5. Example of density-based clustering for wind farms.

Information about wind farms in Canada will be accessed first. Key considerations involve:

a) The distribution density of wind farms in local areas within individual provinces varies.
b) Waste production in different wind farms varies (space-dependent).
c) Waste production in each wind farm is dynamic (time-dependent) due to the limited life span of the blades.
d) In addition, the recycling sites must consider transportation issues (total distances, carbon footprint, cost, etc.), dynamic waste supply from wind farms, possible idle periods and facility relocation problems.

After having the wind farm (waste resources) distributions and time-variant waste production in each wind farm, a clustering algorithm can then be applied to the system for finding the optimized recycling site locations by minimizing the transportation cost, energy cost, carbon footprint etc., subject to the system boundary conditions:

a. All recycling sites should be located within mainland Canada.
b. Specific regions such as residential areas (e.g., downtown areas) should be avoided.

c. The amount of waste accepted by each recycling site should not exceed the maximum site capacity.

d. The annual processing volume of each wind farm should be balanced as much as possible.

The analysis steps for plastic waste management from wind turbine blades in Canada are summarized in **Table 3**. The table outlines a digital infrastructure for the management of turbine blade waste and design for the recycling site locations.

Table 3. Analysis steps for waste management from wind turbine blades in Canada.

#	Step	Comments
1	Data Acquisition	Wind farm distribution in Canada, waste generation in each wind farm
2	Set system boundary conditions	Ensure the effectiveness of the design, for example, clarify the region which is feasible for transportation, storage, and processing of blade waste, before optimizing the recycling site locations, otherwise the results from the LCA will be ineffective or inaccurate.
3	Apply clustering algorithm	Wind farms are set into groups and recycling site locations are determined
4	Results of impacts obtained	Impacts should be low after the optimization
5	Check possible modifications	Possible new recycling sites and site relocations

3.4 Summary and discussion

This research offers a comprehensive framework for the recycling of glass fibers from end-of-life wind turbine blades, including experiments (Phase I), simulations (Phase II) and LCA (Phase III). In Phase I, fibers will be recycled through FFF 3D printing and compression molding process. Corresponding simulation work will be carried out in II. Impact of recycling solutions from Phase I will be evaluated in Phase III. Based on the data obtained from Phases I and III, recycling site selection and optimization will be conducted. The recycling process and associated projects are summarized in **Figure 6**.

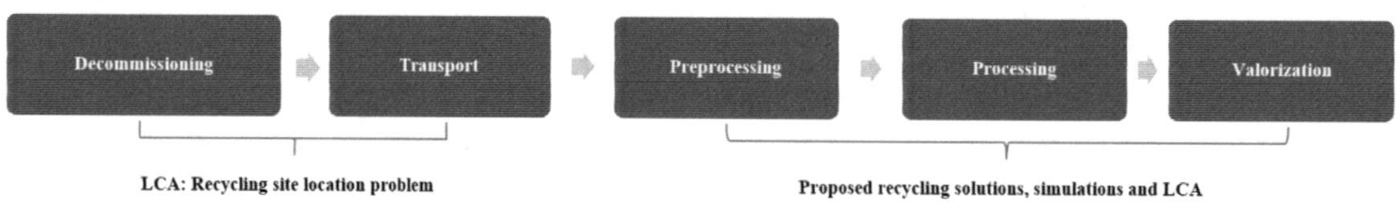

Figure 6. The process of recycling and corresponding projects.

Furthermore, this method uses techniques that can be scaled up to large industrial process. Mechanical grinding machines come in various sizes, thus with larger and multiple grinders, large amounts of fiberglass waste can be decomposed. The sieving process can also be scaled up because the technology for large sieving and sorting operations already existing in the technology of mineral sorting for the mining industry. These sorting machines from the mining industry could be adapted for use in the current fiberglass sorting operation. Finally, in the production of reinforced material for 3D printing, large-scale twin-screw extruders already exist in the polymer processing industry, thus large-scale operations for producing reinforced pellets can be easily developed. These pellets, in turn, can be either used in compression molding industries or fed into industrial-grade filament makers for producing reinforced filament for the 3D printing industry.

It is clear that in this on-going research, not all aspects have been fully developed. The science behind the grinding, sieving, sorting, and production of recycled 3D printing materials is well-advanced and ready for industrial scale automation. However, the first industrial process will be used to recycle a very specific source of similar wind-turbine materials. The industrial process will have to be flexible enough to handle variations in the type of input material (fiber weight percentage, resin type, additives, etc). Life cycle assessment methods will have to take account of these variations and to date, the development of the LCA analysis is not as well defined. This is an area where the research will have to catch up.

4. Expected contributions

The proposed research is expected to result in the following contributions:

- Develop an optimized methodology for transforming the wind turbine blade recyclates into processable pieces for filament extrusion and compression molding processes.
- Develop the optimum process for the extrusion of reinforced thermoplastic filaments as feedstock for 3D printing.
- Develop reinforced FFF 3D printed and compression molded parts through the proper selection of the design and manufacturing process parameters.
- Characterization of the structural properties of 3D printed and compression molded parts made of thermoplastics reinforced with recycled fiberglass.
- Develop novel algorithms and predictive models to facilitate the design process and material assessment of the above recycling solutions.
- Propose life cycle assessments for the above recycling solutions and optimize recycling site locations in Canada.
- Boost wind turbine blade recycling from lab to market.

References

[1] J. Chen, J. Wang, and A. Ni, "Recycling and reuse of composite materials for wind turbine blades: An overview," *Journal of Reinforced Plastics and Composites,* vol. 38, no. 12, pp. 567-577, 2019.

[2] D. S. Cousins, Y. Suzuki, R. E. Murray, J. R. Samaniuk, and A. P. Stebner, "Recycling glass fiber thermoplastic composites from wind turbine blades," *Journal of cleaner production,* vol. 209, pp. 1252-1263, 2019.

[3] L. Mishnaevsky, K. Branner, H. N. Petersen, J. Beauson, M. McGugan, and B. F. Sorensen, "Materials for Wind Turbine Blades: An Overview," (in eng), *Materials (Basel),* vol. 10, no. 11, Nov 9 2017, doi: 10.3390/ma10111285.

[4] H. Albers, S. Greiner, H. Seifert, and U. Kühne, "Recycling of wind turbine rotor blades. Fact or fiction?; Recycling von Rotorblaettern aus Windenergieanlagen. Fakt oder Fiktion?," *DEWI-Magazin,* 2009.

[5] K. Larsen, "Recycling wind turbine blades," *Renewable energy focus,* vol. 9, no. 7, pp. 70-73, 2009.

[6] J. P. Jensen, "Evaluating the environmental impacts of recycling wind turbines," *Wind Energy,* vol. 22, no. 2, pp. 316-326, 2019, doi: 10.1002/we.2287.

[7] K. Ramirez-Tejeda, D. A. Turcotte, and S. Pike, "Unsustainable Wind Turbine Blade Disposal Practices in the United States:A Case for Policy Intervention and Technological Innovation," *NEW SOLUTIONS: A Journal of Environmental and Occupational Health Policy,* vol. 26, no. 4, pp. 581-598, 2017, doi: 10.1177/1048291116676098.
[8] K. Larsen, "Recycling wind," *Reinforced plastics,* vol. 53, no. 1, pp. 20-25, 2009.
[9] S. Job, "Recycling glass fibre reinforced composites–history and progress," *Reinforced Plastics,* vol. 57, no. 5, pp. 19-23, 2013.
[10] Composites UK Trade Association. "End of life options for composite materials." (accessed.
[11] G. Jiang, S. J. Pickering, E. H. Lester, T. Turner, K. Wong, and N. Warrior, "Characterisation of carbon fibres recycled from carbon fibre/epoxy resin composites using supercritical n-propanol," *Composites Science and Technology,* vol. 69, no. 2, pp. 192-198, 2009.
[12] S. Pickering, "Thermal methods for recycling waste composites," in *Management, recycling and reuse of waste composites*: Elsevier, 2010, pp. 65-101.
[13] R. S. Ginder and S. Ozcan, "Recycling of Commercial E-glass Reinforced Thermoset Composites via Two Temperature Step Pyrolysis to Improve Recovered Fiber Tensile Strength and Failure Strain," *Recycling,* vol. 4, no. 2, p. 24, 2019. [Online]. Available: https://www.mdpi.com/2313-4321/4/2/24.
[14] S. Feih, K. Manatpon, Z. Mathys, A. Gibson, and A. Mouritz, "Strength degradation of glass fibers at high temperatures," *Journal of materials science,* vol. 44, no. 2, pp. 392-400, 2009.
[15] A. Nzioka and Y. Kim, "Surface analysis of glass fibres using XPS and AFM: case study of glass fibres recovered from the glass fibre reinforced polymer using chemical recycling," in *Journal of Physics: Conference Series*, 2018, vol. 953, no. 1.
[16] S. Kari, H. Berger, and U. Gabbert, "Numerical evaluation of effective material properties of randomly distributed short cylindrical fibre composites," *Computational Materials Science,* vol. 39, no. 1, pp. 198-204, 2007, doi: 10.1016/j.commatsci.2006.02.024.
[17] B. D. Agarwal, L. J. Broutman, and K. Chandrashekhara, *Analysis and performance of fiber composites*. John Wiley & Sons, 2006.
[18] L. T. Harper, T. A. Turner, N. A. Warrior, and C. D. Rudd, "Characterisation of random carbon fibre composites from a directed fibre preforming process: The effect of fibre length," *Composites Part A: Applied Science and Manufacturing,* vol. 37, no. 11, pp. 1863-1878, 2006, doi: 10.1016/j.compositesa.2005.12.028.
[19] L. T. Harper, T. A. Turner, N. A. Warrior, and C. D. Rudd, "Characterisation of random carbon fibre composites from a directed fibre preforming process: The effect of tow filamentisation," *Composites Part A: Applied Science and Manufacturing,* vol. 38, no. 3, pp. 755-770, 2007, doi: 10.1016/j.compositesa.2006.09.008.
[20] S. Arabnejad and D. Pasini, "Mechanical properties of lattice materials via asymptotic homogenization and comparison with alternative homogenization methods," *International Journal of Mechanical Sciences,* vol. 77, pp. 249-262, 2013.
[21] R. Hill, "Elastic properties of reinforced solids: some theoretical principles," *Journal of the Mechanics and Physics of Solids,* vol. 11, no. 5, pp. 357-372, 1963.
[22] A. Rahimizadeh, J. Kalman, K. Fayazbakhsh, and L. Lessard, "Recycling of fiberglass wind turbine blades into reinforced filaments for use in Additive Manufacturing,"

Composites Part B: Engineering, vol. 175, 2019, doi: 10.1016/j.compositesb.2019.107101.

[23] A. Rahimizadeh, J. Kalman, R. Henri, K. Fayazbakhsh, and L. Lessard, "Recycled Glass Fiber Composites from Wind Turbine Waste for 3D Printing Feedstock: Effects of Fiber Content and Interface on Mechanical Performance," *Materials (Basel),* vol. 12, no. 23, Nov 27 2019, doi: 10.3390/ma12233929.

[24] A. Rahimizadeh, M. Tahir, K. Fayazbakhsh, and L. Lessard, "Tensile properties and interfacial shear strength of recycled fibers from wind turbine waste," *Composites Part A: Applied Science and Manufacturing,* vol. 131, 2020, doi: 10.1016/j.compositesa.2020.105786.

[25] A. Rahimizadeh, J. Kalman, K. Fayazbakhsh, and L. Lessard, "Mechanical and thermal study of 3D printing composite filaments from wind turbine waste," *Polymer Composites,* vol. 42, no. 5, pp. 2305-2316, 2021, doi: 10.1002/pc.25978.

[26] M. Tahir, A. Rahimizadeh, J. Kalman, K. Fayazbakhsh, and L. Lessard, "Experimental and analytical investigation of 3D printed specimens reinforced by different forms of recyclates from wind turbine waste," *Polymer Composites,* vol. 42, no. 9, pp. 4533-4548, 2021, doi: 10.1002/pc.26166.

[27] Z. Yan, A. Rahimizadeh, Y. Zhang, Y. Zhou, and L. Lessard, "A finite element model for 3D printed recycled parts from end-of-life wind turbine blades," *Composite Structures,* vol. 320, 2023, doi: 10.1016/j.compstruct.2023.117177.

[28] H. K. Jani, S. Singh Kachhwaha, G. Nagababu, and A. Das, "A brief review on recycling and reuse of wind turbine blade materials," *Materials Today: Proceedings,* vol. 62, pp. 7124-7130, 2022, doi: 10.1016/j.matpr.2022.02.049.

[29] W. Tian, L. Qi, J. Zhou, J. Liang, and Y. Ma, "Representative volume element for composites reinforced by spatially randomly distributed discontinuous fibers and its applications," *Composite Structures,* vol. 131, pp. 366-373, 2015, doi: 10.1016/j.compstruct.2015.05.014.

[30] A. A. Gusev, P. J. Hine, and I. M. Ward, "Fiber packing and elastic properties of a transversely random unidirectional glass/epoxy composite," *Composites Science and Technology,* vol. 60, no. 4, pp. 535-541, 2000.

[31] S. L. Omairey, P. D. Dunning, and S. Sriramula, "Development of an ABAQUS plugin tool for periodic RVE homogenisation," *Engineering with Computers,* vol. 35, no. 2, pp. 567-577, 2019.

[32] S. H. Owen and M. S. Daskin, "Strategic facility location: A review," *European journal of operational research,* vol. 111, no. 3, pp. 423-447, 1998.

[33] R. Z. Farahani and M. Hekmatfar, *Facility location: concepts, models, algorithms and case studies*. Springer Science & Business Media, 2009.

[34] R. J. Suby, "Facility Layout and Location—An Analytical Approach. By R. L. Francis and J. A. White. (New Jersey : Prentice-Hall Inc., 1974.) [Pp. 468.] Price £8-85," *International Journal of Production Research,* vol. 13, no. 2, pp. 219-219, 1975/01/01 1975, doi: 10.1080/00207547508942989.

IOP Conf. Series: Materials Science and Engineering 1293 (2023) 012005 doi:10.1088/1757-899X/1293/1/012005

Torsional Effects on Wind Turbine Blades and Impact on Field Damages

F. M. Jensen[1], J. O. Arconada[1], M. Werk[1], C. Berggreen[2], J. D. Sørensen[3], G. Zhong[4], W. Haans[4]

[1]Bladena, Banestrøget 13, 2630 Taastrup, Denmark.

[2]Technical University of Denmark, Department of Civil and Mechanical Engineering, Koppels Allé, Building 404, 2800 Kgs. Lyngby, Denmark.

[3]AAU, Department of The Built Environment, Thomas Manns Vej 23, 9220 Aalborg, Denmark.

[4]Shell Global Solutions International B.V., 2596 HR Den Haag, Netherlands.

17th of July

Email: fmj@bladena.com

Abstract.
The purpose of this paper is to present the importance of torsional loads when understanding the significant increase in structural damages on large wind turbine blades. By comparing the stress state of two blades of different length, it is shown that torsional loads grow with an exponent of four (4). Thereby, a function of the length of the blades leading to a new range of challenges that are not met by the current process for design, testing and certification.

The paper presents the study of these new challenges in detail, based upon:
- An experimental perspective, with results from large-scale cyclic loading tests on two blade cut-outs conducted under controlled laboratory conditions in. Both blades were tested to failure.
- A numerical perspective, non-linear geometrical 3D FEM simulations has been performed to study the impact of torsional loads on relevant current structural failure modes on large wind turbine blades.

The conclusion is that the larger torsional loads may explain shear-web disbond, as well as sandwich skin debonding in both the transition zone and in the max chord area of wind turbine blades. In view of the findings, the paper includes proposals for remedies that are expected to mitigate some of the significant failures currently seen in the field.

1. Abbreviations

Leading edge	LE
Trailing edge	TE
Pressure side	PS
Suction side	SS
Cross-Sectional Shear Distortion	CSSD
Root torsional moment	RTM
Root bending moment	RBM
Number of cycles to failure for given cyclic stress of specific material	S-N curves
Trailing edge Towards Leading edge	TTL

Table 1. *List of abbreviations*

2. Introduction

2.1. Background

Publications in recent years have highlighted an increase in structural blade failures (ref.[2.], [5.], [11.]), a fact that has been recognized not only by wind turbine owners, but also by manufacturers and certification bodies (ref.[2.]).

Published articles, see ref. [1.-14.], report reoccurring blade issues on several platforms with large blades on 5MW to 8MW turbines. In some of the reported blade collapses, the operators have shown concern that the critical damage is not a one-off event (ref.[5.]).. Therefore studies have been conducted of whole fleets such as in 2021 at a Norwegian wind farm, see ref.[2] and in 2020 at a wind farm in the USA, see ref.[3]. Furthermore, DNV has recently published a whitepaper acknowledging the relevance of the topic, see ref.[2].

The trend that structural related failures occur early at the large blade's lifetime, suggests that the methodologies used in the development and testing phase of blades need improvements, see ref.[4]. In some instances when similar blade failure happens on the same platform on different wind farms, it can be a clear sign of insufficient testing of the specific blade type, see ref.[5].

The importance of torsional loading is becoming an increasingly significant topic, as found in the studies performed for different blade sizes and presented in this paper. These studies have clearly demonstrated that increasing the length of the blades only adds to the severity of the challenges.

A comprehensive analysis of the importance of the torsional loads for large blades is presented in this paper. According to the current testing standards (ref.[19.]), both edgewise and flapwise load cases are required, but not the operational load case. For large blades, edgewise loading is often the most critical and is therefore used for comparison in this paper. However, neither the edgewise nor the flapwise test load cases include significant torsional loads.

The reason why this topic has been addressed is the rising concerns that out-of-plane panel deformations is a major root cause of critical stresses in adhesive bondlines as well as between the composite layers (interlaminar stresses), especially, the interface between sandwich skin and sandwich core materials. The term sandwich skin is used in this context to refer to the face sheets, thus the outer- or inner-most layers of sandwich panels.

2.2. Torsional Loads Overview

The current design basis considers the loads associated with displacements in the edgewise and flapwise independently and not their combined three-dimensional effect equivalent to what is found under actual operating conditions.

Figure 1 shows the three-dimensional tip deflection in both directions when the blade is in operation. This deflection is a consequence of the combination of flapwise loads due to the aerodynamic impact of the incoming wind, and edgewise loads, due to gravity forces and blade dynamics during rotation. These movements cause forces to act not exactly at the blade's shear centre, thereby creating an arm that generates a Root Torsional Moment (RTM). The direction of this RTM depends on the magnitude of the applied loads and the stiffness of the blade. The load contributions that introduce a root torsional moment on the blade arise from the aerodynamic forces and gravity working on the already deformed blade under operation. As blades increase in length, the tip deflection also increases significantly, leading to notable rise in the RTM. This will be further discussed in the subsequent section of the paper.

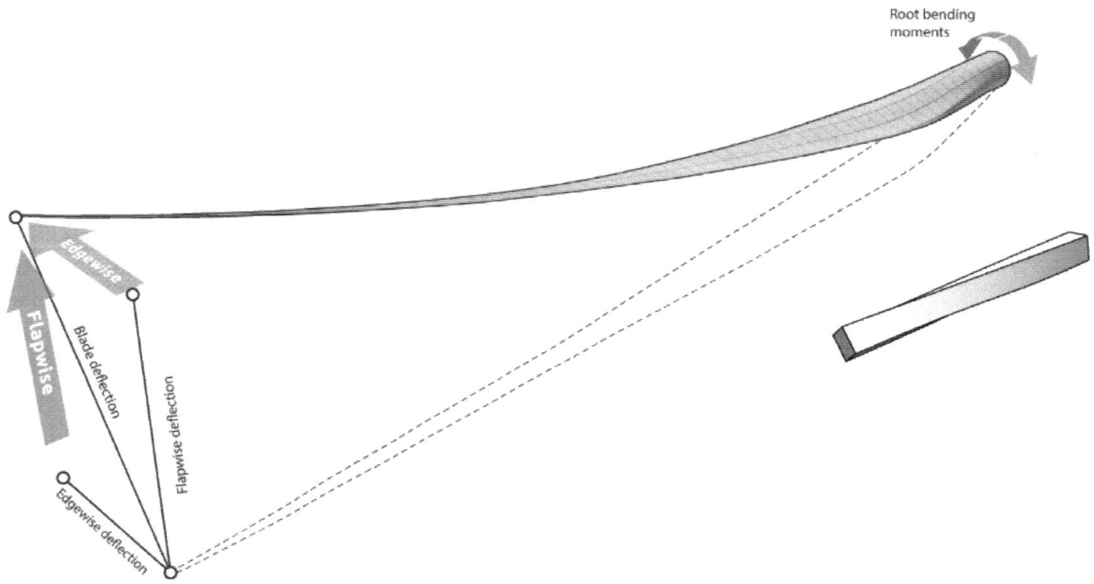

Figure 1. Flapwise and edgewise load components and tip deflection generate Root Torsional Moments (RTM). The orange arrow at the tip represents the edgewise aerodynamic tangential forces and the corresponding root bending moment at the root. The blue arrow at the tip represents the flapwise aerodynamic tangential forces and the corresponding root bending moment at the root. The arrows are representive, the forces act along the whole blade.

Seen from tip towards the root, the edgewise aerodynamic loads working on a flapwise deformed blade twist the blade in a clockwise direction and the flapwise loads generate a counterclockwise twist to the blade. The torsional load components cause localized panel deformations in the transition zone and in the max chord area of the blade (ref.[13.]). As a result of out-of-plane deformation, bending strains in the panels are generated. This occurs mainly in the transition zone and in the max chord area. The deformations also cause peeling stresses in the bondlines of blade designs with an aft shear web.

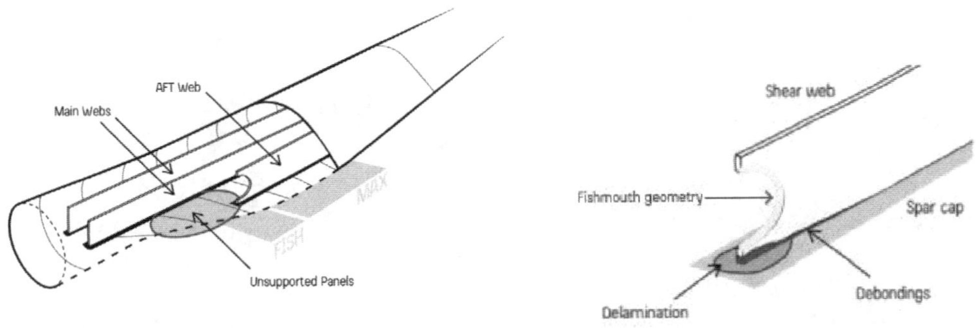

Figure 2. Transition zone of blade with 2 main shear webs and one aft shear web with unsupported panels, the red ellipse illustrating the area where the unsupported panels are located (left). Connection between the aft shear web fishmouth and the spar cap. Highlighted, initiation of delamination and debonding (right).

Localized bending of the sandwich panels in the transition zone (see Figure 3 for further information about the transition zone), often results in peeling in the bondlines of blade designs with an aft shear web. This occurs more frequently due to the peeling stresses caused by the out-of-plane deformation of the unsupported panels, which are carried by the bondlines of the aft shear web. In blades without an aft shear web, the peeling stresses are carried by the bondlines in the trailing edge and at the main shear web.

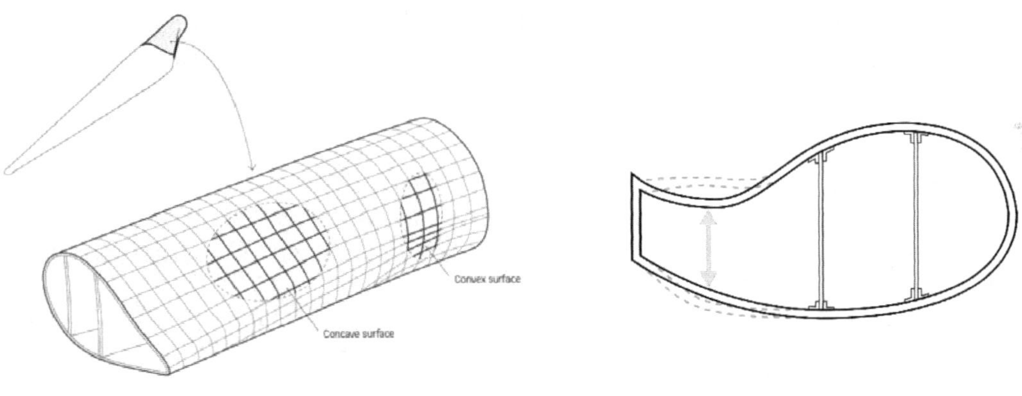

Figure 3. Transition zone of a large blade where the complex geometry can be appreciated, moving from a convex to concave shape, and changing in tapering in 3 directions. On the right, localized bending of the panels is represented, the solid lines represent the un-deformed positions, and the dashed lines present the deformed positions.

Furthermore, the torsional loads cause additional cross-sectional shear distortion (CSSD), which results in additional peeling stresses (ref. [13.]) being transferred to the bondlines, see Figure 4. This phenomenon is in many cases driving structural related failures early in the lifetime of large wind turbine blades, see ref. [1]. In a previous study conducted by Bladena, it was concluded that both cross sectional shear distortion, and especially out-of-plane deformation of the panels in the transition zone, are the two main root causes of shear-web disbonding in the transition zone, see ref.[9].

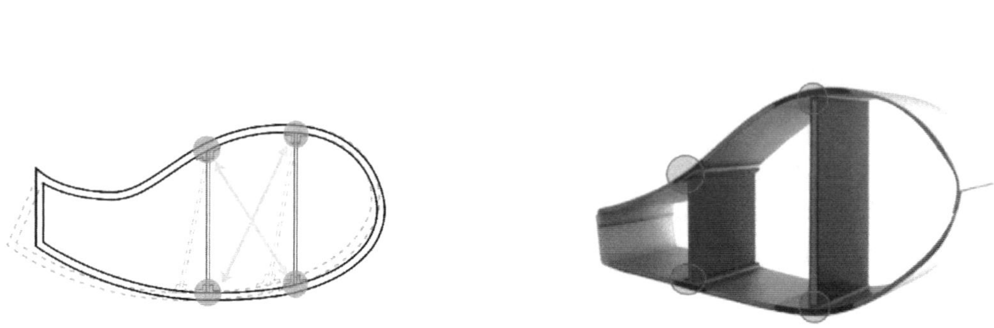

Figure 4. Blade cross-section. On the left, the solid lines represent the un-deformed position wand the dashed lines show the deformed position. On the right, bondlines highlighted in red circles on a blade cross section.

The significance of the torsional loads and the associated deformations is tied to the fact that they increase with a power of 4 in relation with the length of the blades. This will be presented in the paper in more details in section 4. Hence, what was considered an acceptable risk for a certain size of blade may become out of control as the blades increase in length, even leading to failures shortly after the blades have entered into operation.

The paper presents results of non-linear geometric 3D-solid FEM simulations to quantify these phenomena.

3. Experimental analysis: Large-scale testing applying torsional fatigue loads

3.1. Background and experimental methods

A large-scale test campaign was performed by utilizing the strong-floor test facilities. The purpose of the test campaign was to demonstrate the impact of torsional loads on a blade's transition zone region. A detailed description of the test program is presented in ref.[6]. Based on the 3D FEM analysis - as a direct impact of torsional loading - significant out-of-plane panel deformation occurs in the blade sandwich shells on both pressure and suction side. This oscillatory out-of-plane panel deformation is generating peeling stresses in the adhesive bondline connecting the aft shear web to the shells.

The blade type used for this test campaign was a 34m blade manufactured by SSP Technology using a prepreg manufacturing process (pre-impregnated fiber mats without autoclave) (ref. [17.]). High resin content was selected for the laminates close to the PVC foam core and mold to ensure adequate resin quantity According to the manufacturer a modification of the Ampreg 22 epoxy laminating system was used (ref. [18.]). Although, this blade type is considerably smaller than the blades typically discussed when addressing failure in the transition zone and newer blade designs are more likely to use balsa wood or PET foam for the core material. Despite, it is still sufficient to develop the test method and to demonstrate the effect of torsional loads with the use of right setup. A 15m root end section cut-out of the SSP34m blade was utilized with a retrofitted aft shear web in the transition zone. Since the blade originally has a box spar design, a retrofit aft shear web was necessary (ref. [15]). This web was installed 4m from the root towards the tip section, positioned midway between the load carrying box girder and the trailing edge. The position was determined based on previously conducted test results which indicated the highest out-of-plane deformation in this area, see ref. [1.]). The main body of the aft shear web is a sandwich structure consisting of a 10mm core material (PVC foam) and four layers of bi-axial glass fiber. The strong-floor setup and the retrofitted aft shear web are presented on the Figure 5.

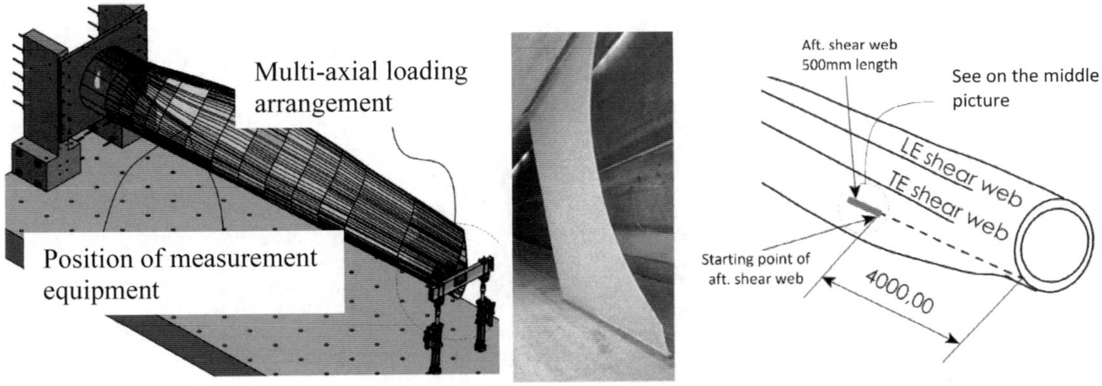

Figure 5. Large-scale strong-floor based test setup, the root section is mounted on a vertical strong-floor and the tip end is mounted onto the load train, the load introduction is explained on Figure 6 (left). Retrofitted aft shear installed in the 2 blade cut-outs (middle). The position of the retrofitted aft shear web illustrated (right).

Since the large-scale test was performed on the inner 15m section containing the transition zone, it is noted that the large-scale test will not be able to address one of the main conclusions explained below in subsection 4.2. Here the deflection of the tip of a large blade gets significantly amplified when both edgewise and flapwise loads are acting simultaneously, creating a large torsional arm that will lead to a high torsion on the structural regions of the blade.

In order to account for this limitation, and to avoid a testing campaign that could last for multiple years using conventional full-scale blade test methodologies, an accelerated fatigue test was carried out using the large-scale test setup and section test configuration. Thus, the loads applied are higher than the tested blade would ever experience in field conditions. Also, the layout and geometrical characteristics of the tested blade differs from current large blade types. Nonetheless, results from the test are illustrative of the field scenario for larger blades.

The torsional cyclic loads which have been applied in the large-scale tests are 50% higher than those that a 1.5MW old blade will meet in the field. The reason is that the objective is to demonstrate the importance of torsional loads on a larger blade. Ideally, a large modern blade should have been tested, but that has not been possible within the framework and budget of this research project.

The loading arrangement consists of two structural actuators which are capable of applying loads at the free end of the root section compromising two degree-of-freedoms: edgewise load along with torsional moment around the longitudinal axis of the blade. The maximum loading used during this fatigue test was 100kN edgewise load component (F_Y) and 100kNm as torsional moment (M_Z) respectively and with a frequency of 0.1Hz. Figure 6 below shows the test setup actuator configuration of the loading at the 15m reinforced cross-section. More details about the test setup can be found in ref. [7]. Figure 6 illustrates the test setup and the actuator specifications with the related coordinate system. During the test campaign various instruments were used, including a Digital Image Correlation (DIC) system, 4 strain gauges mounted on the retrofitted aft shear web, acoustic emission sensors and a wire potentiometer, more details about the instrumentation (e.g., instrument specification, positioning, etc.) see in ref. [7.]. The illustration shows the mounted position of the blade on the test rig and visualizes the applied load components.

IOP Conf. Series: Materials Science and Engineering 1293 (2023) 012005 doi:10.1088/1757-899X/1293/1/012005

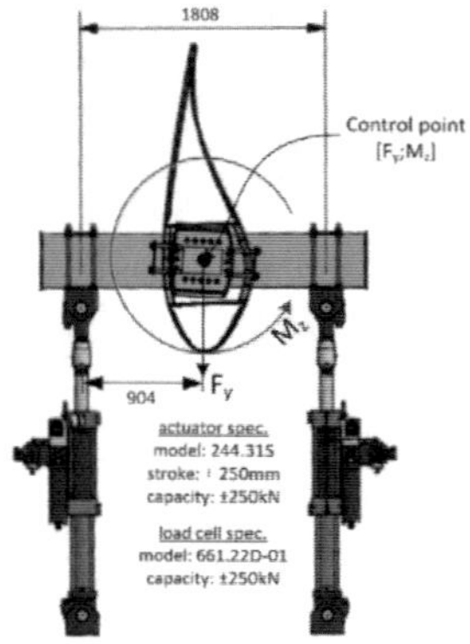

Figure 6. Active coordinate system and notation of the load train. Source: ref. [7]

3.2. Experimental results

The test resulted in a critical damage in the transition zone area on the pressure side panel in the trailing edge region of the blade in approx. 280.000 load cycles. More info about the damage and the test can be found in ref. [7.]. Further testing on the same blade section would have led to a blade collapse due to significant sudden decrease in edgewise stiffness (see Figure 7). This damage was a direct result of the applied torsional fatigue loads, causing large-scale damage on the blade shell in the root transition zone due to delamination from the out-of-plane deformation. The damage is shown on Figure 7. Such damage is considered critical, as it would require an immediate action if it happened during operation; failure to do so could potentially lead to further structural damage to the blade or other areas of the wind turbine. as if no actions were taken it could potentially lead to further structural damages on the blade or on other areas of the wind turbine.

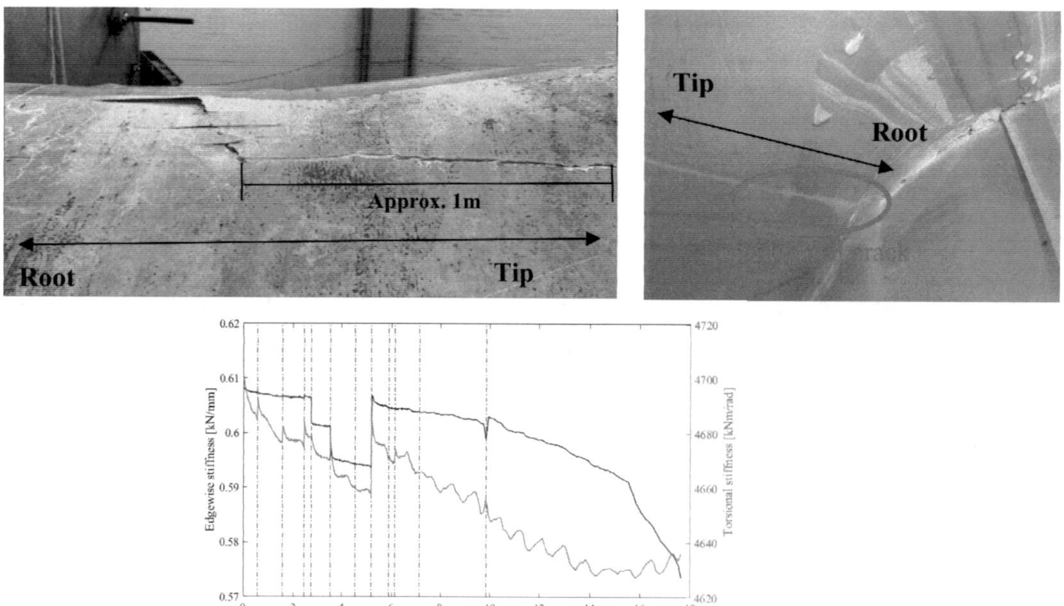

Figure 7. Picture presents the critical damage occurred on the 15m blade section of a 34m blade after 280.000 cycles in fatigue testing, where torsional loads are applied. The external condition of the blade (on the left), the crack observed internally (on the right). The length of the crack both internally and externally is around 1m. Plot presenting the change in edgewise and torsional stiffness (on the bottom).

 doi:10.1088/1757-899X/1293/1/012005

Even though the measurements in the large-scale test focused on the bondlines of the retrofitted aft shear web, limited data is available for the position of the initial damage location (approximately 1-1.5m from the root towards the tip section). However, the initiated damage is highly relevant, as the results show that the application of torsional load components lead to critical damage in the root transition zone on the unsupported panels.

The potentiometer installed 200mm from the front of the aft shear web towards the root (approximately 3.8m from the root has captured out-of-plane deformations in the transition zone. Since the damage occurred further towards the root, the panel deformation is likely significantly higher where the panels are unsupported. At the position where the measurements were made, the aft shear web supported the blade shells which indicates that the out-of-plane deformation of the panels are likely lower in this area. The mounted wire potentiometer is presented in Figure 8 below.

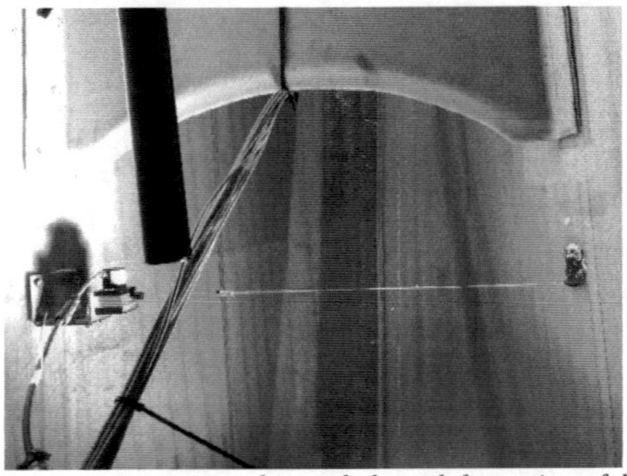

Figure 8. The mounted wire potentiometer measured out-of-plane deformation of the blade panels. The potentiometer was installed 3.8m from the root and 200mm from the retrofitted aft shear web.

The findings suggest that in case the aft shear web is helping to reduce out-of-plane deformation where it is installed, increased "breathing" will occur on the blade further towards the root section which leads to skin delamination and that result in a catastrophic failure mode.

Additionally, a second 15m blade section of a SSP34m blade was tested with a 5cm crack introduced manually into the bondlines on both sides of the retrofitted aft shear web. The test was executed under the same conditions as the first fatigue test (100kN edgewise load component (F_Y) and 100kNm torsional moment (M_Z) with the frequency of 0.1Hz). However, the test was halted after approximately 12000 cycles due to technical issues with the mounting box girder installed onto the tip section of the blade which mounts the blade onto the test rig. During the 12000 cycles significant increase in out-of-plane deformation was observed, from the initial 10mm deformation, it increased to 13mm. as represented on Figure 9, below.

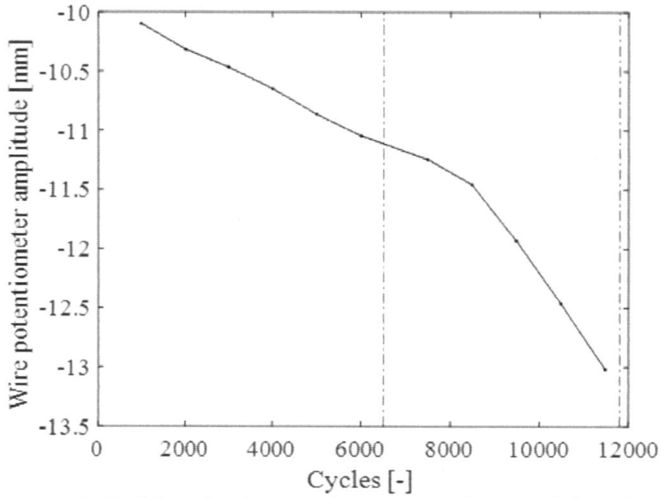

Figure 9. The measurements obtained from the wire potentiometer regarding out-of-plane deformation during the second test.

In Figure 9, a plot is presented of the magnitude of out-of-plane deformations and the nr. of cycles. A sudden drop (a drop in the plot shows an increase in the out-of-plane deformations) is observed on the plot at around 8000 cycles, show that damage is growing faster, hence is unstable. Upon inspecting the aft shear webs bondline, a clear crack propagation of over 10mm was observed on both bondlines of the aft shear web, see Figure 10.

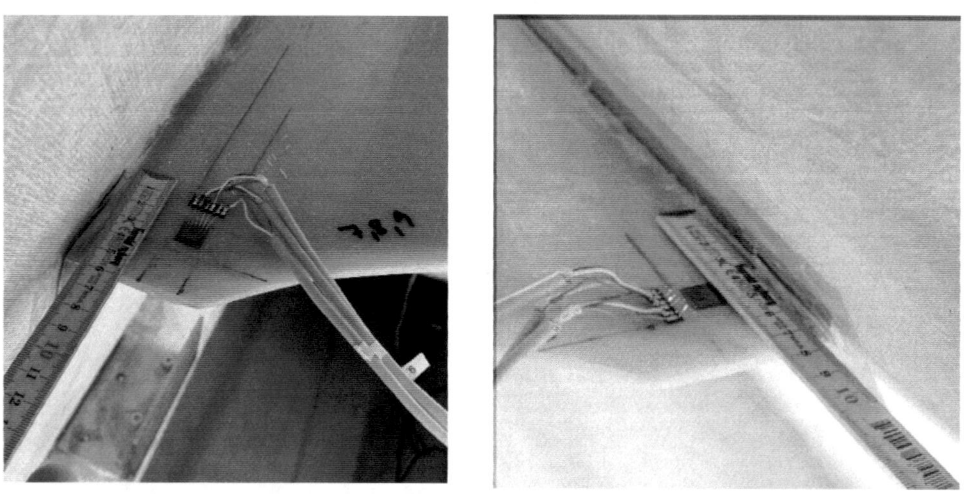

Figure 10. Approximately 10mm crack propagation on the Suction Side (on the left). Approximately 20mm crack propagation on the Pressure Side (on the right).

Figure 10 shows the results of inspection. A 10mm crack and a 20mm crack was observed on the SS and PS, respectively causing a halt to the testing.

In summary, increased "breathing" was observed when torsional load components were applied to the 15m blade cut-outs, leading to a severe damage during large-scale testing. This outcome was encountered during two separate experimental analyses, each revealing the extent of the damage caused by these loads. These tests led to the finding: the impact of torsional loads is not confined to only longer blades but also substantially affects shorter ones. Even a blade with a relatively short length of 34m demonstrated vulnerability to torsional loads. This conclusion emphasizes the importance of considering torsional loads in the design and testing phases of wind turbine blades, regardless of their size.

4. FEM Analysis

Three different analyses have been conducted. Firstly, simulations were performed comparing the additional tip deflection and consequent torsional loads representing actual operational conditions to a scenario with only edgewise loads. This scenario is expressed in several tables below as "Only edge TTL", meaning that the loads are exclusively considered in the edgewise direction, from trailing edge to leading edge. Secondly, torsional moments have been quantified for different blade sizes, obtaining scaling factors for RTM and edgewise and flapwise RBM. Thirdly, 3D FEM analysis have been applied to determine the influence of torsional loads on peeling stresses and bending strains on three hotspots (damage prone regions) of the blade:

1. Max chord area, unsupported panels
2. Transition zone, fishmouth of the aft shear web
3. Transition zone, unsupported panels

Two failure modes are covered for these hotspots: shear web disbonding, and skin debonding.

These hotspots were selected based on the author's in-field experience regarding damage occurrence during operation.

Each hotspot is analysed for the blade position where the highest values were found. Two operational load positions, 90-degree azimuthal position and 270-degree azimuthal position have been used. To clarify this, the azimuthal degree indicates the position of the blade during rotation. As an example, 0-degree represents a blade in a vertical position with the tip pointing up, and 180-degree azimuthal position the blade is also vertical with the tip facing down. For further details, see Figure 11.

The two blade FEM models have been established using state-of-the-art structural and FEM numerical knowledge.

4.1. FEM Technical details

The FEM blades are modelled using 3D layered solids with eight nodes, each possessing composite properties, such as glass epoxy and pultruded carbon profiles in the spar cap, balsa wood in the inner third, and PVC foam in the outer region. These materials, with their biax, triax and unidirectional layers are built up in specified directions and enclosed by a core in the panels.

Regarding the FE model boundary conditions, all nodes are fixed in the root for translations. An MPC-element is used to offset reaction forces/moment at the Origo. The model uses large displacements with a load prescribed following a non-linear geometry analysis. This includes 10 load steps with equal load distribution and a follow-forces option is also used Convergence studies have been performed to ensure the results are reliable, as they guarantee the elements are adequately small.

This modelling approach is very effective at capturing out-of-plane deformations as well as strains and stresses in the lamina. This feature is used to extract strains from the outer and inner layer of panels and calculate bending strains.

The applied loads are calculated by adapting operational loads for specific blades familiar to the author and adjusting them to fit the 80m blade and the 120m blade. In the process, Bladena has gathered different load distributions from various manufacturers for different blade sizes over several years. In addition to this, Bladena has also received information on different Root Bending Moments (RBM) in both edgewise and flapwise direction. Bladena used a preprocessor from MSC-Software (Patran) to calculate the Root Torsional Moment (RTM). First the RBMs are verified to align with calculation from the manufacturer, and subsequently the RTM is extracted from Patran. An analytical calculation of the RTM, obtained by multiplying the edge- and flapwise loads by the "torsional arm" (deformation) across 12 sections along the blade, has been performed. The RTMs from numerical FEM and analytical calculations have been compared and very similar results have been obtained. It is important to note that the flapwise loads are applied at the aerodynamic centre and not at the shear center, as an important contributing to the torsional loads would otherwise be neglected.

4.2. Operational loads cause extra Tip Deflection leading to additional Torsional Loads

The current design basis considers the loads associated with displacements in the edgewise and flapwise independently and not their combined three-dimensional effect equivalent to what is found under actual operating conditions.

This simplification implies a significant underestimation of the actual movements of the blades and thus of the forces/stresses that the blades are exposed to in real life. For the 120m blade the global edge tip deflections are presented in Table 2.

120m wind turbine blade	Edge tip deflection [m]
Only edge TTL	6.6
Operational loads, 90-degree azimuthal position	9.2

Table 2. Edgewise deflection for both operational loads, and only edge load component. 120m wind turbine blade.

Regarding the edgewise direction, blades are significantly stiffer than in the flapwise direction, but the deflection values increase from 6.6m to 9.2m when actual operating conditions are added. The increased edgewise deflections are equivalent to a loss of stiffness compared to the design basis. This effect will not only influence the tip deflection, but also the stresses in the max chord area and in the transition zone.

4.3. Quantification of Torsional moments

As also mentioned in the introduction section, the load contributions that introduce a root torsional moment on the blade arise from the aerodynamic forces and gravity working on the already deformed blade under operation. The aerodynamic forces can be divided in a purely flapwise component as well as a tangential component in relation to the rotor plane, acting as an edgewise load, in relation to the aerodynamic profile of the blade. Additionally, gravitational loads are also acting in the edgewise direction, see Figure 11.

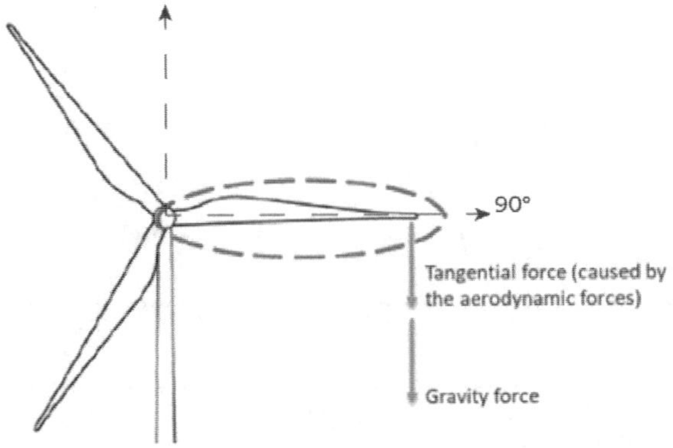

Figure 11. Investigated blade in 90-degree azimuthal position as the blade rotates. The arrows next to the forces indicate direction only, the gravity forces act at the center of gravity of the blade whereas the tangential forces (caused by the aerodynamic forces) act along the blade.

When the blade is loaded and deflected, torsional loads are generated. The edgewise aerodynamic loads working on a flapwise deformed blade twist the blade in a clockwise direction if the blade is seen from the tip towards the root. At the same time aerodynamic flapwise loads work on the edgewise deflected blade and the twist direction generated by this depends on the direction of the edgewise deflection. While the tangential aerodynamic loads (if the aerodynamic loads are split into two, based on directions, we can define normal direction and tangential direction) are always working towards LE, the gravitational load switch direction relative to the blade between the two horizontal positions. Figure 9 shows 90-degree azimuthal position where the aerodynamic tangential forces and gravity works in the same direction. For this position the flapwise loads generate a counterclockwise twist to the blade when seen from tip towards the root. Figure 10 illustrates the resulting torsional moments due to the operational loads.

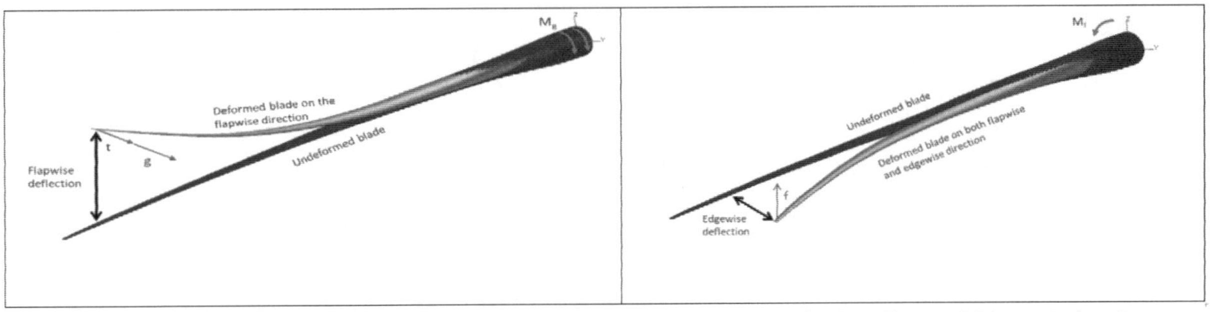

Figure 12. Left: RTM when gravity and tangential forces are in the same direction. Right: Root Torsional Moment induced by flapwise forces. Moment is acting in a different direction than the ones from the edgewise loads. On the figures "t" indicates tangential force and "M" indicates torsional moment.

A scaling study was carried out using two generic blade FEM-models, i.e., a typical 80m blade (corresponding to 7-9MW) and a typical 120m blade (corresponding to 15-18MW).

In the scaling study of the two blades, an 80m blade and a 120m blade, the blade characteristics can be compared. Both blades which are considered large (60m+), share a flatback design with two main shear webs and an aft shear web starting in the transition zone. They also both have carbon in the caps. Despite its large size, the 120m blade is very slender and lacks a clearly defined max chord, instead featuring a max chord region. This means that even though one blade is much larger, they have the same "max chord" profile length of approximately 5m. The sandwich panels are built up with layers of glass fibers (biax, triax, unidirectional).

The three torsional moment contributions and the total moment are listed in Table 3. It is seen that the total root torsional moment is 4 times higher in the 120m blade compared to the 80 m blade.

Torsional load contribution	Generic 80m blade	Generic 120m blade
Gravitational contribution	200 kNm	1600 kNm
Tangential contribution	160 kNm	700 kNm
Flapwise contribution	-160 kNm	-1500 kNm
Total	200 kNm	800 kNm

Table 3. The torsional moment contributions for 80m and 120m blades for 90-degree azimuthal position. A positive moment is defined as the moment that twists the blade clockwise if the blade is seen from the tip towards the root.

Table 4 below shows the result of a study conducted to understand how the RBM and RTM change for different blade sizes. The exponents shown below correspond to a curve fitting performed with the values from this study in which in the x-axis the blade length was presented, and the RBM or RTM in the y-axis. So RBM and RTM are represented as a power function with an exponent x.

$$f(x) = ab^x$$

Where:

- $f(x)$: RTM or RBM
- a: value different to 0
- b: blade length
- x: exponent shown in table below

Load direction	Exponent
Flapwise RBM	2.8
Edgewise RBM	3.4
Torsional RTM	4.0

Table 4. Exponent increase for as a function of blade length for flapwise RBM, edgewise RBM and torsional RTM.

In all cases, magnitudes increase as a function of the blade length, especially in the case of torsional RTM. The mathematical relationship between the three studied variables (Flapwise RBM, edgewise

RBM, and torsional RTM) and its dependency on blade length (exponent shown in Table 4) shows a scaling with a power law with a power of 4 for the torsional RTM.

Finally, Table 5 presents a list of potential failure modes and the impact on them from of torsional loads. Failure modes which are dependent on panel deformation e.g., breathing and CSSD are heavily impacted by the inclusion of torsional loads, whereas damages in the root (T-bolts, root inserts) are not impacted.

Failure mode	IEC 61400-5:2020	IEC 61400-23:2014	DNV-ST-0376:2021	Used in industry	Uncertainty of tools	Impact of torsional loads
Bondlines (Peeling test)	(2)	(1)	(3)	(YES)	MEDIUM	HIGH
Skin debonding from core (Test)	(2)	(1)	(3)	(NO)	MEDIUM	HIGH
Interlaminar failure (Bending test)	(2)	(1)	(2)	(NO)	LOW	HIGH
Global strain (Failure criteria)	(5)	(5)	(5)	YES	LOW	LOW
Shear web disbonding	(2)	(1)	(3)	(YES)	MEDIUM	HIGH
Root failures	(5)	(5)	(5)	YES	LOW	NO

Table 5. Table regarding potential failure modes and the impact of torsional loads on them, ref. [16] p. 46 (1 means there is no reference in the standard, 2 means it is mentioned briefly, 3 means that it is mentioned as "can be considered", and 5 means it is required in standards). The table does not take vibration issues into account. An additional column with the impact of torsional loads has been included. Regarding this column, it must be said that the impact depends on the section of the blade where this is analysed.

4.4 Impact of Torsional Loads on Selected Failure Modes

This sub-section addresses the impact of applying torsional loads in non-linear geometric 3D-solid FEM models for different failure modes.

The objective is to quantify the influence of torsional loads on variables such as peeling stresses, and bending strains, which have a direct relationship with some of the most common structural field damages seen during operational conditions.

The root transition zone of the blade has a complex geometry and contains tapering shell panels in 2 directions double curved in addition to the concave-convex double curved geometry. In addition, depending on the blade model, these panels in the TE box often have a large unsupported panel area in some sections of the blade, since the aft shear web usually starts around the max chord area or at the very end of the root transition zone.

4.4.1. Shear web disbonding

As explained in the introduction, shear web disbonding is a frequently occurring failure mode seen for a variety of larger blades. This failure mode is characterised by disbonds in the aft shear web bondlines, initiating from the spar cap fishmouth region. Thus, special attention has been aimed at blades which have an aft shear web with a fishmouth shape, as well as traces of delamination in the area of the fishmouth foot, see Figure 2.

Consequently, peeling stresses is the main parameter to focus on in the 3D-solid FEM analysis in order to understand the possible initiation of this failure mode.

In the following, results from two different studies will be presented on this topic for two different blade sizes, namely 80m and 120m. The fishmouth region of the aft shear web is located between the trailing edge and the trailing edge shear web and situated at a distance of 3.5m towards the root for the 80m and 9,8m for the 120m blade. These are the positions where the aft shear web starts for the two blades with different length.

1. 80m wind turbine blade

Operational loads have been compared against only edgewise loads.

Results are presented in terms of maximum values. Maximum values show the highest stress or strain calculated by FEM in the selected hotspots. They are considered as being the most likely locations for initiation of damages. An increase of 40% in maximum peeling stresses can be observed in the transition zone due to the impact of torsional loads, presented on Table 6.

It cannot be stated whether this increase in stresses is critical or not, and no literature has been found for reference. Such an evaluation will vary from blade design to blade design. In addition, standard methodologies representing fatigue analysis such as SN curves are not considered to be accurate enough for the analysis of interlaminar strength (ref [21] and [22]). A correlation with field damage would be required in order to estimate peeling stress values that may lead to the initiation of the damage, particularly in the form of shear web disbonding.

However, it must be considered at this point that the transition zone is the main load carrying area of the blade. Thus, damages in this area can influence the structural integrity of the entire blade. Bondlines in this area may suffer from high peeling stresses caused by localised deformations, probably due to the breathing of the panels and the cross-sectional shear distortion of the structure, composed by the spar caps and the shear webs. Increases, like the 40% shown in this paper, may be enough to exceed the stress limit that can be carried by the bondlines.

Figure 13 below shows an example of where the peeling stresses are extracted from the 3D-solid FEM analysis.

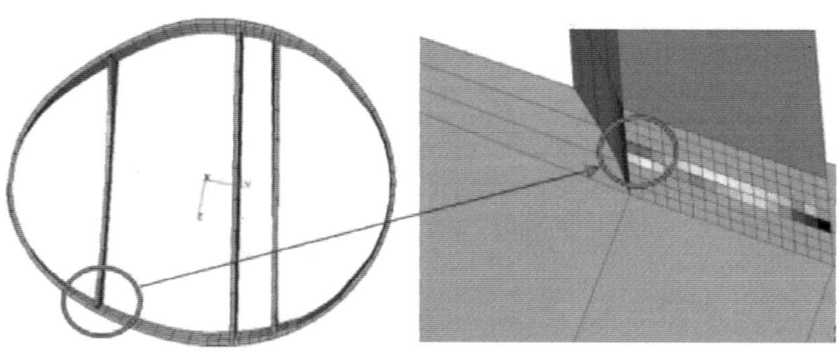

Figure 13. An example of the location where peeling stresses are extracted. The cross section is of a root-transition zone area and the peeling stresses are extracted at the aft shear web fish mouth foot as highlighted in red circle. The illustration is an indication of the increased peeling stresses in the investigated area.

2. 120m wind turbine blade

Again, a comparison between operational loads and edge loads as they are assessed in the current industry design basis has been established. As shown in Table 6, an increase of 23% is deducted from the application of torsional loads.

When comparing results from both the 80m blade and the 120m, see Table 8, it can be observed that, on one side, the values for the 120m blade are between 14% and 30% higher than for the 80m. This result is expected from a theoretical point of view, as the larger torsional arm created in the case of 120m blade, should generate a higher RTM and consequently higher peeling stresses in sensitive structural areas. It may be noted, that for the generation of torque, there must be a force component applied in the perpendicular direction to the line of action of this force from the point around which it is calculated. In other words, if the torsional arm increases, the twisting of the structure will be higher, which will lead to deformation of the blade cross section. As a result, increase in peeling stresses due to both cross section shear distortion and local bending of the unsupported panels between the trailing edge and the aft shear web in the transition zone will be present.

On the other hand, by analysing the percentage increase due to torsion, it can be seen how the impact of torsional loads is increasing faster in the 80m blade than in the 120m blade. The reason for the smaller increase is most likely due to the position of the aft shear web, which is very sensitive.

The most important observation is that for all blade types there is a significant increase in the peeling stresses when torsional loads are applied.

Blade	**Peeling stresses. Operational 90-degree azimuthal position [Mpa]**	**Peeling stresses acc. To current standards [Mpa]**	**Increase due to torsional loads**
80m	1.4	1	+40%
120m	1.6	1.3	+23%
Comparison	+14%	+30%	

Table 6. Comparative table for shear web disbonding. Max values

When comparing the max values for operational load 90-degree azimuthal position between the 80m blade and the 120m, an increase of 14% is found. For the case of pure edgewise load case, the value reaches a +30%. In this case, results are aligned with expectations, showing that peeling stresses increase with the blade size.

Consequently, the reason for the high number of failures due to shear web disbonding in the transition zone is probably due to torsional loads in large wind turbine blades. Only with field data will it be possible to fully validate the results herein. For this purpose, a validation process would be required, in which field and test measurement campaigns would be carried out in specific critical hotspots, measuring the physical response of the blade, and obtaining essential knowledge to validate the FEM models and numerical results obtained. A validation process is considered a key element to reduce risk in large blades. For further information, see ref. [10].

4.4.2. Skin (face sheet) Debonding from Sandwich Core in the Root Transition Zone

This failure mode is also more commonly seen in large blades in operation based on the author's in-field experience. Breathing of the panels due to out-of-plane deformations will potentially initiate skin debonding between the thin skins and the balsa or foam core of the sandwich shell panels. When sandwich panels are subjected to bending there is a risk that the interlaminar strength is not sufficient, causing skin debonding to initiate. If this occurs the bending stiffness of the sandwich panel will decrease significantly, and further crack propagation might develop rapidly. As a result of this, measuring bending strains through FEM will provide an indication of the behavior and possible breathing and bending of the panels which is understood as a possible root cause of this failure mode. Bending strains are calculated as the absolute difference between the outer strain and the inner strain of the panel.

The performed FEM analysis allows for comparison of the bending strains due to only edgewise loads with the operational loads. Results are presented for both the 80m and for the 120m wind turbine blades. For the 80m blade, the bending strains have been measured at 3m from the root and 0.7m from the TE. For the 120m blade, the measured area is around 8m from the root and 1.2m from the TE in the chordwise direction.

The strain value is taken in the surface plane direction transverse of the blade profile (as opposed to longitudinal strains). This direction is best at capturing out-of-plane panel deformation. The complex geometry of the area changing from concave to convex and tapering in two directions is the main explanation for that the out-of-plane deformations are highest in this region, see figure 3.

Results for bending strains will in some cases give negative values, while in others they will be positive. A negative strain is defined as whenever the bending is experienced inwards in relation to the blade cross section, and a positive value indicates that the bending is outwards. Whether positive or negative is not important; the importance is the magnitude, as the focus is on the amount of bending, regardless of whether the bending is inwards or outwards.

1. 80m wind turbine blade

Transition zone. Max values. Results are shown in Table 7.

Bending strains increase due to the increased level of bending and breathing of the panels. The transition zone is the main area of the blade carrying the loads, meaning that even small modifications in the load distribution may have an impact in this structurally critical area.

It can be seen how bending strains are increased 550% under the impact of torsional load. It can also be observed that the direction of strains changes from positive to negative. This sign is not important for the interlaminar stresses between the layers.

One reason that could explain this finding could be that when operational loads are included in the analysis, the panels may suffer an increased bending due to the loss of stiffness in the edgewise direction, as seen in Table 2. Here the tip deflection was compared with and without applied torsional loads for the 120m blade, concluding that operational conditions lead to an extra deflection due to the combined loading scenario. This loss of stiffness experienced for the 120m, is expected to also happen for the 80m blade, as this blade will also suffer from a softer edgewise deflection that could explain the big percentual difference in the impact of torsional loads.

Max chord area. Max values. Results are shown in Table 7.

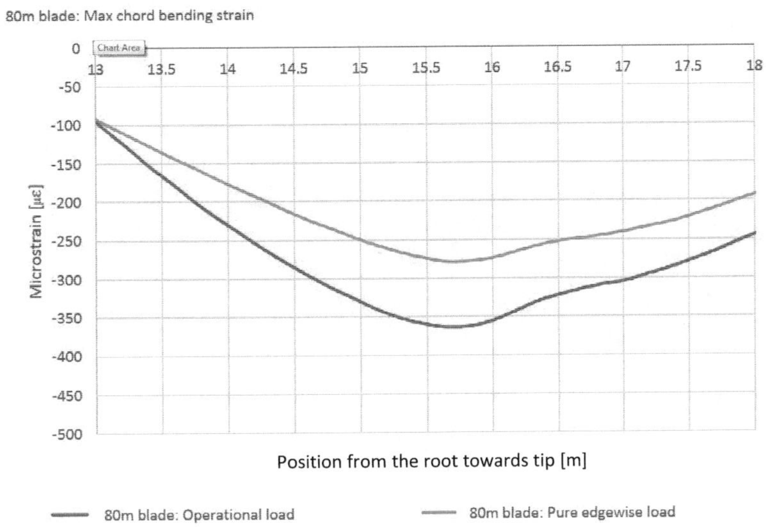

Figure 14. Bending strain comparison for the 80m blade in the max chord area. Operational loads 90-degree azimuthal position vs Only Edgewise loads.

For the max chord area Table 10 shows that bending strains increase around 30% whenever torsional loads are included.

It is also important to highlight that values for operational conditions are lower in the max chord area (360με) than in the transition zone (450με) which strengthens the hypothesis that the transition zone is the most critical and sensitive area from a structural point of view.

2. 120m wind turbine blade

Transition zone. Max values:

The evaluation focuses on the transition zone region, specifically between 8m and 9m along the blade length. The increase in the bending strains due to torsional loads are 120%, as can be seen in Table 7.

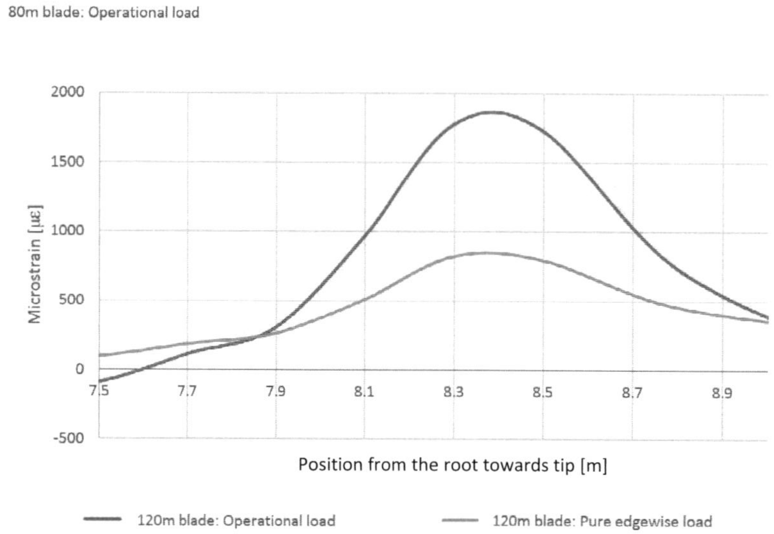

Figure 15. Bending strain comparison for 120m blade in the transition zone PS, operational load 270-degree azimuthal position max values vs Pure edgewise load max values.

Table 11 also indicates that bending of the panels is more severe in a 120m than in the 80m blade, shown in table 9. The difference in bending strains with operational loads reaches a +40% (630με vs 450με).

Consequently, it may be concluded that bending strains in the transition zone could increase linearly to the size of the blade.

Max chord area. Max values, results can be seen in Table 7.

Regarding the max chord area, the increase in the bending strain due to torsional loads is of around +47%.
Summary table for skin debonding with max values is shown in Table 7.

Summary. Bending strains	**Operational [με]**	**Bending strains acc. To current standards [με]**	**Increase due to torsional loads**
80m TZ	-450	69	+550%
80m Max chord	-360	-280	+29%
120m TZ	630	430	+46%
120m Max chord	-1090	-740	+47%

Table 7. Comparative table for skin debonding with max values.

Two main observations and results from the analyses can be highlighted:

- 80m blade vs 120m blade: the bending strains of the larger blade are significantly higher than for the smaller one, which emphasizes the thought that large blades have a much smaller structural reliability than medium or small blades. When values are compared between the 80m blade and the 120m applying operational loads:
    - Max chord area, transverse bending strains: +200% (from -360με to -1090με)
    - Transition zone panels, transverse bending strain: +40% (from -450με to 630με)

- Transition zone vs Max chord area: the transition zone does not always experience higher bending strains than the max chord area.
    - 80m, transverse direction bending strains: +25% (from -360με to -450με)
    - 120m, transverse direction bending strains: -71% (from -1090με to 630με)

5. Impact of torsional loads on lifetime – Discussion of results

The information presented in this paper emphasizes the importance of the torsional load components in order to correctly assess the durability of a given blade. Publicly available information on field damages, experimental analysis, and numerical models all points in the same direction: the larger turbine blades necessitate high maintenance costs (ref.[20.]).

As a result, it seems logical that the wind industry would adapt to the ongoing changes and challenges. This would mean reassessment of the design principles and O&M (Operation and Maintenance) strategies of large modern blades. Moreover, predicting the lifetime of these blades is a complex task, particularly because the traditional S-N curve method may not be accurate enough when it comes to analysing interlaminar strength (ref [21] and [22]). It is a well-established understanding among composite material experts that the S-N curve approach cannot be used to predict the lifetime of bondlines within these materials. Instead, these experts, including the authors of this paper, propose using a fracture mechanics approach. This method offers increased accuracy and allows for the implementation of a damage tolerance approach. The damage tolerance approach is about predicting when damage will occur and understanding how the structure will behave once a damage is present. Our main research project, see ref. [10], over the past two years, has been focused on these aspects.

The experimental analysis resulted in two different failure modes, both of which are considered critical. Meaning that, if damage types like the ones observed (such as delamination, aft shear web disbonding) occur in field, the turbine must be immediately stopped to avoid blade collapse. During testing, the applied loads were approximately 50% higher than what a blade on a 1.5MW turbine would experience; therefore, damage initiation was expected, though there was uncertainty about the specific failure mode. The observed failure modes confirmed that bending of the unsupported panels under applied torsional load components can lead to significant problems. In the case of the first blade, the clear decrease of global stiffness indicated the damage (a delamination leading to longitudinal crack, approximately 1m in length) whereas in the case of the second blade, a sudden increase in out-of-plane deformation indicated the shear web bondline disbond. As previously mentioned, the occurrence of such damages in field could be minimized by following the right strategy.

The majority of our research projects' participants and partners are Wind Turbine Owners whose goals are to minimize operating expenses (OPEX) and maximize the annual energy production (AEP). They are not typically involved in the design process, and therefore have not adopted the classical damage tolerance approach. However, in our views, the traditional methods are viewed from a perspective that is relevant not only to wind turbine owners but, hopefully, to manufacturers and companies involved in service and management. Doing this aims to bring a fresh understanding and approach that benefits all parties involved.

6. Conclusions

The current design and validation testing of large wind turbine blades omit substantial loads on the blades and that these omissions may very well explain the high level of unwanted maintenance costs that are currently found in the industry on large blades; an effect which increases with increasing blade lengths. In comparison with an 80m long blade, the bending strains in a 120m long blade were 200% higher in the max chord region and 40% higher in the transition zone (max chord: from -360με to -1090με; transition zone: from -450με to 630με).

The paper presents non-linear 3D-solid FEM comparisons between different load scenarios as they come out of the current design practice and stresses as they come out when the more realistic 3D non-linear behaviour of the blades is accounted for.

Numerical analysis using non-linear geometric 3D-solid FEM, for the study of shear web disbonding and skin debonding, can conclude that peeling stresses and bending strains are found significant for large blades due to the high torsional loads generated when the blade is in operation.

The results may be summarized as follows:

- The torsional loads that are found under actual operating conditions cause additional and critical out-of-plane deformation of the unsupported panels and also distortion effects such as CSSD and breathing. The localized deformations cause peeling in the bondlines, interlaminar stresses between the composite layers, and in between the sandwich skin and the core material. Layered composite structures and adhesive bondlines do not have a high load carrying capacity for those failure modes.
- The edgewise tip deflection increases by 40% for a 120m blade when the non-linear 3D behavior is accounted for. Amongst others, the larger deflection implies a loss of stiffness which in turn leads to significant increase in bending stresses and peeling stresses.
- Scaling factors have been quantified concluding that RTM's increase with a power of 4 with blade length.
- Experimental testing using large-scale section testing has demonstrated that critical damage in the root transition zone appears when torsional fatigue loads are applied.
- On the second blade shear web disbanding occur and showed an unstable crack development

It should be noted that even though the authors are confident that the results from the FEM models used for the study are valid, it should also be considered that the operational loads used for the analysis may differ from reality, as field measurements for large blades were not available thereby it is not sufficiently covered in this study. Also, layup and geometry are not accurate from any specific blades in the field and therefore, results must be perceived as a trend rather than accurate values.

However, despite any uncertainty of the results, the value of the FEM analysis results, and their conclusions are considered to be significant because the results demonstrate that critical loads are not accounted for in the current industry practice.

7. Recommendations

Taking all this into consideration, some areas have been identified which could potentially help to mitigate the risk of suffering damages from some of the structural damages related to torsional loads. The following recommendations are areas recommended to the wind energy industry to consider. We believe that a joint-venture project concerning the topic would be beneficial for the whole industry.

Suggestion 1. Application of torsional loads during the static blade full scale testing program

The full-scale testing program following IEC-61400-23 standard requires both flapwise and edgewise loads but in separated load cases (ref.[19.]). The static pull is close to the shear centre, hence almost no torsional loads are applied. The importance of both torsional loads, and a combined load case is recognised, but no obligation is imposed.

As an example, the IEC-61400-23 states that "the flap and lead-lag sequence of testing may be performed on two separate blades. However, if an area of the blade is critical due to the combination of flap and lead-lag, then the entire test sequence shall be performed on one blade" (ref.[19.]). The consideration of whether an area of the blade is critical or not due to the combination of loads is open to interpretation, meaning that freedom is given to applying or not a combined loading full scale test.

Another example that emphasizes that torsion is understood as a relevant matter, but not forced to be considered, is the following quote from Annex E in the same IEC-61400-23: "if torsion loads are significant in the structural design of the blade they should be included in the test" (ref.[19.]). Again here, it is subjective the establishment of whether torsion loads are critical or not. And even if they are considered as significant, the word "should" indicate a recommendation, not an obligation.

Therefore, the application of the torsional load components is suggested, as more realistic scenario, but nothing is imposed. As stated also in DNV's whitepaper, see in ref. [2], with the increasing size, the torsional loads and displacement of the shear centre are significant contributors to loading.

If the combination of loads were applied, with an offset, shown on the sketch in Fig. 18, the local deformations and distortion effects (CSSD and breathing) will be representative. To incorporate the combined flap- and edgewise loads, the blade was tilted 30 degrees in the direction of the leading edge toward trailing edge during the full-scale test performed in an earlier executed full-scale blade test study, see ref. [1]. At this full-scale test, the purpose was also to test the buckling capacity of the trailing edge. In this paper the focus is on the structural strength of the max chord region and the root transition zone and therefore the sketch show another load direction, in the positions where the torsional loads are highest.

IOP Conf. Series: Materials Science and Engineering 1293 (2023) 012005 doi:10.1088/1757-899X/1293/1/012005

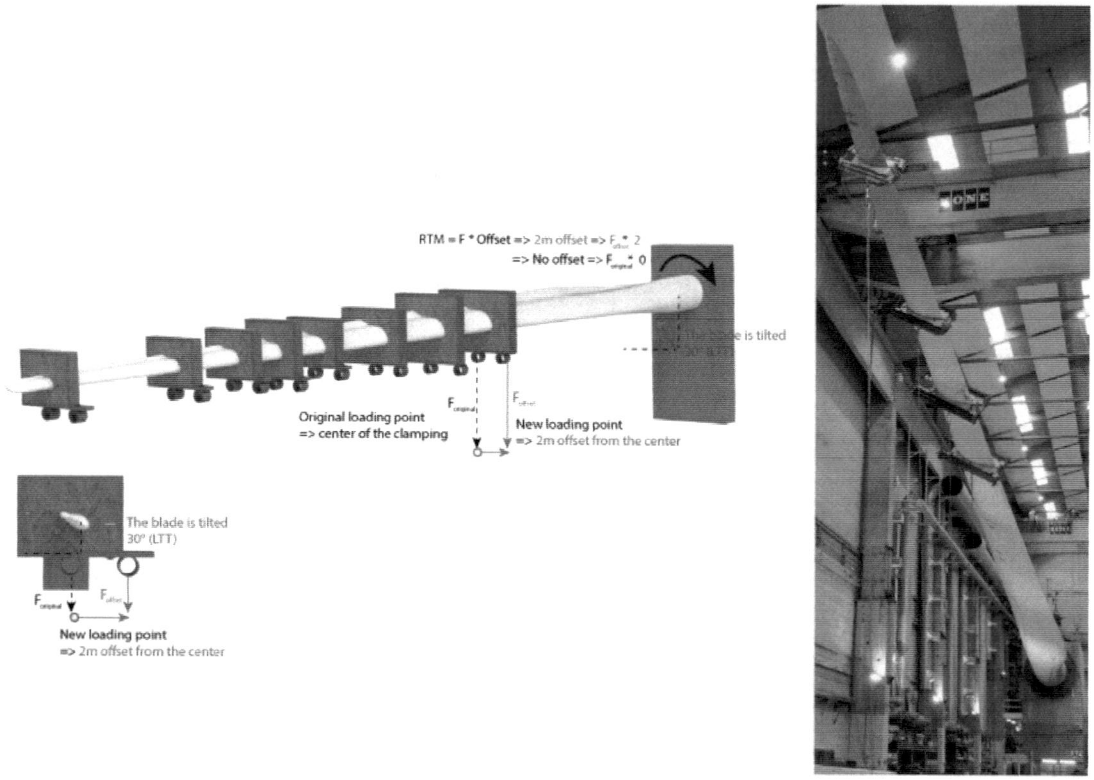

Figure 16: Sketch of a full-scale static test setup with applied torsional load components (left). Photo to the right show a full-scale test of a LM58,7m blade with combined statis loads performed at a commercial test center in collaboration with LM-Windpower and Bladena, see ref. [1] EUDP RATZ project.

This full-scale static test campaign is fast and easy to perform, and the costs are low. Furthermore, the additional load case can be done before the certification fatigue test program is started, and this is important since the measurement, is needed as early, as possible in order to validate FEM-models. When the FEM model is validated the boundaries for of the performed fatigue test sub-component program, can be updated.

More details regarding the needed measurement and the validation program can be found in the following two suggestions. Fig. 18 includes a photo from a full-scale test, Bladena and LM WindPower performed together in a commercial test centre. The static full-scale test with combined load was successfully performed, but at that time the test was performed, the knowledge about size of the torsional loads was known and furthermore the blade was only 58,7m, so the torsional arm offset should not have been as large as illustrated on the sketch.

The calculated torsional arm offset for a 120m blade has been calculated to be approximate 2m offset from the shear centre. 8 load clamps have been estimated, but of course this has been analysed for the actual blade and it can also be considered not to have the same offset for all load clamps, as long the sufficient torsional loads are applied it is not so important how it done.

Suggestion 2. Additional measurements

During a full-scale test campaign, additional measurements, are recommended to be implemented to identify localized deformation in the critical inner third of the blade e.g., cross-sectional shear distortion and breathing. This can be considered as a relevant step prevent from some of the failure modes here studied, as an example, the phenomenon "breathing" of sandwich panels in the trailing edge region, can influence both the formation of transverse cracks due to skin debonding followed by skin fatigue cracks, and the opening of the trailing edge due to high peeling stresses in the adhesive bondlines close to the trailing edge.

The additional measurements would also help to validate the FEM-simulations, so the boundaries for the fatigue sub-component test program for a lifetime estimation can be updated.

Moreover, back-to-back strain gauges on the unsupported panels in the transition zone is highly recommended. By measuring back-to-back strains, the localized bending causing potential critical interlaminar stresses can be determined.

Suggestion 3. Torsional loads during the 3D FEM modelling

Modern aero-elastic load simulations include the torsional loads in the analysis. However, according to the author's knowledge, these loads are not always included in the following non-linear 3D FEM analysis. This will also result in insufficient loading applied to the sub-component fatigue test program.

If the above three suggestions are implemented, potential issues can be found, and design improvements can be performed before the turbine is in operation.

8. Acknowledgement

Bladena would like to acknowledge the Energy Development and Demonstration Program (EUDP) for financial support under the grant number: 64021-1054 CORTIR II Project. Within the frame of the CORTIR II project considerable work has been done by DTU Construct regarding the experimental analysis and Global Wind Service supporting the large-scale test program and visualization work on several illustrations by Kirt X Thomsen.

References

[1] Bladena and Partners, (2019), "Project Report: Root Area Transition Zone RATZ and Reduction O&M cost of WT blades, Energy Development and Demonstration Program (EUDP) RATZ project, 64015-0062,", https://www.bladena.com/uploads/8/7/3/7/87379536/eudp_ratz_-_project_final_report_-_eudp_ref._j64015-0602.pdf

[2] Griffin, D.A., (2023), "The challenges of wind turbine blade durability" DNV Whitepaper, 2023, https://www.dnv.com/Publications/the-challenges-of-wind-turbine-blade-durability-243601

[3] Renews.biz, (2020), "Vestas, EDPR probe blade failure at Ohio wind farm.", Renews 2020 September 9. https://renews.biz/63000/vestas-blade-failure-at-edps-ohio-wind-farm/

[4] Harrison, C., (2019), "Limitations of legacy certification standards and benefits of new standards demonstrated on a100m wind turbine blade", DNVGL, WindEurope Offshore Conference, Copenhagen, 2019 November.

[5] Rajgor, G. (2022). "GE Renewable Energy investigates second Cypress wind turbine failure in two months", Windpower Monthly, 2022 March 16. https://www.windpowermonthly.com/article/1749796/ge-renewable-energy-investigates-second-cypress-wind-turbine-failure-two-months

[6] Bladena and Partners, (2021), "Cost and risk tool for interim and preventive repair (CORTIR) EUDP project 64018-0507 – Final Report", https://www.bladena.com/uploads/8/7/3/7/87379536/eudp_cortir_-_project_final_report_-_ref._64018-0507.pdf
[7] Waldbjørn, J.P., Berggreen, C., Ahmed, S., Nagy, T., "Large-scale fatigue testing of a retrofitted 3rd shear web in a 34m wind turbine blade section" DTU Construct & Bladena, 2023
[8] Bladena, (2022), "Risk considerations on upscaling wind turbine blades", Bladena whitepaper, 2022. https://www.bladena.com/uploads/8/7/3/7/87379536/white_paper_scaling_and_risk_bladena.pdf
[9] Bladena, (2022), "Structural weaknesses of blades in operation" Bladena whitepaper, 2022. https://www.bladena.com/uploads/8/7/3/7/87379536/whitepaper_1_structural_understanding_bladena_13.04.2022_compressed.pdf
[10] Bladena and Partners, (2023), "Cost and risk tool for interim and preventive repair (CORTIR Phase II) EUDP project 64018-0507 – Final Report"
[11] Karlsen, T. (2021), "76 meter langt turbine blad deiset i bakken – kan ikke si hvor utbredt turbinen er i Norge", Teknisk Ukeblad, 2021 November 15. https://www.tu.no/artikler/136-meter-langt-turbinblad-deiset-i-bakken-kan-ikke-si-hvor-utbredt-turbinen-er-i-norge/515102
[12] Brøndsted, P., & Nijssen, R. (Eds.) (2023), "Advances in wind turbine blade design and materials." Woodhead Publishing. Woodhead Publishing Series in Energy No. 47 http://www.woodheadpublishing.com/en/book.aspx?bookID=2548
[13] Jensen, F.M., "Ultimate strength of a large wind turbine blade," Risø DTU, PhD Thesis, 2008. https://backend.orbit.dtu.dk/ws/portalfiles/portal/5434131/ris-phd-34.pdf
[14] Bladena and Partners, (2016), "Torsional Stiffening of Wind Turbine Blades - Mitigating leading edge damages, EUDP project, 64013-0115", https://www.bladena.com/uploads/8/7/3/7/87379536/eudp_lex_project_final_report_september_2016.pdf
[15] Waldbjørn J.P., Buliga A., Berggreen C., Jensen F.M., (2020), "Multi-axial large-scale testing of a 34m wind turbine blade section to evaluate out-of-plane deformations of double-curved trailing edge sandwich panels within the transition zone" DTU and Bladena journal paper, 2020, https://journals.sagepub.com/doi/abs/10.1177/0309524X20978408
[16] Bladena and partners, (2022), "Wind turbine blades - Handbook", Edition 2022 CORTIR, EUDP.
[17] Jensen F.M., Falzon B.G., Ankersen J., Stang H., (2005), "Structural testing and numerical simulation of a 34m composite wind turbine blade", Composite Structures, October 2005, https://www.sciencedirect.com/science/article/abs/pii/S0263822306002480
[18] Gurit, (2020), "Ampreg 22 – Epoxy Laminating System", https://pdf.directindustry.com/pdf/gurit/ampreg-22-epoxy-laminating-system-v17/37817-79061.html
[19] International Electrotechnical Commission, (2015), "International Standard: Wind turbines – Part 23: Full-scale structural testing of rotor blades", IEC 61400-23 Edition 1.0

[20] Wood Mackenzie, (2019), "Unplanned wind turbine repairs to cost industry $8 billion+ in 2019", Wood Mackenzie: One minute read press release https://www.woodmac.com/press-releases/unplanned-wind-turbine-repairs-to-cost-industry-$8-billion-in-2019/
[21] M. Quaresimin, M. Ricotta, (2005), "Fatigue behaviour and damage evolution of single lap bonded joints in composite material", University of Padova, Italy, 2005
[22] A. J. Kinloch, (1997), "Adhesives in engineering", University of London, 1997.

IOP Conf. Series: Materials Science and Engineering 1293 (2023) 012006 doi:10.1088/1757-899X/1293/1/012006

Wind blades circularity - Resins development for improved sustainability

Harald Stecher [1], Francisco Javier Hierro-Olabarria Salgado [2]

1) Siemens Gamesa Renewable Energy A/S, Assensvej 11, 9220 Aalborg, Denmark
2) Siemens Gamesa Renewable Energy S.A., Parque Tecnológico Bizkaia, 48710 Zamudio (Bizkaia), Spain

harald.stecher@siemensgamesa.com, xabier.hierro@siemensgamesa.com

Abstract. Wind turbines play a vital role in the transition to greener energy and to fight climate change. In the coming years the number of installed turbines will increase dramatically. Consequently, the sustainability and especially the circularity of wind turbine components becomes more and more important. Many components of a wind turbine are made from recyclable materials like steel or aluminum, but blades are made from composite and are with that the biggest non-recyclable structure of a wind turbine.

The resin is key for the circularity of wind blades. Glass and carbon fibers offer the strength and stiffness to build long wind blades, core materials like balsa and polymeric foams are applied to reduce weight, and the resin secures fiber straightness, transfers stresses and glues everything together. If it is possible to remove the resin in a mild and efficient way, the other materials can be recovered and reused. This is the reason why the research in the field of composites recycling has intensified over the last years.

Two main directions have been followed in the last years: Discovering a recycling method for blades and in general composite structures based on conventional resin technology or implementation of chemically modified or tailored resins for recyclability by design

1. Wind blades and recycling

Wind turbines are generally perceived as green because they are used for green energy production. But at the same time the public interest in sustainability of wind turbines is growing and several aspects of sustainability and pollution are questioned.

Blades are huge parts, very visible, and made from materials difficult to recycle. But recycling is just one part of the waste management and not the one with the highest impact. As shown in the waste hierarchy established in the waste framework directive [1] (Figure 1) the avoidance of waste is highest priority. The material that is not used has the lowest environmental impact.

Figure 1: Waste hierarchy in five steps [1]

Despite the amount of waste, for wind blades landfill has been the dominant end-of-life option. With the emerging landfill bans other recycling strategies must be identified, and cement co-processing is at the highest readiness level. Pyrolysis is an emerging technology to recover fibers, the resin is decomposed and recovered as an oily residue.

Solvolysis can range from mild conditions using benign chemicals to supercritical conditions and nasty chemicals. The broad range of process conditions is also exhibited in the condition of the recovered fibers and core materials.

A different approach is the manufacture of noise barrier panels for acoustic applications , such as the commercial products offered by Miljøskærm [2]. The composite parts of the blade are converted in to granules, which are then used to manufacture the panels.

In terms of re-use, the application of blades or blade parts for architectural purposes has been shown. Examples are bike sheds from parts of the blade shell, blades as load bearing elements of bridges or blade shell pieces as sunscreens for buildings [3].

1.1. Evaluation of recycling methods

With an increasing number of recycling methods available, it becomes more important to be able to compare the recycling methods based on objective measures. Different measures can be applied and will give a broad and good insight into the characteristics, strengths, and weaknesses of different approaches. Some examples of how to measure and compare recycling methods are: Cost value proposition to evaluate process cost and value of recovered materials, CO_2 footprint of the process and CO_2 saving using the recovered materials [4], [5].

2. Solvolysis of conventional resin systems

The solvolysis of resin systems for wind blades is of high interest. But Solvolysis conflicts with the fact that epoxy amine resin systems have been designed to be strong and chemically resistant. Nevertheless, some interesting processes have been presented over the last few years.

It was shown that a recycling process discovered by Kaufmann et al. at TU Clausthal and MPM Environment Intelligence GmbH. based on boron halides is able to remove the epoxy resin completely. The process is compatible with glass fibers as well as carbon fibers [6].

Another example is the process developed by Catack-H in Korea focussing on the recycling of carbon fiber-based composites. First the decommissioned part is pre-conditioned in a liquid for 24 h, then the actual recycling process takes place [7].

Lately also a two-step recycling process developed in the CETEC project was presented [8]. The first step is performed to separate the resin from the other materials, which can be recovered and used thereafter. The second step is conducted to valorize the remains of the resin coming out of the first step. So far, no details on the two process steps have been disclosed, so it is not possible to judge the feasibility or compare with other processes.

3. Pyrolysis of conventional resin systems.

Pyrolysis intends the separation and recovery of composite materials by thermally decomposing the matrix and recovering the fibers afterwards, due to the higher temperature resistance of carbon and glass reinforcements compared to the cured resins. Alternatively, pyrolysis can be used only to extract energy with the combustion process, sometimes using it in closed loop, or it could be used to recover fibers and degraded matrix in the form of oil. Currently being used at industrial scale for other applications, pyrolysis for blade materials is still at a low technology readiness level.

Pyrolysis has been shown [9] to recover carbon fibers with minimal presence of pyrolytic carbon (Figure 2), organic residues and oxidation, thus retaining a significant level of mechanical properties. The recovery of carbon fibers in these conditions depends mostly on oven atmosphere, temperature and heating rates.

Figure 2: SEM image of carbon fibers recovered by pyrolysis. [9]

When recovering GFRP, the cost balance of the process is much less beneficial, given the higher extent of damage inflicted to the glass fiber surface, lowering the mechanical properties and the lower cost of the glass per kg, which do not justify the large investment and running cost pyrolysis commonly involve. To overcome this limitation, microwave pyrolysis is being investigated [10] to enable faster energy transfer an shorter processing times, which both decreases damage to the fibers and the overall process cost.

Fluidized Bed Reactors are also investigated for blade composites recycling, due to its multi-material versatility and lower process cost. At the moment, FBR systems can only work fed with grinded blade composite residues, thus adding the limitation of only recovering short fibers to the already existing one of degraded properties.

4. Recyclability by design of resins

Recyclability by design with respect to resins means the introduction of chemical groups or additives into the resin. By chemical modification it is possible to enhance the recyclability of resins. The chemical modification can happen in different ways:

1) Additives can be used which are not integrated into the polymer network.
2) Functional groups can be introduced into the polymeric network.

Functional groups that are integrated in the polymer network can either be randomly distributed or located at defined positions in the network. The resin chemistry and especially the mechanism of polymerization is a deciding factor of how controlled the introduction of new functional groups into the polymer network can happen. Polyester and vinylester resins form the network in a chain growth reaction mechanism. Di- or multifunctional oligomers or polymers that contain unsaturated bonds in the main chain as side chains are dissolved in styrene and divinylbenzene or (meth-)acrylic monomers. The reaction is initiated by radical forming initiators (eg. peroxides) and continues in a random fashion by chain extension via reaction of the radicals with double bonds and transfer of the radical to the new chain end. This complicates the directed introduction of functional groups.

Special initiator or catalyst systems allow more controlled polymerization. Examples are RAFT or ATRP polymerization of polyester and vinylester resins. Application of these and similar methods offering living polymerization allows more controlled chain and network formation, and the creation of fully latent resins [11].

Epoxy-amine infusion resins react in a step-growth mechanism. The reaction starts simultaneously in the whole bulk resin. The difunctional epoxy resin reacts with the tetrafunctional amine hardener. With this choice of components the crosslinking happens only through the hardener. As a consequence the modification of the hardener backbone can introduce functional groups into specific and directed locations of the network.

4.1. Introduction of cleavable moieties

The introduction of cleavable moieties has been proven in literature and commercial scale. A good example is Aditya Birla Chemicals´ Briozen product range. The technology builds on the introduction of acetal or ketal groups into the hardener [12] The amine functional hardener can then replace conventional amine hardeners in the formulation of epoxy-amine resin systems. The strength of the technology is, that under neutral conditions the resin is stable and offers the same performance as resins using conventional amine hardeners. First in contact with organic acids, prefentially acetic acid, in aqueous solution and elevated temperature, the cleavage of the acetal or ketal group is catalyzed forming dissolvable polymer chains out of the thermoset network originally formed.

Another example is the recently presented EzCiclo resin technology by Swancor [13]. The epoxy part is modified to introduce a cleavable moiety. In this case the hardener is no different to conventional resins. The recycling process requires a proprietary recycling solution called CleaVer. The composite based on EzCiclo resin is immersed in the CleaVer liquid and heated to 120-140 °C. The cleavable bonds are opened and the resin is dissolved in the Cleaver. The reinforcement fibers can be collected and reused after a cleaning step, the Cleaver solution containing the recyclate can be used as hardener for epoxy resins for example in flooring applications.

4.2. Introduction of reversible bonds

The introduction of reversible bonds into the network is a similar strategy to enhance the recyclability of the resin. But the introduction of reversible bonds adds further features to the resin and the resulting composite. The controlled opening and closing of bonds in the network allow the regulation of the crosslink density of the resin. With a smart positioning of the reversible bonds, the crosslink density of

IOP Conf. Series: Materials Science and Engineering 1293 (2023) 012006 doi:10.1088/1757-899X/1293/1/012006

a thermoset polymer network can be reduced and in the most extreme case shifted to a thermoplastic linear polymer, triggered by an external stimulus. The opening and formation of bonds is a dynamic process and happens simultaneously in an equilibrium state. The bonds are not necessarily reformed at the same position. This results in viscous behavior and enables operations that are normally limited to thermoplastic materials.

A known example for reversible bonds is the Diels-Alder reaction. It is a versatile reaction that forms a six-membered ring from a molecule containing two conjugated double bonds (diene) and another molecule or chain of the same molecule containing one double bond (ene). The reaction is shown in Figure 3 on the example of the dimerization of cyclopentadiene.

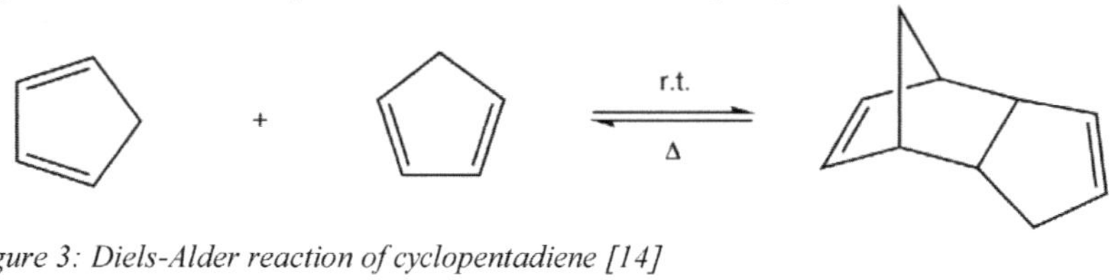

Figure 3: Diels-Alder reaction of cyclopentadiene [14]

At room temperature the equilibrium lies at the 6-membered ring, at increased temperature the equilibrium switches to the open state of diene and ene. Although the concept is well known [15], the application for infusion resins is difficult because the integration of the functional groups into resin and/or hardener result in bulky molecules having a negative impact on the viscosity.

4.3. Introduction of dynamic bonds

A closely related technology to the introduction of reversible bonds is the application of dynamic bonds, the concept is called covalent adaptable network. Two different mechanisms are described in literature: Dissociative covalent adaptable networks, where upon an external trigger, which is normally heat, bonds are broken and then reformed.

Several chemical concepts have been investigated and presented over the last few years. Two prominent examples are resins based on disulfide bond exchange (Figure 4) [16] and resins based on imine exchange (Figure 5) [17]. Both resin technologies behave like conventional thermoset resins until a certain temperature is reached. Above this temperature the active bonds start to rearrange, but the crosslink density is constant. The dynamic bond exchange results in viscous flow, which is enough to enable options, that are normally limited to thermoplastic resins.

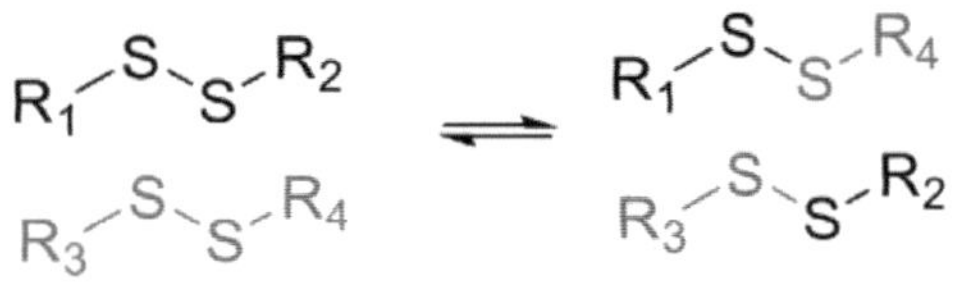

Figure 4: Disulfide bond exchange mechanism

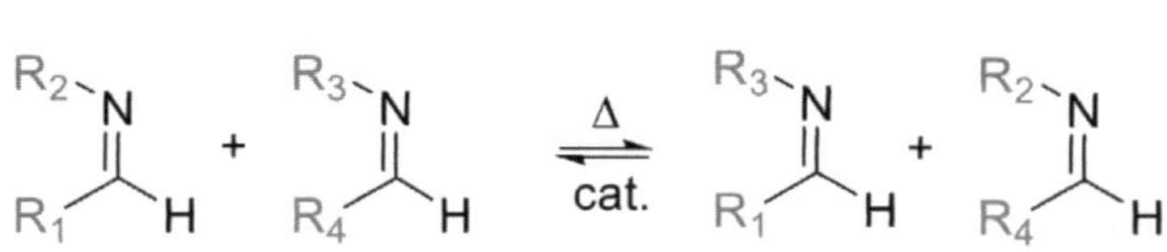

Figure 5: Imine exchange mechanism

The extraordinary behavior also opens up for new recycling techniques. Tailormade solvolysis processes can be applied to conduct a depolymerization of the network under mild conditions.

4.4. Evaluation and comparison of solvolysis process of conventional resins and chemical concepts for enhanced recycling

Like for recycling methods overall it is beneficial to identify measures to evaluate and compare different concepts and technologies. As for the general comparison of recycling methods, the CO_2 footprint and the cost value proposition are important measures. The recycling process conditions are similar resulting in the value of the recovered fibers being almost the same. Differences can be found in the core materials – in which quality is balsa, PET and other core materials recovered.

Given the similar quality of the fibers recovered, the process cost is the determining factor for the fiber price. Process cost is controlled by factors like process temperature and time of the recycling process, cost of chemicals used and simplicity of the process. Similar to the waste hierarchy a classification of recycling processes for polymers has been proposed by Zero Waste Europe (Figure 6) [18] with respect to the recovered material.

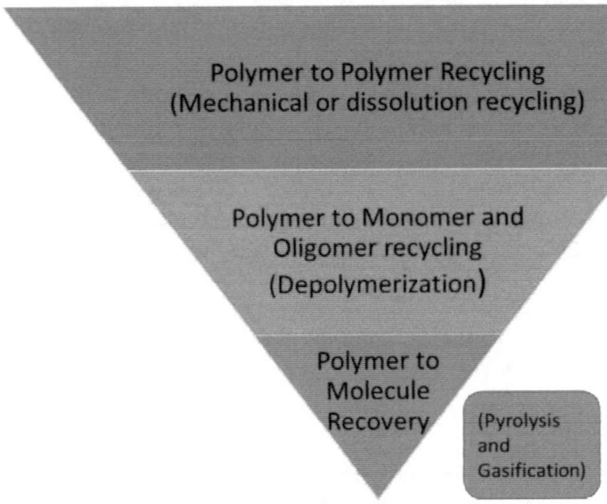

Figure 6: Classification of polymer recycling processes [18]

The solvolysis process degrades the thermoset network of the composite matrix. Depending on the underlying technology the cleavage of the network results in different products. To compare the technologies one aspect is the conservation of functionality and molecular structure, as shown above, and the atom economy. Energy and resources have been consumed in the synthesis of resins and its curing. The more the structure can be conserved and integrated into a valuable product the better.

A cradle-to-cradle recycling process is not automatically the preferred solution over a downcycling process, if the cradle-to-cradle recycling process goes back several steps in the synthetic pathway of the resin components and requires steps of re-synthesis to yield these components again.

The atom economy concept can be adjusted for chemical recycling. The functionality, complexity and purity of the recovered components can be taken into account. The more of the functionality and complexity is recovered at the highest purity, the better. By-products should be avoided as good as possible.

As an example an evaluation of the Aditya Birla Chemicals Briozen resin based on Recyclamine® technology shows, that although the cured resin is converted from a crosslinked thermoset to a thermoplastic resin, almost the full functionality of the resin system is maintained and the thermoplastic resin can be recovered in high purity. Only a small molecule is cleaved off in the recycling process and can be recovered and reused as well.

IOP Conf. Series: Materials Science and Engineering 1293 (2023) 012006 doi:10.1088/1757-899X/1293/1/012006

Another example for comparison is a recycling process, that is able to recover 50 % of the resin system as raw material for the resynthesis of a resin component. The other 50 % are a mixture of different components and can only be used for energy recovery.

Although the latter process offers real cradle-to-cradle recycling for a part of the cured resin, only 50 % of the resin is actually recycled, a poor atom economy. Also the functionality and complexity are reduced coming from the cured resin to the raw material for a resin component. In this comparison of the two technologies, the downcycling appears as the preferred solution over the cradle-to-cradle approach.

5. Performance of recyclable resins on basis of Aditya Birla Chemicals´ Briozen resin

The SGRE RecyclableBlade has been a ground-breaking development towards recyclable blades. The RecyclableBlade has been introduced using Aditya Birla Chemicals Briozen resins which are based on the Recyclamine® technology.

The concept of the Recyclamine® technology allows the synthesis of molecules with defined molecular structure and molecular weight. The components offer similar structural features as the commonly used hardener components.

Acetal groups are already long known in polymer chemistry and find application in polyoxymethylene (POM) or polyvinylbutyrate (PVB) to name some examples.

The Aditya Birla Chemicals Briozen infusion resin is a good example for the smart modification of traditional epoxy amine resins for enhanced recyclability. The general type of resin is still epoxy amine, but the hardener comes with a cleavable group in the backbone. The technology allows the formulation of resins with similar mechanical and process performance. To find suitable resins several different hardeners have been developed by Aditya Birla Chemicals. The hardeners offer different pot life and curing time to have the best fitting solution for each application. Also, the weatherability is not negatively impacted.

5.1. Mechanical properties and interface adhesion

SGRE produces blades in two different ways: IntegralBlades® that are manufactured in a one-shot infusion. So-called "butterfly blades" are manufactured by infusion in several pieces, that are later assembled in a bonding step. The individual pieces of the butterfly blades are smaller in size than the shell of an IntegralBlade®. The infusion is faster, so the pot life can be shorter than for the infusion of an IntegralBlade®.

To have a fitting product for both processes dedicated hardener systems have been developed by Aditya Birla Chemicals which are then combined with the same resin base. Both resin types, the one for IntegralBlades® and butterfly blades, have been thoroughly tested and characterized. The test results showed that throughout the tests the performance is equal to better compared to conventional resins systems. This has been observed for neat resin tests (Figure 7), static (Figure 8) and fatigue tests on laminates level as well as interface tests to hand-lamination repair resins and bonding paste (Figure 9).

IOP Conf. Series: Materials Science and Engineering 1293 (2023) 012006 doi:10.1088/1757-899X/1293/1/012006

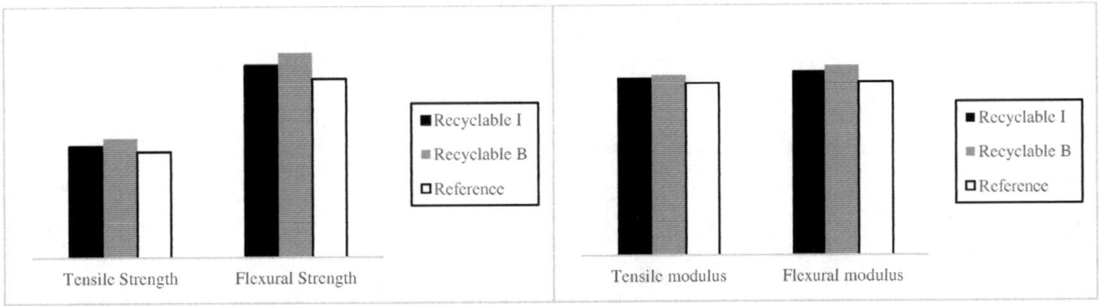

Figure 7: Comparison of neat resin properties of Recyclamine® integral (I), butterfly (B) and reference resin

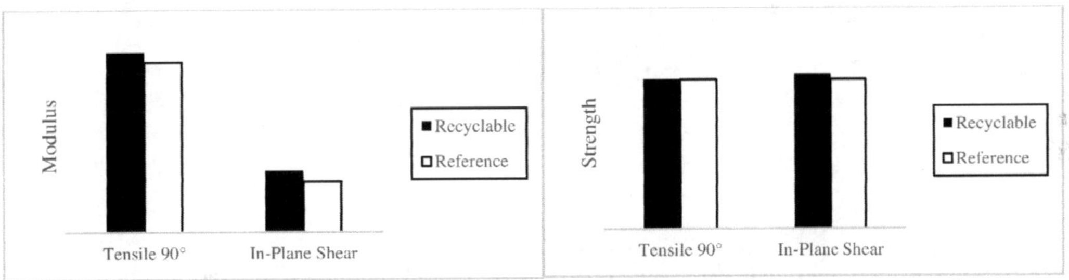

Figure 8: Sample of matrix-dominated property test results for butterfly Recyclamine® system against reference system.

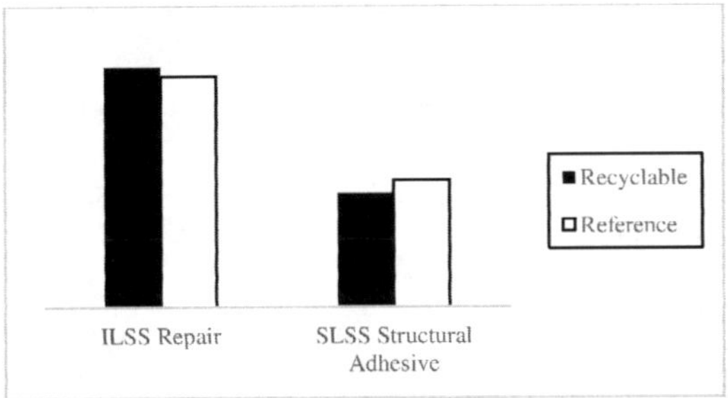

Figure 9: Interlaminar shear strength of Recyclamine® and reference resin laminate to laminate based on conventional epoxy-amine repair resin and single lap shear strength of Recyclamine® and reference resin bonded with structural adhesive (bonding paste)

5.2. *Weathering stability of the cured resin*

Because the blades are exposed to the environment for their whole operational lifetime, the weathering resistance of the resin has been investigated. On one hand the resistance to different acidic conditions was tested in full immersion tests following ISO175. The weight change before (Wiped dry with a cotton cloth) and after drying was checked. The summary of the results is shown in Figure 10.

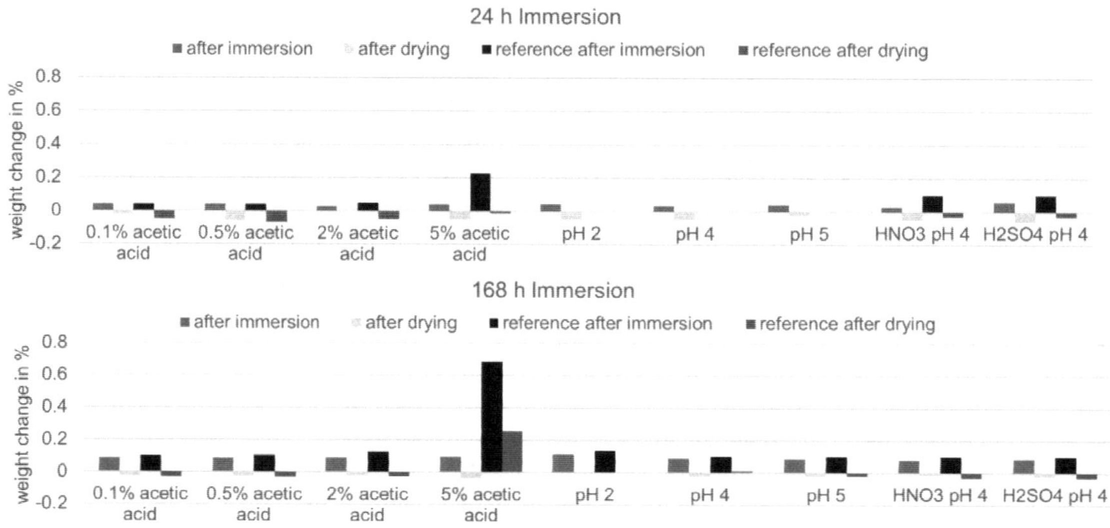

Figure 10: Weight change of Recyclamine® and reference resin based laminate samples after 24 h and 168 h full immersion in different test liquids.

The full immersion tests show little weight change and generally the same trend of the Briozen resin compared to conventional resin used in blades production. Only the 5 % acetic acid solution shows the potential of acetic acid to swell the epoxy network by the weight change of the reference system. No significant effect of the pH, tested with three different buffer solutions, was observed. Acidic rain is an often-mentioned concern. Acidic rain is a mix of various acids typically composed of sulfur- and nitrogen-based acids and a pH of around 4. To mimic these conditions nitric acid and sulfuric acid were diluted to reach pH 4. Both the Briozen resin and the conventional resin did not exhibit any negative effect upon exposure.

Because the chemical nature of the resin is very similar to conventional DGEBA based resins, the UV stability is assumed to be as low as for conventional resin systems. This was validated as part of a weathering test campaign following ISO20340 modified to fit for composites. The test cycle is shown in Figure 11.

Figure 11: Weathering test cycles following ISO20340

The uncoated laminates were tested for weight change and T_G change with the test cycle of three days UV exposure and condensation, three days of salt-spray and one day at -20 C. The results are shown in Figure 12 and Figure 13.

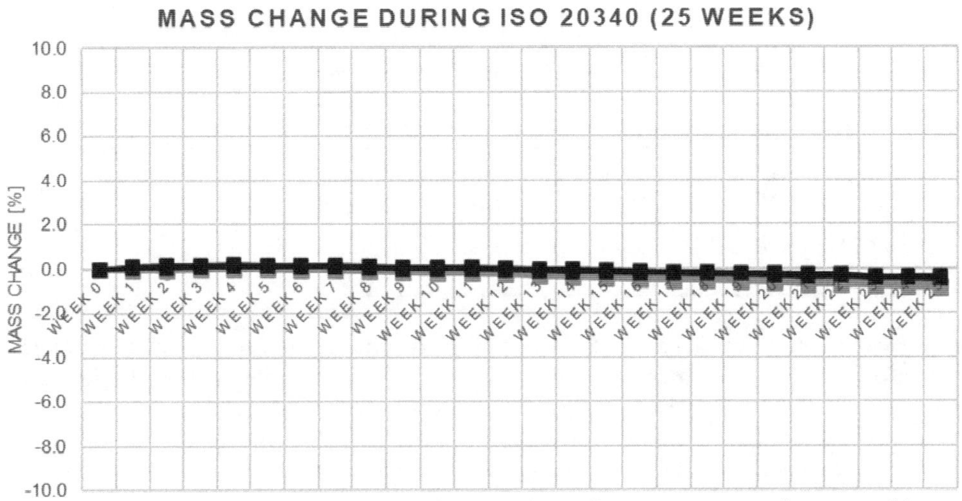

Figure 12: Mass change during weathering test cycles (Recyclamine® resin turquoise, reference purple)

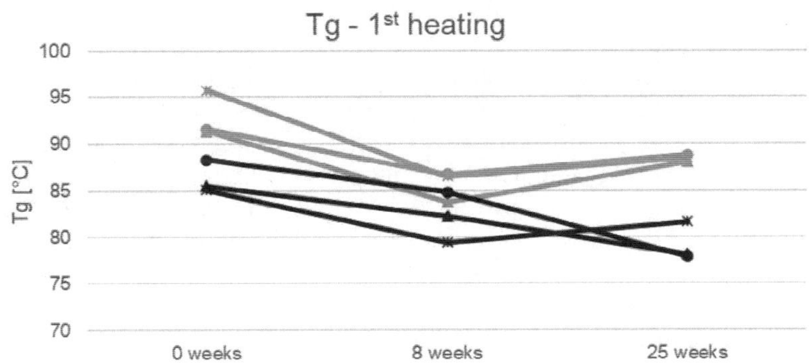

Figure 13: Evolution of T_G before, after 8 and 25 weeks/test cycles (Recyclamine® resin turquoise, reference purple). First heating ramp of DSC is depicted, second heating ramp determining ultimate Tg not shown).

Both resins show slight decrease of the weight during the test, and a non-significant difference of weight change after 25 weeks. Two effects are counteracting: Weight increase by water uptake and weight loss because of degradation under the test conditions. The recyclable resin shows less weight increase in the beginning and slightly more weight loss towards the end. This can indicate a higher water uptake of the reference resin compared to Recyclamine® resin.

5.3. Selectivity of the recycling process

The recycling process is efficient, but selective to the choice of acid and requires elevated temperatures. Representative acids and different conditions have been tested. The results are presented in Table 1 and Table 2.

Table 1: Test of different recycling conditions by weight loss of neat resin samples

Neat resin	HOAc	HOAc	HOAc	HCl
C in % w/w	30	30	30	10
T in °C	80	60	25	60
t in h	2	1	720	2
mass loss in %	75.6	3.8	41.1	0.9

Table 2: Test of different recycling conditions by weight loss of a thin biax glass fabric based laminate

Biax laminate	HOAc	Citric acid	HOAc	Citric acid /DMSO
C in % w/w	30	20	10	20/20
T in °C	80	80	80	60
t in h	2	2	2	5
mass loss in %	24.9	7.9	9.6	0.6

The above shown results demonstrate the selectivity of the recycling process. A strong, but inorganic acid with chloric acid as model substance was tested at 10 wt % concentration and 60 °C. At the same temperature and half the time, the mass loss with acetic acid was more than four times more. This indicates that low pH – this equals a high concentration of H_3O^+ which is the catalytic species for the acetal cleavage – is not enough to cleave the acetal/ketal bonds of the Briozen resin. The same effect is observed with citric acid, which is not able to catalyze the recycling process. So neither inorganic acids nor a standard organic acid are able to catalyze the recycling process efficiently, the addition of DMSO, a good solvent for many organic substances, appears to have no positive effect. Acetic acid seems to offer another special feature other than just generating H_3O^+. A potential explanation can be the property of acetic acid to be able to swell epoxy resins. The excellent swelling behavior is most likely based on the ability to break hydrogen bonds and the character as good solvent.

The effect of the recycling process on other materials was investigated in a series of tests. Sandwich panels based on balsa and PET core were manufactured, composites containing carbon pultruded profiles were prepared. The machined specimens from the panels were immersed in the acetic acid solution (30 wt%) at 80 C. When the whole resin was dissolved from the glass laminate, the core material or carbon profiles was isolated, dried, and weighed. As reference a panel with conventional resin and balsa wood was tested under the same conditions. With the estimated mass loss of the machining the weight change was calculated. The results are presented in Table 3.

Table 3: Mass change of materials subjected to recycling conditions

	PET foam	Balsa wood	Carbon profiles	Reference balsa
Mass change	+5.4 %	+73 %	+7.7 %	+12 %

The results of the balsa wood from the recyclable sandwich laminates show a significant weight increase. A potential explanation is the difficulty to completely remove resin which was taken up by the balsa wood during the infusion process. Also a significant change of the geometry because of swelling was observed.

PET foam was visually unaffected, neither the appearance nor the geometry changed during the recycling process. The weight change was not significant either. This can be explained by the much lower resin uptake of the closed cell structure of PET foam.

The carbon profiles appeared unaffected by the recycling process, but were recovered as one block, the recycling liquid seemed not to be able to penetrate well enough in between the profiles for dissolution of the resin and to allow for the separation of the individual carbon profiles.

The reference sandwich laminate shows a weight change of 12 % of the whole laminate, which is a combination of swelling of the resin and the balsa wood. The reference laminate was still intact, although in some cases the balsa deformed resulting in some delamination.

In a larger test on a representative coated sandwich composite specimen the recycling of a coated structure was investigated. The effect of the recycling process on the blade coating system was observed to be low. Interestingly the coating seems to be susceptible to the diluted acetic acid. At the process temperature the coating starts to blister, and the dissolution of the underlying resin appears to proceed almost unaffected. The coating comes off as a film. This is a good indication, showing that the preparation of the materials for the recycling process does not require a step removing the coating.

6. Conclusion

Recyclability is an important topic in wind industry and many related activities are currently running. Pyrolysis offers interesting potentials, but lately solvolysis has become a strong focus of investigations as well. Nevertheless, solvolysis of conventional resin systems is difficult and requires a sophisticated recycling process.

Resins that are specifically designed facilitate an easier recycling process. The cleavable bonds can be introduced in two different ways: randomly distributed or at defined locations. This influences the product of the recycling process together with the number of cleavable bonds.

A smart choice of the underlying technology allows to manufacture wind blades, that offer the same performance in all aspects as conventional blades, but with enhanced recyclability in a mild solvolysis process. The recovered materials are of good quality and high value, this results in an overall positive business case together with a relevant CO_2 eq. saving potential.

The presented results indicate that recyclability by design of resins is a promising approach and has the potential to replace conventional systems.

References

[1] European Commission, 2023. *environment.ec.europa.eu.* [Online] Available at: https://environment.ec.europa.eu/topics/waste-and-recycling/waste-framework-directive_en#ref-2023-amendment-to-the-waste-framework-directive

[2] Miljoskærm, 2021. *Miljoskærm.* [Online] Available at: https://miljoskarm.dk/en/

[3] André, A. et al. 2020 Re-use of wind turbine blade for construction and infrastructure applications *IOP Conf. Ser.: Mater. Sci. Eng.* **942** 012015

[4] Nagle, A. J., Mullally, G., Leahy, P. G. & Dunphy, N. P. 2022 Life cycle assessment of the use of decommissioned wind blades in second life applications *Journal of Environmental Management* **302**(A) 113994.

[5] Mishnaevsky Jr., L. 2021. Sustainable End-of-Life Management of Wind Turbine Blades: Overview of Current and Coming Solutions *Materials* **14**(5) 1124.

[6] Kaufmann, D. E., Zapolskii, V., Kolbe, P. & Uhde, G., 2020. *Method for complete recycling of inorganic-fiber* reinforced *epoxy composites with boron halides.* World, Patentnr. WO/2020/182484A1.

[7] Catack-H, 2023. *Catack-h.* [Online] Available at: https://catack-h.com/

[8] Teknologisk Institut, 2021. *Project Cetec.* [Online] Available at: https://www.project-cetec.dk/

[9] Meyer, L. O., Schulte, K. & Grove-Nielsen, E., 2009. CFRP-Recycling Following a Pyrolysis Route: Process Optimization and Potentials *Journal of Composite Materials* **43**(12) 1121-1132.

[10] Tavares de Moraes, V. et al. 2019 Microwave-Assisted Recycling Process to Recover Fiber from Fiberglass Polyester Composites *Mat. Res.* **22**(suppl.1).

[11] Cuthbert, J., Wanasinghe, S. . V., Matyjaszewski, K. & Konkolewicz, D. 2021 Are RAFT and ATRP Universally Interchangeable Polymerization Methods in Network Formation? *Macromolecules* **54**(18) 8331-8340.

[12] Pastine, S., 2014. *Sterically hindered aliphatic polyamine cross-linking agents, compositions containing and uses thereof.* World, Patentnr. WO2015/054698A1.

[13] Lin, C.-H.et al., 2022. *Carbonate-containing epoxy resin, manufacturing method thereof, epoxy curable product prepared thereby and method for degrading epoxy curable product.* United States of America, Patentnr. US20230102899A1.

[14] Gunawardena, G., 2023. *chem.libretexts.org.* [Online] Available at: https://chem.libretexts.org/Ancillary_Materials/Reference/Organic_Chemistry_Glossary/Retro_Diels-Alder_Reaction

[15] Briou, B., Améduri, B. & Boutevin, B. 2021 Trends in the Diels–Alder reaction in polymer chemistry *Chem Soc* Rev **50** 11055-11097.

[16] Ruiz de Luzuriaga, A. et al. 2016 Epoxy resin with exchangeable disulfide crosslinks to obtain reprocessable, repairable and recyclable fiber-reinforced thermoset composites. *Materials Horizons* **3** 241-247.

[17] Zhang, W. & Taynton, P., 2015. *Novel covalently cross-linked malleable polymers and methods of use.* World, Patentnr. WO2015138804A1.

[18] Koyuncu, B., Hoffmann, M., Rateau, F. & Vahk, J., 2021. *Chemical Recycling and Recovery,* s.l.: Zero Waste Europe.

IOP Conf. Series: Materials Science and Engineering 1293 (2023) 012007 doi:10.1088/1757-899X/1293/1/012007

A strategic approach to wind turbine blade recycling: Using life cycle assessment to enable data driven decision making

Kyle Pender, Filippo Romoli, Jonathan Fuller

National Composites Centre, Bristol & Bath Science Park, Emersons Green, Bristol BS16 7FS, UK

Kyle.pender@nccuk.com

Abstract. The persistent lack of circular, scalable, and low-impact solutions for decommissioned wind blades remains a challenge for the wind energy sector. With a saturation of proposed technologies, the sector needs to focus efforts on pulling through technologies both for the immediate and longer-term wind blade waste streams. In this work, models for high-profile EoL scenarios have been developed and, using lifecycle assessment, applied to representative onshore and offshore wind blade designs. In their current iterations, not all composite recycling approaches are environmentally beneficial when considering global warming potential (GWP). This is the case particularly for blades consisting of only glass fibre reinforcement, where only cement kiln and mechanical recycling scenarios resulted in GWP lower than landfilling. The introduction of carbon fibre reinforcement leads to reductions in GWP for almost all end-of-life scenarios, the most significant of which coming in the chemical and thermal scenarios. This indicates that to minimise carbon footprint a combination of approaches is required. These findings have been coupled with material circularity indicators and show that, while recycling can be beneficial, it is only one piece of the puzzle. Methods for integrating restorative materials into wind turbine blade designs, as well as finding circular solutions to reduce waste associated with current blade manufacturing practices, are also critical areas of research needed to mitigate the impact of future blade production.

1. Introduction

Most industrialised countries are relying heavily on the expansion of renewable energy generation to meet net zero targets, with wind being at the forefront of the global green energy transition. The amount of electricity generated by wind increased by almost 273 TWh in 2021, which represents the largest growth of all power generation technologies [1]. While wind energy generation can provide electricity with just 1-3% of the Greenhouse Gas (GHG) emissions produced by traditional fossil fuels [2], there remains room for improvement within the industry to help accelerate toward global net zero targets. Although more than 80% of materials used in current wind powerplants are recyclable [3], the persistent lack of circular, scalable, and low-impact solutions for decommissioned wind turbine blades (WTB) remains a challenge for the wind energy sector. Several studies have demonstrated the potential environmental benefits that could be realised by WTB recycling [4][5][6]. These findings point to the development of low impact recycling solutions as a key area of strategic development across the WTB lifecycle with the potential to yield a reduction in the impact of wind energy generation. A strategic and data-driven approach to blade recycling should be followed to ensure a minimal environmental impact and a commercially viable proposition.

The SusWIND programme has developed a life cycle impact approach that incorporates Life Cycle Assessment (LCA) and Material Circularity Indicators (MCI) to give a comprehensive picture of the environmental impact of WTBs from cradle-to-grave. Proprietary models have been developed to assess the environmental impact across a range of WTB end-of-life (EoL) scenarios, which cover all the predominant thermal, chemical, mechanical technologies, in addition to cement kiln co-processing, landfill, and incineration. LCA and MCI analysis has been applied to 1) a 5 MW glass fibre reinforced reference blade produced by the National Renewable Energy Laboratories (NREL) 5 MW [7] and, 2) a 15 MW glass/carbon fibre hybrid reference blade produced by the International Energy Agency (IEA) [8]. This has enabled matching of present and future blade waste streams material compositions with preferred waste treatment options for lowest impact across the WTB lifecycle.

2. Materials and methods

2.1. Reference blades

Two different WTB designs were analysed in this study (Appendix A gives the bill of materials of the two WTB designs):

1) "*GF WTB*": A WTB design reinforced with glass fibre (GF) (i.e. no carbon fibre (CF) reinforcement commonly seen in other WTB models). This design is representative (in terms of material composition) of legacy onshore WTBs comprising the majority of the first wave of WTBs to reach end-of-life (EoL) in the coming decade. The design for the "GF WTB" used in this work is based on the 5 MW reference blade reported by NREL in [7].

2) "*GF/CF hybrid WTB*": A WTB design with GF reinforcement in the shells and shear webs and CF reinforcement in the spar caps. This design is representative of future onshore/offshore EoL WTB waste streams following the widespread use of CF in spar cap regions. The design for the GF/CF hybrid WTB used in this work was developed from the IEA 15 MW reference wind turbine [8] developed as part of the IEA Wind Task 37 by NREL and Technical University of Denmark (DTU). Due to buckling and aeroelastic stability issues of this specific blade design (identified in a thorough analysis performed in [9]), a new round of calculations was conducted, and an optimised blade design was achieved satisfying all the problematic constraints. The design calculations as well as main characteristics of this blade can be found in [9].

2.2. Scenarios

Table 1 gives a summary of the WTB and EoL scenario combinations analysed. Appendix B gives a description of the EoL scenarios and assumptions made in generating the Lifecycle Inventory (LCI) data.

Table 1 Summary of WTB and EoL scenario combinations analysed, ✓- scenario was analysed, X – scenario not analysed

	GF WTB	GF/CF hybrid WTB
Landfill	✓	✓
Incineration	✓	✓
Cement kiln co-processing	✓	X
Mechanical recycling	✓	✓
Fluidised bed recycling	✓	✓
Pyrolysis recycling	✓	✓
Solvolysis recycling	✓	✓
Cement kiln+	X	✓
Mechanical+	X	✓

Figure 1 presents a schematic showing an overview of process and material flows across the various EoL scenarios investigated. For disposal routes (landfill, incineration/Energy from Waste (EfW)), cement kiln, mechanical recycling, thermal recycling (fluidised bed, pyrolysis), and chemical (solvolysis) recycling it is assumed that all WTB substructures (shells, shear webs, spar caps) are processed using the respective EoL treatments. For recycling routes Cement kiln+ and Mechanical+ (shown in Figure 1), it is assumed that the spar cap is isolated from the rest of the blade structure during initial sectioning activities. The spar cap is then recycled using pyrolysis, and all other regions (shear webs, shells – which make up approximately 75% of the WTB mass) are then recycled via cement kiln co-processing or mechanical recycling. Cement kiln+ and Mechanical+ are investigated as they provide an opportunity to extract the valuable CF located in the spar cap, while diverting the other GF rich substructures to (potentially) lower impact recycling routes. Since the GF WTB scenario contains no CF, the Cement kiln+ and Mechanical+ EoL treatment scenarios have not been applied to that blade. Following discussion with stakeholders in the cement production industry, it is believed that CF cannot be processed in the cement kiln. As such, the GF/CF hybrid WTB scenario cannot be processed via cement kiln without prior spar cap removal (given that CF is only located in the spar caps for this design).

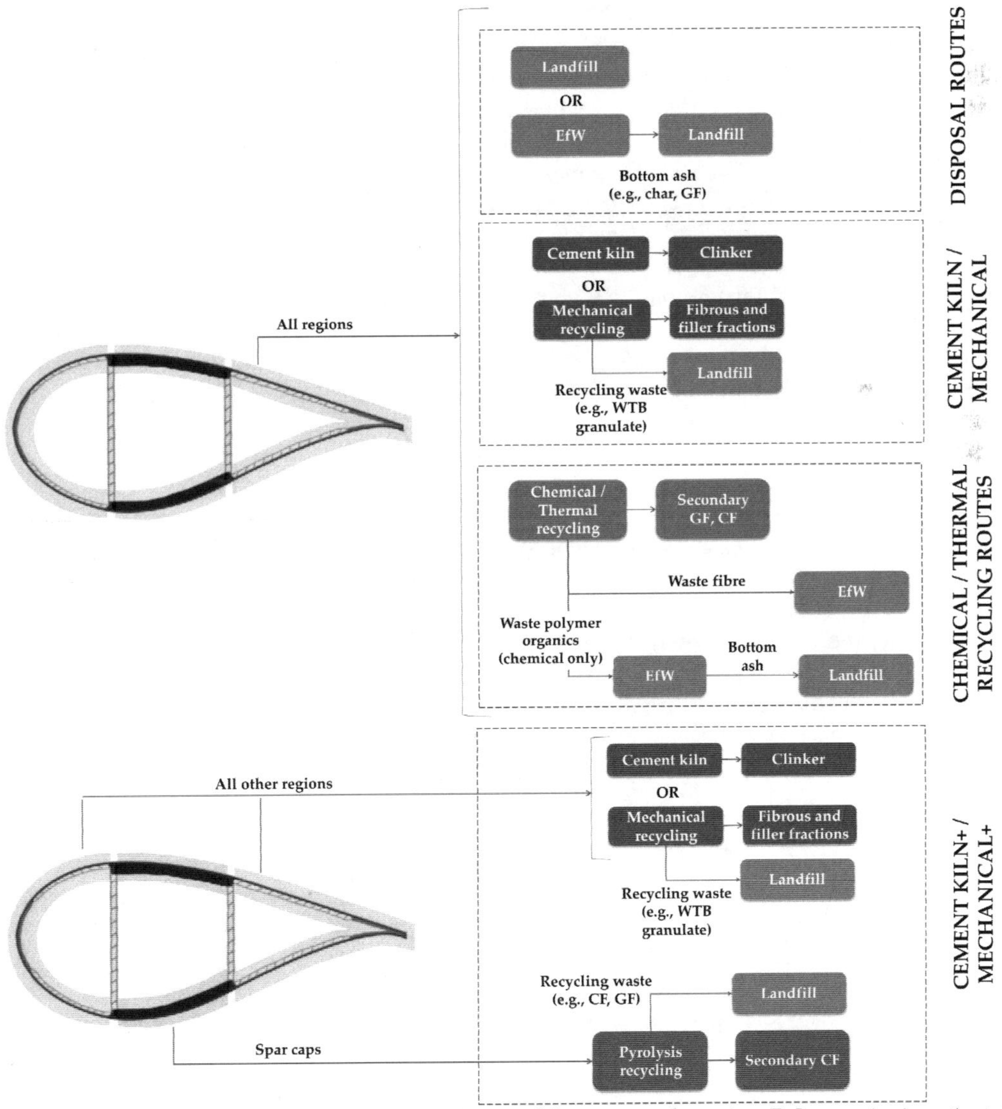

Figure 1 Schematic showing overview of process and material flows across the various EoL scenarios investigated

2.3. Lifecycle assessment

The LCA has been conducted with reference to the methods provided in ISO 14040 related to the preparation of LCA studies. The aim of the LCA was to evaluate the Global Warming Potential (GWP) of various EoL routes that have been proposed for WTBs both with and without CF reinforcement.

The LCA only considers the processes involved in the EoL treatment of the WTB (i.e., secondary material production in the case of recycling routes) and does not consider any processes associated with upstream phases of the WTB lifecycle (such as manufacture, installation, use, decommissioning). Processes considered at EoL are: 1) downsizing the blade using industrial shredding equipment; 2) transporting shredded waste to a waste management facility; 3) disposal / recycling of blade waste using the following processes: landfill, incineration (with energy recovery), cement kiln co-processing, mechanical recycling, thermal recycling (fluidised bed and pyrolysis), and chemical (solvolysis) recycling; 4) transporting recycling waste to a waste management facility; 5) disposing recycling waste. The impact of WTB cutting / sectioning processes performed prior to EoL treatment (including activities to isolate the spar cap in cement kiln+ and mechanical+ scenarios) is expected to be minimal and has not been considered within the LCA.

An avoided burden approach has been used for secondary material production (e.g., recycled GF and CF products from blade recycling), observing the following: 1) the burden for virgin materials used is allocated to the system, 2) the burden of secondary material production is included within the system boundary and is allocated to the system, 3) materials recycled at EoL offset the demand for a quantity of virgin counterpart materials, 4) knockdown in offset rates are included based on loss in performance and the impact this will have on products utilising recycled materials, 5) all blade waste is either recycled or disposed in the system boundary and an avoided burden credit is given to secondary materials produced.

The Lifecycle Inventory (LCI) secondary datasets used in this report have been sourced from the following databases: GaBi Professional Database 2022, Extension database XXII: Carbon Composites 2022, Extension database VII: Plastics 2022. No LCI primary data has been collected or measured. The materials and energy flows needed to manufacture the representative blade have been estimated/calculated based on data and assumptions reported in literature (see Appendix B).

2.4. Material circularity indicators

MCIs were calculated following the methodology outlined in [10]. The "Product-level Methodology" was used as this was deemed the most suitable approach to assessing material circularity on a single WTB level. In this case, the MCI measures the extent to which linear flow has been minimised and restorative flow maximised for the WTB's materials, and how long and intensively it is used compared to a similar industry-average product. The MCI is essentially constructed from a combination of three product characteristics: 1) the mass of virgin raw material used in manufacture, 2) the mass of unrecoverable waste that is attributed to the product, and 3) a utility factor that accounts for the length and intensity of the product's use. The methodology used to calculate the MCI is detailed in [10] and will therefore not be reproduced in this report. A variation on the basic method was used to account for production losses and, when required, additional material inputs to the recycling process (e.g., solvents / chemical inputs in chemical-based recycling processes). The *Multiple Production Steps* approach was also used to account for the different material type inputs used in the blade manufacture [10]. As is common practice within the wind industry, it is assumed that all raw materials input during the WTB production come from virgin sources. In this work it is assumed the blade life-expectancy and use intensity (e.g., annual energy production) is constant across WTB and EoL scenarios analysed for a given blade type. The utility factor was therefore the same across all scenarios and did not influence the MCI calculation. The MCI analysis gives a value between 0 and 1, where higher values indicate higher circularity. A fully circular system has an MCI of 1, which occurs when 1) all feedstock materials are from fully restorative sources, and 2) all waste / output materials are fully restored (e.g., reused, recycled).

3. Results and discussion

3.1. Lifecycle assessment

Figure 2 gives the GWP results from the LCA for each of the WTB and EoL scenario combinations described in Table 1. The various sources of GWP have been categorised in Figure 2 to highlight dominant contributors for each treatment option and to give insight into areas of development that should be pursued to reduce the impact of the processes. A positive GWP indicates a contribution to GWP associated with GHG emissions (either directly or indirectly), whereas a negative GWP indicates a reduction in GWP (through the avoided burden of virgin material or energy production) enabled by the EoL treatment. The "Total recycling GWP" in Figure 2 is the sum of all GWP sources for a given EoL treatment, with a negative total recycling GWP indicating an overall reduction in GWP enabled by the treatment option. While an EoL treatment my increase circularity and reduce landfill burden, it is important to quantify the GHG emissions entailed by recycling processes and assess their contribution to Net Zero targets. Ultimately, optimal EoL solutions for WTB should both increase circularity and result in an overall reduction in GWP.

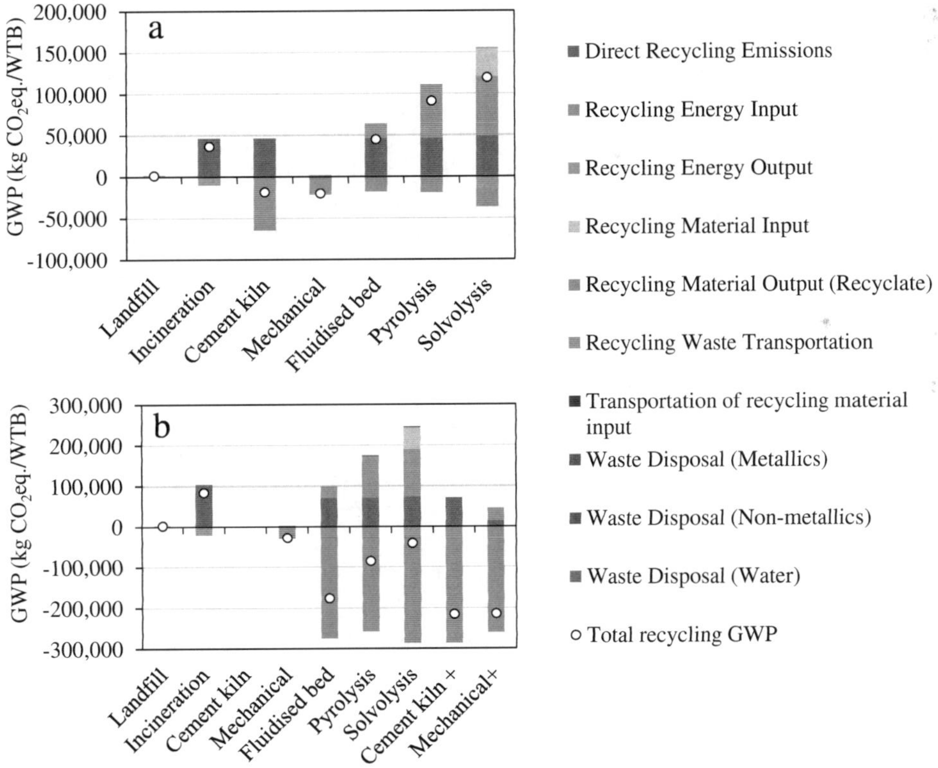

Figure 2 GWP of various EoL scenarios, a) "GF WTB", b) "GF/CF hybrid WTB"

For the GF WTB scenario (Figure 2a), it was found that cement kiln and mechanical recycling are the only EoL treatments able to provide an overall negative GWP. While cement kiln co-processing does result in direct emissions through combustion of the polymer fractions in the WTB, it is assumed that the WTB offsets energy input from petroleum coke which (relative to calorific value) generates more GHG emissions that WTB combustion. Development of green cement is critical for decarbonisation of the construction sector, and WTB scrap as an alternative lower carbon fuel source could play a role. In kilns where alternative fuels, such as solid recovered fuels, are already being used, it is yet to be established if substitution with WTB scrap will be an environmentally preferable alternative. To date, the LCA literature in this area has assumed only fossil fuel replacement (either coal or petroleum coke). It is therefore critical that future studies assess the impact when substituting alternative fuels in cement kilns, which may burn cleaner than fossil fuel counterparts and/or contain higher levels of biogenic carbon, the combustion of which is considered net zero.

Figure 2a shows that the secondary material produced during mechanical, fluidised bed, pyrolysis, and chemical recycling has low opportunity to avoid virgin material burden (as shown by "Recycling Material Output" in Figure 2a). This is for three key reasons, 1) GFs suffer significant mechanical performance loss during recycling which reduces the replacement rate of virgin GF counterparts, and 2) the GWP associated with virgin GF is already low (compared to other blade materials such as resins and CF) meaning that there is a limited amount of GHG emissions that can be displaced by replacing virgin with recycled counterparts, 3) GF is the only secondary material able to offset virgin material production. These factors mean that, for the total recycling GWP to be negative for a given EoL treatment, the impact associated with the recycling processes themselves must also be low to not supersede the limited avoided burden attributed through secondary GF production.

Figure 2a also shows that mechanical recycling is the only recycling technology which can produce secondary GF material and maintain a negative total GWP. This is primarily due to the low energy demand of the process in addition to the lack of direct GHG emissions from the process (which is found in thermal recycling technologies, as well as when incinerating organic residues from chemical recycling). A major barrier to the reuse of mechanical recyclate however is that, while fibre rich fractions can be extracted, resin contaminant persists on the fibre surface which may, 1) alter/limit the adhesion with polymers in secondary composite components, and 2) limit the number of technologies available to manufacturers due to material compatibility with existing processes / practices. It was demonstrated in [11] that the fibrous fractions can replace virgin GF in bulk moulding compound (BMC) production without sacrificing resulting composite strength. It is therefore assumed that the recovered fibrous fractions can offset the production of virgin GF, which would otherwise be used in the production of BMC products.

Figure 2b shows the LCA results for the GF/CF hybrid WTB. Recycling CF materials provides a much greater opportunity to offset GHG emissions by replacing high impact virgin CF production. This means that for all recycling scenarios in Figure 2b, the GF/CF hybrid WTB yields a negative total GWP and are therefore preferable to disposal routes (landfill and incinerations) when aiming to minimise GHG emissions. While mechanical recycling is found to be the lowest impact option for the GF WTB, it is the highest impact recycling scenario investigated for the GF/Cf hybrid WTB. This is because the CF is not being recycled into a format which can compete with virgin CF. It is assumed that fibrous mechanical recyclate from CF composite regions can only replace virgin GF, therefore the avoided burden associated with the process is low, as in the case for the GF WTB. Figure 2b shows that Cement kiln+ and Mechanical+ EoL scenarios provide the lowest GWP for the GF/CF hybrid WTB. This is because the WTB sub-structure material compositions are matched with preferred waste treatment options for lowest overall impact. This approach ensures that the additional GHG emissions associated with thermal or chemical processes (that is required to extract the CFs) are only applied to the CF rich spar caps, while deploying lower impact strategies to the shells and shear webs where there is less opportunity to reclaim valuable, high impact secondary materials.

It should be noted that the results presented in this LCA are a snapshot in time, the outcomes of which will likely change as developments to recycling technologies are realised. In the current iteration, pyrolysis suffers from having both high energy demand and direct GHG emissions from combustion of the polymer pyrolysis products. Future state of pyrolysis technology may enable 1) self-sustained processing through re-circulation and combustion of pyrolysis gases to power the system, and 2) condensation, collection and reprocessing of pyrolysis oils/waxes to produce feedstocks for the chemical industry. These solutions have the potential to reduce both the energy consumption and direct emissions, but also provide additional products from pyrolysis with the opportunity to offset virgin petrochemicals (or similar products). While this work has considered thermo-oxidative recycling using the fluidised bed, these systems can also operate under pyrolytic conditions and therefore has the same potential to reduce GHG emissions as the belt driven pyrolysis process modelled in this LCA. Future state of solvolysis technology may enable 1) a reduction in energy demand through catalysed polymer decomposition, and 2) the reprocessing of recovered organic residues to produce feedstocks for the chemical industry.

3.2. Material circularity indicators

Figure 3 presents a comparison of the WTB MCIs and total recycling GWP (as given in Figure 2) for the various EoL scenarios. The MCI analysis gives a value between 0 and 1, where higher values indicate higher circularity, and a fully circular system has an MCI of 1. EoL solutions existing within the upper left quadrant of the graphs in Figure 3 should be prioritised because they have lower GWP and higher circularity (measured as MCIs) and can better accelerate towards both Net Zero Emissions and Zero Waste to landfill targets. As would be expected, disposal routes (landfill and incineration) have the lowest MCI and are clearly the least favourable EoL strategy for GF/CF hybrid WTBs.

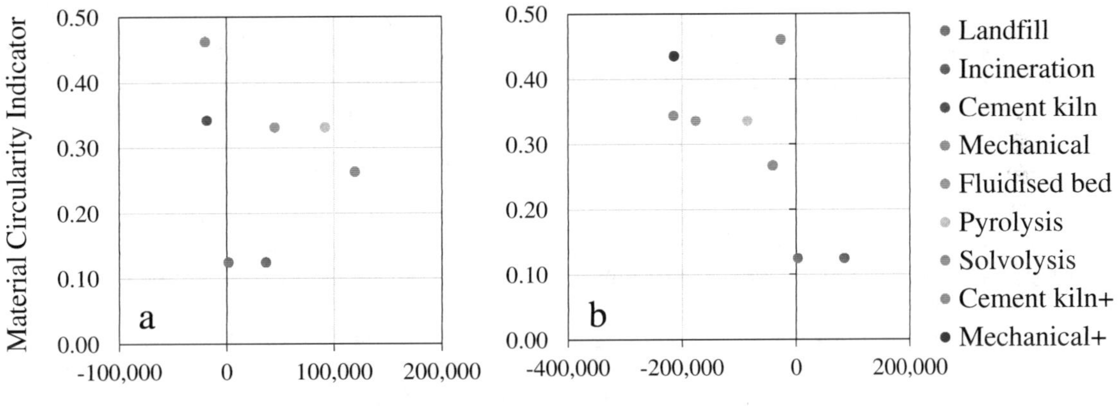

Figure 3 MCI and recycling GWP for the various EoL scenarios, a) "GF WTB", b) "GF/CF hybrid WTB"

Figure 3 shows that mechanical recycling has the highest MCI and lowest GWP for the GF WTB and is therefore identified as the priority EoL strategy to pursue for the first wave of EoL WTBs expected across Europe in the coming decade. Critical in realising economic viability will be the value of mechanical recycled granulates and the products there of. Developing use cases for the materials and strengthening routes to market is needed to ensure that fibrous recyclate can (in some applications) replace virgin GF, as is assumed in this LCA. On the balance of data presented in Figure 3, Mechanical+ is identified as the priority EoL strategy for the GF/CF hybrid WTB scenario. Mechanical+ has the lowest GWP of the EoL scenarios as described above and is approaching the highest MCI (with mechanical recycling alone having a marginally higher MCI).

In all scenarios in Figure 3, the determined MCI values are significantly lower than the theoretical maximum MCI of 1. This is because the MCI methodology also considers the sources of materials entering the product system during manufacture and use phases. As is common practice within the wind industry, it is assumed that all raw materials input during the WTB production come from virgin sources. These are by nature non-restorative and decreases the overall MCI. To enable a circular economy for WTBs, it is therefore important to extend research beyond recycling and focus on methods for integrating restorative material flows (e.g., bio-derived, reused, recycled materials) into blade designs and production. An example that is already done commercially by some turbine OEMs is the use of recycled PET core materials to replace traditional virgin PET material. Additionally, industry standard practices during WTB production result in significant levels of manufacturing waste which is currently not recycled. Identifying circular solutions to these materials will also be critical in transitioning to a circular economy for WTBs.

Recovery of usable matrix polymers from analogous composite structures has been demonstrated using chemical based recycling process for alternative "recyclable epoxy" systems (e.g., Recyclamine

[12]) and standard epoxy systems using targeted catalsyed decomposition [13]. The former requires the use of specific epoxy hardener systems during WTB production and there is currently a lack of published data to conduct independent LCA for the latter, therefore these recycling solutions have not been considered in this analysis. Regardless, both these approaches have the potential to increase WTB circularity by enabling the reuse of the polymer matrix which constitutes 21-23 %wt. of the WTB's bill of materials.

4. Conclusions

This investigation has quantified the GWP of various EoL treatment options for WTB designs with and without CF reinforcement. To compliment the LCA, the circularity across the WTB lifecycle has also been characterised using the MCI methodology for each of the EoL scenarios. Based on the current state of technologies, and the best available data the following conclusion can been made for GF WTBs:

- It was found that cement kiln and mechanical recycling are the only EoL treatments able to provide an overall negative GWP. To date, LCAs investigating cement kiln coprocessing of WTBs have assumed fossil fuel energy replacement. With the growing use of alternative fuels in kilns, it is critical that future studies also assess the impact when substituting alternative fuels which may burn cleaner than fossil fuel counterparts.
- Mechanical recycling provides the highest MCI and is the only recycling technology which can produce secondary GF material and maintain a negative GWP. Developing use cases for the materials and strengthening routes to market is needed to ensure that fibrous recyclate can indeed replace virgin GF as is assumed in this analysis.
- Thermal and chemical recycling routes produce a positive GWP and in their current state of development are not priority routes for WTB recycling. Methods should be pursued which can reduce the energy consumption and direct emissions associated with these processes, in addition to reclaiming resin fractions capable of offsetting feedstock chemicals. As and when these developments are realised, LCA should be used to re-evaluate these technologies and alter recommendations were appropriate.

The following conclusions can be made for GF/CF hybrid WTBs:

- Recycling CF materials provides a much greater opportunity to offset GHG emissions by replacing high impact virgin CF production therefore all recycling scenarios for GF/CF hybrid WTBs yield a negative GWP and are preferable to disposal routes.
- Cement kiln+ and Mechanical+ EoL scenarios provide the lowest GWP, because the WTB sub-structure material compositions are matched with preferred waste treatment options for lowest overall impact. This approach ensures that the additional GHG emissions associated with thermal or chemical processes are only applied to the CF rich spar caps, while deploying lower impact strategies to regions where there is less opportunity to reclaim high impact secondary materials.
- Mechanical+ was identified as the priority EoL strategy for the GF/CF hybrid WTB scenario when considering GWP and MCI environmental metrics. It is therefore critical that method(s) for spar cap extraction are developed in advance of future WTB waste streams where CF in these regions will be ubiquitous (e.g. in the case of offshore WTBs).

Enabling a circular economy for WTB means extending research beyond recycling alone. Methods for integrating restorative materials into WTB designs, as well as finding circular solutions to reduce waste associated with current WTB manufacturing practices, are also critical areas of research needed to mitigate the impact of future WTB production.

References

[1] IEA, "Wind Electricity," 2022. [Online]. Available: https://www.iea.org/reports/wind-electricity.

[2] Schlömer S. *et al.*, "Climate Change 2014: Mitigation of Climate Change. Contribution of Working Group III to the Fifth Assessment Report of the Intergovernmental Panel on Climate Change. Technology-specific Cost and Performance Parameters," 2014. doi: 10.1017/cbo9781107415416.025.

[3] M. Y. Khalid, Z. U. Arif, M. Hossain, and R. Umer, "Recycling of wind turbine blades through modern recycling technologies: A road to zero waste," *Renew. Energy Focus*, vol. 44, pp. 373–389, 2023, doi: 10.1016/j.ref.2023.02.001.

[4] G. Chiesura, H. Stecher, and J. Pagh Jensen, "Blade materials selection influence on sustainability: A case study through LCA," *IOP Conf. Ser. Mater. Sci. Eng.*, vol. 942, no. 1, 2020, doi: 10.1088/1757-899X/942/1/012011.

[5] P. Liu, F. Meng, and C. Y. Barlow, "Wind turbine blade end-of-life options: An eco-audit comparison," *J. Clean. Prod.*, vol. 212, pp. 1268–1281, 2019, doi: https://doi.org/10.1016/j.jclepro.2018.12.043.

[6] A. J. Nagle, E. L. Delaney, L. C. Bank, and P. G. Leahy, "A Comparative Life Cycle Assessment between landfilling and Co-Processing of waste from decommissioned Irish wind turbine blades," *J. Clean. Prod.*, vol. 277, p. 123321, 2020, doi: 10.1016/j.jclepro.2020.123321.

[7] J. Jonkman, S. Butterfield, W. Musial, and G. Scott, "Definition of a 5-MW reference wind turbine for offshore system development," *Contract*, no. February, pp. 1–75, 2009.

[8] E. Gaertner, J. Rinker, L. Sethuraman, B. Anderson, F. Zahle, and G. Barter, "IEA Wind TCP Task 37: Definition of the IEA 15 MW Offshore Reference Wind Turbine," pp. 1–44, 2020, [Online]. Available: https://github.com/IEAWindTask37/IEA-15-240-RWT.

[9] S. Scott, P. Greaves, T. Macquart, and A. Pirrera, "Comparison of blade optimisation strategies for the IEA 15MW reference turbine," *J. Phys. Conf. Ser.*, vol. 2265, no. 3, 2022, doi: 10.1088/1742-6596/2265/3/032029.

[10] Ellen MacArthur Foundation, "Circularity Indicators: An Approach to Measuring Circularity," *Ellen MacArthur Found.*, p. 12, 2015.

[11] J. Palmer, O. R. Ghita, L. Savage, and K. E. Evans, "Successful closed-loop recycling of thermoset composites," *Compos. Part A Appl. Sci. Manuf.*, vol. 40, no. 4, pp. 490–498, 2009, doi: 10.1016/j.compositesa.2009.02.002.

[12] A. D. La Rosa, I. Blanco, D. R. Banatao, S. J. Pastine, A. Björklund, and G. Cicala, "Innovative chemical process for recycling thermosets cured with recyclamines® by converting bio-epoxy composites in reusable thermoplastic-an LCA study," *Materials (Basel).*, vol. 11, no. 3, 2018, doi: 10.3390/ma11030353.

[13] A. Ahrens *et al.*, "Catalytic disconnection of C – O bonds in epoxy resins and composites," no. August 2022, 2023, doi: 10.1038/s41586-023-05944-6.

[14] Tolvik Consulting, "UK Energy from Waste Statistics – 2019," no. June, pp. 1–20, 2019.

[15] A. M. use of municipal solidwaste incineration ash in various building materials: A. B. point of view Joseph, R. Snellings, P. Van den Heede, S. Matthys, and N. De Belie, "The use of municipal solidwaste incineration ash in various building materials: A Belgian point of view," *Materials (Basel).*, vol. 11, no. 1, 2018, doi: 10.3390/ma11010141.

[16] N. A. Shuaib and P. T. Mativenga, "Energy demand in mechanical recycling of glass fibre reinforced thermoset plastic composites," *J. Clean. Prod.*, vol. 120, pp. 198–206, 2016, doi: 10.1016/j.jclepro.2016.01.070.

[17] F. Meng, J. R. McKechnie, and S. J. Pickering, "Environmental and cost analysis of carbon fibre composites recycling," The University of Nottingham, 2017.

[18] F. Meng, E. A. Olivetti, Y. Zhao, J. C. Chang, S. J. Pickering, and J. McKechnie, "Comparing Life Cycle Energy and Global Warming Potential of Carbon Fiber Composite Recycling

Technologies and Waste Management Options," *ACS Sustain. Chem. Eng.*, vol. 6, no. 8, pp. 9854–9865, 2018, doi: 10.1021/acssuschemeng.8b01026.

[19] M. J. Keith, G. Oliveux, and G. A. Leeke, "Optimisation of solvolysis for recycling carbon fibre reinforced composites," *ECCM 2016 - Proceeding 17th Eur. Conf. Compos. Mater.*, no. June, 2016.

Appendix A – WTB bill of materials

	Mass (kg/blade)										
	Gel Coat	Primer	Topcoat	Glass fibre	Carbon fibre	Epoxy	PET (core)	Adhesive	Steel	Aluminium	Total
GF WTB	140	140	140	26,206	0	10,094	5,090	2,357	2,828	139	47,134
GF/CF hybrid WTB	260	260	260	33,746	9,073	16,977	5,799	3,744	4,493	269	74,880

Appendix B – Description of EoL treatments

Landfill

The impact when landfilling the various WTB materials was estimated based on representative GaBi LCI data sets: Glass fibre: "EU-28 - Inert matter (Glass) on landfill"; Carbon fibre, Epoxy, PET, Adhesive, Surface coatings: "EU-28- Plastic waste on landfill".

Incineration

Wind blade waste can be co-processed with solid municipal waste within Energy from Waste (EfW) facilities. The energy produced depends on the waste energy content and the incineration plants efficiency at generating useful energy. The calorific value of the waste is determined by the calorific value of the WTB's constituent materials. EfW facilities vary in efficiency, with typical electrical efficiencies of 15-24%. The average efficiency across all UK energy from waste plants is used in the model. Using data published in [14], the average electrical power and heat efficiencies across all UK was estimated to be 20.7% and 4.3% respectively. Electricity generated is assumed to displace the UK mains electrical supply whereas heat is assumed to offset heat production from natural gas combustion, assuming 80% efficiency [15]. It is assumed that residual fraction (bottom ash) of the WTB is landfilled.

Cement kiln co-processing

WTB waste is already commercially used as both fuel and raw materials in the production of clinker within cement kilns. The production process combines raw material fractions with an energy source (fossil fuel) and heated to around 1450 °C to produce cement clinker; composed of calcium oxide, silica, alumina and iron oxide. Calcium carbonate is a raw material in clinker production which is calcified to calcium oxide within the kiln; it is assumed this can be offset with the calcium oxide present in the glass fibre. Combustion of the organic fraction of the WTB waste within cement kilns can be used to heat the process and offset demand for petroleum coke. In this study it is assumed that WTB waste can replace petroleum coke at an energy equivalence basis. This replacement rate is dependent on the WTB calorific value and varies between WTB scenarios. E-glass with silica, calcium oxide and alumina content of 54, 18 and 15 %wt. respectively was used in modelling the cement kiln co-processing route. It was assumed all minerals will directly replace raw materials on a mass equivalent basis.

Mechanical recycling

Mechanical recycling of WTB waste was modelled using a downsizing and classification process outlined in [11]. WTB waste is first ground using a granulator then classified into four fractions using a zig-zag classifier. The four distinct fractions differ in composition and size, with the process producing two fibre rich fractions, a "coarse" resin rich fraction and a final "powder" fraction. Energy demand for both downsizing and classification stages required for mechanical recycling were modelled. Downsizing energy using a granulator was calculated using the method outlined in [16]. All energy for mechanical recycling is assumed to be electrical and sourced from UK mains supply. It is assumed that the recovered fibre rich fractions can offset the production of virgin GF at a rate proportional to retained strength of GF in the recyclate. It is proposed that the powder fraction may be used as filler and offset the production of calcium carbonate filler material. The coarse, resin rich fraction, is assumed to be a waste product in the study and is assumed to be landfilled.

Fluidised bed recycling
Thermal recycling composite materials in a fluidised bed allows for oxidative decomposition of the polymer matrix, liberating the clean fibre (GF or CF) fractions which are transported by the gas stream out of the reactor and collected. The combustion gases are fully oxidised in an oxidising chamber to remove volatiles then passed through heat exchanger to recover heat to be fed back into the process. Heat is supplied into the system through oxidation of combustibles (e.g. polymers) present in the WTB waste stream, as well as through natural gas oxidation in the oxidiser. The fluidised bed requires fibre length <10 mm to operate efficiently; therefore, pre-treatment of waste WTB involves a secondary downsizing step in a granulator. The energy input requirements have been sourced from literature LCA [17]. Fibre recovery rate is assumed to be 95% [17], with "waste fibres" assumed to be landfilled. It is assumed that all polymer materials in the WTB are fully oxidised during the fluidised bed process. It is assumed that recycled GF and CF can offset the production of virgin counterparts. Offset knockdown factors (proportional to retained strength of GF/CF in the recyclate) have been applied which dictates the replacement rate of virgin fibre counterparts.

Pyrolysis recycling
Thermal recycling of WTB waste using pyrolysis involves decomposition of polymers without (or in low) oxygen at elevated temperatures [18]. This is often a multistage process, requiring different temperatures and/or atmospheric conditions to remove thermally stable char residues in order to reclaim contaminant free fibres. The polymer decomposition is endothermic and requires significant energy input. The sited energy input for pyrolysis recycling of composite materials varies throughout the literature. Energy-related inventory data obtained from a commercial operation from ELG Carbon Fibre (reproduced [18]) was used in this study. It is assumed that all products from the polymer pyrolysis are eventually fully oxidised prior to releasing to the environment. Fibre recovery rate is assumed to be 95%, with "waste fibres" assumed to be landfilled. It is assumed that recycled GF and CF can offset the production of virgin counterparts. Offset knockdown factors (proportional to retained strength of GF/CF in the recyclate) have been applied which dictates the replacement rate of virgin fibre counterparts.

Chemical (solvolysis) recycling
Chemical recycling of WTB waste using solvolysis involves breaking down the polymeric fraction by utilising a solvent, often using elevated temperature and/or pressure. An acetone-based process has been used to model chemical recycling of WTB waste and gather the associated LCI data needed. Data has been extracted from a published LCA on this process using primary data of a lab scale reactor [19]. It is assumed that all polymeric material fractions in the WTB can be broken down during solvolysis, which result in an "organic residue" fraction. It is assumed the organic residue is sent to energy from waste for disposal. Fibre recovery rate is assumed to be 95%, with "waste fibres" assumed to be landfilled. It is assumed that recycled GF and CF can offset the production of virgin counterparts. Offset knockdown factors (proportional to retained strength of GF/CF in the recyclate) have been applied which dictates the replacement rate of virgin fibre counterparts.

IOP Conf. Series: Materials Science and Engineering 1293 (2023) 012008 doi:10.1088/1757-899X/1293/1/012008

Analysis of a small recyclable wind turbine blade

Rafael Carnicero[1], Luis Cano[1], Ignacio Cruz[1], Miguel A Lopez-Manchado[2] and Raquel Verdejo[2 3]

[1] CEntro de Desarrollo de Energías Renovables (CEDER – CIEMAT), Soria, Spain.
[2] Instituto de Ciencia y Tecnología de Polímeros (ICTP – CSIC), Madrid, Spain.

[3]rafael.carnicero@ciemat.es, luis.cano@ciemat.es, ignacio.cruz@ciemat.es, lmanchado@ictp.csic.es, r.verdejo@csic.es

Abstract. The wind energy industry is constantly seeking lower levelized cost of energy (LCOE) and also to improve its environmental and social sustainability. Thermoplastic resins are considered to be key to recycling wind turbine blades in the best possible way to reuse the main raw materials at the end of their life. In this work, a new thermoplastic liquid resin, AKELITE, patented by CSIC, is used to produce a new generation of 100% circular, sustainable wind turbine blades. The aim is to recover the two main components, the fibres and the resin, in optimal conditions to produce a new wind blade. First, the recycling protocol of composite laminates was optimised and the properties of the recycled composite were analysed. Interestingly, the first recycling process resulted in an improvement in the mechanical properties compared to the original laminate, which is ascribed to the presence of a thin layer of resin on the fibre surface that improves the adhesion between the layers. Thus, it is expected that the new blade will have a good performance, using almost 100% of recyclable materials. A 1.9 m recycled wind turbine blade is currently being manufactured to validate the results of this study. The successful development of such blades could have a significant impact on the sustainability of wind energy, providing a path towards circularity and reducing the environmental impact of wind turbine blades at the end of their life cycle.

1. Introduction

Currently, between 85% and 90% of the total mass of a wind turbine can be recycled. Wind turbine blades are the most difficult to recycle due to the composite materials used in their manufacture. Recycling solutions are not yet widely available and cost-competitive [1].

Some solutions are focused on achieving blade circularity. Siemens Gamesa installed in 2022 the world's first turbine equipped with recyclable blades (81 m) in RWE's Kaskasi offshore wind farm [2]. LM Wind Power has manufactured a 62 m blade using Arkema's thermoplastic resin and high-performance glass fabrics from Owens Corning in the ZEBRA project [3]. In 2023, MingYang Smart Energy has launched a 75.7 m turbine blade with over 95% recyclable materials, claiming to be the first Chinese OEM to offer recyclable blades [4].

In this work, a new infusible thermoplastic resin (AKELITE [5]) is processed by in mould polymerisation. A previous study of a 1 m blade made with Akelite showed that the static and centrifugal performance were similar to those of a commercial epoxy [6]. The study also showed the recycling capabilities of the resin through immersion in a suitable solvent.

This paper presents the optimization of the recycling protocol, analysing some parameters such as solvent/composite ratio, dissolution time and dissolution temperature, and the properties of the recycled material. Thus, samples of 75 x 20 cm composite materials have been made, tested, recycled and later re-manufactured in order to test the properties of the second generation composites. These steps are being repeated on a 1.9 m blade, in order to study the scalability for small wind turbine blades.

2. Materials and Methods

Two resins have been used to produce the composites: a thermoplastic resin, Akelite, patented by the CSIC, and a commercial epoxy resin from Sicomin [6]. Akelite is a reactive acrylic resin that undergoes polymerisation through the addition of benzoyl peroxide as an initiator at a temperature of 60 ºC for 2 hours. Its density is 1.18 g/cm^3. The epoxy resin is a two-component system based in an epoxy SR1280 and a fast hardener, SD4772 mixed in a specified ratio of 100:27 parts by weight. It is a very low viscosity system, especially suited for infusion processes. According to the supplier's specifications, the epoxy resin is cured at room temperature for 24 horas, and at 60 °C for 16 hours. Its density is 0.97 g/cm^3. The used fibres were biaxial ±45º carbon fibre fabric 50 K at several areal weights supplied by Mel Composites and an unidirectional glass fibre with an areal weight of 625 g/m^2 supplied by Gavazzi Tessuti Tecnici.

2.1. Manufacturing of composites

Laminates of 75x20 cm were produced using three layers of carbon fibre fabric, one of 200 and two of 600 g/m^2, by vacuum assisted resin infusion (VARI). The fibre content in the laminate is approximately 70 wt.%. The laminate is cut into three pieces of 25x20 cm. The first piece is used to analyse the mechanical properties of the material under bending and inter laminar shear strength (ILSS). The second piece is subjected to a recycling process and a new laminate is made from the recovered fibres and resin. The third piece is subjected to two recycling processes and its mechanical properties are analysed.

2.2. Testing of composites

The mechanical properties of all three composites, original, once and twice recycled, were measured in the longitudinal direction of the fibres using a three-point bending test according to ASTM D790-03 and ASTM D2344, respectively. Specimen dimensions were 50.8 mm long and 12.5 mm wide for flexural tests and 15 mm long and 7.5 mm wide for ILSS tests. An Instron 2204 machine was used for both tests, with a load cell of 1 kN for flexural tests and 50 kN for ILSS tests, at a rate of 1 mm/min. A minimum of seven specimens were tested for each composite.

2.3. Recycling of composites

The laminates were recycled by immersion in a solvent for 18 hours at room temperature without mechanical agitation. Once the fibre was recovered, the resin was extracted from the solvent using a rotary evaporator at 50 ºC.

2.4. Manufacturing of blades

Two 1.9 m blades were manufactured by VARI with a lay-up fibre configuration of satin/45C/0G/sandwich/0G/45C/0G (Figures 1-3). After the vacuum infusion process of the two valves, they were filled with Sicomin PB170+DM03 low density epoxy foam and glued with Sicomin Isobond 735 two-component epoxy adhesive.

Figure 1. Laying glass and carbon fibres on the moulds.

Figure 2. Vacuum infusion process.

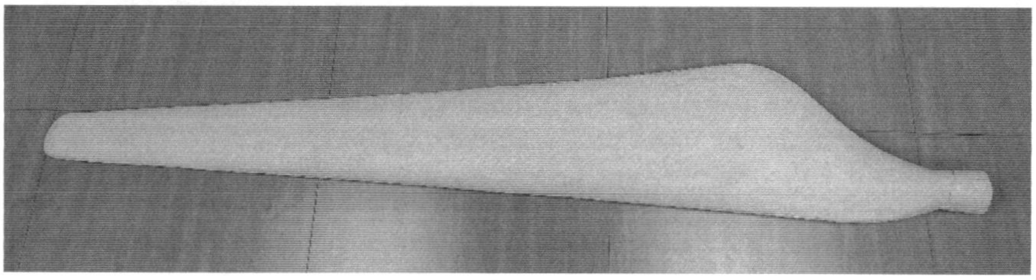

Figure 3. Final aspect of Akelite blade.

2.5. Testing of blades

The wind turbine blades were analysed through static flap-bending test (Figure 4). Static loads were estimated according to the simplified load model of IEC 61400-2 Standard (Small wind turbines). The blade root was secured to a rigid steel test rig using a simple clamp bolted connection. The blade was mounted horizontally (flap-wise), with two winches used to apply the load.

Figure 4. Static test of Akelite blade.

2.6. Recycling of blade

A heating bath circulator has been designed that allows for direct temperature control of a bath tank and indirect heating of the blade immersion solvent, in this case acetone. To achieve this, the blade has been cut into two pieces to immerse them in the bath and reduce the required volume of acetone (as shown in Figure 5). The circulator heats the water in the bath tank, and the heat is transferred to the acetone, allowing it to reach the desired temperature.

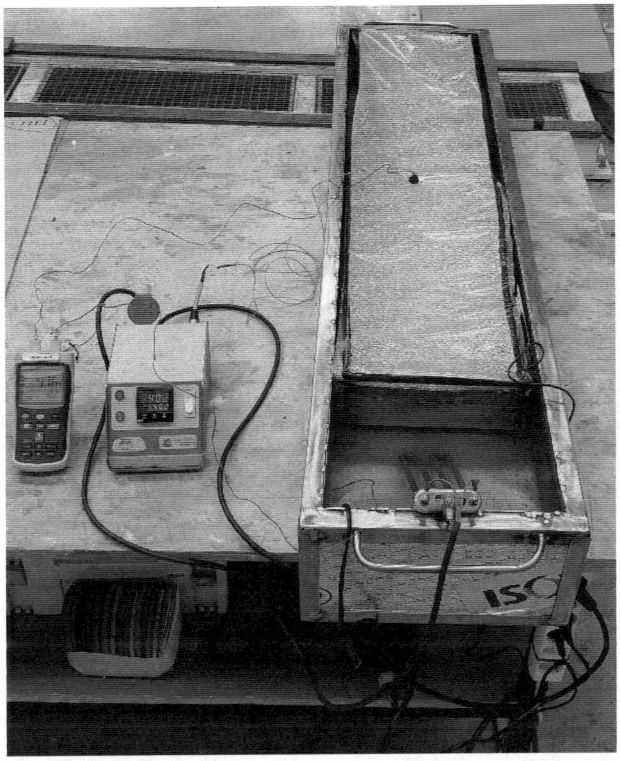

Figure 5. Heating bath circulator for recycling Akelite blade.

3. Results and Discussion

A straightforward and scalable method was used to recycle the laminates. The recycling process involved immersing the laminates in four different solvents - acetone, chloroform, ethyl acetate, and tetrahydrofurane - for a duration of 18 hours at room temperature. A first trial was done with small laminate specimens to select the appropriate solvent. All solvents dissolved the polymer and the laminates were easily separated into the carbon fibres plies by gently pulling by hand. Among the solvents tested, acetone was selected considering the price, toxicity, and environmental impact to continue the study.

A resin:solvent ratio of 1:20 was found to be optimal. After separation, the plies were left to dry at room temperature. The resin is then extracted from the solvent using a rotary evaporator. This allows a complete recovery of the resin and solvent, resulting in a 100% circular process. Approximately, a 95% of the resin is recovered and 2-3% remains on the fibre surface. The recycled fibre and resin are reused to produce a new laminate for testing (Figure 6). This recycling process was repeated twice to analyse the mechanical properties of the recycled composites (Table 1), which were compared to those obtained with the epoxy resin.

Interestingly, the properties of the laminates with the thermoplastic resin are superior to those with epoxy resin, demonstrating the feasibility of this resin for the manufacture of structural materials. The laminate properties after the first recycling (Akelite 1R) tend to improve compared to the original laminate, probably due to the presence of a thin layer of resin on the fibre surface, which acts as a sizing agent and improves the adhesion between the layers. The mechanical behavior slightly deteriorated with the second recycling (Akelite 2R), although the properties were similar or even better than those of the virgin material.

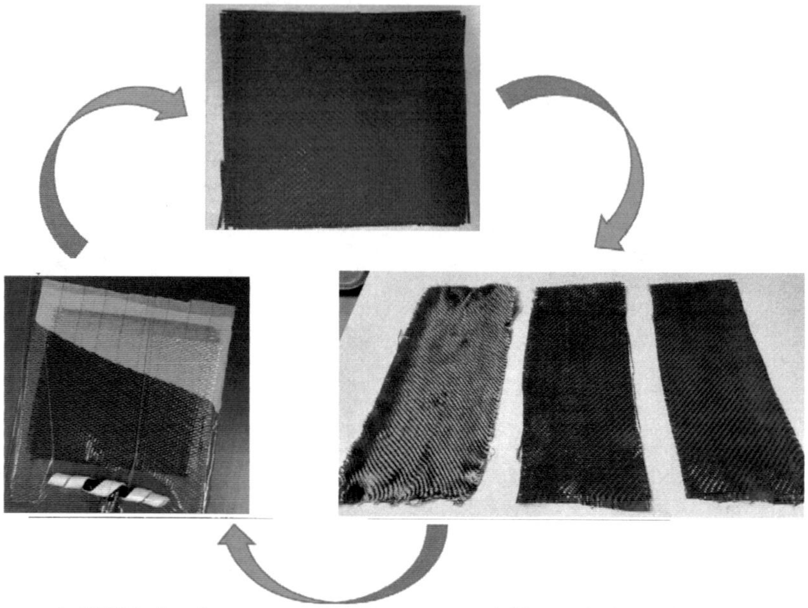

Figure 6. 100% circular process and recovery of fibre reinforced thermoplastics

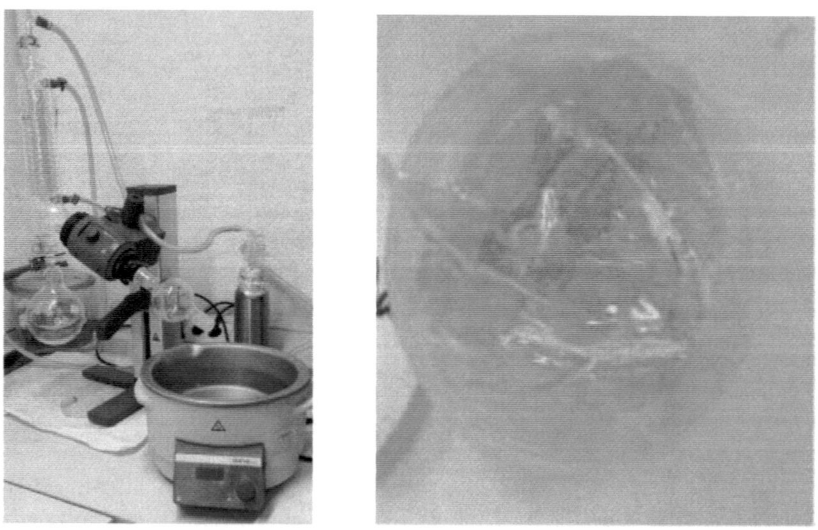

Figure 7. Extraction of resin from acetone using a rotary evaporator

Table 1. Mechanical properties of the laminates after recycling process

	Epoxy	Akelite	Akelite 1R	Akelite 2R
Flexural modulus (GPa)	17.2±1.1	20.6±1.5	22.3±1.7	20.6±1.1
Flexural strength (MPa)	493.3±24.8	492.8±29.7	619.0±24.1	516.5±16.4
ILSS (MPa)	37.9±2.7	45.7±2.6	40.7±2.2	40.2±4.7
Fiber volume fraction (%)	61.2	60.3	61.3	60.2
Thickness (mm)	1.55±0.03	1.48±0.01	1.47±0.01	1.51±0.02

Once the recycling conditions for the laminates were successfully optimised in the laboratory, the current objective is to recycle the 1.9-meter blades and recover the fibres. Thus, 1.9 m blades were manufactured with both the epoxy and Akelite resins and characterised them. Table 2 shows the summary of property tests done to both blades, such as mass, centre of gravity and natural frequencies.

Table 2. Results of property tests

Blade	Length (mm)	Mass (kg)	Centre of gravity (mm)	1ª Flap frequency (Hz)	1ª Lead-Lag frequency (Hz)
Epoxy blade	1886.0	5.3	765.0	13.5	20.4
Akelite blade	1886.0	5.5	769.0	12.5	17.7

For static tests, loads were applied at 0.6 m and 1.4 m from root blade and measured with two load cells. Deformation was measured with a strain gauge glued at 1.0 m from root blade, and tip deflection with a string potentiometer located at blade tip. Tables 3 and 4 show the results of the static test carried out to the epoxy and Akelite blades, respectively. The epoxy resin blade has a tip deflection of 166 mm while the Akelite resin blade has a 10% higher tip deflection of 183 mm. To avoid tower strike, IEC 61400-2 Standard requires that the tip displacement should be less than the clearance provided between the blade tip and the tower, even at worst case loading. In our case, this distance is 303mm, higher than 183 mm tip deflection.

Table 3. Static test of the epoxy resin blade

Winch 1 (N)	Winch 2 (N)	Root moment (Nm)	Deform. (μm/m)	Tip deflection (mm)
28.7	26.2	94.4	0.4	0.0
234.1	210.2	474.9	205.2	-29.0
498.0	372.2	859.6	382.4	-66.0
697.6	612.6	1315.5	640.0	-119.0
974.8	818.5	1769.6	852.0	-166.0

Table 4. Static test of Akelite resin blade

Winch 1 (N)	Winch 2 (N)	Root moment (Nm)	Deform. (μm/m)	Tip deflection (mm)
28.7	26.2	94.4	-3.7	-0.7
279.7	181.4	461.9	173.5	-31.4
496.5	395.8	891.7	409.9	-80.1
731.6	605.0	1325.2	635.3	-140.6
928.7	703.1	1580.5	735.8	-183.2

It is currently carrying out the recycling of Akelite wind blade through a similar immersion protocol. The next step involves separating the resin and acetone using a rotary evaporator. This would enable easy circularity of raw materials, fibers, and resin, making it possible to remanufacture a new blade.

4. Conclusions

It is presented the potential recycling protocol for a wind blade made from a thermoplastic fibre reinforced polymer. Recycling of composite samples revealed materials with improved properties compared to the original thermoplastic composite, which was attributed to the presence of a small percentage of polymer residue. Such residue is likely to act as a sizing agent, improving the interface between the fibre and the matrix. The static test showed small differences in the blade characteristics. In general, variations could be expected from manual blade manufacture.

Again, the Akelite static test was similar to the epoxy static test. This means that thermoplastic resin could be a substitute for epoxy resin for use in the wind energy industry, contributing to the recycling of raw materials at the end of their life. More tests need to be done to scale up the goals and results.

References

[1] Wind Europe. Accelerating Wind Turbine Blade Circularity May 2020.
[2] https://www.siemensgamesa.com/newsroom/2022/07/080122-siemens-gamesa-press-release-recycle-wind-blade-offshore-kaskasi-germany
[3] https://www.jeccomposites.com/news/zebra-project-achieves-key-milestone-with-production-of-the-first-prototype-of-its-recyclable-wind-turbine-blade/
[4] https://www.windpowermonthly.com/article/1817477/mingyang-first-chinese-oem-launch-recyclable-turbine-blade
[5] Verdejo, R. et al. (2021) "Polímeros reforzados con fibras basados en matrices termoplásticas" (Patent N0. PCT/ES2021/070751)
[6] Carnicero, R. et al. (2022) Manufacturing, Testing and Recycling of a small recyclable wind turbine blade, Journal of Physics: Conference Series, 2265 (3), art. no. 32013.
[7] Fiche Technique (sicomin.com) – SR1280/SD47XX: http://sicomin.com/datasheets/product-pdf1169.pdf

IOP Conf. Series: Materials Science and Engineering 1293 (2023) 012009 doi:10.1088/1757-899X/1293/1/012009

Challenges in permeability characterisation for modelling the manufacture of wind turbine blades

Robert S. Pierce

DTU Wind and Energy Systems, Technical University of Denmark, Frederiksborgvej 399, 4000 Roskilde, Denmark

E-mail: rsapi@dtu.dk

Abstract. As wind turbine blades continue to increase in size, and market competition grows, lean manufacturing has become even more important for OEMs. The rapid development of new blade designs, with greater performance, and reduced production waste are driving the need for predictive modelling of the blade infusion process. Such simulations are reliant upon Darcy's Law for the description of fluid flow through porous materials and therefore depend greatly on the permeability properties of the blade preform materials. The characterisation of fabric permeability, although unstandardised, has been well studied in recent years as the focus of numerous international benchmarking efforts. However, the effective permeability properties of infusion consumables, core materials, and pre-cast elements are not so well defined or validated, despite their significance on infusion behaviour. Hence, the great variety of preform materials, stacking configurations, geometric features, and transition regions in wind turbine blades present considerable challenges in terms of permeability characterisation and subsequent modelling. This article reviews some of the challenges, opportunities, and alternatives for characterising permeability in common blade preform materials, along with examples of how these properties have been applied in numerical models to better simulate the resin infusion manufacturing process for wind turbine blades.

1. Introduction

Wind turbine blades are the largest and most complex composite structures being built today. As they continue to grow in size, they become more challenging to manufacture, and more costly to repair when manufacturing goes wrong. Subsequently, the ability to simulate their manufacture and predict potential issues with the manufacturing strategy is becoming more and more valuable. Figure 1 shows the composite layup of a thin sandwich structure representative of the materials involved in the vacuum infusion of a wind turbine blade. Here the composite preform consists of two configurations of glass Non-Crimp Fabric (NCF), unidirecitonal and biaxial, that act as the composite skins either side of a light-weight foam core with a complex arrangement of perpendicular, intersecting, machined grooves that are designed to improve both formability and rein flow. Additionally, consumable layers of resin distribution media, a perforated release film, and peel ply may be placed on top of the composite preform, inside the vacuum bag, to enhance the resin flow and ultimate part quality. The prediction of flow through, and filling of, these materials then becomes highly dependent on the properties of each material and their complex interactions. This article reviews the simulation of both the vacuum infusion process, used to manufacture wind turbine blades, and the important considerations related to material permeability for many of the common blade materials (as shown in Figure 1).

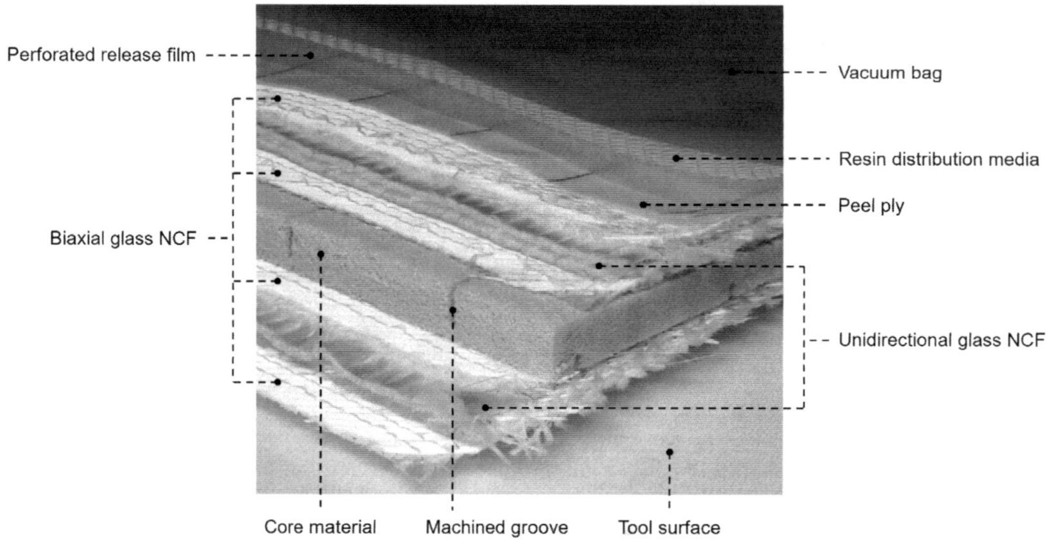

Figure 1. Representative layup of a thin sandwich structure for a wind turbine blade.

2. Vacuum infusion process modelling

2.1. Background

In the absence of consistent naming conventions for Liquid Composite Moulding (LCM) manufacturing processes [1], this article will refer simply to the process used for manufacturing wind turbine blades as 'vacuum infusion'. Other common names for similar processes include Vacuum Assisted Resin Transfer Moulding (VARTM), Resin Infusion under Flexible Tooling (RIFT), Vacuum Assisted Resin Injection (VARI), and Seemann Composites Resin Infusion Molding Process (SCRIMP), just to name a few. The vacuum infusion process fundamentally consists of the layup of reinforcing preform materials against a rigid mould before being sealed under a flexible tooling, typically a thin polymeric film. Then vacuum pressure is applied at an outlet port, such that resin from a higher (often ambient) pressure is driven into the preform, flowing towards the outlet, while the flexible tooling and reinforcing material is held in place by external ambient pressure.

When it comes to modelling this process, LCM simulations generally prioritise the prediction of resin flow behaviour during mould filling in order to avoid catastrophic dry spots. However, such models can also provide a useful estimate of the filling times with different infusion strategies [2, 3, 4], and more sophisticated models can also aim to predict saturation and void content [5, 6], or local fibre volume fractions across a part [7]. These filling simulations can then also be coupled with curing kinetics models [8] in order to better predict residual stresses and part distortion [9]. Independent of the model complexity, the potential for optimising cycle times, reducing costs, and eliminating considerable trial-and-error in manufacturing is therefore particularly valuable to OEMs.

Many commercial and academic software packages exist for LCM simulation. Notably, the use of PAM-Composites/PAM-RTM (ESI Group) [10, 11], RTM-Worx (Polyworx) [2, 12, 13], and LIMS (University of Delaware) [14] have been well demonstrated as tools specifically developed for composites manufacturing applications. While the more general Computational Fluid Dynamics (CFD) packages like ANSYS Fluent (ANSYS, Inc.) [15], COMSOL Multiphysics (COMSOL Inc.) [16], and OpenFOAM (The OpenFOAM Foundation) [11] have also been well demonstrated in this field.

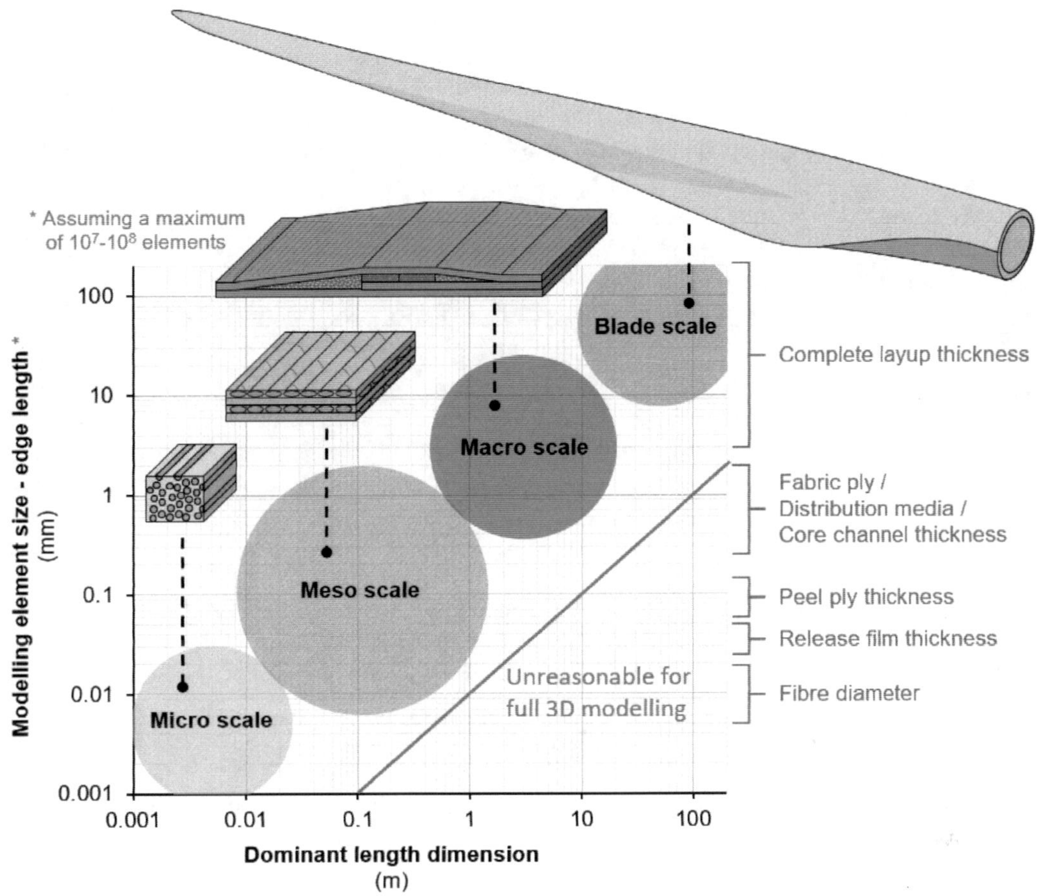

Figure 2. Modelling limitations at increasing simulation scales.

Depending on the desired scale of the infusion simulation, the assumptions and approximations necessary for modelling can vary significantly. Fundamentally, at the microscopic level, resin flow through a porous preform materials can be described by creeping (Stokes) flow since the fluid viscosity is relatively large and its velocity is relatively slow (meaning $Re \ll 1$). However, despite the apparent simplicity of such a fluid dynamics model, the geometric complexity and stochastic nature of these materials makes accurate flow prediction particularly difficult. Micro-scale models, considering the flow between individual fibres in a preform, may simply rely on conventional Computational Fluid Dynamics (CFD) packages to predict 2D or 3D flow, based on simple geometric approximations or real scans from X-ray micro-tomography [17]. As the simulation scale increases though, reasonable limits on the size and number of constitutive elements in the model necessitate significant degrees of homogenisation or geometric simplification. For example, at the meso-scale, where researchers often attempt to simulate fluid flow through stacked reinforcement materials or repeated unit cells, tow bundles are commonly homogenised and treated as a continuous porous or solid material [18, 19, 16]. Then at the macro-scale, details of the fabric weave and architecture are often eliminated in the homogenisation of fabrics to be treated as continuous layers, in order to simulate the full preform behaviour in larger parts [3, 4]. Finally, in the case of blade-scale (or boat hull) simulations, it may be necessary to homogenise the complete preform stack into a single equivalent material [20, 21, 2].

Figure 2 demonstrates the increasing limitations on element sizes as the dimensions of a model increases. Assuming the use of a full 3D model, consisting of a maximum of 10^7 or 10^8 total elements with reasonable aspect ratios, it becomes impossible to model individual plies of reinforcement fabric at the blade scale (let alone the consumable layers that are an order of magnitude thinner). Even at the meso-scale, it is often unreasonable to try to simulate individual fibres. Hence, at these larger simulation scales, where fibre bundles, fabric plies, or the whole layup needs to be approximated by a homogeneous porous material, the volume-averaged flow through these materials is then most commonly described by Darcy's Law, as shown in Eqn. 1 (assuming the fluid is incompressible and Newtonian). Here v is the averaged flow velocity, K is the permeability tensor for the porous material, μ is the viscosity of the fluid, and ∇p is the pressure gradient:

$$v = -\frac{K}{\mu}\nabla p \tag{1}$$

For the purpose of vacuum infusion modelling, the viscosity of the fluid can easily be characterised as a function of temperature using standard viscometers or rheometers. Similarly, in most processes the pressure gradient is able to be monitored or controlled with relative ease. However, the characterisation of permeability properties poses a much greater challenge, as is discussed in detail in Section 3.

A further complication to this process is the well-documented dual-scale nature of the flow through fibrous reinforcement materials [22], where the flow within fibre tows is often driven by capillary forces but flow in the larger gaps between tows is dominated by viscous forces. Depending on the flow velocity, v, viscosity, μ, and ambient surface tension, γ, of the resin (commonly combined as in Eqn. 2 to form the Capillary number, Ca), the significance of viscous or capillary forces in the flow can be assessed. At high Ca, viscous flow drives the infusion as assumed by Darcy's Law, but can result in micro-void formation within the lagging tows. Alternatively at low Ca, capillary forces will lead the infusion through the tows with the potential for macro-void formation between the tows. Hence, an ideal infusion case may attempt to control the resin flow rate such that the capillary and viscous forces are well balanced to minimise void formation throughout the composite [23]. This becomes especially important when high vacuum is not used to drive the infusion (i.e. for conventional Resin Transfer Moulding, RTM [5]) as there is a significant volume of air remaining within the composite mould. In most LCM processes however, the resin viscosity and infiltration velocity tend to result in a capillary number ranging from 10^{-3} to 10^{-1} [22], such that capillary forces can be neglected and flow is assumed to be Darcian and dominated by viscous forces [23].

$$Ca = \frac{\mu \cdot v}{\gamma} \tag{2}$$

2.2. Applications
Across existing literature, there are many examples of LCM process modelling. Early investigations of the Seemann Composites Resin Infusion Molding Process (SCRIMP) accounted for many of the same material systems and features seen in wind turbine blades, including distribution media, peel ply, and a low density core material with machined grooves to improve infusion [24, 25]. These studies also applied Control Volume/Finite Element Models (CV/FEM) for meso-scale filling analysis, observing good experimental agreement. Similarly, Dai *et al* [26] applied an academic software package (RTMSIM) to model the vacuum infusion of foam core composite sandwich panels, comparing the use of either distribution media or core grooves as flow enhancing options. Despite neglecting compaction and deformable bag behaviour, by assuming rigid tool moulds like conventional Resin Transfer Moulding (RTM), these early small-scale models still saw good experimental agreement.

Other work by Poodts *et al* [27] demonstrated the manufacturing and modelling of a 20 mm thick Carbon Fibre Reinforced Polymer (CFRP) component by RTM. This work showed significant asymmetry in the flow behaviour of the real parts due to fibre rearrangement and race-tracking, which could not be predicted in the simulation software without manual manipulation. Such problematic fibre rearrangement was likely the result of the very high injection pressures (3-10 bar). For a vacuum infusion process, where injection pressures are typically 1 bar or less, such fibre rearrangement is therefore not expected to be an issue.

Another study comparing numerical and experimental RTM filling showed good agreement for the prediction of 2D flow around in-mould obstructions, using both PAM-RTM and OpenFOAM [11]. However, for a second case with a narrow gap (around 1 mm) between dissimilar preform materials, the definition and prediction of race-tracking in the open channel was much less reliable, particularly as its geometry is difficult to accurately replicate. Furthermore, it was noted that any such gaps, spaces, infusion channels, or deep grooves cannot be simulated as open channels with dedicated Darcy-based solvers, but instead must be approximated as porous domains, with 100% porosity, and an equivalent permeability [11].

Correia *et al* [14] extended the application of their LIMS RTM software to the simulation of VARTM processes by defining a proportionality constant that accounts for the effects of fibre compaction, fibre volume fraction (porosity), and permeability as a result of using a deformable bag. This approach was then used to highlight filling time differences between RTM and VARTM processes. Further modelling complexity has been introduced by Trochu *et al* [6], who not only considered compaction effects but also investigated saturation and void formation. In this work, 2D and 3D LCM simulations for automotive composite parts were used to demonstrate how controlled flow rates (targeting an optimal capillary number) can improve mechanical performance in the final part by reducing overall void content.

At a larger scale, since part thickness is often much smaller than the width and length dimensions, it is quite common to neglect through-thickness flow and simulate the domain as a 2D shell with 3D curvature, often termed a 2.5D model [28]. Early examples of this modelling approach within RTM-Worx, for the vacuum infusion of wind turbine blade (20 m) and boat hull (16 m) structures, have been introduced by Brouwer *et al* [2], although details of the material definitions were absent from the work.

Other research considering the experimental and simulated vacuum infusion of sandwich assemblies has shown a good prediction of filling times and flow behaviour for core materials, with a complex two-sided and interconnected groove arrangement, for boat hull applications [13]. In order to reduce the geometric complexity of the model for RTM-Worx, two approaches were evaluated. First, a number of horizontal and vertical circular pipe elements were used to approximate rectangular core grooves with a similar cross-sectional area in 2.5D models. Secondly, a further simplification was applied (both for the core material and for the complete sandwich assembly), accounting for the thickness and porosity of each material layer in order to define equivalent properties for a single 2.5D layer [13]. The permeability of this equivalent layer could then be derived empirically from simple 1D flow sandwich infusion experiments. Ultimately, this approach was well validated when applied to a complete boat hull infusion with similar materials, despite the reduction to a 2.5D model. Related work by the same authors showed that the simulated 1D flow behaviour across different sandwich assemblies (in terms of squared flow distance relative to time) is calculated to be somewhat linear, based on the reliance of Darcy's Law [12]. In spite of general agreement between filling time predictions, the experiments exhibit greater non-linearity as a result of through-thickness flow behaviour. Subsequently, considerable error was observed between the simulated and experimental flow front positions during filling, particularly while the flow in the lower laminate lagged behind that of the upper laminate. Hence, the simulation results ultimately resemble a simple linear approximation of the non-linear experimental results.

In order to better account for transverse flow behaviour, a two-layer simulation approach (consisting of separate resin distribution media and reinforcement preform layers) has been proposed [29]. This relies on an approximation of the through-thickness interactions and flow between two layers with very different permeability properties. This method has been applied to the infusion of a 3.5 m wind turbine blade, however the analysis was lacking realistic permeability properties and experimental validation. Alternatively, a multi-layered shell element method for 3D resin flow simulation has been developed in order to better account for through-thickness flow than conventional shell elements or other 2.5D approximations that rely on permeability averaging techniques [21]. This approach was applied to macro-scale components, providing better estimates of the saturation and filling behaviour of different layers in the mould, but did not include experimental validation. Recent work by Yan *et al* [3] has observed good qualitative agreement between PAM-RTM infusion models and the experimental production of foam-core sandwich spoiler structures. Part thickness variations and final filling times were quite well predicted (10-15% deviation), while the models also helped to identify possible causes for manufacturing defects, but the determination of the foam core channel permeability remained unexplained.

Most relevantly, Koefoed [30] performed a comprehensive investigation of vacuum infusion process modelling for the manufacture of wind turbine blades. This work leveraged the fact that vacuum infusion strategies for full blades commonly rely on a linear infusion technique, where span-wise line inlets are used to fill the blade preforms in a chord-wise direction. Subsequently, the models and experiments in this study could be scaled down to prioritise realistic flow through 2D cross-sections from different areas of the blade, rather than attempting to model the full blade infusion. This presents an efficient method of predicting the filling behaviour around complex transition regions in the presence of reinforcement, core, and high permeability distribution media. Furthermore, this approach avoids the need for full 3D models, that would be beyond computational capabilities, or 2.5D shell models that require significant simplifications of the through-thickness flow behaviour.

Ultimately, across all applications and modelling methods, the characterisation and definition of permeability properties remains a key factor to the success and accuracy of any infusion simulation. Hence, the following section will focus on the various aspects of permeability that are most relevant to common wind turbine blade materials.

3. Permeability considerations

3.1. Reinforcement fabrics

3.1.1. Architecture and stacking Inherently, the internal structure and architecture of the reinforcing preform will have a significant effect on its permeability. For example, comparisons of engineering textiles with different weaving patterns, for the same fibre type, tow size, fibre volume fractions, and fibre sizings, have revealed significant differences in permeability [31]. The key observation from this work was the association of permeability properties with the size and frequency of pore spaces: where a higher number of large pore spaces correlated with higher permeability. Evidently, any modification to a reinforcement fabric that may alter its pore sizes or distribution will significantly affect its permeability. Subsequently, meso-scale simulation of flow through fabric repeating unit cells has become a popular method for predicting permeability properties based on fabric architecture [32].

The stitching of fabrics and preforms has also been shown to significantly affect both in-plane and through-thickness permeability, effectively creating channels for fluids to more easily pass through. For example, Talvensaari *et al* [33] observed that a greater number of stitches could increase the in-plane permeability of preforms, but also that stacking multiple stitched preforms together can result in higher in-plane permeability than the same number of plies in a single stitched preform stack. However, other work by Drapier *et al* [34] focusing on transverse (through-thickness) permeability, showed that such changes to the stacking arrangement had no measurable influence on transverse permeability. In this case, the stitching pattern was also determined to have no effect, with only the stitching density being seen to significantly affect the transverse permeability. For woven fabrics, the stochastic nature of nesting between different fabric layers has also been identified as the potential main source of scatter in the permeability behaviour of fabrics [35].

3.1.2. Deformation Another challenge to the characterisation and modelling of fabric permeability behaviour is the relatively compliant and deformable nature of the material. Inherently, as a result of the high performance reinforcing fibres, these materials are resistant to tensile deformation in the fibre directions, but depending on their architecture they are often prone to bending, shearing, and compression. In particular, in-plane shearing and out-of-plane compaction have both been seen to significantly affect permeability behaviour. For example, experimental characterisation of the in-plane permeability for a plain-woven carbon fibre fabric at 40° shearing has shown more than a 2-fold increase in its principal permeability, a 3-fold increase in anisotropy, and a 20° shift in the principal permeability direction from an undeformed state [36]. Numerical investigations have also documented changes of a similar magnitude for glass fabrics [16].

Perhaps more importantly for vacuum infusion processes, that are reliant on ambient pressure across the vacuum bag, the transverse (out-of-plane) compaction of fibrous reinforcement materials can have a significant effect on its filling performance. Evidently, with increasing compaction pressure the fibre volume fraction increases, meaning the size and frequency of pores between tows and fibres will decrease, along with the permeability. Commonly, for the purpose of permeability characterisation, the fibre volume fraction, V_f, is defined by Eqn. 3:

$$V_f = \frac{n \cdot w_A}{t \cdot \rho_f} \tag{3}$$

where n is the number of fabric layers, t is the total thickness of the preform stack, w_A is the fabric areal weight (kg/m^2) and ρ_f is the density of the constituent fibres. However, during infusion, as resin travels through a fibrous reinforcement two important phenomena occur [1]. Firstly, at the advancing flow front the fibres are lubricated, enabling a further reduction in thickness (and increase of local fibre volume fraction, V_f) under the same bag compaction force [37]. Secondly, while the flow front continues ahead, the local pressure will continually increase and the overall compaction force that results from the pressure differential across the vacuum bag will therefore reduce. This results in considerable relaxation of the preform, along with an increase in thickness, decrease in V_f, and therefore also an increase in local permeability. Subsequently, many researchers have noted the importance of modelling dry compaction and wet relaxation properties in order to capture such effects and improve the fidelity of infusion models [10]. This interaction between permeability and fibre volume fraction has been traditionally defined using the Kozeny-Carman equation:

$$K = k \cdot d^2 \frac{(1 - V_f)^3}{V_f^2} \tag{4}$$

where the product kd^2 is an empirical constant determined from experimental fitting. However, recent literature suggests that the use of a power-law model (Eqn. 5) instead of the Kozeny-Carman equation can be more accurate [10], as similar models are used for describing compaction behaviour (Eqn. 6) [38].

$$K = A \cdot V_f^B \tag{5}$$

$$V_f = C \cdot \sigma_f^D \tag{6}$$

where A, B, C, and D are empirical constants and σ_f is the transverse compaction stress applied to the preform. International benchmarking of fabric compaction behaviour has been recently reported, however significant scatter (as much as 50% relative standard deviation) has been observed across the results of different labs using the same materials [39]. Earlier work has also investigated the relationships between textile compaction, injection pressure, porosity, and permeability more broadly for LCM processes [40]. Notably, in the case that injection pressure is lower than the compaction pressure, as is typical for vacuum infusion processes, then the permeability of the fabric should remain independent of the injection pressure.

3.1.3. Other fabric modifications Other, less-studied, influencing factors for fabric permeability are micro-scale modifications, such as fibre sizings or the addition of a tackifier/binder material. Generally the fibre sizing is selected for matrix compatibility and its effect on the ultimate mechanical performance of the composite, however its chemistry will also affect permeability. A thorough study of four different engineering fabrics, both with and without various sizings, using three different test fluids, showed that the removal of sizings from fibres usually improved permeability (by up to 213%) [41]. The exception was a carbon fibre fabric with 68K tows, for which the authors observed no tow gaps. Hence, the explanation for improved permeability in the other fabrics was the clustering of sizing-removed tows that created larger inter-tow gaps. A similar observation of improved permeability (30%), after the sizing was removed from carbon fibres during a recycling process, has also been observed in recent literature [42].

The addition of binder materials, to facilitate preforming and a more efficient layup in large structures, can also modify infusion behaviour. Rohatgi and Lee [43], for example, identified that a thermoplastic binder was more likely to increase flow resistance if it was concentrated between tows rather than inside them, since permeability tends to be largely dependent on the flow behaviour through inter-tow gaps. Similarly, Shih *et al* [44] also noted the significance of the location of an epoxy tackifier, which appeared to shrink the tows from within and increase the size of inter-tow gaps. In the maximum case, at relatively high binder concentration (17%), the permeability was seen to increase by a factor of 4.5 relative to the control preform. However, other research has seen little change in permeability when using a thin thermoplastic film tackifier [45].

Preliminary research at DTU Wind has also demonstrated the significance of a binder material on the permeability properties of a unidirectional NCF, as shown in Figure 3. Here 1D channel flow experiments showed a 75% increase in permeability along the fibre direction for the material with binder, compared to the same fabric without, at a similar fibre volume fraction. This difference is attributed to small scale changes in the pore distribution as a result of the binder pre-consolidation phase (conducted under pressure at elevated temperature).

3.1.4. Characterisation methods Despite extensive study over the last few decades, the characterisation of fabric permeability remains unstandardised. In generally, there are several popular methods for the experimental characterisation of in-plane and through-thickness permeability that have also been the focus of extensive international benchmarking efforts [46, 47, 48, 38]. Fundamentally, experimental methods are commonly classified by their flow dimensions (1D 'channel' or 'through-thickness', 2D 'radial', or 3D), with either saturated (steady-state) or unsaturated (wetting) flow, and are controlled by either a constant flow rate or constant injection pressure. At this stage, 1D channel flow experiments for the characterisation of in-plane permeability have shown the greatest repeatability and reliability in benchmarking [47], demonstrating considerable improvements over a less-prescriptive first round of benchmarking [46].

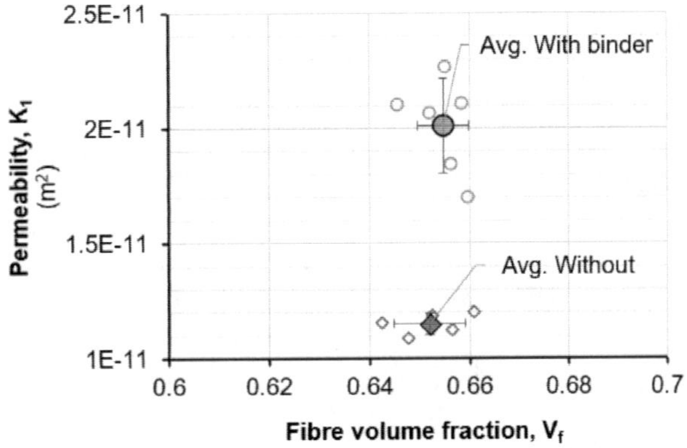

Figure 3. Permeability properties of a unidirectional NCF, with and without binder.

Alternatively, predictive permeability characterisation models are also an attractive area of research in order to overcome many of the tedious and temperamental aspects of the experimental approaches. In addition to the previously mentioned micro- and meso-scale modelling approaches for virtual permeability characterisation [16, 17, 18, 32], there have also been efforts to develop iterative, macro-scale, simulation-enhanced methods for 3D permeability characterisation [49]. These are designed to support experimental procedures for greater accuracy and reliability overall, compared with conventional 3D methods that are based on the theoretical calculations for, and transformations from, an equivalent isotropic flow [50, 51]. Recently, the development of a novel, non-destructive, 3D permeability measurement method has also shown considerable potential by relying on air flow and pressure sensors along with an artificial neural network for the interpretation of results [52]. However the neural network interpreter still needs further development.

Other novel modifications to experimental methods have also been suggested. In one case, permeability values at three different fibre volume fractions have been characterised from a single 1D channel flow experiment by relying on several sequential flow regions with an increasing number of layers for the same cavity height [53]. More recently, characterisation techniques have been proposed to capture preform compaction, expansion, and permeability over a range of fibre volume contents, from a single vacuum-bagged 1D flow experiment representative of the real manufacturing conditions [10]. Despite the obvious efficiency advantages, this approach has only been validated using an acrylic mat material and not the typical engineering textiles that are used as composite reinforcements.

3.2. Core materials and pre-cast elements

Given that wind turbine blades contain significant quantities of materials other than fabric reinforcements, the effect of these materials on infusion behaviour also becomes an important consideration. Pre-cast elements, such as root inserts or pultruded carbon fibre profiles, are solid, impermeable components that are not expected to directly affect resin flow conditions aside from acting as a non-slip boundaries. However, in the layup of these structures there are commonly small gaps between repeated elements, which will act as open channels for fluid flow and facilitate significant race-tracking behaviour. Similarly, foam or balsa core materials for blade structures are commonly machined with various inter-connected grooves that also act as open flow channels during infusion. As was noted in Section 2.1, when modelling at large scales, it is no longer feasible to explicitly simulate these channels and it becomes necessary to develop an approximation or homogenisation of their effect on flow. This is commonly done by developing an equivalent permeability for the channels, or for the whole core material.

Since most dedicated infusion modelling approaches rely on Darcian flow, many early researchers treated channels as a porous medium with an equivalent flow capacity. This requires the definition of an effective permeability (where $V_f = 0$) and the assumption of fully-developed duct flow. Several approaches have been suggested to this end, but the two most common approximations for the effective permeability of a rectangular channel are either to assume a non-slip (zero velocity) condition at the interface with the fabric reinforcement [54], or to consider this interface as permeable and therefore reliant on an empirical slip condition (non-zero boundary velocity) [24]. The latter is more realistic but very difficult to characterise and apply [55]. Other examples of equivalent permeability definitions based on similar principles, and the simulation of LCM for boat hull structures or balsa-core sandwich structures can be seen in literature as well [26, 28]. Hammami *et al* [56] proposed a further simplification of channel flow based on geometric models, independent of channel width, and neglecting transverse flow. However these equivalent permeability approximations were limited in their application and showed relatively poor agreement with experimental testing.

Alternatively, a number of other studies have instead attempted to characterise channel flow behaviour experimentally. Markicevic *et al* [57] performed channel flow experiments to assess the influence of dimensional parameters for both the channel and an adjacent preform material. This work reflected the significance of channel size but also the importance of the preform thickness on the leading velocity of the channel flow, illustrating a strong dependency between the channel and preform properties as they relate to infusion. In support of infusion models for wind turbine blade structures, Koefoed [30] performed vacuum-bagged 1D infusion of a balsa wood core panel alone. The observation of a good linear fit of the squared flow front over time, in this case, supported the use of Darcy's law for the definition of an effective core permeability. Conversely, similar 1D permeability characterisation experiments for foam core materials (with typical perforations and infusion channels, also for wind turbine blades), demonstrated poor fitting with Darcy's description of flow through a porous media [58], suggesting alternative methods for characterising and simulating core materials are necessary.

Additionally, the surface condition of core and pre-cast materials may also affect resin flow, this is particularly true for structural foams that can have vastly different pore structures. Generally these will be closed-cell foams, with minimal resin uptake by design, but the outer few millimetres of these foams will still absorb a considerable volume of resin and influence the flow. Such boundary layer effects are expected to only have a small influence on the creeping flow of an infusion problem, however this may still affect simulation fidelity, and a thorough investigation of these effects is yet to be conducted.

3.3. Consumable materials

In addition to the fabric reinforcement and core materials that make up the structural composite, consumable materials such as resin distribution media (also known as high permeability media or flow mesh/media), peel ply, and release film can significantly affect resin flow. This is particularly the case for the distribution media, as by design it has a much higher permeability than the reinforcing fabrics, in order to enhance overall infusion speeds. Research investigating the infusion of common consumable materials has shown how the permeability of stacked knitted distribution media layers can be 10% lower than that of a single layer, due to nesting effects [25]. The same study also isolated the combination of flow media and peel ply layers (on top of a realistic impermeable preform), which resulted in a 50% reduction in permeability. In spite of this, infusion experiments considering the filling times of similar glass fibre stacks counter-intuitively showed slower filling when the peel-ply was removed. This was again attributed to nesting effects when the distribution media was compacted directly against the glass fibre reinforcement layers. Overall then, the peel ply was considered to have a beneficial effect in separating the distribution media from direct nesting with the reinforcement fabrics, overcoming any potential flow resistance that it might add in the thickness direction [25]. Ultimately mould filling times were found to be highly dependent on the permeability of the distribution media, and relatively insensitive to the preform permeability.

Many researchers have also investigated permeability averaging techniques in order to homogenise the properties of multiple layers, typically including a layer of resin distribution media, for simplified modelling of infusion processes. The simplest approach is to consider a weighted arithmetic average (i.e. the rule of mixtures), however this does not account for transverse flow across different fabric layers. This oversight becomes particularly important in cases where a high permeability distribution media is leading the flow adjacent to a low permeability material like the reinforcing fabrics. Hence, Calado and Advani [59] developed a 'transverse flow' averaging technique that shows significant improvement in the accuracy of flow modelling and allows for dimensional reduction of large flow problems. However, the implementation of this averaging technique requires a numerical solver, and is applicable only in cases where the flow direction aligns with the principal permeability direction and the pressure gradient is therefore assumed to be linear. Bancora *et al* [60] later proposed an extension to this approach by developing a 'generalised effective permeability' scheme that can account for non-uniform ply thickness and flow through off-axis plies. Experimental work has also demonstrated the issues with rule of mixture averaging for materials with very different permeabilities. Specifically, the measured permeability of a distribution media layer was seen to be 1-2 orders of magnitude higher than that of a glass fibre preform, and the apparent/equivalent permeability predicted by the rule of mixtures subsequently deviated from the measured values by an order of magnitude [61]. This work also identified a linear relationship for the apparent lag distance between flow in the distribution media layer and the preform fabric, as the fabric thickness increases. Alternatively, in support of meso-scale flow simulations through a 3D woven fabric unit cell, equivalent permeability calculations have been developed based on an analogy to the conductance of electrical conductors [62]. In this case, an area-weighted arithmetic mean and length-weighted harmonic mean were used for the averaging calculations of materials with different permeabilites in parallel or series respectively. Despite good agreement for some cases, these methods do not account for transverse flows from leading parallel layers, and their results showed unrealistic predictions when the permeabilities of adjacent materials differed by more than a few orders of magnitude.

4. Conclusions

Although some notable cases of infusion simulations for wind turbine blade manufacturing have been demonstrated for many years, infusion process modelling is still not mainstream practice in the wind industry. This is attributed to two main challenges that greatly affect the accuracy and potential of such models: simulation scale and permeability characterisation. In order to model three-dimensional flow through the thick and complex combination of materials in a wind turbine blade, even at the reduced macro-scale for a single panel section, significant homogenisation and simplification of the materials is necessary. However, the characterisation and averaging of permeability behaviour remains unreliable due to many interacting factors like nesting, compaction, and race-tracking. In particular, the characterisation of effective permeability values for core materials, that accurately account for the influence of deep cuts, infusion channels, and perforations is a key challenge to resolve. Similarly, the effects of consumable materials and pre-cast elements, and how to accurately incorporate these into numerical flow models, also remain fundamental challenges. Hence, several on-going projects at DTU Wind aim to investigate and resolve some of these issues, in order to develop more reliable process models for the infusion of wind turbine blade structures.

References

[1] Hindersmann A 2019 *Composites Part A: Applied Science and Manufacturing* **126** 105583 ISSN 1359835X
[2] Brouwer W D, Herpt E C F C V and Labordus M 2003 *Composites Part A: Applied Science and Manufacturing* **34** 551–558
[3] Yan C, Li Y, Su X, Liu Q, Wang Y, Wu K and Wu X 2022 *Materials* **15** 5279 ISSN 1996-1944
[4] Zhao C, Zhang G and Wu Y 2012 *Materials* **5** 1285–1296 ISSN 1996-1944
[5] Ruiz E, Achim V, Soukane S, Trochu F and Bréard J 2006 *Composites Science and Technology* **66** 475–486 ISSN 0266-3538
[6] Trochu F, Ruiz E, Achim V and Soukane S 2006 *Composites Part A: Applied Science and Manufacturing* **37** 890–902 ISSN 1359835X
[7] Shevtsov S, Zhilyaev I, Chang S H, Wu J K and Snezhina N 2022 *Polymers* **14** ISSN 20734360
[8] Ma L, Athreya S R, Mehta R, Barpanda D and Shafi A 2017 *Journal of Reinforced Plastics and Composites* **36** 780–794 ISSN 15307964
[9] Chen W 2019 *An Integrated Flow-Curing Model for Predicting Residual Stresses in Textile Composites in Textile Composites* Ph.D. thesis University of Connecticut
[10] Grofti Z, Croteau-Labouly B, Boukhili R, Causse P, Belouettar S, Perrin H and Trochu F 2022 *Polymer Composites* **43** 3560–3573 ISSN 1548-0569
[11] Grössing H, Stadlmajer N, Fauster E, Fleischmann M and Schledjewski R 2016 *Polymer Composites* **37** 2782–2793 ISSN 15480569
[12] Jhan Y T T, Lee Y J J and Chung C H H 2011 *Journal of Reinforced Plastics and Composites* **30** 533–545 ISSN 07316844
[13] Jhan Y T, Lee Y J and Chung C H 2011 *Journal of Composite Materials* **46** 1417–1430
[14] Correia N C, Robitaille F, Long A C, Rudd C D, Šimáček P and Advani S G 2004 *Journal of Fluids Engineering, Transactions of the ASME* **126** 210–215 ISSN 00982202
[15] Pierce R, Falzon B and Thompson M 2017 *Composites Science and Technology* **149** ISSN 02663538
[16] Loix F, Badel P, Orgéas L, Geindreau C and Boisse P 2008 *Composites Science and Technology* **68** 1624–1630 ISSN 02663538
[17] Syerko E, Schmidt T, May D, Binetruy C, Advani S G, Lomov S, Silva L, Abaimov S, Aissa N, Akhatov I, Ali M, Asiaban N, Broggi G, Bruchon J, Caglar B, Digonnet H, Dittmann J, Drapier S, Endruweit A, Guilloux A, Kandinskii R, Leygue A, Mahato B, Martínez-Lera P, Matveev M, Michaud V, Middendorf P, Moulin N, Orgéas L, Park C H, Rief S, Rouhi M, Sergeichev I, Shakoor M, Shishkina O, Swolfs Y, Tahani M, Umer R, Vanclooster K and Vorobyev R 2023 *Composites Part A: Applied Science and Manufacturing* **167** ISSN 1359835X
[18] Verleye B, Lomov S V, Long A, Verpoest I and Roose D 2010 *Composites Part A: Applied Science and Manufacturing* **41** 29–35 ISSN 1359835X
[19] Ali M A, Umer R, Khan K A and Cantwell W J 2019 *Composites Science and Technology* **173** 99–109 ISSN 0266-3538
[20] Ranjbar R K 2010 *Modeling and Simulation of Flow Pattern and Curing During Manufacturing of Composite Wind Turbine Blades using VARTM process* M.Sc. thesis Wichita State University
[21] Chebil N, Deléglise-Lagardère M and Park C H 2019 *Composites Part A: Applied Science and Manufacturing* **125** 105519 ISSN 1359835X
[22] Michaud V 2016 *Transport in Porous Media* **115** 581–601 ISSN 15731634
[23] Salvatori D, Caglar B, Teixidó H and Michaud V 2018 *Composites Part A: Applied Science and Manufacturing* **108** 41–52 ISSN 1359-835X
[24] Ni J, Li S, Sun X and Lee L J 1998 *Polymer Composites* **19** 818–829 ISSN 1548-0569
[25] Sun X, Li S and Lee L J 1998 *Polymer Composites* **19** 807–817 ISSN 1548-0569
[26] Dai J, Pellaton D and Hahn H T 2003 *Polymer Composites* **24** 672–685 ISSN 02728397
[27] Poodts E, Minak G, Mazzocchetti L and Giorgini L 2014 *Composites Part B: Engineering* **56** 673–680 ISSN 13598368
[28] Han K, Jiang S, Zhang C and Wang B 2000 *Composites Part A: Applied Science and Manufacturing* **31** 79–86 ISSN 1359835X
[29] Young W B 2013 *Applied Composite Materials* **20** 1305–1319
[30] Koefoed M S 2003 *Modeling and Simulation of the VARTM Process for Wind Turbine Blades* Ph.D. thesis Aalborg University
[31] Pearce N R, Guild F J and Summerscales J 1998 *Composites Part A: Applied Science and Manufacturing* **29** 19–27 ISSN 1359835X
[32] Wong C C, Long A C, Sherburn M, Robitaille F, Harrison P and Rudd C D 2006 *Composites Part A: Applied Science and Manufacturing* **37** 847–857 ISSN 1359835X
[33] Talvensaari H, Ladstätter E and Billinger W 2005 *Composite Structures* **71** 371–377 ISSN 02638223

[34] Drapier S, Pagot A, Vautrin A and Henrat P 2002 *Composites Science and Technology* **62** 1979–1991 ISSN 02663538
[35] Hoes K, Dinescu D, Sol H, Parnas R S and Lomov S V 2004 *Composites Part A: Applied Science and Manufacturing* **35** 1407–1418 ISSN 1359835X
[36] Pierce R S, Falzon B G and Thompson M C 2018 *Polymer Composites* **39** 2287–2298 ISSN 15480569
[37] Yenilmez B, Senan M and Murat Sozer E 2009 *Composites Science and Technology* **69** 1710–1719 ISSN 0266-3538
[38] Yong A X, Aktas A, May D, Endruweit A, Advani S, Hubert P, Abaimov S G, Abliz D, Akhatov I, Ali M A, Allen T, Berg D C, Bickerton S, Brauner C, Brütsch D, Caglar B, Caglar H, Causse P, Chiminelli A, Cohades A, Comas-Cardona S, Danzi M, Dittmann J, Dransfeld C, Ermanni P, Fauster E, Garcia-Manrique J A, George A, Graupner R, Grishaev V, Guilloux A, Hancioglu M, Harizi W, Herman T, Huang W, Kabachi M A, Keller A, Kind K, Laspalas M, Lebedev O V, Lizaranzu M, Long A C, Masania K, Michaud V, Middendorf P, Salvatori D, Schubnel R, Sharp N, Sozer M, Thomas J, Trochu F, Umer R, Valette J, Wang J H and Willenbacher B 2021 *Composites Part A: Applied Science and Manufacturing* **148** 106480 ISSN 1359-835X
[39] Yong A X, Aktas A, May D, Endruweit A, Lomov S V, Advani S, Hubert P, Abaimov S G, Abliz D, Akhatov I, Ali M A, Allaoui S, Allen T, Berg D C, Bickerton S, Caglar B, Causse P, Chiminelli A, Comas-Cardona S, Danzi M, Dittmann J, Dransfeld C, Ermanni P, Fauster E, George A, Gillibert J, Govignon Q, Graupner R, Grishaev V, Guilloux A, Kabachi M A, Keller A, Kind K, Large D, Laspalas M, Lebedev O V, Lizaranzu M, Long A C, López C, Masania K, Michaud V, Middendorf P, Mitschang P, van Oosterom S, Schubnel R, Sharp N, Sousa P, Trochu F, Umer R, Valette J and Wang J H 2021 *Composites Part A: Applied Science and Manufacturing* **142** 106243 ISSN 1359-835X
[40] Endruweit A, Luthy T and Ermanni P 2002 *Polymer Composites* **23** 538–554 ISSN 02728397
[41] Sharma S, Siginer D A, Dukipatti R K and Soschinske K A 2009 *Journal of Composite Materials* **43** 1589–1601 ISSN 1530793X
[42] Qazi H, Lin R and Jayaraman K 2021 *Resources, Conservation and Recycling* **169** 105482 ISSN 18790658
[43] Rohatgi V and Lee L J 1997 *Journal of Composite Materials* **31** 720–744 ISSN 00219983
[44] Shih C h, Liu Q and Lee L J 2001 *Polymer Composites* **22** 721–729
[45] Wang C, Ye J, Yue G, Bai G, Liu L and Zhang B 2018 *Journal Wuhan University of Technology, Materials Science Edition* **33** 242–248 ISSN 19930437
[46] Arbter R, Beraud J M, Binetruy C, Bizet L, Bréard J, Comas-Cardona S, Demaria C, Endruweit A, Ermanni P, Gommer F, Hasanovic S, Henrat P, Klunker F, Laine B, Lavanchy S, Lomov S V, Long A, Michaud V, Morren G, Ruiz E, Sol H, Trochu F, Verleye B, Wietgrefe M, Wu W and Ziegmann G 2011 *Composites Part A: Applied Science and Manufacturing* **42** 1157–1168 ISSN 1359-835X
[47] Vernet N, Ruiz E, Advani S, Alms J B, Aubert M, Barburski M, Barari B, Beraud J M, Berg D C, Correia N, Danzi M, Delavière T, Dickert M, Di Fratta C, Endruweit A, Ermanni P, Francucci G, Garcia J A, George A, Hahn C, Klunker F, Lomov S V, Long A, Louis B, Maldonado J, Meier R, Michaud V, Perrin H, Pillai K, Rodriguez E, Trochu F, Verheyden S, Weitgrefe M, Xiong W, Zaremba S and Ziegmann G 2014 *Composites Part A: Applied Science and Manufacturing* **61** 172–184 ISSN 1359-835X
[48] May D, Aktas A, Advani S G, Berg D C, Endruweit A, Fauster E, Lomov S V, Long A, Mitschang P, Abaimov S, Abliz D, Akhatov I, Ali M A, Allen T D, Bickerton S, Bodaghi M, Caglar B, Caglar H, Chiminelli A, Correia N, Cosson B, Danzi M, Dittmann J, Ermanni P, Francucci G, George A, Grishaev V, Hancioglu M, Kabachi M A, Kind K, Deléglise-Lagardère M, Laspalas M, Lebedev O V, Lizaranzu M, Liotier P J, Middendorf P, Morán J, Park C H, Pipes R B, Pucci M F, Raynal J, Rodriguez E S, Schledjewski R, Schubnel R, Sharp N, Sims G, Sozer E M, Sousa P, Thomas J, Umer R, Wijaya W, Willenbacher B, Yong A, Zaremba S and Ziegmann G 2019 *Composites Part A: Applied Science and Manufacturing* **121** 100–114 ISSN 1359-835X
[49] Okonkwo K, Simacek P, Advani S G and Parnas R S 2011 *Composites Part A: Applied Science and Manufacturing* **42** 1283–1292 ISSN 1359-835X
[50] Weitzenböck J R, Shenoi R A and Wilson P A 1999 *Composites Part A: Applied Science and Manufacturing* **30** 781–796 ISSN 1359-835X
[51] Weitzenböck J R, Shenoi R A and Wilson P A 1998 *Composites Part A: Applied Science and Manufacturing* **29** 159–169 ISSN 1359-835X
[52] Abdoli H, Hermann T and Bickerton S 2022 *Composites Part A: Applied Science and Manufacturing* **162** 107167 ISSN 1359-835X
[53] Di Fratta C, Klunker F, Trochu F and Ermanni P 2015 *Composites Part A: Applied Science and Manufacturing* **77** 238–247 ISSN 1359-835X
[54] Bickerton S, Advani S G and A U S 2000 *Polymer Composites* **21** 134–153
[55] Mohan R, Tamma K, Bickerton S, Advani S and Shires D 2001 *Specialized Molding Techniques: Application,*

Design, Materials and Processing vol 2 ed Heim H P and Potente H (William Andrew Publishing) chap 4, pp 193–207 ISBN 9781884207914

[56] Hammami A, Gauvin R and Trochu F 1998 *Composites Part A: Applied Science and Manufacturing* **29** 603–609 ISSN 1359835X

[57] Markicevic B, Nalla A, Navaz H, Heider D, Walsh S and Advani S 2008 *Polymer Composites* **16** 1434 – 1441

[58] Gersborg A R, Jørgensen J E, Gaarde R, Alkemade H, Vanaparthi S K and Patil S 2015 *ICCM International Conferences on Composite Materials* **2015-July** 19–24

[59] Calado V M and Advani S G 1996 *Composites Science and Technology* **56** 519–531 ISSN 02663538

[60] Bancora S P, Binetruy C, Advani S G, Syerko E and Comas-Cardona S 2018 *Composites Part A: Applied Science and Manufacturing* **113** 359–369 ISSN 1359835X

[61] Bian L P, Yang J S and Xiao J Y 2011 *Proceedings of the 18th ICCM International Conferences on Composite Materials, Jeju Island, Korea* (Jeju Island, Korea) pp 21–26

[62] Ali M A, Khan K A and Umer R 2021 *Materials Today Communications* **28** 102565 ISSN 2352-4928

IOP Conf. Series: Materials Science and Engineering 1293 (2023) 012010 doi:10.1088/1757-899X/1293/1/012010

Characterization of a short fiber-reinforced adhesive designed for wind turbine rotor blades

Claudio Balzani, Michael Kuhn

Leibniz University Hannover, Institute for Wind Energy Systems, Appelstr. 9A, 30167 Hannover, Germany

E-mail: research@iwes.uni-hannover.de

Abstract.

Wind turbine rotor blades are usually made of two half-shells and shear webs that are joined together by structural adhesives. The adhesive joints suffer from longitudinal strains from blade bending combined with shear strains from torsion and shear forces acting on the blade. It is thus important to provide an accurate and reliable characterization of the adhesive as a bulk material to account for cohesive failure.

This work focusses on the characterization of a short fiber-reinforced adhesive used in wind turbine rotor blades under uniaxial tension, compression, and shear as well as biaxial tension-compression/shear for both static and fatigue loading. To obtain small scatter in measurement results, a specimen geometry was designed that ensures a clear maximum stress in the test section while minimizing stress concentrations in the transition between the load introduction regions and the test section. The manufacturing process was then designed in such a way that an optimum mixture of resin and hardener was obtained without the inclusion of voids. The latter was verified after manufacturing by means of micro-CT scanning for all specimens. A very extensive test campaign was then carried out in order to quantify stiffness, static strength, fatigue strength, and fatigue-related stiffness degradation. Excerpts of this test campaign are presented in this paper.

1. Introduction

Wind turbine rotor blades are multiaxially loaded structures subjected to loads from aerodynamics, gravitation, operation, gyroscopy, etc. These external loads excite deflections and vibrations of the blades and the entire turbine. The resulting deformations of the blades comprise mainly bending and (geometrically nonlinear) torsion. In the adhesive, a three-dimensional stress state is thus present [1] that is mainly governed by longitudinal strain (or stress) from blade bending and shear stresses from shear forces and torsion.

Failure of adhesive joints may occur due to loss of adhesion (i.e., failure in the interface between the adhesive and the adherend), loss of cohesion (failure of the adhesive itself), or a combination of both. There is a large variety of failure theories and simulation models available for adhesive joints, see e.g. [2] and the references therein. The strength of adhesive joints is normally characterized as a structural member comprising the adherends and the adhesive focussing on the shear strength of the connection, e.g., by means of lap joint testing [3], DCB testing [4], or subcomponent testing [5, 6].

However, adhesive joints in wind turbine rotor blades are designed as thick adhesive layers [7, 8, 9]. It is thus inevitable to characterize the cohesion failure behavior of the respective adhesives as well, which can be done by bulk material characterization tests. As wind turbine rotor blades are subjected to very high ultimate and fatigue loads, the characterization needs to cover both (quasi-)static and fatigue scenarios and loading situations that are typical for the envisaged application. There are numerous specimen types and testing methods available for the multiaxial material characterization, ranging from plates [10, 11] to cruciforms [12, 13, 14] and tubes [15, 16, 17], the choice of which depending on the available test machines, the material properties aimed for, and the material under investigation.

In this work, a paste-like, epoxy-based, short glass fiber-reinforced structural adhesive used in wind energy, namely Hexion Epikote^{TM} Resin MGS^{TM} BPR 135G3 in combination with Hexion Epikure^{TM} Curing Agent MGS^{TM} BPH 137G [18], is considered. A tubular specimen was chosen, as tension/compression and torsion applied to such specimen results in tensile, compressive, and shear stresses in the material. However, available specimen designs from literature [16, 17] had to be modified and the manufacturing processes had to be optimized in order to obtain reliable test results [19]. A very extensive test campaign was carried out including static and fatigue tests and both uni- and biaxial loadings to obtain static strengths and stiffnesses, fatigue strengths, S-N curves, and the fatigue-related stiffness degradation behavior [19, 20, 21, 22]. An excerpt of the respective test results is presented in this paper.

2. Specimen design and manufacturing

Due to the particular loading of an adhesive in a wind turbine rotor blade, it was important to investigate the following stress states in the adhesive: i) uniaxial normal stresses (both in tension and compression), ii) uniaxial shear stresses, and iii) biaxial tension/compression-shear stress combinations.

A tubular specimen was the most suitable choice, as the desired stress states could be introduced with one specimen type and one test machine (walter+bai LFV 100-T2000 [23] with cylindrical specimen grips and independent control in tension/compression and torsion, respectively) for all stress conditions. In this way, consistent material properties could be derived without the need of elaborate considerations on comparability. The specimens additionally needed to provide the following features:

- a distinct test section with an expectable failure position;
- absence of stress concentrations;
- an excellent manufacturing quality.

The first two item points were realized by a thorough finite element-based geometry optimization. The procedure resulted in significantly higher stress ratios between the test section and other positions along the specimen compared to [16, 17] and an almost constant and homogeneous stress distribution along the test section. Moreover, very smooth transitions from the clamping sections to the test section were obtained with optimum curvatures avoiding stress concentrations along the specimen. The final geometry with dimensions is depicted in Figure 1. Note that a clamping length of 43 mm and an outer clamping diameter of 30 mm were design boundary conditions from the testing machine.

The third item point was important to obtain reliable test results with clear mean values and minimized scatter, especially in fatigue. The specimen quality was ensured by optimization of the manufacturing processes, which includes the mixture of the resin with the hardener using a planetary centrifugal vacuum mixer [24] and the development of a 3D-printed injection nozzle avoiding the formation of voids. The absence of voids in the test section was guaranteed by micro-CT scanning of all specimens. For more details on the procedure for the design of the test specimen geometry and the steps in optimizing the manufacturing processes the reader is referred to [19].

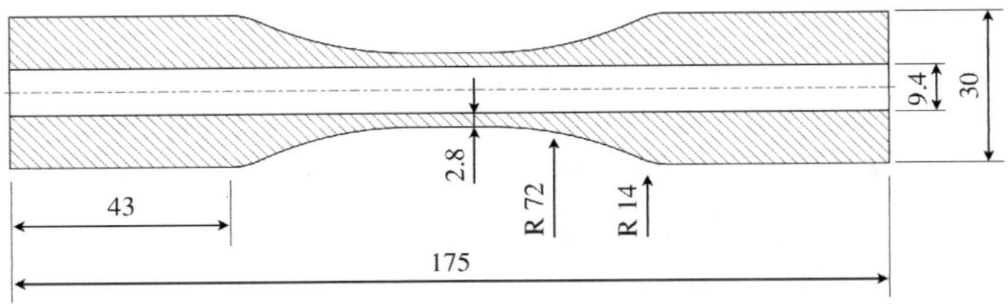

Figure 1. Geometry of the test specimen. All numbers are in mm. Image from [19] (modified, licensed under CC-BY-4.0).

3. Static test results

Static uniaxial tests in tension, compression, and torsion were carried out to determine the respective stiffness, yield point, ultimate strength, and strain to failure. In this paper, the major outcome of the static test campaign is presented. More details can be found in [19, 20].

Strain gauge rosettes were applied to each specimen on opposite sides in circumferential direction in the center of the test section. The load cell of the test machine was calibrated for class 0.5.

The tensile tests have been executed with displacement control at a displacement rate of 1 mm/min. The resulting frequency distribution functions of the ultimate tensile strength for the specimens of this work (both hand-mixed and machine-mixed specimens) are shown in Figure 2 assuming a Gaussian distribution in each case. Test results of other authors [16, 17], who used other specimen geometries and hand mixing-based manufacturing processes are added for comparison.

The test results of the own test campaign [19, 20] are compared first. The tensile strength

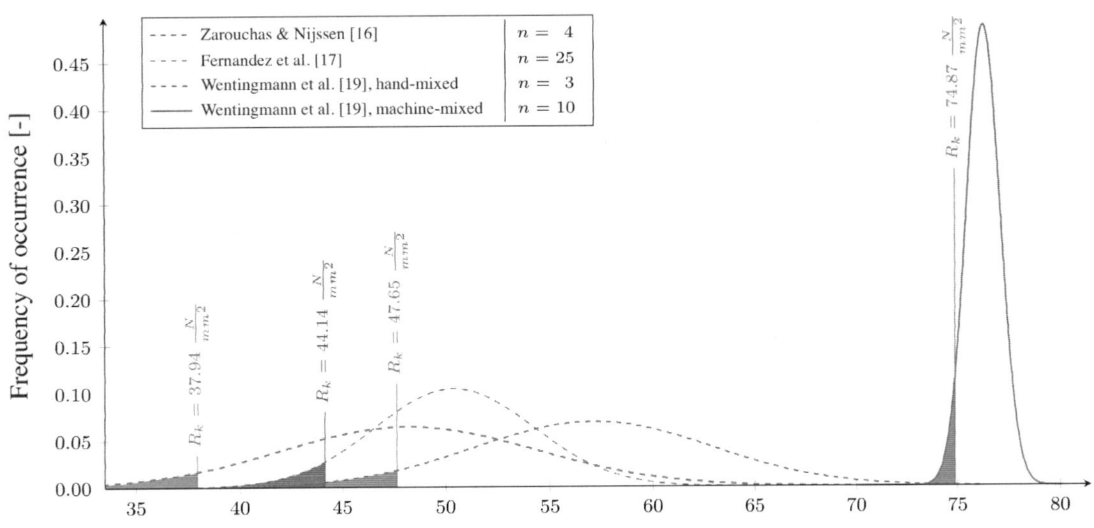

Figure 2. Frequency distributions of the engineering tensile strength for hand-mixed and machine-mixed specimens and two reference specimens from [16, 17] with different geometries and hand-mixing-based manufacturing. The 5% quantile engineering tensile strengths R_k are indicated with vertical lines, n is the number of tested specimens. Image from [19] (modified, licensed under CC-BY-4.0).

of the machine-mixed specimens is much higher than that of the hand-mixed specimens. The reason is that in the hand-mixed specimens a high degree of porosity was present, which reduces the strength significantly, while the machine-mixed specimens were virtually porosity-free. It can also be seen that the standard deviation, is much lower for the machine-mixed specimens. The lower tensile strength and higher standard deviation for hand-mixed specimens is supported by the reference test results from [16, 17].

The high strength obtained in the tests with the machine-mixed specimens is very close to that reported in the product data sheet [18]. It can be concluded that this is the real material property unaffected by defects such as voids. On the contrary, a high porosity content decreases the strength substantially in the hand-mixed specimens, i.e., those specimens suffer from effects of defects. Another reason for the high tensile strength may be a very distinct fiber orientation in longitudinal direction in the machine-mixed specimens, which was identified by high-resolution micro-CT scanning [22]. In the hand-mixed specimens, the voids may have affected the fiber orientation. However, this hypothesis was not yet verified by micro-CT scans. In fact, the tests may in turn have influenced the fiber orientation, e.g., due to plastic deformations. Another difference in test results was that the machine-mixed specimens exhibited a distinct elasto-plastic behavior, whereas the hand-mixed specimens were rather brittle in comparison, see [19].

In the uniaxial compression tests, the specimens failed by buckling, although buckling analyses were carried out in the design of the specimen geometry. However, the strong elasto-plastic material behavior was not expected, as in literature a brittle behavior was reported [16, 17]. The higher failure strain and higher stiffness led to unexpected high stresses and consequently to buckling, so that the uniaxial compression strength cannot be evaluated. However, the transition from the elastic region to the plastic region was at lower strains than the onset of buckling. Hence, the uniaxial tension, compression, and shear tests as well as the biaxial tension-compression/shear tests were employed to derive a yield surface in [20]. The experimentally derived yield surface was compared to well-known criteria from literature, i.e., the criteria of von Mises [25], von Mises-Schleicher [26], and Drucker-Prager [27], respectively. The results are plotted in Figure 3.

The test results imply an elliptical shape of the yield surface with a clear tension-compression asymmetry. An ellipsis fit hence shows very good agreement with the test results. The von Mises criterion cannot capture the tension-compression asymmetry. The von Mises-Schleicher criterion does capture the tension-compression asymmetry, but overestimates the yield strength in shear.

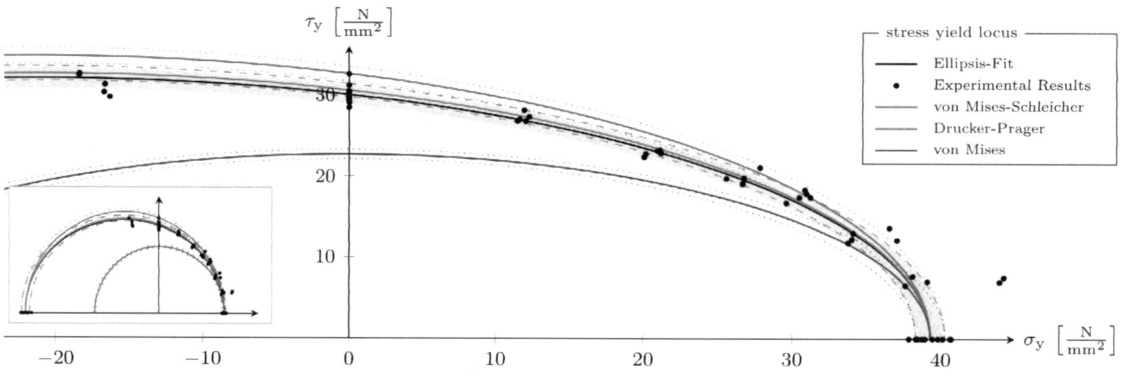

Figure 3. Yield locus on the basis of uni- and biaxial test results (τ_y: shear yield stress, σ_y: normal yield stress). The Drucker-Prager criterion and the ellipsis fit are in very good agreement with the test results. Image from [20] (licensed under CC-BY-4.0).

The Drucker-Prager criterion is a very good fit to the test results, which almost coincides with the ellipsis fit. It is thus recommended to use the Drucker-Prager criterion, as it is well established in other engineering applications.

For more details on the stress-strain curves, the extraction of yield stresses, the required elasto-plastic shear stress correction, the yield surface derivation, and for a thorough discussion of the results with respect to tension/compression bimodularity, viscoelasticity, comparison to existing yield criteria, the reader is referred to [20].

4. Fatigue test results

The fatigue tests were conducted using the same specimen design and testing machine as in the quasi-static tests in [21]. To avoid heating of the specimens, the test frequencies at each load level was adjusted accordingly. The adaptation of the test frequency for different load levels also resulted in a reduction of viscoelasticity-related strain-rate effects. The cyclic loading in tension/compression was applied load-controlled, whereas the torsional load had to be applied displacement-controlled. For the reasoning and the implications in testing and postprocessing see [21]. The determined cycles to failure enabled the extraction of S-N curves, see Figure 4.

Stiffness degradation measurements inspired by [28] accompanied the fatigue tests. Therein, the cyclic loading was stopped at pre-defined intervals, followed by a quasi-static cycle with very small amplitude in tension and torsion. On account of the small amplitudes, the additional

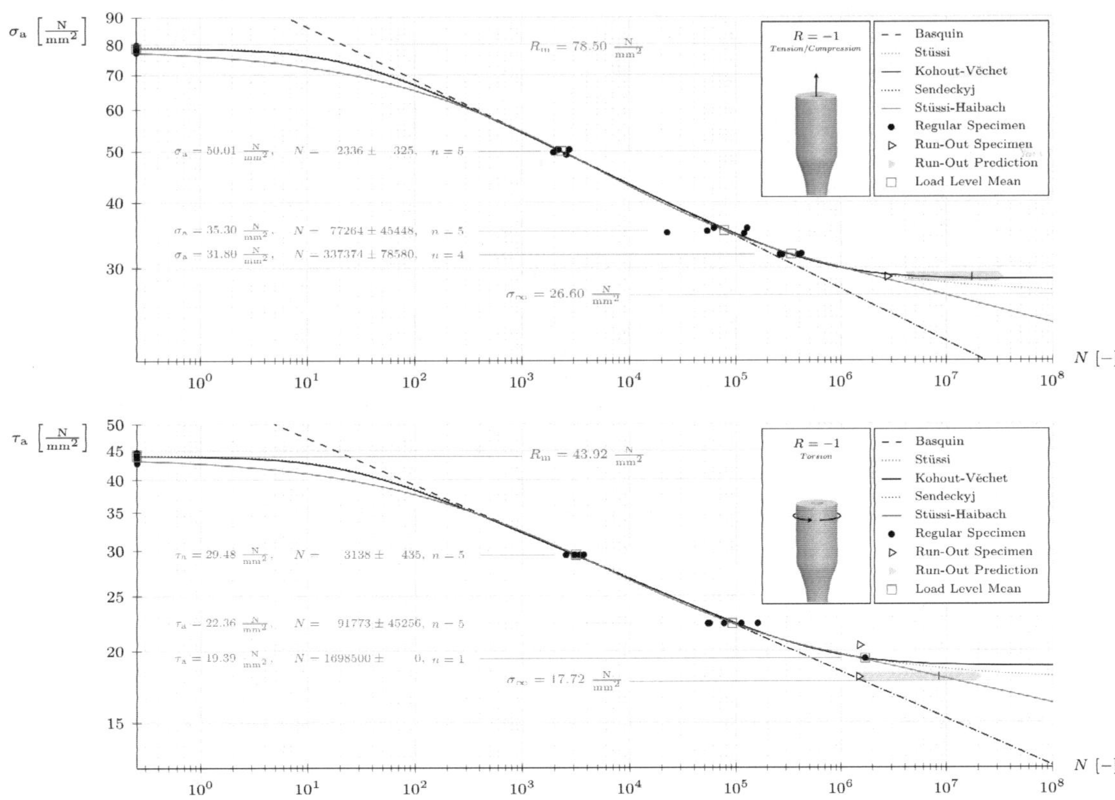

Figure 4. Fatigue test results and S-N curve fits for tension/compression (top, σ_a: axial stress amplitude, $R = -1$) and torsion (bottom, τ_a: shear stress amplitude, $R = -1$). Image from [21] (licensed under CC-BY-4.0).

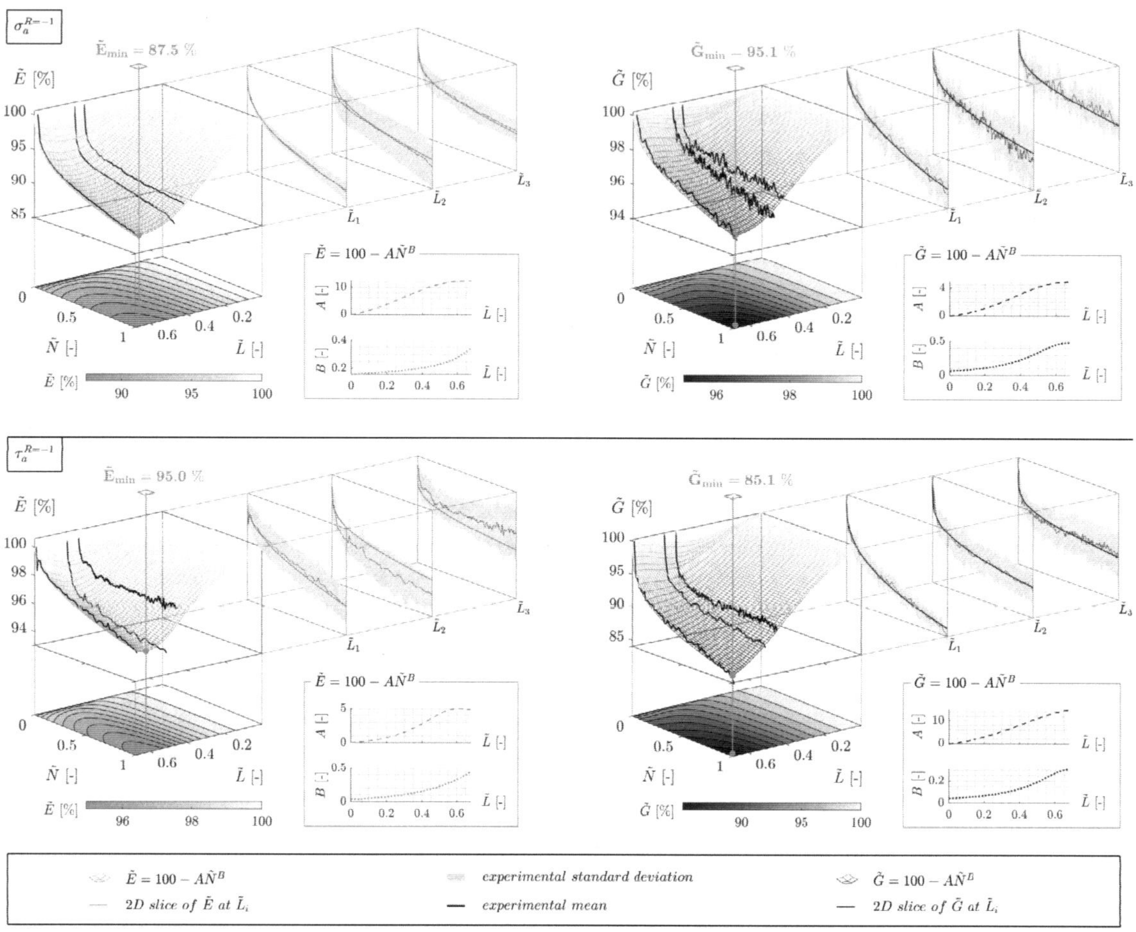

Figure 5. Uniaxial stiffness degradation for tension/compression (top) and torsion (bottom), both for $R = -1$ ($\tilde{E}$: normalized Young's modulus; $\tilde{G}$: normalized shear modulus; $\tilde{N}$: normalized cycle numbers; $\tilde{L}$: normalized load level). The model predictions are shown in green and blue, the test results in black. Image from [21] (licensed under CC-BY-4.0).

cycles of the stiffness degradation measurements had a negligible effect on the S-N results. From these intermediate measurements, the Young's and shear moduli could be extracted, enabling the formulation of moduli as a function of the cycle number and load level, see Figure 5.

The R-ratio, which is defined by the ratio between the minimum and maximum stress, was $R = -1$ and $R = 0$ for tension/compression, and $R = -1$ for torsion. This paper focusses on the results for $R = -1$. The test results for $R = 0$ can be found in [21]. S-N curve models according to Basquin [29], Sendeckyj [30], Stüssi [31], and Kohout & Věchet [32] were fitted to the test results for comparison. The test results together with the S-N curve fits are shown in Figure 4 on a double-logarithmic scale.

The test results form a sigmoidal shape. Three run-out specimens were observed (one in the tension/compression tests, two in the torsion tests). Residual strength tests were carried out with the run-out specimens (black triangles in Figure 4). In combination with the stiffness degradation model, these allowed for a prediction of the fatigue life of the run-out specimens. The results indicate the prescence of a fatigue limit, but due to a lack of data in the very high cycle fatigue (VHCF) regime, a final conclusion is not possible yet.

The Basquin model fails to represent the test results well. It is non-conservative in the low cycle fatigue (LCF) regime and very conservative in the high cycle fatigue (HCF) regime. The Kohout-Věchet and the Stüssi model fit the experimental results very well. However, the Stüssi model is more conservative in the LCF and the HCF regimes. The Sendeckyj model coincides with the Kohout-Věchet model in the LCF regime and with the Basquin model in the HCF regime and is thus very conservative at low amplitudes. The tangential combination of the Stüssi model and the Haibach model [33] for VHCF (approximately half slope of the Basquin model) was proposed earlier [21]. It does not predict a fatigue limit, but is also not as conservative as the Basquin model, especially considering the fatigue life prediction of the run-out specimens. The small scatter in test results on each load level is worth mentioning, and is a result of the very thorough specimen design and manufacturing.

The stiffness degradation test results together with model predictions are summarized in Figure 5. The degradation of the Young's modulus is shown in green, that of the shear modulus in blue. Test restults are indicated by black lines. In the top row, the fatigue load was uniaxial tension/compression with $R = -1$ and in the bottom row the fatigue load was uniaxial torsion with $R = -1$. In Figure 5, the relative Young's modulus $\tilde{E}$ and relative shear modulus $\tilde{G}$ (which are both a function of the cycle number n), the normalized cycle number $\tilde{N}$, and the normalized load level $\tilde{L}$ are defined by the expressions

$$\tilde{E}(n) = \frac{E(n)}{E_0} \cdot 100, \quad \tilde{G}(n) = \frac{G(n)}{G_0} \cdot 100, \quad \tilde{N} = \frac{n}{N}, \quad \tilde{L} = \frac{\sigma_a}{R_m}. \tag{1}$$

Herein, $E(n)$ is the cycle number-dependent Young's modulus, E_0 is the initial Young's modulus, $G(n)$ is the cycle number-dependent shear modulus, G_0 is the initial shear modulus, N is the number of cycles to failure, σ_a is the (normal) stress amplitude (denoted by τ_a for torsion), and R_m is the static ultimate strength, respectively. The modeling approach consists of a power law given by

$$\tilde{E} = 100 - A\tilde{N}^B, \quad \text{with} \quad A = \frac{p_1\tilde{L} + p_2}{\tilde{L}^2 + q_1\tilde{L} + q_2}, \quad B = \frac{p_3\tilde{L} + p_4}{\tilde{L}^2 + q_3\tilde{L} + q_4}. \tag{2}$$

The parameters A, B, p_i, and q_i, where $i = 1, 2, ..., 4$, are fitting parameters. For torsion, $\tilde{E}$ needs to be replaced by $\tilde{G}$ in Equation (2).

The test results show that the stiffness degradation increases with the number of cycles and is more severe for higher load amplitudes. The scatter is quite small in the primary load direction and is bigger in the secondary load direction. The primary load direction corresponds to the applied fatigue loading, i.e., the axial direction for the tension/compression fatigue test and the torsional direction for the torsion fatigue test, respectively. The magnitude of the stiffness degradation is approximately 5% for the secondary load directions, but is as much as 12.5% and 14.9% in the primary load directions tension/compression and torsion, respectively. In general, the modeling approach reflects the experimental results very well, both qualitatively and quantitatively.

For more details on the test setup, the S-N curve models, the experimental results, an engineering approach to very high cycle fatigue, the stiffness degradation model, a comparison with residual strength tests, and a run-out fatigue life estimation, the reader is referred to [21]. Results for multiaxial and non-proportional fatigue tests have recently been published [22].

5. Conclusion

In this paper, an overview of an extensive testing campaign for the characterization of a paste-like, epoxy-based, short fiber-reinforced structural adhesive used in wind turbine rotor blades was presented.

A tubular specimen was chosen in order to carry out uniaxial tension/compression and shear tests as well as biaxial combinations of these using a testing machine with independent control for tension/compression and torsion. In this way, one specimen design could be used for all tests, resulting in a consistent dataset. The specimen was designed in such a way that a distinct test section with an expectable failure position was present in absence of stress concentrations. The manufacturing processes were optimized so that a homogeneous mixture and virtually defect-free specimens were obtained. Micro-CT scanning was used to guarantee a neglectable porosity content.

Static and fatigue tests have been carried out. The very good specimen quality resulted in extraordinarily low scatter of test results, high strength values that were similar to those reported in the manufacturer's data sheet, and a distinct plastic behavior in the static tests. A clear tension-compression asymmetry was present. The Drucker-Prager criterion was in very good agreement with the experimentally derived yield surface. In uniaxial fatigue, an S-N curve model was obtained that was also in very good agreement with the test results for all load levels. The fatigue-related stiffness degradation was also investigated and a cycle number- and load level-dependent degradation model was derived that was in very good agreement with experimental data.

Acknowledgements
This work was supported by the German Federal Ministry for EconomicAffairs and Climate Action (BMWK) in the *ReliaBlade* project (grant numbers 0324335A, 0324335B). The experiments have been carried out at the Fraunhofer Institute for Wind Energy Systems IWES in Bremerhaven, Germany. The authors acknowledge the provided technical support.

References
[1] Noever Castelos P and Balzani C 2016 *Journal of Physics: Conference Series* **753** 062002
[2] Tserpes K, Barroso-Caro A, Carraro P A, Beber V C, Floros I, Gamon W, Kozlowski M, Santandrea F, Shahverdi M, Skejić D, Bedon C and Rajčić V 2022 *The Journal of Adhesion* **98** 1855–1915
[3] American Society for Testing and Materials 2014 Standard test method for determining strength of adhesively bonded rigid plastic lap-shear joints in shear by tension loading Standard ASTM D3163-01R14 West Conshohocken, Pennsylvania, USA
[4] American Society for Testing and Materials 2013 Standard test method for mode i interlaminar fracture toughness of unidirectional fiber-reinforced polymer matrix composites Standard ASTM D5528-13 West Conshohocken, Pennsylvania, USA
[5] Sayer F, Antoniou A and van Wingerde A 2012 *International Journal of Adhesion and Adhesives* **37** 129–135
[6] Zarouchas D S, Makris A A, Sayer F, Van Hemelrijck D and Van Wingerde A M 2012 *Composites Part B: Engineering* **43** 647–654
[7] Subrahmanian K P and Dubouloz F 2009 *Reinforced Plastics* **53** 26–29
[8] Cassano A G, Dev S, Maiaru M, Hansen C J and Stapleton S E 2021 *Journal of Applied Polymer Science* **138** 49989
[9] Noever-Castelos P, Haller B and Balzani C 2021 *Wind Energy Science Discussions* **2021** 1–32
[10] Shewchuk J, Zamrik S Y and Marin J 1968 *Experimental Mechanics* **8** 504–512
[11] George T J, Seidt J, Shen M H H, Nicholas T and Cross C J 2004 *International Journal of Fatigue* **26** 477–486
[12] Makinde A, Thibodeau L and Neale K W 1992 *Experimental Mechanics* **32** 138–144
[13] Demmerle S and Boehler J P 1993 *Journal of the Mechanics and Physics of Solids* **41** 143–181
[14] Smits A, Van Hemelrijck D, Philippidis T P and Cardon A 2006 *Composite Science and Technology* **66** 964–975
[15] Tao G and Xia Z 2008 *Materials Science and Engineering: A* **486** 38–44
[16] Zarouchas D and Nijssen R 2016 *Journal of Adhesion Science and Technology* **30** 1413–1429
[17] Fernandez G, Vandepitte D, Usabiaga H and Debruyne S 2018 *International Journal of Fatigue* **117** 340–351
[18] Hexion^{TM} 2021 Technical Data Sheet - EPIKOTE^{TM} Resin MGS^{TM} BPR 135G-Series and EPIKURE^{TM} Curing Agent MGS^{TM} BPH 134G-137GF
[19] Wentingmann M, Manousides N, Antoniou A and Balzani C 2021 *Materials & Design* **212** 110213

[20] Wentingmann M, Manousides N, Antoniou A and Balzani C 2022 *Polymer Testing* **113** 107648
[21] Kuhn M, Manousides N, Antoniou A and Balzani C 2023 *Fatigue & Fracture of Engineering Materials & Structures* **46** 1121–1139
[22] Kuhn M, Manousides N, Antoniou A and Balzani C 2023 *Fatigue & Fracture of Engineering Materials & Structures* **46** 3161–3178
[23] walter+bai AG 2021 Axial / Torsional Test Systems `https://www.walterbai.com/`
[24] Thinky Corporation 2021 Thinky ARV-310 planetary centrifugal vacuum mixer `https://www.thinkymixer.com/`
[25] von Mises R 1913 *Nachrichten von der Gesellschaft der Wissenschaften zu Götingen, Mathematisch-Physikalische Klasse (in German)* **4** 582–592
[26] Schleicher F 1926 *ZAMM - Journal of Applied Mathematics and Mechanics / Zeitschrift für Angewandte Mathematik und Mechanik* **6** 199–216
[27] Drucker D C and Prager W 1952 *Quarterly of Applied Mathematics* **10** 157–165
[28] Adden S and Horst P 2010 *International Journal of Fatigue* **32** 108–122
[29] Basquin O H 1910 *American Society for Testing Materials* **10** 625–630
[30] Sendeckyj G P 1981 *ASTM STP* **734** 245–260
[31] Stüssi F 1956 *IABSE congress report* **5** 229–234
[32] Kohout J and Věchet S 2001 *International Journal of Fatigue* **23** 175–183
[33] Haibach E 2006 *Betriebsfestigkeit* (Springer-Verlag) 978-3540293637

Circular applications through selection strategies (CATSS): a methodology for identifying reuse applications for end-of-life wind turbine blades

Israel A Carrete*, Jelle J Joustra, A Ruud Balkenende

Faculty of Industrial Design Engineering, Delft University of Technology, 2628 CD Delft, The Netherlands

*I.A.Carrete@tudelft.nl

Abstract. Wind turbines are crucial for the energy transition, but their end-of-life treatment presents a challenge. Most wind turbine blades, made from composites, are currently sent for disposal or recycled through methods that degrade the value of the material. Structural reuse through blade segmentation was introduced as a recovery method that maintains high material value throughout subsequent life cycles. Most recovery attempts focus on thermoset composites, but thermoplastics are becoming more common. Unlike thermosets, thermoplastics can be reshaped through thermoforming processes, which offers the opportunity of adapting the geometry of a blade to new reuse applications. This paper introduces selection strategies to identify secondary applications of reshaped thermoplastic blade sections. A new methodology is proposed based on Landru's selection strategies and the Material Driven Design method (MDD). The Circular Applications Through Selection Strategies (CATSS) methodology proposes understanding a material at different levels to identify applications. Each sectioning level of the blades yields different material characteristics, such as the reshapability, that are then put into Landru's three selection strategies: substitution, selection by function, and inverse selection. Substitution directly supplants other materials with blades in an existing application; selection by function compares material properties and performance indices to derive the most relevant functions (i.e. "light-weight beams"); and inverse selection identifies suitable market sectors. The CATSS method is a systematic approach to exploring the reuse of blade sections across multiple life cycles, taking into consideration the changes in blade geometry introduced by each sectioning level. For example, the second use cycle might use blade segments for infrastructural applications like electrical transmission poles, while 3rd and 4th cycles reuse blade elements or blade units for urban furniture or automotive parts, respectively. Thus, by identifying multiple use cycle applications at various sectioning levels, we introduce structural reuse and reshaping as a long-lasting recovery pathway for decommissioned wind turbine blades. The selection strategies presented on the one hand can help identify new applications for thermoplastic composite products at their end-of-life, while on the other hand they indicate which aspects need to be considered in the original design, thus contributing to more circular practices in the composites industry.

1. Introduction

Wind turbines are one of the most developed sources of renewable energy, making them crucial for the energy transition. The use of wind energy is expected to grow exponentially in the following decades, as will the waste associated with the production and decommissioning of wind turbines. Liu et al suggest that most of the end-of-life waste of wind turbines is derived from the blades. [1] Wind turbine blades (WTBs) are made from fibre reinforced polymers (FRPs). End-of-life (EoL) options for FRPs are limited due to the complexity of separating its constituents. Beauson et al. identified that the reuse, repurposing, and recycling WTBs are a chain of processes where the blades are transformed into new materials in accordance with circular economy principles [2]. In a circular economy, materials and resources flow while preserving their value by looping back into manufacturing processes at different stages [3]. However, due to the complexity of FRPs, landfilling and mechanical recycling are the two most common EoL strategies for WTBs [2], [4]. Recycling is more desirable than landfilling, but it is also the least favourable within a circular economy because all material functionality is lost [3], [5].

A more favourable EoL strategy is *structural reuse*, which is a process that repurposes (partial) components of a blade in sections that retain the load-bearing capabilities of the original material [6], [7]. Joustra et al. propose a method of segmentation that preserves material integrity while expanding the opportunities of reuse to product categories such as urban furniture [8].

The segmentation method was tested on a case study made from a glass fiber/epoxy WTB. However, recent developments in thermoplastic composites has increased their use in WTBs [9]. Unlike their epoxy counterparts, thermoplastic composites can be reshaped with appropriate thermoforming processes. Thus, the geometry of elements from thermoplastic blades can be adapted to different secondary applications than thermoset WTBs.

The structural reuse of reshaped thermoplastic blades in secondary applications has not yet been explored. There is no comprehensive method for identifying suitable secondary applications, either. We aim to establish such a strategy in this work with the Circular Applications Through Selection Strategies (CATSS) method – a systematic approach to identify structural reuse applications.

The CATSS method combines Landru's selection strategies with Karana's Material Driven Design (MDD) method to identify secondary applications for WTBs. Section 2 introduces the existing methods for finding applications for materials. Section 3 presents the CATSS method in detail. Section 4 then reframes applications from literature under the lens of the CATSS method. The two closing sections discuss the strengths and limitations of the methodology at this stage and its potential impact in promoting structural reuse as an EoL strategy for WTBs.

2. Finding Applications for Materials

Materials are typically chosen for a specific application based on a set of constraints and design requirements [10], [11]. In a structural reuse scenario, the logic is inverted – design applications and requirements are derived from a given material. Landru's selection strategies and Karana's MDD method both use this inverted logic to find applications for materials.

Landru et al. propose three strategies to find applications for materials - substitution, selection by function, and inverse selection [12]. In substitution, a new material replaces one already used in a specific application. Substitution facilitates identifying new applications based on what exists in the market. Selection by function uses performance indices to identify functions for the material. Performance indices are combinations of material properties that can be used as a benchmark when comparing material candidates. Material databases such as Ansys Granta EduPack™ can be used to identify the indices that make a material stand out. The results of this strategy include a set of broadly defined functions such as "light-weight load-bearing structures", which is connected to a quantifiable performance index. Inverse selection identifies suitable market sectors based on material characteristics. With the most relevant market sectors identified, this strategy yields the broadest results of the three.

The MDD method is comprised of four steps that begin with understanding a material and end with developing a material demonstrator that emphasizes the material's unique qualities. Only the first step, *Understanding the Material*, is discussed here because it broadens the focus beyond mechanical aspects and includes characteristics such as aesthetics and material experience.

3. CATSS Method

Finding secondary applications for decommissioned WTBs poses the challenge of starting from a material that has been in service for 20-25 years. The two methods mentioned above are the only ones found that find applications from a material. As such, they were used as the basis for the CATSS method, which combines the two in a manner that is relevant for the context of WTBs by introducing a preliminary step of material sectioning.

Since material properties can change after years of service, it is imperative to first understand the residual properties of WTBs before proposing any suitable applications.[13] The following section present the levels of understanding needed to repurpose blades throughout multiple use cycles.

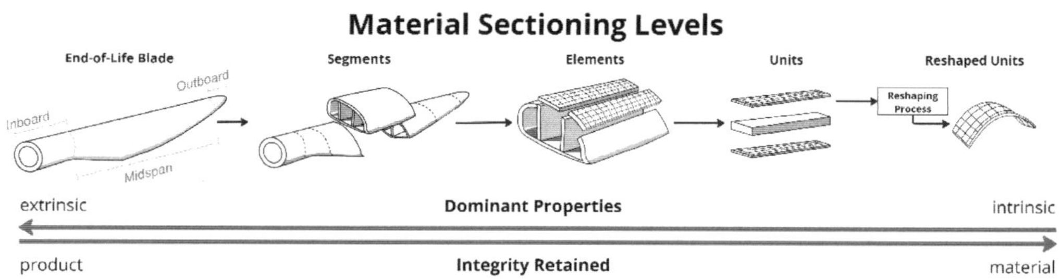

Figure 1 Material sectioning levels for a thermoplastic wind turbine blade

3.1. *Material Sectioning*

The first step of the CATSS methodology emphasizes the importance of understanding the blades' properties. These properties vary depending on the scale of the sections that are used and their location in the blade, so it is worthwhile to explore the levels of *material sectioning* that can be harvested from a blade.

A decommissioned blade is a macrostructure composed of segments like box-beams (composed of 2 shear webs and 2 spar caps) and an aerodynamic shell. Since the properties of these *segments* change

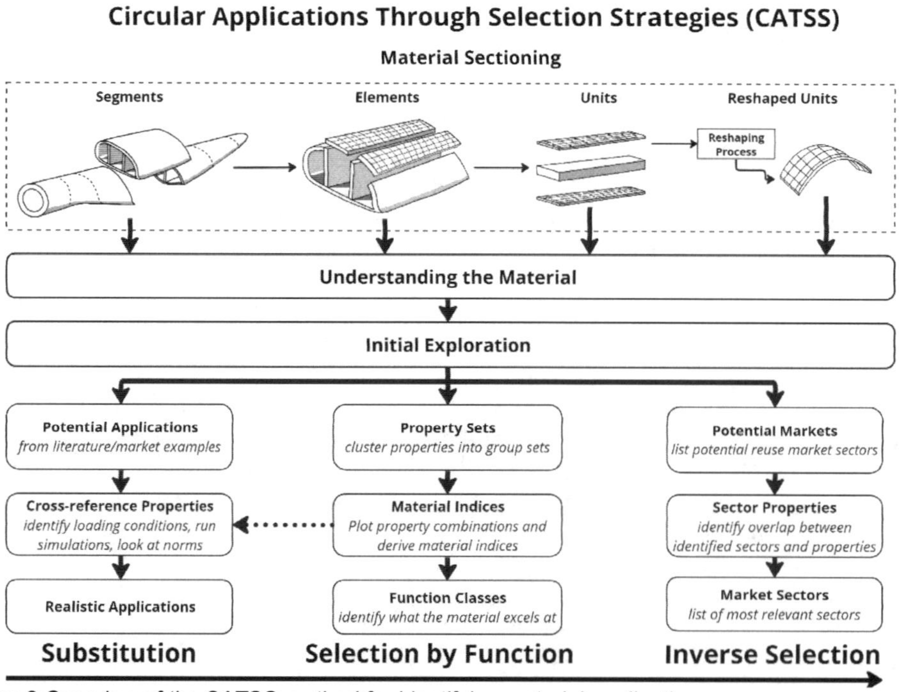

Figure 2 Overview of the CATSS method for identifying material applications

when they are cut, they are governed by dimension-dependent (extrinsic) properties. When these segments are sectioned with longitudinal cuts, they become composite *elements* with different properties. In WTBs, these elements are composed of sandwich structures made from FRP laminates and a core material. When the laminates and core are separated from each other, only the basic *units* remain. Any further separation of the matrix and fibres would lead to recycling rather than structural reuse. The horizontal axis of Fig. 1 shows the gradual changes in material integrity from segments to units, as well as the transition of the dominant properties. Unlike segments, the properties of units are more *intrinsic* and dimension-independent (i.e. density and Young's modulus).[14] Furthermore, the unit sectioning level is where thermoplastic matrices stand out the most because units can be reshaped. The properties of each sectioning level influence which reuse applications are most suitable.

3.2. *Circular Applications Through Selection Strategies (CATSS)*

The second step of CATSS is an adaptation of Landru's selection strategies. Landru's selection strategies can be used independently or in succession, but they all necessitate some understanding of the material. As such, the CATSS method facilitates the use of these strategies with two preliminary steps – *understanding the material* and *initial exploration*. This is especially important since each sectioning level has different characteristics and properties. Once an understanding of the material is achieved, either one of Landru's selection strategies can be pursued, regardless of sectioning level. Fig. 2 gives an overview of the CATSS method when applied to a decommissioned blade.

3.2.1. Understanding the material. As mentioned before, this step is a reference to the MDD method, which encourages a thorough examination of the qualities and constraints that make a material unique. The examination includes classical characterization methods (i.e. empirical tests, literature research, etc) and "material tinkering", which is an explorative process of creation and evaluation with the material.[15] Tinkering is useful in deriving auxiliary material properties such as aesthetics and materials experience, and yields results that are not only focused on mechanical aspects of the material.

Since the segment sectioning level is governed by extrinsic properties, its characterization requires thorough structural analysis. Alshannaq et al. derived the effective bending stiffness of a Clipper C96 wind turbine blade segment [16]. Joustra et al. used Ansys Granta EduPack's software to derive the properties of construction elements cut out from a reference blade. Since this blade included the layup of laminates, that same information is also suitable to characterize the units level.[8] Table 1 provides an overview of the properties for each sectioning level. The exploratory tinkering process was bypassed by referencing the Re-Wind Design Catalog, in which designers identified the auxiliary properties of WTBs [17].

Table 1 Mechanical properties of each sectioning level for decommissioned blades (σ_T *- tensile strength,* σ_c *- compressive strength, and * are actually flexural values)*

Sectioning Level	Size/length	E	G(MPa)	Poisson's ratio	Density	σ_T	σ_C
	m	*GPa*	*GPa*	*[-]*	*kg/m³*	*MPa*	*MPa*
Segments	10 - 20	10 - 27	-	-	-	145-673	145 - 453
Elements	2-10	2-99 *	-	-	300-1650	-	160 - 1330*
Units (GFRP)	-	-	-	-	-	-	-
Unidirectional (UD)	2-10	41.8	26.3	0.28	1920	972	702
Biaxial	2-10	13.6	11.8	0.49	1780	144	213
Triaxial	2-10	27.7	7.2	0.39	1850	700	292
Foam Core	2-10	0.256	0.022	0.3	200	5.6	4.4
CFRP (UD)	2-10	114.5	5.99	0.27	1545	1546	1047

3.2.2. Initial exploration. The overview of mechanical properties and unique qualities of each sectioning level hint at possible uses for the material. The design catalog was used to brainstorm other suitable applications according to the identified properties. This led to the compilation of 60 applications, which act as a primer to think about what each sectioning level has to offer.

The functions of the blade material in these 60 applications were clustered into 11 categories based on the material characteristics that were emphasized in the application. Examples include "chemical resistance", "light-weight", and "load-bearing structures".

Furthermore, the identified applications were categorized according to the market sectors they belong to (i.e. a bridge would be part of "infrastructure"). Here, 16 market sectors were identified in total, but not all sectors were relevant to each sectioning level due to differences in material characteristics. An example of this is explained in Section 4.3. Once these categories are identified, the exploration is complete, and Landru's strategies are employed. The strategies are presented next in order of how specific their outputs are.

3.2.3. Substitution. Substitution is the process in which blades of any sectioning levels replace another material in an existing application. Substitution is composed of three steps: naming potential applications, cross-referencing material properties, and identifying realistic applications. A market analysis and/or state-of-art research into the applications found in section 3.2.2 serve as benchmarks for advances in GFRP reuse. The properties relevant for these applications are cross-referenced with those gathered above to remove anything that is unfeasible. A more elaborate look into the standards and norms of specific applications is appropriate to identify the most realistic applications.

3.2.4. Selection by Function. As a more material-based approach, this strategy consists of three steps: gathering property sets, plotting performance indices, and identifying what the blade sections excel at. Property sets describe combinations of quantifiable properties that, when taken together, describe a function. Recall that the initial exploration stage identified 11 material characteristics. These are composed of relevant material property sets (i.e. density, specific strength, etc). Using a database such as Ansys Granta EduPack, the property sets are plotted into bubble charts that compare various material families. If blade sections outperform other material families within a property set, the corresponding performance index indicates suitable function classes[11]. Alternatively, if the material is close in performance to another material class, then the other material can be used as a starting point for the substitution strategy.

3.2.5. Inverse Selection. The last strategy produces the broadest results. The process consists of identifying potential markets by cross-referencing the material characteristics of a sectioning level with relevant market sectors and evaluating compatibility. Recall that potential market sectors were identified in Section 3.2.2. Each market sector entails specific functions and material characteristics. These properties are cross-referenced with the properties of those of the blade section for compatibility. One way for designers to evaluate the compatibility of a sectioning level with a market sector is by creating a matrix that maps the corresponding market sectors on one axis and material characteristics on the other. Upon each intersection, the relevance of the material characteristic for a specific sector are rated irrelevant to highly relevant (1-5) based on industry knowledge or literature, for example. The sum of all ratings across each market sector is then used to identify the sectors that best emphasize a sectioning level's characteristics. A by-product of this strategy's broad results is that it identifies applications that the other two would miss.

Each selection strategy can be used independently or in conjunction. The CATSS method is a way of exploring a material's compatibility with new applications by deriving relevant criteria, such as material properties, performance indices, and other characteristics relevant to specific sectors.

4. Examples of using the CATSS Method

The CATSS method identifies secondary applications and criteria for feasibility. Despite ongoing efforts to reuse blades, no common approach has been identified. Sections 4.1 and 4.2 take examples from literature that resemble Landru's substitution and selection by function strategies and reframe them under CATSS. The examples illustrate the corresponding strategies as is, but the first two steps of CATSS are added to highlight the advantages of a systematic approach.

4.1. Substitution example. Electrical transmission pole from blade segments

Alshannaq et al. found that WTB segments might be suitable for use as electrical transmission poles because the loading conditions they undergo are similar to those of WTBs in service. Furthermore, FRP poles are entering the market as of late. The compatibility of blade segment properties (identified in a previous step) with those of transmission poles were simulated according to manufacturer data sheets and relevant norms, respectively. The simulations verified that blades abide by the standards used for transmission poles safely. [16] Thus, WTB sections can replace FRP and/or steel poles due to similar loading conditions. The resulting BladePole is an example of substitution, but material data was gathered after finding the application. On the other hand, the first two steps in CATSS gather relevant data beforehand, facilitating the identification of more adequate solutions.

4.2. Selection by function example: Blade element picnic table

Next on the spectrum of specificity, a picnic table was identified as a feasible application for blade elements. To do this, Joustra et al. plotted the flexural properties of blade elements. Figure 3 shows the flexural performance of beams and panels made from materials that optimize performance at minimum mass. The dotted lines correspond to performance indices. For stiffness limited design, the index of $E^{\frac{1}{3}}/\rho$ is used (Fig. 3a). Anything above the line maximizes performance. The recovered blade elements and timber were above this. On the other hand, strength limited design (Fig. 3b) is described by $\sigma^{\frac{1}{2}}/\rho$. The blade elements excelled in this category as well, making them suitable to use in functions that emphasize bending. A picnic table was developed as a demonstrator of this potential. Unlike substitution, the results from selection by function can be applied to any application in which a high specific flexural strength is necessary.

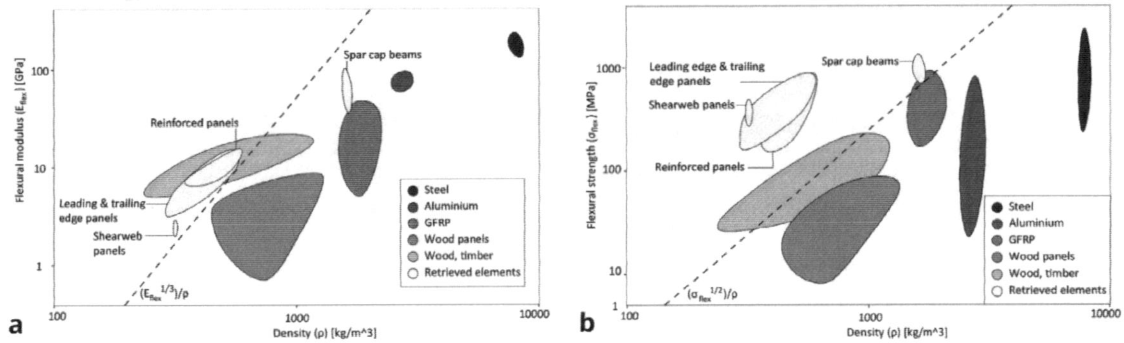

Figure 3 Performance index plots of construction materials for a) stiffness limited design and b) strength limited design (re-printed with the author's permission).

4.3. Inverse selection example: CATSS Exploration

Due to the nature of inverse selection, which yields the broadest results, no concrete examples exist in literature. However, the most suitable market sectors of each sectioning level were already identified during the initial exploration. Taking the segment sectioning level as an example, the affinity between material characteristics and market sectors were evaluated in Fig. 4.

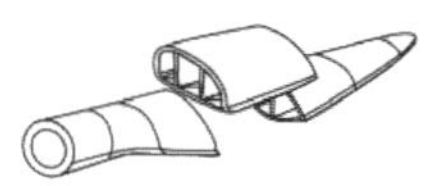

Legend: 1 2 3 4 5

Material Characteristics ⟷ / Market Sectors ↕

Market Sectors	Size	Light-Weight	Load-bearing Capacity	Durability	Environmental Resistance	Chemical Resistance	Toughness	Sound Isolation	Electromagnetic Transparency	Shape	Aesthetics	**Total**
Architecture	5	5	5	4	4	4	3	1	1	3	2	**37**
Infrastructure	5	5	5	4	4	3	4	1	1	3	3	**38**
Environmental Protection	3	2	3	5	5	5	5	2	1	1	1	33
Marine	2	5	3	3	5	4	3	1	1	2	1	30
Arts and Culture	5	2	5	4	4	4	2	1	2	5	5	**39**
Construction	5	5	5	5	4	3	4	1	1	2	1	35
Erosion Resistance	4	3	5	5	5	3	5	1	1	2	1	35
Urban Furniture	2	4	5	5	5	4	5	3	1	3	4	**41**
Acoustics	3	2	1	4	1	3	2	5	3	5	1	28

Figure 4 Sample matrix used to evaluate the market sectors for the segment sectioning level. The relevance of each material characteristic for each market sector is rated on a scale from hardly relevant (1) to highly relevant (5).

The numbers in the matrix indicate the relevance of a material characteristic for each market sector. The market sectors with the highest ratings were considered to make the most of a blade segment's properties and are highlighted in the figure. This process was repeated for all sectioning levels. Table 2 summarizes the top markets per sectioning level. Because the results of this selection strategy are the broadest, they showcase the potential of each sectioning level. CATSS provides a way to identify uncommon applications with logic that is based on the material's characteristics.

Table 2 Most relevant market sectors identified for each sectioning level

	Most relevant market sectors for each sectioning level
Segments	Architecture, Infrastructure, Arts and Culture, Urban furniture
Elements	Urban furniture, Infrastructure, Environmental protection, Arts and Culture, Transportation
Units	Interior design, Sports equipment, Consumer products
Units (reshaped)	Automotive, Interior design, Sports equipment, Transportation, Consumer products

5. Discussion

The CATSS method is an exploratory tool to identify secondary life applications for products at their end-of-life. The material sectioning levels identified can be used as successive use cycles that reduce in size each time. The CATSS method encourages a systematic approach to identifying structural reuse applications. The first two steps of CATSS provide a sufficient understanding of the material to explore a broader solution space than what is usually seen in literature. These steps are followed by three systematic selection strategies that open up application spaces.

The CATSS method is still in development, however. Its strength lies in its exploratory nature, but its benefits to look beyond mechanical properties has not yet been fully explored. There are also limitations to using the method, however. Substitution is reliant on obtaining relevant material properties or applications. Selection by function requires access to a detailed database such as the one presented. Inverse selection requires some previous knowledge of market sectors for the evaluation matrix (which needs further modifying to improve repeatability). It is suggested to explore this method with industry experts to observe how feasible and useful it is in practice.

Each selection strategy has limitations, but combining them strengthens the solution space. The strength of the CATSS method is that it provides a structured approach to identify secondary use applications from different perspectives. All three strategies can be applied to each sectioning level. In doing so, we expect to spark interest in repurposed blade sections and promoting structural reuse as an EoL strategy.

A thorough exploration of CATSS with thermoplastic blades as a case study is in progress, which will expand on the strengths and limitations of the process. This exploration will provide a better understanding of the material characteristics for each sectioning level. With this information, it is also expected that improvements to the original blade design will be identified. This further promotes structural reuse of blades with a broader solution space by informing design decisions. As such, the CATSS method presented can help identify new applications for thermoplastic composite products at their end-of-life, while, on the other hand, they indicate which aspects need to be considered in the original design, thus contributing to more circular practices in the composites industry.

6. Conclusion

A new methodology was proposed to identify relevant secondary applications for wind turbine blades. The Circular Applications Through Selection Strategies (CATSS) method is a systematic approach to identify applications that maximize material integrity at three different sectioning levels of wind turbine blades. The sectioning levels (segments, elements, and units) provide a path toward extending material functionality of WTBs before recycling. Since the unit level disassembles the blades' structural components into laminates, thermoplastic blades can then be reshaped, which opens up the possibilities for market sectors that have yet to be explored. The CATSS methodology was illustrated with examples found in literature that could be reframed as products of Landru strategies. This showed the approach to be promising for further exploration.

References

[1] Liu P 2017 Wind turbine blade waste in 2050 *J. Waste Manag* **62**, 229-240
[2] Beauson J *et al* J 2022 *Renewable Sustainable Energy Rev.* **155**, 111847
[3] den Hollander M C 2018 Design for managing obsolescence *Doctoral Thesis* Delft University of Technology: Delft, The Netherlands
[4] Woo S M and Whale J 2022 *Waste Manag Res* **40**, 1730–1744 (2022)
[5] Ellen MacArthur Foundation 2013 Toward the cicrular economy vol. 1 https://doi.org/10.1162/108819806775545321
[6] Jensen J P and Skelton K 2018 *Renewable Sustainable Energy Rev* **97**, 165–176
[7] Joustra J, Flipsen B and Balkenende R 2021 *RCR Advances* **167**, 105393
[8] Joustra J, Flipsen B and Balkenende R 2021 *JCOMC* **5**, 100137
[9] Alshahrani H, Ahmed A, Kabrein H and Prakash V R A 2022 *Coatings* **12**, 1423
[10] Ashby M F, Bréchet Y and Cebon D 2022 *Adv. Eng. Mater.* **4**, 327–334
[11] Ashby M F 2016 4.2 – The selection strategy. In *Materials Selection in Mechanical Design- 5th Edition* 176-186 (Elsevier).
[12] Landru D, Bréchet Y and Ashby M F 2002 *Adv. Eng. Mater.* **4** 343–349
[13] Alshannaq A A *et al* 2022 *J. Mater. Civ. Eng.* **34** 04022266
[14] Carrete I A *et al* 2023 *Mater. Des.* **225**, 111521
[15] Karana E *et al* 2015 *Int. J. Des.* **9** 35-54
[16] Alshannaq A A *et al* 2021 *J. Compos. Constr* **25** 04021023
[17] Bank L C *et al* 2022 *Current Perspectives and New Directions in Mechanics, Modelling and Design of Structural* 1441-1446

Cohesive zone modelling to predict crack growth under fatigue loading

A. Gomez and C. Berggreen

Technical University of Denmark, DK-2800 Kgs. Lyngby, Denmark

eagme@dtu.dk

Abstract. This manuscript presents the partial results in the development of numerical models based on Cohesive Zone Modelling to analyse quasi-static and fatigue crack propagation in adhesive interfaces. This work applied to adhesive interface between fibre laminates and structural epoxy adhesive of a wind turbine blade 3rd shear-web connection to the trailing edge panels. Fracture characterization experiments using the Double Cantilever Beam with Unequal Bending Moments (DCB-UBM) specimen are performed, and results are used to determine different modelling parameters, such as the mixed-mode cohesive laws and fatigue damage parameters for cohesive crack propagation. The developed numerical model is compared and validated against sub-component testing on two different scale levels, using a Modified Tear Test (MTT) specimen geometry with simplified geometry and loading, and a more complex geometry sub-component specimen with a simplified Shear Web Disbond Test (SBDT). The work is carried out as part of the CORTIR-II project, funded by the Danish Ministry of Energy.

1. Introduction

A wind turbine's blade comprises an outer aerodynamic shell, and shear webs that join the upwind and downwind sections of that shell. These webs are bonded to one another at the leading and trailing edges. Typically, a blade has two shear webs that run almost the entire length of the blade from the transition zone (denoted by the shear web fish mouth).

Due to the increasing global demand for clean and sustainable energy, wind turbines have become more prevalent over the past few decades. This surge in interest has driven manufacturers to develop larger wind turbines that are more efficient and cost-effective than their smaller counterparts. However, the continuous increase in size has resulted in higher loads, which require additional strengthening. This challenge is addressed by adding a third shear web to the blades. Unfortunately, this leads to a failure phenomenon known as shear web disbonding, where cracks in the adhesive interface between the shear web footing and spar cap propagate due to large deflections in the aerodynamic skins near the third shear web. If the crack grows too large this could compromise the integrity of the blade, eventually causing the wind turbine to collapse.

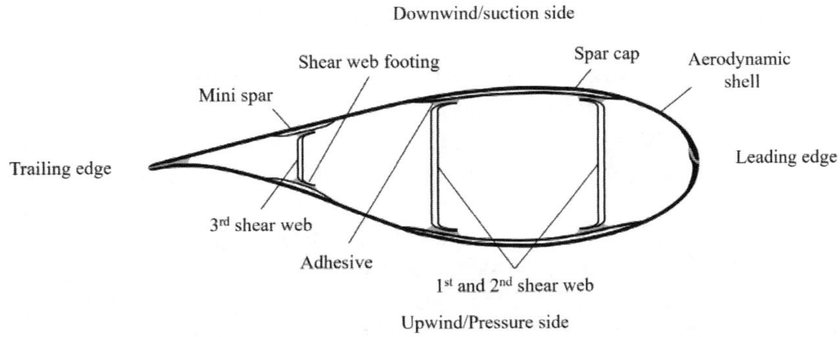

Figure 1. Example of a wind turbine blade section

The Technical University of Denmark (DTU), Aalborg University and Bladena have partnered with wind turbine owners and operators, to initiate the CORTIR-II project. which aims to investigate shear web disbonding in large wind turbine blades. The project is divided into four stages to investigate the issue in detail. It starts with characterisation and moves on to sub-component testing with more detail, before finally conducting large-scale testing.

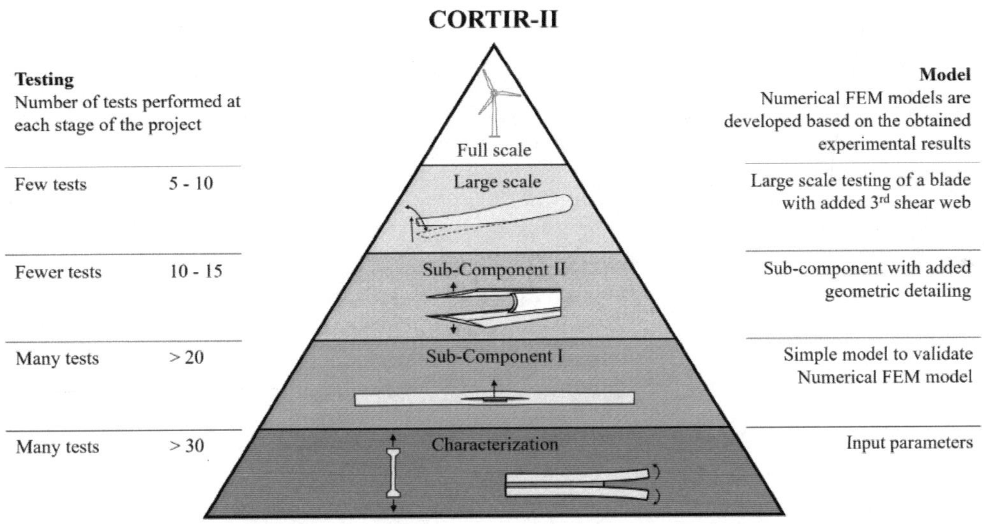

Figure 2 Building block approach for Cortir II project

This work presents the current state of part of this project which is focused on developing numerical models based on Cohesive Zone Modelling to analyze quasi-static and fatigue crack propagation in the adhesive interface between fibre laminates and structural epoxy adhesive. The modelling parameters are determined from results obtained in an experimental campaign using the DCB-UBM test [1], , which is a novel test method for mixed-mode fracture characterization. The developed numerical modelling will be validated using the Modified Tear Test (MTT), and sub-component testing method.

2. Cohesive zone modelling

The fibre bridging phenomenon occurs when the fibres in neighbouring laminate plies bridge the debonding plane, which constitutes an additional mechanism of energy dissipation during crack propagation . As a result, the debonding resistance and inter-laminar fracture toughness are enhanced. Since no purely analytical solution is available to directly determine a bridging/cohesive law it is necessary to perform fracture experiments that allow to compute the J integral. This since the J integral can be used to study non-linear fracture mechanics problems involving large fracture process zones, such as fibre bridging (Sørensen [2]).

Cohesive Zone Modeling (CZM) is a useful tool for predicting and modelling crack initiation and propagation in situations where Linear Elastic Fracture Mechanics (LEFM) cannot be assumed. For this reason and because they are widely implemented in commercial FEA packages over recent decades, researchers have increasingly shown interest in CZM applied in finite element modelling. CZM can incorporate nonlinear behaviour into the material's constitutive response, making it capable of predicting failure in ductile crack problems with significant process zones. The length of the process zone is critical in the fracture process as it can indicate if the fracture is ductile or brittle. If the process zone is much smaller than other characteristic lengths of the specimen such as crack length, specimen length, width, etc., brittle fracture occurs. CZM has previously been applied to the study of various failures in concrete, metals, composites, bi-material interfaces, and adhesive joints.

In the CZM the fracture process zone (FPZ) is substituted with a cohesive region, where the exchange of stress between the two artificial crack surfaces is defined using a relationship between the traction and separation of the FPZ. In practical FEA analysis, the cohesive zone model is commonly applied by a single layer of cohesive elements, which are placed along the expected path of crack propagation. Each element follows a constitutive response which uses a strength-based criterion to predict the onset of damage and the degradation of the material stiffness.

2.1. Traction separation law

Commercial FEA software packages such as Abaqus required a set of parameters that govern the traction-separation law, for example in the case of pure mode I fracture:

- K_{nn}, the stiffness of the cohesive element in the linear elastic regime
- Traction t_n^0 (or separation δ_n^0) for damage onset
- G_n^c, fracture energy release rate
- Damage evolution law

There are multiple traction-separation laws used in literature. Some of the most common bi-linear, exponential, and trapezoidal behaviours although non-standard (tabulated) laws computed directly from experimental tests are also employed.

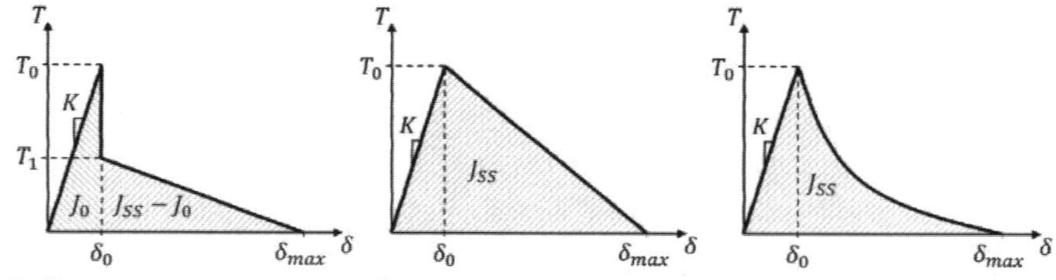

Figure 3. Different traction-separation functions evaluated in the current project. Source: [3]. *Here J_0 represents fracture energy release rate (computed through J integral) for crack onset while J_{ss} represents the fracture energy release rate for steady-state propagation which is also referred as G_c*

CZM also allows studying the fracture process when mode-mixity is present using the power law fracture criterion based on the energy required to propagate a crack under pure mode conditions.

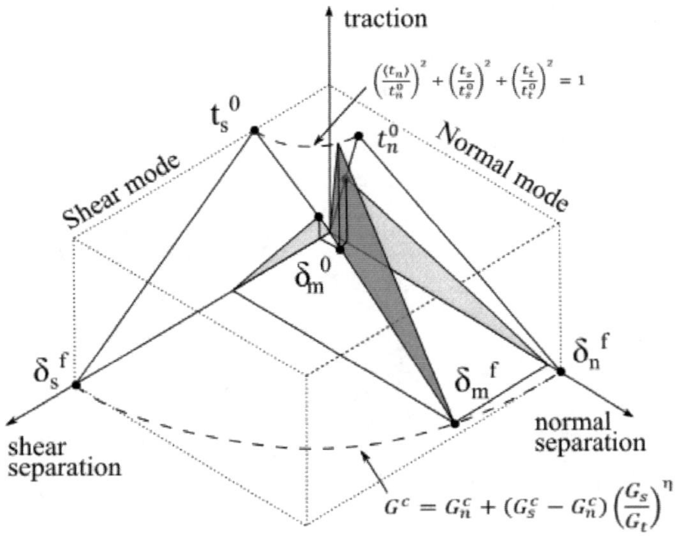

Figure 4. Mixed mode traction-separation and criteria for damage onset and failure

It is generally agreed that the key parameters of traction-separation law previously mentioned should be obtained from experiments although some other studies have used calibration methods based on trial-and-error search or a combination of both.

One of the aims of the present study is to obtain traction-separation law directly from an experimental fracture test following the methodology proposed by [2] which involves the computation of the J integral and the measurement of the crack tip separation in a quasi-static DCB-UBM test. In this method, the traction-separation law is obtained by the partial differentiation of the fracture resistance J_R and the crack tip separation. In pure mode I (opening) for example, the traction-separation law can be obtained by:

$$t_n(\delta_n, \delta_t) = \frac{\partial(J_R(\delta_n, \delta_t))}{\partial(\delta_n)}$$

for model II (shear):

$$t_t(\delta_n, \delta_t) = \frac{\partial(J_R(\delta_n, \delta_t))}{\partial(\delta_t)}$$

2.2. *Modelling fatigue with cohesive zone models*

A novel approach to consider the deterioration of materials caused by fatigue loading is to degrade the traction-separation response as the number of cycles increases (Khoramishad et al. [4]). This model was initially developed for aluminium joints with epoxy adhesive and it was validated against single lap joint (SLJ) and laminated doublers in bending (LDB) fatigue tests. In this approach the cohesive strength (t_i) and critical fracture energy ($G_{c,i}$) decrease through the damage parameter (D_i) which follow a fatigue damage evolution law of the following form:

$$\frac{\Delta D_i}{\Delta N} = \alpha(\varepsilon_{max} - \varepsilon_{th})^{\beta}$$

where ΔD_i is the damage accumulated over ΔN cycles while α, β and ε_{th} are material parameters which must be calibrated against experimental results.

ε_{max} is the maximum overall strain defined in terms of the normal and shear strains in the cohesive element:

$$\varepsilon_{max} = \frac{\varepsilon_n}{2} + \sqrt{\left(\frac{\varepsilon_n}{2}\right)^2 + \left(\frac{\varepsilon_t}{2}\right)^2}$$

In the beginning, the damage parameter D_i for each element within the cohesive zone is set to zero and damage accumulation in the load cycle only occurs if the maximum overall strain is larger than the threshold value ε_{th}. The damage parameter which ranges between 0 and 1 can be computed after each load cycle (or set of load cycles) and update the effective traction and energy release rate used to estimate the crack propagation.

$$t_i = D_i t_0$$
$$G_{c,i} = D_i G_{C,0}$$

Where i denotes the current time step, and t_0 and $G_{C,0}$ is the initial cohesive strength and fracture energy respectively.

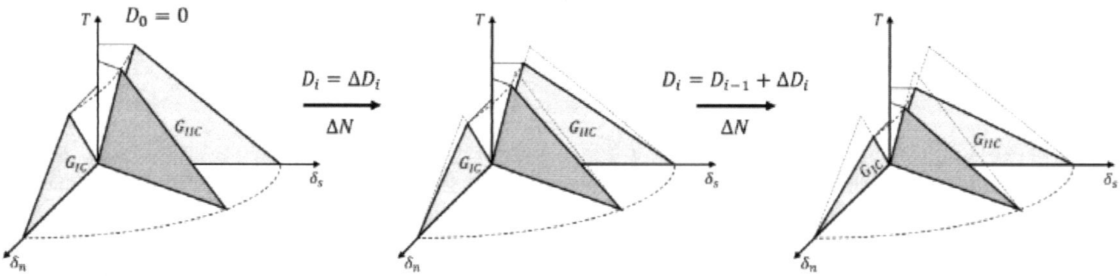

Figure 5. Evolution of the traction-separation law due to fatigue cycles. Source: [3]

3. DCB-UBM tests (Level -0)

DCB-UBM test is the first of three different experimental tests that are carried out to study the quasi-static and fatigue debonding phenomena. This test campaign is aimed to assess the critical fracture parameters such as energy release rates and cohesive damage law under various mixed-mode loading conditions.

For validation purposes, two additional tests are proposed: First, a new test approach, referred to as the Modified Tear Test (MTT), derived from the Sandwich Tear Test (STT), is specifically designed for Sub-component level 1 testing [3]. And second a setup for a simplified shear web geometry (Sub-component II testing) which considers the knowledge acquired from conducting the aforementioned tests and includes external load conditions closer to those encountered in a full-scale blade scenario.

3.1. Test set-up

The DCB-UBM test, as introduced by [2], involves the application of uneven bending moments to the crack flanks' tips in the specimen. By adjusting the ratio of these applied moments, crack propagation can occur under different fracture modes, ranging from pure mode I to pure mode II. Since the load applied to the specimen solely originates from bending moments, both the energy release rate and the mode-mixity angle remain unaffected by the crack length. Another advantage of employing pure bending moments for loading, particularly in cases involving large process zones, is that the applied load remains constant throughout the entire process zone.

The tests were performed in a DCB-UBM servo-hydraulic testing machine developed in-house at DTU. Pure moments are applied over each laminate only on one side of the specimen by two separate servo-hydraulic torsional actuators. To maintain pure moment loading, it is required to minimize any in-plane and out-of-plane forces, so the carriage plates are mounted on raceway shafts, utilizing track rollers.

The DCB-UBM specimens have overall dimensions of 450 mm x 30 mm x 18.2 mm and are made from two GFRP laminates bonded on top of each other (Figure 6). The specimens were manufactured by Global Wind Services by hand layup, manufacturing the laminates separately and then bonding them together by secondary bonding and vacuum bag consolidation. A 50 mm pre-crack is made by inserting a Teflon film during the consolidation process. The top laminate is made from a

triaxial/biaxial fabric with the following stacking sequence $[45/0/-45/\pm45]_8$ and an average cured thickness of 10 mm. The bottom laminate is made from biaxial fabric with a stacking sequence of $[\pm45]_{10}$ and an average cured thickness of 6.8 mm.
The specimen configuration was determined through preliminary studies, utilizing Crack Surface Displacement Extrapolation (CSDE) analysis to calculate mode mixity angles and energy release rates. Due to the significant tip rotations, potential laminate failure and limited fixation options, two 6 mm thick steel reinforcement plates (doublers) were bonded to the top and bottom surfaces of the specimen using an epoxy structural adhesive. Additionally, a clamp was installed to prevent crack initiation between the specimen and plates.

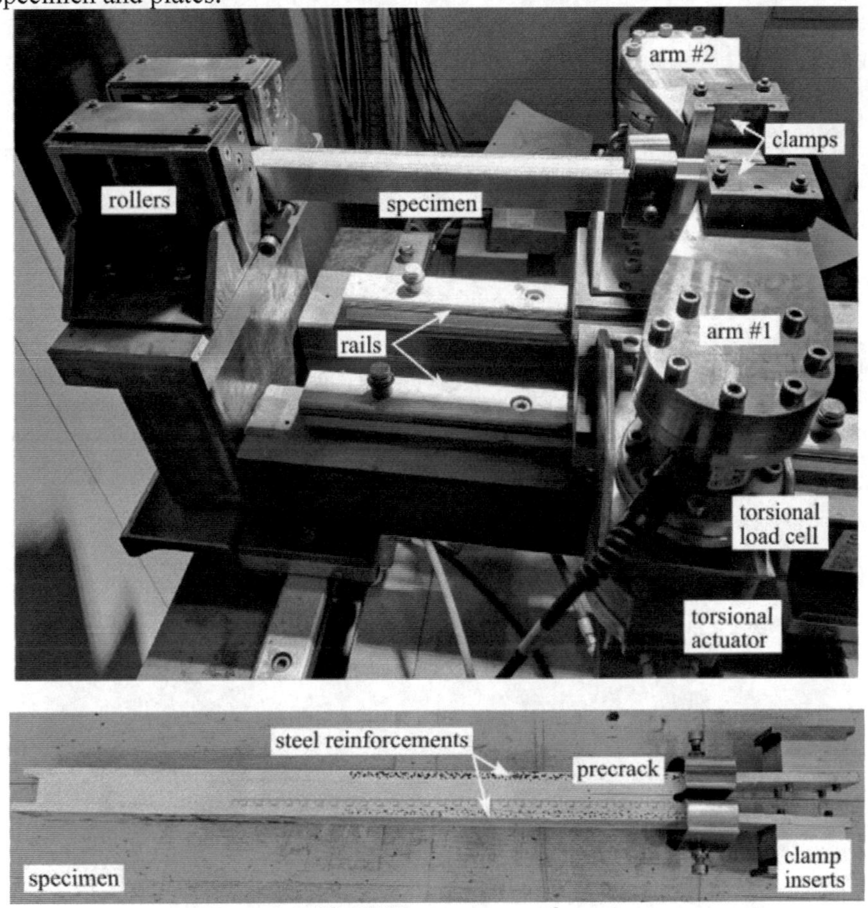

Figure 6. DCB-UBM test set-up and specimen

The moments and rotations of the loading arms are directly obtained from the sensors of the test machine through the MTS data acquisition system. Additional data such as crack length and separation of the crack tip is collected using the iMetrum Non-Contact Video Measurement live-tracking software. The displacements of the tracked crack flanks, which are utilized to determine the phase angle, should be closely monitored near the crack tip and adjacent to the crack flank surface. To ensure precise tracking of crack propagation, it was crucial to accurately define the crack path and initial crack length. Additionally, to enhance the image correlation results, all specimens' surfaces were meticulously sanded and painted white, and speckle patterns of approximately 1mm were applied to the steel plates using a sharp marker. Camera and lightening were adjusted as well as the DIC threshold values until a reliable crack propagation was achieved.
Quasi-static and fatigue tests were conducted to establish the key parameters of the cohesive laws. Three moment ratios ($MR = M_2/M_1$) were tested MR=-1.2, -0.714, and -0.4 corresponding to phase

angles equal to 7.4°, -6.5° and -22.4° respectively. Phase angles larger than 30° (including Mode II) were not considered due to limitations in the load capacity of the testing machine. At least three specimens were tested for each phase angle.

Fatigue tests were performed at three different load levels (M_{1max}) corresponding to 60%, 70%, and 80% of the critical energy release rate (G/G_c) found in quasi-static tests. All tests were performed with a load ratio $R = M_{1min}/M_{1max} = 0.2$ at the same moment ratios (MR) studied in the quasi-static test. The crack extension (Δa) is measured directly from the photo frames at different numbers of cycles in such a way that the crack propagation rate ($\Delta a/\Delta N$) is estimated between two adjacent measurement points. Live-tracking methods were not suitable for this fatigue test due to the relative movement of the specimens and the opening/closing of the crack flanks. Other technical challenges had to be sorted out to perform this test, including a sudden detachment of steel doubler reinforcements and high vibrations. However, the most difficulty was to maintain an accurate moment ratio between the arms when running the experiment at high frequency. In such a case the time delay in the response of both arms made the controller underestimate the moment on arm 2. Thus, the input moment ratio had to be slightly calibrated to match the desired value.

3.2. Tests results

It was discovered that the interface failure was primarily controlled by extensive fibre bridging on a large scale, both during testing under quasi-static and fatigue loading. Fibre bridging was observed across all tested modes, with its greatest impact occurring in the proximity of mode I. The presence of voids within the adhesive, as well as variations in the thickness of the bondline, and internal preload caused by imbalanced composite face-sheets, result in variations in the fracture resistance of each specimen. In some instances, the crack deviated towards the opposite interface from the location of the pre-crack, forming a kink. Besides this, most experiments were successfully performed.

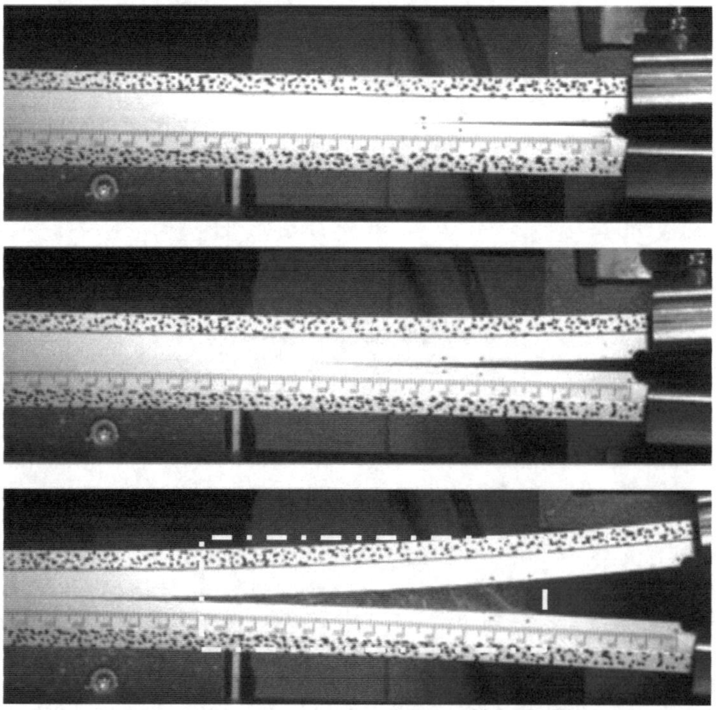

Figure 7. Photograms showing crack propagation in a quasi-static DCB-UBM test (M2/M1=-1.2). Fully developed fibre bridging is visible in the last photogram (yellow)

IOP Conf. Series: Materials Science and Engineering 1293 (2023) 012012 doi:10.1088/1757-899X/1293/1/012012

The obtained data was post-processed in MATLAB to obtain the fracture resistance R-curve by computing the J-integral at each instant taking as input the applied moments (M_1 and M_2). The J-integral formulation presented by [5] was modified to account for the asymmetry due to the difference in the thickness and material properties of the top and bottom laminates. The crack extension (Δa) and tip-end separation (δ_n , δ_t) were also plotted. An example of the post-processed data is shown in Figure 8.

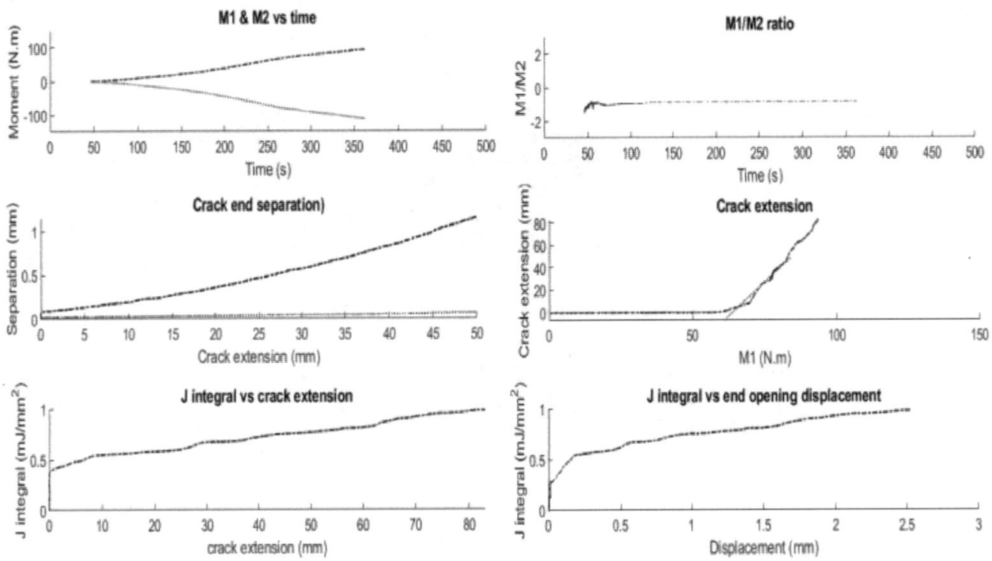

Figure 8. Example of post processed data for MR = -1.2 (Ψ = 7.4°)

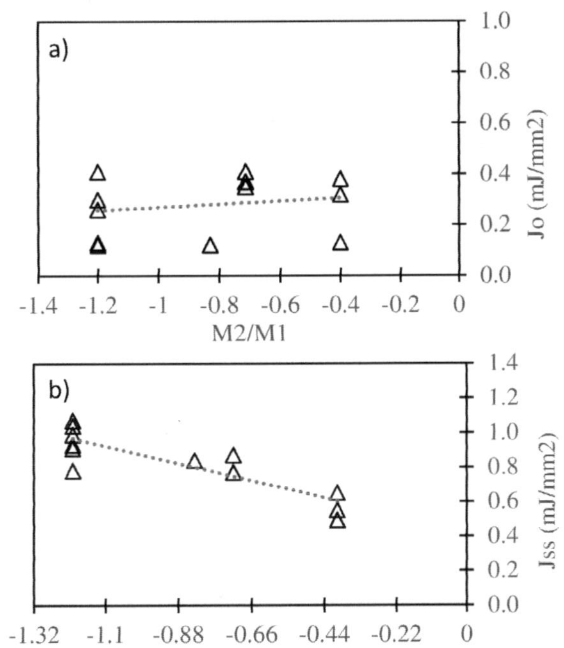

Figure 9 Fracture resistance as a function of moment ratio M1/M2. a) crack initiation, b) steady-state propagation

IOP Conf. Series: Materials Science and Engineering 1293 (2023) 012012 doi:10.1088/1757-899X/1293/1/012012

The fracture resistance for crack initiation (J_o) and steady-state crack growth (J_{ss}) are compared at different phase angles. Results show that J_{ss} is larger than J_o for all studied phase angles which shows the relevance of the fibre-bridging toughening effect, particularly at low phase angles.
To obtain the cohesive traction-separation laws it is convenient to graph the fracture resistance for each test as a function of the normal crack end opening and the tangential crack end opening. Due to the high data scattering a non-linear least square surface polynomial fitting is utilized, yielding a parametric function that enables the estimation of the traction-separation cohesive laws.

$$J_R = a + \left[b\left(x_{sep}y_{sep}\right) + c\right]\left(\sqrt{x_{sep}^2 + y_{sep}^2}\right)^e$$

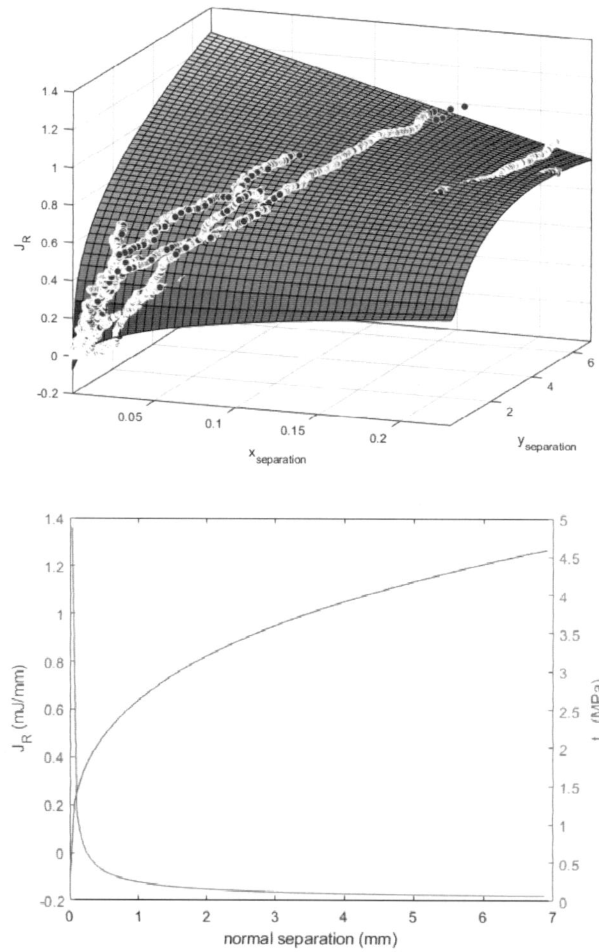

Figure 10. a) Surface polynomial fitting, b) Cohesive traction separation law obtained analytically from quasi-static DCB-UBM tests for pure Mode I,

For the fatigue tests the crack propagation rates (da/dN) were estimated at different load levels (G/G_c) and moment ratios (M_2/M_1). As expected, the crack propagation rate increases with the load level following a power function similar to Pari's law [6](Figure 11 and Figure 12). However large scattering was obtained for da/dN measured despite the load level and moment ratio being the same.

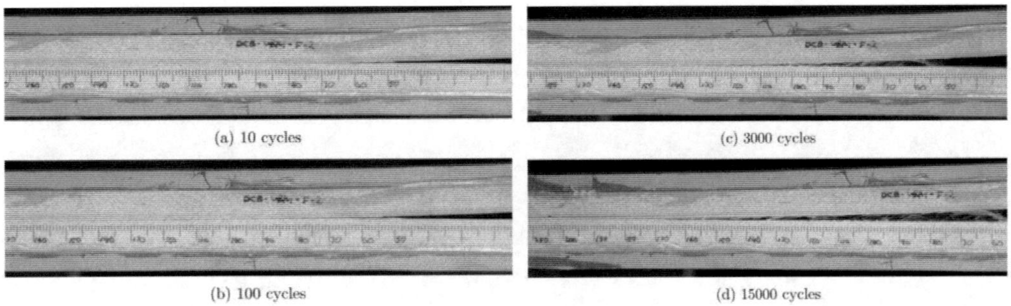

Figure 11. Crack extension at different numbers of cycles. Source: [3]

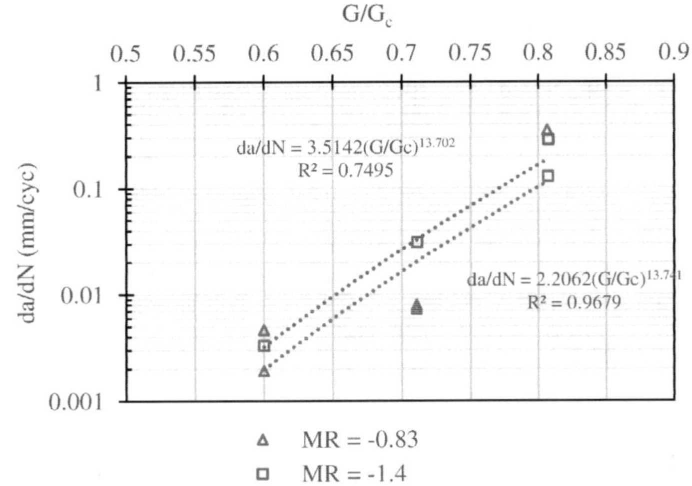

Figure 12 Crack growth rate (da/dN) for different load levels

Through more careful analysis, it was determined that all da/dN measurements were taken at different effective crack lengths. As a result, the size of the bridging zone for each measurement was different, and this could explain the observed data scattering. In fact, crack propagation rates measured at the beginning of the test (when the crack length is small), were much larger than those da/dN measured when the crack has extensively propagated. This suggests that the fatigue crack propagation rate is not only a function of the moment ratio but also a function of the size of the bridging zone (Figure 13). From the same figure, it can be also observed that at large lengths of the bridging zone, the value da/dN asymptotically reaches a constant value which can be related to a fully developed bridging zone and the steady-state critical fracture energy ($G_{cI} = J_{ss}$) determined through the R-curve (section 3.2.)
This dependency has also been previously reported for carbon/epoxy laminates (e.g. L. Yao et al. [7] and G. Murri [8]) and alternative experimental procedures and reduction methods have been proposed. In this regard, a more detailed test campaign is ongoing at DTU Construct to characterize da/dN and account for its dependency on the bridging zone length.

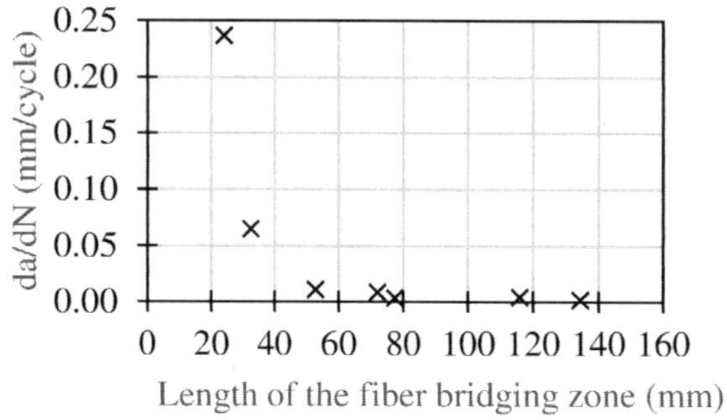

Figure 13 Crack growth rate at different bridging zone lengths. G/GcI = 0.6 and M2/M1 = -1.2

3.3. FEA models of DCB-UBM tests.

An FEA model was created in Abaqus 2021 to replicate the experimental DCB-UBM test explained in the previous subsection. This model uses a cohesive zone model to predict damage initiation and crack propagation both in quasi-static and fatigue loading. Cohesive elements COH2D4 were used together with element deletion to track crack propagation in two alternative ways: when damage initiates (damage variable D>0) and complete failure occurs (D=1).

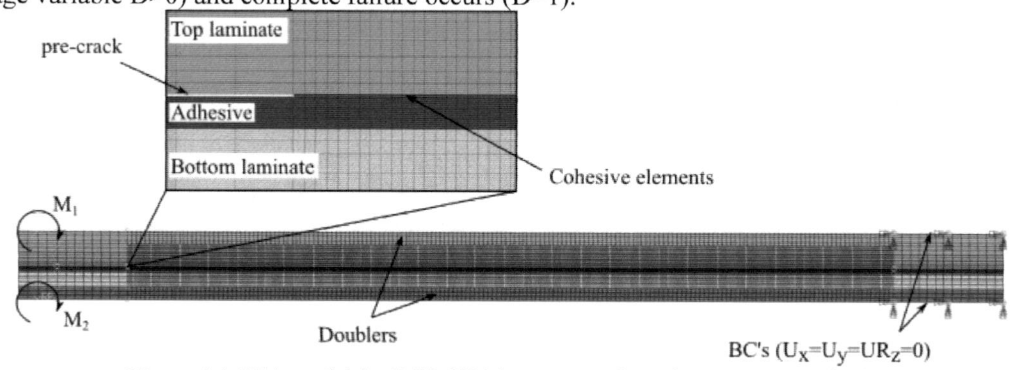

Figure 14. FEA model for DCB-UBM test using the cohesive zone method.

The laminates and adhesive layer are modelled as separate geometric bodies joined at the interface using a tie constraint. The laminates are modelled as orthotropic materials while the adhesive material and the material for the doubler stiffeners an isotropic materials. Material properties for the laminates were obtained by coupon testing during Cortir I project and are summarized in Table 1. The models are mesh using 2D CPE8R, quadratic plane strain elements with reduced integration.

Property	Units	Top laminate	Bottom laminate
E_1	MPa	20159	10067
E_2	MPa	10126	9597
G_{12}	MPa	8705	8962
G_{21}	MPa	9365	9229
Xt_1	MPa	262	119
Xt_2	MPa	90	102
S_{12}	MPa	75	203
S_{21}	MPa	129	206
v_{12}	-	0.44	0.54
v_{21}	-	0.03	0.56

Table 1. Laminate material properties

Two concentrated moments (M_1 and M_2) are applied over reference points located on the left side of the specimen which are coupled to the specimen's external surface by a rigid body constraint. The specimen is restrained on the opposite side simulating the rollers
Quasi-static tests are run using the standard/linear solver while the fatigue test uses an explicit solver. For the fatigue model, the damage criteria described in section 2.2. was implemented in a VUSDFLD Fortran subroutine. This subroutine enables the dynamic adjustment of field variables at a specific material point. These adjustments can be defined as functions of time or any of the material point quantities listed in the "Available output variable keys." This capability is particularly useful for introducing material properties that depend on the solution, as these properties can be conveniently expressed as functions of existing field variables. Since modelling fatigue loading cycle-by-cycle is computationally non-practical a different strategy is adopted in which bundles of multiple load cycles are represented into a single time step.

3.4. FEA model results of DCB-UBM test

The quasi-static cohesive FEA model is primarily used to calibrate and validate the cohesive law estimated analytically from the experimental test results. Due to the large data scattering in the experimental results, the calibration procedure requires a clear measure of the error between experimental and FEA results. Since the primary objective of a cohesive law approach is to be able to predict both crack initiation and crack growth, both measurements are used to define a combined mean squared error of the following form:

$$e_{comb}(\%) = (100\%)\sqrt{\left(\frac{M_1^{FEA} - M_1^{EXP}}{M_1^{EXP}}\right)^2 + \left(\frac{\frac{da}{dM_1}^{FEA} - \frac{da}{dM_1}^{EXP}}{\frac{da}{dM_1}^{EXP}}\right)^2}$$

Where M_1 and $\frac{da}{dM_1}$ are the moment in arm 1 and crack growth rate at crack initiation onset and the superscripts FEA and EXP mean FEA results and experimental results respectively.

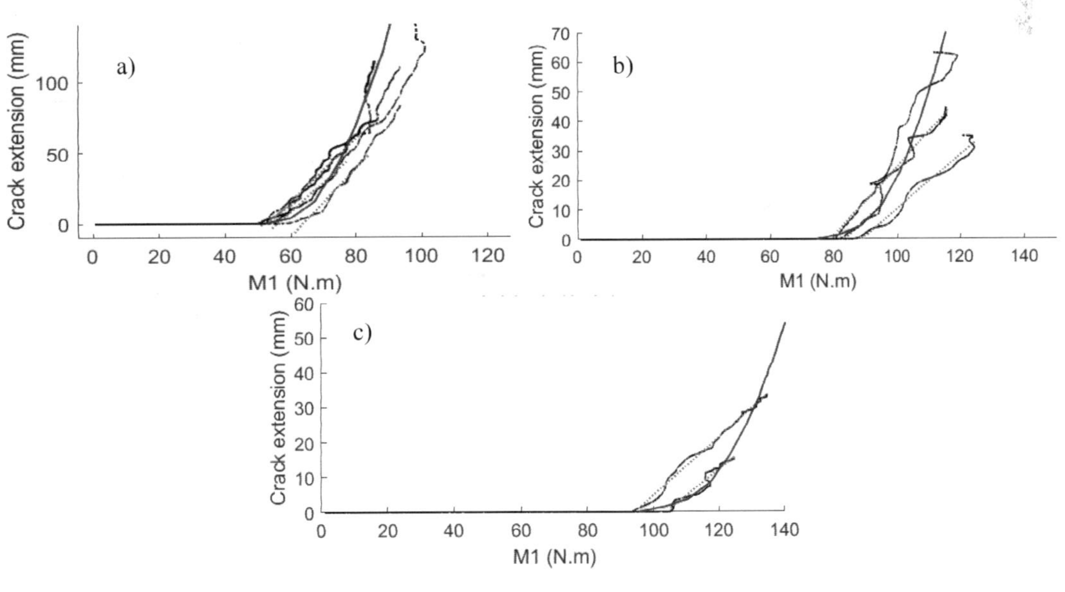

Figure 15. Comparison between FEA and experimental results for a quasi-static DCB-UBM test. Experimental (black), FEA (blue). a) MR = -1.2, b) MR = -0.714, c) MR = -0.4

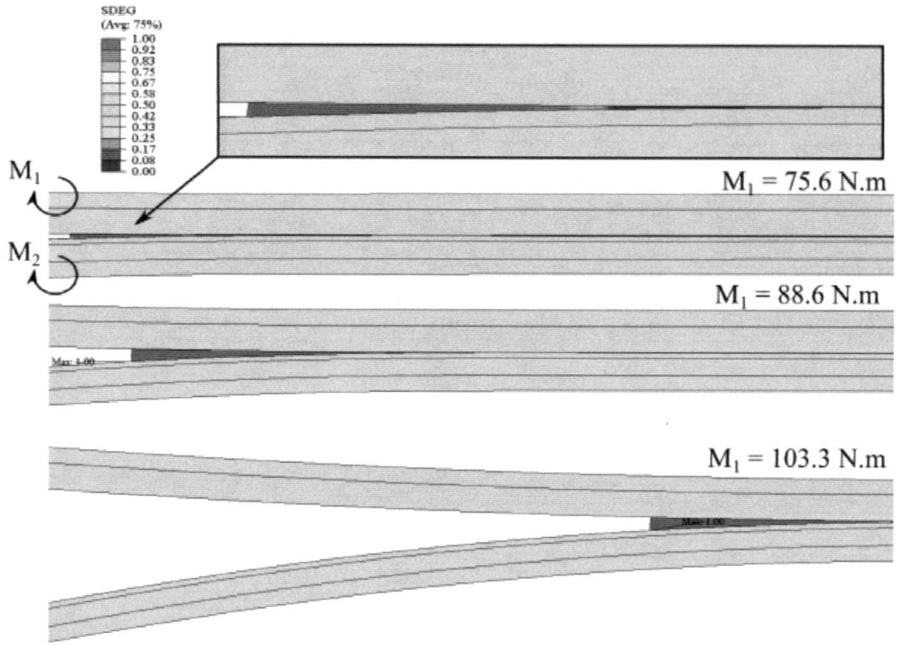

Figure 16. Example of cohesive damage (SDEG) evolution predicted by FEA model FEA (MR = -1.2)

Similarly, the fatigue FEA model is used to calibrate the fatigue damage parameters α, β, ε_{th} against the DCB-UBM test results (section 3.2.). Since the definition of those parameters in the material model is purely phenomenological and cannot be analytically related to any physical measurable quantity it is then necessary to obtain these values through a search method. Numerous attempts were carried out to try to fit these material parameters manually using a trial-and-error technique, however, no successful combination has been found yet. For the sake of computational efficiency, more advanced search algorithms for global optimization are proposed for future analysis including pattern search, multi-start search, genetic algorithms, or surrogate optimization.

Despite this, results from the analysed parameter combinations show that the model can capture the dependency of the crack growth rate (da/dN) with the load level (G/Gc) as well as its increase following a power law. This indicates that the implemented formulation might be suitable to analyse fatigue in fully developed bridging zones.

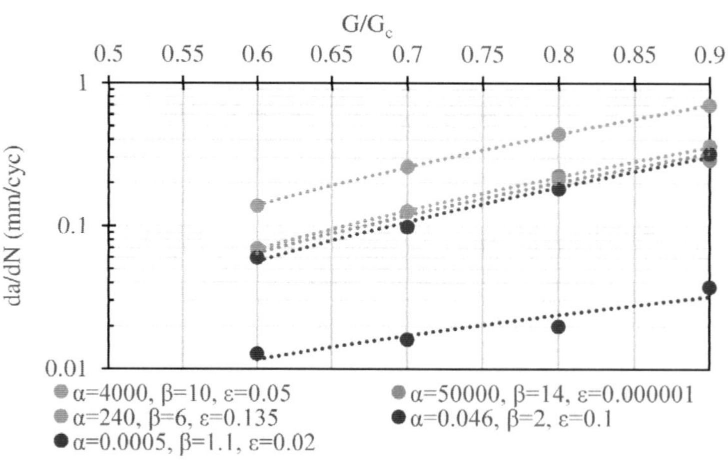

Figure 17. Example of fatigue crack growth estimations predicted by the implemented formulation with different combination of parameters.

Comparing the accuracy of fatigue models with different ΔN jump values is not straightforward because ΔN serves as a prerequisite value in determining other fatigue parameters. Consequently, the fatigue parameters that yield a good fit with the data for a particular ΔN are only applicable to that chosen ΔN. Nonetheless, when comparing models based on time rather than load cycles and using the same fatigue parameters, it can be demonstrated that ΔN has a limited impact on the results.

4. Validation tests and FEA model

Additional tests at higher complexity levels of the test pyramid are currently ongoing at DTU Construct. A brief description of those tests is presented below:

a) MTT tests (Level – 1): The objective of conducting sub-component level I tests is to replicate the real-life occurrence of shear web disbonding using simple test setups and specimens. While it is commonly assumed that shear web disbonding occurs under the mode I condition, the presence of mode mixes can occur due to membrane forces in aerodynamic shells or the torsional shear induced by global loading on wind turbine blades. To explore how these factors might influence the problem, it was necessary to devise a test method in which mode mixity varies during crack propagation. Additionally, this test enables the validation of the cohesive law parameters across different mode mixity angles that are estimated in level-0 of the building block approach. The Modified Tear Test (MTT) is derived from the Sandwich Tear Test (STT) method, which is utilized to examine interface failure between the adhesive and laminates while considering different mixed mode conditions. In this test, the upper laminate is pulled upwards, so the in-plane membrane forces escalate significantly as the test proceeds. This escalation leads to an increase in the mode-mixity angle, transitioning from an initial pure mode I opening condition to mixed fracture modes (opening and shear). A new test set-up was developed and produced, incorporating some components from a previous STT setup while redesigning and manufacturing the majority of them specifically for this test.

Figure 18. MTT test setup. Source: [3]

b) Subcomponent Level II test – Shear Web Disbond Test (SWDT): The purpose of the subcomponent level II test is to study the simplified geometry and loading conditions of a real shear web installed in a wind turbine blade paying particular attention to the bonded interface between the shear web flange and the blade outer shell. The test specimens are a representative simplified version of only half the shear web and a portion of the outer shell to which it is bonded with a pre-crack inserted in the bondline. Four specimens were manufactured by *Global Wind Services*, manufacturing the shear webs and laminates separately by hand layup and vacuum bag consolidation and then joining them by secondary bonding. The top laminate and shear web flange have the same composite layup as those used in level-I and level-II tests.

IOP Conf. Series: Materials Science and Engineering 1293 (2023) 012012 doi:10.1088/1757-899X/1293/1/012012

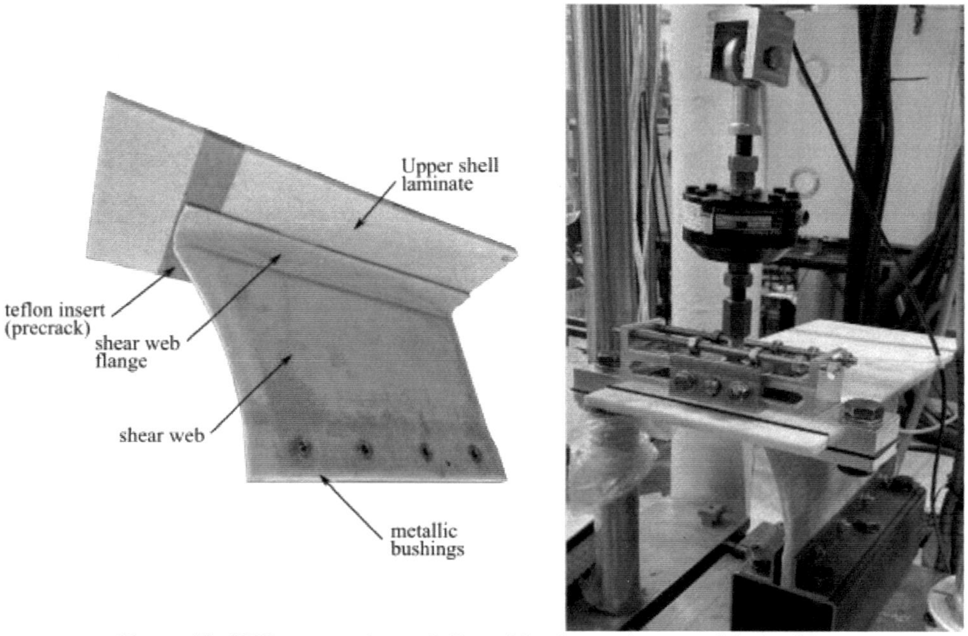

Figure 19. SWD test specimen (left) and load transfer assembly (right)

Conclusions

A building block methodology has been implemented to study crack initiation and propagation in bonded joints of glass fibre laminates and epoxy adhesives, typically encountered in shear webs in wind turbine blades. The phenomena have been studied from an experimental and numerical perspective at three different levels of complexity. This section discusses the partial results of the first-level DCB-UBM characterization.

Experimental study:

Quasi-static DCB-UBM tests show that the crack propagation behaviour is dominated by a large fibre bridging zone in the bondline which acts as a toughening mechanism, particularly at small mode mixites phase angles (mode I, and opening-dominated fracture). Measurements of the fracture resistance (R) through the J integral show a large increase from the crack onset (J_o) to steady-state propagation with fully developed bridging (J_{ss}). The measurements of the fracture energy release rate were highly scattered probably due to differences in the bondline thickness, void content and curvature (due to the asymmetry of top and bottom laminates). Since the manufacturing method, materials and sizes used in the specimen fabrication are representative of those employed in the wind industry for similar applications similar deviations in the fracture resistance could be expected in operating wind turbine blades. It was also possible to determine the traction-separation law directly from the experiments for a range of phase angles by measuring crack tip displacements and applying the data reduction proposed by Sorensen together with nonlinear surface extrapolation. However, due to limitations in the test machine, it was not possible to test phase angles near mode II and therefore the validity of the predicted fracture behaviour in this region is still unknown.

Fatigue DCB-UBM tests were found to be challenging due to large machine vibrations, and difficulties to set the PDI controller, for this reason, a limited number of tests have been performed. Besides this, it was possible to measure the crack growth rate at different loading conditions and results suggest that the crack growth rate (da/dN) follows a power law function which depends on the mode mixity and load level (G/G_c). Due to the large data scattering it was not possible to estimate this dependency quantitatively with enough reliability so additional tests are carried out at this moment to reduce the statistical uncertainty of the curve fitting estimation. Additionally, it was observed that the crack growth rate (da/dN) is highly affected by the length of the bridging zone. A similar dependency of this property to the R-curve has been found by other authors in the delamination of CFRP composites with large fibre bridging zones.

Numerical study:

For characterization (DCB-UBM) tests, 2D finite element analysis (FEA) models have been developed to calibrate the cohesive traction-separation laws obtained in experimental tests. Different traction separation laws have been tested and compared including bi-linear, exponential, and tabulated (based on traction separation curves obtained in experiments). For mode-I tabulated cohesive-law can capture with accuracy the crack onset and crack propagation observed in experimental testing. However, minor calibrations were required for some cohesive law parameters using trial-and-error search methods.

A novel approach to model fatigue crack growth in the bondline was successfully implemented through a material user subroutine in which a fatigue damage law is coupled with the cycle jump technique. The same 2D FEA model was used to calibrate the fatigue damage law parameters using trial-and-error search methods. However, it was not possible to find a combination of material parameters that fully validate the experimental data. Despite this, preliminary results suggest that the proposed approach can capture with some accuracy the crack growth rate of a fully developed bridging zone but further refinement of the experimental test and numerical model is required.

Acknowledgements
The authors would also like to acknowledge the Danish Ministry of Energy for funding this work as well as all the Cortir-II team members for their support during the different project activities, in particular: Find Moholt and Tamas Nagy (Bladena), Yu-Huan Lin (Global wind services), Daniel Søndergaard and Rasmus Thomasen, Ian Rasmussen and Troels Kristensen (DTU Construct).

References

[1] D. Plausinis and J. K. Spelt, "Application of a new constant G load-jig to creep crack growth in adhesive joints," *Int. J. Adhes. Adhes.*, vol. 15, pp. 225–232, 1995.

[2] Sørensen, "Cohesive laws for assessment of materials failure: Theory, experimental methods and application," *Downloaded from orbit.dtu.dk*, 2022, Accessed: Jun. 14, 2022. [Online]. Available: www.risoe.dtu.dk.

[3] D. S. Rasmussen, "Cohesive web joints in wind turbine blades Using cohesive zone modelling to predict crack growth under fatigue loading," 2022. [Online]. Available: www.mek.dtu.dk.

[4] H. Khoramishad, A. D. Crocombe, K. B. Katnam, and I. A. Ashcroft, "Predicting fatigue damage in adhesively bonded joints using a cohesive zone model," *Int. J. Fatigue*, vol. 32, no. 7, pp. 1146–1158, Jul. 2010, doi: 10.1016/J.IJFATIGUE.2009.12.013.

[5] V. Saseendran, C. Berggreen, and L. A. Carlsson, "Fracture Mechanics Analysis of Reinforced DCB Sandwich Debond Specimen Loaded by Moments," doi: 10.2514/1.J056039.

[6] F. Paris, P., and Erdogan, "A critical analysis of crack propagation laws," *J. Basic Eng.*, vol. 85, 1963.

[7] L. Yao, R. Alderliesten, M. Zhao, and R. Benedictus, "Bridging effect on mode I fatigue delamination behavior in composite laminates," doi: 10.1016/j.compositesa.2014.04.007.

[8] G. B. Murri and NASA, "Evaluation of Delamination Growth Characterization Methods Under Mode I Fatigue Loading," 2012.

IOP Conf. Series: Materials Science and Engineering 1293 (2023) 012013 doi:10.1088/1757-899X/1293/1/012013

Complex damage mechanisms and roughness evolution of wind turbine blade surface: Multiphysics and stochastic effect modelling

L. Mishnaevsky Jr. [1*], N. Kuthe[2], A. Tempelis[1], P. Mahajan[2]

[1] Department of Wind Energy, Technical University of Denmark, Department of Wind Energy, Roskilde, Denmark
[2] Department of Applied Mechanics, Indian Institute of Technology, Hauz Khas, New Delhi 110016 India

*lemi@dtu.k

Abstract. Leading edge erosion of wind turbine blades is the most often observed damage mechanism of wind turbines. The surface erosion of blades is influenced by many multiphysics and stochastic factors including humidity and related degradation processes, rough uneven surface and roughness development, random defects in the materials. In this paper, the effects of these factors and possibilities of their computational modelling and prediction are discussed. Competing damage mechanisms in erosion, including debonding and impact damage, as well moisture ingress and weathering are investigated. A predictive model for roughness evolution of leading edge due to the surface damage is presented.

1. Introduction: Modeling of leading edge erosion beyond mechanistic analysis

Leading edge erosion of wind turbine blades is one or the main challenges compromising the development of the wind energy. In order to mitigate the blade erosion, and to find solutions to protect blades, detailed understanding of the erosion mechanisms and predictive models of erosion are necessary [1]-[3].

The main erosion mechanism is largely the material destruction, controlled by small scale impacts (by rain droplet, hail), stress wave propagation and material fatigue processes. That is why the majority of computational models of erosion are based on the methods of continuum mechanics, damage, fracture and fatigue theory [1], starting from the classical Springer model, and up to complex multi-element numerical models, considering liquid/solid contact, 3D stress wave propagation, viscoelastic materials behaviour and random fatigue [2]-[5]. At the same time, leading edge erosion is a complex multiphysics process, which is strongly influenced by the many physical, stochastic effects(see schema Figure 1):

- Materials structure: Micro- and nanoscale structure of the blade coatings, including the polymer chain behaviour, non-linear viscoelastic behaviour, internal structure of coatings,
- Environmental effects: humidity, moisture flow, ultraviolet radiation,
- Stochastic effects: rain scenario, material defects, roughening of the surface.

A solution of the blade erosion problem has to take into account these complex physical processes, going beyond the limits of classical materials mechanics and fracture/fatigue theory. In this paper, computational modelling of some of these effects is discussed.

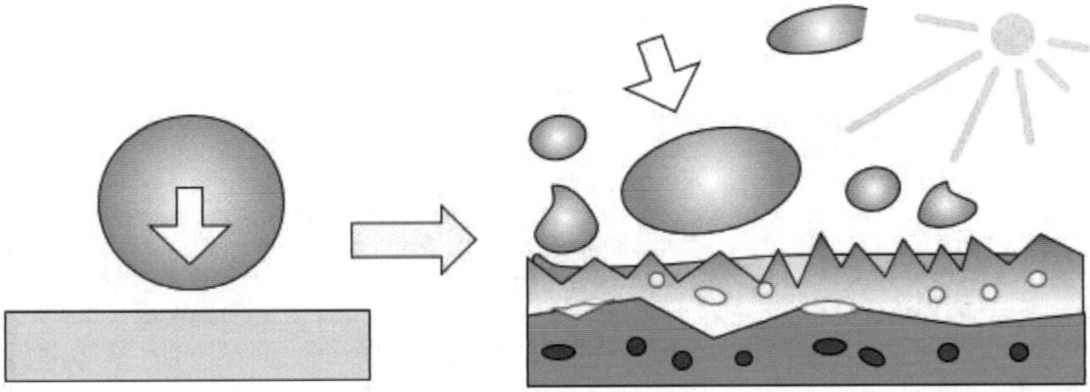

Figure 1. Schema: from idealized model of blade erosion (left) to the analysis taking into account stochastic and multiphysics effects(right).

2. Environmental effects: moisture, temperature variations, radiation

2.1. Effect of environmental loading

Wind turbine blades are subject to both mechanical loading and environmental loading, including high humidity, temperature variation and ultraviolet (UV) radiation. These effects change material properties, interact with mechanical loads and reduce the lifetime of coatings.

Combined mechanical loads, high humidity, and temperature variations can result in residual strain in coatings, hygroscopic swelling in polymers, reducing the materials strength and modulus, and eventually leading to debonding/delamination [6]- [8]. The moisture diffusion in the polymer coatings can create the properties gradient in coatings. UV-radiation can cause photo-oxidation of the polymeric coatings, free radical excitation, coupling reaction, and lead to complicated degradation products, reduction of molecular weight, tensile strength, and impact resistance of polyurethane coating [8]. Under combined UV, water and oxygen loading, polyurethane coatings tend to form blisters on the surface, then the chemical composition of the coatings surface can change (increase of urea and urethane group concentrations; build-up of hydrophilic groups), promoting the water absorption [9].

Fatigue and static strength of polyurethanes depends also on temperature, with decreasing fatigue performance and more cohesive failure at higher temperatures [10]. In particular, degradation of polyurethanes in sea water depends on the degree of cross linking, with crosslinked polyurethanes very resistant to degradation in sea water, and slightly crosslinked polyurethanes only moderately resistant to degradation in sea water [11].

2.2. Humidity effect and coating debonding

In order to simulate the effect of moisture on the leading edge protection degradation, a computational model for moisture diffusion in the multilayered coating system was developed in [12][13].

Several advanced tools were developed to simulate the moisture ingress and its influence on the liquid impact:

- Moisture diffusion and formation of hygroscopic stress field in the coating sample were determined, using the coupled moisture-displacement problem solution, obtained using the thermal-moisture direct analogy in ABAQUS/Standard [14], [15]. Saturation weight gain by the coating at 90% relative humidity and 60°C was approximately 2.5% of the initial weight.
- Coating material changing properties due to moisture ingress is simulated using the Abaqus subroutine User Defined Field. Figure 2 shows von Mises (hygroscopic) stress distribution

at the interfaces due to moisture ingress in multi-layer blade samples. The field variable is changed as a function of local moisture content and changes in turn the local material properties. Potentially, a model of changing material properties due to the external radiation (here, UV radiation) can be also developed, using the User Defined Field subroutine, with external field (with gradient along the vertical axis) and material properties influenced by local radiation level.

- The enhanced interfacial stresses, caused by the polymer swelling and properties gradient due to the moisture diffusion, can lead to the interface debonding of coatings. It was observed in the simulation that stress changes between the polyurethane and putty layers cause additional shear stresses along interfaces caused by the moisture ingress.

High humidity and moisture ingress, as well as thermal and ultraviolet loadings can potentially change drastically the material response, thus, influencing the lifetime and performance of the blade coatings.

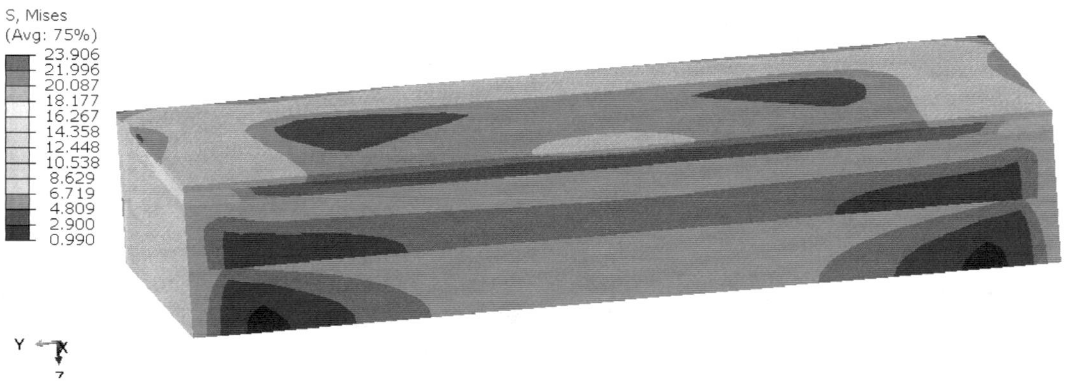

Figure 2. Von Mises (hygroscopic) stress at the interfaces due to moisture ingress in multi-layer blade samples.

Generally, the leading edge erosion process includes competing degradation processes, surface damage near the drop contact surface, and interface damage. A computational model developed allow simulation of both processes, damage due to contact impact and interface damage. It was observed that repetitive raindrop impact of different sizes at arbitrary locations causes damage accumulation and propagation faster at the interface, finally damaging the coating layer. Figure 3 shows the damage distribution on the top surface of the coating (left) and of the interface between coating and putty (right), obtained in the simulation. The validation of this work is underway.

3. Stochastic effects: Rain scenario, defects, roughness

3.1. Small parameter effects and rain scenarios

The surface degradation of blades is strongly influenced by small parameters, factors, which are often out-idealized in continuum mechanical problems. In [16], it was demonstrated that changing the shape of a rain droplet before the impact (from spherical to ellipsoidal) changes drastically the stress field in the coating after the impact, thus, influencing the damage mechanism. The shape of rain droplet changes during its flight, is dependent on the cloud height, wind velocity, and other random factors. Further, availability of small voids or particles in the coating can reduce the coating lifetime by order of magnitude [5]. On the other side, the availability of water film on the surface of a blade can reduce the stresses, thus, extending the coating lifetime [16]. Further, the sequence of rainy periods (strong/weak or vice versa) has strong effect on the blade erosion, according to the results by ORE Catapult team [17]. In this section, some critical stochastic factors, influencing the blade erosion, are listed.

A rain scenario and droplet distributions represent important starting points in all erosion analyses. Many models of rain erosion are based on the Best droplet size distribution [4]. However, Prior and colleagues [18] analysed the precipitation intensity, droplet size distributions (DSD), hydrometeor phase and the wind turbine rotational speed, and concluded that neither Best distribution, nor Marshall Palmer approximation represent realistic atmosphere-relevant drop distribution. Thus, correct, data based rain scenario should be used, when modelling the blade surface degradation.

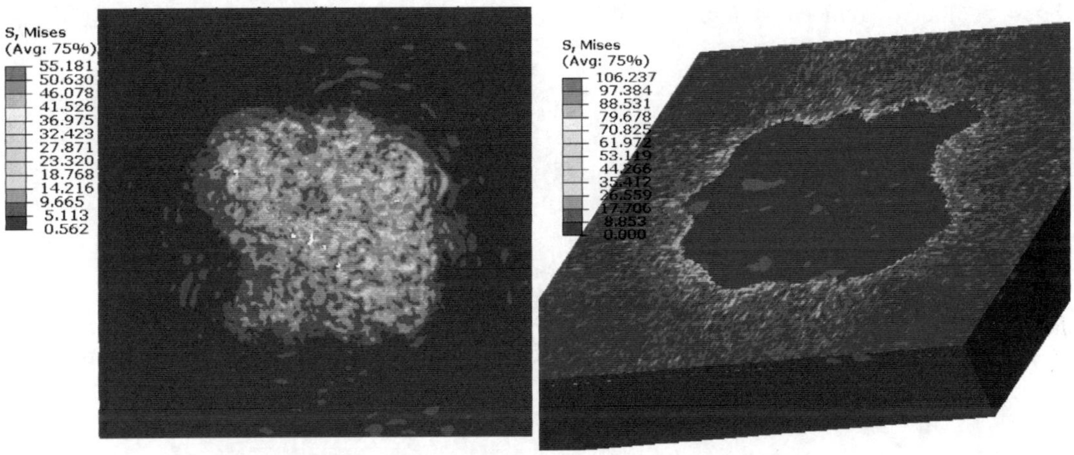

Figure 3. Damage distribution on the top surface of the coating (left) and of the interface between coating and putty (right), obtained in the simulation.

3.2. *Roughness formation and effect of roughness*

The blade erosion leads to the roughening of blade surface. The rough surfaces change the droplet impact conditions, and the roughness increases over time, reducing the aerodynamic performance of the blades. Even initially, the blade surface is not ideally plane, while most of the computational models of the droplet impact still assume the plane surface [4][5]. Figure 4a shows 3D reconstruction of small region of rough blade surface, obtained using the photogrammetry approach from [12]. The digitized rough surface model was introduced into finite element model of liquid impact. Figure 4b shows the stress distribution in the rough surface hit by a rain droplet.

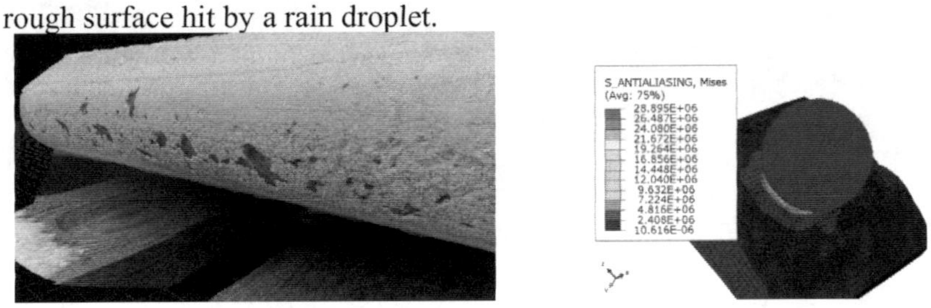

Figure 4. Reconstructed 3D surface of rough blade surface (left) and computational model of droplet impact on rough surface (Reprinted from [12])

In order to predict the roughness evolution of the blade surface, a computational model based on the finite element method, coupled Eulerian-Lagrangian approach, and damage mechanics was developed in [20]. The model uses the stress amplitude calculation method of fatigue damage analysis, and is therefore relatively expensive in computational resources.

Another approach of the roughness is based on simplified model of the material removal. This model simulates material removal by setting a threshold value for the absorbed impact energy of the coating material. Once the absorbed energy from multiple impacts exceeds this threshold, the coating material is removed from that point (see schema Figure 5). A rain event is generated by randomly distributed impact positions of liquid droplets over the surface area. The absorbed impact energy is calculated by summing the contribution of each droplet impact, if the impact occurs close enough to the point. The impact energy is defined as the kinetic energy of the droplet relative to the moving blade. The shape of this area is assumed to be a 3D ellipsoid. If a point of the coating material is within this ellipsoid, then it absorbs an energy amount equal to the droplet's impact energy. Figure 6 shows the examples of the roughness evolution, obtained using the simplified method.

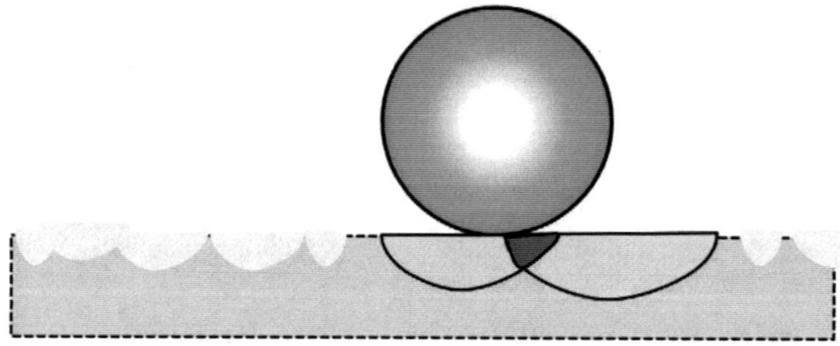

Figure 5. Schema: Roughness formation as successive damage accumulation in regions

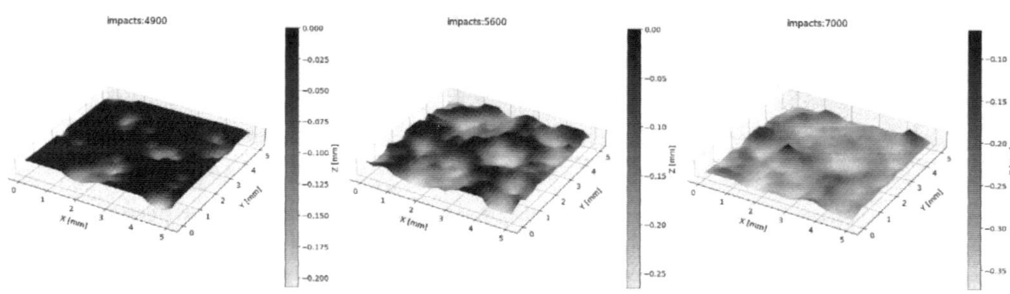

Figure 6. Roughness of the coating layer after different numbers of droplet impacts.

4. Conclusions

Leading edge erosion of wind turbine blades is controlled by complex multiphysics, stochastic processes. These processes include environmental loading, temperature, moisture and load variations, ultraviolet radiation, stochastic variations of mechanical loading and combination of all these factors. In this paper, we consider methods of computational modeling of effects of these effects. Examples of computational simulations of the effect of moisture, roughness evolution, role of defects in blade coatings on the blade erosion are demonstrated.

Acknowledgement: The authors acknowledge the financial support of the Innovation Foundation of Denmark in the framework of project "WiseWind: New generation of sustainable wind turbine blades", No.2079-00004B, EU via Horizon project "Blades2Build: Recycle, repurpose and reuse end-of-life wind blades composites: A coupled pre- and co- processing demonstration plant", grant agreement101096437, and the support of the Ministry of Foreign Affairs of Denmark, via the Danida grant 19-M02-DTU "Maintenance and repair strategy for wind energy development".

References

[1] L. Mishnaevsky Jr., Repair of wind turbine blades: Review of methods, Renewable Energy, Vol. 140, 2019, pp. 828-839
[2] L. Mishnaevsky Jr., C. Hasager, C. Bak, A.M. Tilg, J. I. Bech, S. Doagou Rad, S. Fæster, Leading

edge erosion of wind turbine blades: Understanding, prevention and protection, Renewable Energy 169 (2021) 953e969

[3] M.H. Keegan, D. Nash, M. Stack, Modelling rain drop impact on offshore wind turbine blades. In: ASME Turbo Expo 2012, 2012-06-11 - 2012-06-15.

[4] B. Amirzadeh, A. Louhghalam, M. Raessi, M. Tootkaboni, A computational framework for the analysis of rain-induced erosion in wind turbine blades, part I: stochastic rain texture model and drop impact simulations,. J. Wind Eng. Ind. Aerod., 163 (April 2017), pp. 33-43

[5] S. Doagou Rad, L. Mishnaevsky Jr, Leading edge erosion of wind turbine blades: Computational modelling of multiaxial fatigue, Wind Energy,, 2020, 23/8, pp. 1752-1766

[6] C.S.P. Borges et al. Effect of water ingress on the mechanical and chemical properties of polybutylene terephthalate reinforced with glass fibers. Materials 2021, 14, 1261.

[7] Y. Zhang et al, Effects of moisture ingress on the mesoscale mechanical properties of epoxy adhesives under elevated temperature, Polymer Testing, 94, 2021, 107049

[8] T. Gao et al, Outdoor exposure and accelerated weathering of polyurethane/polysiloxane hybrid coatings, https://www.sciencedirect.com/science/article/pii/S0300944018312657

[9] X.F Yang et al, Weathering degradation of a polyurethane coating, Polymer Degradation and Stability, Volume 74, Issue 2, 2001, Pages 341-351

[10] W.Tan er al, The effects of service temperature on the fatigue behavior of a polyurethane adhesive joint, 2021, International Journal of Adhesion and Adhesives 107(24):102819

[11] M.Rutkowska et al, Degradation of polyurethanes in sea water, Polymer Degradation and Stability, Volume 76, Issue 2, 2002, Pages 233-239

[12] L.Mishnaevsky Jr., A. Tempelis, N.Kuthe, P. Mahajan, Recent developments in the protection of wind turbine blades against leading edge erosion: Materials solutions and predictive modelling (submitted)

[13] N. Kuthe, P. Mahajan, L, Mishnaevsky Jr, Competing and interacting damage mechanisms in erosion of multilayered wind turbine blade coatings: Surface damage, debonding and moisture effect (in preparation)

[14] A. Sasi and P. Gromala, Simulating moisture diffusion in polymers using thermal-moisture analogy, 17th Int Conf Thermal, Mechanical and Multi-Physics Simulation and Experiments in Microelectronics and Microsystems (EuroSimE), 2016, pp. 1-8.

[15] A.S. Abhilash, Sh. P. Joshi, A. Mukherjee, L.Mishnaevsky Jr, Micromechanics of diffusion-induced damage evolution in reinforced polymers, Composites Science and Technology, 71, 3, 2011, pp 333-342

[16] S. Doagou Rad, L. Mishnaevsky Jr. Rain erosion of wind turbine blades: Computational analysis of parameters controlling the surface degradation, Meccanica, 2019, 55, 725–743

[17] K. Dyer, Overview of ORE Catapult activities in leading edge erosion area, Presentation at 4th International Symposium on Leading Edge Erosion of Wind Turbine Blades, 7.2.2023

[18] S. Prior et al, Atmospheric drivers of wind turbine blade leading edge erosion: review and recommendations for future research, Energies 2022, 15(22), 8553; https://doi.org/10.3390/en15228553

[19] L. Mishnaevsky Jr., S. Fæster, L. P. Mikkelsen, Y. Kusano and J.I. Bech, Micromechanisms of leading edge erosion of wind turbine blades: X-Ray tomography analysis and computational studies Wind Energy 2019;1–16.

[20] A. Tempelis, L. Mishnaevsky Jr, Surface roughness evolution of wind turbine blade subject to rain erosion. Materials and Design, 231, 2023, 112011

[21] N. Kuthe, P. Mahajan, S. Ahmad and L, Mishnaevsky Jr, Engineered anti-erosion coating for wind turbine blade protection: Computational analysis, Materials Today Communications, Vol.31, 2022, 103362

IOP Conf. Series: Materials Science and Engineering 1293 (2023) 012014 doi:10.1088/1757-899X/1293/1/012014

Constraining effect of load frames during full-scale rotor blade fatigue testing

F Klaenfoth[1], D Melcher[1], A Nickel[2], T Block[2], and E Petersen[1]

[1]Department Rotor Blades, Fraunhofer Institute for Wind Energy Systems IWES, Am Seedeich 45, 27572 Bremerhaven, Germany

[2]Department of Blade Engineering, Nordex Energy SE & Co. KG, Langenhorner Chaussee 600, 22419 Hamburg, Germany

E-mail: felix.klaenfoth@iwes.fraunhofer.de

Abstract. In rotor blade fatigue tests used for the certification process, load frames are used to introduce the loads to the blade. These load frames have a constraining effect on the cross-sectional deformation. This work investigates these constraining effects using a finite element contact analysis on a current-generation commercial rotor blade. Two different load frame variants are considered: a conventional load frame covering most of the cross-section and a reduced load frame covering only the main spar caps. The interaction between blade and load frame is implemented via contact formulations, which allows pretensioning in combination with a flap and a lead-lag load case. Strains on the outer surface of the blade are evaluated and compared to an artificial reference loading. The area of influence of the load application through the load frame, in which the strains deviate significantly due to clamping effects, mostly corresponds to the 0.75 times the chord length assumed for the certification [1]. The strains in the trailing edge are significantly less affected by the reduced load frame variant than by the conventional one, thus potentially making it possible to consider the trailing edge at the position of the load frame as being properly tested for certification purposes.

1. Introduction

A full-scale certification test of a new rotor blade type is required to validate design assumptions, i.e., to identify relevant failure modes for design details or identify manufacturing details prone to damage initiation. To this end, the design loads occurring during the lifetime of the turbine, i.e., complex loads resulting from the continuous aerodynamic, gravitational, and inertial forces, are simplified and converted into target loads. To achieve these target loads, test loads are introduced at several discrete positions along the blade span via load frames. These load frames locally constrain the deformation of the blade, which alters the stresses in the area around them. As a result, affected areas around the load frames, typically 0.5 to 0.75 times the local blade chord length in both directions, must be excluded from the certification and therefore must not contain any “critical” structural details as described in IEC 61400-23 [1].

Hence, the load frames limit the freedom of test design and introduce artificial stiffening of the blade during testing. However, the excluded areas around the load frames can be reduced if justified by simulation, for example, thereby increasing the test design freedom. This research therefore investigates the effect of the load frames on the structural behavior of the affected area.

The load frames are often modeled in a simplified way using multi-point constraint (MPC) elements that can constrain the profile deformation at a radial position of the blade, as seen in [2]. Chen used contact formulations for modeling load introduction to determine the cause of failure in a full-scale blade test [3].

In the present investigation, the blade section around the load frame, as well as the load frame itself, is modeled in detail using finite elements. The load transmission from the load frame to the blade is realized by means of contact formulations, which enables the analysis of the contact force distribution. The pretensioning process of the load frame is also considered in the simulation.

2. Methods/Approach

Two different load frame variants are investigated: (i) a state-of-the-art load frame encompassing the whole blade profile, and (ii) a reduced load frame which is only in contact with the main spar area and thus allows more deformation of the blade profile. For both variants the area of influence along the blade span in which the load frame has a significant influence on the strains in the blade structure is determined. A load distribution in which the loads are applied directly via MPC elements serves as a reference.

2.1. Load frame design

The load frames investigated are the frames used at IWES for commercial blade tests and consist of a frame made of steel beams fitted around a plywood inlay matching the outer shape of the blade. The two variants are shown in Fig. 1. A 10 mm thick layer of rubber is placed between the inlay and the blade surface to prevent the load frame from slipping and to distribute the load more homogeneously over the surface. The plywood inlay extends 300 mm along the blade.

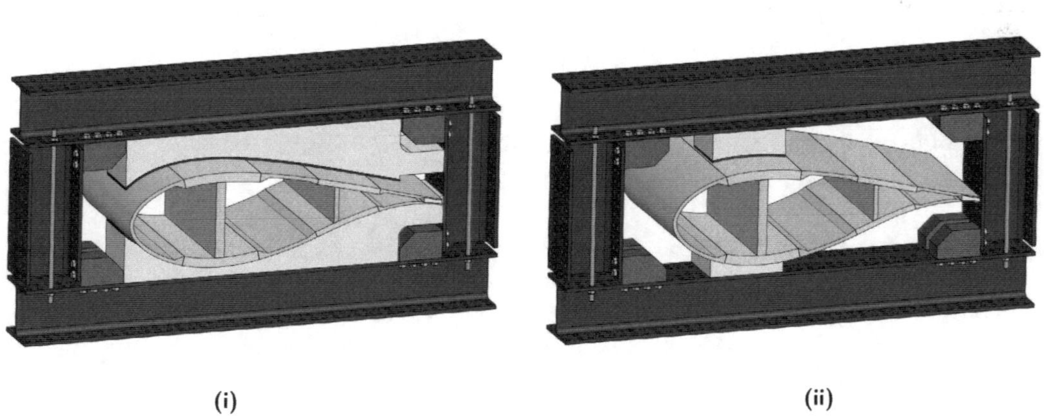

Figure 1. Load frame designs for a generic cross-section; (i) full load frame design; (ii) reduced load frame design

A preload for applying a clamping force is set according to the following procedure: First, the horizontal beams and the plywood inlays are attached to the blade and a pretension is applied via the threaded rods. The ratio of the pretension forces at the leading and trailing edges is set to about 1.8, so that the load is centered around the main spar cap. The pretension forces are kept the same for both load frame variants. The fasteners of the vertical beams are not yet firmly tightened, so that a relative movement between the two horizontal elements is possible and no unwanted constraint force occurs. The vertical beams are then fastened, so that the relative position of the inlays to each other is fixed.

IOP Conf. Series: Materials Science and Engineering 1293 (2023) 012014 doi:10.1088/1757-899X/1293/1/012014

2.2. Blade shell model

A section of a current generation commercial rotor blade, modeled in Ansys® Academic Research Mechanical, Release 22.1, is considered for the investigation presented. The blade section consists of a main spar cap on the suction and pressure sides (SS and PS), connected via a centered web. Another set of spar caps is located in the sandwich panel toward the trailing edge, and also features a web. Shells and web are represented by 4-node `SHELL181` shell elements, the trailing edge adhesive bond is modeled using 8-node `SOLID185` 3-D solid elements. The layup is defined using Ansys Composite PrePost (ACP) [4]. The shells are meshed using predominantly rectangular elements with a side length of 50 mm. In the contact area of the load frame, the element length is reduced to 25 mm.

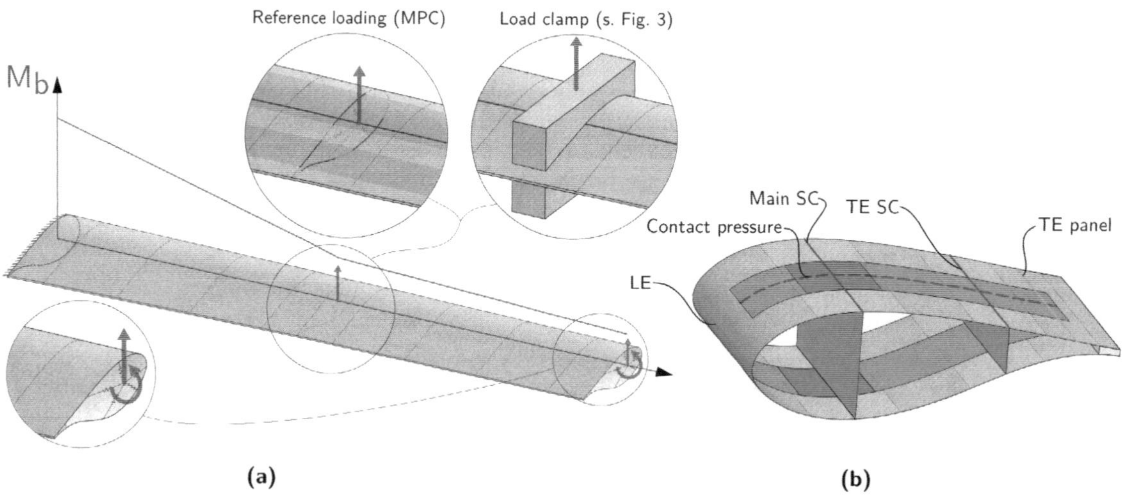

Figure 2. (a) Blade model boundary conditions; (b) Definition of paths for strain and contact pressure evaluation (generic blade cross-section; paths on suction and pressure sides; Leading edge (LE), Main spar cap (main SC), Trailing edge spar cap (TE SC), Trailing edge panel (TE panel), Contact pressure: center path of load clamp)

Two static load cases are considered, which represent cut-outs from global bending moment distributions under simplified lead-lag and flap loading. Additionally, the loads were applied in the opposite directions to determine the strains at both extreme points of a fatigue load. Figure 2 (a) illustrates an overview of the model. The inboard end of the segment is constrained in all degrees of freedom. MPC elements are used to introduce a single force in the shear center of the tip-wise end, which connect to the full circumference of the shells. The shear center position is determined using the software tool BECAS [5]. To offset the applied bending moment, an additional bending moment is applied directly to the circumference of the shells on the tip-wise end. Together with the force acting on the load frame, a bilinear bending moment distribution is formed which approximates the continuous curve of the target loads.

An artificial reference load case is generated to determine the area of influence around the load frame. In this case, the same load is introduced via MPC elements, which act directly on all nodes of the spar caps within the cross-section at the load frame position. A deformable MPC formulation is used to distribute the load over the geometry without creating a rigid connection between the nodes. This method creates a local disturbance of the strains in a limited area around the cross-section which must be considered when interpreting the overloads or underloads near this cross-section. It is assumed that the area of influence in the reference

IOP Conf. Series: Materials Science and Engineering 1293 (2023) 012014 doi:10.1088/1757-899X/1293/1/012014

case is smaller than the one in the other load cases, since the load is only applied directly in a cross-section and without clamping of the blade.

To determine the clamping-induced strain deviations, the lengthwise normal strains in the outermost layer are evaluated along the blade length at different positions, as shown in Fig. 2 (b). Paths are defined in the center of the main SC, the TE SC and in the TE panel, both on the suction and pressure side of the blade. An additional path is placed at the LE of the cross-section. The strains are normalized to the strains from the reference load case, so that values above 1.0 represent a relative overloading of the structure and values below 1.0 a relative underloading.

2.3. Load frame modeling and contact formulation

Solid elements and beam elements are used for modeling the load frames. The load frame inlays and the rubber layer are modeled using 8-node `SOLID185` solid elements. The vertical beams consist of 2-node `BEAM188` elements. The horizontal beams are represented by rigid connections between the beam elements and the surfaces on the top and bottom of the inlays, respectively.

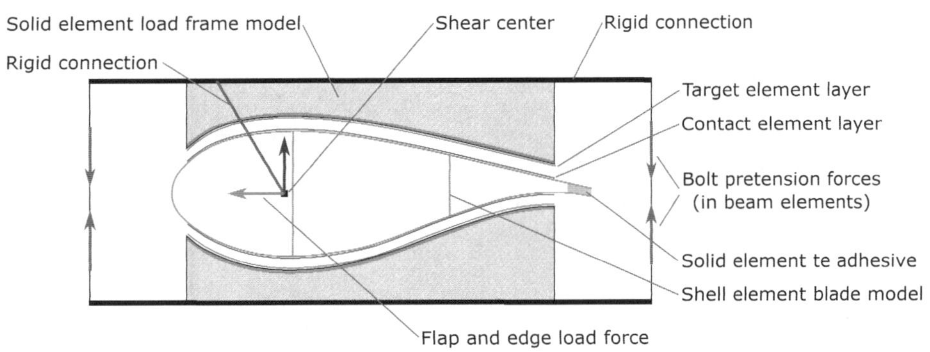

Figure 3. Load frame model (generic cross-section)

The bolt forces are applied in a simplified way via the bolt pretension `PRETS179` elements integrated in Ansys. The vertical beams are used to apply the forces in a first load step. The position of the beams is then locked throughout the next load steps, simulating the fixing of the vertical beams to the horizontal ones. In the last load step the loads on the blade are applied as described above.

The contacts between the blade and the inlays are implemented using 3-D surface-to-surface contact element pairs consisting of `CONTA174` and `TARGE170` element types. A small study was performed to identify suitable parameters for the contact formulation. Asymmetric contact behavior was chosen, with the target elements on the rubber surface of the load frame inlays, and contact elements on the blade shell surface. The contacts are modeled as frictional contacts with a friction coefficient of 0.4 using a pure penalty formulation with a normal stiffness factor of 1.

3. Results and Discussion

Figure 4 shows the pressure distribution on the interface surface for both load frame variants, with pretension and no additional load applied. The pressure is normalized to the largest occurring pressure value. The distribution is inhomogeneous with concentrations of pressure at the position of the webs and regions of low pressure in between. In the longitudinal direction of the blade, the pressure is evenly distributed since the geometry and the layup at this point of the blade change only slightly over the width of the load frame. For load frame variant (ii), the pressure distribution is similar to that of the corresponding area within variant (i), but of roughly twice the magnitude.

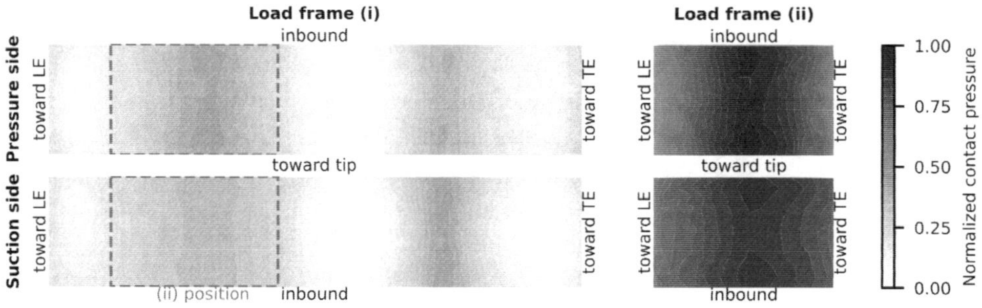

Figure 4. Normalized contact pressure distribution due to pretension

The applied flap and lead-lag loads are superimposed with the pressure resulting from the pretension. In Fig. 5 the pressures are evaluated on a path through the center of the load frames for the two different variants (i) and (ii) and both flap and lead-lag load cases. The flap loads result in a reduction of the contact pressure on one load frame side, and an increase on the other side while the lead-lag loads lead to a shift of the pressure maxima in the main SC area in the opposite direction of the applied force on both sides of the blade.

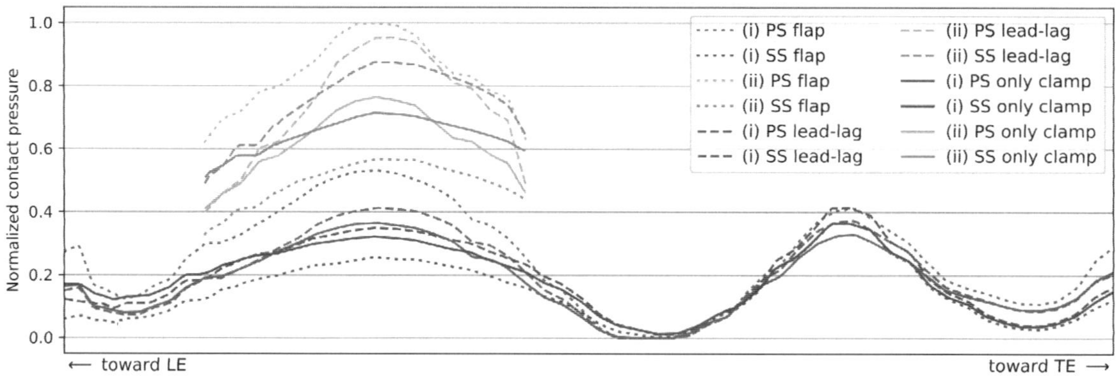

Figure 5. Contact pressure on load frames along center path

Figure 6 shows the strain distribution in the outermost layer of the SS blade shell along different paths from Fig. 1, normalized as described in subsection 2.2. The pretension of the load frame creates compressive strain in all areas covered by the plywood inlay and a tensile strain in the longitudinal direction of the blade. In the leading edge area of the cross-section the effect is reversed, with a tensile strain maximum in the area under contact. These loads are superimposed on the bending loads applied to the blade.

Depending on the path considered, two different strain patterns can be identified: Either an overload occurs in the clamping area under contact and an underload around it, or vice versa. The width of the region in which a significant overload or underload occurs also varies.

Under flap loading, an overload in the clamping area and an underload at around three quarters of the chord length around the load frame can be identified for both spar caps. The overload in the main SC is higher for load frame (ii), resulting from the higher load concentration

IOP Conf. Series: Materials Science and Engineering 1293 (2023) 012014 doi:10.1088/1757-899X/1293/1/012014

due to the smaller contact area. A local overload can also be observed in the TE SC, although it is not covered by the load frame. Strains in the leading and trailing edges are not evaluated for the flap load case, as they are very small and therefore do not make a significant contribution to the fatigue exposure.

For the lead-lag case, a high overload in the main SC area with a relatively high underload in the surrounding area can be seen. For load frame (ii), this overload increases, while the affected area stays the same, spanning around three quarters of a chord length in both directions. At the TE SC as well as the TE panel, an underload can be observed for load frame (i). The strains for the reduced load frame only show a small overload in a small area along the blade. The strains at the LE of the blade profile show a strong underload for both load frames together with a small overload in the surrounding area spanning an area of about 1.2 chord lengths in both directions.

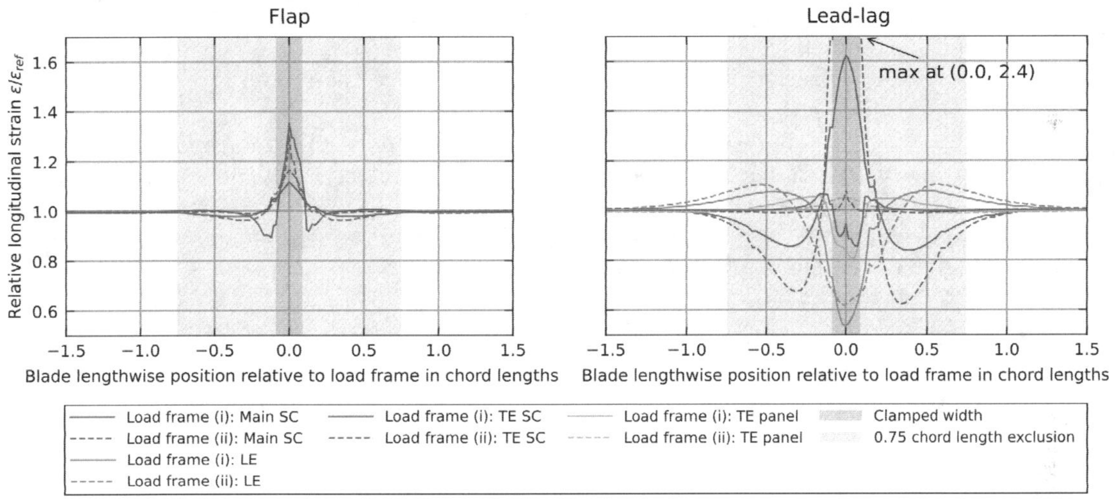

Figure 6. Relative strains evaluated in lengthwise direction

The strains plotted in Fig. 6 show the blade clamping has a clear influence on the loads in the area surrounding the load frames. The influence is mainly limited to the area of three quarters of a chord length around the cross-section where the load is applied, except for the LE profiles, where the area of influence is significantly wider.

Of particular interest are the strains in the trailing edge for the lead-lag loading. In the TE panel as well as at the TE SC, the strains are only slightly affected by load frame variant (ii). The amount of over- and underload is comparable to the amount prevailing at the position of 0.75 chord lengths for the other paths.

The strains were also evaluated from the load cases, where the flap and lead-lag loads are applied in the opposite direction representing the other extremum within a fatigue test cycle. The qualitative pattern of relative strain is similar, but with the opposite sign, so that overloads become underloads and vice versa. Only in the case of the path on the TE SC with load frame variant (ii), no sign change can be observed. This behavior could result from the loading via the MPC elements in the reference load case. The area affected by the load frames and the relative strain amplitudes are similar or smaller for the opposite load cases than for the load cases shown in Fig. 6. Since the relative strain deviates in the opposite direction between the two loading directions, the deviation caused by the clamping affects mostly the mean value of the strains and not the strain ranges.

4. Conclusions

The aim of this study was to identify constraining effects of load frame designs in rotor blade fatigue tests by nonlinear numerical simulation including contact formulation. The results demonstrate that with conventional designs, the pressure and thus the load transfer is not evenly distributed on the surface of the load frame.

By reducing the lateral length of the load frame, the over- and underload of the trailing edge can be reduced significantly for the lead-lag load case. It might be conceivable to place a load frame at a lengthwise position where a critical detail exists on the trailing edge that needs to be included in the certification, providing additional flexibility in the design of the test setups.

Further research needs to be undertaken on the required pretension forces in order to minimize the constraining effect. However, it is necessary to ensure that the load frames do not move under alternating loads. The preload force is kept the same for both load frame variants. This results in a higher contact pressure for the reduced load frame and may also result in a higher buckling load on the shear web. The higher load in the load application area and buckling of the shear web could be compensated by local reinforcement layers. This is in accordance with the certification guidelines; however, it might be necessary to prove that these layers do not affect the loading of the trailing edge.

In this work, overload is determined by lengthwise strain only. To obtain a deeper understanding of overloads, more complex static and fatigue failure criteria should be applied which take into account the complex stress state of the laminate at all positions. Likewise, the local failure of the blade due to the increased contact pressure of the load clamp could be investigated further, especially the buckling and failure of the webs and the failure of the adhesive bonds.

Acknowledgments

We acknowledge the support provided by the German Federal Ministry for Economic Affairs and Climate Action (BMWK) within the SmarTestBlade project (03EE2037).

References

[1] IEC 2014 *IEC 61400-23 - Wind Turbines Part 23: Full-scale Structural Testing of Rotor Blades* 1st ed (Geneva, Switzerland: International Electrotechnical Commission)
[2] Noever-Castelos P, Haller B and Balzani C 2022 *Wind Energy Science* **7** 105–127 URL `https://wes.copernicus.org/articles/7/105/2022/`
[3] Chen X 2017 *35th Wind Energy Symposium* p 1166
[4] Ansys® Academic Research Mechanical 2022 *ACP User's Guide*
[5] Blasques J P A A 2012 *User's Manual for BECAS: A cross section analysis tool for anisotropic and inhomogeneous beam sections of arbitrary geometry* (Denmark: Risø DTU – National Laboratory for Sustainable Energy)

Cure characterisation and prediction of thermosetting epoxy for wind turbine blade manufacturing

Jesper K. Jørgensen, Lars P. Mikkelsen, Tom L. Andersen

DTU Wind, Technical University of Denmark, 4000 Roskilde, Denmark

E-mail: jkjjo@dtu.dk

Abstract. In the wind industry, the increasing length of turbine blades and the use of thick composite sections make it vital to understand the thermoset and the underlying curing process. In this study, the curing of a specific thermoset epoxy for application in wind turbine blades is characterised by differential scanning calorimetry. The thermoset is analysed under dynamic and isothermal conditions to determine the complex cure behaviour under these various conditions. The data is fitted to determine firstly; the cure kinetics of the material through a kinetic model. Secondly; the phenomenon of the glass transition temperature T_g is studied to predict the development of T_g as a function of the degree of cure. Thirdly, a Diffusion-Kinetic model was utilised to make more accurate cure predictions based on the interaction between kinetic chemical reactions and diffusion control affecting the reaction rate through the incorporation of the glass transition temperature. A simple experimental framework for validating the diffusion-kinetic model was used to make a final validation of the model predictions on a larger scale. A set of final parameters is presented for application in a full diffusion-kinetic model for cure predictions and simulations beyond this study.

1. Introduction

For many years, the study of thermoset curing has been an important aspect of understanding and ensuring good composite performance. Residual stresses play an important role with respect to curing. Previous studies have shown a correlation between the level of residual stresses resulting from different curing profiles and the tensile fatigue lifetime of laminates for turbine blades [1]. However, the linking of curing and residual stresses can benefit industries such as; automotive, aerospace, sport and maritime industries.

A principle for determining the characteristics of thermosets has been well established through the application of heat-flow-based Differential Scanning Calorimetry (DSC) [2]. In this configuration, DSC is capable of both measuring the heat flow coming from the thermoset and the heat transferred to the thermoset to maintain the desired temperature in the sample and chamber. It is possible to determine the extent of cure from the thermoset based on the measured heat flow from the DSC device.

Furthermore, the concept of heat capacity can be easily studied with DSC. For many years, the DSC has been used to measure the glass transition temperature, which is easily observed by a change in the heat capacity of the polymer studied. The reason for the clearly observed transition is that the glass transition is a transition dependent on thermodynamical equilibrium [3]. Below the glass transition temperature, the thermoset is not in thermal equilibrium, resulting in the compaction of the molecular chains. As the glass transition is reached the material becomes more saturated and thus increasing the distance between the molecular chains. This can be seen as a clear step change in the heat going into the thermoset when scanned with DSC. This has resulted in very clear definitions of the glass transition temperature range and how it should be found [4].

Over the years, a number of phenomenological models for defining the curing of thermosets have been developed. Early work by M. R. Kamal and S. Sourour presented that the reaction of thermosets, and especially epoxy, follows an autocatalytic reaction [5]. The authors presented a model with a term for the accelerating nature in the early stage of curing and deaccelerating for higher degrees of cure i.e., the model was autocatalytic. The model was based on a simple power law and combined with the Arrhenius reaction equation. The authors found a good fit with the experimental data for both the epoxy and polyester systems. Later, another study showed that predicting polyester behaviour could be overcome with another modified model with an additional term, making the model more accurate at higher conversions [6].

As the glass transition temperature is a thermodynamical principle, A. T. DiBenedetto [7] proposed a nonlinear relation between the glass transition temperature and the degree of cure. In which, the nonlinearity was determined by the ratio of heat capacity in the uncured and fully cured states. Further developments were made to replace this rather less definable ratio with a fitting factor [8]. Ultimately, the modifications of the relation proposed by A.T. DiBenedetto give a relationship between the glass transition temperature and the degree of cure easily applicable. The glass transition temperature is a phase transition between the glassy and rubbery states. Therefore, it is important to note that the material is thermodynamically unsaturated in the glassy state, i.e. it has less molecular mobility; thus, the material appears more rigid. The mobility increases as the material becomes increasingly saturated when reaching the glass transition temperature[3]. The sink in mobility when the material enters a glassy state is also known as vitrification. Therefore, to accurately predict the curing of a thermoset where mobility is at least in part prohibited by the glass transition temperature, modifications of kinetic models are necessary. The mobility-reduced behaviour is commonly referred to as diffusion-control [9].

A simple approach to creating a diffusion-modified model is implicitly including the glass transition temperature through a critical degree of cure [10]. This was later modified with a linear interpretation of the critical degree of cure to compensate for the changes in the glass transition temperature with increasing process temperature [11]. Different approaches have been applied to define the diffusion control. One approach was to take the actual measured reaction rate and divide it with the model predictions, only accounting for the kinetic part of the reaction [11]. The response should show that the model and the data agree as long as diffusion control is not dominant, i.e. when the measured reaction rate and the kinetic model prediction coincide. Then, as the measured reaction rate becomes lower than the model prediction, increasingly prohibited molecular mobility appears. Another approach is to use the critical degree of cure function and correlate that with the equation that describes the glass transition temperature as a function of the degree of cure [1]. In this case, the relation proposed by A.T. DiBenedetto [7] was used and the degree of cure was found by setting typical cure temperatures equal to the glass transition temperature and isolating the critical degree of cure. The critical degree of cure evolution was fitted from the critical degree of cure at the different cure temperatures with a good fit. To monitor the behaviour of epoxy systems different approaches exist [12, 13]. These studies have compared embedded sensor technology with predictions from cure kinetic models. However, these typically neglect the presence of diffusion control in the applied cure models.

This study investigates the curing of thermoset epoxy with DSC utilising a simple and robust experimental procedure. The investigation involves the fitting of DSC data and the selection of

IOP Conf. Series: Materials Science and Engineering 1293 (2023) 012015 doi:10.1088/1757-899X/1293/1/012015

a diffusion-kinetic model in order to better predict the curing of a larger-scale resin experiment with an embedded sensor.

2. Theory

2.1. Cure kinetics

To study the behaviour of an epoxy thermoset resin this section will present the relevant theory applied in the study. First, the amount of reacted polymer is defined by the degree of cure, X, determined by the ratio between the reacted enthalpy, and total enthalpy of reaction,

$$X(t) = \frac{H(t)}{H_T}. \tag{1}$$

Here, $H(t)$ is the accumulated heat released from the curing reaction up to a certain time t. If the accumulated amount is unknown, one can determine the remaining amount (residual) instead. The total enthalpy of the reaction, H_T, is the total enthalpy released by the exothermal nature of the reaction. Thus, the degree of cure X, will be in the range $X \in [0; 1]$, with $X = 0$ for uncured and $X = 1$ for fully cured, respectively. To capture the kinetic reaction mechanisms, the following relation [5, 6, 11], is used,

$$\begin{aligned} \frac{\mathrm{d}X}{\mathrm{d}t} &= k(T) \times f(X) \\ k(T) &= A \exp\left(-\frac{e_a}{RT}\right) \\ f(X) &= X^m(1-X)^n. \end{aligned} \tag{2}$$

This model involves the temperature-dependent Arrhenius reaction relation $k(T)$. The relation consists of A as the preexponential factor, e_a as the activation energy, and R as the universal gas constant. For the power law function, $f(X)$, n and m are power law coefficients.

The activation energy can be found using the Kissinger's method [12, 14] in

$$\ln \frac{\beta}{T_p^2} = \ln \frac{AR}{e_a} - \frac{e_a}{RT_p}, \tag{3}$$

where β is the heating rate and T_p is the temperature at the peak of the heat flow.

2.2. Glass transition temperature

The glass transition temperature, T_g, is defined in the range from an onset, over a midpoint, to the end of the transition [4]. This study defines the glass transition as the midpoint value. The DiBenedetto relation [7] shown in (4) is used for relating the degree of cure to the glass transition temperature.

$$T_g(X) = T_{g0} + \frac{\xi X(T_{g\infty} - T_{g0})}{1 - (1-\xi)X} \tag{4}$$

Here T_{g0} is the glass transition temperature for the uncured state, $T_{g\infty}$ is the glass transition temperature for a fully cured state. Furthermore, ξ is a fitting parameter given by the ratio of specific heat capacity $\Delta C_{p\infty}/\Delta C_{p0}$. The heat capacities $\Delta C_{p\infty}$ and ΔC_{p0} are the difference in heat capacity for a fully cured and uncured sample at the crossing of the glass transition temperature.

2.3. Diffusion-controlled effects on cure kinetics

A model accounting for the interaction between T_g and the cure evolution [11] is applied to incorporate the prohibited molecular mobility effect by diffusion control into the cure predictions

$$\frac{\mathrm{d}X}{\mathrm{d}t} = \frac{k(T)X^m(1-X)^n}{1+\exp\left[C(X-X_c(T))\right]}. \tag{5}$$

The model is based on (2) modified by dividing with a term that is an exponential function including a diffusion constant C. It is related to X and the critical degree of cure X_c. The critical degree of cure is the level of cure where diffusion will dominate. In this study, a linear function, of X_c dependent on temperature T is applied [11], where, X_{cT} is the critical degree of cure increase and X_{c0} is the initial critical degree of cure

$$X_c(T) = X_{cT}T + X_{c0}. \tag{6}$$

3. Experimental

3.1. Materials

In the presented study, an industrially available thermoset epoxy resin is investigated. The resin is a conventional diglycidyl ether of bisphenol-A (DGEBA) resin. The hardener is a modified cyclo-aliphatic- and aliphatic-amine. A mixing ratio by weight was base:hardener; 100:31, was used according to supplier guidance.

3.2. Sample preparation and storage

A small batch of approximately 200 g was mixed to prepare the samples. The mixing was carried out in a small bucket over 5 minutes to ensure sufficient mixing. The mixture was degassed and pipetted into smaller 5 ml vials that were immediately quenched in a Kryogen bottle filled with liquid nitrogen. Liquid nitrogen storage serves two purposes: preventing premature curing and acting as a purge gas to protect the vials from moisture contamination during storage. DSC testing was carried out over a total time period of 4 weeks. In this period, the total enthalpy of reaction was monitored at regular intervals to verify that the material had not been reacting while storing.

The sample preparation procedure was performed systematically: A vial was taken from the Kryogen container and allowed to thaw for 20 minutes before 18-20 mg of epoxy was taken from the vial and transferred to a Netzsch Concavus® pan.

3.3. DSC measurements

The DSC measurements were performed using the Netzsch™ Polyma 214. For the experiments, a protective purge gas of nitrogen was used in the main chamber of the DSC. The flow rate was set to 40 ml/min to ensure sufficient sample protection and reasonable heat conductivity.

For the purpose of investigating the curing of thermosets accurately, it is necessary to determine the curing behaviour based on both dynamic and isothermal conditions. Differential Scanning Calorimetry can with dynamic scans accurately determine; enthalpy of fusion, the peak temperature of the reaction with constant heating rates. This makes a dynamic scan ideal for fitting of Kissinger's method in (3) to determine the activation energy, e_a [12]. Furthermore, the dynamic scan can also determine the glass transition temperature, T_g. But better results are achieved by applying temperature-modulated DSC (TM-DSC) as the glass transition temperature becomes much clearer with this approach [2].

The isothermal scan with DSC is applied to determine the reaction rate and degree of cure at various cure temperatures. This approach is applied to make sure that the cure models are fitted over a varied range of common isothermal cure temperatures. The isothermal scan can be relatively long in order to capture the transformation from a high rate of cure to a slower rate that occurs for higher levels of cure. Thus, it makes it important to use both dynamic and isothermal conditions to determine the evolution of the cure.

However, it is also a great benefit to vary the time period of the isothermal scan and in combination with a TM-DSC scan, it creates a great method for determining the evolution of T_g with X. This method will be utilised in Section4 for the fitting of the DiBenedetto relation in (4).

Two techniques exist for putting samples into the DSC: The first involves the sample being mounted at room temperature and then scanned. For dynamic and TM-DSC scans, this approach has been applied. The second is to preheat the chamber before mounting the sample for scanning[15]. For isothermal scans, the second approach was applied.

4. Results and discussion

4.1. DSC cure characterisation

To get the total extent of the reaction, samples were cured with dynamic DSC scans with heating rates of 2, 5, 10, 15 and 20 K/min. The lower temperature limit was chosen as -40°C for the temperature ramp. This was chosen to ensure that the temperature ramp starts far away from any onset of reaction. The upper limit was 275°C as this was found to be below any material degradation.

Figure 1 shows the 5 scans plotted with heat flow in W/g as a function of temperature in K. In all five cases, a second ramp was performed to check for any residual cure. For the 5K/min case, the signal for this second heating ramp has been included as a dashed line in Figure 1. The second heating ramp shows that no additional cure has occurred as the signal is more or less flat. The slight bump observed on the included second ramp is the glass transition temperature. Table 1 shows all the corresponding peak temperatures T_p, as well as the individual heating rates denoted β. The five scans are integrated over time from a point where the response is fairly flat, i.e. before the reaction starts and until the ramp-up ends at 275°C. This integration is interpreted

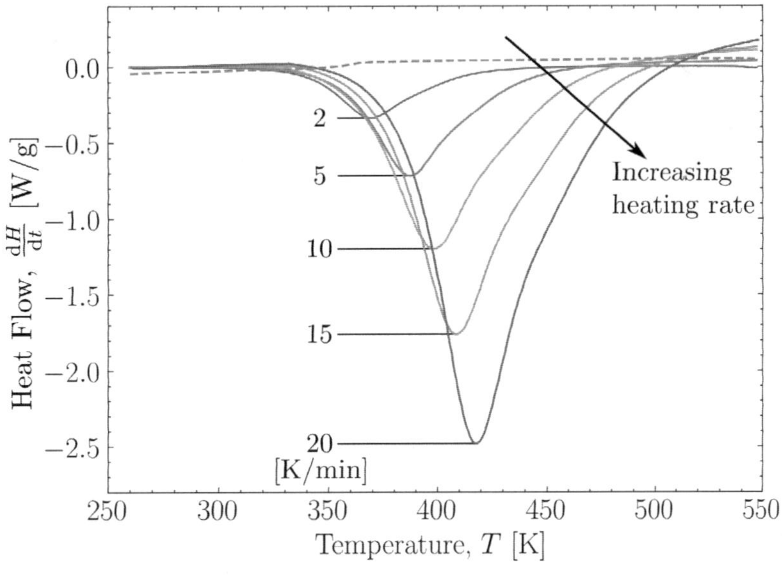

Figure 1. Dynamic scans of uncured samples at different heating rates as solid line and the DSC signal from a second heating ramp at 5K/min to check for any residual cure - dashed line.

Table 1. Heating rate β, Peak temperature T_p, total enthalpies of reaction H_T and final glass transition temperatures $T_{g\infty}$.

β K/min	T_p [°C]	H_T [J/g]	$T_{g\infty}$ [°C]
2	96.3	471.8	89.3
5	113.2	464.2	88.9
10	129.7	471.0	88.4
15	136.0	472.3	89.4
20	144.6	470.3	89.0
Avg.	-	470 ± 3	89.0 ± 0.4

as the total enthalpy of reaction H_T. A TM-DSC scan has been performed as a second dynamic scan to determine, what is taken as the final glass transition temperature $T_{g\infty}$. Table 1 gives the H_T values at the different heating ramps and the average value is $H_T = (470 \pm 3)$J/g. Thus, a consistent value, independent of the heating rate. It is worth noting, that these values of H_T have been obtained throughout the whole test period of 4 weeks and therefore, the utilised storage facility was ideal. Similarly, the average value $T_{g\infty} = (89.0 \pm 0.4)$°C obtained, was found to be with good accuracy and well representing.

Kissinger's method from equation (3) is used to estimate the activation energy. To do so, $\ln(\beta/T_p^2)$ is plotted as a function of $1/T_p$. Figure 2 shows the scatter of measurement points with linear regression. The slope of the linear curve then represents the term e_a/R. The fitted activation energy becomes $e_a = (56.23 \pm 0.36)$kJ/mol.

4.1.1. Cure kinetic fitting The isothermal scans were performed for $T_c = [40, 50, 60, 70, 80, 90, 100]$°C. The scans were run in time periods from 3.5 h up to 20 h, as lower T_c requires a longer cure time. As H_T has been defined, $H(t)$ is found by integrating the heat flow, as the area under the

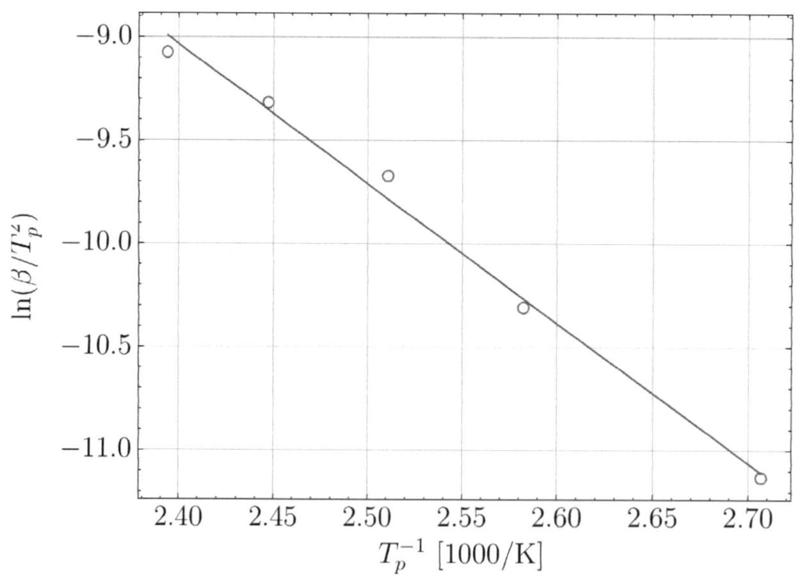

Figure 2. Kissinger's plot to fit the activation energy based on dynamic scans.

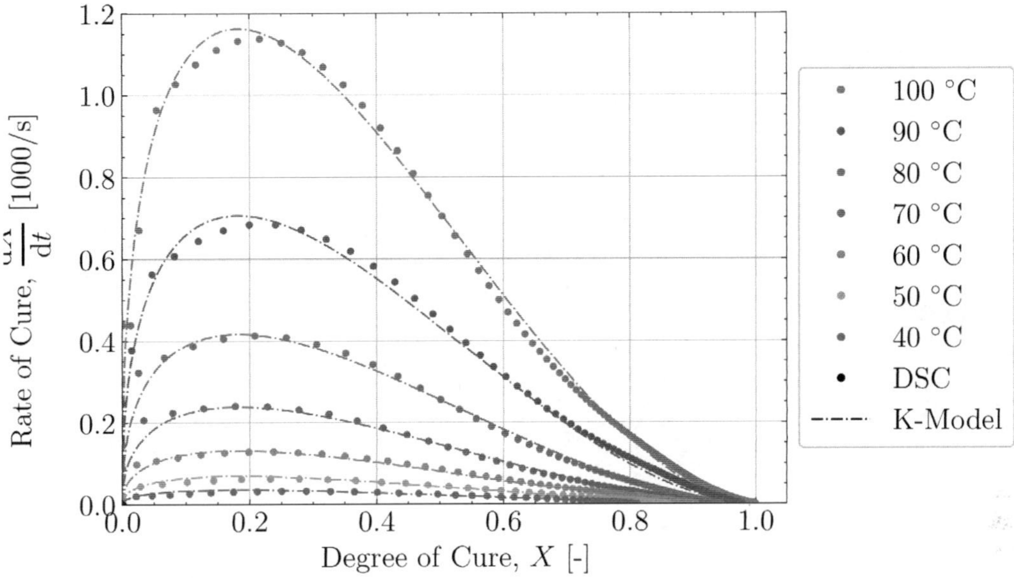

Figure 3. Least squares fit of the kinetic model (K-model) with the experimental values as dots and full lines for the fit.

curves coming from each individual scan at different T_c. The approach is to use a horizontal integration baseline, starting at the end of the isothermal cure where the heat flow has stabilised. The integration is then done for the settled cure response and back to the start of the measurement. Thus, through (1) the degree of cure X and the derivative $\mathrm{d}X/\mathrm{d}t$, can be found for each scanning.

From (2) a least-squares fit is performed with initial guessed parameters for A, n and m, whereas the e_a found by Kissinger's method from the slope of Figure 2 was used as the initial guess for the activation energy.

Figure 3 shows the experimental values, from the DSC at different T_c as points with different colours for each T_c and the data fitting as dashed lines. Overall, the fitting is reasonable by a qualitative comparison of the experimental data and the model.

Table 2 compiles the parameters found by least-squares fit for the K-model from (2). The activation energy found for the isothermal fitting agrees well with the value found by Kissinger's method. This shows that predictions can be expected to perform well under both dynamic as well as isothermal conditions.

As the K-model in (2) is an ordinary differential equation (ODE) it can be integrated numerically, knowing the fitted parameters from Table 2. The model is integrated with respect to time and temperature from the experimental data. The K-model can then be plotted as $X(t)$ with the experimental data for comparison of prediction.

Figure 4 shows the predictions made with (2) based on the fitting of the DSC measurements.

Table 2. Fitting parameters for the kinetic model.

A [s^{-1}]	e_a [kJ/mol]	n [-]	m [-]
$2.50 \cdot 10^5$	56.24	1.83	0.41

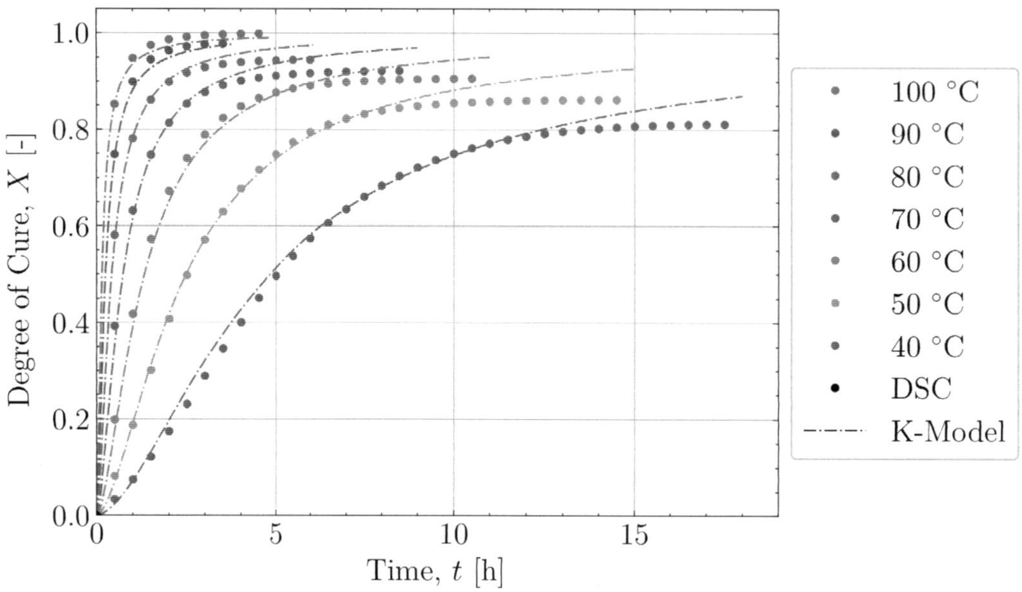

Figure 4. Predictions based on fitted parameters from the kinetic model with DSC measurements for different T_c.

For the different T_c measured, the model seems to predict well up to around 80% of curing. For lower T_c, the model somewhat overpredicts the experimental observations except for the case of 90°C and 100°C, where the same deviation is not observed. This indicates that for $T_c < 90$°C a slowdown in the chemical reaction may be due to elements not considered.

This is likely to be due to the reaction mechanism changing to a diffusion-controlled regime. As this becomes important when the glass transition temperature is greater than the cure temperature, it is important to determine the evolution of T_g with cure.

4.1.2. Glass transition temperature fitting Section 3.3 described that for the determination of T_g the isothermal scan followed by a TM-DSC scan was used. Appropriate T_c were chosen as [60, 80 and 100]°C in time periods from 2 min to 2 h. Furthermore, T_g was found, based on the data used to fit the K-model, and was also included in the fitting of T_g.

The DiBennedetto relation (4) was used to fit the measured data, with the lowest T_g measured at approximately −29°C as it was not of interest to measure lower values than this. The model was fitted with T_{g0} and ξ as fitting parameters and $T_{g\infty}$ was the known parameter in Table 1.

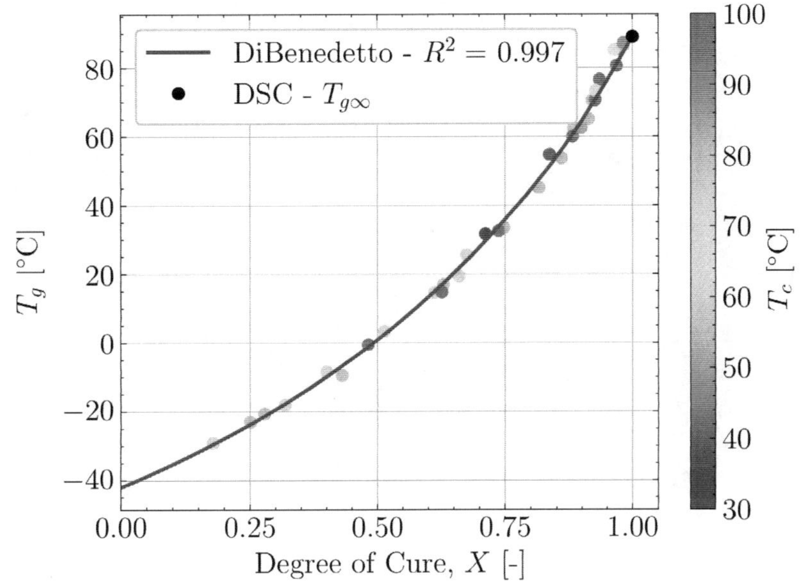

Figure 5. Fitting of the DiBenedetto relation to predict the glass transition temperature development with curing. The colorbar shows the cure temperatures reflected in the colour of the dots.

Figure 5 shows the measured glass transition temperatures as dots. Every measurement point is mapped with a colourmap marking at which curing temperature T_c the T_g has been measured.

The fitting of the DiBenedetto relation has a R^2 value of 0.997 related to the fitted data, which generally confirms a good fit.

Table 3 summarises the fitted values of the model where the fitted parameters can be for $T_{g0} = -42.0$°C and $\xi = 0.487$. As the development of T_g with X has been fitted, it can be used to relate the potential diffusion-controlled behaviour into the cure kinetics.

Table 3. Parameters used in the DiBenedetto equation to predict the development of the glass transition temperature and the fitting coefficient.

T_{g0} [°C]	ξ [-]	$T_{g\infty}$ [°C]	R^2 [-]
-42.0	0.487	89.0	0.997

4.1.3. Determination of diffusion characteristics for curing Determining the change from a cure kinetic model reaction without diffusion behaviour to a diffusion-controlled cure kinetic model is described by (5) [11]. This model will be used to modify the predictions from the K-model. The model involves the term X_c, the critical degree of cure. The method of determining this crossover, from one molecular mobility mechanism to another, is to set $T_g = T_c$ and, through (4), find the corresponding degree of cure [1]. The value of X found is then interpreted as the critical degree of cure X_c at the given T_c.

Figure 6 illustrates the determined values of X_c as a function of T_c from 293 K to 353 K i.e., 20°C to 80°C. The line is the fitted linear regression of (6) and the overall fit is good. Table 4 lists the fitting parameters. The slope corresponds to X_{cT} and takes into account an increase or

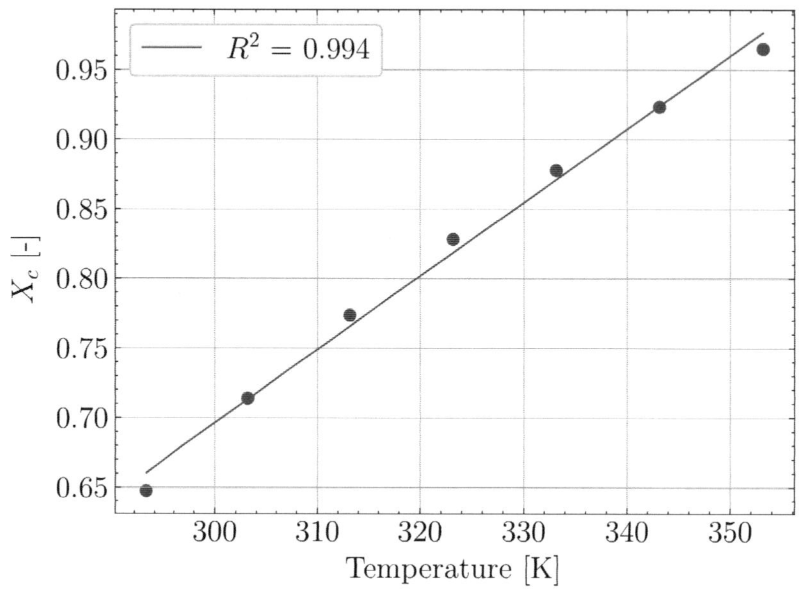

Figure 6. Fitting of the critical cure development based on T_c.

decrease in X_c with the cure temperature. The initial critical value $X_{c0} = -0.885$ is invalid for temperatures below -105°C as X is non-physical below zero. The remaining unknown in (5) is the diffusion coefficient C. By fixing the parameters found for the K-model in Table 2 and the critical degree of cure function in Table 4 the diffusion coefficient C, is found. The diffusion coefficient is found to be $C = 43.7$.

Figure 7 shows the Diffusion-Kinetic model (DK-model) based on (5). The predictions by the DK-model seem to better adapt the DSC measurements for levels of conversion above $X = 0.8$. For T_c at 40°C to 60°C the prediction at high conversion follows the DSC data well. For T_c at 70°C and 80°C, the model slightly overpredicts compared to the data. However, the predictions are still improved compared to the K-model predictions in Figure 3. For 90°C and 100°C, the predictions seem to be unaffected by the diffusion term, since molecular mobility is not significantly prohibited at these cure temperatures. It is relevant to compare the kinetic model (K-model) in (2), and the diffusion-modified (DK-model) from (5). To justify a good comparison, one should look at isothermal curing for longer time periods.

Figure 8 shows the development of curing at the same T_c as in Figure 7, over a logarithmic time period of up to 200 h. Figure 8 visualises that for curing at lower temperatures, the K-model seems to over-predict X compared with the measurements from the DSC. However, the DK-model predicts significantly slower speeds as the reactions slows down due to the prohibited molecular-mobility by the glass transition temperature. Thus, it is infeasible to expect the

Table 4. Fitting parameters found for the evolution of the critical degree of cure with the glass transition temperature.

X_{cT} [K^{-}1]	X_{c0} [-]	R^2 [-]
$5.27 \cdot 10^{-3}$	-0.885	0.994

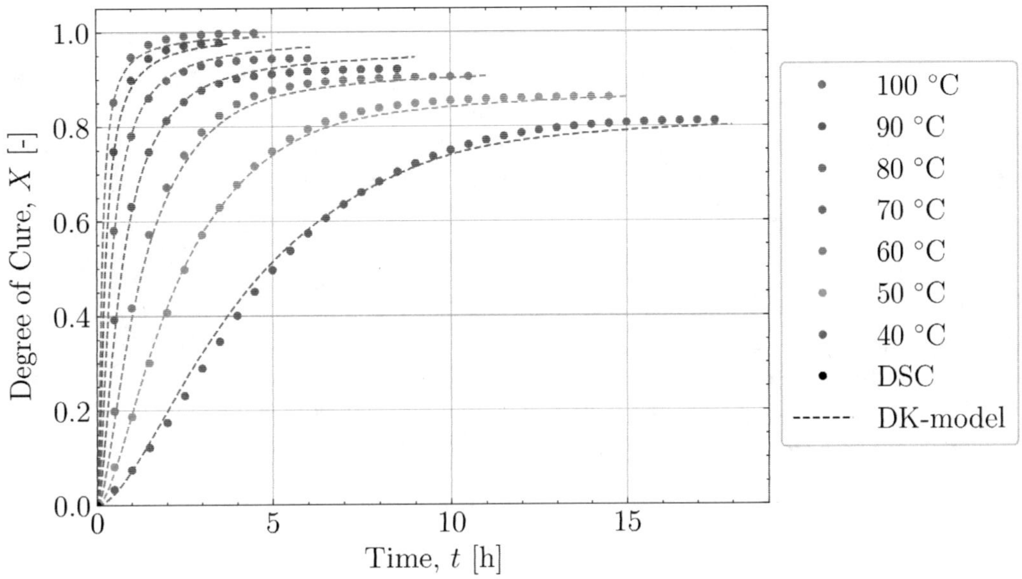

Figure 7. Predictions based on fitting of the diffusion modified model.

material to fully cure at an isothermal of 40°C held for a very long time.

The two models tend to agree more as curing is performed at higher T_c as the molecular mobility is not so prohibited at these cure temperatures. Overall, it is important to apply the DK-model for either or both long cure times and low cure temperatures.

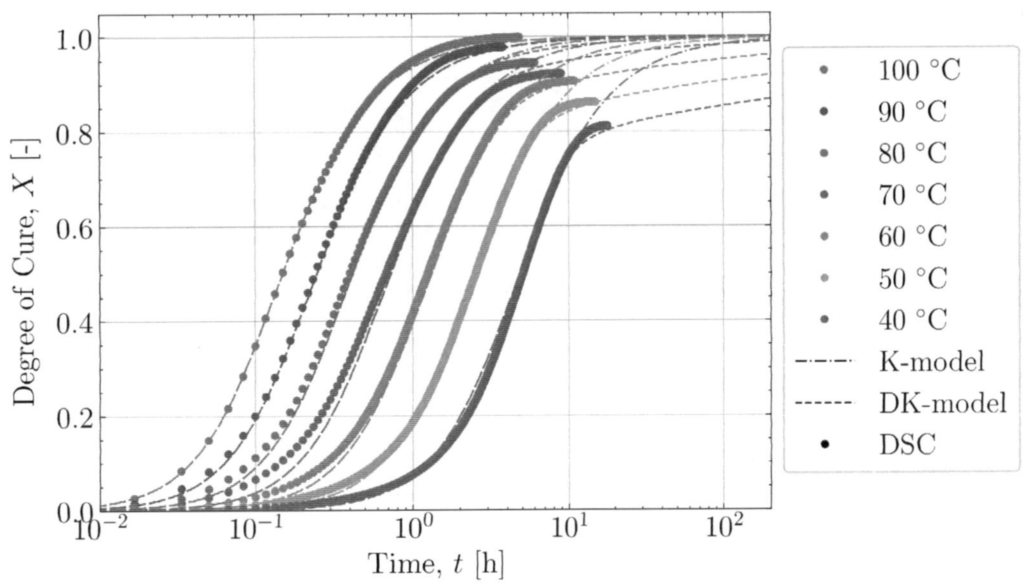

Figure 8. Predictions of the kinetic model and DK-model for relatively long cure times.

4.2. Prediction of cure in neat resin experiment

The previous section dealt with the analysis and fitting of measurements performed with Differential Scanning Calorimetry. This section will deal with the in situ monitoring of a neat resin plate manufactured with the industrially available resin system. The kinetic model (K-model) and diffusion-kinetic model (DK-model) will be applied and validated on this in situ monitored neat resin plate to investigate their applicability on larger components.

4.2.1. Experimental procedure The neat resin plate is cured in an industrial-scale universal oven. The neat resin plate is consist of a zip-bag of 150 x 200 mm. The plate is placed in the middle of the oven and a thermocouple is placed so it will measure the temperature of the thermoset in situ. At the top and bottom of the oven, thermocouples are placed approx. 100 mm from the oven surfaces to monitor the temperature variation across the oven.

Just before starting the experiment, the thermoset is mixed, following the procedure in Section 3.1. The thermoset is degassed and infused into the bag with a syringe. The thickness of the neat resin plate is approximately 4 mm at the end of cure.

In addition to the neat resin plate, a concavus® aluminium DSC pan is also placed in the oven, filled with 20.0 mg of the remaining thermoset. This DSC sample was cured simultaneously and thus has experienced the same cure profile. This DSC sample served as a validation sample, since T_g and the possible residual enthalpy H_R could be determined immediately after the cure ended. Table 5 tabulates the parameters used for the cure profile applied.

To assist the cure profile listed in Table 5 a letter annotation is adopted for illustration purposes for explaining the cure profile steps.

- Ⓐ Isothermal - Pre-cure.
- Ⓑ Heat-up before post-cure.
- Ⓒ Isothermal - Post-cure.
- Ⓓ Cool-down after post-cure.

Figure 9 shows the temperature, as a function of time, registered by the thermocouple inside the resin, for which, the predictions of X will be based. Thinner dotted lines have been used to illustrate the temperature of the air in the oven around the neat resin plate. Figure 9 illustrates both models; the kinetic (K-model) from (2) as a green dashed line, and the diffusion-modified kinetic model (DK-model) as a green dash-dotted line based on (5). Similarly, two orange curves illustrate the T_g predictions based on the two models. At the end of the predictions, two larger dots mark the measured DSC values from the DSC sample cured simultaneously with the neat resin plate. Both models in Figure 9, agree at the early stages of X and start to deviate after approximately 8 hours. The DK-model slows down compared to the K-model, which is also reflected in the glass transition temperature development. This behaviour is expected as it was confirmed in Figure 8 that for longer isothermal cure time the K-model is less conservative with respect to X compared to the DK-model. As the temperature is elevated during Ⓑ, both models predict an increase in reaction rate. Furthermore, the speed is higher for the DK-model, as it almost catches up with the kinetic model as the temperature enters Ⓒ. This is expected

Table 5. Cure profiles applied for the sample cured.

Heat rate T [K/min]	Pre-cure [h @ °C]	Heat rate [K/min]	Post-cure [h @ °C]	Total cure time [h]
1	12 @ 40	1	10 @ 80	23

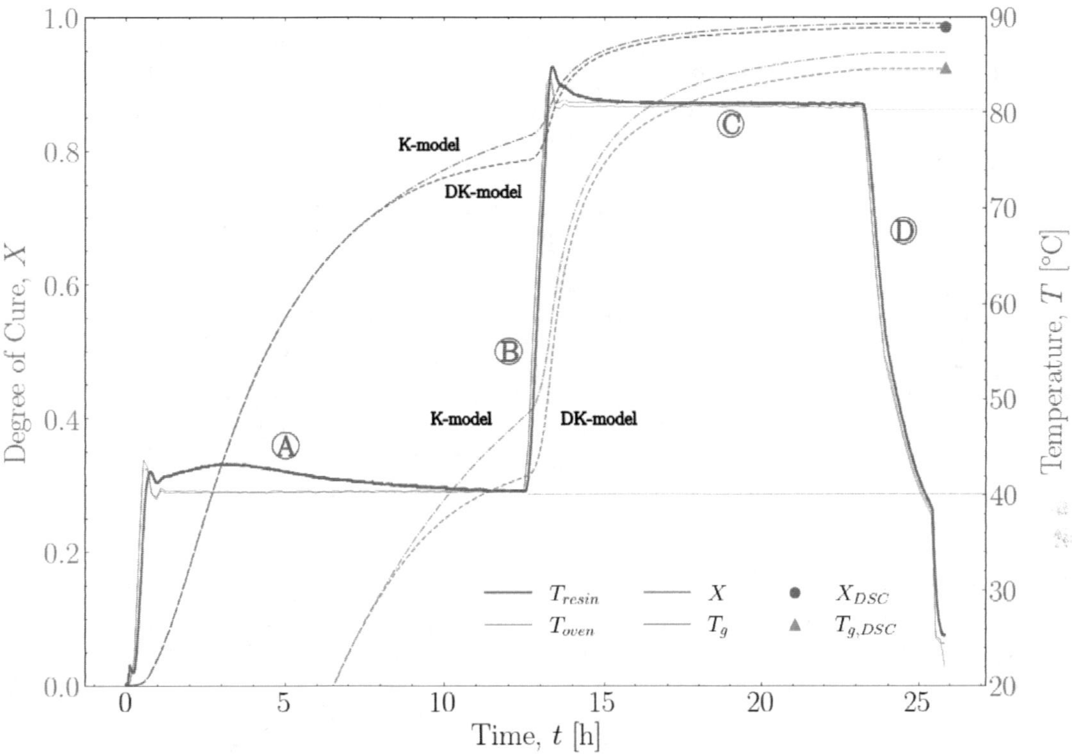

Figure 9. Prediction of X and T_g as a function of time from the in situ measurement of temperature and time. The temperature profile measured inside the neat resin plate is shown, together with the temperature in the oven, above and below the plate, indicated by the two thin lines. Predictions are based on both models investigated and include the values of the DSC sample at the end of the cure.

as the T_c crosses T_g and the DK-model becomes more dominated by kinetic reactions. The glass transition temperature is naturally affected by the development of X and the fact that $T_g(X)$ is a non-linear relation. The difference in the predicted X by the two models has a great effect on T_g. Figure 9 illustrates this as T_g predicted during Ⓐ shows a significant difference between the two models. During the heat-up at Ⓑ, the difference becomes small as observed and as the temperature enters the next isotherm Ⓒ the two predictions start to deviate again. Although, the difference in X is not significant, it affects the glass transition temperature achieved at the end of the cure.

The DSC sample, cured along the neat resin plate was scanned with TM-DSC and the residual enthalpy of reaction H_R and T_g was found. The found H_R was used to determine X_{DSC} through (1) and plotted together with $T_{g,DSC}$ at the end of the curing.

Table 6 compiles the values found by scanning the DSC sample together with the final values predicted by the two models at the end of the cool-down Ⓓ where the thermoset has returned to a temperature near room temperature. The values measured with the DSC, are not far away from either of the predictions. As the H_R found with the DSC is attached with some uncertainty due to the low response measured with the DSC. This makes it difficult to determine where the limits for integrating the residual enthalpy should be. Thus, the glass transition temperature is a more stable entity to determine as it is easily observed through the TM-DSC response. The $T_{g,DSC}$ found was very close to the predictions from the DK-model, confirming that under these

Table 6. Summarised final values of X and T_g for the two predictions of the neat resin panel as well as the control sample measured with DSC.

	DSC measurement	K-model prediction	DK-model prediction
X	0.985	0.990	0.984
T_g	84.7	86.3	84.5

conditions, this model gives the best prediction.

5. Conclusion

This study presents a procedure for analysing the curing and glass transition temperature of a thermoset epoxy system. Based on the data analysis, predictions from a simple kinetic model were made. This model was found to lack some accuracy when the thermoset was cured isothermally for a long time. To accommodate this, the glass transition temperature development was investigated and a diffusion-modified kinetic model was proposed to take into account the glass transition temperature. This diffusion-modified kinetic model showed that accounting for the diffusion control for cure temperatures below the glass transition was important to make reasonable predictions for long cure times. The model also showed that when curing at lower temperatures relative to the glass transition temperature this behaviour became more dominant, and thus the model was also found to be relevant at low cure temperatures. Based on the two models, an experimental validation was made by curing a neat resin panel and predicting in situ the degree of cure and glass transition temperature based on thermocouple data over time. The predictions were verified by a DSC sample cured simultaneously with the neat resin plate. The result of scanning the DSC sample immediately after the experiment showed that the value of the glass transition temperature, as midpoint values, was in good agreement with the diffusion kinetic model prediction.

Thus, this study has shown a relatively simple experimental procedure. The study demonstrated several techniques to fit the data measured with DSC and in the end, presented a diffusion-modified kinetic model able to predict the extent of cure in the studied thermoset with good results. Table 7 lists a set of recommendable parameters for future predictions of an epoxy thermoset for application in wind turbine blade manufacturing.

Table 7. Final parameters recommended for simulation of X and T_g of the resin system.

A [s^{-1}]	e_a [kJ/mol]	n [-]	m [-]	C [-]	X_{cT} [K^{-1}]	X_{c0} [-]	T_{g0} [°C]	ξ [-]	$T_{g\infty}$ [°C]
$2.50 \cdot 10^5$	56.24	1.83	0.41	43.7	$5.27 \cdot 10^{-3}$	-0.885	-42.0	0.487	89.0

Acknowledgments
Jesper Kjær Jørgensen acknowledges financial support from Horizon Europe, the European Union's Framework Programme for Research and Innovation, grant Agreement No. 101058054: Towards tURbine Blade production with zerO waste (TURBO). Furthermore, Anette Pedersen and Jonas Kreutzfeldt Heininge are acknowledged for their technical assistance with the experiments.

Data availability
The data used in this study is available [16]. The data given is time, temperature and measured degree of cure. Hence; t, T, X.

References
[1] Maduro M A M 2021 *Influence of Curing Cycle on the Build-up of Residual Stresses and the Effect on the Mechanical Performance of Fibre Composites* Ph.D. thesis DTU URL `https://orbit.dtu.dk/en/publications/influence-of-curing-cycle-on-the-build-up-of-residual-stresses-an`
[2] Schick C 2002 *Handbook of Thermal Analysis and Calorimetry* ed Cheng S Z D (Ohio: Elsevier) chap 16, pp 713–802
[3] McKenna G B and Simon S L 2002 *Handbook of Thermal Analysis and Calorimetry* ed Cheng S Z D (Ohio: Elsevier) chap 2, pp 49–104
[4] ISO Central Secretary 2020 Plastics - differential scanning calorimetry (dsc) - part 2: Determination of glass transition temperature and step height Standard (ISO 11357-2:2020) International Organization for Standardization Geneva, CH URL `https://www.iso.org/obp/ui/en/#iso:std:iso:11357:-2:ed-3:v1:en`
[5] Kamal M R and Sourour S 1973 *Polymer Engineering and Science* **13**(1) 59–64 ISSN 15482634
[6] Lee S, Chiu M and Lin H S 1992 *Polymer Engineering and Science* **32**(15) 1037–1046 ISSN 15482634
[7] Dibenedetto A T 1987 *Polymer Science: Part B: Polymer Physics* **25** 1949–1969
[8] Pascault J P and Williams R J 1990 *Journal of Polymer Science Part B: Polymer Physics* **28** 85–95
[9] Rabinowitch B E 1936 *Faraday Society* **32** 1225–1381
[10] Chern C S and Poehlein G W 1987 *Polymer Engineering and Science* **27**(11) 788–795 ISSN 15482634
[11] Cole K C, Hechler J J and Noel D 1991 *Macromolecules* **24** 3098–3110
[12] Hardis R, Jessop J L, Peters F E and Kessler M R 2013 *Composites Part A: Applied Science and Manufacturing* **49** 100–108 ISSN 1359835X
[13] Mikkelsen L, Jørgensen J, Mortensen U and Andersen T 2023 *Zenodo* [**Pre-print**] URL `http://doi.org/10.5281/zenodo.7868314`
[14] Kissinger H E 1956 *Helv. Chim. Acta* **28** 73–78
[15] Hale A 2002 *Handbook of Thermal Analysis and Calorimetry* ed Cheng S Z D (Ohio: Elsevier) chap 2, pp 49–104
[16] Jørgensen J K and Mikkelsen L P 2023 Isothermal and dynamic dsc dataset for an epoxy resin system for wind turbine blade applications URL `https://doi.org/10.5281/zenodo.8113838`

IOP Conf. Series: Materials Science and Engineering 1293 (2023) 012016 doi:10.1088/1757-899X/1293/1/012016

Defect detection in carbon fiber-reinforced composites using directional dark-field imaging and tomography

Azat Slyamov[1], Adriaan van Roosmalen[1], Mads A. Pedersen[1], Kenneth K. Nielsen[1], Ole V. Ferguson[2,3], Lars P. Mikkelsen[2], Erik Lauridsen[1]

1. Xnovo Technology ApS, 4600 Køge, Denmark
2. DTU Wind, Technical University of Denmark, 4000 Roskilde, Denmark
3. LM Wind Power, 6000 Kolding, Denmark

E-mail: elauridsen@xnovotech.com

Abstract. This paper proposes the use of circular X-ray grating interferometry as an effective technique for defect detection with potential applications for in-line inspection of carbon fiber-reinforced pultruded profiles used inside the load-carrying spar caps of wind turbine blades. A fuzzball defect in the pultruded profile is characterized as a demonstration. The method allows for large field-of-view quantification of local fiber alignment and relative fiber volume fraction. A two-dimensional through the thickness averaged distribution of the fiber orientation, the mean scattering, and fractional anisotropy are determined. Based on this, it is possible to determine the size of the defect as well as quantify the severity of the defect.

1. Introduction

Renewable energy technology, such as wind turbines, plays a critical role in reducing greenhouse gas emissions and combating climate change [1]. The efficiency of wind turbines depends largely on the design of their components, particularly the blades, which convert the kinetic energy of wind into mechanical energy [2]. One important component of wind turbine blades is the spar cap, which provides structural support and stiffness to the blade. Carbon fiber reinforced polymer (CFRP) is a widely-used material for manufacturing spar caps due to its high strength-to-weight ratio and fatigue resistance.

A new generation of wind turbine blades use spar caps manufactured from carbon fiber-reinforced pultruded profiles. These profiles are manufactured with fixed geometry and used as building blocks in the spar cap design. The pultrusion process involves a continuous resin impregnation of carbon fibers which is subsequently pulled through a die, resulting in a composite material with a tunable cross-sectional shape [3].

The orientation of the carbon fibers within the pultruded profile significantly affects the mechanical performance [4]. In particular, the alignment of the fibers with respect to the loading direction can greatly influence the strength and stiffness, as well as its fatigue resistance. Therefore, it is crucial to carefully control the fiber orientation during the pultrusion process to achieve the desired mechanical properties. This requires precise control of the fiber placement and orientation within the die, as well as optimization of the resin system to ensure complete impregnation of the fibers.

Common manufacturing defects in CFRPs include voids, porosities, and inconsistent fiber placement, which results in non-uniform mechanical properties. The pultrusion process is ideal for mitigating the occurrence and severity of these types of defects and thus yields more consistent properties. A more random and localized manufacturing defect in pultruded profiles includes fuzzballs, which are an accumulation of broken fibers in the carbon fiber yarn. The broken fibers form a small ball that is trapped in between the yarns and subsequently fed into the pultrusion die. Fuzzball defects can act as a foreign inclusion and thus affect the alignment of the surrounding fibers, which can have a detrimental effect on mechanical strength and, ultimately, the structural integrity of the wind turbine blade. While quality control measures are employed during the pultrusion process to minimize the occurrence of defects, their complete elimination remains a challenge. The use of advanced monitoring and inspection techniques is necessary to identify defects.

To characterize fiber misalignment in CFRP pultruded profiles, various techniques have been developed, including X-ray radiography, computed tomography (CT), ultrasonic testing, and optical microscopy [5]. Ultrasonic testing can be used to determine the orientation of fibers and the presence of delamination, but its accuracy is limited by the size of the fibers and the presence of other defects. Optical microscopy can provide high-resolution images of the composite material, but it is limited to surface analysis and cannot detect defects embedded within the material without preparation of a cross-section of the specimen. X-ray radiography and CT provide a non-destructive way to visualize the internal structure of composite materials and identify defects such as voids, resin-rich areas, and fiber misalignment. Conventional characterization of Fiber-Reinforced Composites (FRC)s using X-rays is performed in the so-called absorption mode. This approach requires sufficient contrast to differentiate between the composite matrix and reinforcement materials as well as direct resolution of individual microfibers, which drastically limits the imaged field-of-view (FOV). Radiography images and tomographic reconstructions of composite materials are usually followed by either segmentation of individual fibers or statistical analysis of the fiber orientations [6, 7]. In the meantime, X-ray dark-field or scattering mode enables accessing anisotropic microstructure due to the local small-angle X-ray scattering. Scattering from FRCs is not limited by the resolution of the imaging system as it occurs on multiple length-scales ranging from nanostructures up to individual fibers and fiber bundles.

Grating interferometry is one of the effective techniques to measure the dark-field signal from FRCs. Design of the interferometer can be chosen to optimize the sensitivity to a certain length scale without restricting the overall FOV. With a conventional interferometer system for compact X-ray systems consisting of a set of linear gratings, the same sample area is measured multiple times in order to retrieve the dark-field signal. Moreover, due to the dependence of the signal on the relative orientation of the grating lines to the alignment of the microstructure, the optical elements are usually rotated in plane to provide orientation sensitivity [8]. A novel design of the grating comprised of an array of circular unit cells allows single-shot measurement capabilities for omni-directional orientation sensitivity, significantly relaxing the instrumental complexity and total measurement time per image. This is particularly crucial for in-line inspection applications as well as for 3D tomographic fiber characterization that is obtained from a collection of 2D images [9, 10, 11, 12].

Here we present an application of directional dark-field imaging using circular gratings on industrial samples of 5 mm thick pultruded carbon fiber profile based on 7 µm thick carbon fibers used in the load carrying pultruded profiles in large wind turbine blades. We show the potential of this method for rapid and robust in-line inspection for sub-millimeter scale characterization of fiber misalignment and realtive fiber density in CFRP pultruded profiles over a wide area.

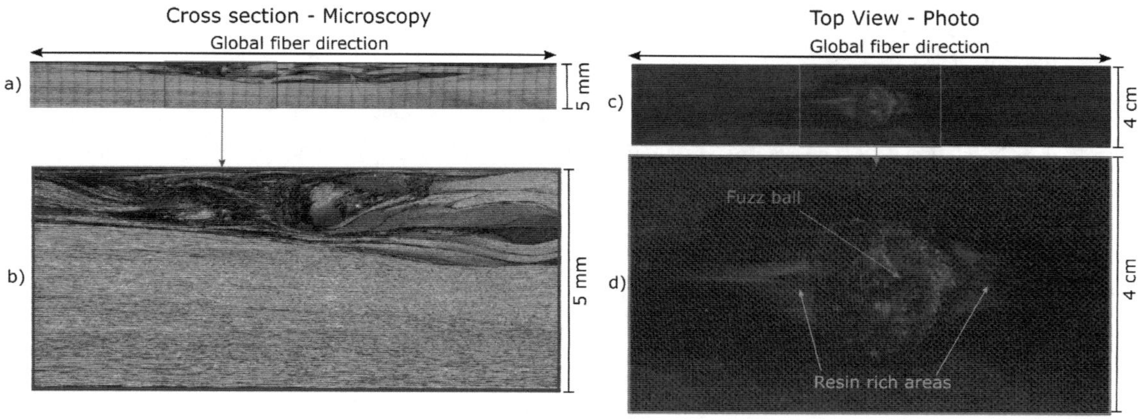

Figure 1. Carbon fiber-reinforced pultruded profile with a fuzzball defect at the surface. a) Microscopy image of fuzzball defect cross-section where carbon fibers appear white and the resin dark gray. Resin rich areas within the defect are black. b) Zoom of microscopy image with clear out-of-plane fiber misalignment w.r.t. the global fiber direction. c) Top view of 10 cm wide sample cut aimed for mechanical testing. d) Zoom of fuzzball defect with annotated defect characteristics.

2. Methods

2.1. Sample

The current study is based on a carbon fiber-reinforced polymer matrix pultruded profile, where a fuzzball defect has purposely been introduced in the pultrusion process line with the intention of studying the effect on mechanical properties. The fuzzball defect is manufactured by adding a ball of accumulated broken carbon fibers in between the carbon yarns in the pultrusion line. The fuzzball forces the surrounding carbon fibers to compact and misalign relative to the global fiber direction, ultimately changing the local fiber-matrix structure (Fig. 1.a and b). Fuzzballs can, in reality, occur in any position in the pultruded profile but have been placed near the surface in order to visually inspect the defect characteristics (Fig. 1.c and d). The fuzzball defect presented in this work is approximately 16 mm in diameter prior to manufacturing the pultruded profile. In reality, fuzzball defects can have any size and are only limited by the dimensions of the pultrusion die. Smaller fuzzballs may be hidden, while larger fuzzballs are typically visible on the surface.

2.2. Imaging setup

A schematic overview of the measurement setup at Xnovo Technology is shown in Fig. 2. A fiber-reinforced polymer component is positioned between the X-ray source and detector, and a circular grating optics is placed at a specific location depending on the desired image resolution. The probed FOV, length scale, and resolution of the setup can be adjusted accordingly. The unit cell size of the optical element determines the smallest pixel size of the reconstructed image, which is typically few orders of magnitude larger than the fiber diameter. For the experiment, a microfocus X-ray source was operated at 40 kV and 20 W. A large FOV CMOS detector with 50 µm pixel size was used to capture X-ray images. A circular grating optics prototype with 222.75 µm unit cell size was placed equidistantly between the source and detector providing a $6 \times 4\,\mathrm{cm}^2$ effective FOV. The sample was raster scanned (via single scan line) in the proximity of the grating via step-scan mode with 10 seconds per single FOV. To extract the contribution of the scattering signal from fibers, a reference image was captured at the beginning of the

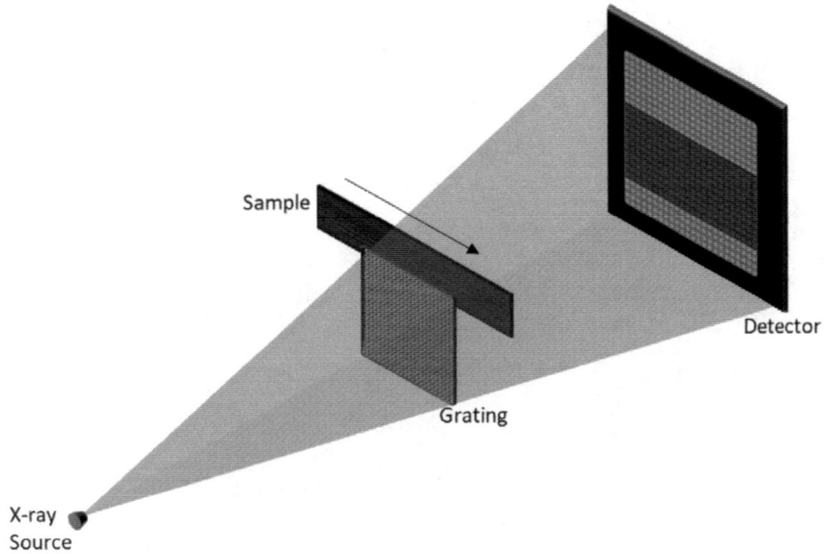

Figure 2. Schematic representation of the experimental setup for omni-directional dark-field imaging. The X-rays from the source illuminate the grating consisted of an array of circular unit cells. The sample is scanned at different positions along the scanning axis shown as black arrow. Reference grating fringes are measured by the detector prior to the placement of the sample under X-rays.

scan, followed by a series of images of the sample at different positions of the scan line. For comparison, a typical absorption contrast tomography of a CFRP sample with 1 µm pixel (voxel) size over $2 \times 2 \times 2\,\text{mm}^3$ volume takes between 10-20 hours of scanning time [6, 12].

2.3. Signal processing

A detailed description of the directional scattering signal retrieval algorithm is presented in [9, 10]. For each unit cell in the grating, the algorithm produces a 2D scattering distribution within angular bins uniformly spread over 180 degrees. The scattering distribution can be subsequently redefined by three eigenvectors v_1, v_2, v_3 and the corresponding eigenvalues $\lambda_1, \lambda_2, \lambda_3$ via principal component analysis. The eigenvector with the lowest eigenvalue is considered as the preferential local fiber orientation, because the scattering is weakest along the fiber orientation. The mean scattering (MS) can provide an estimate of the relative local density of fibers when defined as

$$\text{MS} = \frac{\lambda_1 + \lambda_2 + \lambda_3}{\max(\lambda_1 + \lambda_2 + \lambda_3)}. \tag{1}$$

where $\max(\lambda_1 + \lambda_2 + \lambda_3)$ indicate the volumetric maximum of the sum of the eigenvalues.

Fractional anisotropy (FA) is a metric commonly used in diffusion tensor imaging to quantify the degree of directionality, or anisotropy, of water diffusion within tissues, including those with fiber structures such as white matter in the brain. FA reflects the organization and coherence of the fibers, with higher FA values indicating greater directionality and alignment of the fibers. Thus, the more directional the fiber structures, the higher the FA value. The FA value is computed as [13]:

$$\mathrm{FA} = \sqrt{\frac{3}{2}} \frac{\sqrt{(\lambda_1 - \overline{\lambda})^2 + (\lambda_2 - \overline{\lambda})^2 + (\lambda_3 - \overline{\lambda})^2}}{\sqrt{\lambda_1^2 + \lambda_2^2 + \lambda_3^2}}, \tag{2}$$

where $\overline{\lambda} = (\lambda_1 + \lambda_2 + \lambda_3)/3$. Areas with a high volume fraction of well-aligned fibers will exhibit both high FA and high MS, whereas regions with a high volume fraction of randomly distributed fibers will have low FA but high MS. In the case of unidirectional CFRP pultruded profiles, local changes in MS and FA values are useful for indicating defects that alter the density and orientation of the fibers, respectively.

3. Results

The measured omni-directional fiber orientation data is visualized using a color wheel, where each color corresponds to an orientation within a half-circle, and the brightness represents the degree of anisotropy. The overview of the fiber orientation (see Fig. 3.a) reveals predominant unidirectional alignment across the majority of the pultruded profile. However, a distinct region is observed where the local fiber orientation deviates from the main orientation, indicating the presence of a fuzzball defect that extends approximately 20 mm along the pulling direction. To

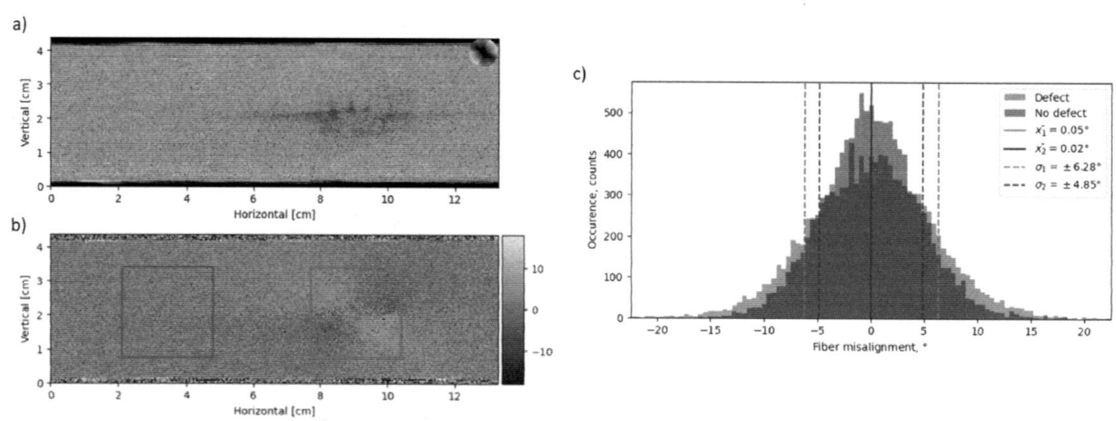

Figure 3. a) omni-directional fiber orientation, b) fiber orientation misalignment wrt pulling direction, c) histogram of fiber misalignment within boxes indicated in (b): around the fuzzball defect and non-defected area as red and blue boxes, respectively .

illustrate the impact of this defect on fiber misalignment, an angular deviation map is presented (see Fig. 3.b), showcasing the increased variation in fiber orientation around the defect. The mean value of fiber misalignment within this region (see Fig. 3.c) is close to zero indicating a symmetrical shape of the defect, while the standard deviation of fiber misalignment is larger than in the non-defect region.

In a conventional absorption contrast (Fig. 4.a), the fuzzball defect doesn't exhibit sufficient contrast relative to the rest of the profile due to the small difference in material density of fibers and resin. However, the map of MS (Fig. 4.b) provides more insights into the effect of the fuzzball on local fiber density. Notably, within the affected region, the MS values are elevated compared to the rest of the pultruded profile, suggesting an increased local fiber density. Simultaneously, the FA in the same area (Fig. 4.c) is lower, indicating a more randomly distributed local fiber orientation. Orthogonal line profiles across the defect area (Fig. 4.d-e) reveal dimensions of the defect for different data modalities. Both the mean scattering and fractional anisotropy exhibit higher definition of the defected area via line profiles than conventional absorption contrast. This can ultimately lead to more reliable metrological studies of fuzzball defects.

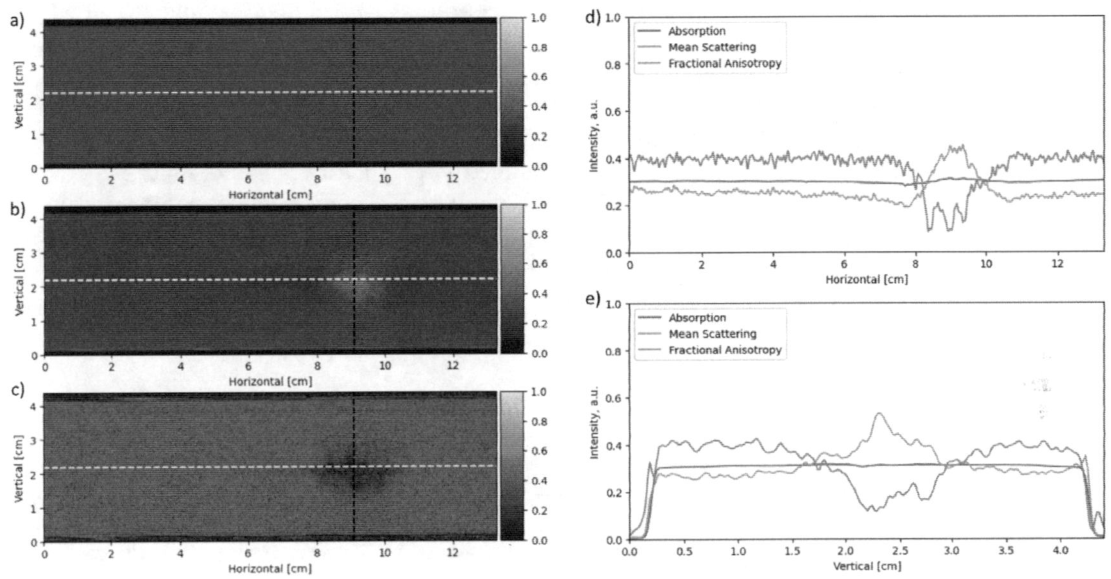

Figure 4. Maps of absorption (a), mean scattering (b) and fractional anisotropy (c). Horizontal (d) and vertical (e) line profiles indicated in (a, b, c) as white and black dashed lines, respectively.

To demonstrate the potential of using circular-grating based optics for in-line inspection of CFRP pultruded profiles, a fuzzball defect segmentation from mean scattering is shown in Fig. 5. Here, the segmented defect is overlaid with the greyscale mean scattering map.

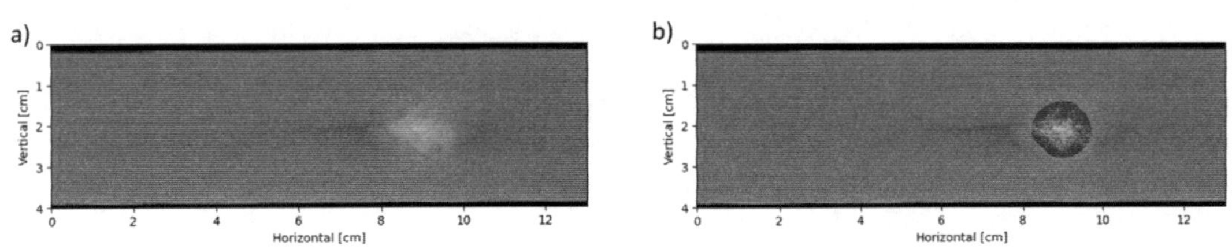

Figure 5. Example of defect segmentation based on dark-field image data: a) greyscale image of mean scattering, b) overlaid mask of the defect

4. Conclusion

The fiber orientation in carbon fiber-reinforced pultruded profiles plays a critical role in determining their mechanical properties and overall performance in wind turbine blades. Various factors during the manufacturing process, such as resin cure rate, temperature, and pulling speed, can affect fiber orientation and ultimately influence the performance of the composite material. Optimization of the pultrusion process and post-processing techniques can result in improved mechanical properties and ultimately lead to more efficient wind turbines, contributing to a more sustainable future. Detecting and minimizing defects that cause fiber misalignment, is also crucial for ensuring optimal performance of the spar cap. This paper demonstrates that advanced techniques such as dark-field X-ray microscopy using circular gratings yield promising results in rapid and reliable detection of defects and characterization of fiber misalignment.

The findings of this study have implications for the design and optimization of wind turbine blades. The fast and accurate characterization of fiber orientation enables the development of more precise numerical models and simulations, which can aid in predicting the mechanical behavior of the spar caps under different loading conditions. This, in turn, facilitates the design of blades with improved structural integrity, enhanced energy capture, and increased overall efficiency. It is important to note that while this study is focused on carbon fiber-reinforced pultruded profiles, similar principles and considerations apply to other manufacturing methods and composite structures. The role of fiber orientation in determining the performance of composites is a fundamental aspect across various industries, including aerospace, automotive, and marine applications.

Acknowledgments

Xnovo Technology acknowledges European Union EIC Accelerator grant agreement no. 946611. Ole V. Ferguson acknowledges grant no. 1044-00124B from the Innovation Fund Denmark. Lars P. Mikkelsen acknowledges European Union grant no. 101073040, MSCA Doctoral Network 2021, Reliance.

References

[1] Saidur R, Rahim N A, Islam M R and Solangi K H 2011 *Renewable and sustainable energy reviews* **15** 2423–2430
[2] Sarkar A and Behera D K 2012 *International Journal of Scientific and Research Publications* **2** 1–5
[3] Volk M, Yuksel O, Baran I, Hattel J H, Spangenberg J and Sandberg M 2022 *Composites Part B: Engineering* 110135
[4] Vedernikov A, Safonov A, Tucci F, Carlone P and Akhatov I 2020 *Journal of Composite Materials* **54** 4081–4117
[5] Márquez F P G and Chacón A M P 2020 *Renewable Energy* **161** 998–1010
[6] Jeppesen N, Mikkelsen L, Dahl A, Christensen A and Dahl V 2021 *Composites Part A: Applied Science and Manufacturing* **149** 106541
[7] Emerson M J, Jespersen K M, Dahl A B, Conradsen K and Mikkelsen L P 2017 *Composites Part A: Applied Science and Manufacturing* **97** 83–92
[8] Revol V, Plank B, Kaufmann R, Kastner J, Kottler C and Neels A 2013 *Ndt & E International* **58** 64–71
[9] Kagias M, Wang Z, Villanueva-Perez P, Jefimovs K and Stampanoni M 2016 *Physical Review Letters* **116** 093902
[10] Kagias M, Wang Z, Birkbak M E, Lauridsen E, Abis M, Lovric G, Jefimovs K and Stampanoni M 2019 *Nature communications* **10** 5130
[11] Kim J, Kagias M, Marone F and Stampanoni M 2020 *Applied Physics Letters* **116** 134102
[12] Kim J, Slyamov A, Lauridsen E, Birkbak M, Ramos T, Marone F, Andreasen J W, Stampanoni M and Kagias M 2022 *Composites Part B: Engineering* **233** 109634
[13] Basser P J and Pierpaoli C 2011 *Journal of magnetic resonance* **213** 560–570

IOP Conf. Series: Materials Science and Engineering 1293 (2023) 012017 doi:10.1088/1757-899X/1293/1/012017

Design of an element test specimen for fatigue delamination growth of a thick laminate

R. Erives, B.F. Sørensen, S.S. Samareh-Mousavi, A. Bangaru, X. Chen, X.-Y. Miao, M. McGugan

Technical University of Denmark, Department of Wind and Energy Systems
Risø campus, Frederiksborgvej 399, 4000 Roskilde, Denmark

riea@dtu.dk

Abstract. Following the observation of multiple delaminations growing from a tunnelling crack in the spar cap of a wind turbine rotor blade subjected to cyclic loading, an element test specimen is proposed. Design of the element test specimen to study fatigue delamination of a thick laminate is presented. Complexities inherent to a full structure e.g., curvatures, ply drops, manufacturing defects, etc. are removed to study the fatigue delamination damage mechanism. The element test specimen is made of unidirectional layers with two embedded artificial delamination cracks connected by a tunnelling crack. In this paper, analytical and numerical (2D) models are used to design the element test specimen.

1. Introduction

The current design tools and certification procedure of modern wind turbine blades rely heavily on testing at different scales. Numerous versions of the so-called "testing pyramid" are available in the literature [1] [2]. At the base of the pyramid there are the coupon tests needed to obtain material properties, in the middle there are typically sub-structures and/or sub-component testing specimens, while full-scale structure testing is at the top. The number of tests carried out at each of these blocks is decreasing as their complexity as well as the associated cost increase i.e., many coupon tests and a few full-scale tests. With the complexity of the specimens the statistical uncertainty also increases. This concept was developed in the aerospace industry, where tests are carried out bottom up with the final aim of ensuring structural integrity.

It has been noted that the full pyramid testing is very costly. A value of 300 million dollars has been estimated for the state-of-the-art carbon fibre epoxy composites used in aerospace [3]. Such costs are preventive for materials innovation. It has been proposed [3] that in the future, numerical models of structures ("digital twins") should be so accurate in the prediction of damage evolution that the models can be trusted, and thus full-scale testing and sub-structure testing can be avoided. To reach this state, it is essential that the high accuracy of predictive models is demonstrated convincingly. This brings in element test specimens. Then, the purpose of conducting element testing will be to study (to document) the evolution of at least two interacting damage types in generic specimens relevant for critical details in structures. In parallel, the damage evolution of the element test specimens should be predicted using materials laws (cohesive laws, crack growth laws, etc.) measured independently by materials testing.

By comparing the observed and predicted damage evolution, the accuracy of the numerical models can be assessed.

In the wind energy industry, it is a well-documented fact that wind turbine blades develop damage (e.g., delamination) during their lifecycle [4,5]. As such, it makes more sense to have a pyramid testing focused on monitoring stable damage growth which more closely resembles the real service life of a wind turbine. Moreover, this would enable the development and validation of predictive damage propagation models based on a given current damage state.

With our current tools for life service prediction not fully matured, full-scale testing of composite blades is still required to certify structural integrity. This is the closest representation of the real structure subject to realistic loads. However, these tests are expensive, lengthy, and complicated, which is why very few of these tests are carried out in the conventional test pyramid. Sub-structures allow the test of smaller sections of the blade, while sub-components deal with general design details such as beams, flanges, webs, and connections [6]. While these are simplifications from the full-scale tests, complexities inherent to structures, e.g., curvatures, ply-drops, and manufacturing defects remain in both (i.e., sub-structures and sub-components). These can lead to multiple damage mechanisms interacting with each other making it difficult to understand each damage mechanism. To address this, the idea of element tests has emerged with the intention of studying the evolution of interacting damage mechanisms in a wind turbine blade (e.g., spar cap delamination) allowing the investigation of scales larger than coupon tests (e.g., thick laminates), and to check the accuracy of predictive models.

Following the observation of delamination growth in the spar cap of a wind turbine rotor blade subjected to cyclic loading [7], an element test specimen is proposed here to investigate the delamination crack growth starting from a tunnelling crack in thick laminates under cyclic loading. The design of such an element subject to fatigue delamination is outlined in detail. The current study involves the design of the element test using numerical and analytical tools.

The present article is organized as follows; an introduction with a motivation for the study is presented in section 1 followed by the problem definition in section 2. Analytical and numerical (using FEM) approaches are presented in section 3. The results are given in section 4 followed by the discussion and conclusion in section 5 and 6 respectively.

2. Problem definition

The design of the present element test specimen comes after the observation of delamination growth in the spar cap of a wind turbine rotor blade subject to cyclic loading [7]. In the study, an artificial defect in the form of a tunnelling crack and two delaminations was embedded into the spar cap of the blade, consisting of a central layer surrounded by two additional layers on each side. The blade was tested under flap-wise cyclic bending induced by transverse force with a maximum strain range of 3400 $\mu\varepsilon$ at a load ratio of approximately $R = -1$. Under the cyclic loading, fatigue damage in the form of two delaminations evolved from the embedded defect. The delamination crack growth was monitored via visual inspection, acoustic emission sensors, and surface thermography techniques.

In the present study we are designing an element test specimen that enables documentation of the damage growth for two interacting damage types, a central tunnelling crack and two delaminations under cyclic loading, the damage types observed in the blade [7]. The element test specimen proposed here is essentially a 2D specimen that should be relative cheap to manufacture. The tunnelling crack and starter cracks for two delaminations will be made by embedding a slip foil during specimen manufacturing. A schematic of the geometry of the element test specimen is shown in Figure 1. Unlike the full-blade test [7], the present element test specimen has been designed for tension-tension loading with a loading ratio of $R = 0.1$ to avoid buckling since the element test specimen will be unsupported.

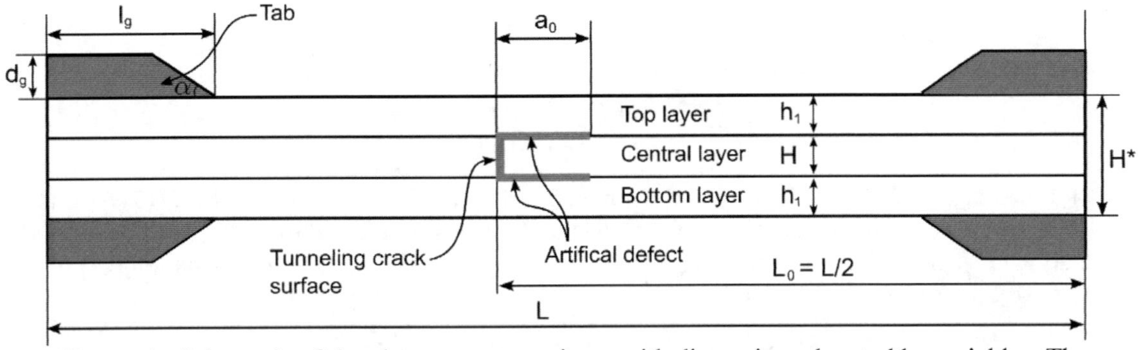

Figure 1. Schematic of the element test specimen with dimensions denoted by variables. The artificial defects are shown with the thick red lines.

3. Modelling

In the following, we present two models; an analytical model for the prediction of the steady-state energy release rate for long cracks, disregarding friction, and a numerical finite element model (FEM), which can calculate the energy release rate for short cracks. The effect of uneven crack lengths and friction between the crack faces can be assessed with the numerical model.

3.1. Analytical model

An analytical model can be established to calculate the path-independent J-integral around the fracture process zone. The problem analysed, depicted in Figure 2, is a symmetric multilayer where the central layer has a stress-free surface (the tunnelling crack) and is delaminated from the remaining layers. The central layer has a Youngs modulus in the x_1-direction, E_0, and a thickness, H. Two layers, placed outside the central layer, have a Young's modulus in the x_1-direction E_1 and E_2 and a thickness h_1 and h_2, respectively. When their crack length exceeds a few times the layer thickness, H, the problem becomes a steady-state problem that can be analysed analytically by the J-integral. Then, the strain component ε_{11} in all layers will be uniform (i.e., independent of x_1 and x_2) far ahead of and far below the fracture process zone and will be denoted ε^+ and ε^-, respectively.

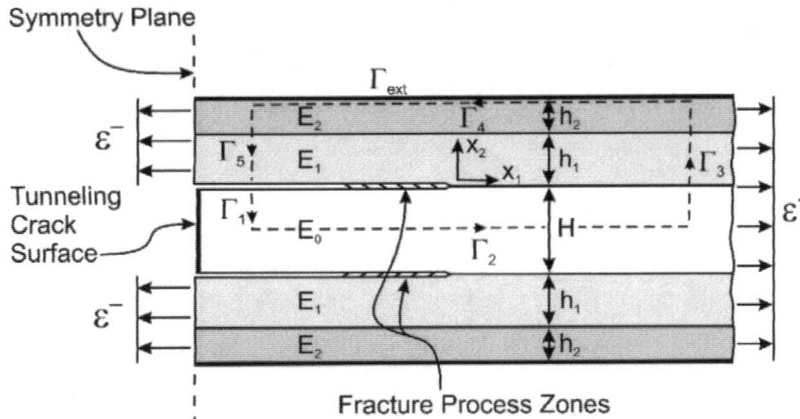

Figure 2. The problem analysed by the analytical model: Delamination of a middle layer in a symmetric five-layer structure. The applied strain in the uncracked part of the structure is ε^+.

Introducing the following non-dimensional parameters:

$$\Sigma_1 = \frac{E_1}{E_0} \qquad \Sigma_2 = \frac{E_2}{E_0} \qquad \eta_1 = \frac{h_1}{H} \qquad \eta_2 = \frac{h_2}{H}, \tag{1}$$

the strain far below the fracture process zone is found via force balance to be (plane stress):

$$\varepsilon^- = \frac{[1 + 2(\Sigma_1\eta_1 + \Sigma_2\eta_2)]}{2(\Sigma_1\eta_1 + \Sigma_2\eta_2)}\varepsilon^+. \tag{2}$$

For plane strain, the Young's modulus, E, of each layer should be replaced with $E/(1 - \nu_{13}\nu_{31})$ where ν_{13} and ν_{31} are the Poisson's ratios for each layer (here, subscripts indicates materials directions). We analyse the upper half of the problem (with only one fracture process zone) by the path-independent J-integral [8], defined as

$$J = \int_\Gamma W\,dx_2 - \sigma_{ij}n_j\frac{\partial u_i}{\partial x_1}\,dS\,, \tag{3}$$

where W is the strain energy density, σ_{ij} is the stress tensor, u_i is the displacement vector, and n_j is the outwards unit normal vector to the integration path, Γ, that runs from the lower traction-free crack surface to upper crack face in a counter-clockwise direction. Eq. (3) is written in index notation (i.e., indices i and j takes the values 1, 2, 3, and the Einstein summation rule applies for repeated indices). The integration path for the half-problem, Γ_{ext}, is chosen along the external boundaries, far ahead and far below the fracture process zone and along the mid-plane. This is advantageous since far away from the fracture process zone, the stress state (plane stress) is uniaxial stress, only stress component σ_{11} is non-zero. The integration path Γ_{ext} is split into five straight line segments, denoted Γ_1, Γ_2, Γ_3, Γ_4, and Γ_5, as shown in Figure 2.

Far below the fracture process zone the middle layer is stress free, $\sigma_{ij} = 0$. Then, along Γ_1, $W = 0$, and with $\sigma_{ij} = 0$ we get $J_1 = 0$ (here, subscript indicates integration path). Along Γ_2, $dS = dx_1$ so that $dx_2 = 0$ and with $n_j = \{0, -1, 0\}$, we get $\sigma_{ij}n_j\,\partial u_i/\partial x_1 = -\sigma_{i2}\,\partial u_i/\partial x_1 = 0$, since (for $i = 1$) $\sigma_{12} = 0$ and (for $i = 2$) $\partial u_2/\partial x_1 = 0$ (Γ_2 is a symmetry plane), so that $J_2 = 0$. Along Γ_3, we have $W = E/2\ (\varepsilon^+)^2$ (with E being the relevant Young's modulus), $dS = dx_2$, $n_j = \{1, 0, 0\}$, $\sigma_{11} = E\varepsilon^+$ and $\partial u_1/\partial x_1 = \varepsilon_{11} = \varepsilon^+$ so we get

$$J_3 = -\frac{(\varepsilon^+)^2}{2}\left\{\int_{-\frac{H}{2}}^{0} E_0 dx_2 + \int_0^{h_1} E_1 dx_2 + \int_{h_1}^{h_1+h_2} E_2 dx_2\right\}, \tag{4}$$

or

$$J_3 = -\frac{E_0 H}{4}[1 + 2(\Sigma_1\eta_1 + \Sigma_2\eta_2)](\varepsilon^+)^2. \tag{5}$$

Along Γ_4, $dx_2 = 0$ and with $n_j = \{0, 1, 0\}$, so that $\sigma_{ij}n_j\,\partial u_i/\partial x_1 = \sigma_{i2}\,\partial u_i/\partial x_1 = 0$ ($\sigma_{12} = \sigma_{22} = 0$ since Γ_4 is a traction-free surface), we get $J_4 = 0$. Along Γ_5, the J-integral contribution becomes:

$$J_5 = \frac{E_0 H}{8}\frac{[1 + 2(\Sigma_1\eta_1 + \Sigma_2\eta_2)]^2}{(\Sigma_1\eta_1 + \Sigma_2\eta_2)}(\varepsilon^+)^2. \tag{6}$$

Now summing the J-integral contributions, we get

$$J_{ext} = J_3 + J_5 = \frac{E_0 H}{8}\frac{[1 + 2(\Sigma_1\eta_1 + \Sigma_2\eta_2)]}{(\Sigma_1\eta_1 + \Sigma_2\eta_2)}(\varepsilon^+)^2. \tag{7}$$

The analytical model presented above gives values of the J integral for the steady-state situation only. The steady-state situation is anticipated for long cracks. Since we wish to design our specimen such that the cracking will be in the steady-state situation, we will conduct finite element modelling (FEM) with different crack length to explore how long the initial crack lengths must be to ensure that the subsequent cracking occurs in steady-state.

3.2. Finite Element model

A 2D finite element model of the element test specimen was created in the commercial finite element program ABAQUS. The model was created with total length $L = 1800$ mm, and variable total thickness $H^* = 6$ mm - 30 mm, and variable H and h_1 between 2 mm and 10 mm. Figure 1 and Table 1. Mechanical properties of unidirectional layers manufactured from glass fabric and epoxy were assigned to the geometry as linear elastic orthotropic with transversely isotropy with $E_1 = 42$ GPa, $E_2 = 12.6$ GPa, $\nu_{12} = 0.3$, and $G_{12} = 3.89$ GPa. The left edge of the model is fixed in the x_1-direction (u_1), while a tensile force is applied to the right edge which induces με all over the model and far away from the crack tips. Vertical displacement (u_2) of the middle nodes in left and right edges are constrained to prevent rigid body displacements and convergence difficulties. Figure 3 shows the applied boundary conditions schematically.

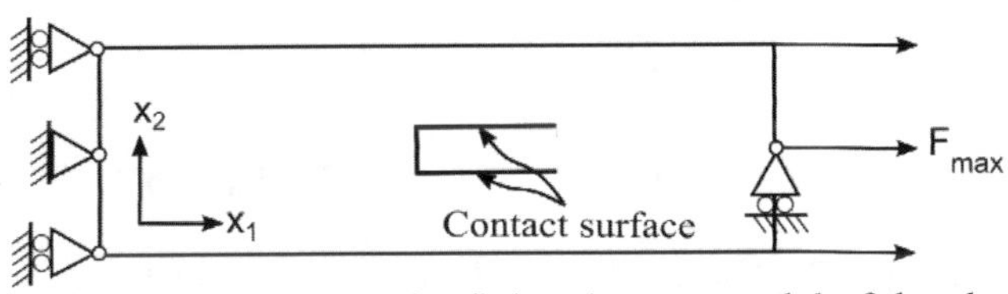

Figure 3. Boundary condition applied to the finite element model of the element test specimen.

A continuous seam, which is a region in the model that can open during the analysis, was considered for two crack surfaces and the connecting tunnelling crack. At seam-assigned lines, two overlapping nodes are created which allows the creation of two surfaces. Contact interaction was considered between crack surfaces to prevent surface interpenetration and simulate coulomb friction. Hard contact in normal direction and different values of friction coefficient, μ, was defined in the tangential direction for the contact properties. 8-node quadratic plane stress elements (S8) were used for the model. The model was partitioned in two circles around crack tips and an oriented seed was used to generate a gradually finer mesh in the radial direction to the crack tips. In addition, to meet the sharp crack tip condition and create stress singularity, the elements located at the crack tip collapsed and the middle node of elements moved closer to the collapsed nodes. A mesh sensitivity analysis was done to find the maximum allowed element size in the whole model and around the crack tip. The length of the smallest element around the crack tip is $0.002H^*$ to $0.01H^*$ in all FE models. Figure 4 shows a zoomed view of the mesh around the crack tips. J-integral for the cracks were calculated through 10 paths. Three paths close to the crack tip were inconsistent due to stress singularity, while for farther paths the J-integral converges to constant values. The converged values will be compared with the analytical model for further analyses.

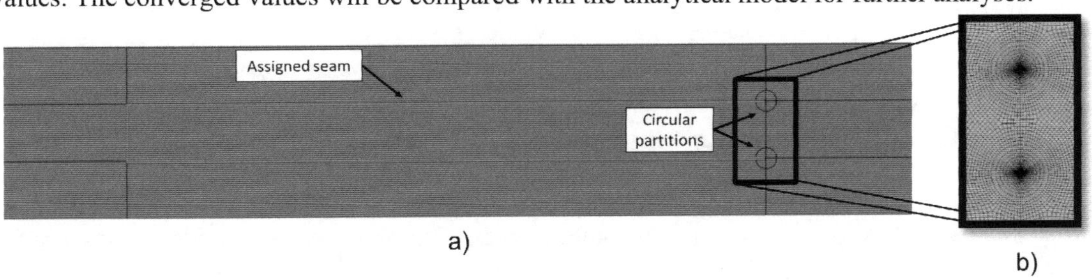

Figure 4. Crack modelling a) assigned seam on crack surfaces, b) generated mesh

4. Results

4.1. Analytical and FEM results

The steady-state energy release rate from the analytical model is compared with the J-integral obtained from FEM results for equal crack lengths ($a_1 = a_2 = 120$ mm) in Figure 5. It is seen the J-integral decreases with increasing h_1 and increases with increasing H. It can be noted that the FEM simulation values are very close to the analytical values with a maximum difference of less than 2%. The J-integral results from the FEM are plotted as a function of the crack length (equal crack lengths, $a_1 = a_2$) in Figure 6. For small crack lengths, the energy release rate is high, with increasing crack length the energy release rate decreases to an asymptotic value, the steady-state energy release rate, J_{SS}. It is seen that a crack length is needed to ensure a steady state has been attained for the three different H values shown in Figure 6. Moreover, the results from the FEM showed that the Mode I stress intensity factor was vanishingly small in comparison with the Mode II stress intensity factor, indicating that the delamination takes place under pure Mode II.

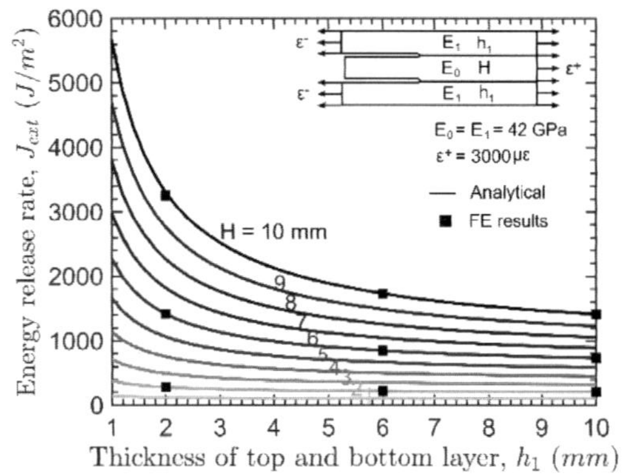

Figure 5. Steady-state energy release rate plotted for different values of top layer thickness, h_1. The square markets are the FEM values, and the curves are the analytical results.

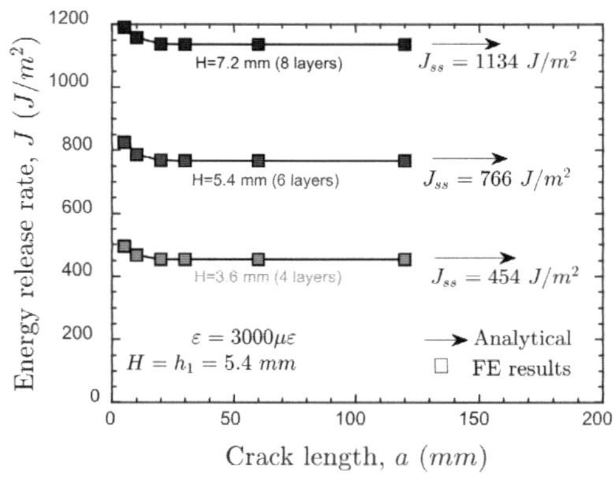

Figure 6. Energy release rate from the FEM analysis plotted as a function of the crack length (equal crack lengths, $a_1 = a_2$) for different ply thickness values showed with square markers.

4.2. Design methodology of the element test specimen

In this section, we will describe the procedure followed to determine the dimensions of the element test specimen shown in Figure 1.

1. The length of the element test specimen is determined based on the maximum distance between the grips of the test machine. In this case, we chose $L = 1300$ mm.
2. The maximum load, F_{max}, of the test machine is 100 kN and the design strain is 3000 με, assuming a longitudinal modulus (E_1) of a unidirectional glass-epoxy laminate of 42 GPa, we obtain the maximum cross-section area, $A = F_{max}/(E_1\varepsilon^+)$, as 794 mm^2. Next, to have a larger width than thickness, we set the width as $w = 50$ mm, which then results in a maximum total thickness of H^*_{max}, of 15.8 mm.
3. We aim for a maximum crack growth rate $(da/dN)_{max}$ of 10^{-2} mm/cycle. Using data from the literature (Fig. 6(d) of [9]), the J-integral of a glass-epoxy laminate loaded in Mode II for the targeted crack growth rate would be approximately $J_{max} = 1000$ J/m^2.
4. It was decided to omit the outer (top and bottom) layers depicted in Figure 2 since bi-axial layers are normally used, which are prone to off-axis tunnelling cracks that may obstruct the visualisation of the main delaminations.
5. Different combinations of H and h_1 can produce J values of 1000 J/m^2 (See Figure 7). However, with a UD layer thickness of 0.9 mm [10], both H and h_1 must be multiples of 0.9 mm. Here we have

chosen $H = 5.4$ mm (6 UD layers) and $h_1 = 3.6$ mm (4 UD layers). Now, the total thickness of the element specimen is smaller than the maximum total thickness, $H^* = H + 2h_1 < H^*_{max}$.

6. The initial length of the artificial defect a_0 is determined such that the crack growth will be in steady state. From FE analysis (Figure 6), it can be observed that an a_0 of 60 mm ensures the crack growth is in steady state.

7. The position of initial crack (slip foil) is chosen to be at the centre of the test specimen, $L_0 = L/2$, since it was found that during the blade test [7], cracks grew in both directions from the slip foil.

Table 1 summarizes the chosen dimensions of the element test specimen.

Table 1. Element specimen geometry design parameters [a].

	Symbol	value	Unit
Initial delamination length	a_0	60	mm
Central layer	H	5.4	mm
Top and bottom layer thickness	h_1	3.6	mm

[a] Additional geometry parameters are $L = 1300$ mm, $d_g = 2$ mm, $l_g = 150$ mm, $\alpha = 2°$, $w = 50$ mm

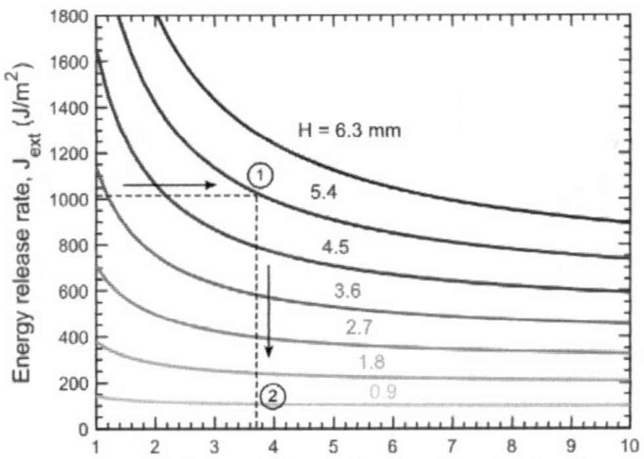

Figure 7. J-integral is shown as a function of top and bottom layers thickness for different central layer thicknesses. The steps for determining H and h_1 for $J_{max} = 1000$ J/m² are indicated with the dashed lines.

4.3. Issues disregarded in design process: different crack lengths and friction between crack faces

The previous figures are obtained for a frictionless, symmetric configuration (i.e., top and bottom cracks with equal initial lengths, $a_{0-1} = a_{0-2}$). An asymmetric configuration with a fixed bottom crack length of 120 mm, and a varying top crack was modelled using FEM. The energy release rate obtained from such an asymmetric configuration is shown in Figure 8. The top crack, which is shorter, has a higher energy release rate, while the bottom crack (the longer crack) has a lower energy release rate. This means that for an asymmetric configuration, under cyclic loading, the shorter crack would grow faster until they reach the same length i.e., approaching the situation of equal crack lengths.

Friction between the crack surfaces is not considered in either of the two plots shown above. However, in Figure 9, it can be observed that friction influences the computed energy release rate. In Figure 9, the J-integral computed from FEM is plotted using different values of friction coefficient, μ. The inclusion of friction in the FEM model results in lower energy release rate as it can be expected. In the case of glass/epoxy composites, where friction coefficient for dry conditions and low normal forces typically ranges between 0.2 and 0.5 [11], a reduction of 10-30% of the J-integral is expected.

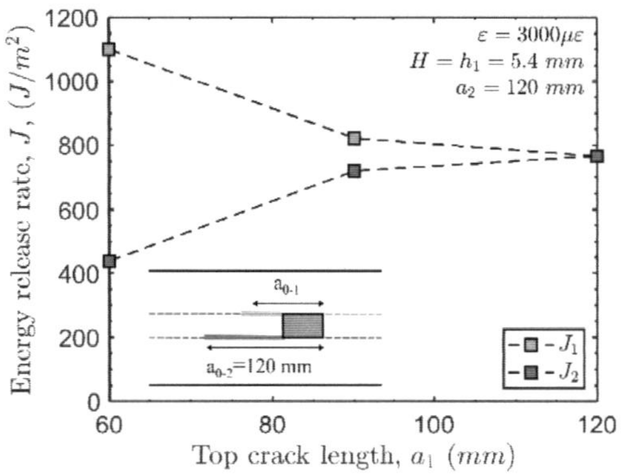

Figure 8. J-integral plotted against the crack length with the top and bottom cracks marked with green and purple squares respectively.

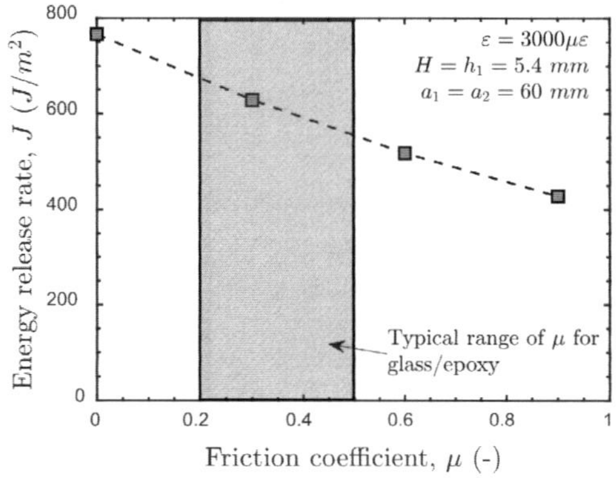

Figure 9. J-integral plotted against the friction coefficient.

5. Discussion

5.1. Limitations of proposed element test specimen and predictive models

As mentioned, the analytical model is restricted to steady-state cracking, i.e., for long cracks. The model assumes symmetry about the horizontal midplane, and thus equal crack lengths. However, since the steady-state result is independent of crack length, it is anticipated that the analytical result would correspond to the average value of the energy release rate of the two cracks if the two-crack length differed, as indicate in Figure 8.

It is appropriate to discuss the differences between the element specimen and the situation of testing a full blade with similar, designed cracks. In comparison with a full blade, element testing offers several advantages. First, it is possible to observe the crack at both faces of the element test specimen; for a full blade, the crack can be seen from the outside, but it might be difficult to see the crack inside a full blade. Secondly, there are limitations on how high the blade can be loaded – a higher load could induce other undesired damage types. An element test specimen can be loaded higher. Thirdly, the manufacturing, instrumentation and testing of a full blade is several orders of magnitude larger than for an element test specimen. It is therefore possible to manufacture and test several nominally identical element test specimens, and get data for specimen-to-specimen variation or to explore effects of one or more geometric parameters. A drawback of the present element specimen is that it is restricted to tension loading only – a compression loading could cause buckling. A full blade includes one or more shear webs that prevents buckling, and therefore damage in a main spar can undergo tension-compression loads. It is, however, also unclear what crack growth data there can be used for a crack tip that undergoes reverse loading (i.e., the crack tip goes from positive to negative shear stresses at the crack tip).

Friction between crack faces is not considered in the analytical model. Without friction, the J-integral value evaluated around the external boundaries, J_{ext}, will be equal to the energy release rate of the crack tip. In case of friction, stress would be transferred across the crack faces, unloading the crack tip, as shown in Figure 9. In the analytical model, friction could approximately be modelled as a constant frictional shear stress (as opposed to the Coulomb friction modelled in FEM), see e.g., [12]. Note that disregarding friction provides with conservative predictions, as friction further contributes to energy dissipation as mentioned previously.

The FEM model relies on linear elastic fracture mechanics, which assumes a small-scale fracture process zone. This means that fibre bridging is not considered. However, if there were significant fibre bridging it would have an effect decreasing the cyclic growth rate [12,13]. The analytical model could model crack bridging in a simple way, e.g., as a constant shear stress acting along the bridging zone [12]. Furthermore, the present models only consider the choice of literature data for the cyclic crack growth rate (using Mode II data).

The present element test is designed under tension-tension cyclic load since the analytical model cannot handle compression loads. In addition, a specimen under uniaxial compression loading is vulnerable to undesired fibre micro buckling and global buckling, which needs extra constraints to be prevented. As mentioned, the proposed element test specimen is robust towards different crack lengths in that a shorter crack would be loaded higher and would thus tend to grow faster and catch up with the longer crack under cyclic loadings.

5.2. Prognosis and perspective

It is clear from the previous work that delamination is highly dependent on geometry, e.g., depth and thickness. Damages identified during inspection of wind turbine blades are assessed based on their perceived risk to the remaining operational requirement of the turbine. Information about the damage (e.g., geometry), provided by visual inspection alone often needs to be supplemented with sub-surface details provided by ultrasonic scans if a confident assessment is to be made. Highly accurate automated prognostic tools will be required within the digital twin representations of very large offshore wind turbine blades in the future [15].

The work described in this paper suggests the development of a damage predictive framework where measured crack growth rate parameters from fracture mechanics tests can be used as input to models that predict the delamination crack length as a function of the number of cycles in the element specimen. Such an approach can provide the validated test data necessary for automated prognostic assessment of damages, and thus service life, in composite structures based on known geometries and parameters.

6. Concluding remarks

The major outcomes from the present study are the following:

- Establish a design methodology for element test specimen to study delamination crack growth initiated from a tunnelling crack. This methodology could be generalised to other element test specimens intended to study other damage types.
- The designed element specimen is a steady state specimen, which means that the energy release rate is independent of the crack length.
- The designed element test specimen can be used to check the accuracy for predicting delamination crack growth.
- The designed element test specimen is robust in terms of handling different lengths of top and bottom delamination crack lengths.

References

[1] R. Wanhill, 2018, "Fatigue requirements for aircraft structures," *Aircraft sustainment and repair*, Butterworth-Heinemann, 17-40.

[2] MIL-HDBK-17-1F, 2002, "Composite Materials Handbook: Volume 1. Polymer Matrix Composites Guidelines for Characterization of Structural Materials," *U.S Department of Defense.*

[3] Douglas S. Cairns, 2023, Private communication, Department of Mechanical and Industrial Engineering, Montana State University.

[4] K. Branner, A. Ghadirian, 2014, "Database about blade faults," DTU Wind Energy.

[5] S. Ataya, M.M. Ahmed, 2013, "Damages of wind turbine blade trailing edge: Forms, location, and root causes," *Eng. Fail. Anal,* 35, 480-488.

[6] Laustsen, S., Lund, E., Kühlmeier, L. and Thomsen, O.T., 2014, "Development of a High-fidelity Experimental Substructure Test Rig for Grid-scored Sandwich Panels in Wind Turbine Blades," *Strain,* 50(2), 111-131.

[7] S. Mousavi, X. Chen, Semenov S. P., McGugan M. and K. Branner, 2023, "Fatigue delamination growth identification in a wind turbine blade with embedded delamination defect using thermography and acoustic emission,".

[8] J. Rice, 1968, "A path independent integral and the approximate analysis of strain concentration by notches and cracks," *Journal of Applied Mechanics,*379-388.

[9] M. Kenane, 2009, "Delamination Growth In Unidirectional Glass/Epoxy Composite Under Static And Fatigue Loads," *Physics Procedia,* 2(3),1195-1203.

[10] R. Erives, B. F. Sørensen and S. Goutianos, 2023, "Extraction of mix-mode cohesive laws of composites undergoing delamination with large scale bridging," *Composites Part A: Applied Science and Manufacturing,* 165.

[11] J. Alotaibi, A. Eid Alajmi, G. Mehoub and B. Yousif, 2021, "Epoxy and Polyester Composites' Characteristics under Tribological Loading Conditions," *Polymers,* 13(2230).

[12] B. F. Sørensen, 2020, "Microscale testing and modelling for damage tolerant composite materials and structures," *IOP Conference Series: Materials Science and Engineering,* 942(012004).

[13] J. Holmes, L. Liu, Sørensen, B.F. and S. Wahlgren, 2014, "Experimental approach for mixed-mode fatigue delamination crack growth with large-scale bridging in polymer composites," *Journal of Composite Materials,* 48(25), 3111-3128.

[14] Yao, L., Cui, H., Sun, Y., Guo, L., Chen, X., Zhao, M., Alderliesten, R.C., 2018, "Fibre-bridged fatigue delamination in multidirectional composite laminates," *Composites Part A: Applied Science and Manufacturing,* 115, 175-186.

[15] K. Branner, M. A. Eder, H. K. Danielsen, X. Chen and M. McGugan, 2021, "Towards more smart, efficient and reliable wind-turbine structures," *DTU International Energy Report 2021: Perspectives on Wind Energy,* 115-124.

Development of wind turbine blade recycling baselines in the United States

Matthew Korey[a], Evan Sproul[b], Mitchell L. Rencheck[a], Brandon L. Ennis[b]

a. Manufacturing Science Division, Oak Ridge National Laboratory, Oak Ridge, TN 37831
b. Sandia National Laboratories, P.O. Box 5800, Albuquerque, NM 87185

Corresponding author's e-mail address: koreym@ornl.gov

Abstract. Over the past several years, the wind energy industry has received scrutiny regarding wind turbine blade (WTB) recycling due to the landfilling of WTBs caused by a lack of industrially viable recycling solutions. The amount of WTBs that will need to be recycled is set to increase in the United States as the deployment of wind energy is expected to rapidly grow to meet the nation's energy goals by 2035. While significant progress has been made worldwide, it is still unclear which WTB recycling solutions would be the most cost and energy effective within the United States for the existing fleet of wind turbines. To guide researchers and industry with a clear path forward, a range of options for WTB recycling in the United States are modeled through development of baseline scenarios and the use of formal life cycle assessment (LCA). Model data have been collected through literature review, industry engagement, and expert opinion regarding current end of life practices and considerations surrounding equipment, labor, and logistics. A detailed baseline for WTB decommissioning processes has been developed and used to assess alternative approaches, such as on-site shredding to compare the impacts on greenhouse gas (GHG) emissions. The developed LCA model and baseline scenarios for WTB recycling is used to assess the current WTB decommissioning practices in the United States along with emerging recycling pathways, including cement kiln co-processing and pyrolysis. Initial findings indicate that there are different approaches to decommissioning WTBs in the United States, each of which has unique implications for recycling. In light of this finding, additional results from the modeling will be used to better understand decommissioning practices and assist in making educated decisions on recycling pathways for the future. Throughout the analysis, focus was given to where international efforts might differ from the United States. WTB recycling is occurring worldwide, and different countries have different drivers for creating markets for recycled WTB materials. The contrasts and similarities between the United States and other countries offer insight to areas of opportunity that the United States could investigate and areas that can be readily transferred from existing solutions. By modeling and characterizing the current decommissioning practices and potential recycling solutions for the United States, a clearer vision will be created for pathways forward as to how to handle end of life WTBs to enable more efficient and cost-effective opportunities for material recovery from end-of-life WTBs.

1. Introduction

to grow by 50% by 2050 amounting to 45 trillion kWh [1]. In the U.S. alone, electrification of heating and the transportation industries alone will cause around a 50% increase in electricity demand by 2050. Growing energy demand is significant where, in the US in 2019, a quarter of all greenhouse gas (GHG) emissions were from the burning of fossil fuels for electricity alone [2]. Climate scientists overwhelmingly agree that the burning of fossil fuels, producing GHGs, is significantly contributing to the growing environmental crisis associated with global warming [3]. Even so, our global society is becoming ever more dependent on energy to meet basic human needs, such as health, food, mobility, and communication. To combat this phenomenon, there has been a tremendous global effort to utilize renewable energy sources – such as solar energy, geothermal energy, hydropower, wind energy, and many others – which can provide energy without simultaneously generating GHGs as with fossil fuels [4]. Of these, wind energy is the largest source of renewable power in the U.S., is one of the fastest growing sources of renewable power (15% growth per year), is showing tremendous job growth, and is one of the most cost-effective sectors (0.01-0.02 USD/kWh) [5, 6]. The United States already has over 85,000 wind turbines in 50 states and its offshore wind industry is projected to grow rapidly [7].

While wind turbines can generate electricity during their functional use without the use of GHG emitting fuels, they create an additional challenge at end of life. Wind turbines must be decommissioned after their 20–30-year life[8, 9] and many reach their end of life sooner due to financial incentives, defects in manufacturing, and damage from operation in the field [10-14]. With increasing climate concerns and need for sustainability, the industry is turning towards new technologies with net zero environmental impacts and developing circular economies of domestic materials. Today, a wind turbine is made up of the blades/rotor, nacelle and drivetrain, and tower. Most of the turbine is recyclable other than the blades/rotor, which include a mix of fiberglass, balsa wood, polymer resins, carbon fiber, and metals from the lightning protection system [9]. Currently, in the U.S. wind turbine blades are most commonly landfilled after being decommissioned [8, 15]. According to a recent study by Cooperman et al. published in 2020, assuming a 20-year turbine lifetime, the cumulative blade waste in 2050 is approximately 2.2 million tons [8]. Cooperman et al. reviewed several processes to recover energy and materials from wind blades, finding that none are at cost-parity with landfilling – the current cost of disposing of blades in large segments or through grinding was relatively low compared to life cycle cost of energy. Ultimately, step-changes were needed in terms of recycling technology, new blade material and design, in addition to policy to achieve true circular economy of wind blades.

There are several recycling technologies that have been developed which are capable of processing the materials within wind turbine blades. These include mechanical recycling [16-18], thermal or thermochemical recycling [19-23], chemical or solvent recycling [24-26], and enzymatic recycling [27-29]. While these technologies all exist, not all are available at scales relevant for the wind energy industry. Further, there is yet to be a robust analysis of the benefits/limitations of each technology along with life cycle assessment (LCA) to compare the technologies side-by-side within the existing infrastructure available in the U.S. for plastics recycling. In previous work by Sproul et al. similar analysis was performed for future recycling technologies including chemical recycling and microwave pyrolysis, with little attention paid to the decommissioning pathway utilized. However, in this work, the team strives to establish a set of robust technical baselines for current end of life processing technologies for wind turbine blades in the U.S. available at scales capable of processing wind turbine blade waste including the decommissioning step within the analysis. Recycling technologies included within this analysis include landfilling, cement kiln co-processing, pyrolysis, and mechanical recycling all of which are available and can process wind turbine blade waste in the U.S. today.

2. Methods

Data in the results section of this work were collected through U.S. energy industry stakeholder engagement and interviews. Stakeholders throughout the supply chain were involved in this process. This included manufacturers, owner/operators, local stakeholders, policymakers, decommissioners, recyclers, repurposers, and others from both the distributed or "small" and utility wind industries.

Stakeholders were engaged in regular meetings throughout the project to ensure analysis and results were in-line with industrial perspectives. Their thoughts and input were summarized in the narrative below.

Process model methods were developed using the procedure outlined by Sproul et al [30]. The model system used in this study was a wind turbine in Texas, U.S. which had all the metal removed, was decommissioned (disassembled/cut/shredded) on-site, and then transported to the landfill, cement kiln, or recycler. Inputs to the model were collected from peer-reviewed literature in combination with the knowledge from industry discussions. Models were developed using several peer-reviewed publications [31-35] and input from industry partners.

3. Results

3.1. Decommissioning

Wind turbine blades are designed for a 20 to 30-year lifecycle. A tremendous amount of research and development has gone into designing blades to maximize their lifetime. As such, the materials within them are highly engineered to minimize costs and maximize functional life. These materials can include a mix of fiberglass, balsa wood, thermoset resins (either epoxy, polyester, and/or vinyl ester), carbon fiber in the main structural member, foams, and metals from the lightning protection system. Owner/operators want blades that survive in the field for as long as possible and require as little maintenance as possible. As such, the market is competitively pushed to highly engineered materials with maximized lifetimes. Limiting the lifetime of functional use of these blades are weathering and wear/tear on blades seen during their functional use – especially in the blade tip Conversely, wind turbine owners and operators indicate that as new wind turbine technologies are developed it is more difficult to compete on the market with older machines having smaller blades. As such, manufacturers are rapidly designing longer blades and releasing them into the market to capitalize on this opportunity. To remain competitive in the market, and maximize profit, owners/operators must decommission older but still functional blades and replace them with new blades (known as re-powering) approximately every 7-9 years (after only 23-45% of the design life of the materials in the blade). Repowering can be

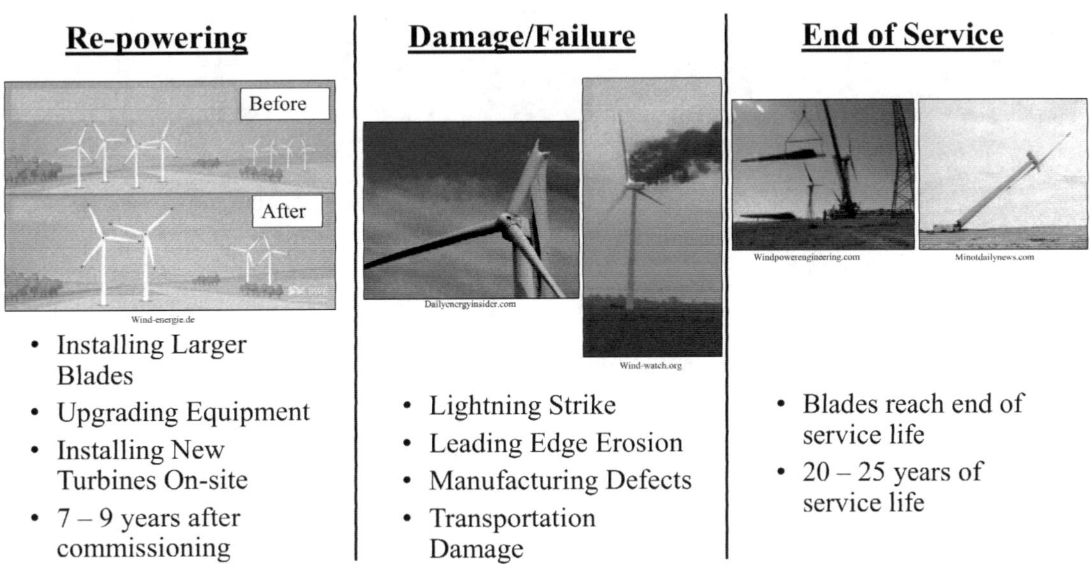

Figure 1: Causes for wind turbine blade decommissioning as provided by industrial partners in the U.S. Figures from Wind-energie.de, Dailyenergyinsider.com, Wind-watch.org, Windpowerengineering.com, and Minotdailynews.com

seen in Figure 1. Although blades removed through this method are likely still functional, owner/operators still see these materials as waste that need to be removed from their site.

Another source for premature blade decommissioning results from damage and failure of the blade. Common sources of blade failure include lightning strike, leading edge erosion, manufacturing defects, and damage during transportation. Although there is growing interest and significant development in advanced, non-destructive characterization methods for finding and repairing damage in blades, it is often difficult to spot in the field. For example, damage in the inside of the blade from lightning strikes can be impossible to detect using current technology without someone physically climbing onto the blade. As a result, damage is often missed by owner/operators until non-repairable failure occurs. Currently, lightning protection systems are being created which can detect damage in real time and some are being deployed along with drone-mounted damage detection devices. However, these systems have been found to have their own failure mechanisms and do not operate with perfect reliability in the field.

Another reason for blade decommissioning is that materials reach the end of their design life at which point they are often decommissioned, though in the U.S. some can remain up and non-functional for months to years before being brought down. This often occurs between 20-30 years after the blades are installed and when the materials are no longer useful in this application.

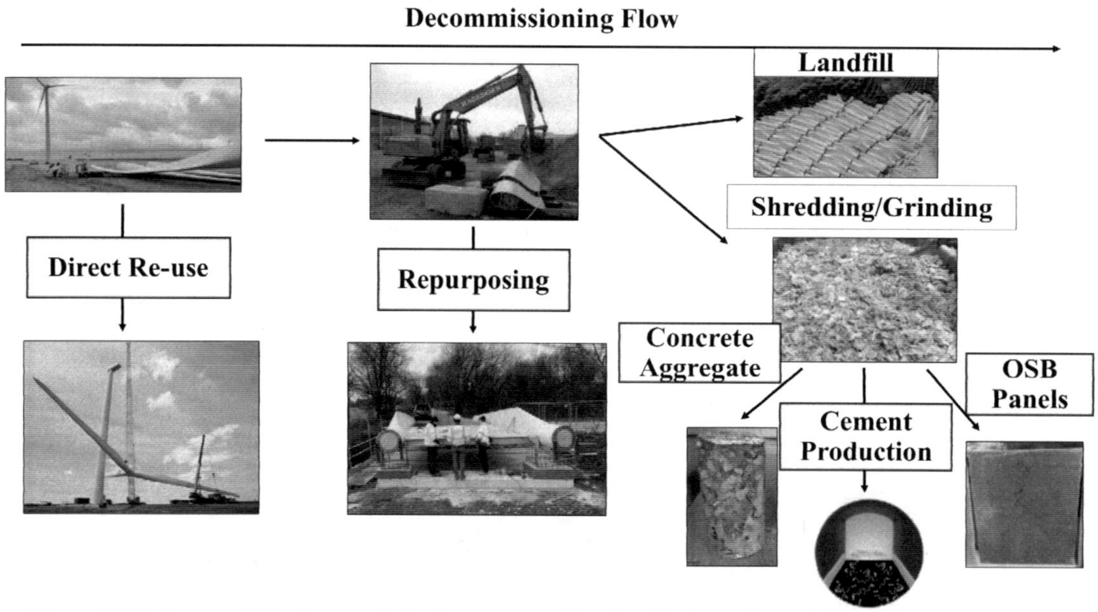

Figure 2: Each step of the decommissioning process visualized. Images from GE.com, wind-energy-the-facts.org, echidna.com, compositeworld.com, TPI Composites, and WSJ.com

The typical decommissioning process flow for the U.S. is displayed in Figure 2. As the further processing (left to right in Figure 2) occurs the recycling pathway options begin to decrease. The highest value recovery of the blade is for direct re-use in another turbine for sustainable energy generation. Industry partners indicate this is not often done for utility scale blades in the U.S. but is more common in the distributed or small wind industry. Additionally, wind blades at end of life can be repurposed into semi-structural or structural applications, such as bridges. However, most re-purposing efforts are not industrialized processes at this time and the structures would need to be recycled at the end of the repurposed life, which may be made more complex by the newly designed/built structure.

In the U.S., end of life blades are most commonly landfilled [8, 15]. However, in some cases they are shredded/ground into small particles which can be used as feedstock for concrete aggregate, cement production (most common), or as filler for oriented strand board panel alternatives.[36] It is difficult to estimate volumes for the different applications provided in this analysis, but industrial partners also suggest that most of the material at end of life is currently being landfilled at this time. Decommissioners indicated there have been recyclers who have agreed to recycle material but went out of business before successfully recycling all of the material in their possession.

Although often overlooked, decommissioning is an important step in the end-of-life management of wind turbine blades in relationship to the downstream material application. As shown in Figure 2, decommissioning significantly impacts the method through which the materials can be re-used or recycled impacting the recycled value of the blade materials. For example, re-powered blades at year 7-9 of their life could be valuable for 10-20 more years in service, though likely not in their original location. It could be possible to re-use those blades in a different location if it is cost competitive to do so and logistically possible to relocate them. However, due to the present lack of knowledge about the number of blades on the second hand market and where they are at any given time, and lack of predictability of this stream, it is not clear if such reuse strategies will be adopted in the U.S. market.

3.2. Decommissioning Analysis

Using data gathered through literature review and consultation with industry experts, an LCA was performed to compare the greenhouse gas emissions of different recycling approaches [30]. Within that LCA, decommissioning, size reduction, and transportation of a 48.7m long wind turbine blade were considered. To compliment that work, decommissioning and transportation scenarios were developed in this paper to compare different methods of size reduction and transportation by truck. The primary considerations of these scenarios were the method of size reduction (on-site versus at the recycling facility), the packing efficiency of down sized material onto the truck relative to its available payload, and the transportation distance.

In six of the scenarios the method of downsizing material is limited to cutting material into 10-meter segments. In the remaining two scenarios, the same cutting operation occurs, but the material is then fed into an on-site mobile shredder. The way in which a blade is cut or shredded directly impacts how efficiently it can be loaded onto a truck for transportation. The scenarios developed in this LCA include consideration of four different packing efficiencies on the truck based on a range of values from expert opinion. The first three packing efficiencies are 25%, 50%, and 75% relative to the available payload. These efficiencies represent how much of the truck's 16,000 kg mass payload can be filled by blade mass. A low efficiency of 25% represents a scenario in which the blade segments cannot be packed efficiently and the volumetric constraints of the blade limit how much blade material is loaded on the truck. A higher efficiency of 75% represents a scenario in which blades are cut and stacked in an efficient manner on the truck, reducing the volumetric constraints and increasing the amount of mass that can be loaded on the truck. For shredding, a 100% efficiency is used to represent the likely scenario in which material can be condensed and will take up the full mass payload of the truck.

Once material is loaded on a truck, the final consideration is the distance of transportation. The current challenge in evaluating transport distance is that few blade recycling locations exist within the U.S. Therefore, there are no historical distances that can be used to gauge the impacts of transporting the blades. As a result, two generic transportation distances were considered in these scenarios to represent a likely upper and lower bound. The first distance of 200km represents a nearby end-of-life destination, more likely associated with landfill or cement co-processing. The second distance of 1000km represents transportation to a theoretical dedicated wind turbine blade recycling facility. While this distance remains uncertain, the 1000km scenario is in line with current distances of transport to some pilot scale facilities.

Figure 3 shows the greenhouse gas emissions of the size reduction and transportation scenarios considered. Across all scenarios truck transportation has the largest share of greenhouse gas emissions. Consequently, the packing efficiency and transportation distance have a significant impact on greenhouse gas emissions. However, this impact is far more pronounced for longer distances on an absolute scale. This means that on-site shredding has the most potential to meaningfully reduce greenhouse gas emissions if recycling facilities are located further away. Evaluating these emissions in combination with economic considerations would help identify the trade-offs between different size reduction approaches and identify an optimal solution for a given location.

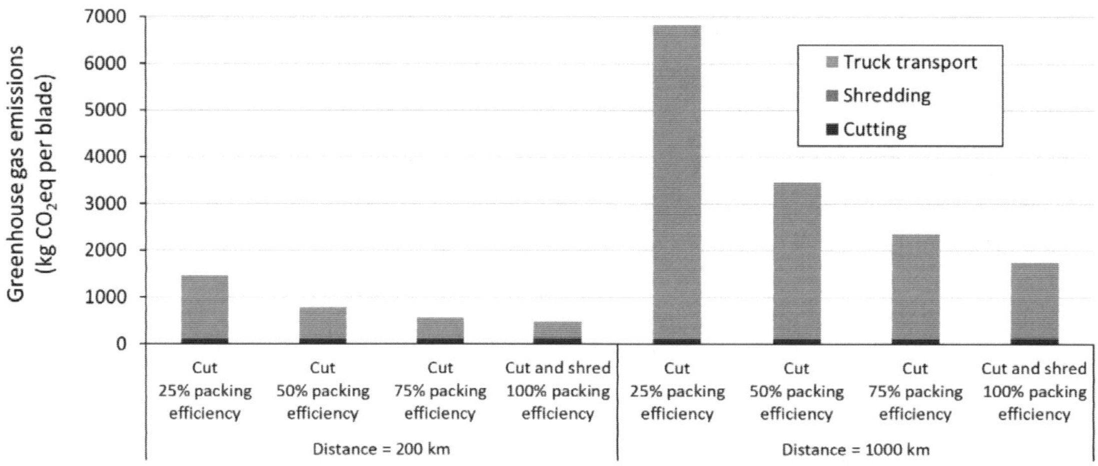

Figure 3: Greenhouse gas emissions from eight different downsizing and transportation scenarios.

3.3. Current Industrialized End of Life Strategies in the U.S. for Processing Wind Turbine Blades

A robust analysis of the state-of-the-art end of life strategies for wind turbine blades within the last 5 years is available in previous literature [37-41]. For this work, we are summarizing the key processes below for purposes of the analysis performed. It is important to note that recyclers are financially motivated in this process and many of the technologies to recover wind turbine blades are low margin, high volume, meaning throughput of material is the limiting factor. Maximizing throughput of systems and therefore product generation is critical to enabling the recycling pathway.

3.3.1. Landfill. Landfilling is the most common pathway for end-of-life blades due to the low cost in the U.S. (35-75 USD/ton) [42, 43]. However, this method sets the cost target for other recycling technologies. Because blades are so large and so tough, they can be challenging to shred. Further, maintenance of a shredder is expensive. Thus, often they are sliced into smaller pieces and then buried without further size reduction (minimizing costs and logistics).

3.3.2. Mechanical Recycling. Mechanical recycling is the process through which material is downsized using industrial equipment to feedstock which can be reprocessed as-is using different recycling technologies. For mechanical recycling, wind turbine blades present a unique challenge given their size during functional use. Because of this, they require several steps to become small enough to be remanufactured. This process traditionally uses heavy machinery with large saws and then manual labor to cut the blade into parts small enough to go into an industrial-scale shredder system [8, 9, 38, 44]. This material is then ground up into granulate which is small enough to then be remanufactured using several recycling technologies. Costs of shredding, labor hours, and maintenance are the limitation to this process, as shredders are expensive to maintain and cutting large fiber reinforced composites is difficult. There are several companies in the U.S. employing this technology at pilot scale. This includes GreenTex Solutions, LLC who uses granulate mixtures as filler for new composite pieces including flooring and counter tops [36]. RiversEdge has a similar technology where they are making 4'x8' panels filled with chopped wind turbine blades for lower cost boards for the marine industry [45].

3.3.3. Cement Kiln Co-Processing. In this process, wind turbine blades are used simultaneously as feedstock and fuel for cement clinker production. The blades must be shredded down to an adequate size to feed into the cement kiln. Inside the kiln, the organic components within the blend are burned and used as fuel to partially power the process. Although it is not yet clear from an academic standpoint whether cement kiln co-processing should truly be considered recycling, the European Composites Industry Association considers cement kiln as an option because of the e-glass component being fully recycled into the cement product (~50% of the blade waste) [46]. This process is often limited by the cost of shredding and other downsizing processes, as large scale shredders and size reduction technologies are expensive to maintain and use. The partnership between General Electric and Veolia in the U.S. is the best-known example of cement kiln co-processing of wind turbine blade waste. Estimates showed one 7-ton blade in the U.S. offsets 5 tons of coal and resulted in a 27% and 13% reduction in CO_2 emissions and water consumption, respectively [47]. However, it is uncertain what the maximum weight percent of end of life blade in the cement kiln co-processing fuel mixture can be before causing negative composition changes in the resulting clinker. Furthermore, as carbon fibers are more readily utilized in wind turbine blades cement kiln co-processing as an end of life wind turbine blade recycling pathway may become more difficult as fuel mixtures containing carbon fiber are often rejected.

3.3.4. Pyrolysis. Pyrolysis is a thermal decomposition process performed in an inert gas atmosphere (in the absence of oxygen) above 600°C [48-50]. During pyrolysis the polymer matrix and other constituent materials in the blade can be removed from the fiber reinforcement, to generate bio-oil, gas, and char which are recovered separately from the fiber [51]. The fibers recovered from this process, unlike in cement kiln, can be re-compounded back into new polymer composites for downcycling into other high-volume applications, such as automotive plastics. Carbon Rivers' (based in Knoxville, TN) G2G technology is a well-known example of this approach who is capable of recovering chopped glass fiber and already has industrially available products for new applications. Another example currently being installed in the U.S. is thermolysis [52]. Both technologies are similar and have been demonstrated for use for recovery of materials from wind turbine blades at pilot scale.

3.4. Existing Recycling Technology Analysis

In addition to size reduction and transportation, the overall results of the LCA can be compared across existing approaches [30]. **Figure 4** displays the baseline net greenhouse gas emissions of landfilling, mechanical recycling, cement co-processing, and pyrolysis. In the landfill baseline, blades are cut into segments, loaded onto a truck at 57% packing efficiency, and transported 100km to a sanitary landfill. In cement co-processing the blades are shredded on-site, loaded at 100% efficiency, and transported 200km to a cement plant. In mechanical recycling and pyrolysis blades are also shredded on-site and packed at 100% efficiency, but then are transported 500km to a dedicated recycling facility. The transportation distances for each approach were selected to capture the likely differences between end-of-life destinations in the U.S. In this approach it is estimated that landfilling locations are abundant, requiring minimal transportation distance, while cement plants and theoretical recycling locations would on average be located further away. Given that there are currently limited cement co-processing locations and no full-scale recycling locations in the U.S., the distances selected are hypothetical and are based on a conservative estimate of potential future locations. The baseline size reduction and transportation estimates result in variations in greenhouse gas emissions which could be important for low emissions approaches such as landfill and mechanical recycling. However, these emissions are relatively small compared to the recycling emissions for cement co-processing and pyrolysis.

The recycling emissions shown in Figure 4 are the result of converting wind turbine blade materials into recycled products. For cement co-processing the recycled material is cement clinker, while in mechanical and pyrolysis the primary recycled material is recovered fiberglass. In addition to the primary recycled product, both mechanical recycling and pyrolysis produce additional co-products. For mechanical this co-product is powder from grinding, while in pyrolysis co-products include powder and oil. Substitution credits are given for the recycled materials of each process with a negative emission.

The results of adding these substitution credits and the positive process emissions are the net greenhouse gas emissions, shown by the diamond markers.

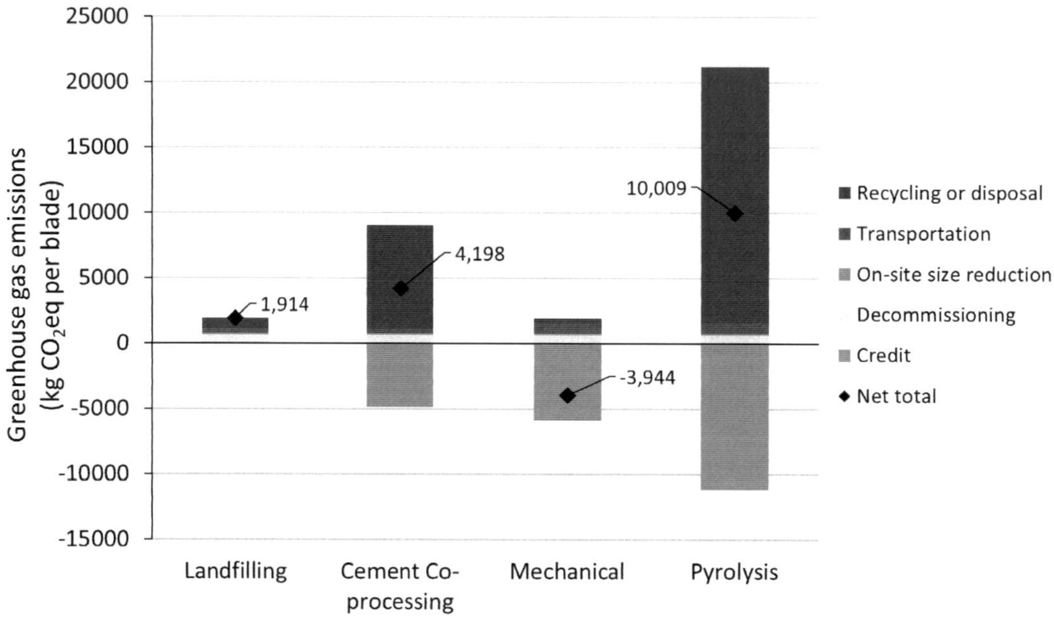

Figure 4: Greenhouse gas emissions of end-of-life approaches for wind turbine blades in the U.S [30].

Comparing the net greenhouse gas emissions across processes highlights some key findings. First, the simplicity of mechanical recycling makes it the only process with the potential for net negative emissions. However, this finding comes with the caveat that the markets and corresponding value for mechanically recycled fibers and powders remains uncertain. Further research is needed to understand these markets and update the credit accordingly. Another finding from the baseline comparison is that cement co-processing and pyrolysis both result in higher net greenhouse gas emissions than landfilling. In both cement co-processing and pyrolysis, the major contributor to greenhouse gas emissions is thermal energy consumption. Within cement co-processing this thermal energy comes from incinerating the organic blade material, which results in direct carbon dioxide emissions. In pyrolysis this thermal energy is largely supplied by natural gas, which also results in carbon dioxide emissions at the point of combustion. Reducing thermal energy demand or changing to alternative sources of thermal energy could help reduce the emissions of pyrolysis. However, cement co-processing is limited by the origin of the organic content in the blade. As a result, lifetime emissions of the materials including cement co-processing are unlikely to be significantly reduced unless the organic material of future blades is sourced from bio-based resources.

4. Discussion

Presently, the majority of wind turbine blade waste being generated in the U.S. is being landfilled. Owners and operators have indicated the reason for this is because they are motivated to choose the cheapest option which in the current U.S. market is landfilling material at a cost of 35-75 USD/ton, which varies based on location. Until costs of recycling are less than that of landfilling, there will be little incentive for owner-operators to choose to pay more to recycle materials. Several manufacturers, notably TPI Composites and GE Vernova, have indicated they are now working with owners to ensure their products do not end up going to landfill. It is not clear at this time if this will be a long-term solution to minimize blade waste going to landfill or merely a stepwise solution. In addition, because new blades are commonly decommissioned prior to the designed lifetime, more towers are being repowered than previous analysis indicated [8] resulting in more blade waste going to landfill in the U.S. Until manufacturers are incentivized to design blades for the known re-powering lifecycle, or designs become so optimized that there is little financial incentive to repower, there will be more end-of-life blade material to handle than is representative of the energy production of the design life.

Another major challenge facing this industry is that disassemblers/decommissioners are not incentivized to prepare the material for the most optimized recycling pathway. Disassembling a wind turbine is an engineering challenge involving many steps. Participants in this study indicate the businesses who do this are incentivized to get the material ready to be recycled as quickly as possible. The goal is to recover as much metal as possible for recycling separate from the non-recyclable streams, while maximizing packing density of those streams which cannot be recycled (including wind turbine blades). Because so many recyclers have promised to recycle these materials and then went out of business, at least one such company interviewed indicate they are reluctant to pursue other recycling technologies for blades until they are properly vetted by industry partners. This step has been engineered so that a full turbine can be taken down in just a few days. Until this portion of the supply chain for end-of-life materials is fully understood, and cost structures for recycling pathways are well developed, this portion will not be efficiently designed with the end of life of the blade in mind and will continue to be done as quickly as possible with metal recovery in mind.

From the recyclers' perspective the reason they are not recycling more material is that although the technologies have been proven at pilot scale, several of them are not available at production capacity. Currently, companies interviewed indicate promising recycling technologies have been funded by U.S. Federal agencies through demonstrations at the pilot scale and therefore to technology readiness level (TRL) 7-8. However, beyond demonstrations these recyclers cannot scale up their technology without private investment to industrially relevant scales (*i.e.,* 100 MT/yr, etc.) for large scale recycling industries such as wind turbine blades. However, private investors will not invest in scaling-up processes without contracts being in place by customers (tier 1-2 suppliers, chemical manufacturers, original equipment manufacturers (OEMs)) for those industries. These customers will not put contracts in place without the recycler being able to process material at the scale relevant to industry. Right now, small business recyclers in the U.S. are caught in the middle of this. If it was possible to enable high-impact, high-likelihood of success companies to reach industrially relevant scale beyond pilot, this would significantly de-risk the adoption of that technology nationwide by enabling contracts to be put in place and therefore prompting private investment. As it stands, recycling approaches which are being employed are using technologies that are already available at industrially relevant scale without further investment (cement kiln, landfill, etc.) which may limit the adoption of new recycling technologies in the U.S. market.

In addition, participants indicated that more robust analysis of the cost and functional performance of recycled materials from commercially available recycling technologies is needed in the U.S. from reputable, non-biased sources. As it stands for the recycling industry in general, partners have indicated the demand for recycled materials (most importantly demand for recycling material for the automotive industry) is significantly higher than the current production rates available for them in the U.S. and it is unlikely the volume of material from recycled wind turbine blades alone will be enough to meet the demand for recycled glass fiber. It will likely require pulling similar materials from other waste streams (i.e. marine, vehicle, and aerospace scrap/waste) into the recycling stream in order to meet demand. This lends itself to a consistency, cost, and performance question which is limiting the adoption of these materials to higher performing applications. As the cost, performance, and consistency/quality of the materials recovered from recyclers become better understood it is likely there will be significantly more industrial pull for the recycled materials. Further, it is likely this will enable further optimization of the cost of recyclate and therefore more widespread adoption of the materials.

5. Conclusions

The unique analysis in this paper suggests that optimizing packing density in the truck is critical during the decommissioning phase, and that transportation over long distances is the driver for carbon intensity leading up to recycling facilities. It is likely that there will exist fewer companies doing plastic recycling than landfilling in the U.S. for quite some time, and thus these materials may need to be transported greater distances than otherwise to be re-processed. However, in the meantime developing techniques to reach 100% packing efficiency (i.e. shredding, maximizing stacking, etc.) can reduce carbon intensity of the blade decommissioning and transportation process by over 50% alone and should be considered. The results of this analysis suggest as well the promise of mechanical recycling as a decarbonization pathway for composite materials. However, the limitation of this analysis is largely that the value and performance of recovered materials are not currently well understood. Further research is needed to unpack the cost structures associated with the existing recycling infrastructure in the U.S. and determine the consistency and functional performance of the recovered materials.

6. Acknowledgments

This research has been funded by the U.S. Department of Energy Wind Energy Technologies Office through the Wind Technology Recycling Assessment project. Oak Ridge National Laboratory is operated by UT-Battelle, LLC. under Contract No. DE-AC05-00OR22725 with the U.S. Department of Energy. Sandia National Laboratories is a multi mission laboratory managed and operated by National Technology & Engineering Solutions of Sandia, LLC, a wholly owned subsidiary of Honeywell International Inc., for the U.S. Department of Energy's National Nuclear Security Administration under contract DE-NA0003525. This paper describes objective technical results and analysis. Any subjective views or opinions that might be expressed in the paper do not necessarily represent the views of the U.S. Department of Energy or the United States Government. The US government retains and the publisher, by accepting the article for publication, acknowledges that the US government retains a nonexclusive, paid-up, irrevocable, worldwide license to publish or reproduce the published form of this manuscript, or allow others to do so, for US government purposes.

References

[1] L. Capuano, "International Energy Outlook 2020," U.S. Energy Information Administration, Washington, DC, October 14, 2020 2020.

[2] U. S. Environmental Protection Agency, "Inventory of U.S. Greenhouse Gas Emissions and Sinks: 1990-2019," Environmental Protection Agency, Washington, District of Columbia, U.S., 2021.

[3] J. Cook *et al.*, "Consensus on consensus: a synthesis of consensus estimates on human-caused global warming," *Environmental Research Letters,* vol. 11, no. 4, 2016, doi: 10.1088/1748-9326/11/4/048002.

[4] P. A. Owusu, S. Asumadu-Sarkodie, and S. Dubey, "A review of renewable energy sources, sustainability issues and climate change mitigation," *Cogent Engineering,* vol. 3, no. 1, 2016, doi: 10.1080/23311916.2016.1167990.

[5] U. D. o. Energy, "Wind Vision Detailed Roadmap Actions," US Department of Energy, Springfield, VA, USA, 2017.

[6] R. Wiser *et al.*, "Land-Based Wind Market Report," U.S. Department of Energy, Washington, D.C., August 2022 2022.

[7] D. Keyser and S. Tegen, "The Wind Energy Workforce in the United States: Training, Hiring, and Future Needs," National Renewable Energy Laboratory, Golden, CO, July 2019 2019. [Online]. Available: https://www.nrel.gov/docs/fy19osti/73908.pdf

[8] A. Cooperman, A. Eberle, and E. Lantz, "Wind turbine blade material in the United States: Quantities, costs, and end-of-life options," *Resources, Conservation and Recycling,* vol. 168, p. 105439, 2021/05/01/ 2021, doi: https://doi.org/10.1016/j.resconrec.2021.105439.

[9] R. Martini and G. Xydis, "Repurposing and recycling wind turbine blades in the United States," *Environmental Progress & Sustainable Energy,* vol. 42, no. 1, 2022, doi: 10.1002/ep.13932.

[10] L. Mishnaevsky, Jr., "Sustainable End-of-Life Management of Wind Turbine Blades: Overview of Current and Coming Solutions," *Materials (Basel),* vol. 14, no. 5, Feb 27 2021, doi: 10.3390/ma14051124.

[11] L. Mishnaevsky, Jr., "Root Causes and Mechanisms of Failure of Wind Turbine Blades: Overview," *Materials (Basel),* vol. 15, no. 9, Apr 19 2022, doi: 10.3390/ma15092959.

[12] L. Mishnaevsky, "Repair of wind turbine blades: Review of methods and related computational mechanics problems," *Renewable Energy,* vol. 140, pp. 828-839, 2019, doi: 10.1016/j.renene.2019.03.113.

[13] A. M. Jadali, A. Ioannou, K. Salonitis, and A. Kolios, "Decommissioning vs. repowering of offshore wind farms—a techno-economic assessment," *The International Journal of Advanced Manufacturing Technology,* vol. 112, no. 9-10, pp. 2519-2532, 2021, doi: 10.1007/s00170-020-06349-9.

[14] A. Jezierski, C. Mańkowski, and R. Śpiewak, "Energy Savings Analysis in Logistics of a Wind Farm Repowering Process: A Case Study," *Energies,* vol. 14, no. 17, 2021, doi: 10.3390/en14175452.

[15] A. Alshannaq, D. Scott, L. Bank, M. Bermek, and R. Gentry, "Structural Re-use of De-commissioned Wind Turbine Blades in Civil Engineering Applciations," Georgia Technical University, Atlanta, GA, 2019.

[16] F. A. Cruz Sanchez, H. Boudaoud, M. Camargo, and J. M. Pearce, "Plastic recycling in additive manufacturing: A systematic literature review and opportunities for the circular economy," *Journal of Cleaner Production,* vol. 264, p. 121602, 2020, doi: 10.1016/j.jclepro.2020.121602.

[17] E. Moreno *et al.*, "Technical Evaluation of Mechanical Recycling of PLA 3D Printing Wastes," in *The First International Conference on "Green" Polymer Materials 2020*, 04 November 2020 2020: MDPI.

[18] Y. Zhang, Y. Pontikes, L. Lessard, and A. Willem van Vuure, "Recycling and valorization of glass fibre thermoset composite waste by cold incorporation into a sustainable inorganic polymer matrix," *Composites Part B: Engineering,* vol. 223, p. 109120, 2021/10/15/ 2021, doi: https://doi.org/10.1016/j.compositesb.2021.109120.

[19] S. D. Anuar Sharuddin, F. Abnisa, W. M. A. Wan Daud, and M. K. Aroua, "A review on pyrolysis of plastic wastes," *Energy Conversion and Management,* vol. 115, pp. 308-326, 2016, doi: 10.1016/j.enconman.2016.02.037.

[20] D. Coughlin, C. Ludwig, S. Ozcan, D. Hartman, and R. Ginder, "Controlled Pyrolysis: A Robust Scalable Composite Recycling Technology," Institute for Advanced Composites Manufacturing Innovation, Knoxville, Tennessee, 2021.

[21] M.-x. Xu *et al.*, "The pyrolysis of end-of-life wind turbine blades under different atmospheres and their effects on the recovered glass fibers," *Composites Part B: Engineering,* vol. 251, p. 110493, 2023/02/15/ 2023, doi: https://doi.org/10.1016/j.compositesb.2022.110493.

[22] P. S. Stelzer *et al.*, "Experimental feasibility and environmental impacts of compression molded discontinuous carbon fiber composites with opportunities for circular economy," *Composites Part B: Engineering,* vol. 234, p. 109638, 2022/04/01/ 2022, doi: https://doi.org/10.1016/j.compositesb.2022.109638.

[23] P. E. Caltagirone *et al.*, "Substitution of virgin carbon fiber with low-cost recycled fiber in automotive grade injection molding polyamide 66 for equivalent composite mechanical performance with improved sustainability," *Composites Part B: Engineering,* vol. 221, p. 109007, 2021/09/15/ 2021, doi: https://doi.org/10.1016/j.compositesb.2021.109007.

[24] G. W. Coates and Y. D. Y. L. Getzler, "Chemical recycling to monomer for an ideal, circular polymer economy," *Nature Reviews Materials,* vol. 5, no. 7, pp. 501-516, 2020, doi: 10.1038/s41578-020-0190-4.

[25] A. Rahimi and J. M. García, "Chemical recycling of waste plastics for new materials production," *Nature Reviews Chemistry,* vol. 1, no. 6, p. 0046, 2017/06/07 2017, doi: 10.1038/s41570-017-0046.

[26] J. Zhao, P. Liu, J. Yue, H. Huan, G. Bi, and L. Zhang, "Recycling glass fibers from thermoset epoxy composites by in situ oxonium-type polyionic liquid formation and naphthalene-containing superplasticizer synthesis with the degradation solution of the epoxy resin," *Composites Part B: Engineering,* vol. 254, p. 110435, 2023/04/01/ 2023, doi: https://doi.org/10.1016/j.compositesb.2022.110435.

[27] J. Kaushal, M. Khatri, and S. K. Arya, "Recent insight into enzymatic degradation of plastics prevalent in the environment: A mini - review," *Cleaner Engineering and Technology,* vol. 2, 2021, doi: 10.1016/j.clet.2021.100083.

[28] N. Mohanan, Z. Montazer, P. K. Sharma, and D. B. Levin, "Microbial and Enzymatic Degradation of Synthetic Plastics," (in English), *Frontiers in Microbiology,* Review vol. 11, no. 2837, 2020-November-26 2020, doi: 10.3389/fmicb.2020.580709.

[29] C.-C. Chen, L. Dai, L. Ma, and R.-T. Guo, "Enzymatic degradation of plant biomass and synthetic polymers," *Nature Reviews Chemistry,* vol. 4, no. 3, pp. 114-126, 2020/03/01 2020, doi: 10.1038/s41570-020-0163-6.

[30] E. Sproul, M. Williams, M. L. Rencheck, M. Korey, and B. Ennis, "Life Cycle Assessment of Wind Turbine Blade Recycling Approaches in the United States," presented at the 43rd Risø International Symposium on Materials Science: Composite for wind energy: Manufacturing, operation and end-of-life, Roskilde, Denmark, 2023.

[31] M. Rani, P. Choudhary, V. Krishnan, and S. Zafar, "A review on recycling and reuse methods for carbon fiber/glass fiber composites waste from wind turbine blades," *Composites Part B: Engineering,* vol. 215, p. 108768, 2021/06/15/ 2021, doi: https://doi.org/10.1016/j.compositesb.2021.108768.

[32] J. Yang, J. Liu, W. Liu, J. Wang, and T. Tang, "Recycling of carbon fibre reinforced epoxy resin composites under various oxygen concentrations in nitrogen–oxygen atmosphere," *Journal of Analytical and Applied Pyrolysis,* vol. 112, pp. 253-261, 2015/03/01/ 2015, doi: https://doi.org/10.1016/j.jaap.2015.01.017.

[33] A. M. Cunliffe, N. Jones, and P. T. Williams, "Recycling of fibre-reinforced polymeric waste by pyrolysis: thermo-gravimetric and bench-scale investigations," *Journal of Analytical and Applied Pyrolysis,* vol. 70, no. 2, pp. 315-338, 2003/12/01/ 2003, doi: https://doi.org/10.1016/S0165-2370(02)00161-4.

[34] P. Liu, F. Meng, and C. Y. Barlow, "Wind turbine blade end-of-life options: An eco-audit comparison," (in en), *Journal of Cleaner Production,* vol. 212, pp. 1268-1281, 2019/03/01/ 2019, doi: 10.1016/j.jclepro.2018.12.043.

[35] "Roadmap to Carbon Neutrality," Portland Cement Association, 2021.

[36] J. Holmes, "Greentex Solutions, LLC Executive Summary," presented at the Institute of Advanced Composites Manufacturing and Innovation Members Meeting, Knoxville, TN, July 25, 2018, 2018.

[37] M. Leon, Jr., "Recycling of wind turbine blades: Recent developments," *Current Opinion in Green and Sustainable Chemistry,* vol. 39, 2023, doi: 10.1016/j.cogsc.2022.100746.

[38] H. Jani, S. Kachhwaha, N. Garlapati, and A. Das, "A brief review on recycling and reuse of wind turbine blade materials," *Materials Today: Proceedings,* vol. 62, 02/01 2022, doi: 10.1016/j.matpr.2022.02.049.

[39] C. Psomopoulos, K. Kalkanis, S. Kaminaris, G. Ioannidis, and P. Pachos, "A Review of the Potential for the Recovery of Wind Turbine Blade Waste Materials," *Recycling,* vol. 4, no. 1, 2019, doi: 10.3390/recycling4010007.

[40] N. Rathore and N. L. Panwar, "Environmental impact and waste recycling technologies for modern wind turbines: An overview," (in eng), *Waste Manag Res,* vol. 41, no. 4, pp. 744-759, Apr 2023, doi: 10.1177/0734242x221135527.

[41] J. P. Jensen and K. Skelton, "Wind turbine blade recycling: Experiences, challenges and possibilities in a circular economy," *Renewable and Sustainable Energy Reviews,* vol. 97, pp. 165-176, 2018/12/01/ 2018, doi: https://doi.org/10.1016/j.rser.2018.08.041.
[42] B. Alves, "Price of landfilling municipal waste in the U.S. 2020-2021," Statistica, 2023.
[43] M. L. Rencheck, M. Korey, L. Ghobrial, D. Berry, and S. Ozcan, "The Institute for Advanced Composites Manufacturing and Innovation Circular Economy for Wind Workshop Report," Washington, D.C., 2021.
[44] M. Korey *et al.*, "Recycling polymer composite granulate/regrind using big area additive manufacturing," *Composites Part B: Engineering,* vol. 256, p. 110652, 2023/05/01/ 2023, doi: https://doi.org/10.1016/j.compositesb.2023.110652.
[45] N. Rohan. https://www.riversedgecomposites.com/ (accessed June 28, 2023.
[46] A. Jacob, "Composites can be recycled," *Reinforced Plastics,* vol. 55, no. 3, pp. 45-46, 2011/05/01/ 2011, doi: https://doi.org/10.1016/S0034-3617(11)70079-0.
[47] B. North, C. Griffiths, and S. Guendoul. "GE Renewable Energy Announces US Blade Recycling Contract with Veolia." Bloomberg. https://www.bloomberg.com/press-releases/2020-12-08/ge-renewable-energy-announces-us-blade-recycling-contract-with-veolia (accessed.
[48] J. Zhang, V. S. Chevali, H. Wang, and C.-H. Wang, "Current status of carbon fibre and carbon fibre composites recycling," *Compos Part B-Eng,* vol. 193, 2020, doi: 10.1016/j.compositesb.2020.108053.
[49] X. Zhao *et al.*, "Plastic waste upcycling toward a circular economy," *Chem Eng J,* p. 131928, 2021, doi: 10.1016/j.cej.2021.131928.
[50] X. Zhao *et al.*, "Plastic waste upcycling toward a circular economy," *Chemical Engineering Journal,* vol. 428, p. 131928, 2022/01/15/ 2022, doi: https://doi.org/10.1016/j.cej.2021.131928.
[51] L. Giorgini *et al.*, "Pyrolysis of fiberglass/polyester composites: Recovery and characterization of obtained products," *FME Transaction,* vol. 44, no. 4, pp. 405-414, 2016, doi: 10.5937/fmet1604405G.
[52] D. Coughlin, C. Ludwig, S. Ozcan, D. Hartman, and R. S. Ginder, "Controlled Pyrolysis: A Robust Scalable Composite Recycling Technology," The Institute for Advanced Composites Manufacturing Innovation, U.S. Department of Energy, 2021.

Effect of span/thickness ratio on fatigue properties of thick quasi-isotropic CFRP laminates subjected to three-point bending loading

M Otsubo[1], A Hosoi[1,2], T Sakurai[3], T Yamatogi[3] and H Kawada[1,2]
[1] Department of Applied Mechanics and Aerospace Engineering, Waseda University, 3-4-1, Okubo, Shinjuku-ku, Tokyo 169-8555, Japan
[2] Kagami Memorial Research Institute for Materials Science and Technology, Waseda University, 2-8-26, Nishiwaseda, Shinjuku-ku, Tokyo 169-0051, Japan
[3] Nakashima Propeller Co., Ltd, 688-1, Joto-Kitagata, Higashi-Ku, Okayama City, Okayama 709-0625, Japan

E-mail: hosoi@waseda.jp

Abstract. In this study, three-point bending fatigue tests were conducted to investigate damage growth behavior of thick quasi-isotropic carbon fiber reinforced plastic (CFRP) laminates. The span/thickness ratio *L/h* was varied for both static tests and fatigue tests to determine failure mechanisms. In the static tests, the failure mode changed from delamination to compressive buckling as span/thickness ratio increases. The apparent interlaminar shear strength was higher for the specimens with lower span/thickness ratios, and the shear failure was likely to be caused by oblique crack in 90° plies. In the fatigue tests, the specimens with span/thickness ratio *L/h*=7 caused delamination regardless of stress level. For the specimens with span/thickness ratio *L/h* =15, three types of failure modes were obtained depending on stress level; delamination near the neutral plane, buckling at compressive area and matrix crack at tensile area. During the fatigue tests, stiffness of specimens with different span/thickness ratios which fail in shear was compared using digital image correlation (DIC) method. The specimens with span/thickness ratio 4 suddenly decreased in stiffness without increasing normal strain over cycles. Both specimens with span/thickness ratio 4 and 11 finally failed by delamination but they showed different failure process in how normal or shear strain developed.

1. Introduction

In recent years, carbon neutral society is widely aimed and carbon fiber reinforced plastics (CFRPs), with high corrosion resistance, have been adopted for several marine structures including blades for tidal turbines and ship propellers as well. Applying to ship propellers is said to contribute to reduction of fossil fuels use [1]. The application to ship propellers is subjected to very high cyclic loadings, resulting in 10^9 cycles. In addition, bending load will increase at the root of blades where thicker materials are used. Thus, it is necessary to accurately evaluate the bending fatigue properties for thick CFRP laminates to ensure long-term reliability.

The damage growth behaviors of composites by three-point bending test have been widely investigated in previous works. Makeev [2] studied shear properties under three-point bending cyclic loading for unidirectional carbon/epoxy composites with 3.8 mm thick. A sharp increase in shear strain was found just before the failure and a likely reason for this result was concluded that accumulation of non-visible damage increased in matrix. It was found that undetectable micro matrix cracking caused a sharp increase in shear strain just before the failure. He and Makeev [3] performed three-point bending tests of unidirectional polymer matrix composites with different span/thickness ratios. They showed good agreement between numerical and experimental shear strength properties when the nonlinear shear properties of the coupon materials were considered in the finite element simulation. May and Hallett [4] also investigated initiation of damage for unidirectional carbon/epoxy laminates with 2.55 mm thick. They showed progressive internal damage growth prior to shear failure of the specimens indirectly as a result of fatigue testing by X-ray computed tomography.

Currently, both static and fatigue bending properties have been widely studied for thinner samples [5] whereas research of thick laminates bending behavior is limited. Moreover, some specimens have shown brittle failure under fatigue bending loading with no prior indication [6], but the failure mechanisms have not been revealed for thick quasi-isotropic CFRP laminates. Previous researches [7][8] pointed the effect of span/thickness ratio on failure modes as well. However, little research has been completed to examine how it happens, resulting in the overestimate of ship propeller blades safety. In this paper, both static and fatigue three-point bending tests were conducted for 16 mm thick quasi-isotropic CFRP laminates to evaluate the effect of span/thickness ratio on failure modes and damage mechanisms.

2. Materials and methods

2.1. Material and specimen

In this research, the material was made from T700 carbon-fiber (Toray Co., Ltd) and amine-cured epoxy resin (DIC Co., Ltd.), and manufactured by vacuum-assisted resin transfer molding (VaRTM). An 80-ply quasi isotropic layup ($[(+45/-45)/(0/90)]_{10s}$) was adopted as specimens for all tests and they are about 16 mm thick, and 20 mm wide. The length of the specimens is determined depending on span/thickness ratio *L/h*. The fiber-volume fraction is about 59.4%. Three-point bending test was done as shown in figure 1 where specimens were put symmetrically between two supports and loading nose.

2.2. Static three-point bending test

Static tests were conducted to obtain shear and bending strength with a universal testing machine Autograph AG-Xplus. 9 types of specimens with different span/thickness ratios *L/h* = 4, 5, 7, 9, 11, 13, 15, 20 and 30 were prepared to investigate the effect of *L/h* and were subjected to three-point bending. Test configuration is shown as figure 1 and a loading nose with 15 mm diameter was adopted to avoid excessive stress concentration in the vicinity of loading nose. Each specimen was loaded at 1.0 mm/min until failure. Shear and bending strength were calculated based on classical beam theory. Failure modes were obtained after tests for each specimen. Several specimens were used to observe the crack using replica, stopping the test at the severities of 20%, 40%, 60%, 80%, 90% relative to failure load. To capture the images of failure behavior, a digital microscope (KEYENCE VHX5000) was used with the resolution of 1600×1200 pixels, at the magnification of 200 ×.

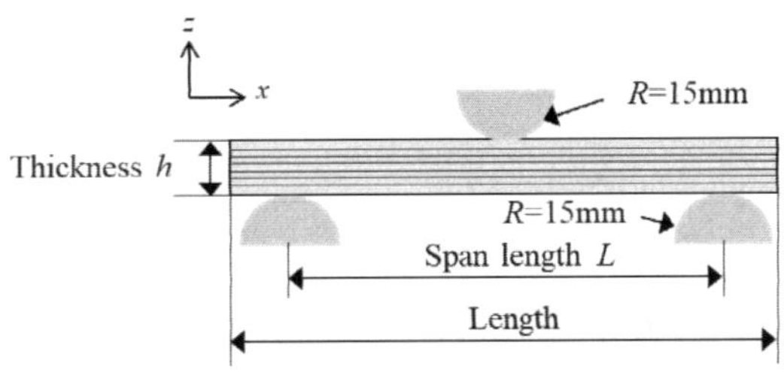

Figure 1. Specimen and support rollers geometry of three-point bending test.

2.3. Fatigue three-point bending test

Fatigue three-point bending tests have been carried out under load control to assess failure cycles and failure mechanisms of specimens. Test configuration is the same as figure 1. A servo hydraulic test machine was employed, setting the stress ratio $R = 0.1$. The frequency of the applied stress was flexibly changed in the range 0.8-2.0 Hz depending on maximum deflection of specimens and the performance of machines. The maximum stress of the fatigue tests was set at the severities of 50%, 60%, and 70% relative to static shear strength τ_b and the tests were stopped when specimens failed by shear or compressive buckling and exceeded its displacement limit in this study. The span/thickness ratio L/h of specimens was varied in the range 4-15 to determine failure mechanisms. After each test, the failure cycle was recorded and used to produce *S-N* curve. DIC software VIC-2D (Correlated Solutions Inc, version 7) was also employed in this work for assessment of shear strain distribution on the surface of tested specimens.

3. Results and discussion

3.1. Static three-point bending test

Figure 2 shows the result of static three-point bending test for thick quasi-isotropic laminates, and the relationship between span/thickness ratio and shear or bending strength. 9 types of tested specimens with different span/thickness ratios were divided into two groups according to failure modes. The shear or bending strength was calculated based on the classical beam theory. By the static tests, it was revealed that the failure modes changed from delamination to bending buckling at ratio 15. There was an obvious correlation between span/thickness ratio and strength. Whereas the bending strength indicates almost the same, the apparent shear strength decreased with the increase in span/thickness ratio. The previous research [9] mentioned that thinner specimens with smaller span/thickness ratios are influenced by nonlinearity of shear modulus, which is likely to be the same as thicker samples used in this study. Double notch shear (DNS) test was also done [10] and obtained 42.5MPa (standard deviation: 6.41) as interlaminar shear strength (ILSS), which is shown in figure 2 by a red dashed line. ILSS exceeds the shear strength of samples with span/thickness ratio = 7 or more in three-point bending tests.

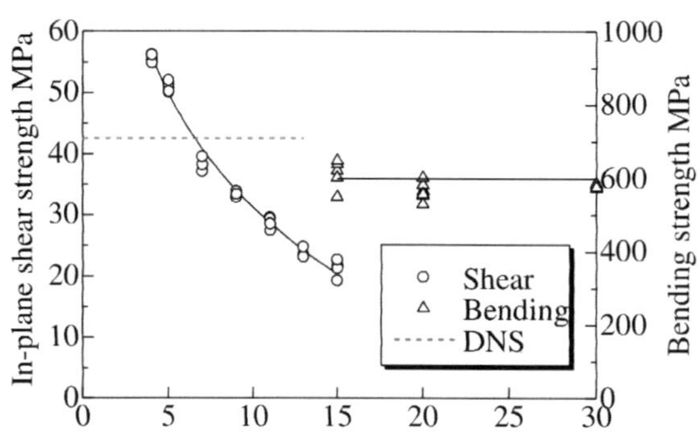

Figure 2. Relationship between shear or bending strength and span/thickness ratio.

The onset of shear failure was also studied through static tests with replica using the specimen with span/thickness ratio 11. The test ended with a typical delamination failure in the vicinity of neutral plane. Figure 3 compares the crack of specimen depending on the percentage of the applied load relative to the failure load. From the observation of failed sample, oblique crack was confirmed in 90° ply directly above the delamination. It is likely to originate the initial defects on the surface of a sample and stress concentration there could cause sudden delamination of specimen failure. As explained above, the crack in 90° ply could initiate delamination between 0°/90° plies. Previous research, aimed at 0° unidirectional composites, accounted for the impossibility of predicting the onset of damage but it can be said that thick quasi-isotropic laminates show different damage mechanism from that.

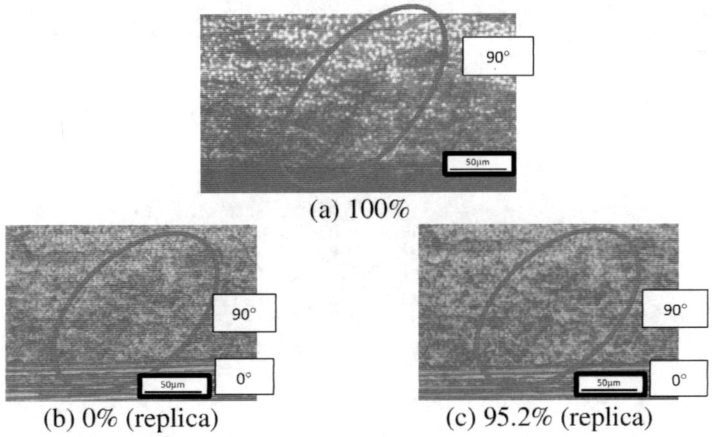

Figure 3. Image of transverse crack in 90° ply at each loading point

3.2. Fatigue three-point bending test

Figure 4 shows *S-N* diagram of specimens with different span/thickness ratios generated from the result of fatigue tests. This diagram confirmed that the specimens with larger span/thickness ratios are likely to fail by buckling at compressive area. Furthermore, about 14 MPa was referred as a boundary of the failure modes between shear and bending in accordance with maximum loaded shear stress.

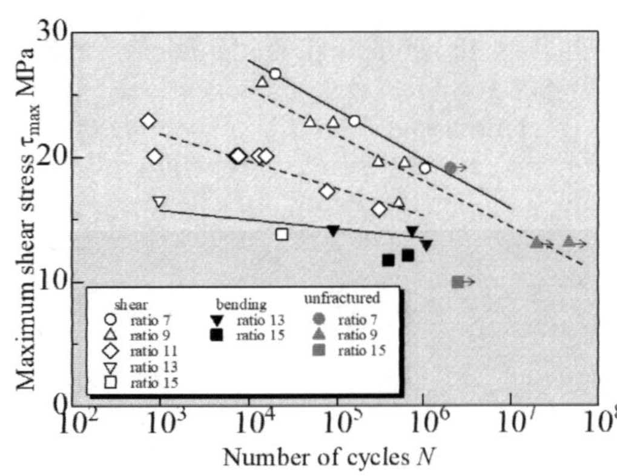

Figure 4. *S-N* diagram of specimens with different span/thickness ratio

Figures 5-7 compared the result of failure behavior for specimens with span/thickness ratio 15 tested at the different stress amplitude. Stress level was defined as τ_{max}/τ_b, where τ_{max} is the maximum shear stress of the specimen during fatigue tests and τ_b is shear strength obtained by static tests for respective span/thickness ratio specimens (values in figure 2). When loaded at higher stress, stress level τ_{max}/τ_b = 70%, specimens failed by delamination in the vicinity of neutral plane as figure 5. In the case of stress level τ_{max}/τ_b = 60%, buckling occurred at compressive area in a stable manner. Contrary to those two types of failure modes, only matrix crack in 90° ply and delamination were observed at the stress level τ_{max}/τ_b = 50%. This is attributed to the fact that the compressive strength of 90° ply basically exceeds the tensile strength of 90° ply. Tests at lower stress level induced only damage at tensile area, therefore, stress concentration here could cause the final tensile bending failure to happen.

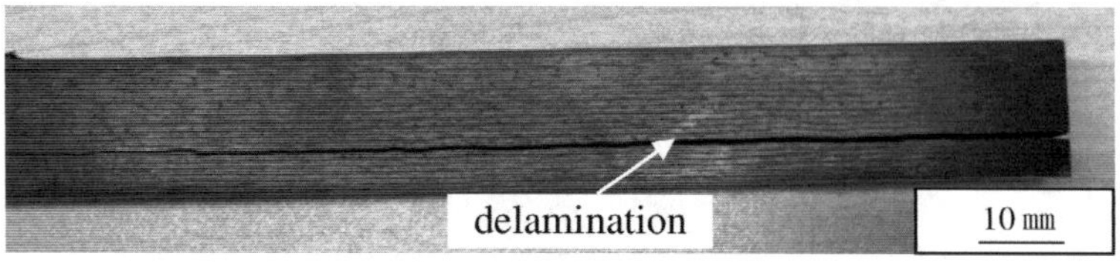

Figure 5. Shear failure (15:1, $\tau_{max}/\tau_b = 70\%$, $N_f = 2.4 \times 10^4$ cycles)

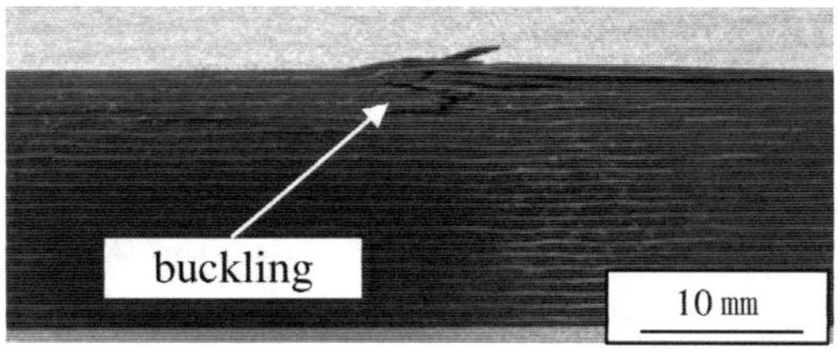

Figure 6. Buckling failure (15:1, $\tau_{max}/\tau_b = 60\%$, $N_f = 4.0 \times 10^5$ cycles)

Figure 7. Crack at a central bottom layer (15:1, $\tau_{max}/\tau_b = 50\%$, $N_f = 2.5 \times 10^6$ cycles)

Stiffness, normalized by the first cycle value, is plotted against number of cycles in figure 8 (a) and (b) for the specimens with span/thickness ratio 4 and 11, respectively. Both specimens failed in a sudden manner by delamination and showed a gradual decrease in stiffness during tests. However, there was a significant difference in how they decreased their stiffness and while the former specimen caused its delamination at about 90% stiffness, the latter specimen's stiffness gradually decreased, resulting in its delamination at about 60%. The slope of the curve in figure 8 (b) clearly changed after stiffness 80%, showing a sharper decrease before failure.

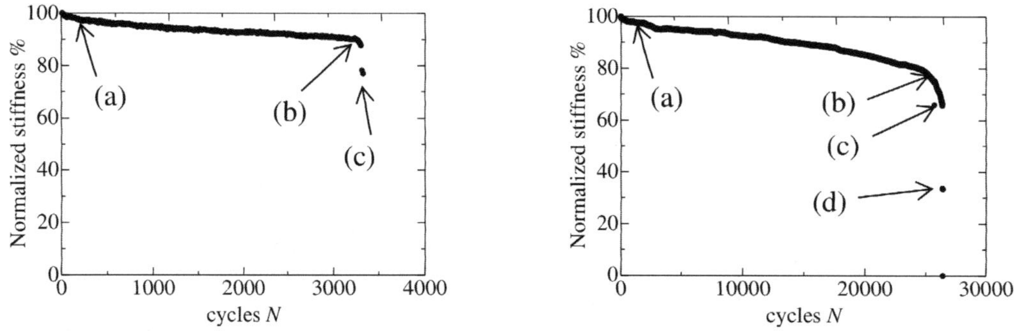

(a) Specimen with span/thickness ratio 4 (b) Specimen with span/thickness ratio 11

Figure 8. Normalized stiffness vs fatigue cycles ($\tau_{max}/\tau_b = 70\%$)

Figures 9 and 10 illustrate shear strain distribution histories at the side face of specimens made from a DIC measurement at the peak stress of each cycle during fatigue tests. Local shear and bending strains are also numerically shown in these figures. Each cycle (a)-(d) in figures 9 and 10 corresponds the same symbol in figure 8 as an observation cycle. As these figures show, delamination occurred at the point where the shear strain along through-thickness was almost maximum. The specimen with span/thickness ratio 4 failed by shear in the vicinity of neutral plane with shear strain concentration whereas the specimen with span/thickness ratio 11 had the same behavior below the neutral plane. From the DIC measurement in figure 10, shear strain concentration occurred approximately at the center for longitudinal direction, at which delamination was surely initiated. This shear strain concentration seemed to be affected by fiber buckling at the compressive area almost directly below the loading nose. Since the several layers at the top failed and were no longer able to transfer the load, the shear strain concentration was likely to occur at through-thickness lower area. While the both specimens showed a steady increase in shear strain with cycles, normal strain's (ε_{xx}) behavior was a little different according to span/thickness ratio. For the specimen with span/thickness ratio 4, normal strain at the compressive area was almost the same ε_{xx} = -0.6 % over cycles but the specimen with span/thickness ratio 11 showed its increase from ε_{xx} = -0.83 % to -6.2 %, which seems attributed to the difference of stress distribution by the effect of span/thickness ratio. It has to be cared that the value -6.2 % is the apparent strain due to fiber buckling before delamination. Furthermore, comparing (b) and (c) in figure 10, a significant increase in normal strain was obtained between 24500 and 26410 cycles. Seen from figure 8(b), there was a sharp decrease in stiffness during these cycles, therefore, the correlation was confirmed between the value of normal strain and specimen's stiffness. From these results, it can be seen that the specimen with larger span/thickness ratios shows buckling failure prior to shear failure as both normal and shear strain increase during fatigue tests, followed by final delamination. However, the specimen with smaller span/thickness ratios does not clearly develop its normal strain and shear failure happens before even the increase in the effect of the bending strain. As the above results demonstrated, although the specimens show the same failure behavior by delamination, the failure process depends on span/thickness ratio and normal or shear strain developed in a different manner.

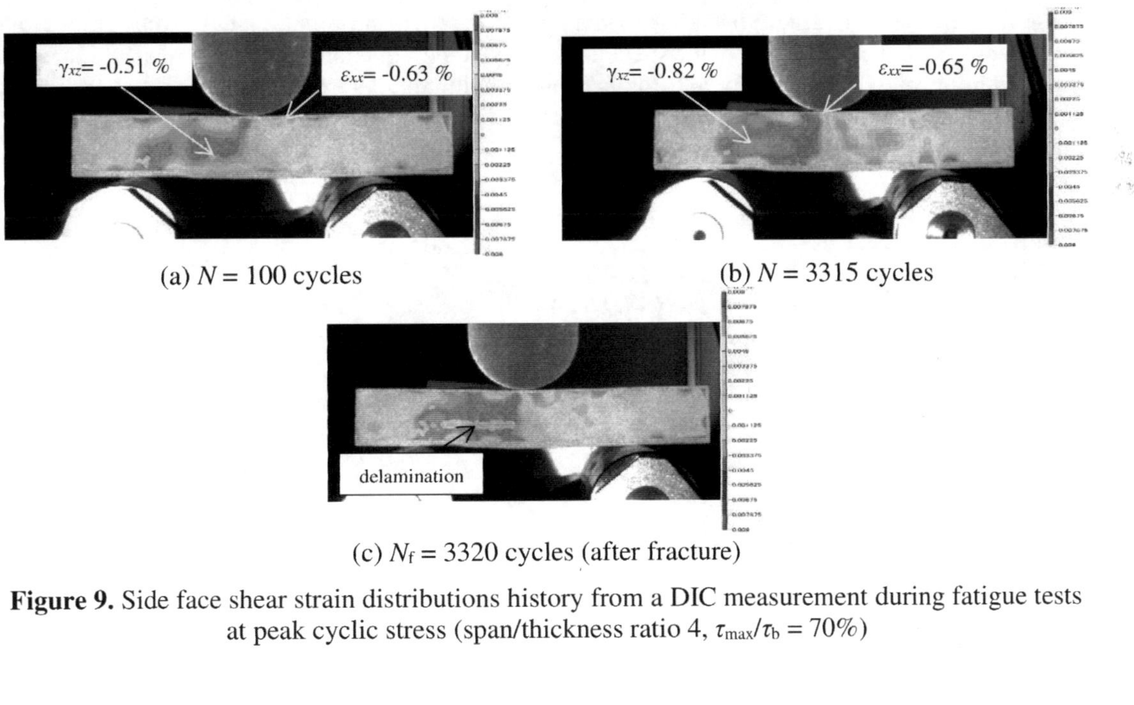

(a) N = 100 cycles (b) N = 3315 cycles

(c) N_f = 3320 cycles (after fracture)

Figure 9. Side face shear strain distributions history from a DIC measurement during fatigue tests at peak cyclic stress (span/thickness ratio 4, τ_{max}/τ_b = 70%)

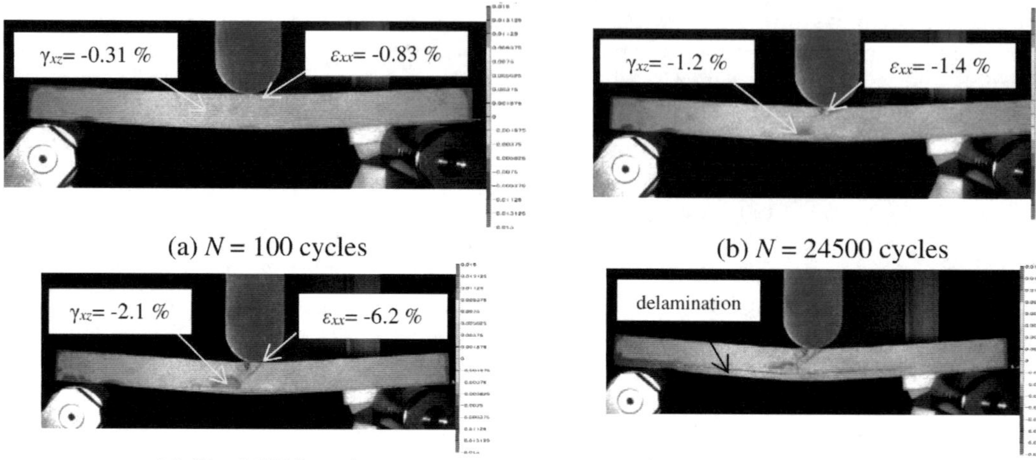

(a) $N = 100$ cycles (b) $N = 24500$ cycles

(c) $N = 26410$ cycles (d) $N_f = 26412$ cycles (after fracture)

Figure 10. Side face shear strain distributions history from a DIC measurement during fatigue tests at peak cyclic stress (span/thickness ratio 11, $\tau_{max}/\tau_b = 70\%$)

4. Conclusion

The effect of span/thickness ratio of the specimens on bending fatigue properties was investigated using thick CFRP laminates in this research. The results of three-point bending test show that the apparent interlaminar shear strength and failure modes depend on the span/thickness ratio. The DIC method was also used to study shear fatigue behavior of the specimens with different span/thickness ratios over cycles. It confirmed that even though the specimens showed the same failure by delamination, shear and normal strain developed differently and the specimens with smaller span/thickness ratios exhibit no increase in bending strain. New experiments will be performed to investigate the detailed relationship between span/thickness ratio and the effect of shear/bending stress, focusing on the stress state of each span/thickness ratio.

References

[1] Marsh G 2004 *Reinf. Plast.* **48** 34-38
[2] Makeev A 2013 *Compos. Sci. Technol.* **80** 93-100
[3] He Y and Makeev A 2014 *Int. J. Solids. Struct.* **51** 1263-73
[4] May M and Hallett S R 2010 *Compos. Part A-Appl. S.* **41** 1263-73
[5] Backe D, Balle F and Eifler D 2015 *Compos. Sci. Technol.* **106** 93-99
[6] Djabali A, Toubal L, Zitoune R and Rechak S 2019 *Compos. Sci. Technol.* **183** 107815
[7] Solyaev Y and Babaytsec A 2021 *Compos. Part B-Eng.* **224** 109145
[8] Xie M and Adams D F 1995 *Composites.* **26** 653-59
[9] He Y, Makeev A and Shonkwiler B 2012 *Compos. Sci. Technol.* **73** 64-71
[10] Otsubo M, Hosoi A, Sakurai T, Yamatogi T and Kawada H 2023 *Proc. 14th Japanese Conference on Composites Materials (JCCM-14)* **3B-13** (in Japanese)

IOP Conf. Series: Materials Science and Engineering 1293 (2023) 012020 doi:10.1088/1757-899X/1293/1/012020

Experimental testing method to characterise the drapability of UD non-crimp fabrics used in wind turbine blades

Rajnish Kumar*, Kaveh Rashvand, Anthony Fraisse, Ali Sarhadi and Tom L. Andersen

Division of Wind Energy Materials and Components, Department of Wind and Energy Systems, Technical University of Denmark, 4000 Roskilde, Denmark

*E-mail: rakum@dtu.dk

Abstract. This scientific article presents a novel approach for characterising the drapability of fabrics used in wind turbine blade production. This study defines drapability as an intrinsic property of fabric to shear. Specifically, it refers to the potential of the rovings to slide with respect to each other. The evolution of wrinkles has been quantified by the ratio of height-to-width corresponding to a shear angle. The growing industrial interest in binder fabrics, for their preforming ability and improved handling leading to faster blade production, has motivated this study. In this research, two types of non-crimp fabrics, with and without binder, were analysed to study the evolution of wrinkles concerning applied shear angles. A state-of-the-art 3D blue light scanning technique is employed to accurately measure the aspect ratio (height/width) of wrinkles at various shear angles, including 0°, 4°, 6°, 8°, 12°, and 16°. A wrinkle having an aspect ratio of 1/10 was determined to correlate with an applied shear angle of 9° for non-binder fabrics, and 3° for binder-based fabrics. The findings clearly demonstrate the influence of binders on fabric drapability, reducing it by a factor of three. These results provide valuable insights into the influence of different parameters on wrinkle formation, aiding in controlling these factors to avoid manufacturing defects in wind turbine blades.

1. Introduction

Drapability is a fundamental characteristic that describes a fabric's ability to conform to the surface of moulds [1]. The existing standards for drapability, such as BS 5058:73 [2] and AFNOR G07-109 [3], adequately apply to apparel, upholstery, and non-woven fabrics, there is a need for better adaptation to address drapability capabilities in composite preforms [1]. For engineering textiles like woven fabrics, unidirectional or biaxial non-crimp fabrics, the draping behaviour is mostly affected by in-plane shear and bending characteristics [4]. To characterise the shear deformation behaviour of fabrics, picture frame tests [5,6] and uniaxial bias extension tests [7,8] are commonly used. Several researchers have worked to characterise and analyse of wrinkling behaviour of fabrics with hemispherical [9], square box [10] and double-dome [11] punching preforming methods.

Wind turbine blades possess a complex geometry optimised for aerodynamic efficiency [12] and structural integrity [13]. Among the various manufacturing techniques employed, vacuum-assisted resin transfer moulding is widely used to produce longer blades exceeding 55 meters in length. This technique involves placing fabrics in a mould, applying vacuum pressure, and injecting resin into the cavity. The resin fills the space between the fibres, and the resulting component is cured through heat [14]. Currently, the most economically feasible approach to wind blade manufacturing involves manual fabric manipulation by workers to ensure a layup without wrinkles in the mould [15]. Preventing out-of-plane deformation during the fabric layup process is crucial. Such deformation occurs when certain fibres are forced to span shorter distances than their parallel counterparts in the fabric, leading to compression and buckling, ultimately forming wrinkles [16]. This is especially important in double-curved areas, such as the root section.

A wide range of non-destructive testing (NDT) methods exists for detecting wrinkles in fibre-reinforced composites, including ultrasonic techniques, linear phase FIR-filtered ultrasonic array data, laser and X-ray tomography [17], [18], [19]. Griffin and Malkin [20] from the DNV have identified fibre wrinkles as a significant manufacturing defect leading to catastrophic failures of wind turbine blades, compromising their reliability and cost-effectiveness. Several studies have reported substantial reductions in strength and stiffness in laminates containing wrinkles [21], [22]. Nartey et al. [23] investigated the influence of fibre wrinkles on composite laminate's mechanical properties and failure mechanisms, revealing a significant drop of up to 21% and 37% in tensile and compressive strength, respectively, for the specimen with wrinkle angle of 7.8°.

The use of binder fabrics has gained industrial interest due to their superior handling characteristics compared to binder-free fabrics. Binder fabrics offer the potential for faster wind turbine blade manufacturing by preparing fabric stacks outside the mould and subsequently placing them inside. Automation of the blade manufacturing process has also gained considerable attention, with attempts made to equip robotic arms with standard layup tools to replicate the manual movements of workers during fabric seating in the mould. These robotic arms employ sensory input to apply appropriate pressure to manipulate the fabric effectively. Controlling wrinkle formation during automated fabric installation is critical for enhancing manufacturing quality.

Despite extensive research on the detection of fibre wrinkles and their impact on the mechanical performance of fibre-reinforced composites, there remains a notable research gap concerning the assessment of fabric wrinkle formation during composite manufacturing. This paper presents a method for characterising fabric drapability, defining it as an intrinsic property related to the shearing potential between rovings without the formation of wrinkles. The study focuses on analysing two types of non-crimp fabrics, one with a binder and one without, investigating the evolution of wrinkles in response to applied shear angles. A 3D blue light scanning technique is employed to measure the out-of-plane movement of the fabrics and quantify the aspect ratio of wrinkles for various shear angles (0°, 4°, 6°, 8°, 12°, and 16°).

2. Materials and methods

2.1. Materials

The two investigated non-crimp fabrics, with and without binder, have a total glass fibre areal weight of 1382 g/m^2. Each fabric consists mainly (96 wt %) of H-glass fibres with a fibre diameter of 17 to 24 μm oriented at 0°. In addition, there are 4 wt % E-glass backing fibres, with a fibre diameter of 9 μm, oriented at ±80°. The fibre angles are given relative to the warp direction of the fabric. The stitch to keep fibres in place is a combined tricot-chain polyester thread stitch with an areal weight of 15 g/m^2. The preform fabrics have an additional 15 g/m^2 of a soluble polyester binder dispersed on the roving side of the fabric. Three fabric specimens of each fabric type with a width of 400 mm have been tested.

2.2. Methods

2.2.1. Experimental setup. The experimental setup for the drapability test is depicted in Figure 1. The setup consists of an Instron universal testing machine (UTM) fitted with a 1 kN load cell, depicted on the left-hand side of Figure 1. A parallelogram fixture, similar to the previously referred method [24] forms the subsequent component of the setup. While one side of the parallelogram fixture is affixed to the frame, the other side can move freely in the in-plane direction. A metallic wire links the top grip of the UTM to the movable side of the parallelogram fixture through a pulley mechanism. The test fabric is then placed on the fixture and secured using bolts. A constant torque of 7 Nm is applied to the bolts utilising a torque wrench.

The upward movement of the top UTM grip initiates a shear action, causing the fabric to form wrinkles. To monitor and analyse the fabric's behaviour during shearing, the top surface of the fabric is scanned at different shear angles (0°, 4°, 6°, 8°, 12°, and 16°) using a blue light 3D scanner (EviXscan 3D).

Before the scans, the 3D scanner's projector is configured for brightness and distance, and the camera's exposure is calibrated. The initial state of the fabric is set as the zero height reference. Subsequently, the acquired 3D scans are processed and analysed using the Geomagic Design X software developed by 3D Systems. Figure 2 provides an illustrative example of a 3D scan obtained during the drapability test.

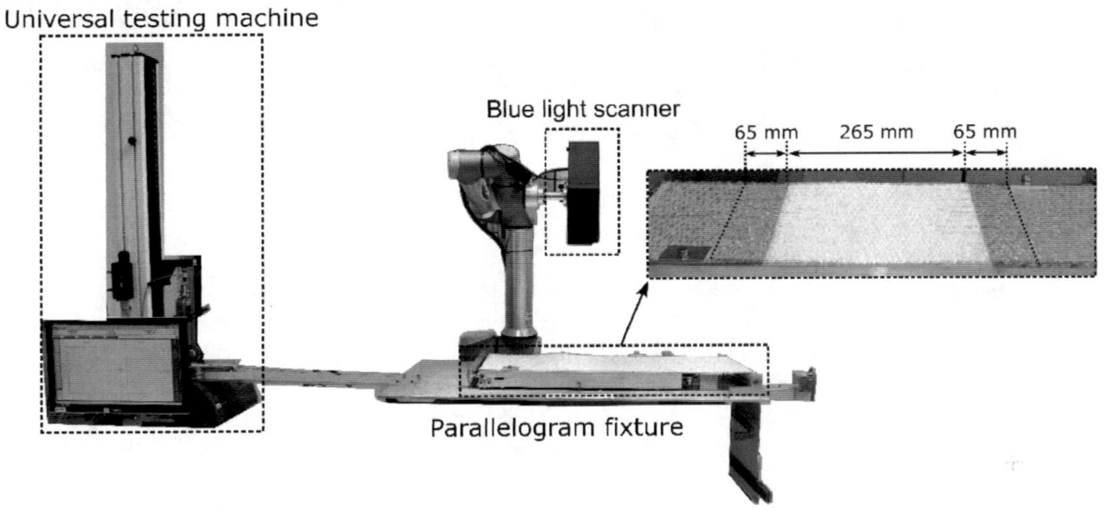

Figure 1. Experimental drapability set-up enabling fabric shear and wrinkle detection by 3D blue light scanning.

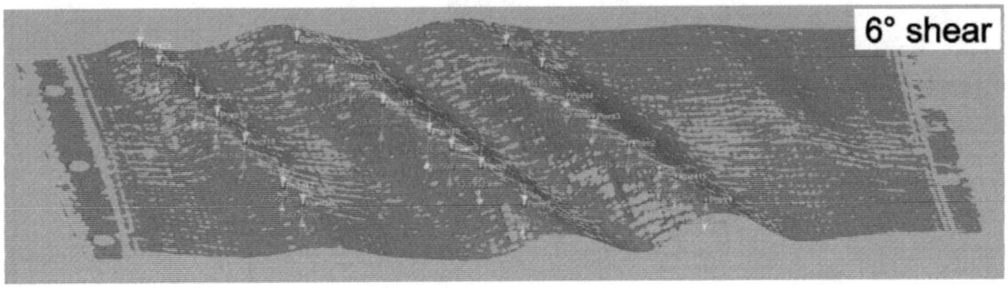

Figure 2. Example of a 3D scan showing wrinkling of the non-crimp fabric at 6° shear.

2.2.2. Testing configurations. In developing this novel test method, two different configurations representing distinct boundary conditions of the test setup have been investigated (see Figure 3). In configuration 1, the fabric is completely constrained to the frame using a rectangular metal plate and bolts. This configuration simulates fully constrained boundary conditions. It is similar to the bias-extension test to investigate the in-plane shear kinematics of a fabric [25]. The main concern for this configuration is that the boundary conditions might lead to the full locking of the fabric, especially preventing its shear and instead testing the backing fibres in tension. Therefore, this configuration does not represent a fabric layer laid out in the wind turbine blade mould. In configuration 2, the fabric is partially constrained to the frame at the diagonal positions. It gives freedom to the fabric to move around, replicating the fabric placement scenario in a wind turbine blade. The analysed area of the fabric is 65 mm away from the clamps, following the Saint Venant principle [26] to leave the boundary conditions far enough from the measurement area (see Figure 1).

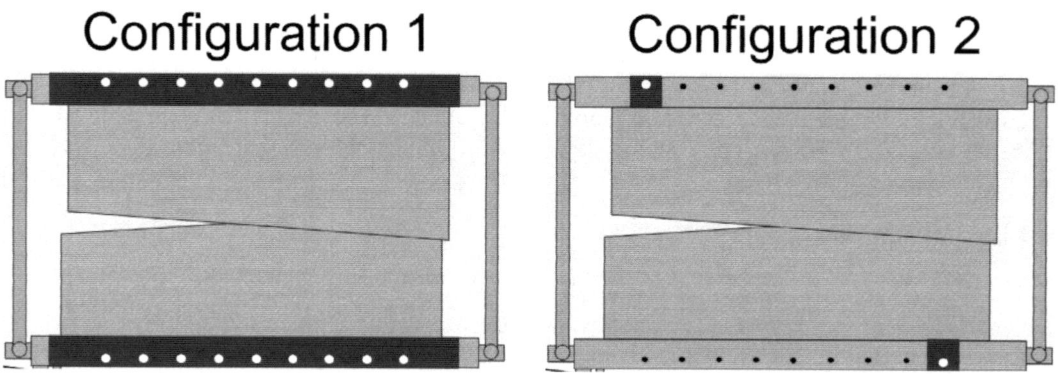

Figure 3. Two configurations of the drapability tester set-up. The left side represents a completely constrained configuration, whereas the right represents a partially constrained configuration.

3. Results and discussion

Figure 4 shows the load-displacement curve for the same non-binder fabric tested with two configurations. In *configuration 1*, the resultant load is higher than in *configuration 2*. In *configuration 1,* the backing fibres oriented at ±80° are stretched due to the fully constrained boundary conditions. Therefore, the measured load is not describing the shear load necessary to shear the fabric but also the tensile load carried by the backing fibres. *Configuration 2* shows a lower applied load than *configuration 1*, representing the shear load needed to shear this specific fabric. In this configuration, shearing between the unidirectional (UD) rovings is thus possible, leading to wrinkle formation after a certain angle. Therefore, the rest of the study used *configuration 2* of the drapability tester setup. The decision has been taken to observe the wrinkle height and width as a function of the applied shear angle rather than monitoring load and displacement, as it is more relevant and practical for wind turbine blade manufacturing applications.

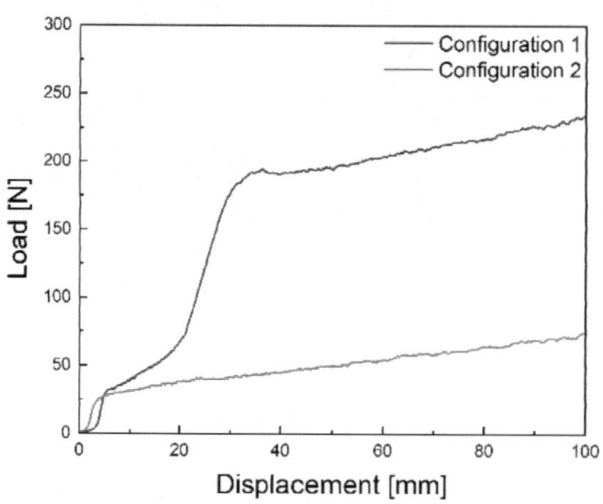

Figure 4. Load vs displacement curves of non-binder fabrics tested with two configurations of the drapability tester set-up.

This study defines drapability as an intrinsic property of fabric to shear. Specifically, it refers to the potential of the rovings to slide with respect to each other without creating wrinkles. Two non-crimp fabric types, with and without binder, have been analysed in terms of the evolution of wrinkles with respect to the applied shear angle. Figure 5 shows an example of wrinkle evolution with increasing shear angles of 0°, 4°, 6°, 8°, 12°, and 16° for a non-binder fabric.

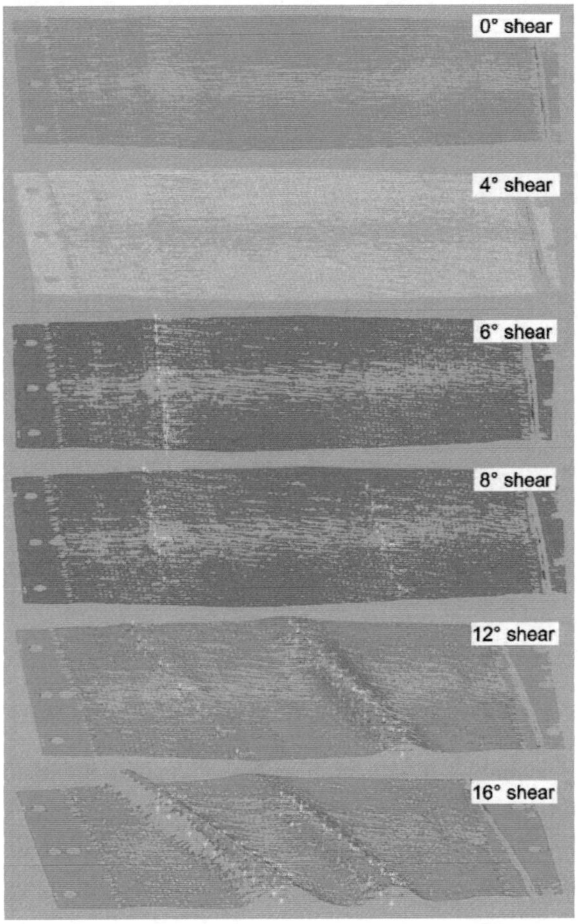

Figure 5. Example of the evolution of wrinkles in the non-binder fabric with increasing shear angle.

Figure 6A shows the relationship between the average maximum height of fabric wrinkles and the applied shear angle (°) for a standard non-binder fabric. The "average maximum height" refers to the arithmetic mean of maximum height values for all the observed wrinkles. Notably, no wrinkles were observed at a shear angle of 4°. This observation indicates that at low shear angles, the fibre rovings possess the ability to slide, thereby avoiding wrinkle formation. This finding aligns well with production and design data. After 4°, the backing fibres are stretched, creating a locking phenomenon on the UD rovings. The UD rovings are no longer capable of sliding next to each other, and as the displacement continues to increase, the fabric wrinkles. In other words, wrinkles appear when the shear is prevented, and Poisson's effect becomes dominant. The non-binder fabric shows an increasing trend in the average maximum height of the wrinkle with an increasing shear angle. Figure 6B shows the average width of fabric wrinkle at the maximum height of the wrinkle. The width of the wrinkle also shows an increasing trend with an increasing shear angle.

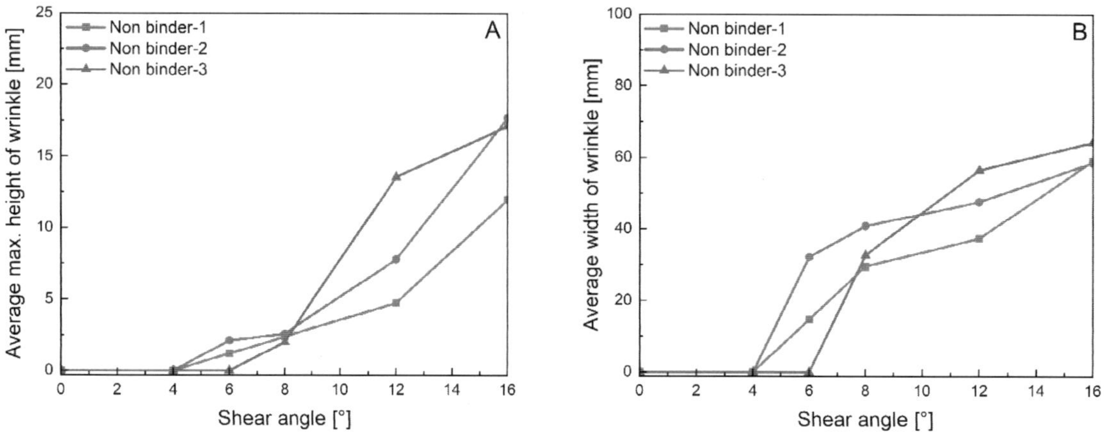

Figure 6. A) The average of maximum wrinkle height (in mm) for the non-binder fabric as a function of the applied shear angle B) The average wrinkle width (in mm) at the point of maximum wrinkle height.

Figure 7A presents the average maximum height of fabric wrinkle as a function of the applied shear angle for a binder fabric. Wrinkles were observed at a shear angle of 4°, with an average maximum height of 10 mm. It is highly likely that wrinkles begin to appear at a shear angle lower than 4°. The binder fabric demonstrated an increasing trend in the average maximum height of wrinkles as the shear angle increased. However, at a shear angle of 12°, the height of the wrinkles reached a constant value. Figure 7B shows the average width of fabric wrinkle at the maximum height of the wrinkle. The width of the wrinkle shows an increasing trend until a shear angle of 6°, then decreases with increasing shear angle. In comparison to the non-binder fabric, the binder fabric exhibited lower drapability. The binder material acts as a glue between the UD rovings and prevents their movement relative to each other. Due to this phenomenon, coupled with the locking effect from the backing bundles, binder fabrics exhibit a higher average wrinkle height and width than standard fabrics with no binder.

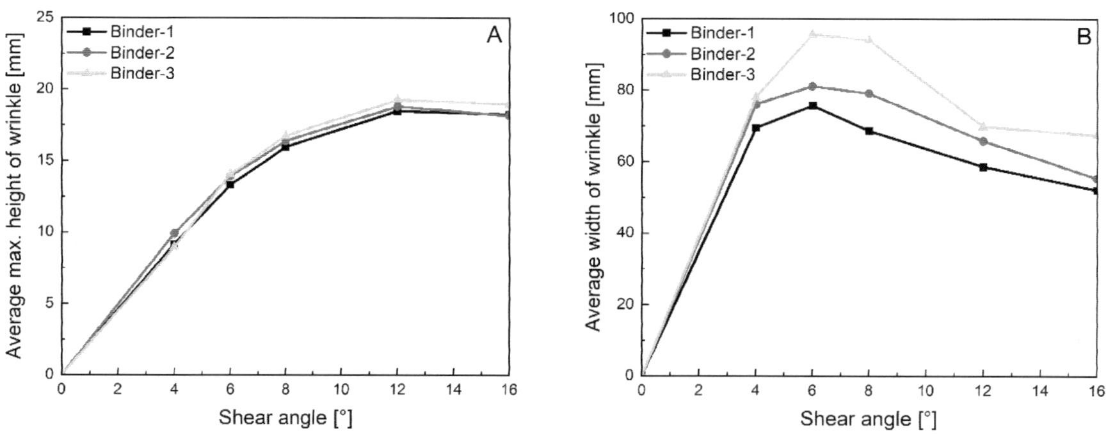

Figure 7. A) The average of maximum wrinkle height (in mm) for a binder fabric as a function of the applied shear angle B) The average wrinkle width (in mm) taken at the point of maximum wrinkle height.

Figures 8A and 8B show the height-to-width ratio of the wrinkle for non-binder and binder fabric types, respectively. The severity of a wrinkle can be measured by the aspect ratio of its height and width [27]. Mendonça et al. [28] evaluated the effect of wrinkle defects on the stiffness response of composite laminates for wind turbine blades. According to their definition of aspect ratio (height/half-width), a wrinkle with a 1/5 aspect ratio was the most severe one. Compared to flat composite laminate, the laminate with wrinkle showed a knockdown of 54% in stiffness properties. In this study, on the fabric level, the aspect ratio used by Mendonça et al. corresponds to the aspect ratio of 1/10. This ratio is also plotted in the dashed line in Figure 8.
In this study, an aspect ratio of 1/10 corresponds to an applied shear angle of 9° and 3° for non-binder fabric and binder fabric, respectively. These results clearly showed the effect of binder on the fabric, reducing the drapability by a factor of 3. The aspect ratio measured on the fabric level will not directly translate to the same aspect ratio on a composite level, given the fact that the fabric layers will undergo the vacuum infusion process, potentially reducing the overall size of the wrinkle.

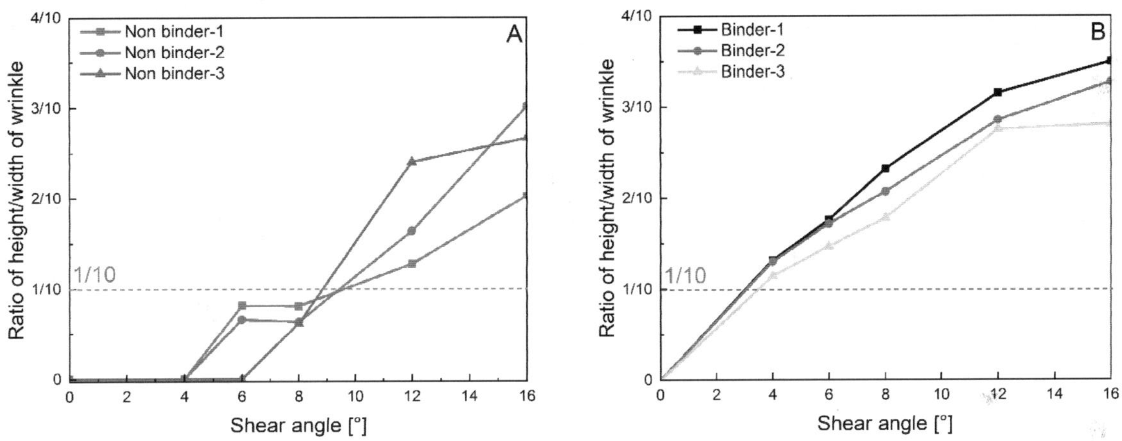

Figure 8. The height-to-width ratio of the wrinkles for A) non-binder fabric and B) binder fabric.

There is a scope to study the characteristics of wrinkles prior to and after infusion stages. The purpose is to bridge a link between a wrinkle on a fabric to the wrinkle on the composite laminate level, drawn up to the structural level. Manufacturers implement various procedures to detect and repair these defects. Therefore, avoiding these defects and ensuring a high level of manufacturing quality and productivity is vital. This study provides a basis for identifying the potential zones for repair exceeding the permissible limit. The results show that to resolve challenges associated with the upcoming fabric types and avoid wrinkles forming, it would be necessary to define and qualify drapability more precisely.

4. Conclusion

A novel methodology has been introduced for characterising fabric drapability as an intrinsic property associated with the shear angle. In this study, two types of non-crimp fabrics, with and without binder, have been examined to investigate the evolution of wrinkles in response to varying shear angles. A state-of-the-art 3D scanning technique was employed to quantify the aspect ratio at shear angles ranging from 0° to 16°. For the non-binder fabric, a critical aspect ratio of 1/10 corresponded to a shear angle of 9°, while for the binder fabric, it corresponded to a shear angle of 3°. The findings clearly demonstrated the detrimental effect of binders on fabric drapability, resulting in a three-fold reduction. These results provide valuable insights into the impact of binders and other parameters on wrinkle formation, enabling better control of these factors to avoid manufacturing defects in wind turbine blades. Consequently, this novel experimental method could assist the industry in selecting suitable fabrics according to their drapability for specific blade designs, thereby minimising post-production repairs. However, this study also highlights challenges, including the need for a precise and universally accepted definition of drapability to prevent wrinkle formation, particularly in regions with highly curved geometry. Future investigations will focus on refining the aspect ratio measurement at lower shear angles and establishing the relationship between fabric wrinkles and laminate wrinkles, as well as their influence on the mechanical properties of laminates at a structural level.

Acknowledgements
The MADEBLADES project, funded by the Energy Technology Development and Demonstration Program, Grant No. 64019–0514, supported the work. This support is gratefully acknowledged. The authors also would like to acknowledge Steffen Rasmussen for his work leading to the development of this methodology, Mariana Rojas for her help in providing the fabrics for the test, and Vagn Jensen for his work on the fixture modification.

References
[1] Chennamsetti R 2015 *Structural Health Monitoring of composite structures* vol 95
[2] British Standards Institution 1974 *The assessment of Drapeability of fabrics, B.S Handbook No 11* (London)
[3] Anon 1980 *Method for determination of the drape of a woven or knitted fabric*
[4] Schirmaier F J, Weidenmann K A, Kärger L and Henning F 2016 Characterisation of the draping behaviour of unidirectional non-crimp fabrics (UD-NCF) *Compos Part A Appl Sci Manuf* **80** 28-38
[5] Harrison P, Clifford M J and Long A C 2004 Shear characterisation of viscous woven textile composites: a comparison between picture frame and bias extension experiments *Compos Sci Technol* **64** 1453-65
[6] Lomov S V, Barburski M, Stoilova Tz, Verpoest I, Akkerman R, Loendersloot R and Thije R H W ten 2005 Carbon composites based on multiaxial multiply stitched preforms. Part 3: Biaxial tension, picture frame and compression tests of the preforms *Compos Part A Appl Sci Manuf* **36** 1188-206
[7] Härtel F and Harrison P 2014 Evaluation of normalisation methods for uniaxial bias extension tests on engineering fabrics *Compos Part A Appl Sci Manuf* **67** 61-9
[8] Bel S, Boisse P and Dumont F 2012 Analyses of the Deformation Mechanisms of Non-Crimp Fabric Composite Reinforcements during Preforming *Applied Composite Materials* **19** 513-28
[9] Viisainen J V and Sutcliffe M P F 2021 Characterising the variability in wrinkling during the preforming of non-crimp fabrics *Compos Part A Appl Sci Manuf* **149** 106536
[10] Bai R, Chen B, Colmars J and Boisse P 2022 Physics-based evaluation of the drapability of textile composite reinforcements *Compos B Eng* **242** 110089
[11] Ali H, Noor S, Huiqi S, Jinhua J and Nanliang C 2020 Characterization and analysis of wrinkling behavior of glass warp knitted non-crimp fabrics based on double-dome draping geometry *J Eng Fiber Fabr* **15** 1-17
[12] Mishnaevsky L, Branner K, Petersen H N, Beauson J, McGugan M and Sørensen B F 2017 Materials for Wind Turbine Blades: An Overview *Materials* **10** 1285
[13] Gonzaga P, Toft H, Worden K, Dervilis N, Bernhammer L, Stevanovic N and Gonzales A 2022 Impact of blade structural and aerodynamic uncertainties on wind turbine loads *Wind Energy* **25** 1060-76
[14] Summerscales J and Searle T J 2005 Low-pressure (vacuum infusion) techniques for moulding large composite structures *Proceedings of the Institution of Mechanical Engineers, Part L: Journal of Materials: Design and Applications* **219** 45-58
[15] Deisenroth D 2011 Development of an Automated Fabric Deformation System for Composite Wind Turbine Blade Manufacturing

[16] Bloom L D, Wang J and Potter K D 2013 Damage progression and defect sensitivity: An experimental study of representative wrinkles in tension *Compos B Eng* **45** 449-58
[17] Larrañaga-Valsero B, Smith R A, Tayong R B, Fernández-López A and Güemes A 2018 Wrinkle measurement in glass-carbon hybrid laminates comparing ultrasonic techniques: A case study *Compos Part A Appl Sci Manuf* **114** 225–40
[18] Ma T, Li Y, Zhou Z and Meng J 2023 Wrinkle Detection in Carbon Fiber-Reinforced Polymers Using Linear Phase FIR-Filtered Ultrasonic Array Data *Aerospace* **10**
[19] Smith R A, Nelson L J, Mienczakowski M J and Wilcox P D 2018 Ultrasonic Analytic-Signal Responses From Polymer-Matrix Composite Laminates *IEEE Trans Ultrason Ferroelectr Freq Control* **65** 231–43
[20] Griffin D and Malkin M 2011 Lessons Learned from Recent Blade Failures: Primary Causes and Risk-Reducing Technologies *49th AIAA Aerospace Sciences Meeting including the New Horizons Forum and Aerospace Exposition* Aerospace Sciences Meetings (American Institute of Aeronautics and Astronautics)
[21] Potter K, Khan B, Wisnom M, Bell T and Stevens J 2008 Variability, fibre waviness and misalignment in the determination of the properties of composite materials and structures *Compos Part A Appl Sci Manuf* **39** 1343–54
[22] El-Hajjar R F and Petersen D R 2011 Gaussian function characterization of unnotched tension behavior in a carbon/epoxy composite containing localized fiber waviness *Compos Struct* **93** 2400–8
[23] Nartey M, Zhang T, Gong B, Wang J, Peng S, Wang H and Peng H X 2020 Understanding the impact of fibre wrinkle architectures on composite laminates through tailored gaps and overlaps *Compos B Eng* **196** 108097
[24] Pourtier J, Duchamp B, Kowalski M, Wang P, Legrand X and Soulat D 2019 Two-way approach for deformation analysis of non-crimp fabrics in uniaxial bias extension tests based on pure and simple shear assumption *International Journal of Material Forming* **12** 995–1008
[25] Krogh C, Kepler J A and Jakobsen J 2021 Pure and simple: investigating the in-plane shear kinematics of a quasi-unidirectional glass fiber non-crimp fabric using the bias-extension test *International Journal of Material Forming* **14** 1483–95
[26] Saint-Venant 1855 Memoire sur la Torsion des Prismes, Mem. Divers Savants14, pp. 233–560
[27] Wang J, Potter K D, Hazra K and Wisnom M R 2011 Experimental fabrication and characterization of out-of-plane fiber waviness in continuous fiber-reinforced composites *J Compos Mater* **46** 2041–53
[28] Mendonça H G, Mikkelsen L P, Chen X, Bode J, Mortensen F, Haselbach P U and Branner K 2022 Methodology to predict stiffness knock-down in laminates for wind turbine blades with artificial wrinkles *Wind Energ. Sci.* **7** 2513–25

Fatigue S-N curve approach for impact loading of hyper- and visco-elastic leading edge protection systems of wind turbine blades

T W Hamers[1*], K M Jespersen[1], L P Mikkelsen[1], J I Bech[1], N F-J Johansen[1]

[1]Department of Wind Energy, Technical University of Denmark, Frederiksborgvej 399, 4000 Roskilde, Denmark

E-mail: twiha@dtu.dk*

Abstract.
Computational evaluation of leading edge erosion remains challenging due to the high-strain rate loading conditions caused by impact of the wind turbine blade leading edge with rain droplets and other environmental particles. Here, a methodology is proposed for obtaining an S-N curve which can be used for impact fatigue evaluation of hyper- and viscoelastic leading edge protection systems for wind turbine blades, in the relevant strain rate domain. Two material systems (hard and soft polyurethane (PU)) are characterised experimentally by dynamic mechanical analysis (DMA) and static tensile tests. Time-temperature superposition is applied to the raw DMA data in order to obtain the material's mastercurve, describing its visco-elastic behaviour in an expanded strain rate domain. The Yeoh (hyperelastic) and prony series (viscoelastic) material model parameters are calibrated and form the input for a 2D-axisymmetric finite element model, in which Single Point Impact Fatigue Test (SPIFT) testing conditions are simulated. The stress field experienced by the coating during SPIFT testing is obtained and combined with the experimental measurements, allowing the determination of the material systems S-N curve, in the relevant strain rate domain. Results for a hard and soft PU coating system are compared with rain erosion test (RET) data. The RET data shows higher lifetime for the hard PU systems, a tendency that can be predicted when comparing the S-N curve for the hard and soft PU system. This methodology can be utilised in computational lifetime evaluation of leading edge coating systems. Furthermore, the methodology has the potential to partly alleviate the need of RET in the development and comparison of next-generation leading edge protection systems.

Keywords: Wind turbine blades, Leading edge erosion, Impact fatigue, Viscoelasticity, High strain rate, S-N curve

1. Introduction

Leading edge erosion (LEE) constitutes the degradation of a wind turbine blade's leading edge due to impact with environmental particles. Causing AEP losses of up to 5% and leading to expensive maintenance campaigns, LEE is one of the biggest challenges in the wind industry [1]. This is further emphasised by the tendency to larger and more powerful turbines [2], increasing the blade length, potentially leading to higher tip-speeds. This means more erosion resistant and durable leading edge protection systems will be required in the future.

Different solutions have been proposed such as erosion shield, tapes and protective coatings. Of these, elastomeric coating systems are the most common practise in industry today [3, 4]. The mechanical properties of these materials show dependency on strain rate and should therefore be evaluated in the relevant strain rate domain. Different studies have been performed in which the lifetime of such systems has been predicted [5, 6, 7]. However, only limited studies exist on the evaluation of impact fatigue performance in the relevant strain rate domain.

Here, a methodology is proposed to obtain an S-N curve, describing the materials fatigue properties in the relevant strain rate domain. Two polyurethane (PU) systems, a soft and a hard system, are characterised based on dynamic mechanical analysis (DMA) and static tensile tests. Both systems are experimentally evaluated using the single point impact fatigue test (SPIFT). The experimental results are combined with a numerical evaluation of the SPIFT conditions [8] in order to obtain the material's stress based damage criteria in the form of a S-N curve. The relative material's performance is validated by comparison with the performance in rain erosion whirling arm test conditions.

2. Material Characterisation

Two PU systems are investigated for this study. They are distinguished by their respective hardness with one system relatively soft and one relatively hard to the touch. Furthermore, the nitrile rubber used as the projectile material in SPIFT experimental testing is discussed. The material systems are characterised by Yeoh model (hyperelastic) and prony series (viscoelastic), calibrated based on experimental results which are described below.

2.1. Materials

The basic material properties of both PU systems are displayed in Table 1. Furthermore, the properties of the projectile material used in SPIFT (nitrile rubber) are shown. The Poisson's ratio at high strain rate is taken from [8]. The density of the systems are based on measurements according to Archimedes principle [9].

Table 1: Basic assumed material properties used in current study

	Young's modulus [MPa]	Poisson's ratio [-]	Density $\left[\frac{g}{cm^3}\right]$
Soft polyurethane	Hyper-viscoelastic	0.40	1.140
Hard polyurethane	Hyper-viscoelastic	0.40	1.130
Nitrile rubber	Hyper-viscoelastic	0.49	1.087

2.2. Tensile test

Tensile tests for the soft and hard PU were carried out in accordance to ISO 527-2 (specimen 1A) [10] at test speeds corresponding to a strain rate of $\dot{\epsilon} = 10^{-4}/s$. The test results are visualised in Figure 1 and serve as one of the inputs for the Yeoh hyperelastic material model calibration.

2.3. Dynamic Mechanical Analysis (DMA)

DMA testing was carried out in order to obtain the material systems visco-elastic properties, expressed by the storage (G') and loss (G") shear modulus. Furthermore, the loss factor (Tanδ) is determined according to Equation 1.

$$Tan\delta = \frac{G''}{G'} \tag{1}$$

A constant amplitude, cyclic shear loading frequency sweep ([10^{-2},10^{3}] Hz) was performed at constant, discrete temperature levels ranging from -70°C to +140°C. These frequency sweeps are post processed by utilising the time-temperature superposition (TTS) principle. This allows

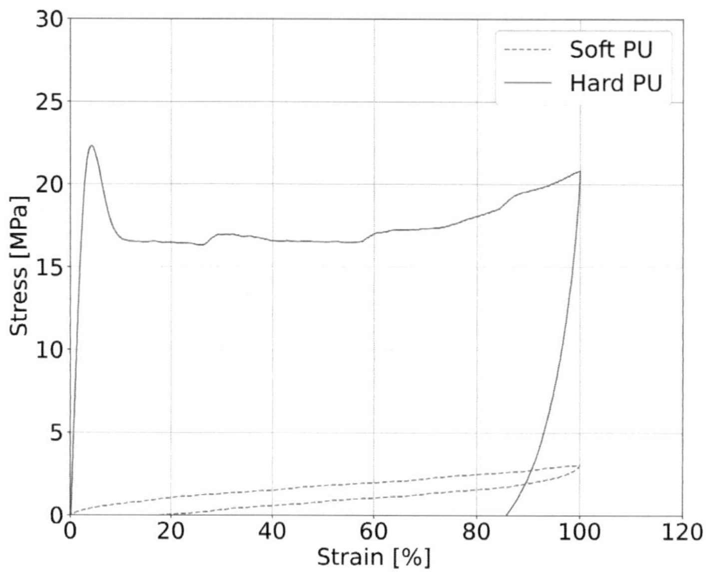

Figure 1: Tensile test experimental data for the hard and soft PU systems, corresponding to a strain rate of $\dot{\epsilon} = 10^{-4}/s$

the construction of the material's mastercurves describing the visco-elastic properties over an increased ($[10^{-5}, 10^{5}]$ Hz) frequency range, for a given reference temperature. For this purpose, a python based post-processing script was developed which calculates horizontal shift factors based on the overlapping range of the storage modulus data of adjacent temperature level's frequency sweeps. For adjacent frequency sweeps, the shift factor is determined according to Equation 2. Here, a_T represents the shift factor of the frequency sweep for temperate T_i, while ω_{Ti+1} and ω_{Ti} represent the frequency at which equal storage modulus is found for respective temperature levels T_{i+1} and T_i.

$$a_T = \frac{\omega_{Ti+1}}{\omega_{T_i}} \tag{2}$$

Figure 2 visualises the working mechanism of this tool as well as illustrates the increase in frequency range when comparing the raw DMA data with the mastercurve. Figure (a) shows the raw DMA data for the soft PU system, as well as its corresponding mastercurve, while Figure (b) visualises the horizontal shift factors applied to the respective temperature levels frequency sweeps. In the extreme temperature ranges, the material's behave increasingly unpredictable, resulting in a large scatter in the data. This data is disregarded from the analysis and are not included in the mastercurve.

2.4. Material model calibration

The software MCalibration by PolymerFEM [11] is used to calibrate the Yeoh-Prony hyper-viscoelastic material models. The calibration is carried out based on the storage and loss modulus mastercurves and the static tensile test data. The software utilises the tensile data at the slow strain rate and the mapped strain amplitudes of the DMA data covering a larger strain range to calibrate the hyperelastic material model. The viscoelastic part of the material

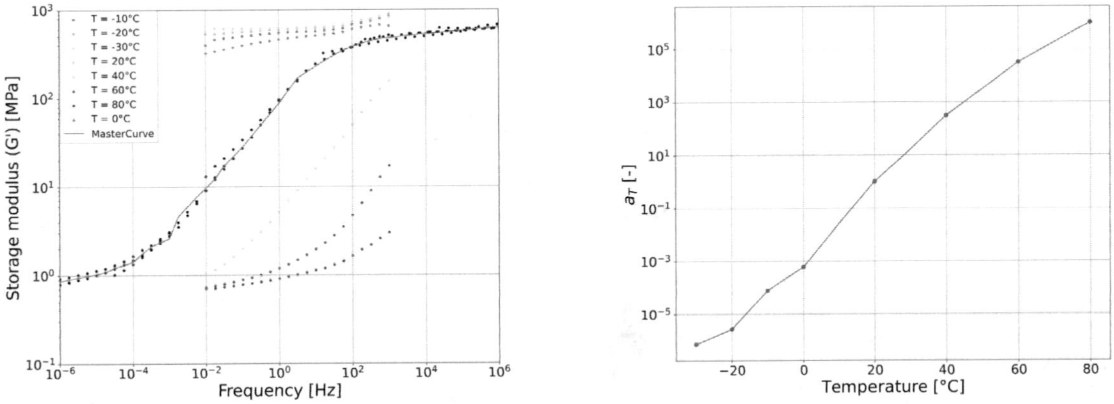

(a) (b)

Figure 2: DMA storage modulus experimental data before and after mastercurve construction (a) and shift factor a_T as a function of temperature used in mastercurve construction (b) for the soft PU system with $T_{ref} = 20°C$.

model is calibrated based on the master curves for the storage and loss modulus as a function of frequency. The calibration of the hyper- and viscoelastic material model are carried out simultaneously since the respective model parameters are dependent of each other.

3. Methodology

Here the methodology for determination of the materials impact fatigue S-N curve is discussed. The experimental evaluation of the PU systems is elaborated upon in subsection 3.1, while subsection 3.2 discusses the numerical evaluation of the testing conditions. Both results are combined, allowing for the construction of the material systems impact fatigue S-N curve as discussed in subsection 3.3

3.1. SPIFT: V-N curve

The SPIFT is a electro-pneumatic projectile firing device as described in [12] [13]. The main goal of testing with the SPIFT is to establish V-N curves (Velocity - Number of impacts to failure). As can be seen on Figure 3 the SPIFT allows for a wide array of instrumentation i.e. high-speed, optical and thermal cameras. In order to generate V-N curves, a suitable impact rate must be chosen. This is done using the thermal imaging system (Optris Pi 640 recording 120 Hz with a 0.1 °C temperature resolution) to empirically determine an impact interval that allows for the dissipation of the thermal energy imparted by each impact [13] at a given impact speed. The impact velocity (V_{impact}) is set by adjusting the air pressure via the regulator and the Optical speed trap is used to measure the mean impact velocity. The impact rate and desired maximum number of impacts is then input into the controller and the test is started. During the test the surface of the sample is monitored for damage using the inline microscope camera (M7915MZTL long working distance USB microscope from Dino-lite) with 3.1 Megapixel (2048 × 1534 @10 Hz). A continuous video of the sample is made and the test is terminated once a significant damage is observed on the sample. After the conclusion of the test, the recorded video is analysed, and the time to a given failure state as defined in [13] is measured. The number of impacts is calculated in Equation 3, with N_0 as the number of impacts until failure, t as the testing time and R_i as the impact rate.

$$N_0 = t \cdot R_i \tag{3}$$

After the V-N data is recorded the data is fitted following ASTM E739 − 10[14], using N_0 as the dependent parameter

$$N(V_{impact}) = C \cdot V_{impact}^{-m} \tag{4}$$

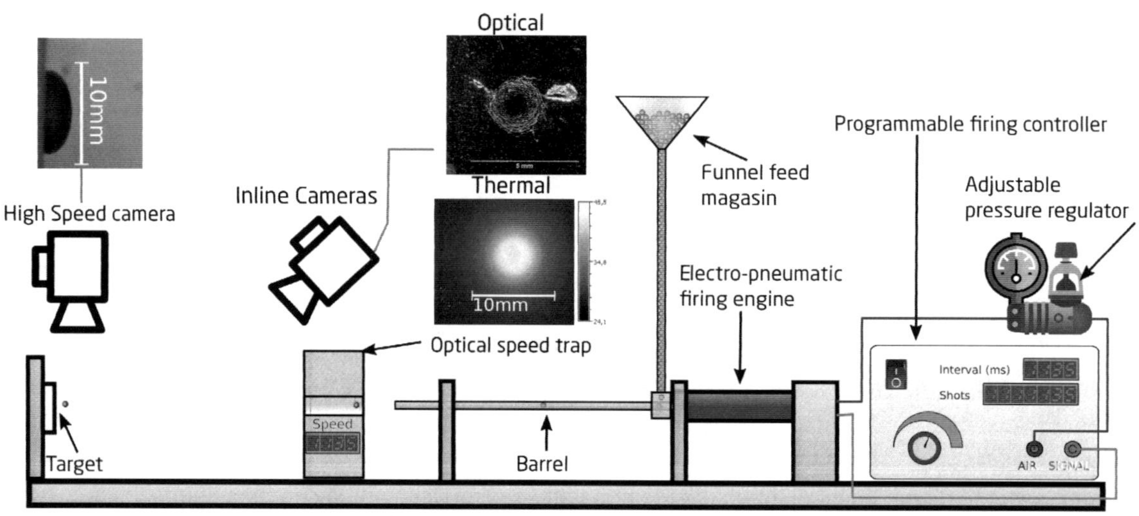

Figure 3: Illustration of the SPIFT, with instrumentation and control systems relevant for the conducted experiments.

3.2. SPIFT: FE model evaluation

SPIFT testing conditions are modelled in an Abaqus/Explicit axisymmetric finite element model, from which the stress states experienced during SPIFT testing are obtained. The model is schematically shown in Figure 4 (a) and described in [8]. The calibrated material models describing the hard and soft PU systems are used as input and the model is evaluated for $V_{impact} \in [50, 190]\ m/s$ with a step size of 10 m/s. The impact material experience a stress wave moving away from the centre of impact. Figure 4 (b) visualises such a stress state in terms of the Von Mises stress criteria. The maximum stress amplitude experienced by one of the nodes is extracted and used for subsequent analysis. The stress criteria considered is the signed von mises (SVM) stress criteria.

3.3. V-N to S-N curve

The experimental SPIFT results provide a V-N (Velocity vs. Number of impacts) relation for the coating system, while the numerical evaluation provides a S-V (Stress amplitude - Velocity) relation. Both results are combined and the following relation is fitted to the newly established S-N relation:

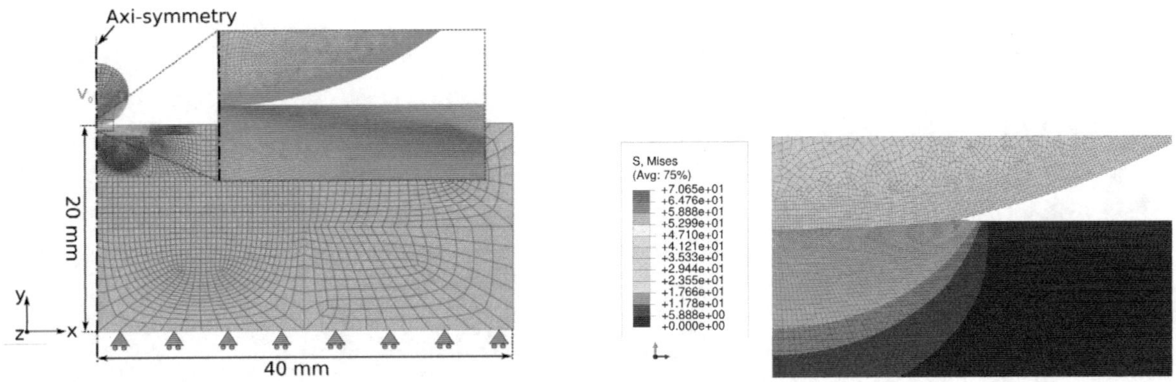

(a) (b)

Figure 4: Model geometry and boundary conditions for simulating SPIFT testing conditions with V_0 as the impact velocity indicated on the ball (a) and stress field in terms of the Von Mises stress criteria for the soft PU system, with V_{impact} = 110 m/s approximately 1 μs following impact initiation (b)

$$S = A \cdot \ln N + B \tag{5}$$

Here, S represents the maximum stress amplitude experienced by the coating system following impact, N the number of impacts until failure. A and B are fitted to the relation and describe the coating systems stress based damage criteria.

4. Results and discussion

The material characterisation and corresponding model parameters used for subsequent analysis are discussed in subsection 4.1. The S-N curve established for the hard and soft PU system is discussed and validated in subsection 4.2. A discussion on the methodologies and limitation of this work is included in subsection 4.3.

4.1. Material Characterisation

The master curves for the storage and loss modulus as well as Tanδ for both material systems is visualised in Figure 5. These form the basis for the material model calibration. The resulting Yeoh-prony model parameters are displayed in Appendix A, Table 2. The relevant strain rates for rain erosion (($>10^4$/s) [8] can be linked to the high frequency domain ($>10^4$ Hz) of the mastercurves. Here, the soft PU system shows a higher degree of energy dissipation, as compared to the hard system.

4.2. S-N curve

Figure 6 visualises the SPIFT experimental results in the form of the materials VN curve (a) and maximum stress amplitude, expressed by the SVM stress criteria (b). The soft PU system shows higher lifetime in SPIFT when compared to the hard system. Furthermore, it can be seen that the soft system experiences lower stress amplitude over the analysed impact velocities, indicating a higher degree of energy dissipation during the test. This can be recognised in the respecive materials master curves.

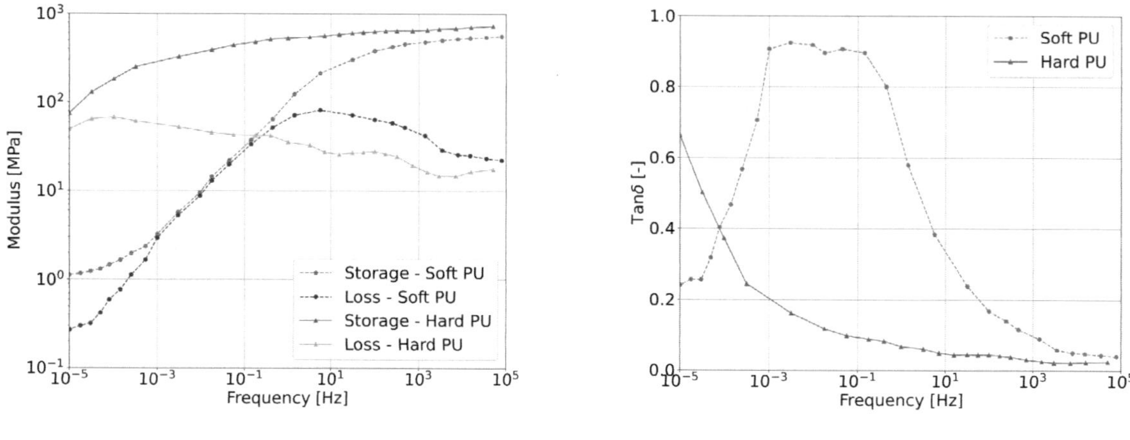

(a) (b)

Figure 5: Mastercurves for $T_{ref} = 20°C$, for hard and soft PU material system, in terms of the storage and loss modulus (a) and Tanδ (b)

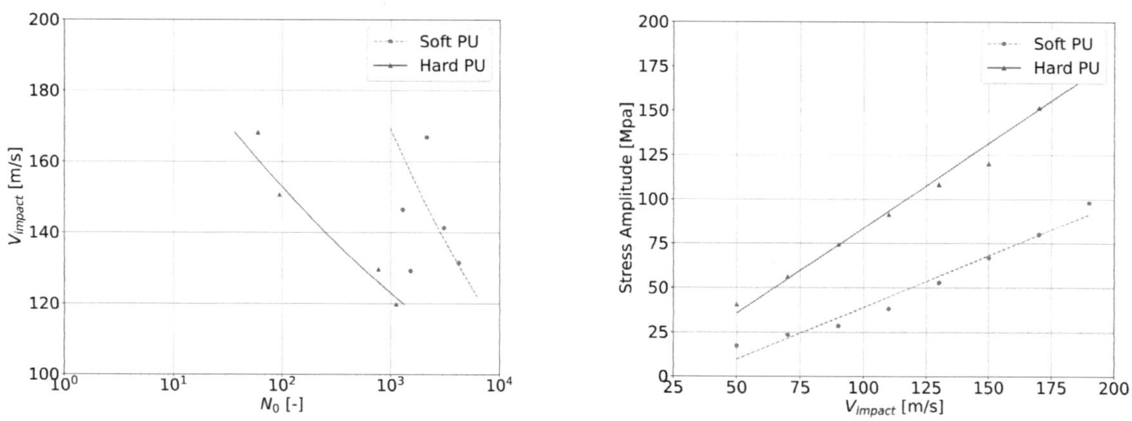

(a) (b)

Figure 6: SPIFT V-N curve (a) and Signed von Mises stress maximum amplitude as obtained by FEM model (b)

Combining these results leads to the materials S-N curve, visualised in Figure 7, (a). It can be seen under similar loading conditions, the soft PU system shows lower lifetime then the hard PU system. This behaviour is opposite to the results found in the experimental SPIFT analysis, confirming the necessity to consider the stress states in the material. To get an insight into the relative material performance in terms of impact fatigue resistance, Figure 7, (b) shows RET experimental data for both material systems. It can be seen that the relative performance of both systems corresponds to the relative position of respective material systems S-N curve.

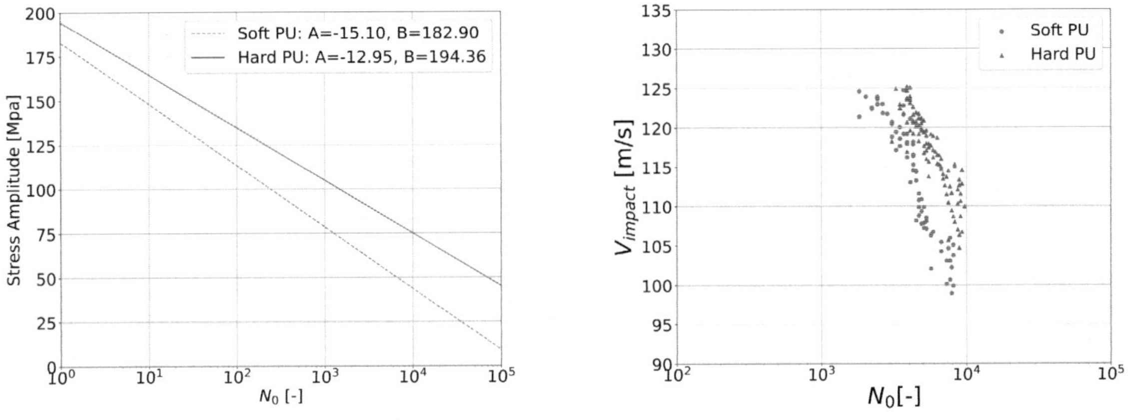

(a) (b)

Figure 7: S-N curve relation (a) and RET data (b) for the hard and soft PU system

4.3. Discussion and limitations

The S-N curves for the material systems can be used for computational lifetime evaluation of impact fatigue problems. Direct validation with RET data is not valid since the stress states during impact fatigue loading, as experienced in RET, are dependent on the visco-elastic material properties. Furthermore, the same equivalent stress criteria (SVM) should be used when expressing the loading history. Alternatives such as the Von Mises, Tesca or Principle stress (maximum, minimum, absolute maximum) could be used. Here the choice for SVM provides a simplistic approach while considering the directionallity of the loading conditions. Due to the nature of the loading (multi-axial, non-proportional), the equivalent stress damage accumulation approach might provide overestimated lifetime results.

5. Conclusion

A methodology is presented for the determination of a S-N curve, relevant for impact fatigue (high strain rate) loading conditions. A hard and soft PU system are characterised based on tensile test and DMA analysis. The master curves indicate the soft PU system to show relatively high energy dissipation during impact loading, leading to lower amplitude stress waves in the coating system. However, when combining the SPIFT experimental and numerical evaluation of both systems, the respective S-N relations shows that, under the same stress conditions, the hard PU system will experience a longer lifetime. This tendency is recognised with RET data in which the same relative material performance is seen. Indications are seen that the relative material performance can be predicted by the presented methodology, however further analyses is required to conclusively validate the results.

Acknowledgments

The authors gratefully acknowledge the financial support of the Innovation Foundation of Denmark in the framework of the Grand Solutions project DURALEDGE, Durable leading-edges for high tip speed wind turbine blades, File nr.: 8055–00012A. We warmly thank our collaborating partners from Covestro in Leverkusen, Germany for formulating polyurethane resins and providing DMA and tensile test data and PU targets.

Appendix A

Table 2: Yeoh-Prony model parameters for the Hard and Soft PU system

Yeoh Parameters		
	Hard PU	**Soft PU**
C10 [MPa]	338.52	93.2
D1 $[\frac{1}{MPa}]$	0.0001	0.011
Prony Parameters		
τ_i **[ms]**	g_i **[Pa]**	g_i **[Pa]**
1.00E+11	0.000177	
1.00E+10	0.002656	
1.00E+09	0.009392	
1.00E+08	0.071452	
1.00E+07	0.095845	
1.00E+06	0.104812	0.00173
1.00E+05	0.213968	0.004841
1.00E+04	0.105051	0.020505
1.00E+03	0.085177	0.08461
1.00E+02	0.079553	0.23238
1.00E+01	0.050308	0.241676
1.00E+00	0.06829	0.257243
1.00E-01	0.026673	0.045852
1.00E-02	0.032427	0.102229
1.00E-03	0.052281	6.78E-06

References

[1] Bakırcı M and Yılmaz S 2018 *Engineering Science and Technology, an International Journal* **21** 1128–1142 ISSN 2215-0986 URL `https://www.sciencedirect.com/science/article/pii/S2215098618301381`
[2] IRENA 2021 Renewable power generation cost in 2021 Tech. rep. International Renewable Energy Agency, Abu Dhabi
[3] Sareen A, Sapre C and Selig M 2012 *Wind Eng.* **36** 525–534 ISSN 0309524X
[4] Herring R, Dyer K, Martin F and Ward C 2019 *Renew. Sustain. Energy Rev.* **115** ISSN 18790690
[5] Doagou-Rad S, Mishnaevsky Jr L and Bech J I 2020 *Wind Energy* **23** 1752–1766
[6] Amirzadeh B, Louhghalam A, Raessi M and Tootkaboni M 2017 *Journal of Wind Engineering and Industrial Aerodynamics* **163** 33–43
[7] Amirzadeh B, Louhghalam A, Raessi M and Tootkaboni M 2017 *Journal of Wind Engineering and Industrial Aerodynamics* **163** 44–54
[8] Jespersen K M, Eftekhar M, Johansen N F J, Bech J I, Mishnaevsky-Jr L and Mikkelsen L P 2022 *Submitted 2023*
[9] ASTM 2013 Standard Test Methods for Density and Specific Gravity (Relative Density) of Plastics by Displacement1 ASTM D792-8
[10] International Organization for Standardization 2012 Plastics — Determination of tensile properties ISO 527-2 Retrieved from https://www.iso.org/obp/ui/fr/iso:std:iso:527:-2:ed-2:v1:en
[11] Bergstrom D J and Elabbasi D N 2020 *PolymerFEM.com*
[12] Fraisse A, Frost-Jensen Johansen N, Bech J I, Borum K K, Fedorov V, McGugan M, Mishnaevsky L and Kusano Y 2018 *Renewable Energy* **126** 1102–1112 ISSN 18790682 URL `http://production.datastore.cvt.dk/filestore?oid=5b0e7bcc5010df026c075232&targetid=5b0e7bcc5010df026c075236`
[13] Johansen N F J, Mishnaevsky L, Dashtkar A, Williams N A, Fæster S, Silvello A, Cano I G and Hadavinia H 2021 *Coatings* **11** ISSN 20796412
[14] ASTM 2012 *Annual Book of ASTM Standards* **i** 1–7 ISSN 5935522004

Generative AI and image based numerical mechanics in wind blade adhesive composites

AW Khan, C Balzani

Gottfried Wilhelm Leibniz University Hannover, Institute for Wind Energy Systems, Appelstr. 9A, 30167 Hannover, Germany

E-mail: abdul.wasay.khan@iwes.uni-hannover.de

Abstract. Numerical modelling of adhesive composites in wind energy is complicated in part due to material heterogeneity. Microstructural CT scan fibre composite patterns or representative elements, which play a major role in defining the mechanical behaviour of these adhesive structures, are both difficult to characterize as well as hard to numerically simulate. With advances in deep learning based generative AI, new ways of predicting the mechanical behaviour of heterogeneous materials is now possible. Here we put forward a data driven method to relate input composite adhesive microstructures with field data using deep learning and generative AI based methods. The prediction of mechanical stress or strain fields or other similar patterns and combining them as a function of boundary conditions, fibre composite microstructure and material models is achieved and the models are trained such that they closely approximate computationally expensive simulations based on numerical FE techniques and would have the ability to generalize. We also create a dataset of wind energy adhesives with their numerical mechanics based FE simulations subject to different boundary conditions and material models for further deep learning based composite studies.

1. Introduction

Establishing numerical models that capture the mechanical behavior of wind blade adhesives considering the fibres at a microscale is an ardent but important task. Traditionally when structures are solved numerically as a Boundary-value problem, two types of equations are combined, equations based on Conservation laws, which are inherently uncertainty-free and equations derived from physical modelling and observations which are empirical and uncertain. The classical numerical paradigm is based on making use of the observed experimental data and then use the calibrated material model in other calculations. Often times the method is expensive and time consuming while being empirical in nature [1–3]. A new direction based on data driven techniques using machine learning architectures opened up new methods for computational solid mechanics [4–9]. Off late generative AI is making news and disrupting and revolutionizing various industries. It is based on generative deep learning methods with the ability to leverage both supervised and unsupervised learning approaches, with the ultimate goal of taking some training inputs from a distribution and then learning a model that characterizes that distribution. One such class of Generative AI methods are the Generative Adversarial Networks (GANs) [10]. In this work we will explore on how to use Generative Adversarial Networks for stress predictions in a wind energy fibre composite adhesive microstructure.

2. Fibre Adhesive Microstructure and Numerical FE Modelling

Short fibre adhesives are composites used in the bondlines of wind turbine blades. Various orientations and alignments are reported in prior experiments where dependencies are based on type and characteristics of adhesive flow during the joining process. This anisotropy then effects the structural properties of the composite. Several CT scans were conducted at various magnifications and the investigations reveal approximately 8-9% fibre volume fraction. One such CT- Scan of the adhesive at a magnification of 10X conducted with Zeiss Xradia 410 Versa is shown in Figure 1. The scans depict white fibres (lighter) embedded in an adhesive (dark) matrix.

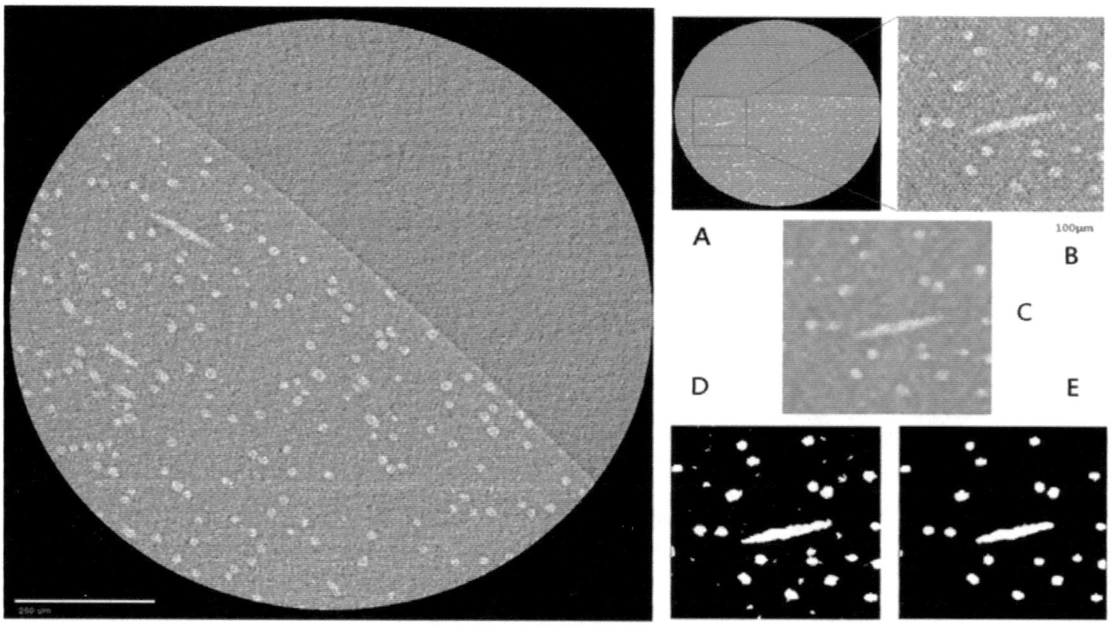

Figure 1. Left: Original Computed Tomography Scan. Right:(A) Photo after rotation (B) Cut and enlarged image (C) Filtered image with non-local mean de-noising (D) Binarized image with OTSU's method (E) Image after using open and close operation.

Segmentation of the fibres and creation of representative microstructures from the adhesive matrix CT Scans are important prerequisites that have to be met to perform subsequent analysis. Most images however due to the reasons of image acquisition and various other operational modalities contain noisy fluctuations of varying pixel values which force us to define a segmentation approach. Since the current emphasis is on the extraction and segmentation of fibres which have reasonably well defined boundaries or edges thus high intensity methods are employed. Small subsets of microstructures are cropped and marker controlled watershed algorithms are applied. In some cases, pixels below a certain threshold are removed and then ellipses around the amorphous regions are fitted. The segmented regions are then replaced with these fitted ellipses to complete the segmentation. Noise removal strategies are also sought since excess noise even after binarization can alter the results for further processing (Figure 2). The approach is inspired by the reconstruction approach used for tomographic data of heterogeneous fibre systems [11].

After the segmentation is done and a collection of diverse binarized microstructures are obtained with scanned fibre directions, we do a set of finite element simulations where we

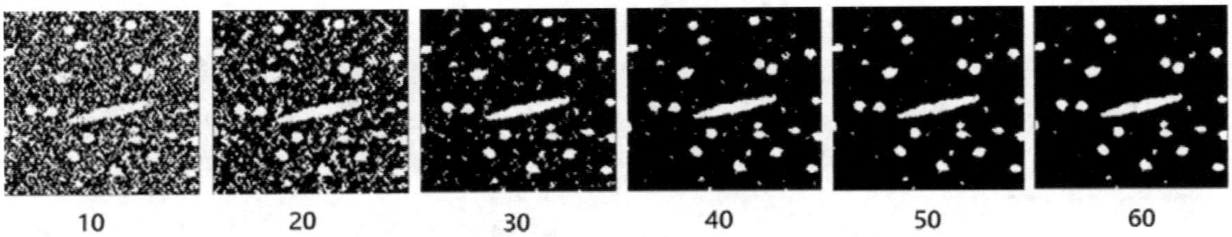

Figure 2. Iterative test results with different de-noising strengths.

employ the input CT binarized patterns to inform the heterogeneous fibre matrix distribution of the domain and subject it to equibiaxial extension. For achieving this goal, the binary bitmaps are converted to two different materials of meshed regions. The segmented fibre matrix coordinates are translated to a mesh with two distinct subdomains using Python-based pygmsh. Quadratic triangular elements are chosen for the mesh. Once the microstructures are meshed, a compressible Neo Hookean material model is chosen, defined by a free energy function W and mathematically representing the deformation gradient as $\boldsymbol{F}$ and the right Cauchy Green Tensor as $\boldsymbol{C}$. Then the free energy function could be written as:

$$W = \frac{\mu}{2}(I_c - 3) - \mu \ln(J) + \frac{\lambda}{2}\left(\frac{1}{2}(J^2 - 1) - \ln(J)\right) \tag{1}$$

Where $I_c = \text{trace}(C) = \text{trace}(F^T F) = F : F$ and $J = \det(F)$, and μ and λ are the Lamé's parameters respectively. The values of the material parameters are chosen from previous studies on fiber reinforced adhesives in wind energy [12, 13]. Specifically, moduli of 79050 MPa for fiber and 2798 MPa for matrix are used respectively. The Poisson's ratio for fiber is taken as 0.22, and for the matrix, the value is set as 0.40. A perfectly bonded interface behavior between the matrix and fiber is assumed, and numerically it is handled by defining subdomains inside the domain. A finite element database with varying boundary conditions, magnifications, and loading conditions is created. For this study, however, only biaxial extensions at 2% and 4% are considered. The Frobenius norm of the Cauchy stress maps is obtained for each of the 90 microstructures for each of these load conditions, creating a small but reasonably diverse dataset. The study would use this as the training-testing dataset with a split ratio of 80% − 20% respectively.

3. Generative Modeling

Once the training dataset is obtained, a strategy is sought for generating corresponding stresses using an artificial intelligence-based approach. The objective of the proposed neural network will be thus to output a stress map, given a new microstructure. This could be thought of as a paired image-to-image problem (I2I) where an input image x_a from the microstructural domain $\boldsymbol{a}$ is being translated to an output image in the stress domain $\boldsymbol{v}$. Mathematically, we need to train a mapping $G_{(a \to v)}$ such that it could generate an image $x_{av} \in v$, which is identical to the target image in the stress domain $x_v \in v$, given the input microstructure image $x_a \in a$. Mathematically, we can model it as follows:

$$x_{av} \in v : x_{av} = G_{(a \to v)}(x_a) \tag{2}$$

Generative modelling answers questions on how to achieve this mapping between the two image domains. Generally, in generative modelling, it is assumed that the creation of data is related to a distribution and usually that distribution is described by parametric and non-

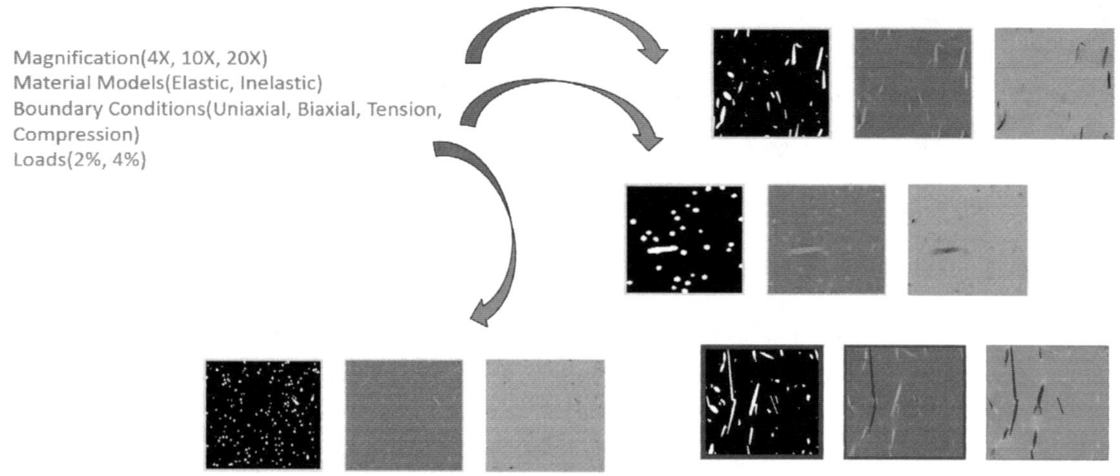

Figure 3. Creation of a FEA database with diversity.

parametric variants, respectively [14,15]. The goal in generative modeling is then to approximate this underlying distribution with various algorithms and techniques.

Our work is based on Generative Adversarial Networks (GANs) [10], which consist of a generator (G) and a discriminator (D), two neural networks which are in competition with one another. In the most classic setting, the generator turns random noise (z) into an imitation of the data, trying to trick the discriminator. The discriminator tries to discriminate real data from fake ones which are created by the generator. The generator continuously improves with its imitation of the data and tries to fool the discriminator. The objectives are thus adversarial, hence the name. A stage comes where the generator is able to output data so close to the real ones that the discriminator can no longer differentiate them from the real ones, a state of Nash equilibrium.

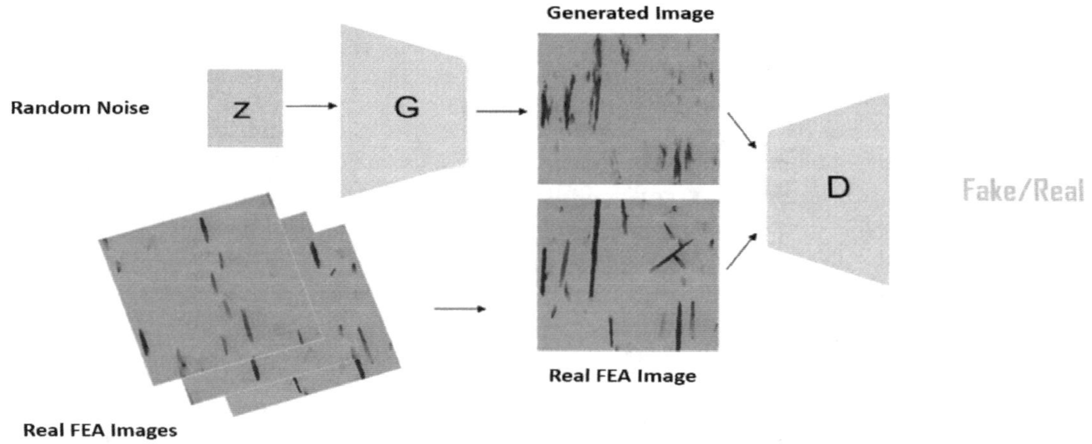

Figure 4. Unconditional GAN where G and D denote the Generator and Discriminator respectively.

As shown in Figure 4, the generator G inputs a random noise z sampled from the model's prior distribution to generate a fake image $G(z)$ to fit the distribution of real data as much as possible. Then, the discriminator D randomly takes the real sample x from the dataset and the fake sample $G(z)$ as input to output a probability between 0 and 1, indicating whether the input is a real or fake image. In other words, D wants to discriminate the generated fake sample $G(z)$ while G intends to create samples to confuse D. Consequently, the objective optimization problem is as shown below:

$$\min_G \max_D L(D,G) = \mathbb{E}_{x\sim p(x)}[\log D(x)] + \mathbb{E}_{z\sim p(z)}[\log(1 - D(G(z)))] \tag{3}$$

The only shortcoming with this configuration though is that the user has no control over the final output because the only input is random noise. It was suggested that additional information could be added with random noise to generate the final output [16]. In other words, conditioning the input with labels, text and attributes of the images. Using this conditional variable c, we are able to achieve control over the generated output image. This is something called Conditional GANs where in addition to the random noise we also have another factor or label to condition for [17–19], in the present scenario the microstructure image.

Thus the problem that we deal with falls under the category of supervised image to image problem, wherein we have a microstructure and the corresponding stresses in the form of paired data. The problem could be thought of as single modal if there is no time dependency and multimodal if time also becomes a variable i.e. when we want to see the evolution of stresses in the microstructure with time. For the case of simplicity, in this paper we will only focus on single modal problems. We will use the Pix to Pix architecture as illustrated in figure (5) with minor changes to fit our problem of stress predictions from a composite microstructure in wind energy [20].

The generator and discriminator models in Pix2Pix GAN use standard Convolution-Batch Normalization- ReLU blocks of layers. The generator uses a popular encoder-decoder network "U-Net" architecture with skip connections [21]. A patchGAN network is used by the discriminator which instead of predicting the whole image as fake or real, takes an image patch and then subsequently makes a prediction for every pixel in that patch for its authenticity. The final objective then becomes as follows:

$$G^* = \min_G \max_D \mathbb{E}_{x,c}[\log D(x|c)] + \mathbb{E}_{z,c}\log(1 - D(G(z|c))) + \lambda \mathbb{E}_{x,z,c}[||x - G(z|c)||_1] \tag{4}$$

$$G^* = A + \lambda B \tag{5}$$

where,

$$A = \min_G \max_D \mathbb{E}_{x,c}[\log D(x|c)] + \mathbb{E}_{z,c}\log(1 - D(G(z|c))) \tag{6}$$

$$B = \mathbb{E}_{x,z,c}\left[||x - G(z|c)||_1\right] \tag{7}$$

Here c represents the conditional variable, x denotes the real FEM data, z characterizes the random noise vector, and $G(z|c)$ are the fake samples generated by the generator G under the control of the conditioning variable c. The quantity $D(x|c)$ specifies the probability that the discriminator's conditioned input is real. The adversarial loss A controls whether the generator in the network has the power to produce images that are plausible in the target FEM domain. The $L1$ loss B regulates the generator to output images that are a possible translation of the image from the source domain. The parameter λ gives the user control as to how much importance needs to be given to the $L1$ loss. Thus it is kind of a weighting factor between the adversarial loss and the $L1$ loss.

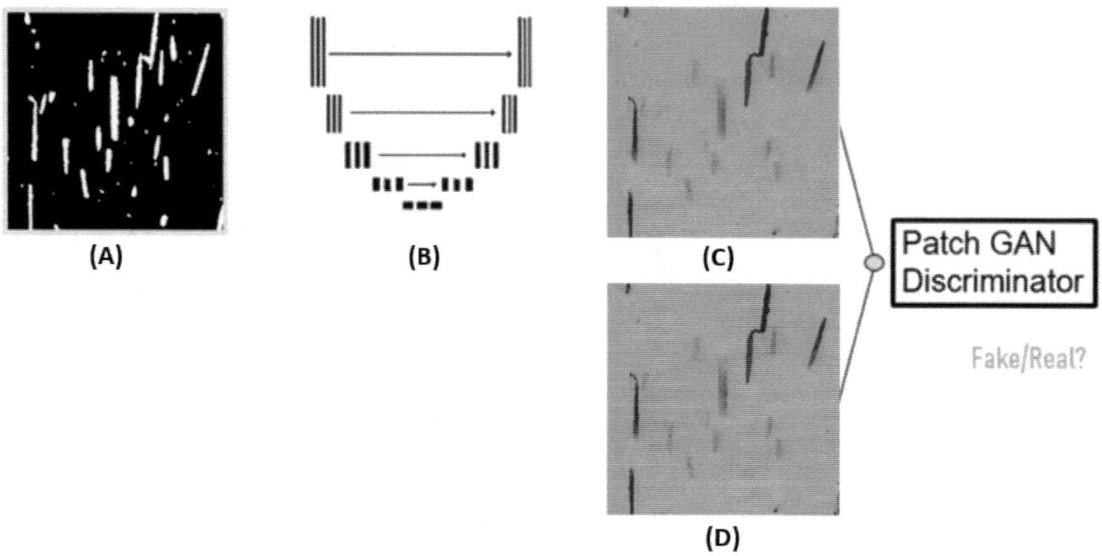

Figure 5. Conditional GAN for stress predictions using Pix to Pix Architecture. (A) Conditioning Map (B) Segmentation Network (C) Generated AI Stress Map (D) FEA based Stress Map.

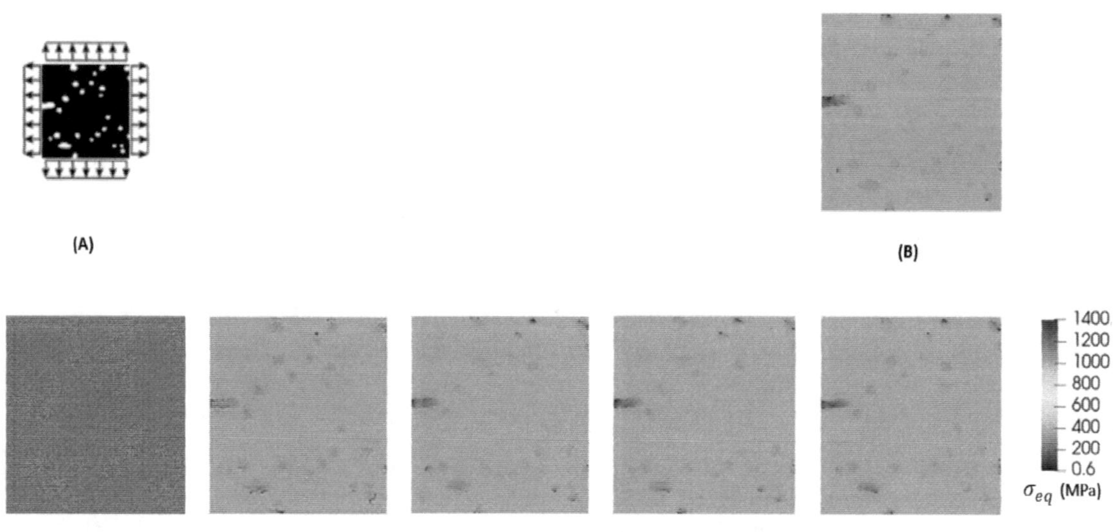

Figure 6. Training process of the conditional GAN. (A) Composite microstructure under biaxial extension (B) FEA based prediction acting as ground truth (C) The iteration process, notice the predictions become better with advancing iterations, finally reaching to the tolerance value where the discriminator no longer can differentiate between the real and the fake predictions.

4. Results and Discussions

The deep learning model used and proposed was evaluated on biaxial extension both for the case of 2% and 4% loading to check for its performance in adhesive composites and the related mechanical behaviour. The microstructure image is the input, while the Frobenius norm of the Cauchy stress tensor σ_{eq} is the label target. Error Maps are calculated using Mean Absolute Error as a metric to test for performance and to compare the results between FEM ground truth and the conditional GAN (AI) based predictions.

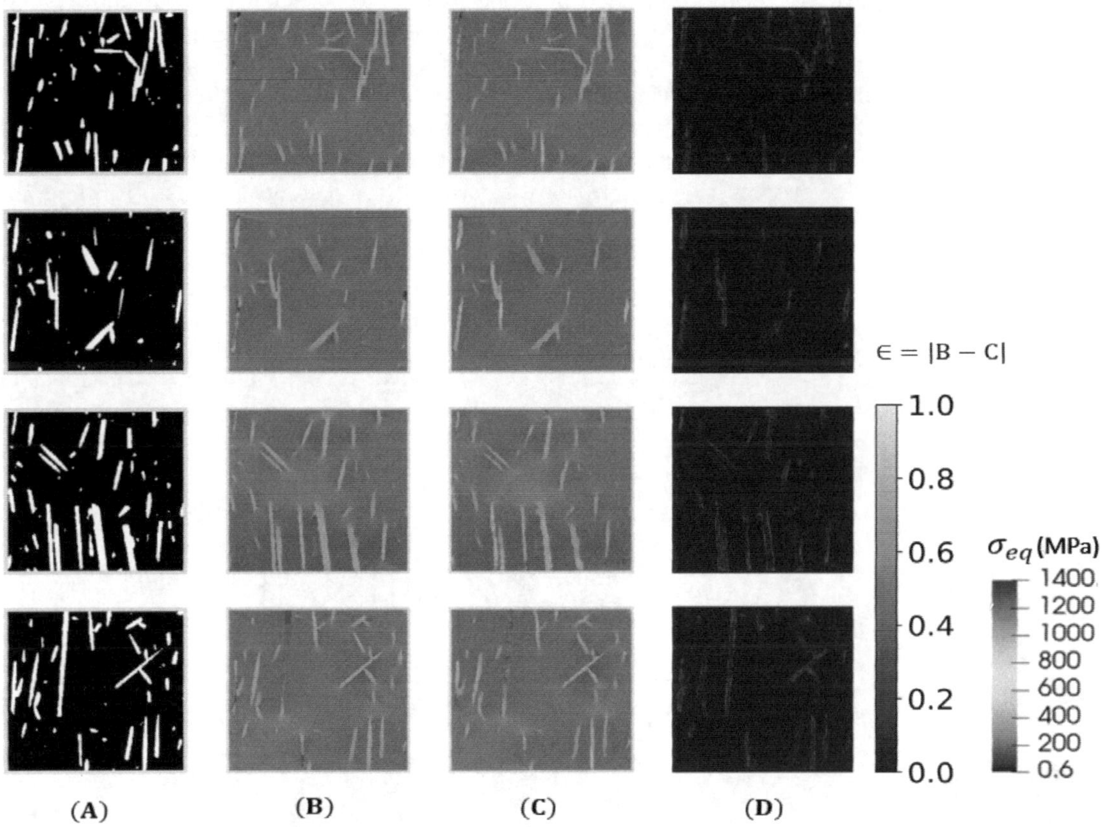

Figure 7. Equivalent stress field predictions for an adhesive microstructure under biaxial extension of 2%.(A) Microstructure of the adhesive composite, where the white colour represents the fibre and the black colour the adhesive matrix respectively (B) FEM predictions (Ground Truth) (C) Conditional GAN (AI) based predictions (D) Mean Absolute Error between the FEM results and the AI based predictions. Please note that whilst the composite image could be acquired at different magnifications but the overall image resolution is 256 pixels × 256 pixels in each case.

Figures (7), (8) show that the results of using conditional GAN's on data out of the training dataset and it could be seen that even with very limited training data, we have good quality predictions. We have some deviations between the ground FEM and the conditional GAN based predictions near the boundaries of the fibres especially in highly loaded cases but nonetheless the predictions are promising considering how frugal the dataset was and the complexity of the problem. The variation in fibre shapes, small sizes and different volume fractions combined with the stresses they carry which could be very different for each probable case makes it quite a complicated task. For a deep learning algorithm to be able to predict the fibre stresses opens up broad applicability and transferability to other use cases. For more trustworthy results and confidence in our predictions, it's important to have a much more diverse dataset and this is something which is left for future work. If the dataset covers the whole spectrum of microstructures used in wind adhesives then in future, we can immediately know the mechanical behavior of the adhesive composite if we have the microstructure. Further extensions include incorporation of further material properties of composites as this will enlarge the scope of application. Also, the study is reporting results only from Generative Adversarial Networks as the base algorithm. Further research needs to be done to see how other generative models like Auto-encoders and Diffusion models behave in these situations. The latest and more powerful text to image models can be adapted where the conditioning prompt is not the text but rather can be taken as an embedding of the input adhesive microstructure from which the model then using a learning strategy is trained to predict stress values or other field data. This is also left as future work. Another very important direction as part of the whole data driven strategy is the effective, cheap and representative generation of artificial microstructures as obtaining the real computed tomography is usually quite expensive. Generative AI can also be used for this purpose. The work could also be well extended to cracks and studies involving crack propagation in wind turbine blades. Also porosity and other defects can be incorporated. These effects can be employed with something called transfer learning where if a model is repurposed for a similar but repurposed activity. The main idea is to use the knowledge that has been attained in one task to enhance generalization in another. So instead of learning from scratch, we start with useful knowledge attained before is repurposed for a similar but repurposed activity.

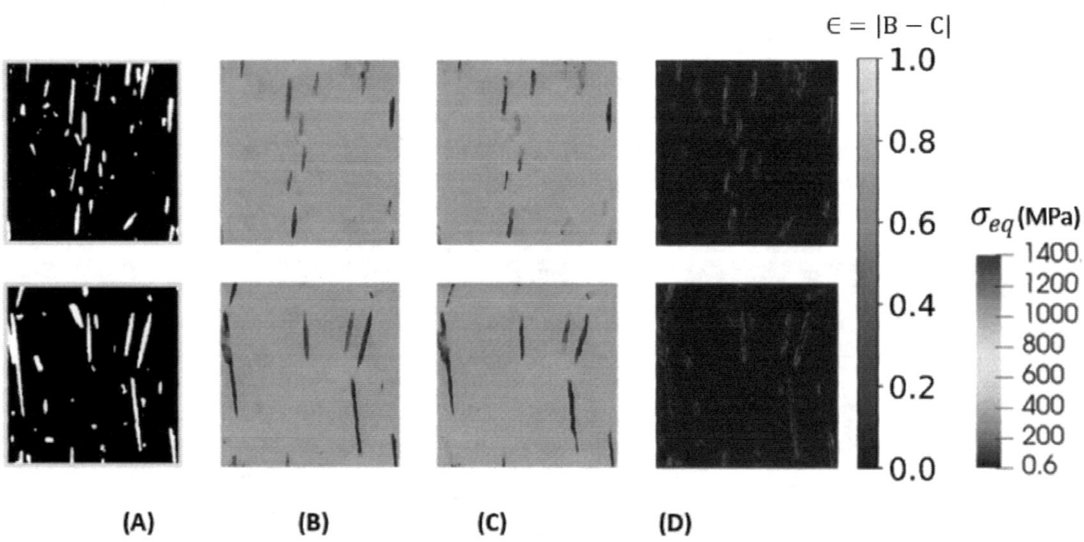

Figure 8. Equivalent stress field predictions for an adhesive microstructure under biaxial extension of 4%.(A) Microstructure of the adhesive composite, where the white colour represents the fibre and the black colour the adhesive matrix respectively (B) FEM predictions (Ground Truth) (C) Conditional GAN (AI) based predictions (D) Mean Absolute Error between the FEM results and the AI based predictions. Please note that whilst the composite image could be acquired at different magnifications but the overall image resolution is 256 pixels × 256 pixels in each case.

5. Conclusions

In this work, Generative modelling specifically a conditional GAN is used to predict the stress fields from the 2D microstructures slices of a fibre reinforced adhesive used in wind energy blades. The microstructures were obtained from computed tomography. The conditional GAN model was trained on finite element based simulations and a database with varying boundary and load conditions was created. Here it is postulated that a conditional GAN model can predict stresses even with frugal data especially on fibres which encourages the use of Generative AI. However, the results at the edges of the borders need further improvement in accuracy. Further methods and configurations in Generative Deep Learning for example Conditional GANs and Diffusion models need to be investigated. Also more randomness need to be incorporated along with big datasets to make the model more generalizable.

References

[1] Karapiperis, K., Ortiz, M. and Andrade, J.E., 2021. Data-Driven nonlocal mechanics: Discovering the internal length scales of materials. Computer Methods in Applied Mechanics and Engineering, 386, p.114039.

[2] Yagawa, G. and Oishi, A., 2021. Computational mechanics with neural networks. Springer International Publishing.

[3] Kumar, S. and Kochmann, D.M., 2022. What machine learning can do for computational solid mechanics. In Current trends and open problems in computational mechanics (pp. 275-285). Cham: Springer International Publishing.

[4] Ghaboussi, J. and Sidarta, D.E., 1998. New nested adaptive neural networks (NANN) for constitutive modeling. Computers and Geotechnics, 22(1), pp.29-52.

[5] Hashash, Y.M.A., Jung, S. and Ghaboussi, J., 2004. Numerical implementation of a neural network based material model in finite element analysis. International Journal for numerical methods in engineering, 59(7), pp.989-1005.

[6] González, D., Chinesta, F. and Cueto, E., 2019. Learning corrections for hyperelastic models from data. Frontiers in Materials, 6, p.14.

[7] Chun, S., Roy, S., Nguyen, Y.T., Choi, J.B., Udaykumar, H.S. and Baek, S.S., 2020. Deep learning for synthetic microstructure generation in a materials-by-design framework for heterogeneous energetic materials. Scientific reports, 10(1), p.13307.

[8] Ghavamian, F. and Simone, A., 2019. Accelerating multiscale finite element simulations of history-dependent materials using a recurrent neural network. Computer Methods in Applied Mechanics and Engineering, 357, p.112594.

[9] Huang, D., Fuhg, J.N., Weißenfels, C. and Wriggers, P., 2020. A machine learning based plasticity model using proper orthogonal decomposition. Computer Methods in Applied Mechanics and Engineering, 365, p.113008.

[10] Goodfellow, I., Pouget-Abadie, J., Mirza, M., Xu, B., Warde-Farley, D., Ozair, S., Courville, A. and Bengio, Y., 2014. Generative adversarial nets Advances in neural information processing systems. arXiv preprint arXiv:1406.2661.

[11] Agyei, R.F. and Sangid, M.D., 2018. A supervised iterative approach to 3D microstructure reconstruction from acquired tomographic data of heterogeneous fibrous systems. Composite Structures, 206, pp.234-246.

[12] Holst, T., Antoniou, A., Englisch, N., Manousides, N. and Balzani, C., 2023. On the Estimation of Short Fiber Orientation in a Filled Epoxy Adhesive and its Effect on the Tensile Strength. In AIAA SCITECH 2023 Forum (p. 1922).

[13] Wentingmann, M., Manousides, N., Antoniou, A. and Balzani, C., 2022. Yield surface derivation for a structural adhesive based on multiaxial experiments. Polymer Testing, 113, p.107648.

[14] LeCun, Y., Chopra, S., Hadsell, R., Ranzato, M. and Huang, F., 2006. A tutorial on energy-based learning. Predicting structured data, 1(0).

[15] Oussidi, A. and Elhassouny, A., 2018, April. Deep generative models: Survey. In 2018 International conference on intelligent systems and computer vision (ISCV) (pp. 1-8). IEEE.

[16] Mirza, M. and Osindero, S., 2014. Conditional generative adversarial nets. arXiv preprint arXiv:1411.1784.

[17] Wang, C., Zheng, H., Yu, Z., Zheng, Z., Gu, Z. and Zheng, B., 2018. Discriminative region proposal adversarial networks for high-quality image-to-image translation. In Proceedings of the European conference on computer vision (ECCV) (pp. 770-785).

[18] Wang, T.C., Liu, M.Y., Zhu, J.Y., Tao, A., Kautz, J. and Catanzaro, B., 2018. High-resolution image synthesis and semantic manipulation with conditional gans. In Proceedings of the IEEE conference on computer vision and pattern recognition (pp. 8798-8807).

[19] AlBahar, B. and Huang, J.B., 2019. Guided image-to-image translation with bi-directional feature transformation. In Proceedings of the IEEE/CVF international conference on computer vision (pp. 9016-9025).
[20] Isola, P., Zhu, J.Y., Zhou, T. and Efros, A.A., 2017. Image-to-image translation with conditional adversarial networks. In Proceedings of the IEEE conference on computer vision and pattern recognition (pp. 1125-1134).
[21] Ronneberger, O., Fischer, P. and Brox, T., 2015. U-net: Convolutional networks for biomedical image segmentation. In Medical Image Computing and Computer-Assisted Intervention–MICCAI 2015: 18th International Conference, Munich, Germany, October 5-9, 2015, Proceedings, Part III 18 (pp. 234-241). Springer International Publishing.

IOP Conf. Series: Materials Science and Engineering 1293 (2023) 012023 doi:10.1088/1757-899X/1293/1/012023

How epoxy cure profiles (may) impact performance of non-crimp fabric composites

Ulrich A. Mortensen, Tom L. Andersen, Lars P. Mikkelsen (✉)

DTU Wind Energy Systems, Technical University of Denmark, Risø, 4000 Roskilde, Denmark

E-mail: `lapm@dtu.dk`

Abstract. Glass fibre reinforced polymers made from quasi uni-directional non-crimp fabrics have recently been shown to suffer from a specific damage mechanism. This damage mechanism is dominated by off-axis cracks in the off-axis backing bundles of the fabrics. In addition, research results have indicated that the onset and propagation of such off-axis cracks may be influenced by the residual stresses that arising from the manufacturing process used to produce the Glass fibre reinforced composites. The current article investigates the potential effect of residual stresses on the fatigue life through an experimental set-up that involves tailored cure profiles and image based evaluation of tensile test experiments.

1. Introduction

Non-crimp fabric reinforced polymer matrix composites are used in many structural applications due to a low material cost, good stiffness properties and high resistance to fatigue degradation. These properties makes it beneficial to use non-crimp fabrics in a quasi uni-directional (quasi-UD) form, especially in structures subjected to loading conditions where one loading direction is particularly dominant. An example of such an application with high stiffness and fatigue requirement is the load-carrying laminates in the spar cap of a modern large wind turbine blade, in which UD Non-crimp fabrics are heavily used.

Brøndsted et al. [1] found that the most critical loading condition for Non-crimp UD fabrics was tension/tension fatigue. For tension/tension fatigue, Zangenberg et al. [2] postulated a fatigue damage evolution mechanism based on an extensive microscopy study of a quasi-UD non-crimp glass fiber fabric reinforced composites. This fatigue damage evolution mechanism which was also observed by Jespersen et al. [3, 4]. Furthermore, a study using Large-Field optical microscopy by Mortensen et. al [5] showed both qualitative and quantitative results that supports the existence of this mechanism. In these studies, the fatigue damage in tension/tension has been found to be controlled by the off-axis tunnel cracks. Tunnel cracks which was found to appear rather early during the fatigue life testing [5] and which was resulting in fiber breakage located in alignment with those. Therefore, a delay of these off-axis cracks may have the potential of a better fatigue resistance of the composite materials.

In addition to the fiber-bundle architecture, the fatigue resistance of polymer matrix composites has recently been found to be influenced by the chosen cure profile [6, 7]. This study aims to combine recent developments knowledge about epoxy cure profiles [8, 9], and experimental procedures for quantifying the crack-density of off-axis cracks in tension/tension

Table 1: Fabric properties.

Classification	Angle [°]	Area Weight [g/m^2]	Linear Density [tex]	Filament [μm]
UD	0	1133	2400	17
	-45	50	200	16
Backing	90	17.6	200	16
	+45	50	200	16
Stitch Thread	-	9.2	8	-

fatigue tests of UD Non-crimp fabrics [10]. The goal of the study is to investigate the potential effect of residual stresses on off-axis cracks in backing bundles of UD Non-crimp fabrics.

2. Methodology

2.1. Materials and Manufacturing

The glass fibre composite specimens were manufactured using Vacuum Assisted Resin Transfer Moulding (VARTM). The details of the fabric is presented in Table 1, showing that the area weight of the fabric is dominated by axially oriented UD fibre bundles, resulting in high axial stiffness of the final composite. The focus in this study is , however, the off-axis cracks that form in the backing bundles of the fabric. Non-crimp fabrics of this sort are sometimes referred to as quasi-UD fabrics in the literature [4] because only around $\approx$ 10 percent of the fabrics area weight is made up from off-axis bundles. Quasi-UD non-crimp fabrics are commonly used in the wind turbine industry where handling of the fabrics are aided by the backing bundles, while stiffness in the axial direction of the blade remains high [11]. The backing bundles of a non-impregnated fabric sheet are depicted in Figure 1a where the axial orientation is aligned horizontally, and the backing bundle is presented. It is worth noting here that even though quasi-UD non-crimp fabrics contains off-axis bundles, the lay-up notation for a quasi-UD composite often ignores the presence of the backing bundles. A lay-up of 10 non-crimp quasi-UD fabrics could be described with a notation like this $[0°]_{5s}$, where the subscript s refers creating a symmetric lay-up (2 time 5) wrt. to whether the backing side faces up or down . For the lay-ups in this work a notation that is similar to the one used by Jespersen [4] will be applied. The axial direction will be noted by the 0° angle, an individual backing layer will be noted by /b$^{\theta}$ where the /b represents that a backing layer is part of a fabric sheet, and the superscript denotes the angle orientation. To mark the separation of two non-crimp fabric sheets a semi-colon ';' will be used.

The crack-counting method, described in further details below, counts cracks in a plane field of view with no perception of depth. The plane view means that only a single layer of backing in each of the backing orientation may be on the field of view for crack counting. If more than one layer of backing is present in the material, then no information is available in the images to confirm to which layer the crack belongs. Manufacturing a symmetric composite plate with only a single layer of backing is not possible without manipulating the fabric sheets. Figure 1b shows the lay-up modification made in order to have a gauge area in the plate that is symmetric yet only have one layer of each of the backing orientations. The symmetric lay-up with only one layer of backing bundles has been build by defining a gauge area of roughly 10 cm in width, where the backing from one fabric sheet has been removed. Stacking the modified fabric sheet on top of another (un-modified) fabric sheet results in a composite plate lay-up that is slightly asymmetric in most of the plate - with a lay-up of $[\,0°\ /b^{+45}/b^{90}/b^{-45}]_2$ - but symmetric in the

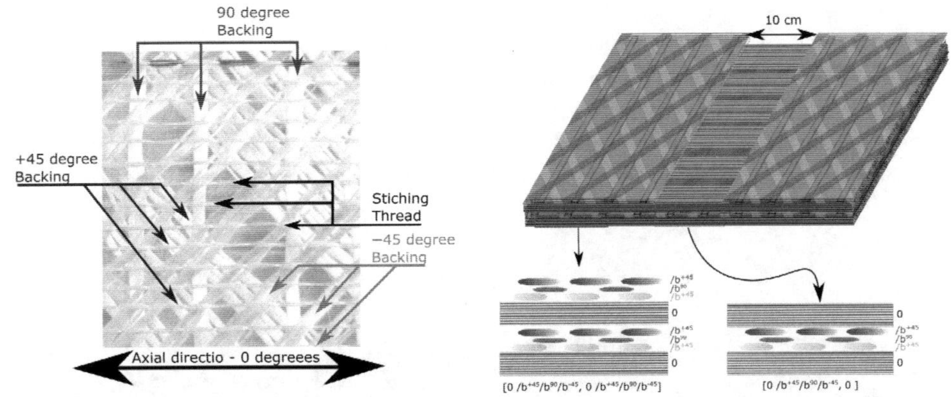

(a) Structure of non-crimp fabric backing showing the +45° , 90° , −45° fabric bundles and the stiching thread.

(b) Lay-up used for manufacturing of composite plates with only a single layer of backing.

Figure 1: Illustrations of the fabric used in the laminate lay-up, and the special gauge-area with a single layer of non-crimp fabric.

center strip with a lay-up of [0° $/b^{+45}/b^{90}/b^{-45}$; 0°]. The area where the plate is symmetric coincides with the area specified as the gauge-area in the test specimens described later. An extra benefit of the described lay-up modification is that the curvature that forms in the asymmetric part of the plate can be used as a macroscopic measure to the internal stresses that are formed in the plate as a consequence of the polymer curing during manufacturing.

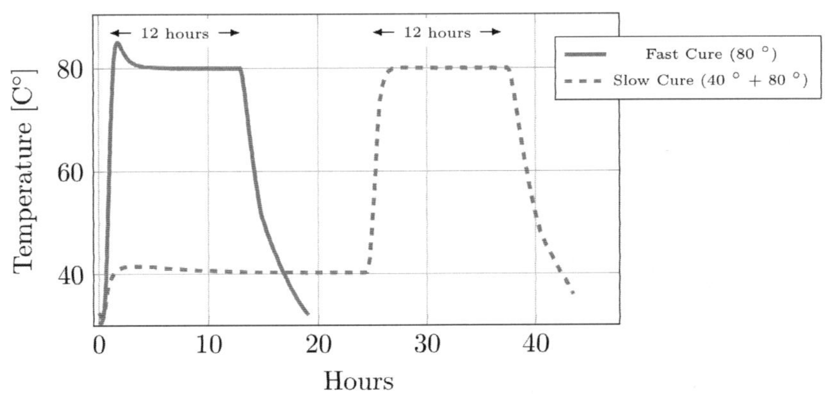

Figure 2: Temperature profiles used for plate manufacturing.

The polymer matrix used in the VARTM process was a commercially avaible epoxy (Huntsman Araldite LY 1568 / Aradur 3489) described in the literature by Mortensen et al. [9]. The ultimate glass transition temperature of the epoxy is $T_g^{ult} = 84°C$. As mentioned previously the plates that tested in this work was manufactured using VARTM. Two plates were made with the same manufacturing set-up, but with different temperature profiles applied for curing. The two temperature profiles - a one step (fast) and a two-step (slow) - was used, and temperature data from embedded thermocouples are shown in Figure 2. The temperature profiles were based on research papers presented by Mortensen et al. [8] and Mortensen et al. [9] which suggests that the magnitude of residual stresses induced from curing the epoxy matrix can be substantially lowered by using a two-step cure profile instead of a one-step cure profile. The outcome of using a two-step cure profile, which requires significantly longer time for curing, compared to a one-step cure profile, is a significant reduction in cure induced residual stresses. The cure kinetic model presented by Mortensen et al. [9] confirms that the cure profile for both plates ensures that the epoxy matrix in the composite plates have reached full cure.

2.2. Testing

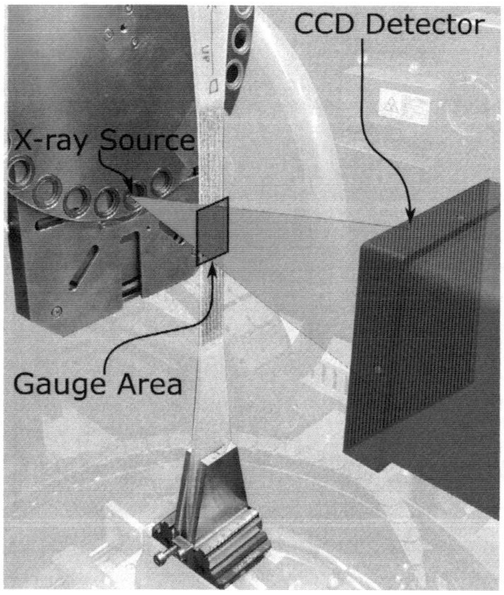

(a) X-ray tomography ste-up used for sinlge image exposure of gauge area.

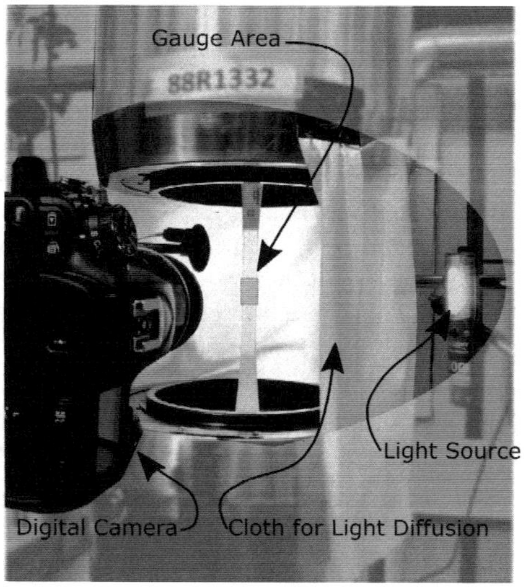

(b) Tensile testing set-up with image capturing digital camera

Figure 3

Testing of the composite material was performed by cutting the plates and mounting tabs material using the improved tensile fatigue specimen geometry from DTU Wind Energy [12] of 410 mm length. All tests were conducted in load control with R=0.1. The load were determined from an initial measurement of specimen stiffness conducted with two extenso-meters. The frequencies used for the specimens were varied based on the maximum strain at which the specimens were tested. Varying the frequencies allowed the strain rate of the specimens to be kept at identical levels for all specimens. The strain rate was set to 14 $\frac{1}{s}$ based on a reference test with 0.5% maximum strain at 5 Hz. During testing a digital camera was used to capture images of specimen gauge areas, which were later used as data in an automated crack counting scheme. The image series captured were taken such that roughly[1] 100 images were captured for each log-scale decade. All captured images had a 10 ms exposure time, and were captured at the peak load of the sinusoidal load curve exerted by the testing machine. The test machine and camera set-up are displayed in Figure 3b.

[1] Extra images were captured if large changes in cross-head displacement were recorded

2.3. Automated Crack Counting and X-ray image masking

Images captured during testing have been used for Automated Crack Counting (ACC) using the procedure described by Glud et al [10] and used by Jespersen et al [4] for crack counting in a quasi-UD non-crimp fabric material system. In a material system like the one investigated in this work the off-axis cracks that can be counted by ACC is limited to the areas where off-axis backing bundles are present. Jespersen et al [4] used 3D X-ray tomography to extract the fibre bundle structure that makes up this area. A similar approach to defining the appropriate area was used by Jespersen et al. [4], except only a single x-ray exposure was used rather than the full x-ray scan applied by Jespersen.

The single exposure X-ray images were captured using an Xradia Versa 520 μCT scanner with a Low Energy filter. An exmaple of the test set-up used for capturing an x-ray image exposure is shown in Figure 3a. An example (shown horizontally) of a single exposure image is shown in Figure 4 where it is combined with an image by the digital camera during tensile testing. The backing bundles in the middle layer of the specimen are visible in the x-ray images as areas that are darker than the surrounding areas of the images. The x-ray images are darker in the regions where there is backing because fewer gamma particles penetrates these areas during the x-ray exposure, causing fewer detections on the x-ray CCD detector, and eventually resulting in dark areas of the image. The optical image and the x-ray image can ultimately be scaled and translated such that the images covers the same physical image space. The pixel size p - with units $\frac{\text{mm}}{\text{pixel}}$ - for the x-ray images are given by the tomography equipment, while the pixel size for the optical images are calibrated based on the measured width of the specimen and the width observed in the optical image. The scaling of the x-ray image is performed automatically based in the known pixel sizes of each image, while manual positioning of the x-ray image wrt. to the reference optical image have been performed base on the small paint-marks - *not* pen-marks - that are visible in the x-ray and optical reference image. The manual position of the x-ray image wrt. the optical image is only needed for the first image - the reference image - in used in ACC because the algorithm automatically aligns all subsequent images wrt. to the reference image position.

As portrayed in Figure 1a, and documented in Table 1, the backing bundles are sown onto the

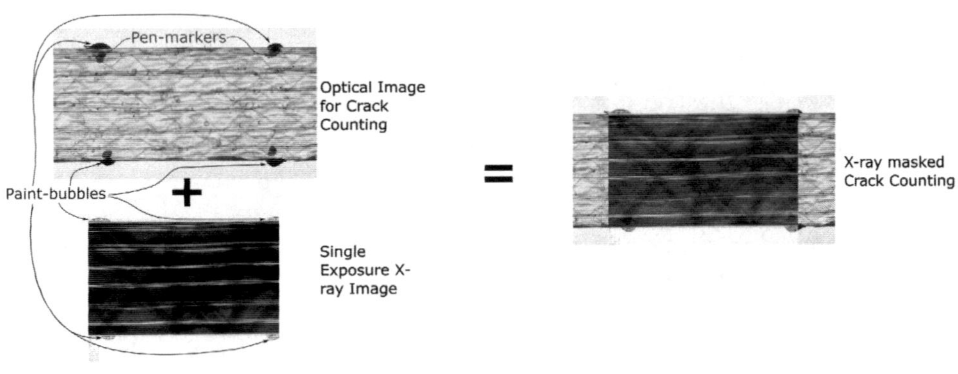

Figure 4: Combination of image obtained with optical digital camera and X-ray exposure of gauge area. The overlay of the x-ray exposure allows a to be overlayed the optical image for crack counting.

main UD bundles in three distinct orientations: +45°, 90° and -45°. These three distinct orientations can not be automatically extracted from the x-ray images. Thus the backing areas from the single exposure x-ray images are segmented manually and the area related to a particular orientation θ are computed as the product of the pixel size p and the number of foreground pixels n_{mask} in the manual segmentation, as written in eq. (1). The relation for the crack density ρ_θ for a specific orientation θ is printed in eq. (2), and can be calculated from the sum of lengths of all cracks in the θ-direction divided by the backing area of the θ-direction.

$$A_\theta = n_{\text{mask}} \cdot p^2 \tag{1}$$

$$\rho_\theta = \frac{\sum_{i=1}^{N_c} L_i}{A_\theta} \tag{2}$$

3. Results

3.1. S-N curve and tensile specimens failure modes

The results of the tensile test in terms of final failure is shown in Figure 5. The test series with the fast cure have lower fatigue resistance than the test series with the slower two-step cure. The relative difference in performance is relatively low, and for the tests conducted at a 1.0 % strain level the specimens failed at nearly identical amount of load cycles. There are, however, no instances in which the specimens cure with the fast cure can sustain more loading cycles than the slow cure.

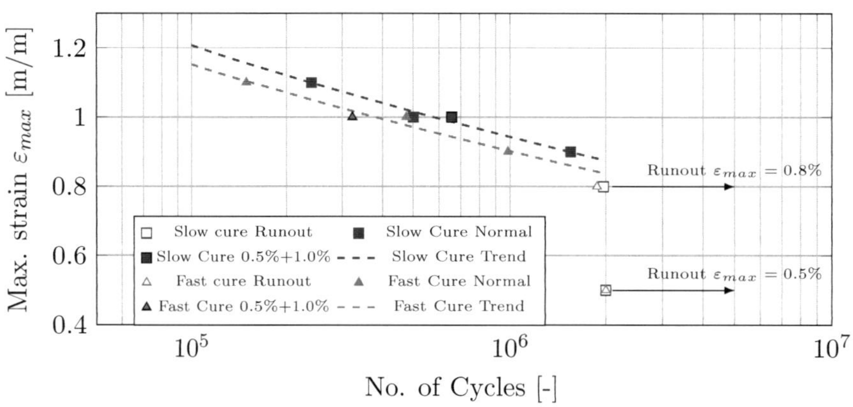

Figure 5: S-N Curve for fast and slow cure

3.2. Crack Counting Results

A representative example of cracks detected in an image is displayed in Figure 6. The opaque red areas in the image marks the areas of the gauge ares where 90° is present and the red lines mark the detected cracks. Similarly, the opaque blue areas and opaque green areas marks the areas of the gauge ares with +45° and 45° is present. The cracks are colored in the same representative colours. The masked areas functions both as a representation of the area, and as mask that removes cracks that are falsely detected outside the area with backing bundles in the same direction as the detected crack. Figure 7 compares the development of the crack densities in the specimens manufactured with the fast one-step cure cycle and the slow two step cure cycle. Before addressing the differences that arise as a result of the cure cycles it is worth addressing how the crack density development is influenced by the off-axis backing fibre bundle orientation. While no ±45° cracks were observed for the 0.5% strain level, the general trend is that the crack density develops at a much faster rate in the 90° backing bundles than in the ±45° backing bundles. The difference is not surprising giving that the fracture mode for the 90° cracks is pure mode I, while the fracture mode for the ±45° is mixed mode, and dominated by mode II, which means that crack growth in the 45° direction requires a higher fracture energy to propagate. The stress state in the ±45° bundles are likely to cause the cracks to open less than

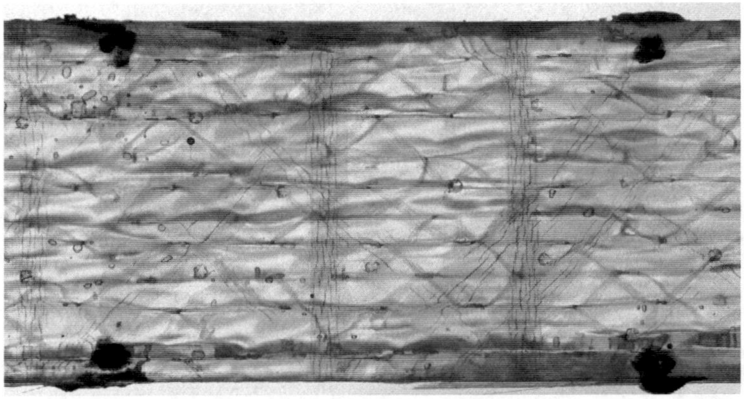

Figure 6: Cracks counted in image from tensile test with 1.0 % maximum strain after 100 000 loading cycles. The areas marked in green, red and blue are the segmented backing areas A_θ from each respective backing orientations. The cracks are the lines in the accordingly colors.

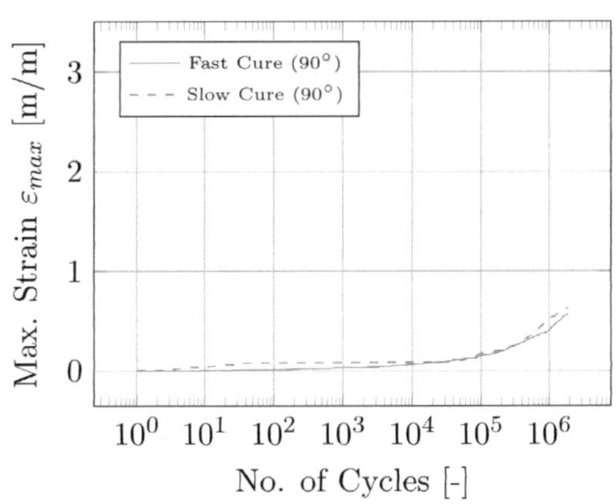

(a) Max strain 0.5 percent

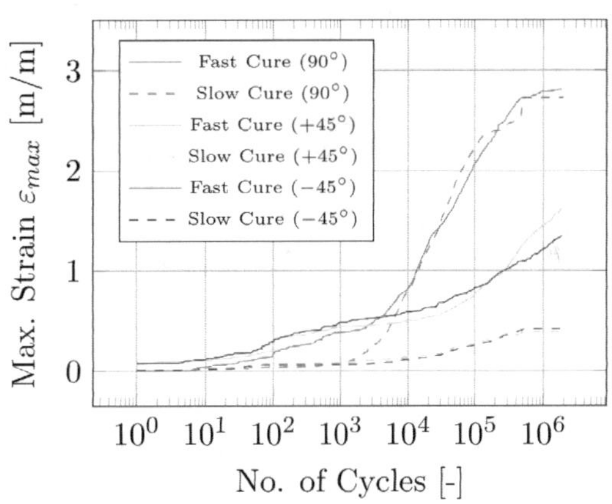

(b) Max strain 0.8 percent

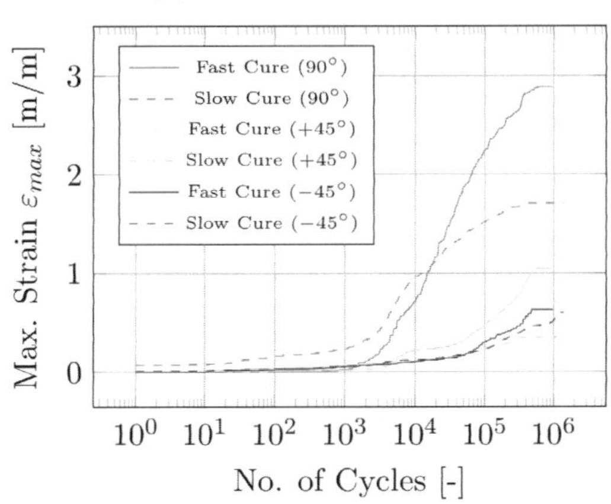

(c) Max strain 0.9 percent

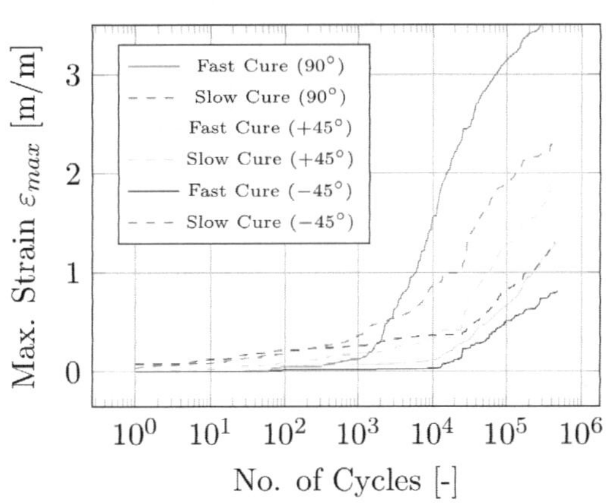

(d) Max strain 1.0 percent

Figure 7: Curvature of unsymmetric laminates

for the 90° bundles. Furthermore, the crack density of the ±45 backing bundles are expected to be less sensitive to the residual stresses, and the actual effect of residual stresses on the ±45° bundles are not as clear as for the 90° bundles that are loaded in pure mode I.Therefore the impact of residual stresses will be based on the crack density in the 90° backing layers.

For strain levels $\varepsilon = 0.5$ and $\varepsilon = 0.8$ there is no significant difference between the development of the crack densities for the 90° backing bundles. For strain levels of $\varepsilon = 0.9$ and $\varepsilon = 1.0$ the crack density for the 90° backing bundles grows faster for the fast cure, and importantly achiieves levels of crack density that is 1.5 to 2 times higher than what is reached for the slow cure.

4. Conclusion

A crack counting scheme was applied to measure the development of crack densities in two test series of quasi UD NCF composites. The two test series had been manufactured with the same manufacturing technique, but with two different cure cycles, designed to create variation in the magnitude of residual stresses in the plates from which the test series specimens were cut. The S-N curve for the two test series confirmed a relative performance difference between the two test series. The expected faster development of the crack density for the fast cure, with the high magnitude of residual stresses, was observed for tests conducted at two specific test strain levels ($\varepsilon = 0.9\%$,$\varepsilon = 1.0\%$). For two other strain levels the observed effect was small or non-existent.

References

[1] P. Brøndsted, H. Lilholt, and A. Lystrup, "Composite Materials for Wind Power Turbine Blades," *Annual Review of Materials Research*, vol. 35, pp. 505–538, aug 2005.

[2] J. Zangenberg, P. Brndsted, and J. John W Gillespie, "Fatigue damage propagation in unidirectional glass fibre reinforced composites made of a non-crimp fabric," *Journal of Composite Materials*, vol. 48, no. 22, pp. 2711–2727, 2014.

[3] K. M. Jespersen and L. P. Mikkelsen, "Three dimensional fatigue damage evolution in non-crimp glass fibre fabric based composites used for wind turbine blades," *Composites Science and Technology*, vol. 153, pp. 261 – 272, 2017.

[4] K. M. Jespersen, J. A. Glud, J. Zangenberg, A. Hosoi, H. Kawada, and L. P. Mikkelsen, "Uncovering the fatigue damage initiation and progression in uni-directional non-crimp fabric reinforced polyester composite," *Composites Part A: Applied Science and Manufacturing*, pp. –, 2018.

[5] U. Mortensen, L. Mikkelsen, and T. Andersen, "Observation of the interaction between transverse cracking and fibre breaks in uni-directional non-crimp fabric composites subjected to cyclic bending fatigue damage mechanism," *Journal of Composite Materials*, vol. 56, no. 23, pp. 3511–3527, 2022.

[6] J. Hther and P. Brndsted, "Influence of the curing cycles on the fatigue performance of unidirectional glass fiber reinforced epoxy composites," *IOP Conference Series: Materials Science and Engineering*, 2016.

[7] M. Miranda Maduro, *Influence of Curing Cycle on the Build-up of Residual Stresses and the Effect on the Mechanical Performance of Fibre Composites.* PhD thesis, Denmark, 2021.

[8] U. A. Mortensen, T. L. Andersen, J. Christensen, and M. A. M. Maduro, "Experimental investigation of process induced strain during cure of epoxy using optical fibre bragg grating and dielectric analysis," *Proceedings of the 18th European Conference on Composite Materials*, 2018.

[9] U. A. Mortensen, T. L. Andersen, and L. P. Mikkelsen, "Pragmatic approach to cure profile enhancement for improved fatigue performance of thermoset matrix composites," *IOP Conference Series: Materials Science and Engineering*, vol. 388, no. 1, p. 012014, 2018.

[10] J. Glud, J. Dulieu-Barton, O. Thomsen, and L. Overgaard, "Automated counting of off-axis tunnelling cracks using digital image processing," *Composites Science and Technology*, vol. 125, pp. 80 – 89, 2016.

[11] K. Jespersen, *Fatigue Damage Evolution in Fibre Composites for Wind Turbine Blades.* PhD thesis, Technical University of Denmark, Denmark, 2017.

[12] J. Hansen and R. Østergaard, *The effects of fibre architecture on fatigue life-time of composite materials.* PhD thesis, Technical University of Denmark, Denmark, 2013.

IOP Conf. Series: Materials Science and Engineering 1293 (2023) 012024 doi:10.1088/1757-899X/1293/1/012024

J integral strain-curvature approach for multilayer fracture specimens with unknown thickness and stiffness of layers

B F Sørensen[1] **and H L Toftegaard**[1]

[1]Technical University of Denmark, DTU Wind and Energy Systems, Risø Campus Frederiksborgvej 399, DK-4000 Roskilde, Denmark

E-mail: bsqr@dtu.dk

Abstract. In the coming years, many wind turbines will reach their planned design life, and it is of great interest to investigate if their service life can be extended so that the turbines can produce a larger amount of electricity before decommissioning. Such an assessment can include the measurement of the fracture mechanics properties of materials interfaces, e.g., for trailing edge bondlines, web foot bondlines and interfaces between layers in the load-carrying main spars. It would then be preferable to conduct the fracture mechanics testing and data analysis by approaches that give accurate results with the smallest amount of testing. In the present work we propose an approach for measurement of mixed mode fracture resistance (and mixed cohesive laws) that does not require knowledge of the stiffness properties and the elastic centre of the laminates of the test specimens. This is advantageous since the layer stiffness, layer thickness and the lay-up of a laminate cut from an old blade may not be known. Furthermore, the elastic properties of polymer resins of bondlines and matrix materials in fibre composites can have changed (aging) due to the blades long-time environmental exposure.

1. Introduction

When the use-time of a wind turbine in a wind farm is close to the projected service life (typically, 20 to 25 years), it is relevant to consider if the service life of the turbines can be extended in order to get more energy produced before the turbine is taken out of service. The process of assessing the extension of the use-life of old wind turbine rotor blades in a wind turbine park can involve the characterization of the mechanical properties of the materials and interfaces by testing specimens cut from a single representative blade. The fracture mechanics properties of the blade can then be used to assess the growth of damages in the blades of the remaining turbines. The primary load-carrying parts of a wind turbine blade is made of layers of composites materials (typically, made of glass and/or carbon fibres in a polymer matrix) and therefore it would be relevant to make fracture mechanics testing of interfaces between layers in laminates (e.g., within the main spar) and laminate/adhesive interfaces (e.g., web-foot and trailing edge bondlines). It would be convenient to cut fracture mechanics specimens of regular shape from a blade, so that established fracture mechanics testing procedures can be used. There are well-established fracture mechanics testing methods for linear elastic fracture mechanics characterisation, both for pure Mode I and mixed mode [1, 2] (valid for small scale fracture process zone) and J integral test specimens (valid for both small-scale fracture process zones and large scale bridging problems) [3, 4].

However, the elastic properties, thickness and orientation of each layer may not be known for fracture mechanics test specimens cut from an old wind turbine blades. Therefore, it would be convenient to use fracture mechanics testing methods for which the calculation of the fracture mechanics properties (mixed mode fracture resistance or mixed mode cohesive laws) can be done without the need for knowing the elastic properties and thickness of the individual layers. For Mode I Double Cantilever Beam (DCB) specimens loaded with wedge forces or bending moments, there are known J integral equations that are independent of the elastic properties [5, 6, 7]. These approaches have been generalized to DCB specimen mixed mode specimens loaded by transverse forces [8] and DCB specimen loaded with axial forces and bending moments [9]. For beam-like specimens loaded by axial forces and bending moments the J integral solution can be expressed in terms of the applied axial forces, bending moments as well as the curvature and the axial strain of the elastic centres of the specimen beams [9]. Neither the elastic properties of the individual layers nor the elastic properties of the beams are required. The approach is fairly general; it is applicable for multilayered and graded specimens and, as a J integral solution, remains valid for both small scale fracture (linear elastic fracture mechanics conditions) and for large-scale bridging problems. The approach requires, however, knowledge of the elastic centres for each of the three beams of the DCB-specimen. Although this can be determined experimentally [9], it would be preferable if these additional measurements were not required. Only in the absence of axial forces (i.e., pure bending), the approach does not require knowledge of the elastic centre.

The purpose of the present paper is to establish a J integral equation where forces, moments, curvatures and axial strains are defined from the geometric mid-plane instead of the elastic centre. It is found that while the bending moments and axial strains are different from those defined from the elastic centres, the J integral equation takes the exact same mathematical form. This enables a novel fracture mechanics testing approach, which only requires the overall dimensions (thickness and width) of each beam as well as continuous measurements of the applied axial forces, moments, curvatures and axial strains of the geometric mid-plane of each beam. The strains and curvatures can be measured experimentally by the use of the digital image correlation technique, e.g. by measurement of axial strains at the top and bottom of the three beams, i.e. six strain measurements. For the measurement of the fracture resistance to be accurate, it is necessary that these strains are measured with sufficient accuracy. It is therefore also important to consider how these strains can be measured. Strain gauges are known to be very accurate but could be difficult to place on beams running parallel to each other. Therefore, the use of digital image correlation (DIC) [10] should be considered. However, since strain measurements by DIC are not as accurate as strain measurements by strain gauges, this paper considers how to check, if the accuracy is sufficient to enable J integral measurement from curvature measurements obtained from DIC.

The paper is organised as follows. First, in Section 2 we define the problem. In Section 3, we prove mathematically that the J equation of Toftegaard and Sørensen [9] holds true irrespective of what coordinate system is used to define forces and moments, and set up equations for estimation of the uncertainty of the J integral value accounting for uncertainties associated with DIC strain measurements. Section 4 presents results in the form of tables for the estimated uncertainties. Section 5 discusses the advantages and disadvantages of the proposed approach, and Section 6 contains the main conclusion of the study.

2. Problem description

The problem we consider is a beam-like fracture specimen consisting of three beams, as shown in figure 1. The specimen has a constant width, B and the beam height is denoted H_1, H_2 and H_3, for Beam 1, Beam 2 and Beam 3, respectively. The specimen can consist of a single, homogenous, linear-elastic material or of multiple layers made of orthotropic materials oriented parallel to the plane of the fracture process zone (the x_1-x_3-plane) or the elastic properties can be graded in the x_2-direction, but must be independent of the x_1-position. For a layered structure, the elastic properties and the layer thicknesses must be independent of the x_1-coordinate. The specimen is loaded with a combination of axial forces and bending moments. The axial forces, acting at the elastic centres, are denoted by N_1, N_2 and N_3 (subscript indicates beam number), and the bending moments are denoted M_1, M_2 and M_3. The forces and moments are taken to be positive as indicated in figure 1. The distance from the ends (where the forces and moments apply) to the fracture process zone is several times the beam thickness. The fracture process zone may not necessary be small, it can include large-scale fibre bridging. Therefore, we will analyse the specimen by the path-independent J integral [11], not by linear elastic fracture mechanics.

As mentioned above, this problem has been analysed by Toftegaard and Sørensen [9]. The J integral evaluation around the external boundaries can be written as

$$J_{ext} = \frac{N_1\varepsilon_1^0}{2B} + \frac{N_2\varepsilon_2^0}{2B} - \frac{N_3\varepsilon_3^0}{2B} + \frac{M_1\kappa_1}{2B} + \frac{M_2\kappa_2}{2B} - \frac{M_3\kappa_3}{2B}, \tag{1}$$

where ε_1^0, ε_2^0 and ε_3^0 are the axial strain component, ε_{11}, at the elastic centres of Beam 1, Beam 2 and Beam 3, and κ_1, κ_2 and κ_3 are the (elastic centre axis) curvatures of Beam 1, Beam 2 and Beam 3. Thus, by measuring the strain at the elastic centres and the curvatures along with the applied axial forces and moments, the J integral value can be determined without needing to know anything about the elastic properties of the beams, except that they are linearly elastic.

But this approach requires the knowledge of the position of the elastic centre for each of the three beams. In the following, we will express (1) in terms of axial forces and moments associated with the geometric mid-height of each beam instead of the elastic centre of each beam.

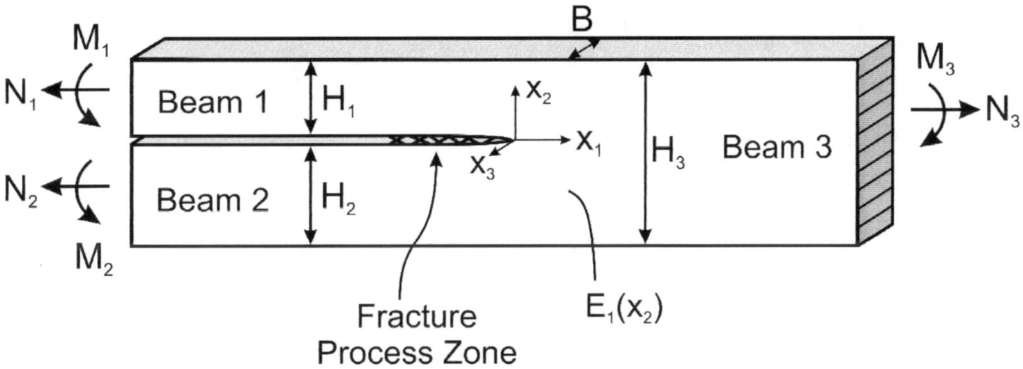

Figure 1. The problem under investigation: A beam-like specimen loaded with axial forces and bending moments remote from the fracture process zone. The elastic properties of the beams are independent of the x_1 coordinate but can vary layer wise or graded in the x_2-direction.

3. Mathematical derivation

3.1. J integral result

In the following, we will analyse the J integral contribution from Beam 2 with reference to two coordinate systems, one located at the elastic centre and another coordinate system located elsewhere (e.g., at the geometric mid-height), as shown in figure 2. The J integral is evaluated away from the fracture process zone, where the stress and strain fields are uniform, i.e., independent of the x_1-position [9]. Define the local coordinate system with origin in the elastic centre by $(\bar{x}_2, \bar{y}_2, \bar{z}_2)$, and the other coordinate system by (x_2^*, y_2^*, z_2^*) (for both coordinate systems, subscript 2 indicates Beam 2). In the global x_i-coordinate system, the x_2-coordinate of the elastic centre is denoted by δ_2 and the x_2-coordinate of the origin of the other coordinate system is denoted ξ_2 (again, for both symbols, subscript 2 is used to denote Beam 2). The axial force and moment defined with reference to the (x_2^*, y_2^*, z_2^*)-coordinate system is denoted N_2^* and M_2^*.

From (1) it follows that the J integral contribution for an integration path from top to bottom of Beam 2 is (subscript 2 is used to indicate Beam 2)

$$J_2 = \frac{N_2\varepsilon_2^0}{2B} + \frac{M_2\kappa_2}{2B}. \tag{2}$$

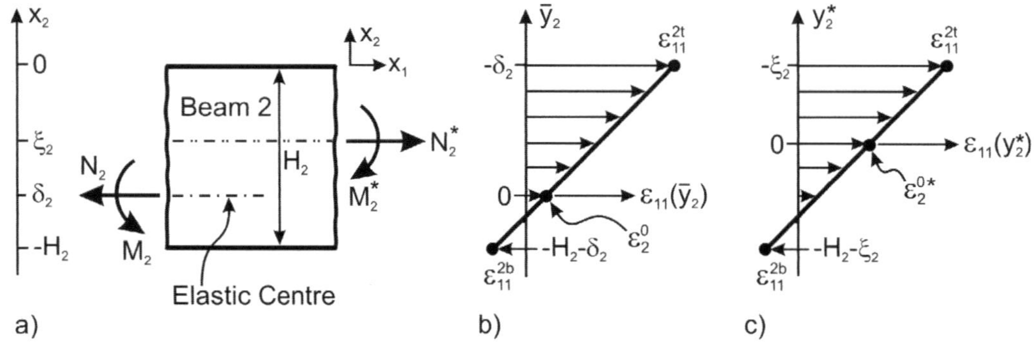

Figure 2. The definition of the axial force, moment, strain and curvature with respect to the two coordinate axes, $\bar{y}_2$, and y_2^*.

With respect to the original $\bar{y}_2$-coordinate, the normal strain in Beam 2 can be written as

$$\varepsilon_{11}(\bar{y}_2) = \varepsilon_2^0 + \kappa_2 \bar{y}_2 \, , \tag{3}$$

where $\varepsilon_2^0 = \varepsilon_{11}(x_2 = \delta_2)$. In the (x_2^*, y_2^*, z_2^*)-coordinate system the strain distribution would be written as

$$\varepsilon_{11}(y_2^*) = \varepsilon_2^{0*} + \kappa_2 y_2^* \, , \tag{4}$$

where $\varepsilon_2^{0*} = \varepsilon_{11}(x_2 = \xi_2)$. Note that the curvature κ_2 is the same for both descriptions. Since $x_2 = \delta_2 + \bar{y}_2 = \xi_2 + y_2^*$, the relation between $\bar{y}_2$ and y_2^* is

$$\bar{y}_2 = y_2^* + \xi_2 - \delta_2. \tag{5}$$

We proceed by expressing ε_2^0 in terms of ε_2^{0*}. By setting $y_2^* = 0$ (and thus by (5), $\bar{y}_2 = \xi_2 - \delta_2$), we get from (3) and (4) using (5):

$$\varepsilon_2^0 = \varepsilon_2^{0*} - \kappa_2(\xi_2 - \delta_2). \tag{6}$$

Next, from force equilibrium we find:

$$N_2 = N_2^* \qquad \mathit{and} \qquad M_2 = M_2^* + N_2^*(\xi_2 - \delta_2)\,. \tag{7}$$

Finally, we insert N_2 from (7), ε_2^0 from (6) and M_2 from (7) into (2):

$$J_2 = \frac{N_2^*}{2B}[\varepsilon_2^{0*} - \kappa_2(\xi_2 - \delta_2)] + \frac{\kappa_2}{2B}[M_2^* + N_2^*(\xi_2 - \delta_2)]\,, \tag{8}$$

and tidying up we notice that the terms with ξ_2 and δ_2 cancel out, so that eq. (8) reduces to

$$J_2 = \frac{N_2^* \varepsilon_2^{0*}}{2B} + \frac{M_2^* \kappa_2}{2B}\,. \tag{9}$$

The J integral equation expressed in the (x_2^*, y_2^*, z_2^*)-coordinate system, eq. (9), takes exactly the *same mathematical form* as the J integral expression with the normal force and moment defined from the elastic centre, eq. (2), with N_2 replaced by N_2^*, ε_2^0 replaced by ε_2^{0*} and M_2 replaced by M_2^*. It follows, that we do not need to define the axial force and moment from the elastic centre of the beam. Irrespective of what coordinate system is used when defining the axial force and moment, the same J integral equation applies.

Similar results can be found for Beam 1 and Beam 3 (since the integration path of Beam 3 is opposite of the integration path of Beam 1 and Beam 2, the J integral contributions become negative), so that the J integral evaluated around the external boundaries of the specimen can be written as

$$J_{ext} = \frac{N_1^* \varepsilon_1^{0*}}{2B} + \frac{M_1^* \kappa_1}{2B} + \frac{N_2^* \varepsilon_2^{0*}}{2B} + \frac{M_2^* \kappa_2}{2B} - \frac{N_3^* \varepsilon_3^{0*}}{2B} + \frac{M_3^* \kappa_3}{2B}. \tag{10}$$

where N_1^* and M_1^* and N_3^* and M_3^* are the axial force and moments applied to Beam 1 and Beam 3, respectively, being defined from conveniently chosen local (x_1^*, y_1^*, z_1^*)- and (x_3^*, y_3^*, z_3^*)-coordinate systems for the two beams. Eq. (10) is our main result.

3.2. Measurements of ε_2^{0*} and κ_2 from surface strains

In practice, the variables ε_1^{0*}, κ_1, ε_2^{0*}, κ_2 as well as ε_3^{0*} and κ_3 must be measured during fracture experiments. Focussing again on Beam 2, we will now describe an approach from which we can determine ε_2^{0*} and κ_2 from measurements of the axial strain at the top and bottom of the beam. Denote the axial strain at the top of the beam by ε_{11}^{2t} and the axial strain at the bottom of the beam by ε_{11}^{2b} (see figure 2). From (4) we get the strains as:

$$\varepsilon_{11}^{2t} = \varepsilon_{11}(y_2^* = 0 - \xi_2) = \varepsilon_2^{0*} - \kappa_2 \xi_2 , \tag{11}$$

and

$$\varepsilon_{11}^{2b} = \varepsilon_{11}(y_2^* = -H_2 - \xi_2) = \varepsilon_2^{0*} - \kappa_2 (H_2 + \xi_2) . \tag{12}$$

Solving the two equations (11) and (12) for ε_2^{0*} and κ_2 gives

$$\varepsilon_2^{0*} = \varepsilon_{11}^{2t}\left(1 + \frac{\xi_2}{H_2}\right) - \varepsilon_{11}^{2b}\frac{\xi_2}{H_2} \tag{13}$$

and

$$\kappa_2 = \frac{\left(\varepsilon_{11}^{2t} - \varepsilon_{11}^{2b}\right)}{H_2}. \tag{14}$$

Note that (14) is independent of the choice of ξ_2.

For the special case that the coordinate system is chosen at the geometric mid-plane of the beam, $\xi_2 = -H_2/2$, eq. (13) reduces to the expected result

$$\varepsilon_2^{0*} = \frac{\left(\varepsilon_{11}^{2t} + \varepsilon_{11}^{2b}\right)}{2}. \tag{15}$$

Similar equations apply for Beam 1 and Beam 3, i.e., for ε_1^{0*}, κ_1, ε_3^{0*} and κ_3.

To sum up, the width B should be measured before the experiment, and the following variables should be measured during the experiment: N_1^*, M_1^*, ε_{11}^{1t}, ε_{11}^{1b} (Beam 1); N_2^*, M_2^*, ε_{11}^{2t}, ε_{11}^{2b} (Beam 2); N_3^*, M_3^*, ε_{11}^{3t}, ε_{11}^{3b} (Beam 3).

3.3. Equations for sensitivity analysis

Before using the proposed approach experimentally, it is useful to contemplate the accuracy of the approach of measuring axial strains and curvatures instead of measuring the extension and bending stiffness properties of the three beams, Beam 1, Beam 2 and Beam 3. In the following we will present some simple estimates of the uncertainties on the J integral contributions. Our focus will be on the sensitivity of the measurement of curvature from strain measurements in accordance with (14). For simplicity we will consider a symmetric (i.e., identical beam heights, $H_1 = H_2$), homogenous DCB-specimen loaded with symmetric moments, $M_1 = -M_2$ (with $M_2 > 0$), with $N_1 = N_2 = N_3 = 0$ and $M_3 = 0$. In an experiment, the relevant moments and strains (M_2, ε_{11}^{2t} and ε_{11}^{2b}) would be measured. In this sensitivity analysis we instead assume the material to have a certain stiffness (Young's modulus) and set up some equations to establish consistent sets of moments and strains as a function of J_{ext} and H_2.

For a symmetric, homogenous DCB-specimen loaded with symmetric moments (Mode I), the J integral result (10) (or equivalently (1)) reduces to

$$J_{ext} = \frac{M_2 \kappa_2}{B}. \tag{16}$$

Assuming that Beam 1 and Beam 2 are homogeneous beams, it follows that $\xi_1 = \delta_1 = H_2/2$ and $\xi_2 = \delta_2 = -H_2/2$. With the moment of inertia of the rectangular cross section of Beam 1 and Beam 2 given by [12]

$$I = \frac{B(H_2)^3}{12}, \tag{17}$$

and the Young's modulus in the x_1-direction denoted by E_{11}, the moment is given by [12]

$$M_2 = E_{11} I \kappa_2 , \tag{18}$$

we find by inserting κ_2 (expressed in terms of M_2) from (18) into (16)

$$J_{ext} = \frac{(M_2)^2}{B E_{11} I}. \tag{19}$$

From (19), we can find the moment required for a given J integral value:

$$M_2 = \sqrt{J_{ext} B E_{11} I}\,. \tag{20}$$

Eq. (20) gives M_2 as a function of H_2 (since I depends on H_2, see (17)). With M_2 determined, we can derive equations for ε_{11}^{2t} and ε_{11}^{2b}. With $N_2 = 0$, ε_2^{0*} must be zero for a homogenous specimen. An equation for Beam 2, similar to (14), gives

$$\varepsilon_{11}^{2b} = -\,\varepsilon_{11}^{2t}, \tag{21}$$

so that an equation for Beam 2 similar to (15) gives

$$\kappa_2 = \frac{2\varepsilon_{11}^{2t}}{H_2}, \tag{22}$$

or

$$\varepsilon_{11}^{2t} = \frac{H_2 \kappa_2}{2}. \tag{23}$$

Now we can set up calculations to assess the uncertainly calculations. Since, in some test configurations, the bending moment is generated by a force couple [3], we will consider the uncertainty of the moment in terms of both the applied force, P, and the moment arm, ℓ_2, where the moment is given by

$$M_2 = P\ell_2\,. \tag{24}$$

Then, by (22) and (24), the J integral equation (16) becomes (dropping the subscript of J_{ext})

$$J = \frac{2P\ell_2 \varepsilon_{11}^{2t}}{BH_2}. \tag{25}$$

The uncertainty of J, denoted by ΔJ, can now be estimated using the general formula for error propagation [13], as follows:

IOP Conf. Series: Materials Science and Engineering 1293 (2023) 012024 doi:10.1088/1757-899X/1293/1/012024

$$(\Delta J)^2 = \left(\frac{\partial J}{\partial P}\Delta P\right)^2 + \left(\frac{\partial J}{\partial \ell_2}\Delta \ell_2\right)^2 + \left(\frac{\partial J}{\partial B}\Delta B\right)^2 + \left(\frac{\partial J}{\partial H_2}\Delta H_2\right)^2 + \left(\frac{\partial J}{\partial \varepsilon_{11}^{2t}}\Delta \varepsilon_{11}^{2t}\right)^2, \quad (26)$$

where ΔP, $\Delta \ell_2$, ΔB, ΔH_2 and $\Delta \varepsilon_{11}^{2t}$ are the uncertain of measured values of the applied force, P, the moment arm, ℓ_2, the specimen width, B, the specimen height, H_2 and the strain at the top of the beam, ε_{11}^{2t} , respectively. Partial differentiation of (25) gives

$$\frac{\partial J}{\partial P} = \frac{J}{P}, \quad \frac{\partial J}{\partial \ell_2} = \frac{J}{\ell_2}, \quad \frac{\partial J}{\partial B} = -\frac{J}{B}, \quad \frac{\partial J}{\partial H_2} = -\frac{J}{H_2}, \quad \frac{\partial J}{\partial \varepsilon_{11}^{2t}} = \frac{J}{\varepsilon_{11}^{2t}}. \quad (27)$$

Inserting (27) into (26) gives:

$$\left(\frac{\Delta J}{J}\right)^2 = \left(\frac{\Delta P}{P}\right)^2 + \left(\frac{\Delta \ell_2}{\ell_2}\right)^2 + \left(-\frac{\Delta B}{B}\right)^2 + \left(-\frac{\Delta H_2}{H_2}\right)^2 + \left(\frac{\Delta \varepsilon_{11}^{2t}}{\varepsilon_{11}^{2t}}\right)^2. \quad (28)$$

Taking the square root of (28) gives the relative uncertainty of J, $\Delta J/J$.

4. Results

4.1 Input parameters for sensitivity study

Although our analysis, Eq. (28), includes the sensitivity of all parameters, we will in the following focus on the effect of strain accuracy for curvature measurements. This is motivated by the fact that different strain measurement techniques have different accuracy, and it is important to choose ~~a~~ strain measurement techniques that lead to a sufficient accuracy result for the J integral value.

The uncertainty of strain measurements by foil strain gauges bonded to the surface is estimated to be ± 5 µstrain by Carlsson *et al.* [14]. Hannah and Reed [15] discuss many of the parameters that can influence the output from strain gauge measurement but wrote that, with reasonably care, one can realize strain readings accurate to ±2 % with a threshold strain sensitivity of 5 µstrain. Similar values can be found in other text books on strain measurements [16].

The uncertainty of strain measurements by DIC has been estimated to be about 10^{-4} (100 µstrain) [17]. In cases where the strain is uniform, the strain values for 100 images can be averaged and accuracy of the averaged value can be about 3 x 10^{-6} (3 µstrain) [18]. However, Reu *et al.* [19] found, comparing results from 11 different DIC codes, that for a varying strain field (Sample 14 in that study) that by averaging 50 rows of data, the standard deviation in the strain was between 172 x 10^{-6} to 875 x 10^{-6} (172 - 875 µstrain), with the majority in the range of 300 - 600 x 10^{-6} (300 - 600 µstrain).

In the following, we use these values of typical DCB test specimens for glass fibre composites: B = 30.0 mm, H_2 in the range of 2 to 16 mm and E_{11} = 40 GPa [4]. We use a moment arm with ℓ_2 = 86.5 mm. We investigate the situation of characterising a rather weak interface (fracture energy of 10 J/m^2), as well as a value corresponding to onset of crack growth of typical glass fibre composites (200 J/m^2) and a relative high fully-developed (steady-state) fracture resistance value (1000 J/m^2) [4].

We assume the following uncertainties: For the geometric dimensions we assume some uncertainties in measurement, ΔB = 0.1 mm, ΔH_2 = 0.2 mm and $\Delta \ell_2$ = 0.3 mm, while for the load, we take a load cell to be accurate within one percentage (twice the value claimed by a load-cell manufacturer [20]), thus we set $\Delta P/P$ = 0.01. Furthermore, with B and ℓ_2 fixed, we have $\Delta B/B$ = 3.33 x 10^{-3} and $\Delta \ell_2/\ell_2$ = 3.47 x 10^{-3}. For the strain measurement, we explore four values: A relative uncertainty of 2%, corresponding to the order of magnitude of uncertainty of strain gauges [15], uncertainties of 100 x 10^{-6} (100 µstrain), 300 x 10^{-6} (300 µstrain), and 600 x 10^{-6} (600 µstrain), the two last values taken to be representative of DIC [19].

Table 1 lists the squared uncertainty terms that are kept fixed in the present study. They constitute some of the parameter to go into (28). The remaining terms are explored next.

Table 1. Fixed uncertainty terms.

$\left(\frac{\Delta P}{P}\right)^2$	$\left(\frac{\Delta \ell_2}{\ell_2}\right)^2$	$\left(-\frac{\Delta B}{B}\right)^2$
100 x 10^{-6}	12 x 10^{-6}	11 x 10^{-6}

Specifying the fracture resistance, and thus the applied J integral value, J_{ext}, the stiffness, E_{11}, as well as the width and height of the beams, B and H_2, eq. (20) then gives the required moment, M_2. Next, κ_2 is calculated from (18), and from (23) we get ε_{11}^{2t}. Inserting these values together with the estimated uncertainty terms into (28) gives the relative uncertainty of J.

4.2. Estimated uncertainties

Results for uncertainty terms are shown in Table 2 for H_2 = 2 mm (rather thin DCB specimens for wind turbine blade materials [4]), Table 3 for H_2 = 8 mm (a representative DCB test height [4]) and Table 4 for H_2 = 16 mm (thicker specimens). The squared terms shown under the light green background colour enter eq. (28).

Table 2. Estimated uncertainty terms for beam height H_2= 2 mm calculated as a function of applied J value and various uncertainties of strain measurements. The strain uncertainty values in the fourth column from the left are absolute values ($\Delta\varepsilon_{11}^{2t}$) except for relative values ($\Delta\varepsilon_{11}^{2t}/\varepsilon_{11}^{2t}$) marked with #.

J (J/m^2)	P (N)	κ_2 (mm^{-1})	$\Delta\varepsilon_{11}^{2t}$ (-)	$\left(-\frac{\Delta H_2}{H_2}\right)^2$	$\left(\frac{\Delta\varepsilon_{11}^{2t}}{\varepsilon_{11}^{2t}}\right)^2$	$\left(\frac{\Delta J}{J}\right)$
10	5.66	0.612 x 10^{-3}	0.02#	0.01	0.40 x 10^{-3}	0.10
10	5.66	0.612 x 10^{-3}	100 x 10^{-6}	0.01	27 x 10^{-3}	0.19
10	5.66	0.612 x 10^{-3}	300 x 10^{-6}	0.01	0.24	0.50
10	5.66	0.612 x 10^{-3}	600 x 10^{-6}	0.01	0.96	0.98
200	25.3	2.74 x 10^{-3}	0.02#	0.01	0.40 x 10^{-3}	0.10
200	25.3	2.74 x 10^{-3}	100 x 10^{-6}	0.01	1.3 x 10^{-3}	0.11
200	25.3	2.74 x 10^{-3}	300 x 10^{-6}	0.01	12 x 10^{-3}	0.15
200	25.3	2.74 x 10^{-3}	600 x 10^{-6}	0.01	48 x 10^{-3}	0.24
1000	56.6	6.12 x 10^{-3}	0.02#	0.01	0.40 x 10^{-3}	0.10
1000	56.6	6.12 x 10^{-3}	100 x 10^{-6}	0.01	0.27 x 10^{-3}	0.10
1000	56.6	6.12 x 10^{-3}	300 x 10^{-6}	0.01	2.4 x 10^{-3}	0.11
1000	56.6	6.12 x 10^{-3}	600 x 10^{-6}	0.01	9.6 x 10^{-3}	0.14

Several points are noteworthy. Since it is not uncommon to experience a scatter of ±20% in fracture experiments [21], we choose an acceptance criterion of maximum 20% relative uncertainty on J (but other acceptance levels could also be argued). Then from Table 2 (H_2 = 2 mm) it is seen that for J = 10 J/m^2, a strain resolution of 100 μstrain is required. The expected resolution of DIC (300 μstrain) leads to an uncertainty of 50% in J, which, with the chosen acceptance criterion, would be unacceptable high for H_2 = 2 mm. Consulting Table 3 and Table 4, we note that an increasing value of H_2 (with all other parameters fixed) would result in an even higher uncertainty in in J.

Table 3. Estimated uncertainty terms for beam height H_2= 8 mm calculated as a function of applied J value and various uncertainties of strain measurements. The strain uncertainty values in the fourth column from the left are absolute values ($\Delta\varepsilon_{11}^{2t}$) except for relative values ($\Delta\varepsilon_{11}^{2t}/\varepsilon_{11}^{2t}$) marked with #.

J (J/m²)	P (N)	κ_2 (mm⁻¹)	$\Delta\varepsilon_{11}^{2t}$ (-)	$\left(-\frac{\Delta H_2}{H_2}\right)^2$	$\left(\frac{\Delta\varepsilon_{11}^{2t}}{\varepsilon_{11}^{2t}}\right)^2$	$\left(\frac{\Delta J}{J}\right)$
10	45.31	76.5×10^{-6}	0.02#	0.63×10^{-3}	0.4×10^{-3}	0.03
10	45.31	76.5×10^{-6}	100×10^{-6}	0.63×10^{-3}	0.11	0.33
10	45.31	76.5×10^{-6}	300×10^{-6}	0.63×10^{-3}	0.96	0.98
10	45.31	76.5×10^{-6}	600×10^{-6}	0.63×10^{-3}	3.8	1.96
200	202.6	0.342×10^{-3}	0.02#	0.63×10^{-3}	0.4×10^{-3}	0.03
200	202.6	0.342×10^{-3}	100×10^{-6}	0.63×10^{-3}	5.3×10^{-3}	0.08
200	202.6	0.342×10^{-3}	300×10^{-6}	0.63×10^{-3}	48×10^{-3}	0.22
200	202.6	0.342×10^{-3}	600×10^{-6}	0.63×10^{-3}	0.19	0.44
1000	453.1	0.765×10^{-3}	0.02#	0.63×10^{-3}	0.4×10^{-3}	0.03
1000	453.1	0.765×10^{-3}	100×10^{-6}	0.63×10^{-3}	1.1×10^{-3}	0.04
1000	453.1	0.765×10^{-3}	300×10^{-6}	0.63×10^{-3}	9.6×10^{-3}	0.10
1000	453.1	0.765×10^{-3}	600×10^{-6}	0.63×10^{-3}	38×10^{-3}	0.20

Table 4. Estimated uncertainty terms for beam height H_2= 16 mm calculated as a function of applied J value and various uncertainties of strain measurements. The strain uncertainty values in the fourth column from the left are absolute values ($\Delta\varepsilon_{11}^{2t}$) except for relative values ($\Delta\varepsilon_{11}^{2t}/\varepsilon_{11}^{2t}$) marked with #.

J (J/m²)	P (N)	κ_2 (mm⁻¹)	$\Delta\varepsilon_{11}^{2t}$ (-)	$\left(-\frac{\Delta H_2}{H_2}\right)^2$	$\left(\frac{\Delta\varepsilon_{11}^{2t}}{\varepsilon_{11}^{2t}}\right)^2$	$\left(\frac{\Delta J}{J}\right)$
10	128.15	27.1×10^{-6}	0.02#	0.16×10^{-3}	0.40×10^{-3}	0.03
10	128.15	27.1×10^{-6}	100×10^{-6}	0.16×10^{-3}	0.21	0.46
10	128.15	27.1×10^{-6}	300×10^{-6}	0.16×10^{-3}	1.9	1.39
10	128.15	27.1×10^{-6}	600×10^{-6}	0.16×10^{-3}	7.7	2.77
200	573.1	0.121×10^{-3}	0.02#	0.16×10^{-3}	0.40×10^{-3}	0.03
200	573.1	0.121×10^{-3}	100×10^{-6}	0.16×10^{-3}	11×10^{-3}	0.10
200	573.1	0.121×10^{-3}	300×10^{-6}	0.16×10^{-3}	96×10^{-3}	0.31
200	573.1	0.121×10^{-3}	600×10^{-6}	0.16×10^{-3}	0.38	0.62
1000	1281.5	0.271×10^{-3}	0.02#	0.16×10^{-3}	0.40×10^{-3}	0.03
1000	1281.5	0.271×10^{-3}	100×10^{-6}	0.16×10^{-3}	2.1×10^{-3}	0.05
1000	1281.5	0.271×10^{-3}	300×10^{-6}	0.16×10^{-3}	19×10^{-3}	0.14
1000	1281.5	0.271×10^{-3}	600×10^{-6}	0.16×10^{-3}	77×10^{-3}	0.28

From the tables we note that an increase in J increases both P and κ_2 (this is also expected from eq. (16)) and increases ε_{11}^{2t} (as expected from (25)). The higher strain means that the uncertainty contribution associated with strain measurement decreases. For example, from Table 3 it can be inferred that for J = 200 J/m², a value of H_2 = 8 mm gives a relative uncertainty of J of 22% (i.e., exceeding the acceptance criterion) for a strain uncertainty of 300 µstrain. But (still for H_2 = 8 mm) for the highest J value (1000 J/m²) even the highest strain uncertainty (600 µstrain) would give a relative uncertainly of 20% on J, just within the acceptance criterion set here.

With other parameter fixed, an increase in H_2 leads to an increase in P and a decrease in κ_2, resulting in an increasing uncertainty in J for an uncertainty in strain resolution of 300 µstrain (corresponding to the expected uncertainty of DIC).

Insight can be gained by comparing the magnitude of the squared uncertainty terms (showed under the light green colors in Tables 1 to 4) associated with the various parameters. The contributions from P, ℓ_2 and B are in the order of 10^{-5} to 10^{-4} (Table 1). The squared uncertainty terms associated with H_2 and ε_{11}^{2t} are (in nearly all cases) larger. The squared uncertainty term from H_2 is in the range of approximately 10^{-4} to 0.01, decreasing with increasing H_2. The squared uncertainty contribution from ε_{11}^{2t} is in the order of magnitude of 10^{-4} to 1, increasing with increasing H_2. Comparing the tables, it is seen that in most cases, the uncertainty contribution from ε_{11}^{2t} is the largest contributor to the combined uncertainly of J. Optimization of specimen geometry could include reducing the term $(\Delta\varepsilon_{11}^{2t}/\varepsilon_{11}^{2t})^2$ to the same magnitude as the other terms.

5. Discussion

5.1. Limitation of the approach

Tables 2 to 4 show that for J = 10 J/m^2, a strain resolution of 300 μstrain gives an uncertainty of J that is larger than 20%. This suggests that DIC would not be suitable for measurement of strains for curvature for materials and interfaces having low fracture energy values. Measurement using strain gauges would most likely be needed.

In practice, it would be difficult to bond foil strain gauges to the bottom of Beam 1 and top of Beam 2 (the cracking plane). It is therefore highly desirable to be able to obtain the strain measurements from the frontal face of the beams using DIC. A broader sensitivity study could be made to determine the best specimen dimensions following the lines of the small example presented here.

The approach is not limited to the type of results given in Table 2 to 4. The formulas in Section 3.3 can for example be used to determine the strain resolution necessary to obtain a desired relative uncertainty (e.g., $\Delta J/J$ = 20% or maybe $\Delta J/J$ = 10%) for different levels of H_2 and J.

5.2. Practical measurement of ε_{11}^{2t} and ε_{11}^{2b}

As described earlier, for a beam segment loaded by an axial force and a bending moment, the strain component ε_{11} varies linearly across the beam height in accordance with (3) and (4), being independent of the x_1-position. Therefore, the averaging approach proposed by Wang and Tong [18] can be used for the present specimen remote from the fracture process zone. More precisely, the strain component ε_{11} could be averaged along the length for the same x_2-position (or as an example, y_2^*-position for Beam 2).

Furthermore, the fact that the strain varies linearly across the beam in accordance with (4), enables precise measurement of ε_{11}^{2t} and ε_{11}^{2b} (or κ_2 and ε_2^{0*}) as follows. First, as shown in figure 3a), the strain component ε_{11} is determined for many points along a line at a given x_2-position (the approach of Wang and Tong). These strain values are then averaged (see Fig. 3b). Likewise, the average value of ε_{11} is determined for points along another line at a smaller x_2-position. This can be done for several lines along the x_1-direction, creating a data set of $\varepsilon_{11} = \varepsilon_{11}(x_2)$. A straight line can be fitted to the $\varepsilon_{11}(x_2)$-data and the parameters κ_2 and ε_2^{0*} can be obtained by linear regression in accordance with (4). Such an approach might give a strain measurement accuracy better than the range of 300 - 600 μstrain obtained by Reu *et. al.* [19].

6. Conclusions

The J integral contributions for beams of beam-like specimens subjected to axial forces and bending moments can be calculated from measurement of applied loads and resulting deformation (strains and curvature) of three beams. No stiffness data are required. The approach eliminates the need to make additional tests to determine the tensile and bending stiffnesses.

Such a testing approach is particularly useful for fracture mechanic characterisation of materials with unknown elastic properties, and specifically for analysis of substructure cut from old blades where the elastic properties and thickness of layers are not readily known.

The proposed approach of using digital image correlation for strain and curvature measurements for fracture mechanical characterisation by J may be associated with large uncertainties for materials and interfaces that have a low fracture energy (here exemplified by a fracture energy of 10 J/m^2). But for materials and interfaces with higher fracture resistance (fracture energy in the order of 200 J/m^2 or higher), the proposed approach seems feasible.

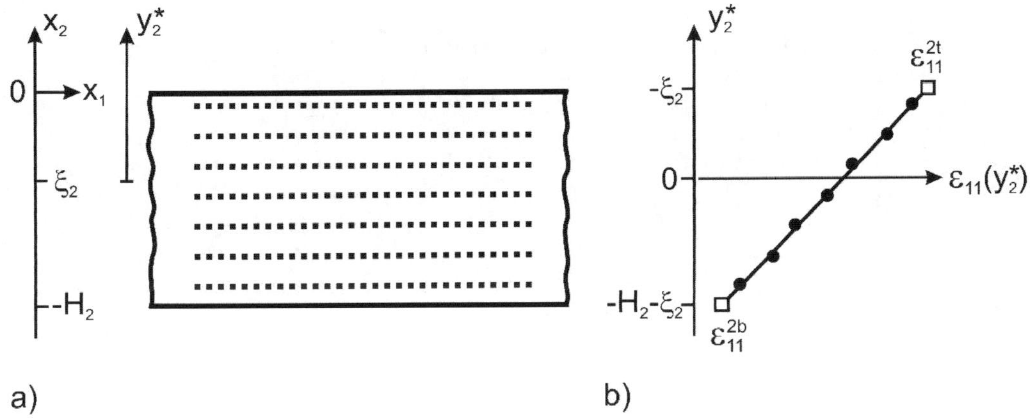

Figure 3. An approach for measurement of strain variation across a beam segment subjected to uniform extension and bending by the DIC technique. (a) measurement of strains of many points along several lines parallel to the beam (the x_1-directions), (b) for each line average points are anticipated to lie on a straight line, which allows extrapolation to identify ε_{11}^{2t} and ε_{11}^{2b}.

Acknowledgements
This works was made as a part of the ReLife project ("Methodology for assessment of remaining life time of wind turbine blades based on damage state") funded by the Energy Technology Development and Demonstration Programme (EUDP) under the Danish Energy Agency (grant no. 64019-0543).

References
[1] ASTM D5528-01 (2001); "Standard test for Mode I interlaminar fracture toughness of unidirectional fiber-reinforced polymer matrix composites", American Society for Testing and Materials, West Conshohocken, Pennsylvania.
[2] ASTM D6671/D6671M-22 (2022), "Standard Test Method for Mixed Mode I-Mode II Interlaminar Fracture Toughness of Unidirectional Fiber Reinforced Polymer Matrix Composites", American Society for Testing and Materials, West Conshohocken, Pennsylvania.
[3] Sørensen B F, Jørgensen K, Jacobsen T K and Østergaard R C 2006 *Int. J. Fract.* **141** 159
[4] Sørensen B F and Jacobsen T K 2009 *Comp. Sci. Techn.* **69** 445
[5] Paris A J and Paris P C 1988 *Int. J. Fracture* **38** R19
[6] Olsson P and Stigh U 1989 *Int. J. Fracture* **41** R71
[7] Rask M and Sørensen B F 2012 *Engr. Fracture Mechn.* **96** 37
[8] Sarrado C, Turon A, Renart J and Costa J 2015 *Comp. Sci. Techn.* **117** 85
[9] Toftegaard H L and Sørensen B F 2019 *Engr. Fracture Mechn.* **217** 106500
[10] Sutton M A, Cheng M, Peters W H, Chao Y J and McNeill S R 1986 *Image Vis. Comput.* **4** 143
[11] Rice J R 1968 *J. Appl. Mech.* **35** 379
[12] Beer F P and Russell Johnston E Jr 1992 *Mechanics of Materials* (London: McGraw-Hill), 2 edit.
[13] Taylor J R 1982 *An introduction to error analysis. The study of uncertainties in physical measurements* (Sausalito, University Science Books, second edition)
[14] Carlsson L A, Adams D F and Byron Pipes R 2014, *Experimental characterisation of advanced composite materials* (Boca Raton: CRC Press, Taylor & Francis Group), 4th ed., Appendix C.

[15] Hannah R L and Reed S E 1992 *Strain gage users' handbook* (Dordrecht: Elsevier Science Publishers) p. 137
[16] Dally J W, Riley W F and McConnell K G 1984 *Instrumentation for engineering measurements*, John Wiley and Sons p. 193, p. 205
[17] Smith B W, Li X and Tong W 1998 *Experimental Techniques* **22** 19
[18] Wang Y G and Tong W 2013 *Optics and Lasers in Engineering* **51** 30
[19] Reu P L, Toussaint E, Jones E, Bruck H A, Iadicola M, Balcaen R, Turner D Z, Siebert T, Lava P and Simonsen M 2018 *Experimental Mechanics* **58** 1067
[20] Instron 2527 Series Bi-axial Dynacell Load Cells, https://www.instron.com/en/products/testing-accessories/load-cells/dynamic-fatigue.
[21] Sørensen B F and Jacobsen T K 1998, *Comp. A.* **29A** 1443

IOP Conf. Series: Materials Science and Engineering 1293 (2023) 012025 doi:10.1088/1757-899X/1293/1/012025

Impact of shell structure stiffness on aero-structural coupling in wind turbine rotor blades

Claudio Balzani, Julia Gebauer

Leibniz University Hannover, Institute for Wind Energy Systems, Appelstr. 9A, 30167 Hannover, Germany

E-mail: research@iwes.uni-hannover.de

Abstract.

Wind turbine rotor blades are heavily loaded composite structures that experience a mixture of aerodynamic, inertial, gravitational, and gyroscopic forces during their operation life. Due to the high loads, the cross-sections of the blades are subjected to in-plane and out-of-plane deformations. The out-of-plane deformations are referred to as shear warping while the in-plane deformations are also called blade breathing. Blade breathing depends on the magnitude of the mechanical loads, which are expressed by means of internal forces and moments, and the stiffness of the blade shell.

In this work, the relationships between in-plane cross-sectional deformations and internal loads are investigated. For the quantification of the deformation, a reference blade is studied via 3D finite shell element simulations for different loading scenarios. The cross-section of interest is located at the radial position of maximum chord. To compare the shape of the cross-sections in the undeformed and the deformed configurations, a procedure is proposed to relate the positions of nodes associated with the cross-section of interest in both configurations to a joint coordinate system. The shape of the deformed cross-section is then extracted and compared with the undeformed configuration. The comparison is executed for the individual internal forces and moments, namely flapwise and edgewise bending moments, normal force, shear forces, and torsion moment, respectively. The deformation patterns are discussed and it is addressed how these may influence the aerodynamic behavior of the cross-section under consideration.

1. Introduction

The performance of a wind turbine is governed by the aeroelastic characteristics of the rotor to a large extent. The aerodynamic characteristics of the rotor are defined by the characteristics of the airfoils used in the aerodynamic design. The lift-to-drag ratio as a measure of the airfoil efficiency is determined by the overall shape of the airfoil, which can be reduced to the relative thickness and the camber of the airfoil. A thin airfoil has a high lift-to-drag ratio, and the camber further increases the lift. Hence, a blade would consist only of thin airfoils, if it was designed just with respect to aerodynamic considerations. However, from a structural point of view, thick airfoils are beneficial for the higher loaded sections close to the blade root, as an increase of thickness increases the cross-section's moment of inertia and with it the structural features of the blade (i.e., stiffness, strength, natural frequencies, etc.).

Cross-sectional deformations are normally subdivided into out-of-plane deformations called shear warping, and in-plane deformations. In the general case, in-plane cross-sectional deformations (IPCD) in a blade correspond to a variation of the airfoil shape and the aerodynamic characteristics on the one hand, and a variation of the cross-section's moment of inertia and the structural characteristics on the other. Consequently, cross-sectional deformations may compromise the aeroelastic performance of the blade, especially for very large and slender blade structures.

The effect of cross-sectional deformations on thin-walled shell-like structures was already demonstrated in [1]. It was mentioned under the name *Brazier effect* as a geometric nonlinearity in wind turbine blades in [2]. A solution for the calculation of cross-sectional deformations in symmetrical multi-cell airfoils was provided in [3]. A simplified approach for the analysis of the Brazier effect in composite beams was proposed in [4]. Cross-sectional deformations have been investigated in the context of damage analyses [5]. They have also been considered to calculate peel stresses in a trailing edge adhesive joint [6]. Buckling as another geometrically nonlinear effect was found to be a possible driver for trailing edge damages [7]. However, to the best knowledge of the authors, there is no publication available on the effect of IPCD on the aerodynamic or aeroelastic behavior of the blade.

In this paper, the cross-sectional deformation is investigated for a reference wind turbine blade at the maximum chord position. A shell element model is used for 3D finite element (FE) simulations of the structural response. The blade is considered clamped at the root, and the force-like boundary conditions are applied in such a way that the internal force and moment distributions from loads analyses are approximated. The impact of the normal force, flapwise and edgewise shear forces, torsion moment, and flapwise and edgewise bending moments are analyzed independent of each other. Although this is unrealistic, as a blade in real life is always subjected to a combination of these loads, it allows to estimate what internal force or moment contributes in which manner to the IPCD.

To identify the shape of the deformed cross-section, a plane is identified considered the cross-sectional plane in the deformed configuration. The nodal positions of the nodes associated with the cross-section under investigation in the deformed configuration are projected onto that plane. The method retains the assumptions according to Timoshenko beam theory, i.e., planarity of cross-sections and constant shear angles, but contains detailed information of the 3D model (e.g., blade compliance). The cross-sectional deformation contributions of each internal force and moment are described, both qualitatively and quantitatively, and conclusions are drawn related to their criticality in the context of the aeroelastic behavior of the blade.

2. Calculation of cross-sectional deformations

For the cross-sectional deformations, the following assumptions were made:

- The cross-sections of interest undergo a translational and rotational deflection in space.
- The cross-section of interest in the undeformed and the deformed configuration is planar.
- The cross-section of interest is subjected to out-of-plane and in-plane warping, the latter of which called IPCD in this paper.

The aim was to extract the deformed cross-section from the solution of the 3D FE simulation and to compare it with the undeformed configuration. Both cross-sections thus had to be related to a joint coordinate system, which was the (global) blade root coordinate system. There, $\boldsymbol{z}$ was the basis vector pointing along the pitch axis of the blade, i.e., the director vector of the blade root cross-section, $\boldsymbol{x}$ was the in-plane basis vector perpendicular to the rotor plane (i.e., in wind direction), and $\boldsymbol{y}$ was the in-plane basis vector parallel to the rotor plane forming a right-hand Cartesian coordinate system, see also Figure 1.

The centroid of the undeformed and deformed cross-sections were calculated by the mean of all position vectors of the associated nodes in the undeformed and the deformed configuration,

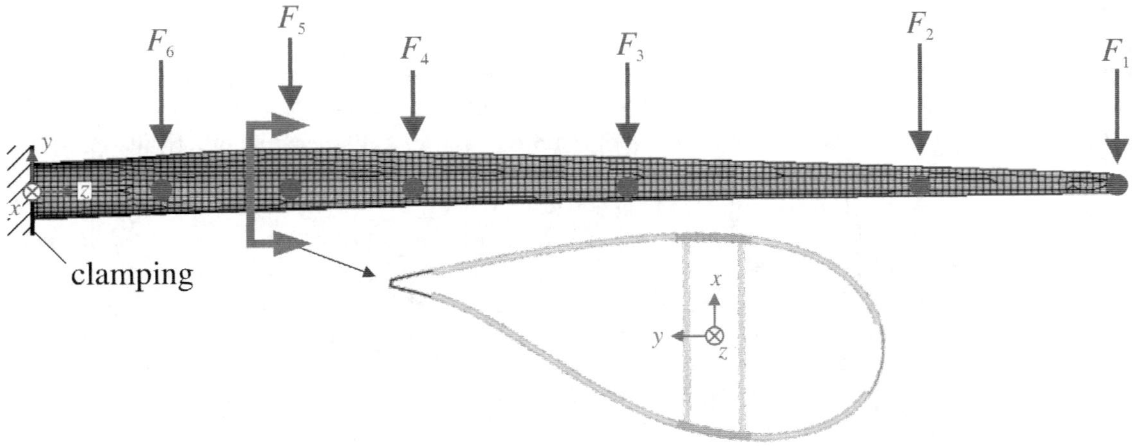

Figure 1. Planform of the blade's 3D FE model, visualization of the cross-section at the marked maximum chord position, coordinate systems, and exemplary load boundary conditions. The blade model was created with MoCA [12], the in-house model creation and analysis tool (for visibility the mesh is coarser than that used for the analyses). The cross-section visualization was created with BECAS [13]. The red dots show the load points. The external single forces F_i approximate the internal forces and moments. The global blade coordinate system is plotted in red, the local cross-section coordinate system in blue.

respectively. The associated nodes were then translated by the centroid vectors. The centroid was thus positioned in the origin of the blade root coordinate system for both configurations.

For the undeformed cross-section, the (local) chord coordinate system was used[1] to set up a rotation matrix. The rotation matrix was then utilized to transform the nodal coordinates so that the cross-section plane was aligned with the blade root coordinate system. Note that this step is only necessary if the blade exhibits prebend and/or presweep. Otherwise, the plane of the undeformed cross-section is parallel to the blade root coordinate system and the rotation matrix becomes the identity matrix.

The translational motion of the deformed cross-section was already accounted for by the translation of the cross-section centroid to the blade root center. For the rotational motion, the orientation of the cross-sectional plane in the deformed configuration had to be found. In the general case, the nodes in the deformed configuration are not located on a plane due to out-of-plane warping that is non-constant due to changes in shell stiffness around the circumference of the airfoil. However, we assumed that the deformed cross-section is planar (see the second bullet point above). It was thus necessary to identify a plane that is considered the cross-sectional plane. In other words: The error of nodal positions not matching the cross-sectional plane had to be minimized. The mean value of the rotations of nodes associated with the cross-section of interest was calculated for setting up a rotation. This rotation matrix was then used for the coordinate transformation from the deformed to the undeformed cross-sectional plane. The nodal coordinates in the setting that was already translated to the blade root were then transformed via the rotation matrix and subsequently projected onto the plane parallel to the blade root. As a result, the shape of the deformed cross-section in the blade root coordinate system was obtained.

[1] The in-house FE model generator *MoCA* that was used for modeling the blade calculated the chord coordinate system of each nodal position along the blade to define cross-sections in the undeformed configuration including the inclusion of twist angle, prebend and/or presweep.

3. Rotor blade and load cases

The rotor blade under consideration was that of the reference wind turbine model developed in the IEA Wind TCP Task 37 [8]. The turbine has a rated power of 15 MW, a rotor diameter of 240 m, and a blade length of 117 m. As the turbine exhibits a cone angle of -4° and a prebend with a tip amplitude of 4 m, the cross-section plane in the undeformed configuration is not parallel to the blade root cross-section and the joint coordinate axes. Hence, the rotational transformation described in section 2 is necessary.

A loads assessment was executed for the design load case *normal operation* [9] at a mean wind velocity of 10 m/s, which is approximately the rated wind speed. OpenFast [10] was used with the ElastoDyn module [11] for the blades and the tower. The highest bending moment distribution was identified and the internal forces and moments were extracted for that time instance in the time series.

For creating the FE model, the in-house model creation and analysis tool called MoCA [12] was used. A sketch of the blade model in the planform view is shown in Figure 1. The coarse mesh depicted there is for visualization purposes only. The FE model used in the simulations consisted of 26,267 shell elements with linear shape functions and 25,839 nodes. Ansys® Mechanical APDL [14] was used for the solution of the FE problem. The cross-section of interest was chosen to be located at the maximum chord position, as the largest cross-sectional deformations were expected there. A sketch of the cross-section of interest is included in Figure 1, which was created with BECAS [13] for visualization via a MoCA2BECAS interface. The internal forces and moments extracted from the loads assessment at this position are given in Table 1. Note that a positive flapwise shear force or bending moment points in positive x direction, a positive edgewise shear force or bending moment in positive y direction, and a positive normal force or torsion moment in positive z direction on the positive cutting edge[2].

Table 1. Internal loads in the cross-section of interest (maximum chord position).

Load description	Unit	Minimum	Maximum
Normal force (axial)	kN	501	2,091
Shear force (flapwise)	kN	104	1,500
Shear force (edgewise)	kN	-572	382
Torsion moment	kNm	-832	236
Bending moment (flapwise)	kNm	-204	69,621
Bending moment (edgewise)	kNm	-12,828	19,463

All 6 degrees of freedom were fixed in each node at the blade root, i.e., the blade was clamped. The force-like boundary conditions consisted of external forces and/or moments applied at a finite number of positions along the blade named load points. The load point positions and the magnitude of the forces and/or moments were chosen in such a way that the internal force and moment distributions from the loads analysis were approximated, with a focus on the cross-section of interest. In the load points, a single force or moment was applied to a master node that was linked to the nodes in the cross-section at the same radial position via multiple point constraints. The internal forces and the torsion moment were thus approximated as piecewise constant distributions with jumps in the load point positions, whereas the bending moments

[2] The positive cutting edge is defined by the cutting edge where the z axis points away from. It is thus the cutting edge on the root end of the blade.

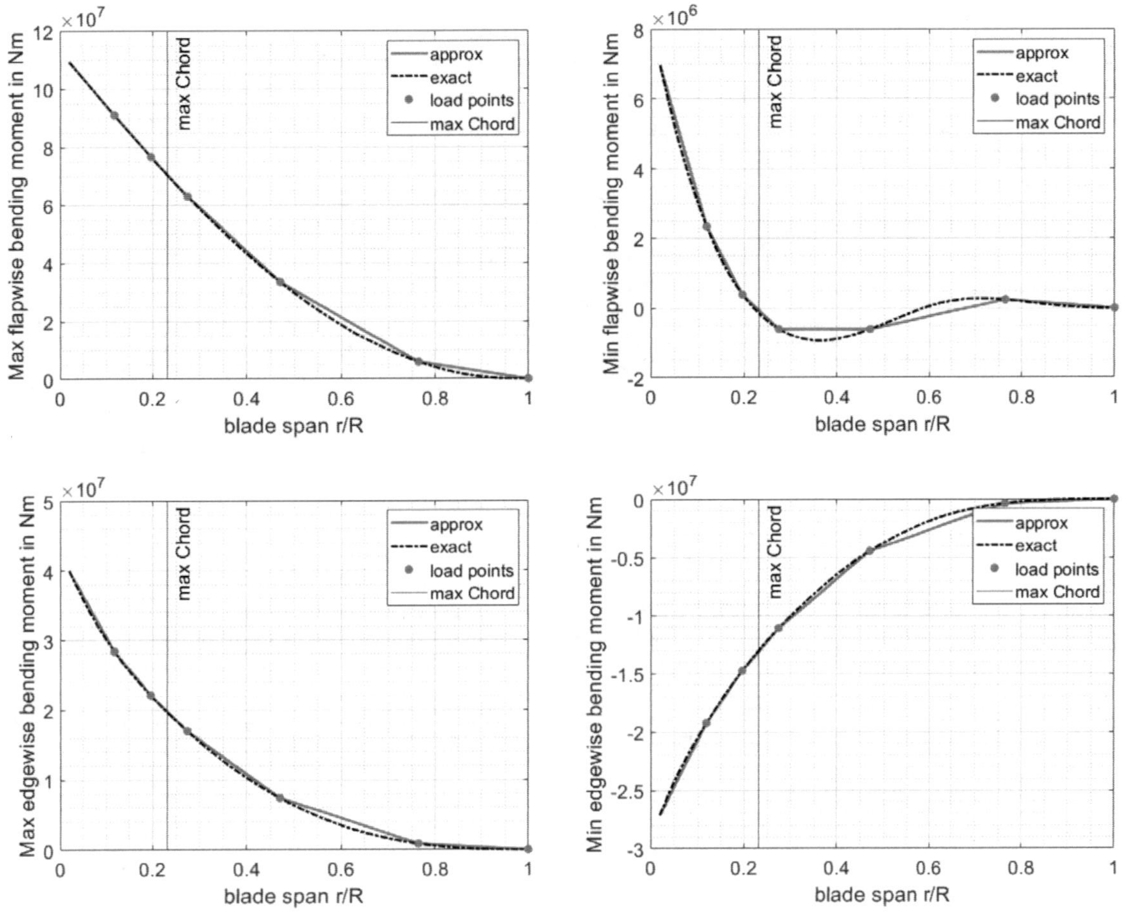

Figure 2. Bending moment distributions along the blade. Distributions from load assessments (exact) and the approximation from force-like boundary conditions (approx) are shown. Flapwise bending moments are plotted at the top, edgewise bending moments at the bottom. Maximum bending moments are plotted on the left, minimum bending moments on the right. Red dots are the load points where the external loads are applied. The maximum chord position is indicated by the vertical lines.

were approximated by continuous and piecewise linear distributions with kinks in the load point positions. The load points and the single force distribution for the maximum edgewise bending moment (i.e., compression in the trailing edge) is depicted in Figure 1.

Figure 2 exemplarily shows the minimum and maximum flapwise and edgewise bending moment distributions. As the blade root is clamped it cannot contribute to cross-sectional deformations. The load values at the blade root, i.e., at $r/R = 0$, are thus omitted in Figure 2. The convex graphs of the maximum bending moments as well as the concave graph of the minimum edgewise bending moments are well approximated. There is only a certain inaccuracy in the outboard region between load points 5 and 6 (counted from the root to the tip). This inaccuracy should be far enough away from the cross-section of interest to not influence the results. The minimum flapwise bending moment oscillates around zero, which is not well approximated between the load points 4 and 6. Here, an optimization of the load point positions would help to improve the solution. In any case, the bending moments in the cross-section of interest are all captured well. The convexity of the bending moment distribution determines whether the applied load distribution is an over- or an underestimation (i.e., bending moments in convex distributions are slightly overestimated, those in concave distributions are underestimated).

Recall that we intended to have only the desired internal force or moment in the cross-section of interest in order to be able to evaluate their impact on the IPCD separately. Hence, for the normal force, the shear force, and the torsion moment distributions the external force and moment were applied to the fourth load point, which is the load point next to the cross-section of interest towards the tip. For the normal force, the load point was moved inside the respective cross-section so that the bending moment induced by the prebend and the offset to the elastic center was compensated as much as possible, and a counter-moment was further applied to have a bending-free cross-section of interest. A similar strategy was followed for the shear forces, where, due to a small offset between the load point and the shear center, an additional counter-acting torsion moment was applied so that the cross-section of interest was torsion-free. Note that shear forces and bending moments are not independent of each other, so that the distance between cross-section of interest and load point results in a bending moment in the cross-section of interest when a shear force is applied to the load point. A counter-moment was thus applied in the load point In order to obtain a bending-free cross-section of interest that is only loaded by shear forces.

4. Results

In this section, the results of the IPCD for the different load cases are presented and discussed. The order of load cases is normal force, flapwise shear force, edgewise shear force, torsion moment, flapwise bending moment, and edgewise bending moment. For each, deformations due to both minimum and maximum load is presented, respectively.

Figure 3 shows the IPCD for the applied internal forces. The nodes in the undeformed configuration are plotted by circles, and the arrows indicate the nodal displacements. IPCD for normal forces are plotted at the top, those for shear forces in the flapwise direction in the middle, those for shear forces in the edgewise direction at the bottom; maximum forces are on the left and minimum forces on the right.

For the normal force, there is no significant IPCD. Only the leading edge panel (between the leading edge and the left shear web) on the pressure side slightly deform inwards for the maximum normal force. However, the magnitude of deformation is negligible, given that the deformations are amplified by a scaling factor of $s = 100$.

For the maximum shear force in flapwise direction, the leading edge panel on the suction side is deformed inwards, that on the pressure side outwards. The trailing edge panel (between the right shear web and the trailing edge) on the suction side is mainly deformed inwards and that on the pressure side mainly outwards. The spar caps that are located in the shells between the shear webs are deformed upwards. Hence, the shear webs are translated upwards, but additionally deform partially towards the leading edge. The deformations in the leading edge panels are larger in magnitude than those in the trailing edge panels and the deformations in the left shear web are larger than those in the right shear web. The deformation pattern for the minimum shear force in flapwise direction looks qualitatively very similar, but is quantitatively smaller. This is because the minimum shear force has the same sign as the maximum shear force, but is two orders of magnitude smaller, see Table 1.

For the maximum shear force in edgewise direction, the leading edge panel on the suction side and the trailing edge panel on the pressure side deform outwards. All other parts of the cross-section hardly show any deformation. The deformation at the trailing edge is larger than that

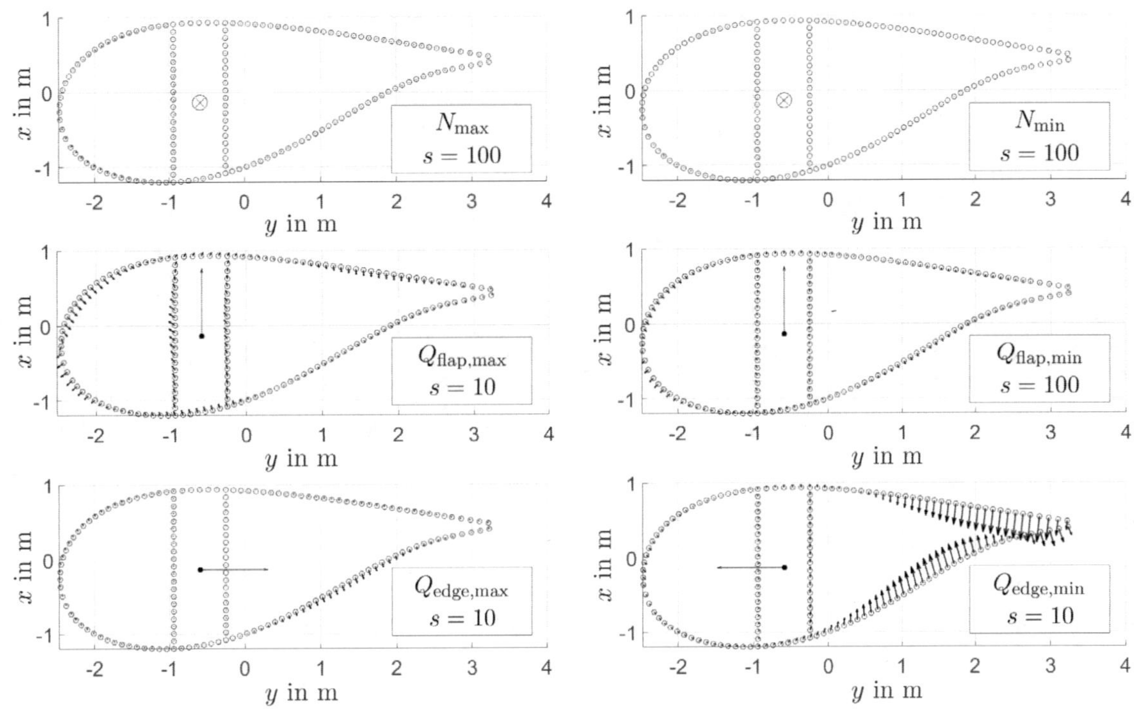

Figure 3. In-plane cross-sectional deformations (IPCD) for internal forces: Normal forces N (top), flapwise shear forces Q_{flap} (middle), edgewise shear forces Q_{edge} (bottom), maximum forces (left), minimum forces (right). The circles show the nodes in the undeformed configuration and the arrows indicate the nodal displacement vectors. The scaling factor s for the deformation vectors and the direction of the forces are given in the chart areas. View direction is in positive z direction, i.e., from towards the blade tip.

at the leading edge. For the minimum shear force in edgewise direction, the trailing edge panels (both at the suction and the pressure side) show deformations inwards. The spar caps around the right shear web deform upwards, so that the right shear web is translated upwards. The rest of the cross-section hardly deforms. The magnitude of the trailing edge deformations is orders of magnitude larger than those in the rest of the cross-section. Note that in this particular plot the scaling factor is $s = 100$, while the scaling factor in the other shear force-related plots is $s = 10$. The results thus must be treated with caution when it comes to quantitative comparisons.

Figure 4 shows the IPCD for the torsion and bending moments. Torsion moments are at the top, flapwise bending moments in the middle, and edgewise bending moments at the bottom. The maximum moments are on the left and the minimum moments on the right. The circles depict the nodes in the undeformed configuration, and the arrows are the nodal displacement vectors.

The deformation due to the maximum torsion moment is generally very small, given the amplification factor of $s = 100$. Qualitatively, the leading edge panel on the suction side deforms outwards, that on the pressure side outwards. In the trailing edge panels it is the other way round: That on the suction side deforms outwards and that on the pressure side inwards. The spar (spar caps and shear webs) hardly show any deformation. The deformation pattern for the minimum torsion moment is qualitatively the opposite, but has a larger amplitude. The reason is that the torsion moment has the opposite sign, but the magnitude is about 4 times larger.

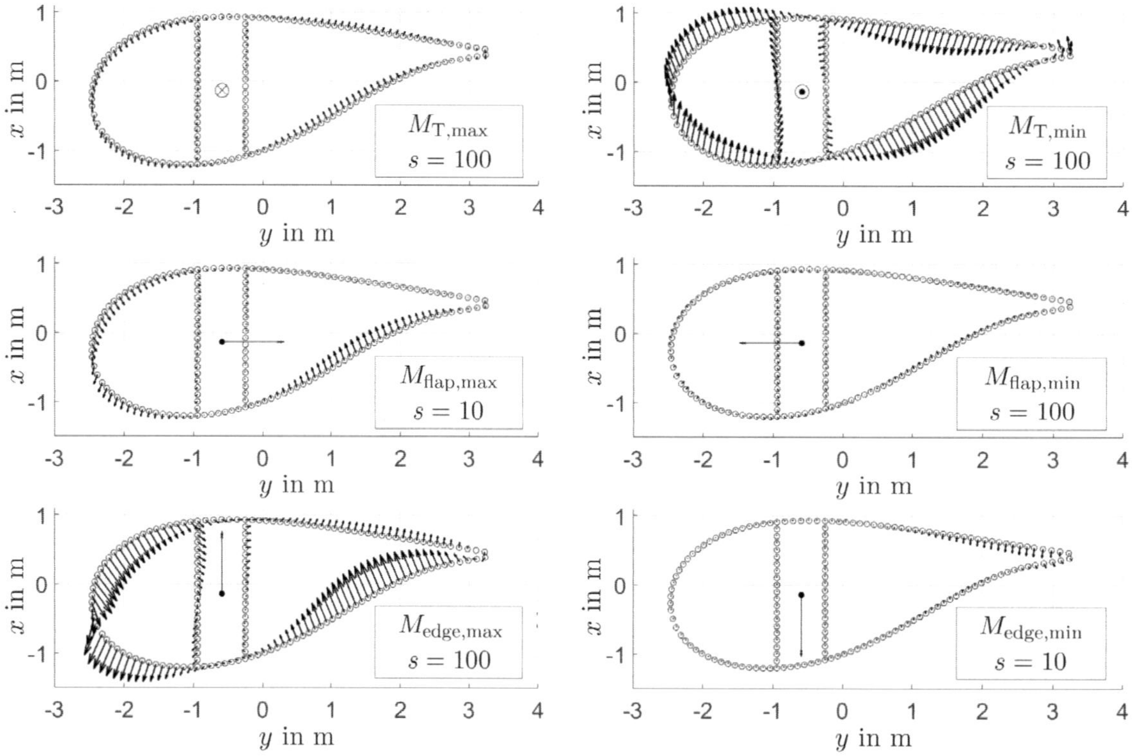

Figure 4. In-plane cross-sectional deformations (IPCD) for internal moments: Torsion moments M_T (top), flapwise bending moments M_{flap} (middle), edgewise bending moments M_{edge} (bottom), maximum moments (left), minimum moments (right). The circles show the nodes in the undeformed configuration and the arrows indicate the nodal displacement vectors. The scaling factor s for the deformation vectors and the direction of the moments are given in the chart areas. View direction is in positive z direction, i.e., from towards the blade tip.

In the case of the maximum flapwise bending moment, the leading edge panel on the suction side and the trailing edge panel on the pressure side deform inwards, while the leading edge panel on the pressure side deforms outward. The trailing edge panel on the suction side does not show any significant deformation. The spar (spar caps and shear webs) stay also in place. The deformation pattern for the minimum flapwise bending moment looks qualitatively more or less like the negatively scaled version of that for the maximum flapwise bending moment, but with a much smaller amplitude, especially given the amplification factor that is one order of magnitude higher. The reason is that the minimum flapwise bending moment in the cross-section of interest is almost zero, while the maximum flapwise bending moment is quite high (two orders of magnitude higher than the minimum flapwise bending moment, see Table 1).

For the maximum edgewise bending moment, the deformation of the leading edge panel on the suction side is inwards, as well as that of the trailing edge panel on the pressure side. The leading edge panel on the pressure side and the trailing edge panel on the suction side deform outwards. Note that the shear webs are bended and rotated clockwise. It looks like the spar is undergoing a certain portion of torsion. This may be due to some bend-torsion coupling or an offset between the external forces and the shear center of the respective cross-sections resulting in an unintentional torsion moment. Generally, the deformation amplitudes are comparably high. For the minimum edgewise bending moment, the cross-section hardly deforms. Only the trailing edge is moved downwards, which results in a deformation inwards in the trailing edge panel on the suction side and a deformation outwards on the pressure side.

5. Discussion

The cross-sectional deformation of the investigated blade at the maximum chord position is generally small, given that the deformation had to be scaled by a factor of at least $s = 10$ to make it visible. However, a decrease in stiffness especially of the shell structure would result in more pronounced blade breathing. It is thus likely that for very large and slender blades, where the panels have a smaller local bending stiffness, suffer more from cross-sectional deformations.

The effect of normal forces on the cross-sectional deformations is in any case negligible. Even with a scaling factor of $s = 100$ the cross-sectional deformation is hardly visible. A decrease of the blade structure in such extent that the cross-sections show significant in-plane deformations would very likely result in loss of structural integrity. As the exterior shape of the airfoil is unaffected by normal forces, the aerodynamic performance is unaffected as well.

The cross-sectional deformations due to flapwise shear forces may become significant if the shell bending stiffness is reduced (or the length of the cross-section is increased). The deformation pattern shows a deflection of the leading edge towards the pressure side, which in turn affects the camber line (increase in camber). This may result in higher lift forces and thus in an increase of power output. However, an increase in lift force also leads to an increase of extreme loads and fatigue load amplitudes, and thus to a reduction of the fatigue life. The shape change at the leading edge and the trailing edge panel on the suction side may also be detrimental for the angle of attack at stall, which should be further investigated.

Edgewise shear forces mainly influence the shape of the trailing edge panels (deflection inwards or outwards, depending on the sign of the shear force). This can have an impact on the camber line and on the stall behavior and also deserves further attention in the future.

Naturally, a torsion moment triggers an elastic twist around the blade axis. This elastic twist will affect the angle of attack, a key parameter for the determination of aerodynamic forces. However, a torsion moment also affects the shape of the camber line significantly, as can clearly be seen in Figure 4. For the maximum torsion moment, the camber is increased (higher lift at the same angle of attack in the linear range of the lift coefficient). However, for the minimum torsion moment, the camber is reduced. Depending on the stiffness of the shell panels, the camber may be reverted in sign, which completely changes the aerodynamic behavior of the airfoil.

In case of bending, the ovalization appears only in the shell panels, as the shear webs are normally designed stiff enough to hold the spar caps in position. However, the deformation can affect the camber line and the airfoils shape significantly, in both the leading edge region and the trailing edge region. This holds independent of the direction of bending, i.e., for both flapwise and edgewise bending.

It is clear that the cross-sectional deformation depends on the magnitude of the respective internal forces and moments, as can clearly be seen when comparing Figure 3 and Figure 4 with Table 1. However, flapwise shear forces, torsion, and bending (both flapwise and edgewise) seem to be the most critical loads in the context of aerostructural coupling of cross-sectional deformations.

From the results obtained so far, we suspect an impact of the cross-sectional deformations on the aerodynamic airfoil characteristics, as discussed above. Due to a camber change, the lift coefficient in the attached flow regime may be shifted and the angle of attack at stall may be modified as well. Depending on the particular operation state of the turbine, this can have a beneficial or detrimental effect on the power production and the load distribution along the blade. However, a detailed evaluation requires cross-sectional airfoil polars for the undeformed and deformed configuration. The calculation of those is work in progress and a comparison of the aerodynamic performance will be published in the near future.

The aim of this paper was to analyze the impact of internal forces and moments individually in order to highlight the importance of each with respect to IPCD. However, a blade in real life is subjected to all loads simultaneously. Hence, the results of this paper cannot be used directly to draw conclusions for a blade in operation, as the cross-sectional deformations are geometrically non-linear in nature and cannot be superimposed. Furthermore, the internal loads not only appear simultaneously, but also vary with time at different frequencies. This induces multiaxial non-proportional fatigue in the blade's subcomponents, e.g., the adhesive joints, which is currently subject of extensive research [15, 16]. Hence, conclusions on the relations between cross-sectional deformations and fatigue damage evolution is very difficult and may be addressed in future work.

6. Conclusions

In this paper, in-plane cross-sectional deformations of a wind turbine blade were investigated. A methodology has been established for the extraction of the shape of a deformed cross-section from a 3D finite element simulation. This methodology has then been employed for the investigation of in-plane cross-sectional deformations of a 15 MW reference wind turbine blade at one particular position of interest. The position of interest was the maximum chord position, as the largest cross-sectional deformations are normally expected there due to large shell panel dimensions. The deformed cross-sections were compared with the undeformed cross-sections, and it was discussed how the deformation may influence the aerodynamic performance of the corresponding airfoil.

It was found that the deformations were principally quite small, as a magnification factor was required to make the deformations visible. However, the deformation amplitude depends on the bending stiffness of the shell panels. Hence, in very large rotor blades, where the local bending stiffness of the panels decreases due to very large panel dimensions, the cross-sectional deformations may become significant. Generally, the airfoil shape and consequently the camber line is modified by the cross-sectional deformations. The relative thickness is hardly affected. The results give implications that the flapwise shear forces, torsion moments, and bending moments (both flapwise and edgewise) have the most significant impact on the cross-sectional deformations. However, the impact of cross-sectional deformations on the aerodynamic performance have to be investigated in more detail, which is work in progress.

Acknowledgements

This research was funded by the Deutsche Forschungsgemeinschaft (DFG, German Research Foundation) as part of the Collaborative Research Center 1463 *Integrated Design and Operation Methodology for Offshore Megastructures* (SFB1463 – Project-ID 434502799). The authors acknowledge the financial support.

References

[1] Brazier L 1927 *Proceedings of the Royal Society A* **116** 104–114
[2] Chen X and Eder M A 2020 *IOP Conference Series: Materials Science and Engineering* **942** 012001
[3] Cecchini L and Weaver P 2005 *AIAA Journal* **43** 2252–2258
[4] Damkilde L and Lund B 2009 A simplified analysis of the brazier effect in composite beams *Proceedings of the Twelfth International Conference on Civil, Structural and Environmental Engineering Computing* Paper 207 ed Topping B, Costa Neves L and Barros R (Civil-Comp Press, Stirlingshire, UK)
[5] Eder M and Bitsche R 2015 *Thin-Walled Structures* **93** 1–9
[6] Rosemeier M, Gebauer T and Antoniou A 2020 *IOP Conference Series: Materials Science and Engineering* **942** 012026
[7] Miao X Y and Chen X 2023 *Composite Structures* **308** 116680

[8] Gaertner E, Rinker J, Sethuraman L, Zahle F, Benjamin A, Barter G, Abbas N, Meng F, Bortolotti P, Skrzypinski W, Scott G, Feil R, Bredmose H, Dykes K, Shields M, Allen C and Viselli A 2020 Definition of the iea 15-megawatt offshore referencewind Tech. Rep. NREL/TP-5000-75698 National Renewable Energy Laboratory Golden, Colorado, USA URL `https://www.nrel.gov/docs/fy20osti/75698.pdf`

[9] International Electrotechnical Commission 2019 Wind energy generation systems - part 1: Design requirements, standard iec 61400-1:2019

[10] 2022 Openfast v3.1.0 URL `https://doi.org/10.5281/zenodo.6324288`

[11] Jonkman J M 2013 The new modularization framework for the fast wind turbine cae tool *Online Proceedings of the 51st AIAA Aerospace Sciences Meeting including the New Horizons Forum and Aerospace Exposition* (Grapevine, Dallas/Ft. Worth Region, Texas, USA: American Institute of Aeronautics and Astronautics)

[12] Noever-Castelos P, Haller B and Balzani C 2022 *Wind Energy Science* **2021** 105–127

[13] Blasques J P A A 2012 Users manual for becas a cross section analysis tool for anisotropic and inhomogeneous beam sections of arbitrary geometry Report Risoe-R No. 1785(EN) Risø DTU National Laboratory for Sustainable Energy Denmark. Forskningscenter Risoe

[14] Ansys® Academic Research Mechanical APDL, Release 2020 R2

[15] Kuhn M, Manousides N, Antoniou A and Balzani C 2022 *Fatigue & Fracture of Engineering Materials & Structures* **46** 1121–1139

[16] Kuhn M, Manousides N, Antoniou A and Balzani C 2023 *Fatigue & Fracture of Engineering Materials & Structures* 1–18

Large-scale fatigue testing of a retrofitted 3rd shear web in a 34m wind turbine blade section

J. P. Waldbjørn[1], C. Berggreen[1], S. Ahmed[2] and T. Nagy[2]

[1]Technical University of Denmark, Department of Civil and Mechanical Engineering, Koppels Allé, Building 404, 2800 Kongens Lyngby, Denmark
[2]Bladena ApS, Banestrøget 13, 2630 Taastrup, Denmark

Email: jpwa@dtu.dk

Abstract. Utilizing a fatigue-rated multi-axial strong-floor based test rig, the effect of a retro-fitted fish-mouth third shear-web geometry detail located within the double-curved trailing edge sandwich panels are evaluated in an inner 15m root section from a 34m wind turbine blade manufactured by SSP Technology. From previous research, a load configuration is identified, capable of triggering the breathing/pumping deformations in the trailing edge panels within the root and transition zone, which will drive the propagation on the shear-web disbonding mechanism. Investigation and evaluation of the shear web disbonding and associated mitigation are acquired through strain gauges, wire potentiometer and digital image correlation (DIC) measurements inside the root section.

1. Introduction

Wind turbines are progressively used as a substitute to fossil fuel energy resources, enhancing the demand for larger and more efficient wind turbine blades [1]. With an increasing size of wind turbine blades a growing need is emerging to address structural challenges due to higher utilization of the structural capacity – especially within the root, transition zone and max. chord region [2]. Among one of the increasingly encountered in-field damages found in wind turbine blades, regardless of blade make and model, is transverse cracks within the transition zone and max chord region [3]. This hypothesis is from an operational perspective, supported by the wind turbine owners (WTOs), who are reporting a gradually increasing amount of transverse cracks. Believed to be the root cause of these transverse cracks are, among others, the out-of-plane deformations in the large double curved trailing edge aerodynamic sandwich panels on the pressure side within the max-chord region and towards the root – referred to here as breathing [4]. These deformations, often referred to as 3D longitudinal out-of-plane bending, leads to critical bending stresses in the area, where the trailing edge panels are connected/kinked into the stiff cylindrical root section. To mitigate these critical bending stresses in the trailing edge panels, a retro-fitted fish-mouth third shear-web geometry detail is installed in the inner 15m section of a 34m wind turbine blade designed and built by SSP Technology – referred to here as the root section, connecting the inner skin of the trailing edge pressure side (PS) and suction side (SS) sandwich panels through an adhesive bond line.

The scope of this paper is to investigate and evaluate the overall structural integrity of the retro-fitted third shear-web along with the mitigation associated with the shear web disbonding mechanism triggered by breathing of the trailing edge panels. This breathing is primary triggered by i) the edgewise dominated loading governed by gravity which have to be carried by the large curved sandwich panels combined with ii) a torsional moment around the longitudinal axis of the wind turbine blade, generated by the flap-wise deformation, which acts as an in-plane eccentricity to the edgewise load [5]. Utilizing a fatigue rated multi-axial test rig, this dual degree-of-freedom (dof) load configuration is applied to the free end of the root section through a servo-hydraulic loading arrangement. Magnitude and phase shift of the dual dof load configuration capable of triggering the maximum breathing deformations in the trailing edge panels within the root and transition zone, which will drive the propagation on the shear-web disbonding mechanism, are identified in accordance to [6]. Investigation and evaluation of the shear web disbonding and associated mitigation are acquired through internally mounted strain gauges, wire potentiometer and digital image correlation (DIC) measurements.

2. Test specimen and installed third shear-web

The 34m wind turbine blade used in this application was used on a 1.5MW wind turbine. General blade dimensions and properties are outlined in table 1.

Table 1: Overall blade dimensions

Type	SSP34
Blade net weight	4500 [kg]
Weight of the truncated blade	2800 [kg]
Original length of blade	34 [m]
Length of the truncated blade	15 [m]
Diameter at the root	1.8 [m]
Max chord position from blade root	7.0 [m]
Chord position from blade root	2.7 [m]
Structural configuration	Box bar construction
Sandwich panel locations	Leading edge (LE) and trailing edge (TE)

The blade is produced using a high quality production method including pre-impregnated fibre mats without autoclave [7]. To ensure adequate resin quantity towards the core and mold a high resin content (HRC) is selected for the laminates next to those regions. The resin system used is unknown to the authors, but according to SSP Technology a modification of the Ampreg 22 epoxy laminating system [8] was used. At a distance of 4m from the root (z = 4m) a retrofitted fish-mouth third shear-web is installed connecting the inner skin trailing edge PS and SS sandwich panels through a single lap adhesive bond line according to figure 1. The main body of the shear-web consists of a sandwich structure with 10mm core material covered by 4 layers of bi-axial (±45°) glass fibre on both sides (equivalent to a skin thickness of approximately 2mm).

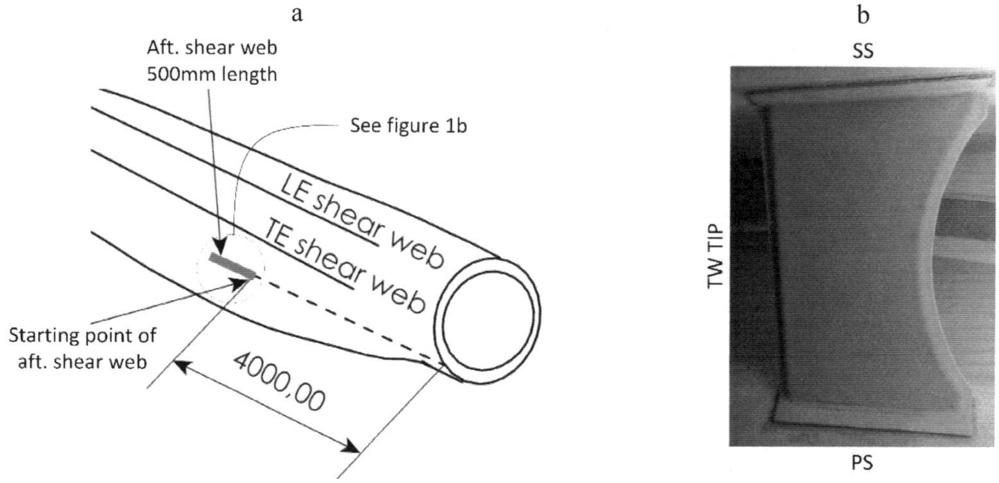

Figure 1: Retro-fitted 3rd shear-web including: a) overall position and b) photographic illustration of installation

3. Experimental test setup

A fatigue rated multi-axial test rig for structural assessment of the root section has been established on the strong floor [6]. A 3D illustration of the test setup including the loading arrangement and clamped support is presented in figure 2.

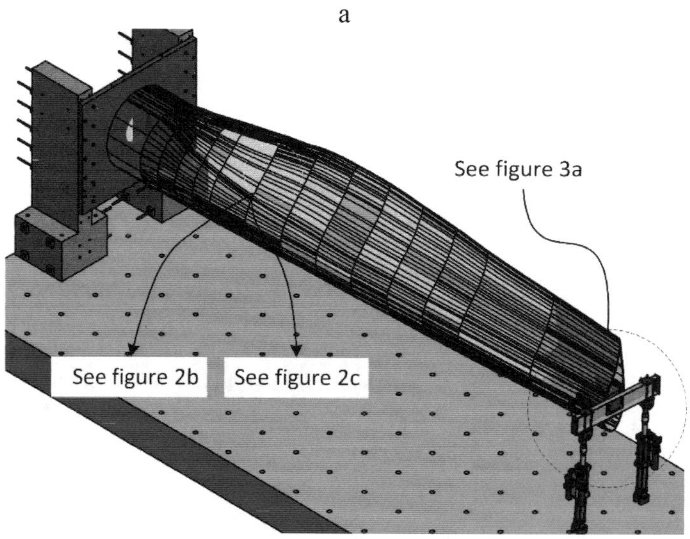

b c

Figure 2: Test setup including: a) the multi-axial loading arrang and b) photographic illustration of PS bondline monitored by DIC and c) photographic illustration of aft – mounted wire potentiometer

The fatigue rated clamped support of the root section is achieved using a mobile strong wall consisting of i) two strong floor mounted concrete towers and ii) steel plate attached to the concrete towers. The root section is center mounted on the steel plate. All connections are established using pretensioned threaded bars. Pretensioning levels of critical connections are monitored with "donut" load cells of the type: K-181/N550-G31 by Lorenz Messtechnik GmbH.

The complete loading arrangement – consisting of two structural actuators – is capable of applying discrete loads at the free end of the root section comprising two dofs including: edgewise load (F_y) along with a torsional moment (M_z) around the longitudinal axis of the blade. Actuator specifications are provided in figure 3b. A servo hydraulic control system is operated in load control through a Proportional Integral Derivative (PID) controller of the type MTS FlexTest using the MTS 793 software suite. The coupling between the control point, located in the center of the load carrying box girder and the corresponding force of each actuator, are defined through the MTS 793.15 Degree of Freedom Control software following the assumption of rigid body motion. Related coordinate system and notation is presented in figure 3b.

a b

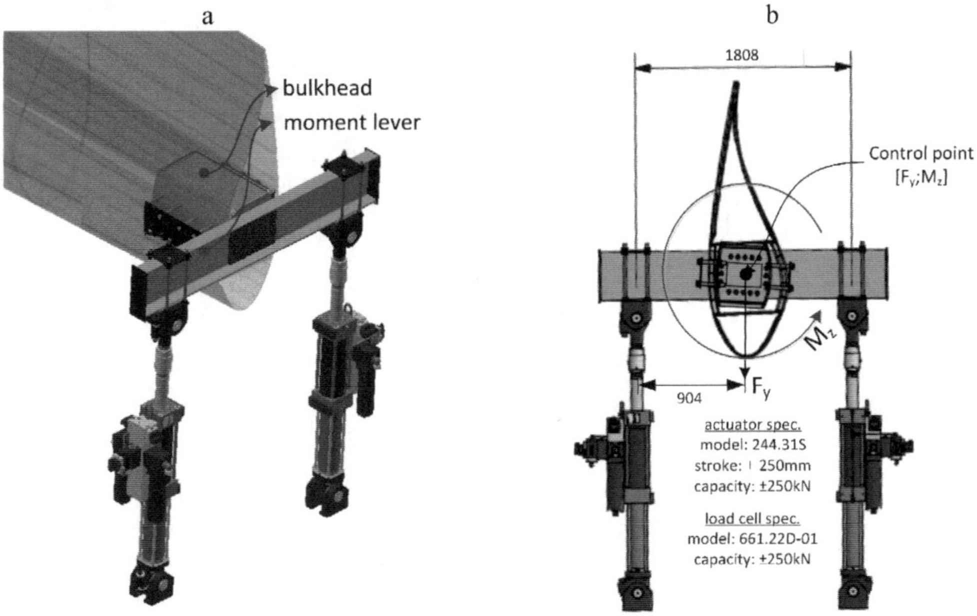

Figure 3: Load train including: a) detailed 3D illustration of multi-axial loading arrangement and b) active coordinate system and notation

The load is transferred to the free end of the root section through a steel bulkhead, which is extending 750mm into the free end of the load carrying box girder, and fixed to the inner surface of both spar caps using glue and threaded bars. To avoid critical peeling stresses in the adhesive bondline connecting the trailing edge (TE) and leading edge (LE) panels with the spar caps, the free end of the root section is fully constraint against in-plane distortion by closing the cross section with plates of over laminated plywood. The tip end region – covering approximately 2m towards the root – is over laminated using GFRP fabrics. This over-lamination is solely added to avoid structural damages generated by the localized loading induces trough the steel box girder. The zone of interest – referred to here as the gauge zone – is located around the trailing edge panels at the transition zone. According to a numerical study further described in [6], no influence on the breathing response were detected within the gauge zone. The loading arrangement and clamped support presented in figure 2a constitutes a fatigue rated structural test setup with an edgewise load capacity of 100kN and moment capacity of 100kNm around the longitudinal axis.

Investigation and evaluation of the retrofitted fish-mouth third shear-web disbonding and associated mitigation is monitored through strain gauges, wire potentiometer and DIC. Strain gauge rosettes are applied on all four corners of the shear-web to monitor the overall transfer of shear loads from the connected trailing edge panels. These strain gauge rosettes are applied at the surface towards the trailing edge of the shear-web and labelled according to figure 4a. The breathing of the large double curved trailing edge aerodynamic sandwich panels are monitored through a wire potentiometer located 200mm aft of the shear-web according to figure 2c. The out-of-plane and in-plane deformation of the PS single lap adhesive bonde-line relative to the adjacent trailing edge sandwich panel (see figure 4b) is monitored through full-field 3D DIC system of the type: ARAMIS 12M supplied by Gesellschaft für Optische Messtechnik GmbH (GOM GmbH). The camera setup and performance are presented in table 2.

Table 2: Setup and performance of the 3D-DIC system

Technique used	3D DIC system
Subset	32x32px
Shift	6px (81% overlap)
Camera	Dalsa Falcon2 FA-80-12M1H with an 12-bit, 4096x3072pixel CMOS (6.00µm pitch) censor and 24mm
Field of View	lens
Measurement points	380x305mm (4050x2882px)
	9531
Displacement	
Spatial resolution	1.62mm, 17px
Resolution	
In plane	0.045mm, 0.48px
Out of plane	0.090mm, 0.96px

By tracking the out-of-plane (z-direction) and in-plane (xy-direction) deformation of the three points located at the PS single lap adhesive bonde-line (named: shear-web) and adjacent pressure side trailing edge panel (named: ps panel) through DIC, the relative peak/valley deformation is derived.

IOP Conf. Series: Materials Science and Engineering 1293 (2023) 012026 doi:10.1088/1757-899X/1293/1/012026

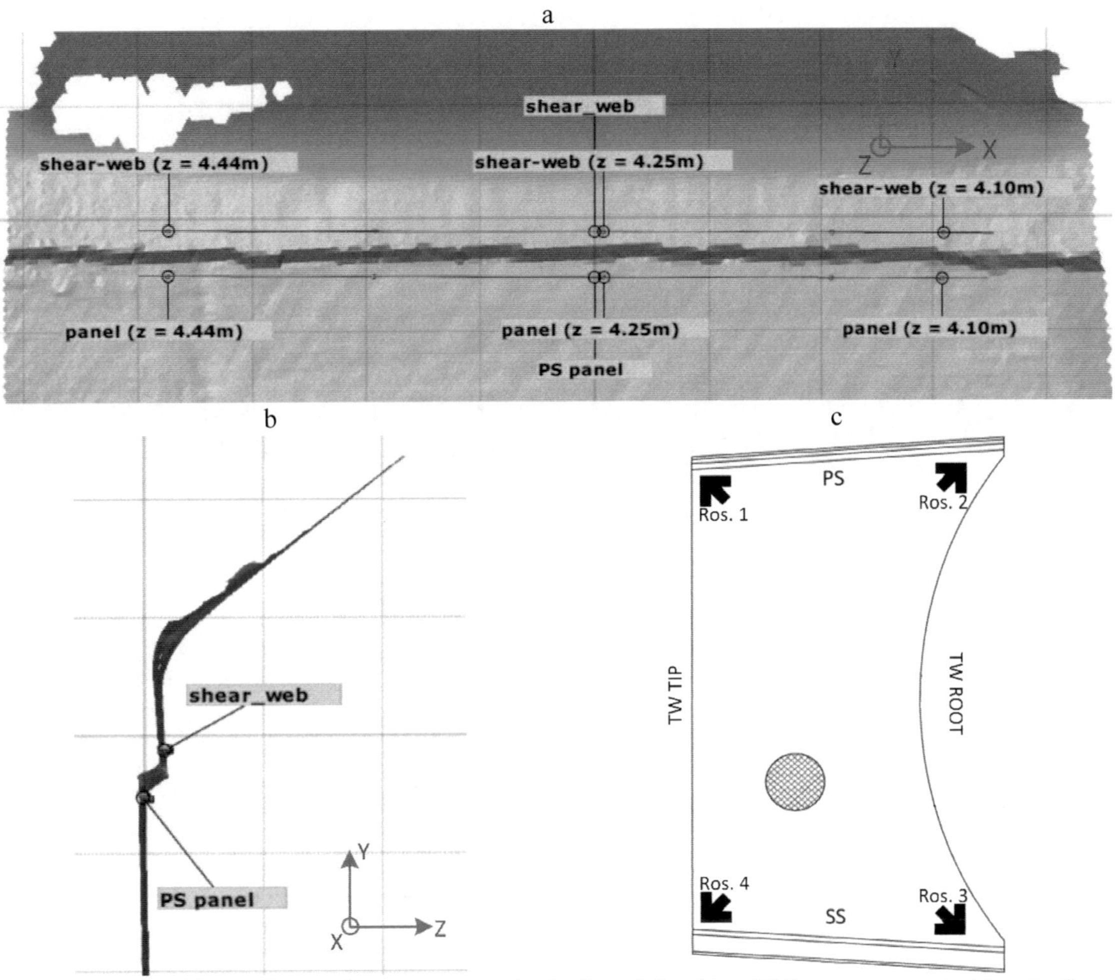

Figure 4: Labelling of internal measuremens including a) in-plane DIC measurements, b) out-of-plane DIC measurements and c) strain gauges

4. Results and discussions

The root section is loaded in a fatigue test campaign comprising approximately 280.000 load cycles with a combined load configuration of two dofs including F_y and M_z. With a sinusoidal waveform, execution frequency of 0.1Hz and phase shift of 90 degrees – F_y and M_z is applied a max load of 100kN and 100kNm respectively both with R=-1 according to figure 5. This load configuration is according to [6] found to generate the highest magnitude of breathing of the large double curved trailing edge aerodynamic sandwich panels. As the retrofitted 3rd shear web is installed to mitigate these deformations, that load configuration is established with the highest possible load magnitude to redice the number of cycles required for design validation process – even though the shear forces are different and higher than the in-service loads of the full-scale blade. A stable phase shift error of 0.7 sec between F_y and M_z were detected throughout the entire testing campaign. This phase shift error could not be lowered further with the available compensators.

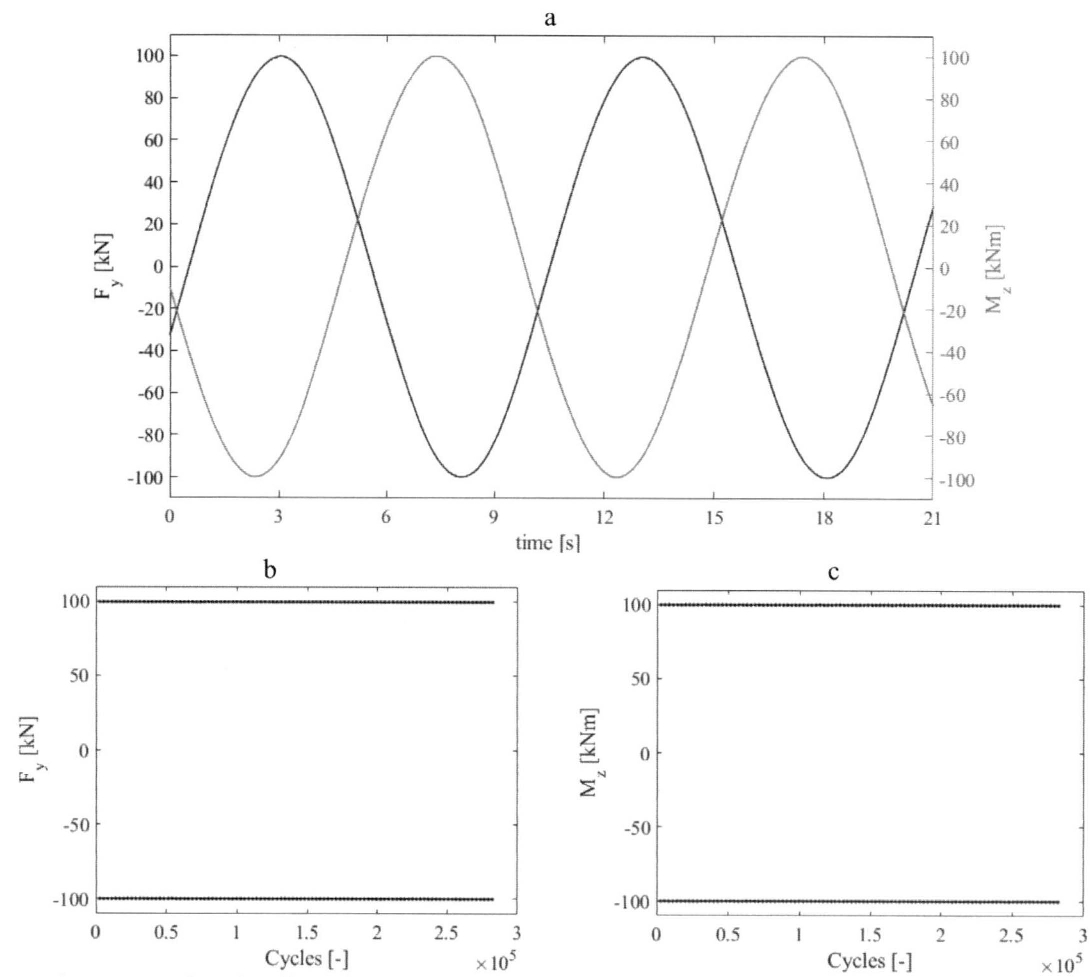

Figure 5: Fatigue loading of root section including a) timed loading sequence, b) peak/valley edgewise load and c) peak/valley torsional load

The wire potentiometer, monitoring the breathing of the PS and SS trailing edge sandwich panels, are stable throughout the entire test campaign with an amplitude of approximately 7mm.

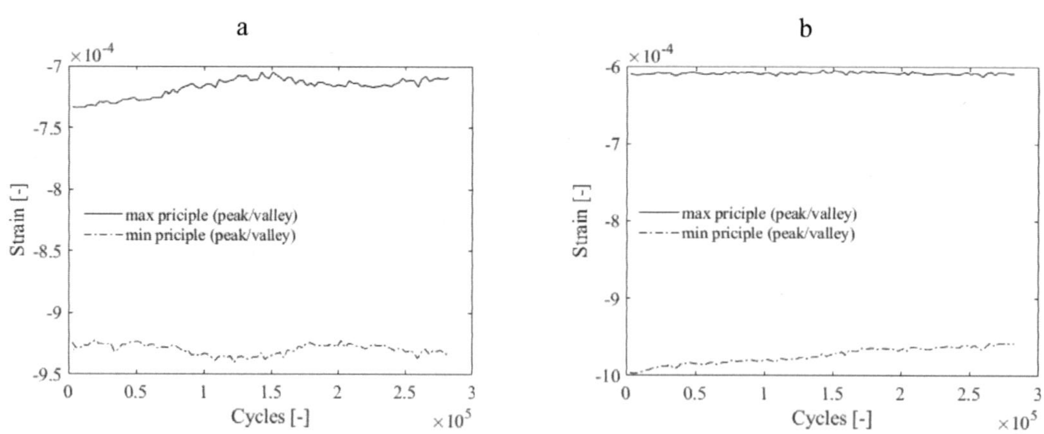

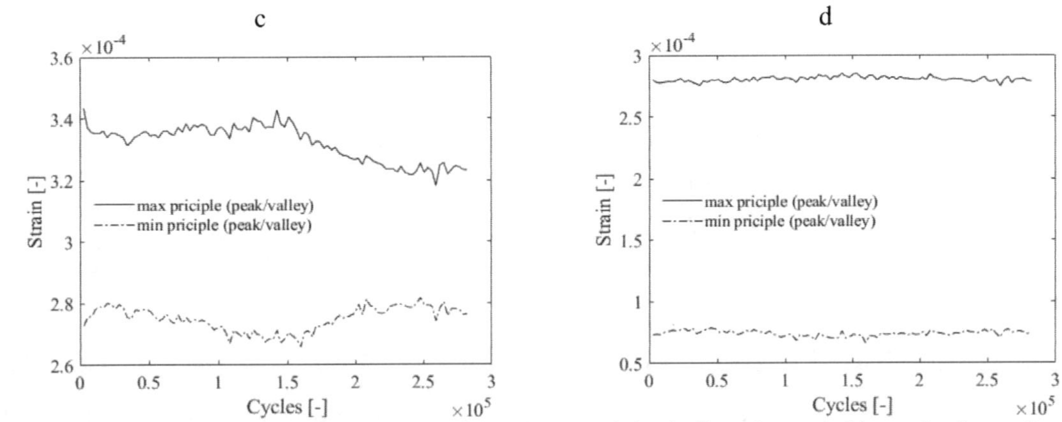

Figure 6: Principle strain at each corner of the shear-web including a) ros. 1, b) ros. 2, c) ros. 3 and d) ros. 4

From the max- and min principle strain provided in figure 6, no significant and consistent variation of the magnitude is detected throughout the fatigue test campaign indicating that the peak/valley strain field throughout the third shear-web is unchanged during the fatigue test campaign. Furthermore, no rotation of the principal axis is detected during the fatigue test campaign.

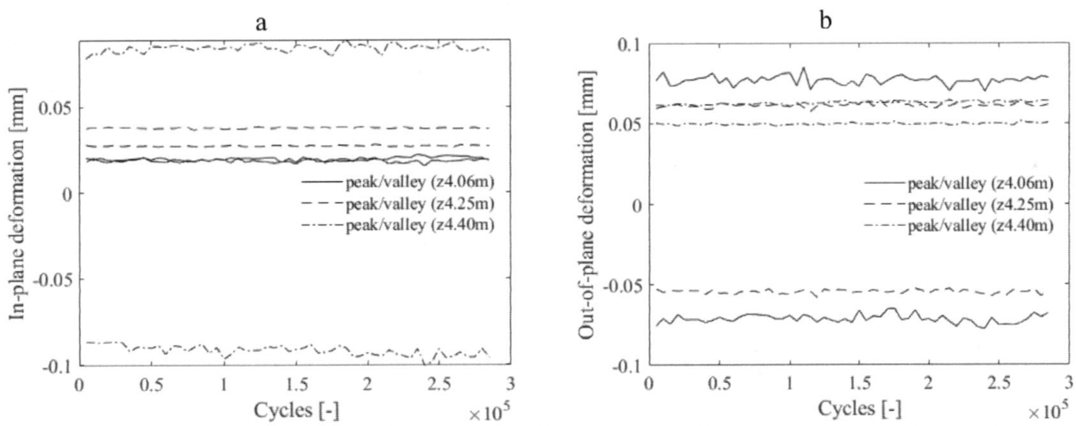

Figure 7: Relative deformation of PS single lap adhesive bonde-line and adjacent trailing edge panel including a) in-plane and b) out-of-plane

From the DIC measurements of the peak/valley in-plane (resulting deformation in the xy-plane) and out-of-plane (deformation in the z-direction) deformation of the PS single lap adhesive bonde-line and adjacent trailing edge panel is stable throughout the entire fatigue test campaign. From all the above measurements on the third shear web and visual inspection of both the shear-web and SS and PS bondline, it is concluded that the adhesive connection between the trailing edge sandwich panels and single lap bondline is undamaged.

5. Conclusion and further work

To investigate and evaluate the overall structural integrity of the retro-fitted third shear-web along with the mitigation associated with the shear web disbonding mechanism triggered by breathing of the trailing edge panels, a fatigue test campaign was conducted covering 280.000 load cycles. Monitored by strain gauges, a wire potentiometer and DIC, the structural behavior of the third shear-web was evaluated. Here it was found that the shear-web and single lap bondline connecting the shear web with the trailing edge was undamaged at this point. To accelerate the process, it is intended to initiate a damage/crack in the bondline between the third shear-web and adjacent trailing edge sandwich panels.

6. Acknowledgement

The authors would like to acknowledge the Energy Development and Demonstration Program (EUDP) for financial support under the grant number: 64021-1054 CORTIR II project. Within the frame of the CORTIR II project, considerable work has been done by Global Wind Service supporting the large-scale test program.

References

[1] S. Singh, "Wind power: Future lies within," in *7th India International Conference on Power Electronics*, Patiala, 2016.

[2] G. Sieros, P. Chaviaropoulos, J. D. Sørensen, B. H. Bulder and P. Jamieson, "Upscaling wind turbines: theoretical and practical aspects and their impact on the cost of energy," *Wind Energy,* vol. 15, no. 1, pp. 3-17, 2012.

[3] F. M. Jensen, M. Werk, A. Buliga, T. Pardalakis, C. Berggreen, J. Waldbjoern and J. D. Sørensen, "Root Area Transition Zone – RATZ and Reduction O&M cost of WT blades," Bladena A/S, Taastrup, 2019.

[4] F. M. Jensen, "Ultimate Strength of Large Wind Turbine Blade," Department of Civil Engineering, Technical University of Denmark, Kgs. Lyngby, 2008.

[5] F. M. Jensen, M. Werk, A. Buliga, C. Berggreen, J. P. Waldbjoern, J. D. Sørensen, Y. Yang and E. S. Jensen, "Root and Transition zone (RATZ) and reduction O&M cost of WT blades," Bladena A/S, Roskilde, 2019.

[6] J. P. Waldbjørn, A. Buliga, C. Berggreen and F. M. Jensen, "Multi-axial large-scale testing of a 34m wind turbine blade section to evaluate out-of-plae deformations of double-curved trailing edge sandwich panels within the transition zone," *Wind Engineering,* p. 0309524X2097840, 2020.

[7] F. M. Jensen, B. G. Falzon, J. Ankersen and H. Stang, "Structural Testing and Numerical Simulation of a 34m Composite Wind Turbine Blade," *Composite Structures,* vol. 76, no. 1-2, pp. 52 - 61, 2006.

[8] "Ampreg 22 - Epoxy Laminating System," Gurit, [Online]. Available: https://pdf.directindustry.com/pdf/gurit/ampreg-22-epoxy-laminating-system-v17/37817-79061.html. [Accessed 8 June 2023].

Life cycle assessment of wind turbine blade recycling approaches in the United States

Evan Sproul[1], Michelle Williams[2], Mitchell L. Rencheck[3], Matthew Korey[3] and Brandon L. Ennis[1]

[1] Wind Energy Computational Sciences Department, Sandia National Laboratories, Albuquerque, NM, 87123, USA
[2] Wind Energy Design and Experimentation Department, Sandia National Laboratories, Albuquerque, NM 87123, USA
[3] Manufacturing Science Division, Oak Ridge National Laboratory, Oak Ridge, TN 37831, USA

Corresponding author email: esproul@sandia.gov

Abstract. Most wind turbine blades reaching end-of-life are sent to landfill where embedded cost, energy, and materials are lost. To avoid landfilling future blades, a broad range of recycling and material recovery approaches have been proposed as solutions in the U.S., each with benefits, challenges, and varying levels of technical maturity. The approaches include 1) cement co-processing, 2) mechanical recycling, 3) pyrolysis, 4) microwave pyrolysis and 5) solvolysis. While these approaches are all capable of recovering various forms of materials for use in secondary markets, there are trade-offs between material circularity, reducing harmful environmental emissions, and cost-effectiveness for the U.S. market. Life cycle assessment (LCA) is a critical step needed to compare these trade-offs and determine where future research and development should be focused. As a result, some previous LCA has been performed on recycling approaches. However, attempts to quantify and compare greenhouse gas emissions across a broad range of technologies have been limited, particularly within the U.S. market where landfill availability and costs do not hinder disposing of wind blades. This work addresses this limitation by presenting a detailed comparison of LCA greenhouse gas emissions and material yields from a range of wind turbine blade recycling approaches in the U.S. The LCA presented in this work includes baseline results, as well as a variety of sensitivity and scenario analyses that look at the impact of process modelling uncertainty, future energy mixes, and other critical input parameters. Overall, results show that mechanical recycling and microwave pyrolysis have the lowest net greenhouse gas emissions. However, the value of mechanically recycled materials is highly uncertain, as mechanical recycling generates a mixed feedstock that may underperform compared to virgin materials. Cement co-processing has higher net emissions than mechanical recycling or microwave pyrolysis but does generate a value-added feedstock that offsets virgin material from mining for cement production. Other advanced thermal and chemical recycling methods such as pyrolysis and solvolysis have higher net emissions due to increased energy consumption but are also highly sensitive to thermal energy sources within the model.

1. Introduction

Wind turbine blades are constructed primarily using fiber reinforced polymer composites due to their light weight and durability, allowing blades to survive 20-30 years in extreme conditions and variable climates. However, the material properties that provide these advantages also present challenges when blades reach the end of their service life. The primary challenge is that thermoset polymers are difficult to separate from fiber reinforcement. As a result, most blades currently reaching end-of-life are sent to landfill where the embedded energy and costs of those blades are left unrecovered[1]. Recognizing this issue, researchers and industry have started to propose a range of approaches for recycling wind turbine blades[2]. While many of these options can recover materials or energy from the wind turbine blades, each approach has a unique set of economic and environmental trade-offs. Evaluating these trade-offs at an early stage of development is critical to ensure that solutions are optimized to increase the overall sustainability and circularity of wind turbine blades.

Addressing the trade-offs of different wind turbine blade recycling approaches requires quantifying a range of metrics. One set of metrics that are especially important are the emissions released to the environment during recycling. Quantifying these emissions requires performing a detailed life cycle assessment (LCA) of each recycling approach. Previous efforts have performed LCA on select recycling approaches specifically for wind turbine blades[3-5]. Other efforts have performed LCA in the more general context of recycling fiber reinforced polymers[6-8]. However, there is still a research gap where additional detailed LCA is needed across multiple recycling approaches. Furthermore, the majority of LCA that has been performed is focused on Europe, where current and future laws regarding landfilling are accelerating the need for near-term solutions. The large amount available landfill space and relatively low cost associated with landfilling in the U.S. poses a unique challenge and has specific implications for performing LCA on blade recycling in the U.S.

This paper highlights an ongoing effort to perform a U.S. specific LCA of multiple wind turbine blade recycling approaches. This effort has consisted of three major phases to date. First, a broad literature review was conducted to gather qualitative and quantitative data about existing recycling approaches[9]. Second, data from the literature review and discussion with industry experts were used to develop detailed process models which track all materials and energy required by each recycling approach. Third, the process modelling data were combined with life cycle inventory (LCI) databases to perform a baseline gate-to-gate LCA of cement co-processing, mechanical recycling, pyrolysis, microwave pyrolysis, and solvolysis, as well as landfilling for comparative purposes. Following this baseline assessment, a series of scenario analyses were performed to quantify uncertainty and identify the impact of different energy mixes on greenhouse gas emissions.

2. Methods

2.1. Process Modelling

Process models for each recycling approach were developed to track energy and mass entering and leaving the system boundary. Process modelling data were sourced from published literature, as well as discussions with industry and fellow researchers regarding the current standard practices within the U.S.[9]. Using this information, the baseline scenario was developed around a land-based turbine being decommissioned in Texas, U.S with the properties shown in Table 1. Each wind turbine blade is disassembled, downsized on-site, and transported using the parameters shared in Table 2. Although not explicitly defined in the modelling, any metal components are assumed to be removed from the blade during the cutting and shredding steps to avoid contamination in downstream recycling processes.

The packing density of blade segments or shredded materials greatly impacts the efficiency of transportation. The current practices around downsizing and transportation logistics vary widely across industry, leading to a range of possible outcomes for packing and transportation efficiency [9]. Based on this range, a middle estimate of packing efficiency was selected to represent a conservative improvement of on-site shredding, while avoiding an overly optimistic result. In the baseline, shredded material fills 100% of a standard truck's payload. Meanwhile, blade segments are assumed to have a packing density of 50% due to volumetric constraints. In addition to packing density considerations, the transportation distance has a significant impact on the process modelling. Since few commercial scale options for recycling blades exist in the U.S., estimates for varying transport distances were selected to capture the potential differences across approaches, while maintaining an overall conservative approach.

Table 1. Wind turbine blade properties used in analysis.

Property	Value	Unit
Turbine capacity	1.7	MW
Blade length	48.70	m
Total blade mass	9000	kg
Blade fiberglass mass	5283	kg
Blade resin mass	2401	kg
Blade balsa mass	416	kg
Blade gelcoat mass	180	kg
Blade adhesive mass	450	kg
Blade metal mass	270	kg

Table 2. Key parameters for disassembly, on-site size reduction, and transportation of blade.

Parameter	Value	Unit
Disassembly with crane	2	hours/blade
Cutting with saw	1.6	hours/blade
Shredding energy	0.27	MJ/kg
Shredder material loss	2%	kg/kg
Truck payload	16,000	kg
Shredded material on truck	16,000	kg
Cut blade segments on truck	8,000	kg
Landfill distance	100	km
Cement co-processing distance	200	km
Recycling distance	500	km

After transportation, the blade materials arrive at a landfill, cement co-processing facility, or recycling facility. Select process modelling parameters associated with each of the approaches are displayed in Table 3. The two simplest process models developed for this stage were landfilling and mechanical recycling. In landfilling all blade material is treated as mixed plastic waste at a sanitary landfill. For mechanical recycling there are a broad range of potential options including downsizing for reuse as aggregate in construction materials or simple grinding to recover resin rich fibers and powders. In this process model, simple grinding was selected to represent mechanical recycling due to the higher likelihood of market adoption.

In contrast to landfilling and mechanical recycling, cement co-processing represents the most complex process model developed. In the model, a wind turbine blade is assumed to be further downsized, then mixed into a cement kiln where the resin is burned as fuel and the remaining constituent materials from fiberglass are used as feedstock for cement clinker. The key materials in the blade required for clinker are CaO, SiO_2, Al_2O_3, and Fe_2O_2. These materials must be balanced with typical cement clinker feedstock materials such as limestone, sand, clay, and ore. In addition to these mass balance considerations, the kiln requires a significant amount of thermal energy that comes from different fossil fuels such as coal, petroleum coke, and natural gas, as well alternative fuels such as the organic materials of wind turbine blades considered in this analysis[10].

Table 3. Key parameters used in process modelling of recycling approaches.

Cement co-processing parameters	Value	Unit
Blade mass feedstock substitution	18%	kg/kg
Clinker CaO content	70%	kg/kg
Clinker SiO_2 content	20%	kg/kg
Clinker Al_2O_3 content	7%	kg/kg
Clinker Fe_2O_2 content	3%	kg/kg
Fiberglass CaO content	35%	kg/kg
Fiberglass SiO_2 content	55%	kg/kg
Fiberglass Al_2O_3 content	8%	kg/kg
Fiberglass Fe_2O_2 content	1%	kg/kg
Total kiln energy consumption	3.14	MJ/kg
Portion of supplemental energy from natural gas	25%	MJ/MJ
Portion of supplemental energy from coal/petroleum coke	75%	MJ/MJ
Mechanical recycling parameters	**Value**	**Unit**
Energy consumption	0.27	MJ/kg
Fiber yield	42%	kg/kg
Powder yield	58%	kg/kg
Pyrolysis parameters	**Value**	**Unit**
Energy consumption	21.2	MJ/kg
Solids yield	72%	kg/kg
Solids fiber content	80%	kg/kg
Solids non-fiber content	20%	kg/kg
Oil yield	16%	kg/kg
Gas yield	12%	kg/kg
Oil energy content	35	MJ/kg
Gas oil content	38	MJ/kg
Microwave pyrolysis parameters	**Value**	**Unit**
Energy consumption	2.52	MJ/kg
Fiber yield	61%	kg/kg
Oil yield	28%	kg/kg
Gas yield	21%	kg/kg
Oil energy content	35	MJ/kg
Gas oil content	38	MJ/kg
Solvolysis parameters	**Value**	**Unit**
Energy consumption (not including solvent recovery)	15.36	MJ/kg
Solvent to composite ratio	400%	kg/kg
Acetone portion of solvent	80%	kg/kg
Water portion of solvent	20%	kg/kg
Solvent recovery efficiency	99.50%	kg/kg

Pyrolysis is also a relatively complex process, especially considering the range of potential operating conditions and corresponding mix of output products. As a result, the baseline process model was developed using data from multiple publications [2, 5, 11, 12], as well as feedback from industry experts. In pyrolysis, downsized composite material enters a reactor where pressure and temperature are increased to thermally remove the epoxy from the fiber. Recovered fiber represents the primary output product of the pyrolysis system, while co-products of the system include the oils and non-fiber solids. Gases generated through pyrolysis are assumed to be recirculated and combusted within the system to offset a portion of the thermal energy demand. Given the energy demands of pyrolysis, several alternative approaches have been proposed to reduce the energy demand to recover fiber. One promising approach is microwave pyrolysis, which uses microwave radiation in place of traditional thermal energy sources. The microwave pyrolysis model was primarily based upon reports from a previous project[13, 14]. Since all energy provided to the microwave pyrolysis is electrical, any oil and gas are treated as co-products that would be sold to external customers.

When modeling solvolysis there are a broad range of options relating to the type of solvents, temperature, pressure, and other operating parameters. In this work the solvolysis model was developed based upon a previous publication with experimental results[15]. In this model a combination of acetone and water are used as solvent. The solvent is raised to an elevated temperature via natural gas heating to expedite dissolution. At the end of the process solvent is recovered and recirculated through distillation also heated by natural gas, which corresponds to a full scale, continuous process. After dissolution, the fiber is recovered while the resin and a small fraction of spent solvent are removed from the system and combusted to produce thermal energy that offsets a portion of natural gas demand.

2.2. Life cycle Assessment and Scenario Analyses

LCA was performed in accordance with guidance from ISO14040 and 14044[16, 17]. Based on this guidance the four standard phases of LCA are 1) goal and scope definition, 2) LCI analysis, 3) life cycle impact assessment (LCIA), and 4) interpretation and decision making. Within this specific LCA the current goal is to compare the greenhouse gas emissions of multiple wind turbine blade recycling approaches. In the future, additional environmental impacts such as human toxicity, ecotoxicity, particulate matter, and water consumption should be considered to obtain a full set of environmental trade-offs. The system boundary is held consistent across all recycling approaches and is considered "gate-to-gate". It begins when a blade reaches end-of-life on a wind turbine and ends when the blade has been transformed into a new material intended for remanufacturing. To enable a reasonable scope for this paper, life cycle stages outside of the system boundary such as blade manufacturing and remanufacturing of recycled materials have been excluded from the LCA. The primary functional unit considered in this analysis is one kilogram of recycled material leaving the system boundary. However, since landfilling does not result in any recycled material, comparisons with landfill are shown on a per blade basis. Primary LCI data were sourced from the process modelling. Secondary LCI data were sourced from Ecoinvent 3.9.1[18]. When possible, Ecoinvent LCI data were selected based upon specific geographic regions within the U.S. If these data were unavailable, "rest-of-world" or "global" options were selected to represent a best possible approximation. Within Ecoinvent only "cut-off unit process" data were selected to avoid any duplicate accounting for recycling processes.

To account for the benefit of recycled materials, substitution credits were taken for any mass of cement clinker, fiberglass, powders, oils, and combustible gases produced during recycling. Table 4 displays the credit taken for one kilogram of each material. The amount of credit given to a recovered material was based upon the properties of the recovered material compared to the properties of a similar virgin material. Cement clinker was assumed to meet all requirements for standard quality clinker and receives a credit equal to the emissions of producing and equivalent amount of standard cement clinker. Recovered fiberglass was given a credit based upon the emissions of producing virgin fiberglass. However, that credit was weighted by the estimated tensile strength of recovered fiberglass compared to the tensile strength of virgin fiberglass. Recovered oil or gas were given credit based on standard production of light fuel oil and natural gas. These credits were assumed proportional to their higher heating value. Recovered powder was assumed to have properties equivalent to mortar used in construction processes and therefore received a credit equal to the emissions of producing virgin mortar.

Once all LCI data were collected, LCIA was performed using climate change emissions factors from the U.S. Environmental Protection Agency's Tool for Reduction and Assessment of Chemicals and Other Environmental Impacts (TRACI v2.1)[19]. As mentioned previously, the scope of this paper was limited to assessing the greenhouse gas emissions (kg CO_2eq). These results were compared across a range of scenarios to understand the impact of different process modelling assumptions and parameters. The first two scenarios considered were an optimistic and conservative scenario. These scenarios were developed to capture the simultaneous effect of adjusting all process model input parameters to their best or worst value. While these scenarios are unlikely to represent real world outcomes, they are useful in defining likely upper and lower bounds of uncertainty for each technology. The next two scenarios considered were based around lower carbon energy sources for both electricity and thermal energy. The first scenario looks at the potential emissions reductions based on electricity being supplied by a projection of the 2035 U.S. electrical grid mix[20]. The second scenario considers this same electrical grid mix, but also analyzes the impact of introducing a low emissions fuel source such as green hydrogen or biomethane for all thermal energy demand. Based on data from Ecoinvent, this fuel source is estimated to have emissions of 0.015 kg CO_2eq per MJ of heat delivered, which is also in line with optimistic estimates for green hydrogen production via renewable energy[21].

Table 4. Credit given for production of one kilogram of clinker, fiber, filler, oil, and gas.

Approach	Recovered Material	Credit Material [kgCO_2eq/kg]	Criteria	Credit Applied		
				Optimistic	Baseline [kgCO_2eq/kg]	Conservative
Cement Co-processing	Clinker	Clinker [0.95]	All properties equal	100%	100% [0.95]	100%
Mechanical	Fiberglass	Virgin Fiberglass [2.57]	Tensile strength	80%	50% [1.29]	30%
Mechanical	Filler	Construction Mortar [0.25]	All properties equal	100%	100% [0.25]	80%
Pyrolysis	Fiber	Virgin Fiberglass [2.57]	Tensile strength	90%	80% [2.06]	50%
Pyrolysis	Filler	Construction Mortar [0.25]	All properties equal	100%	100% [0.25]	80%
Pyrolysis	Oils	Light Fuel Oil [0.73]	Heating value	83%	80% [0.58]	77%
Solvolysis	Fiber	Virgin Fiberglass [2.57]	Tensile strength	90%	80% [2.06]	50%
Microwave Pyrolysis	Fiber	Virgin Fiberglass [2.57]	Tensile strength	90%	80% [2.06]	50%
Microwave Pyrolysis	Oils	Light Fuel Oil [0.73]	Heating value	82%	82% [0.60]	82%
Microwave Pyrolysis	Gases	Natural Gas [0.0015]	Heating value	97%	72% [0.0011]	60%

3. Results

Figure 1 displays the blade mass yields for each approach considered. All five recycling approaches lose 5-6% of material due to system inefficiencies. These include dust lost to the environment during downsizing and the metal that is removed from the blade before recycling. Four of the approaches include incineration of organic material within the system boundary. This includes initial incineration of the resin when it is heated, as well as recirculation of certain oils or gases for secondary combustion to provide thermal energy back into the process. In addition to this incineration, pyrolysis and microwave pyrolysis also result in an export of some oil or gas, which is not utilized within the system boundary. The remaining solid recycled materials are broken into two categories. One is the primary solid recycled material, which consists of either recovered fiber or cement clinker. The other category is secondary solid material, which consists of powders resulting from the recycling processes.

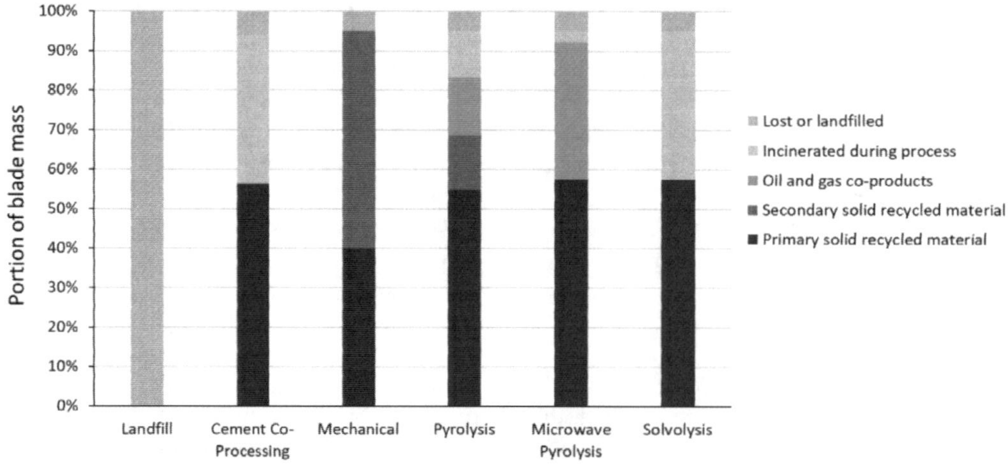

Figure 1. Wind turbine blade mass yields through different approaches.

The baseline results for greenhouse gas emissions of recycling approaches and landfilling are displayed in Figure 2. These results show the positive greenhouse gas emissions from decommissioning, size reduction, transportation, and recycling or disposal. They also show an estimate for credits due to substitution of recycled materials for cement clinker, virgin fiberglass, mortars, light fuel oil, and natural gas. The sum of the positive emissions and the negative credit represents the net greenhouse gas emissions of each recycling approach. To enable comparison of net emissions to landfill, the results are shown per one wind turbine blade entering the system boundary.

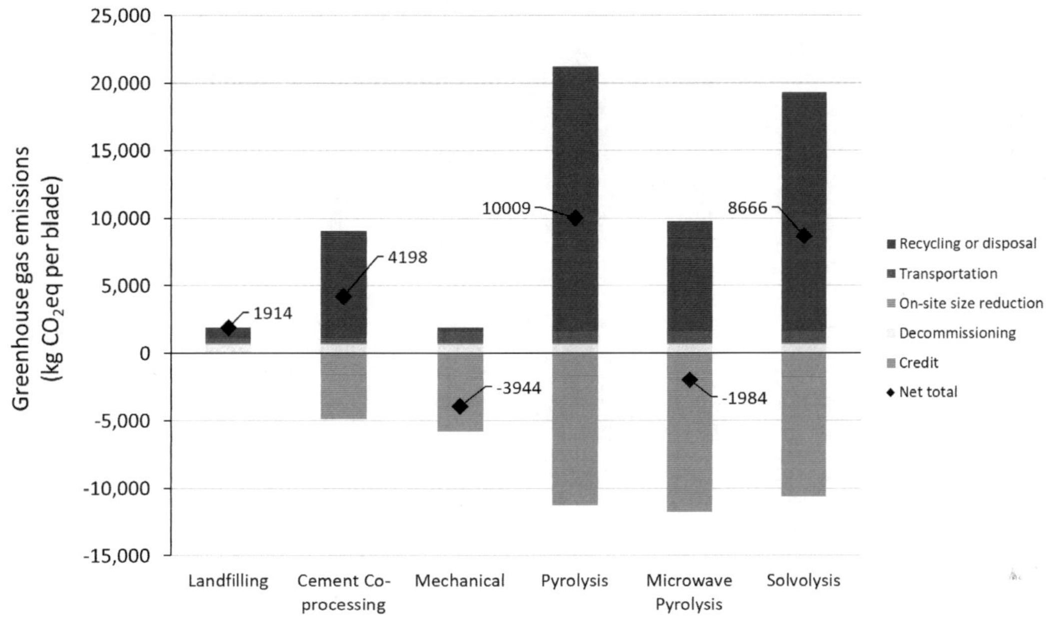

Figure 2. Greenhouse gas emissions of recycling approaches compared with landfilling.

The net greenhouse gas emissions of mechanical recycling and microwave pyrolysis are negative, meaning the credit taken for producing recycled materials is greater than the emissions of the recycling process itself. All other processes result in net positive greenhouse gas emissions, meaning that the emissions of recycling a wind turbine blade are greater than the credit for recycled material. In four out of six approaches the greenhouse gas emissions of decommissioning, on-site size reduction, and transportation are small compared to the recycling process. Across the approaches there is a small but notable difference in transportation emissions due to truck packing efficiency and transportation distance.

Net greenhouse gas emissions results from the optimistic and conservative scenarios are shown in Figure 3, with a functional unit of one kilogram of solid recycled material. These results show the total cumulative uncertainty of net greenhouse gas emissions for each recycling approach, and inherently account for the mass yield of solid recycled materials. Within the figure, all approaches show a wide range of uncertainty due to the breadth of different approaches available in literature. While the general trends across recycling approaches remain similar between baseline, optimistic, and conservative scenarios, some findings do change dependent upon the scenario considered. For example, microwave pyrolysis shows a net-negative emissions result in the baseline and optimistic scenario, but net-positive in the conservative scenario. Based on the wide uncertainty bands, a sensitivity analysis was also performed to identify which individual input parameters are most sensitive to uncertainty. Figure 4 displays the top six most sensitive parameters to uncertainty for pyrolysis and solvolysis. In both approaches, energy consumption and fiber yield are highly sensitive to uncertainty.

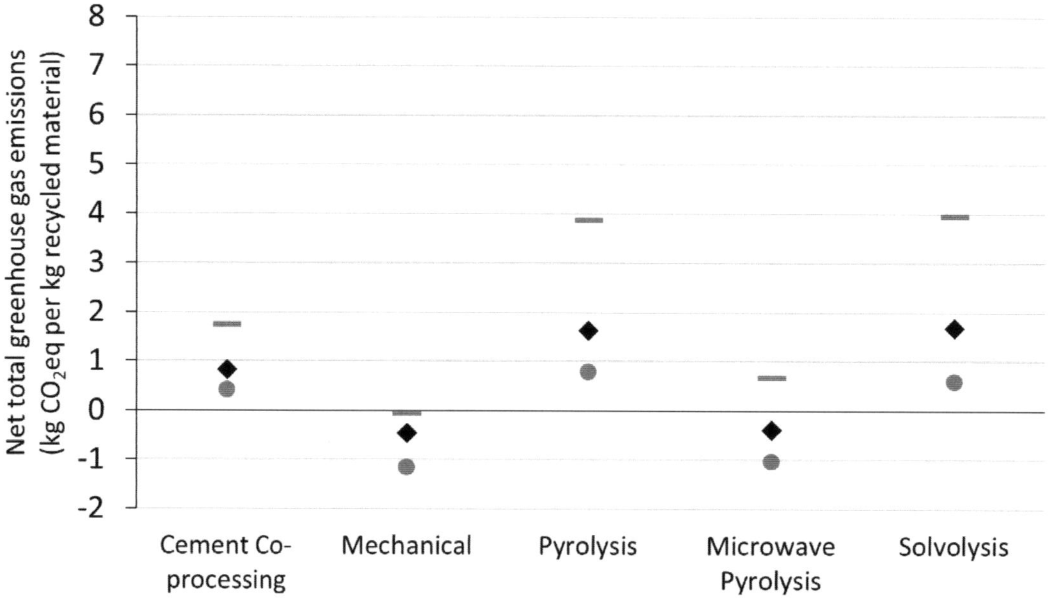

Figure 3. Optimistic and conservative scenario net greenhouse gas emissions compared with the baseline results.

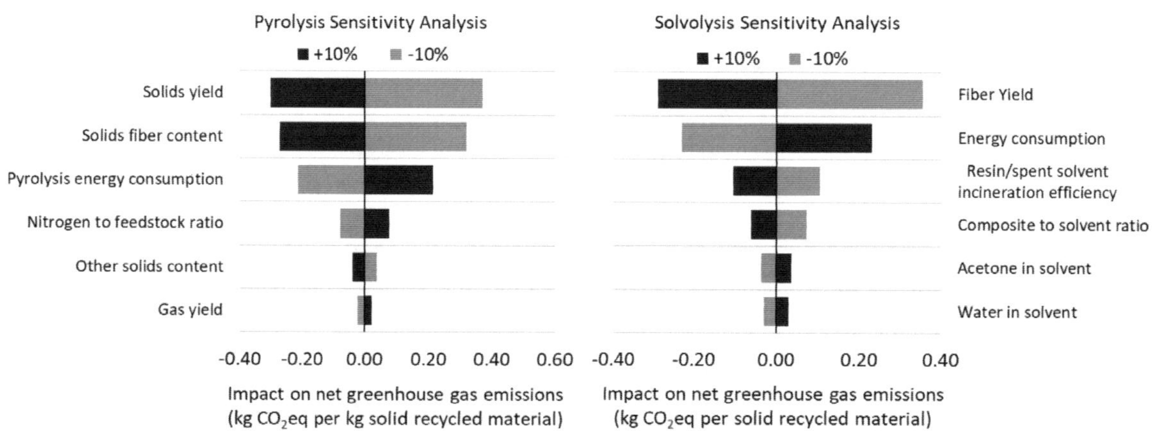

Figure 4. Sensitivity analysis of adjusting individual input parameters by 10%.

Figure 5 shows the impact of a 2035 electrical grid mix that is mostly decarbonized and low emission fuel sources. The 2035 electrical grid reduces greenhouse gas emission from 1%-23% across different recycling approaches. The largest impacts are on mechanical recycling and microwave pyrolysis which see reductions of 12% and 23%, respectively. Simultaneously adding a low emissions fuel for thermal energy results in total reduction of emissions from 8%-50% across all recycling approaches. The largest reductions are seen in solvolysis and pyrolysis which are reduced by 40% and 50%, respectively. Across all approaches the decommissioning, on-site size reduction, and transportation steps are estimated to operate via standard diesel power and are not impacted by the 2035 grid or low-emissions fuel scenarios.

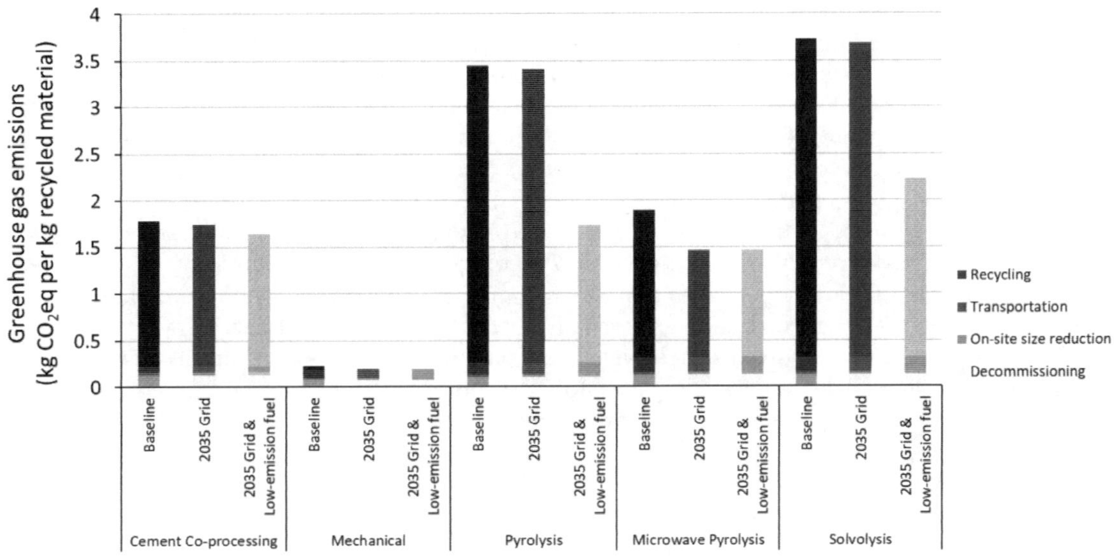

Figure 5. Greenhouse gas emissions for different recycling approaches with a 2035 grid mix and low-emission fuel source for thermal energy.

4. Discussion

Recovery of wind turbine blade materials at their end of life is currently challenged by increased cost and emissions relative to landfilling in the U.S. However, there are recycling approaches that show promise as a responsible alternative to disposal. For example, this paper has identified promising results for mechanical recycling in terms of both circularity and greenhouse gas emissions in the baseline scenario. This is largely due to modelling a simplified grinding process, which does not incinerate any material, and requires a relatively small amount of electrical energy to recover fibers and powder. However, it is important to note that the resin rich fiber recovered through this process will likely have significantly degraded properties compared to virgin fiberglass. While the credit taken for this recovered fiber does account for a 50% reduction in tensile strength, it does not consider the market demand for this reduced strength material. As a result, more work is needed to research potential markets for mechanically recovered fiber and assess actual value beyond a ratio of tensile strength. Microwave pyrolysis also shows promise in terms of low net greenhouse gas emissions. In this case, the microwave pyrolysis benefits from both the efficiency of the microwave pyrolysis process, as well as the source of energy being electricity. Not only does this result in low baseline emissions, but it also opens the potential to see significant emissions reductions with future U.S. electrical grid mixes. The current limitation of microwave pyrolysis is that it is an early-stage technology for recycling wind turbine blades at scale. While there have been pilot scale efforts outside the U.S., most research in the U.S. is still at laboratory scale, meaning unforeseen challenges may arise in the future and commercialization is not near-term.

Following mechanical recycling and microwave pyrolysis, the next lowest emitting approach is cement co-processing. Given that that cement co-processing is already occurring in the U.S., it seems like a sensible near-term solution for avoiding wind turbine blades in landfills. However, by looking at the results from this analysis, some limitations appear with respect to minimizing future greenhouse gas emissions. First, cement co-processing burns the epoxy resin of the wind turbine blade for energy. While this process does capture embedded energy that would otherwise be lost to landfill, it is also a direct source of CO_2 emissions. Second, cement co-processing of wind turbine blades currently benefits from avoiding the direct carbon emissions associated with calcination of standard feedstock materials. Given the current focus on decarbonizing U.S. cement production emissions with carbon capture or alternative feedstocks[10, 22], the future benefit of co-processing wind turbine blades may be reduced.

Pyrolysis and solvolysis have the largest net greenhouse gas emissions in the baseline case. However, they also have significant uncertainties and the greatest potential for reduction if low emissions fuel sources can be implemented. Ongoing research on pyrolysis of wind turbine blades in the U.S. is likely to further optimize processing and decrease the overall greenhouse gas emissions. Efficient use of oil and gas byproducts from pyrolysis will be especially important to this reduction. One other major consideration for pyrolysis is the value of the recovered fibers. Research on cleaning or recovering the properties of fibers after pyrolysis will be critical to enabling this approach. Currently, this step is often the most proprietary part of the pyrolysis process and gaining additional data for future modelling may prove challenging. Like pyrolysis, solvolysis can also benefit from further research, however the added complexity of using harsh solvents at elevated temperatures is a challenge for use with existing thermoset resins. Future thermoplastic or reversible thermoset materials creates a greater opportunity for effective use of less abrasive solvents at lower temperatures.

It is important to note that the scope of this analysis was limited to the fiberglass composites that will make up most near-term blade waste in the U.S. Therefore, it does not explicitly consider the greenhouse gas emissions associated with processing other materials, most notably carbon fiber. If carbon fiber was considered it would have significant impacts on the overall results. Specifically, the credit received for recovered fiber would be higher, as the greenhouse gas emissions of producing virgin carbon fiber are much higher than that of virgin fiberglass. As a result, pyrolysis, microwave pyrolysis, and solvolysis would all see reduced net greenhouse gas emissions when processing carbon fiber due to the increased credit. In contrast to these approaches, the presence of carbon fiber in materials destined for cement co-processing is unacceptable, as it cannot be processed in most cement kilns. As a result, any increase of carbon fiber content in future wind turbine blades will pose an added challenge for cement co-processing. With mechanical recycling, carbon fiber would have a mixed impact on greenhouse gas emissions. Energy loads and the time for size reduction would likely increase due to carbon fiber's increased strength. At the same time, the approach may benefit from a higher value of mixed-fiber recovered material. For mechanical recycling, as well as the other approaches, understanding the net impact of carbon fiber on greenhouse gas emissions requires its own dedicated analysis.

The overall results of this paper indicate that there are multiple approaches for wind turbine blade recycling that could result in net greenhouse gas emissions similar to or below that of landfilling. The results also show that there are many ways to reduce the net greenhouse gas emissions of these approaches, including cleaner electricity grid mixes and low-emission fuels for thermal energy. While this analysis helps enable a comparison of different recycling approaches, there are remaining research and market challenges to be addressed. First, the value of recovered products is critical to this analysis and requires more detailed investigation for representative secondary markets to increase certainty. This includes understanding potential U.S. markets for recycled products, as well as the physical characteristics that are important in those markets. Second, many of these approaches are at a low technology readiness level and have only been demonstrated at laboratory or pilot scale. Optimizing these approaches while scaling them up will be critical for their success. Last, greenhouse gases are just one of many important metrics that must be considered when assessing the trade-offs of different recycling approaches. Ultimately, more analysis is needed to quantify and compare the costs, circularity, and other environmental impacts of each approach to ensure future sustainability.

Acknowledgements

This research was supported by the Wind Energy Technologies Office of the U.S. Department of Energy (DOE) Office of Energy Efficiency and Renewable Energy through the Wind Technology Recycling Assessment project. Sandia National Laboratories is a multimission laboratory managed and operated by National Technology & Engineering Solutions of Sandia, LLC, a wholly owned subsidiary of Honeywell International Inc., for the U.S. Department of Energy's National Nuclear Security Administration under contract DE-NA0003525. Oak Ridge National Laboratory is operated by UT-Battelle, LLC. under Contract No. DE-AC05-00OR22725 with the U.S. Department of Energy. This paper describes objective technical results and analysis. Any subjective views or opinions that might be expressed in the paper do not necessarily represent the views of the U.S. Department of Energy or the United States Government. The US government retains and the publisher, by accepting the article for publication, acknowledges that the US government retains a nonexclusive, paid-up, irrevocable, worldwide license to publish or reproduce the published form of this manuscript, or allow others to do so, for US government purposes.

References

[1] A. Cooperman, A. Eberle, and E. Lantz, "Wind turbine blade material in the United States: Quantities, costs, and end-of-life options," (in en), *Resources, Conservation and Recycling,* vol. 168, p. 105439, 2021/05/01/ 2021, doi: 10.1016/j.resconrec.2021.105439.

[2] M. Rani, P. Choudhary, V. Krishnan, and S. Zafar, "A review on recycling and reuse methods for carbon fiber/glass fiber composites waste from wind turbine blades," (in en), *Composites Part B: Engineering,* vol. 215, p. 108768, 2021/06/15/ 2021, doi: 10.1016/j.compositesb.2021.108768.

[3] A. J. Nagle, E. L. Delaney, L. C. Bank, and P. G. Leahy, "A Comparative Life Cycle Assessment between landfilling and Co-Processing of waste from decommissioned Irish wind turbine blades," (in en), *Journal of Cleaner Production,* vol. 277, p. 123321, 2020/12/20/ 2020, doi: 10.1016/j.jclepro.2020.123321.

[4] A. J. Nagle, G. Mullally, P. G. Leahy, and N. P. Dunphy, "Life cycle assessment of the use of decommissioned wind blades in second life applications," *Journal of Environmental Management,* vol. 302, p. 113994, 2022/01/15/ 2022, doi: https://doi.org/10.1016/j.jenvman.2021.113994.

[5] P. Liu, F. Meng, and C. Y. Barlow, "Wind turbine blade end-of-life options: An eco-audit comparison," (in en), *Journal of Cleaner Production,* vol. 212, pp. 1268-1281, 2019/03/01/ 2019, doi: 10.1016/j.jclepro.2018.12.043.

[6] F. Meng, E. A. Olivetti, Y. Zhao, J. C. Chang, S. J. Pickering, and J. McKechnie, "Comparing Life Cycle Energy and Global Warming Potential of Carbon Fiber Composite Recycling Technologies and Waste Management Options," *ACS Sustainable Chemistry & Engineering,* vol. 6, no. 8, pp. 9854-9865, 2018/08/06 2018, doi: 10.1021/acssuschemeng.8b01026.

[7] R. J. Tapper, M. L. Longana, A. Norton, K. D. Potter, and I. Hamerton, "An evaluation of life cycle assessment and its application to the closed-loop recycling of carbon fibre reinforced polymers," *Composites Part B: Engineering,* vol. 184, p. 107665, 2020/03/01/ 2020, doi: https://doi.org/10.1016/j.compositesb.2019.107665.

[8] S. Karuppannan Gopalraj, I. Deviatkin, M. Horttanainen, and T. Kärki, "Life Cycle Assessment of a Thermal Recycling Process as an Alternative to Existing CFRP and GFRP Composite Wastes Management Options," *Polymers*, vol. 13, no. 24, doi: 10.3390/polym13244430.

[9] M. Korey, Sproul, E., Rencheck, M.L., Ennis, B.L., "Development of Wind Turbine Blade Recycling Baselines in the United States," presented at the 43rd Riso International Symposium on Materials Science, Roskilde, Denmark, September 2023, 2023.

[10] "Roadmap to Carbon Neutrality," Portland Cement Association, 2021.

[11] J. Yang, J. Liu, W. Liu, J. Wang, and T. Tang, "Recycling of carbon fibre reinforced epoxy resin composites under various oxygen concentrations in nitrogen–oxygen atmosphere," *Journal of Analytical and Applied Pyrolysis,* vol. 112, pp. 253-261, 2015/03/01/ 2015, doi: https://doi.org/10.1016/j.jaap.2015.01.017.

[12] A. M. Cunliffe, N. Jones, and P. T. Williams, "Recycling of fibre-reinforced polymeric waste by pyrolysis: thermo-gravimetric and bench-scale investigations," *Journal of Analytical and*

Applied Pyrolysis, vol. 70, no. 2, pp. 315-338, 2003/12/01/ 2003, doi: https://doi.org/10.1016/S0165-2370(02)00161-4.
[13] "Recycling of waste glass fiber reinforced plastic with microwave pyrolsis - final technical report," Stena Metall, 2012.
[14] "Recycling of waste glass fiber reinforced plastic with microwave pyrolsis - Layman report," Stena Metall, 2012.
[15] M. J. Keith, Oliveux, G., Leeke, G.A., "Optimisation of Solvolysis for Recycling Carbon Fibre Reinforced Composites," presented at the 17th European Conference on Composite Materials, Munich, Germany, 2016.
[16] *ISO 14040:2006 Environmental management - Life cycle assessment - Principles and framework*, I. O. f. Standardization, 2006.
[17] *ISO 14044:2006 Environmental management - Life cycle assessment - Requirements and guidelines*, I. O. f. Standardization.
[18] G. Wernet, C. Bauer, B. Steubing, J. Reinhard, E. Moreno-Ruiz, and B. Weidema, "The ecoinvent database version 3 (part I): overview and methodology," *Int J Life Cycle Assess,* vol. 21, no. 9, pp. 1218-1230, 2016/09/01 2016, doi: 10.1007/s11367-016-1087-8.
[19] J. Bare, "Tool for the Reduction and Assessment of Chemical and Other Environmental Impacts (TRACI) TRACI version 2.1 User's Guide," United States Environmental Protection Agency, July 2012 2012.
[20] W. Cole, Carag, J.V., "2021 Standard Scenarios Report: A U.S. Electricity Sector Outlook," National Renewable Energy Laboratory, November 2021 2021.
[21] K. de Kleijne, H. de Coninck, R. van Zelm, M. A. J. Huijbregts, and S. V. Hanssen, "The many greenhouse gas footprints of green hydrogen," *Sustainable Energy & Fuels,* 10.1039/D2SE00444E vol. 6, no. 19, pp. 4383-4387, 2022, doi: 10.1039/D2SE00444E.
[22] "Industrial Decarbonization Roadmap," U.S. Department of Energy, 2022.

Non-metallic lightning strike protection coating for wind-turbine blades

V Kumar[1,4], S Saha[1], S Millen[2], A Murphy[2], B Seta[3], J Spangenberg[3], J Tobin[1], B Knouff[1], A Hassen[1], V Kunc[1]

[1]Manufacturing Science Division, Oak Ridge National Laboratory, Oak Ridge, TN 37830, United States

[2]Queen's University Belfast, School of Mechanical and Aerospace Engineering, UK

[3]Department of Civil and Mechanical Engineering, Technical University of Denmark, Kgs. Lyngby, Denmark

[4]E-mail: kumarvi@ornl.gov

Abstract. The increasing use of Carbon Fiber Reinforced Polymer (CFRP) composites in the wind energy industry presents a challenge concerning lightning strike protection (LSP). Due to their low electrical conductivity, these materials are inferior to metals in dissipating electrical currents generated by lightning strikes, potentially leading to catastrophic damage. The current LSP system for wind turbine blades involves metallic lightning arrestors, which may only sometimes be effective due to the accumulation of debris and salt on other parts of the blade. Other commercially available products, such as diverter strips and nano-filler coatings, are expensive and impractical for use throughout the entire blade. Recently, researchers have proposed using electrically conductive polymeric coatings as a potential solution. These coatings are easy to apply and can be spray-coated, painted, or manufactured via automation. In this study, newly developed polymeric coating solutions were experimentally tested and compared to traditional metallic-based LSP systems. The experiments showed promising results in dissipating the current generated by lightning strikes. In addition, the thicker coatings reduced catastrophic damage, including puncture, fiber breakage, and resin evaporation, compared to thinner coatings. Overall, the study highlights the potential of polymeric coatings as a viable solution for lightning strike protection in the wind energy industry.

1. Introduction

In recent years, fiber-reinforced polymer (FRP) composites have gained popularity and are already dominating the aerospace and wind energy industries. Carbon FRP incorporation in future wind turbine blades is also growing [1]. Although FRPs can replace their metal counterparts in applications where specific strength is desired, but not in applications where other properties, such as high electrical conductivity, are required. Due to low electrical conductivity, CFRP structures are inferior to their counterpart metal structures in the event of a lightning strike [2]. Lightning strikes on wind turbine structures are severe due to substantial repair costs and downtime associated with lightning strike damages. The location of installed wind turbine blades (areas with the most probability of strong winds) also makes them highly vulnerable to lightning strike damages [3]. During a lightning strike, Carbon/Glass Fiber Reinforced Polymer C/GFRP structures can be destroyed due to the extreme heat produced by resistive heating or Joule's heating.

Wind energy has been a great source of clean, renewable energy; therefore, the United States of America is expected to increase wind turbine blade deployment by 5X. The target for offshore energy production is 30 GW by 2030, which will pay the path for 110 GW of total energy production via offshore wind turbines. The offshore turbine will deliver 10% of the USA's energy need by 2050.

With this exponential expansion, it has become critical to safeguard these tall structures from extreme weather and increase their credibility. Keeping these wind farms running without properly mitigating weathering problems will cost significantly. Therefore, a new lightning strike protection system is needed. Many research groups and industries are investigating novel LSP systems as the current LSP systems are inadequate.

The current lightning strike protection system of wind turbine blades includes a metallic lightning arrestor, which creates a lightning attachment point on the tip of the blades and is connected to a metal rod (down conductor) placed underneath the C/GFRP structure [4]. However, the current technology has yet to be effective during a lightning strike, as the accumulation of debris and salt on the other blade areas can also act as the lightning attachment points. In such events, there is no down conductor available to dissipate the current safely to the ground, and the C/GFRP structure suffers potentially catastrophic damage in the form of puncture, fiber breakage, resin evaporation, etc. Some of these damages grow over time and can ultimately put the blade out of commission [5]. Diverter strips and nano-filler coatings are other commercially available products for lightning strike protection of wind turbine blades. Both technologies depend on the arc jumping mechanism, where a dielectric material separates the two closely placed conductive electrodes. These expensive diverter strips cannot be used throughout the blade, leaving many unprotected surfaces exposed.

A promising solution being extensively researched is the development of a nano-engineered multifunctional coating of polymer/ceramic composite that could create an in-situ 3D conductive network to dissipate the lightning current. Painting or advanced automation manufacturing techniques can apply the coating to the wind turbine. In addition, it can operate under harsh conditions of lightning strikes and severe wear during the operation of onshore/offshore wind turbine blades.

Serbescu et al. studied the silver-based conductive coating as a scalable lightning strike protection for composite aircraft. Silver-coated carbon fibers (SCCF) with various binders were used, including epoxy and PEDOT:PSS conductive ink. All SCCF-based LSP exhibited around 45% - 50% retention of flexural strength, which was lower than the commercial Cu-mesh LSP, which exhibited an 83% retention of flexural strength [6]. Zhang et al. studied the thin, flexible graphene surface coating as lightning strike protection of carbon fiber epoxy prepreg laminate. They reported that the damaged area and volume in the graphene-coated laminate were reduced by 94% and 96%, respectively, compared to the laminate without the coating [7]. In another study, Zhu et al. applied a dual conductive network of nickel-coated carbon fiber woven fabrics (Ni-CFWF) on CFRP. They investigated both the indirect and direct lightning strike protection response of the composites. This study showed that the maximum lightning strike-induced damage depth and area-protected composites were reduced by 21.59% and 6.11% under a simulated lightning current of 100 kA, respectively [8]. Wang et al. designed a film reinforced by nickel-coated carbon fibers/carbonyl iron powders to provide a conductivity framework (NCF/CIPCF) for lightning protection of CFRP laminates [9]. Yokozeki and his group have been working on intrinsic conductive polymer, i.e., Polyaniline (PANI) based polymer, to make lightning strike protection coatings [10–12]. Their multiple studies focused on the coating performance at various lightning intensities and on optimizing the thickness of the coatings for optimal performance. The current work is also inspired by PANI based system.

In the current work, we embedded conductive PANI in the polyurethane-based system to represent the coatings used in wind applications. For the reference, expanded metal (Cu) foils were also tested for comparison.

2. Materials and Method

2.1 CFRP substrate

A commercial CFRP specimen (230 mm × 230 mm × 3.17 mm, McMaster Carr) was used as a substrate control specimen, and various conductive coatings were applied on top of it to prepare a PANI-PU LSP coating on top of the CFRP. Coated and uncoated CFRP panels were compared against lightning strikes. The following information about the CFRP was provided by the supplier (PAN-based woven fabric, tensile strength ≈ 827-1200 MPa, compressive strength ≈ 517-880 MPa, flexural strength ≈ 613-1200 MPa, and service temperature up to 82.2 °C).

2.2 Synthesis of conducting coating solution

The flow chart below shows a detailed experimental procedure to make the PANI-coating.

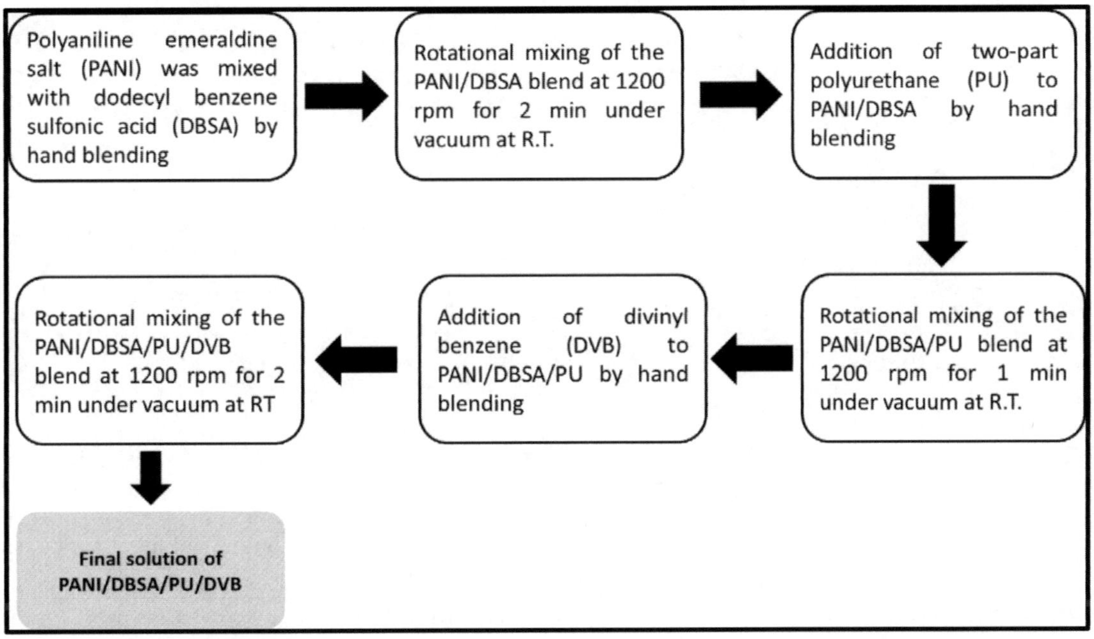

Figure 1. Synthesis of PANI/DBSA/PU/DVB blend for conductive coating layer

2.3 LSP- coating on CFRP

Before applying a coating, the specimen's outermost layer was lightly sanded for better bonding between the PANI-based LSP system and composites. Further, different PANI-based resin weights were applied over the composite panels to achieve the different thicknesses of LSP coating. Finally, the PANI-based LSP coating was cured in the oven. The heating rate was 10°C/min, from 23 to 115°C, followed by 5 mins of a dwell period at 130°C.

2.4 Reaction mechanism

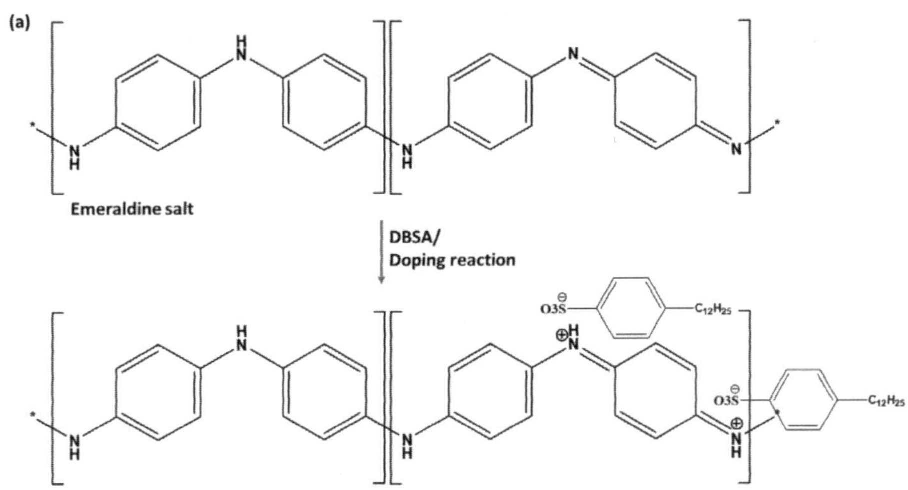

IOP Conf. Series: Materials Science and Engineering 1293 (2023) 012028 doi:10.1088/1757-899X/1293/1/012028

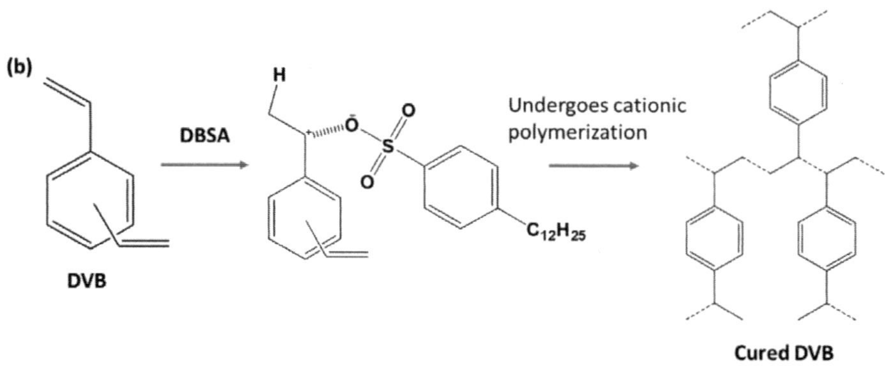

Figure 2. (a) PANI-DBSA doping and (b) DVB curing mechanism.

2.5 Lightning testing

LSP testing was conducted at Wichita State University's National Institute for Aviation Research Center as per SAE ARP 5416 standard. The actual waveform varies for various specimens, but the standardized waveform is shown below. The benchmarking (unprotected and Cu-mesh) samples were tested for all the three waveforms, while thin-PANI and Thick-PANI were tested only for the waveform A. This research aimed to understand the PANI-based coating performance against lightning strike, rather than the full certification, therefore only the modified waveform A was used for PANI-based samples.

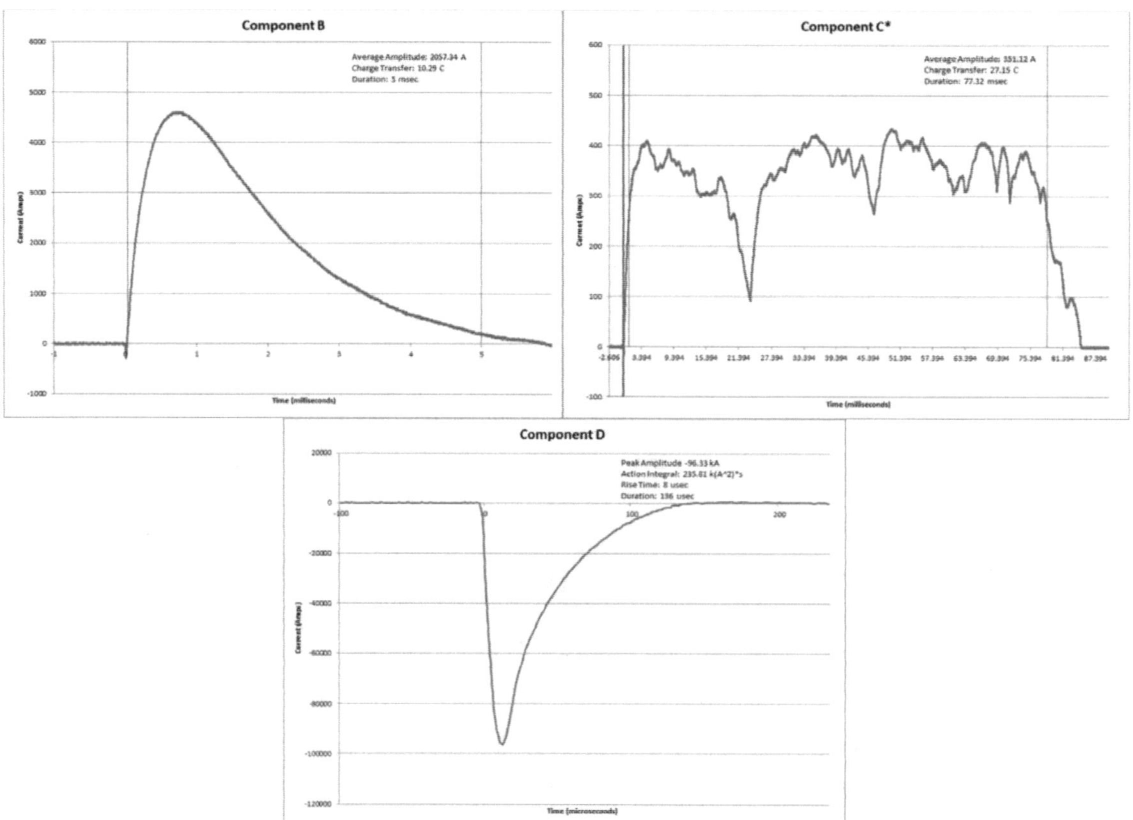

Figure 3. Applied current waveforms as per SAE ARP 5416 standard.

Table 1. Applied lightning waveforms on the CFRP samples.

Sample	Component A/D				Component B			Component C		
	Peak Amplitude (kA)	A/I (AAs)	Rise Time (µs)	Duration (µs)	Average Amplitude (kA)	Charge Transfer (C)	Duration (ms)	Average Amplitude (kA)	Charge Transfer (C)	Duration (ms)
Unprotected	-96.33	2.36E+5	8	136	2.06E+3	10.29	5	351.12	27.15	77.32
Cu-Mesh	-93.60	2.14E+5	8	128	2.06E+3	10.29	5	336.02	28.47	84.72
Thin-PANI	-59.47	7.6 E+4	10	94	-	-	-	-	-	-
Thick-PANI	-58.27	7.2 E+4	8	90	-	-	-	-	-	-

3. Results

Benchmarking: In the first experiment, one bare CFRP specimen without any protection was subjected to a lightning waveform, as shown in Table 1. Lightning damaged the specimen significantly, and fiber breakage, resin evaporation, and ply delimitation were observed. The affected area of was around 150 mm in length and 140 mm in width. Secondary attachment to a nearby place was also observed.

In the second benchmarking experiment, a CFRP protected with a 76 g/m^2 weight Cu-Mesh was also tested against the lightning waveform shown in Table 1. Lightning energy evaporated the Cu-Mesh to 115 mm long and 110 in width. However, the damage to the CFRP substrate was reduced to 35 mm in length and width. The fiber damage was significantly minor compared to the unprotected CFRP specimen, confirming that the Cu-Mesh provided Faraday cage protection to the CFRP substrate until its failure (evaporation). Once the Cu-mesh is evaporated, the lightning enters the CFRP specimen damaging it. This benchmarking testing confirmed that even aerospace-grade LSP does get damaged and needs repair to be functional over a long period, which is a big challenge for wind blades, where repair cost is colossal.

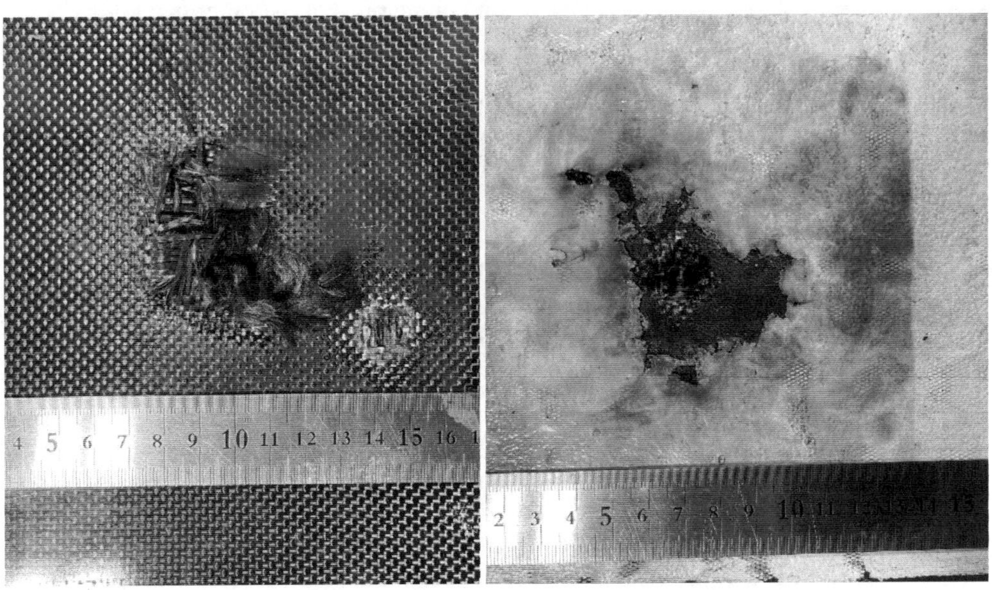

(a) Unprotected (b) Cu-mesh protected

Figure 4. (a) CFRP Specimen without any protection (b) CFRP with 76 GSM CU-mesh protected LSP after waveforms (A=100 kA, B= 10 C, and D= 20C charges).

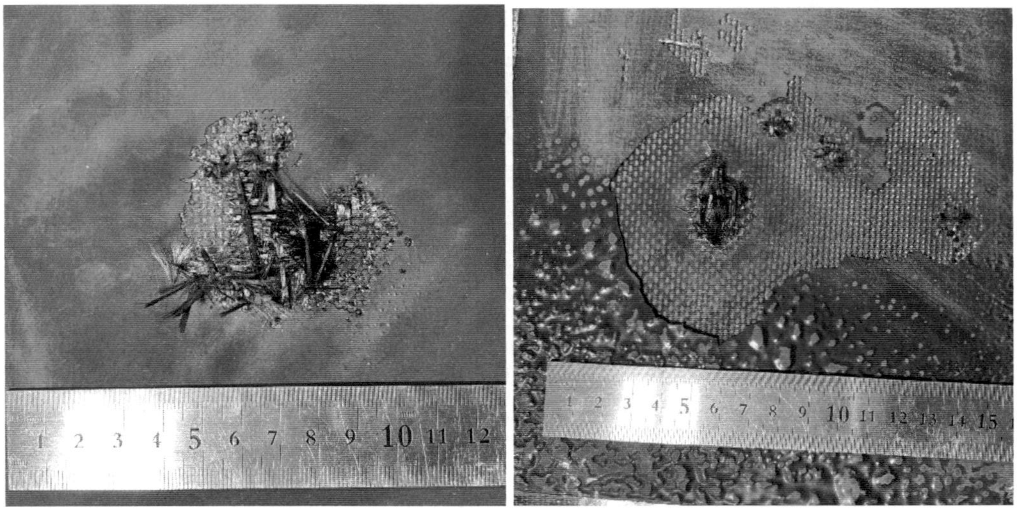

(a) Thin PANI-PU LSP (b) Thick PANI-PU LSP

Fig. 5 (a) CFRP Specimen without thin PANI-PU coating, and (b) CFRP with thicker PANI-PU LSP after waveform (A=60 kA).

In this experiment, we applied 20 gm and 30 gm of PANI-PU coating over a 280 mm × 280 mm CFRP specimen, which added roughly 255 GSM (thin coating) and 382 GSM (thick coating) of weight to the CFRF specimen. The lightning testing parameters for both panels are also shown in Table 1. The thinner PANI-PU coating showed less effectiveness against lightning, and the damaged area was around 100 mm × 100 mm. The damage is smaller than the bare uncoated CFRP specimen, but that could be due to different lightning current intensities in both cases. On the other hand, thicker PANI-PU coating showed a significant reduction in damage of around 28 mm × 25 mm in length and width, respectively. The damage is even more minor compared to the Cu-Mesh LSP. However, it is cautioned to draw a direct comparison as both samples had different lightning intensities. Nevertheless, the thicker PANI-PU coating performance is very promising for developing all-polymeric LSP solutions for wind blades. It is also worth mentioning that repairing a polymeric coating would be much more feasible than metal-mesh-type LSP repairs.

4. Conclusion

In conclusion, electrically conductive polymeric coatings have emerged as a promising solution for lightning strike protection. These coatings offer easy application methods such as spray-coating, painting, or automated manufacturing. Through experimental testing and comparison with traditional metallic-based LSP systems, newly developed polymeric coating solutions demonstrated their effectiveness. The results indicated that these coatings effectively dissipated electrical currents generated by lightning strikes and successfully safeguarded G/CFRP structures against severe damage, including puncture, fiber breakage, and resin evaporation. Consequently, polymeric coatings present a practical and cost-effective alternative for lightning strike protection in wind turbines. This study underscores the potential of polymeric coatings as a viable solution in the wind energy industry.

Acknowledgments

This manuscript has been authored by UT-Battelle, LLC under Contract No. DE-AC05-00OR22725 with the U.S. Department of Energy. The United States Government retains and the publisher, by accepting the article for publication, acknowledges that the United States Government retains a non-exclusive, paid-up, irrevocable, world-wide license to publish or reproduce the published form of this manuscript, or allow others to do so, for United States Government purposes. The Department of Energy will provide public access to these results of federally sponsored research in accordance with the DOE Public Access Plan (https://www.energy.gov/doe-public-access-plan).

References

[1] A. Cooperman, A. Eberle, E. Lantz, Wind turbine blade material in the United States: Quantities, costs, and end-of-life options, Resour. Conserv. Recycl. 168 (2021) 105439. doi:10.1016/J.RESCONREC.2021.105439.

[2] H. Che, M. Gagné, P.S.M. Rajesh, J.E. Klemberg-Sapieha, F. Sirois, D. Therriault, S. Yue, Metallization of Carbon Fiber Reinforced Polymers for Lightning Strike Protection, J. Mater. Eng. Perform. 27 (2018) 5205–5211. doi:10.1007/s11665-018-3609-y.

[3] V. Peesapati, I. Cotton, Lightning protection of wind turbines - A comparison of real lightning strike data and finite element lightning attachment analysis, 1st Int. Conf. Sustain. Power Gener. Supply, SUPERGEN '09. (2009). doi:10.1109/SUPERGEN.2009.5348122.

[4] S. Yokoyama, Lightning protection of wind turbine blades, Electr. Power Syst. Res. 94 (2013) 3–9. doi:10.1016/J.EPSR.2012.07.017.

[5] D. Li, S.C.M. Ho, G. Song, L. Ren, H. Li, A review of damage detection methods for wind turbine blades, Smart Mater. Struct. 24 (2015). doi:10.1088/0964-1726/24/3/033001.

[6] A. Serbescu, D. Brassard, J. Langot, E. Gourcerol, K. Chizari, M. Lapalme, A. Desautels, F. Sirois, D. Therriault, Silver-based conductive materials for lightning strike protection of aircraft composite structures, ICLP 2022 - 36th Int. Conf. Light. Prot. (2022) 250–255. doi:10.1109/ICLP56858.2022.9942506.

[7] B. Zhang, S.A. Soltani, L.N. Le, R. Asmatulu, Fabrication and assessment of a thin flexible surface coating made of pristine graphene for lightning strike protection, Mater. Sci. Eng. B Solid-State Mater. Adv. Technol. 216 (2017) 31–40. doi:10.1016/j.mseb.2017.02.008.

[8] H. Zhu, L. Chang, K. Fu, Y. Li, Dual conductive network of nickel-coated carbon fiber woven fabric for indirect and direct lightning strike protection of carbon fiber reinforced polymer composites, Polym. Compos. 43 (2022) 4437–4446. doi:10.1002/pc.26703.

[9] F. Wang, X. Ma, Y. Zhang, S. Jia, Lightning damage testing of aircraft composite-reinforced panels and its metal protection structures, Appl. Sci. 8 (2018). doi:10.3390/app8101791.

[10] S. Manomaisantiphap, V. Kumar, T. Okada, T. Yokozeki, Electrically conductive carbon fiber layers as lightning strike protection for non-conductive epoxy-based CFRP substrate, J. Compos. Mater. (2020) 002199832093594. doi:10.1177/0021998320935946.

[11] Y. Zhou, S.N.V. Raghu, V. Kumar, T. Okada, T. Yokozeki, Simulated lightning strike investigation of CFRP comprising a novel polyaniline/phenol based electrically conductive resin matrix, Compos. Sci. Technol. 214 (2021) 108971. doi:10.1016/J.COMPSCITECH.2021.108971.

[12] V. Kumar, T. Yokozeki, T. Okada, Y. Hirano, T. Goto, T. Takahashi, A.A. Hassen, T. Ogasawara, Polyaniline-based all-polymeric adhesive layer: An effective lightning strike protection technology for high residual mechanical strength of CFRPs, Compos. Sci. Technol. 172 (2019) 49–57. doi:10.1016/j.compscitech.2019.01.006.

IOP Conf. Series: Materials Science and Engineering 1293 (2023) 012029 doi:10.1088/1757-899X/1293/1/012029

Manufacturing methods for assessing the impact of wrinkles in wind turbine blades

Heloisa Guedes Mendonça[1,2], Sergei Semenov[2], Lars Pilgaard Mikkelsen[2], Kim Branner[2]

[1]Vestas Wind Systems, Hedeager 44, 8200 Aarhus, DK

[2]Department of Wind and Energy Systems, Technical University of Denmark, Frederiksborgvej 399, 4000 Roskilde, DK

E-mail: lapm@dtu.dk

Abstract. Wrinkles are defects prone to occur during the manufacturing process of wind turbine blades. Wrinkles can emerge at various locations of the structure, exhibiting a range of diverse shapes. Different manufacturing steps can lead to different defect types. This work presents different manufacturing methods to embed artificial wrinkle defects in laminates representing a typical defect that can be found in a wind turbine blade. Several methods are tested at coupon scale to design a critical wrinkle defect that can lead to the blade structural knock-down on the mechanical performance during high cycle fatigue operation. Following the selection of the defect types for investigation at the test coupon scale, the corresponding defect types are embedded while manufacturing a 12.6 m wind turbine blade.

1. Introduction

In the production of large structures, such as wind turbine blades, manufacturing procedures can lead to different sources of defects. Wrinkles are one such defect that can initiate failure mechanisms, including fiber fracture and delaminations [1]. Different steps in the manufacturing process can lead to different types of wrinkles. It can arise from the poor attachment of plies due to low interlaminar shear stress affecting compaction during layup, geometric changes on tooling surfaces, errors during operation, flaws in the layup process, distorted fibers, abrupt change in ply thickness, among other sources [2,3]. Wrinkles can emerge at various locations assuming diverse configurations, of which out-of-plane waviness appears at a higher frequency [4]. Several configurations are mentioned in the literature, divided into two major groups: uniform and non-uniform ply waviness [5]. The embedded uniform pattern is a configuration replicated in labs making it feasible to examine laminates with such defects [6]. Additionally, folded wrinkles are considered an extreme case in which the wrinkle is folded back to itself with a maximum deviation of the fiber misalignment [4,7]. Concerning the presence of gaps in a laminate, it can create regions with a high tendency of resin accumulation after curing [8], referred to as resin pockets. Different methods are adopted when it comes to producing artificial defects in a lab [9,10,11,12]. Wrinkles are introduced in a laminate during vacuum-assisted resin transfer molding (VARTM). A wrinkled shape is placed at the top of a vacuum bag, and pressure is manually applied to it after the infusion process but before the resin is fully cured [9]. An alternative method is adopted to induce wrinkles in a sandwich laminate by introducing a plastic rod between a face sheet and a balsa core, followed by applying a vacuum to hold the defect shape. Subsequently, the vacuum is released, the plastic rod is removed, and the vacuum is reapplied with the resin infusion carried controlled at an angled table [10]. A different approach involves the utilization of 90-degree thin strips of pre-impregnated fibers (prepreg) positioned at specific locations of the laminate to induce the formation of wrinkles without adding foreign material [11].

The present work outlines various manufacturing methods leading to out-of-plane wrinkles of different severity types for a glass fiber reinforced (GFR) composite laminate. The severity types elected are submitted to a screening fatigue test at R=-1 to define a defect type that falls in the criteria for assessing its impact on realistic cycle conditions for fatigue failure of a wind turbine blade. Within the field of mechanics of materials, a screening test is a preliminary test performed to identify characteristics or properties of a certain structure. After selecting the severity types through the coupon test, the same types of wrinkles are embedded in a 12.6 m wind turbine blade.

This work proposes a protocol for selecting wrinkle types likely to occur in a wind turbine blade. The protocol integrates a method chain comprising defect design, manufacturing procedures, testing programs, and qualitative evaluation of fracture characteristics. The ultimate purpose is the selection of manufacturing procedures to create out-of-plane wrinkles similar to those found in a wind turbine blade that can potentially lead it to failure at high cycle fatigue. The protocol is implemented by upscaling the defect manufactured at the test coupon level to the structure level. The method chain is presented in section 2.

2. Methodology

2.1. Types of Wrinkles

Feasible wrinkle profiles for glass fiber wind turbine blades serve as the basis for the adopted defect configurations evaluated in this work. The five wrinkle types illustrated in Figure 1 were produced using three distinct manufacturing processes: *parameter control*, *non-controlled parameter*, and *resin pocket*. The methods *parameter control* and *resin pocket* use a cured UD insert for the central wrinkle, while the *non-controlled parameter* creates a fold based wrinkle. Wrinkle types one to four were created using a layup of $[+45/-45/0/0/0/0/0]s$, while a layup of $[0/0/0/0/0]s$ was applied for defect type five.

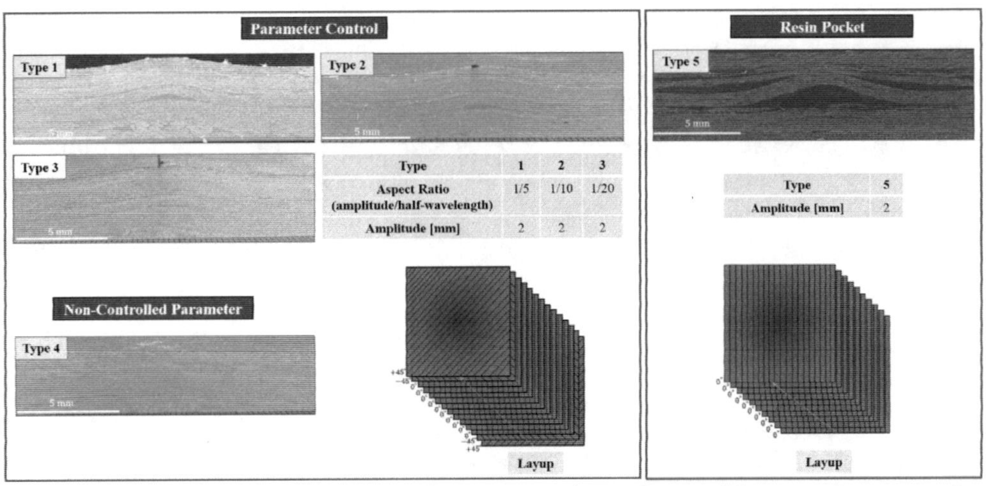

Figure 1: Cross-section of different types of wrinkles defects obtained from various manufacturing methods with an assigned layup and geometrical parameters.

The root section, a structurally thicker region, exhibits a higher sensitivity to wrinkles formations when compared to other sections. To ensure accurate representation of the defects being examined, it is crucial that the thickness of the test coupon specimens replicates the equivalent section thickness of the laminate, which imitates a region near the root section of the wind turbine blade. In addition, a critical factor in selecting the defect parameters was to ensure the defect size was sufficiently small to keep it in the blade during operation without repair. The specimens were produced and sliced in 25 mm wide samples resulting in uniform volumes measuring $400 \times 25 \times 11\,\text{mm}^3$. The manufacturing procedures are detailed in section 2.2.

2.2. Manufacturing Methods

The present investigation has explored three distinct manufacturing methods, which are depicted and identified in Figure 2, Figure 3, and Figure 4, respectively. One of the limiting factors in the manufacturing of test coupons with artificial defects is the need to ensure consistent replication of the geometrical parameters within a given batch of samples.

The *parameter control* method (Type 1, 2 and 3) ensures that predefined geometrical parameters, such as amplitude and wavelength, are maintained during production. The process is divided into two distinct phases. Phase 1 entails manufacturing an insert that conforms to the pre-defined geometrical parameters through the utilization of a cast mold. The insert comprises two unidirectional (UD) layers arranged on the aluminum cast mold in a stacked configuration. The defect amplitude and aspect ratio are defined according to the specifications outlined in Figure 1. The wrinkle defect is introduced transversely to the orientation of the UD fibers. The vacuum-Assisted Resin Infusion (VARI) process is used to impregnate the UD plies with resin. The Hexion epoxy resin system uses a mixing ratio of 100/28 (by weight). The curing process involves treating the resin-impregnated insert to a temperature of 40 °C at 12 h, followed by subsequent heating at 80 °C at 10 h. The objective of Phase 2 is to integrate the UD insert into the symmetry axis of the layup sequence $[+45/-45/0/0/0/0/0]s$ along with the remaining plies. The second Biax (biaxial) ply is folded to align with the center of the insert containing the wrinkle defect. The folding process emulates the undulation that may occur during the layup process for the blade and, therefore, can be the reason for causing wrinkles. Additionally, the folded wrinkle facilitates increased compaction of the material during the infusion process, reducing the resin content between the undulated plies. The same resin mixing ratio and curing cycles implemented during Phase 1 are applied to Phase 2. The manufacturing procedure is repeated for the classification of wrinkles type 1, type 2, and type 3, as depicted in Figure 1.

The *non-controlled parameter* method (Type 4) comprises a single phase as indicated in Figure 3. The only wrinkle present is the result of the fold at the second biax layer in the composite layup. Subsequently, the layers are assembled with the same layup and undergo the same resin infusion procedure while adhering to the equivalent curing cycle employed in the *parameter control* Methodology. The laminate contains the wrinkle classification type 4, resulting in a less pronounced wrinkle and a reduced potential for retention of resin pockets.

The *resin pocket* method (Type 5) follows two phases, as pointed out in Figure 4. The initial phase is focused to produce a two-layer UD insert by applying the identical manufacturing principle adopted in the initial phase of the *parameter control* method. Subsequently, in phase 2, the aforementioned insert is embedded at the center of a ply stacking sequence comprising $[0/0/0/0/0]s$. The stacked configuration is confined between two aluminum plates placed at an angled position for a controlled resin infusion process. This approach effectively contains the formation of wrinkles in the pre-cured UD insert, thereby mitigating its propagation to adjacent plies in the stacking sequence. By not implementing the biax fold underneath the insert, the space between the flat ply and the wrinkle insert is filled by resin during the infusion process. The current approach adopts identical curing cycle and temperature conditions as those described in the previous methods.

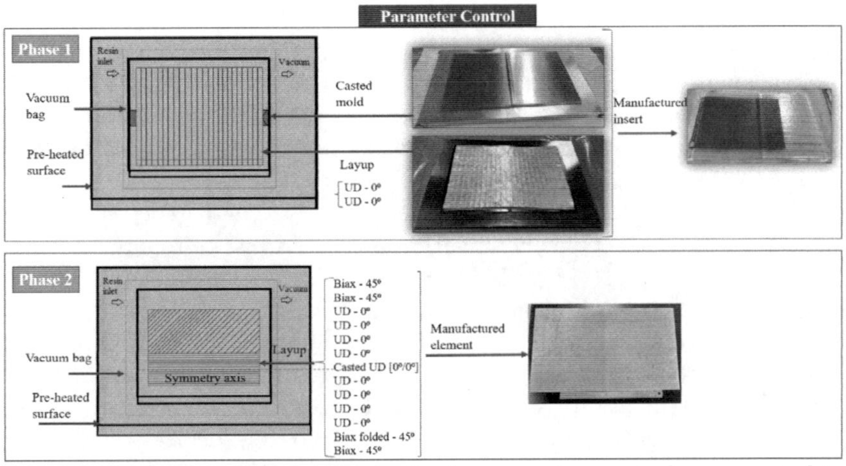

Figure 2: Schematics of the manufacturing method *parameter control* represented in two phases.

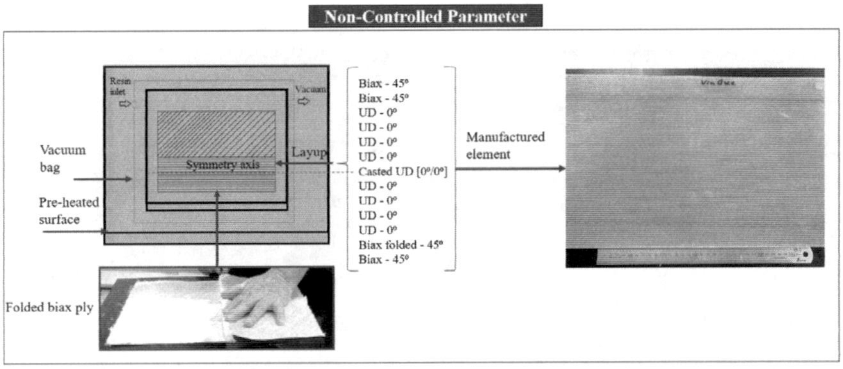

Figure 3: Schematics of the manufacturing method *non-controlled parameter* represented in a single phase.

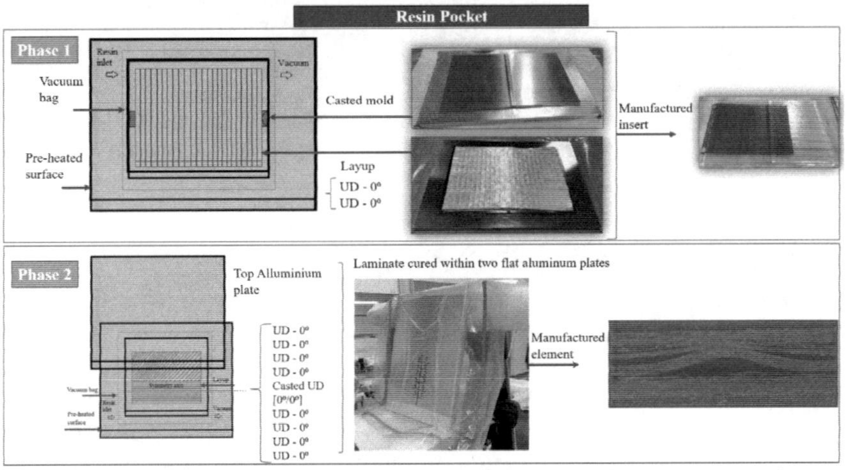

Figure 4: Schematics of the manufacturing method *resin pocket* represented in two phases.

2.3. Testing Procedures

The purpose of the experimental tests is to select a manufacturing process and wrinkles size to be embedded into a DTU designed 12.6 m blade [13, 14] to investigate delamination growth initiated from wrinkle defects during a fatigue test campaign. The DTU blade is described in references 13, 14. Screening fatigue experimental tests assess the structural performance of the produced laminates with diverse wrinkle classifications. The tests aim to identify a wrinkle classification that could persist within the blade structure undetected, ultimately resulting in high cycle fatigue damage. A wrinkle classification is defined by the manufacturing method and the geometrical features.

The experimental protocol is performed for fatigue with $R = -1$, normally defined as load or stress ratio with $R = \varepsilon_{min}/\varepsilon_{max}$. The fatigue test is carried out on a single representative sample for each wrinkle classification. Whereby considered representative of the full batch of samples within the same defect classification. Figure 5 show the test setup incorporating consumer-grade Nikon D7500 cameras to capture the cross-sectional and outer surfaces of the specimens. The cameras track the crack initiation and propagation during the test. Additionally, a thermal camera is installed to track the temperature gradient associated with the fracture progression. Extensometers are positioned at a distance of 40 mm from the center of the wrinkle to enable strain measurement.

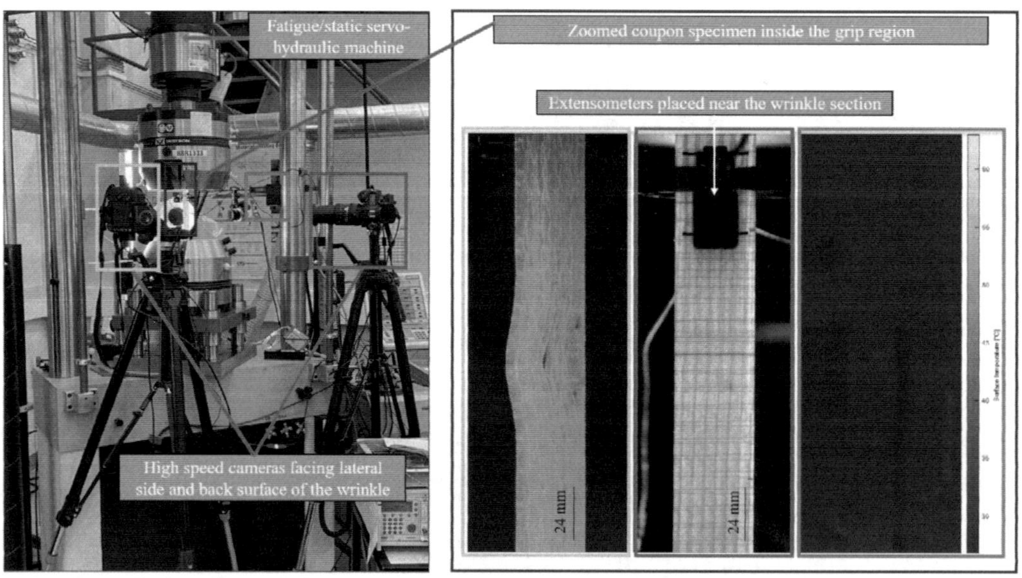

Figure 5: Experimental setup for testing coupon specimens embedded with various classifications of wrinkle defects. Yellow is the side view, blue is the surface view and red is the thermal camera.

The order of tests follows the indication in Table 1 and Figure 6. The experimental program starts with the sample manufactured with the *resin pocket* method for wrinkle classification type 5. The linear strain amplitude is established to its maximum limit avoiding off-axis buckling. It is kept constant throughout the fatigue test. As the strain amplitude applied drove the representative sample to low cycle failure, the subsequent test was performed using a reduced strain amplitude compatible with what is intended to be used for the mid-span area of the 12.6 m blade. The representative samples produced with the *parameter control* method and classification type 1 and type 2 were tested with high cycle failure observed. As the latter two classification types fulfilled the requirements for the defect choice, defect classification type 3 was not tested further. Considered less critical by the smooth wrinkle configuration, the characteristic sample for non-controlled parameter classification type 4 was tested in a staircase approach. The test strain amplitude is initially incremented from classification type 2 and gradually increased if crack propagation remains undetected. The latter test resulted in a runout without failure detected after 4 million cycles.

Table 1: Testing order and corresponding strain amplitude for each defect classification.

Order of test	Method (Specimen ID)	Strain amplitude [%]	Cycles to Failure
1	Resin Pocket (Type 5)	0.324	42,639
2	Parameter Control (Type 1)	0.200	250,048
3	Parameter Control (Type 2)	0.225	581,711
4	Non-controlled Parameter (Type 4)	staircase: 0.227-0.260-0.280-0.300-0.330	runout

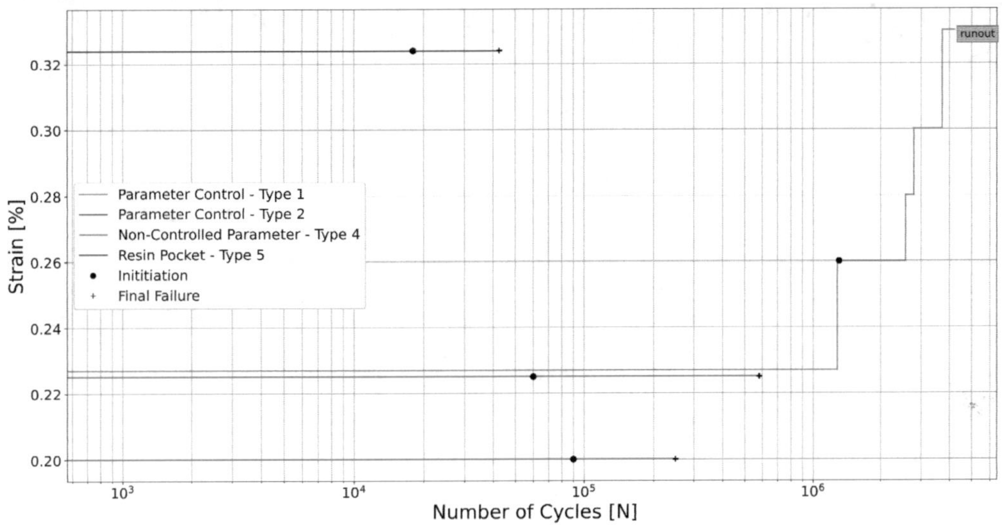

Figure 6: Testing program depicting the number of cycles versus percentage strain with regions outlined for crack initiation and final failure.

2.4. Fracture Characteristics

After evaluating the test images, it was determined that the dominant fatigue failure mechanism is represented by a critical crack at the maximum amplitude wrinkle, resulting in multiple delaminations along the interfaces and a kink band as the final failure. The failure sequence for each of the samples tested for the various manufacturing processes are depicted in Figure 7. The characteristic samples classification type 1, type 2, and type 5 experience a critical crack length around the wrinkle insert during the fatigue test. After testing the sample resultant from the *non-controlled parameter* method, it was noticed that a crack initiates around the fold ply. However, propagation is not observed along the interface. The specimens manufactured through the *parameter control* method passed the initial assessment adhering to the failure criteria in high cycle fatigue. Additionally, the wrinkle defects embedded through the latter method are considered sufficiently fine to remain within a thick blade section despite surface repair. Regarding manufacturing, the latter method is equally feasible for embedding in the blade. Defect classification type 5 is ruled out due to difficulties controlling the through-thickness wrinkle shape propagation during blade manufacturing. Therefore, the *parameter control* method is selected to assess wrinkles impact on the blade undergoing fatigue loads. Wrinkle classification type 1 and type 2 are chosen to compare the impact of two severity levels in the blade. These selected wrinkle defects are selected for embedding in the DTU 12.6 m blade to evaluate crack growth in wrinkle defects under high cycle fatigue conditions.

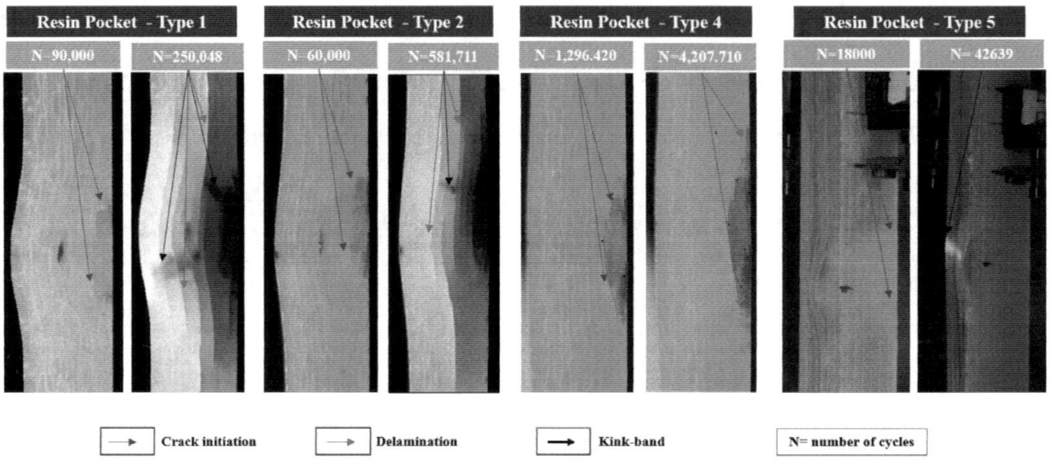

Figure 7: Crack initiation and propagation detected during fatigue tests for various defect configurations.

2.5. Defect in the Blade

The defect classifications selected at the coupon scale are embedded in a 12.6 m blade. The defect is embedded in the spar cap on the pressure side mid-span within a region where the strain amplitude can drive crack growth for wrinkle type 1 and type 2, as shown in Figure 8. The *parameter control* manufacturing method is adjusted to accommodate the structural up-scaling. Wrinkle type 1 and type 2 are manufactured with the same method, applying cast molds designed to meet the pre-defined geometric parameters unique for each wrinkle type in phase 1. The method is changed at phase 1 by producing an insert with one UD layer submitted to evenly spaced cut-outs a the fiber direction, performed before the infusion. Subsequently, the cut-out spaces are filled in with dry UD fibers. This procedure is necessary for the viability of the resin infusion process in the blade. Manufacturing trials conducted for a thick laminate using the insert produced in phase 1 in its pristine version proved that it restrains the resin infusion, resulting in a laminate with the accumulation of dry spots. After modification of the method, the defects are effectively embedded in the blade without dry spots.

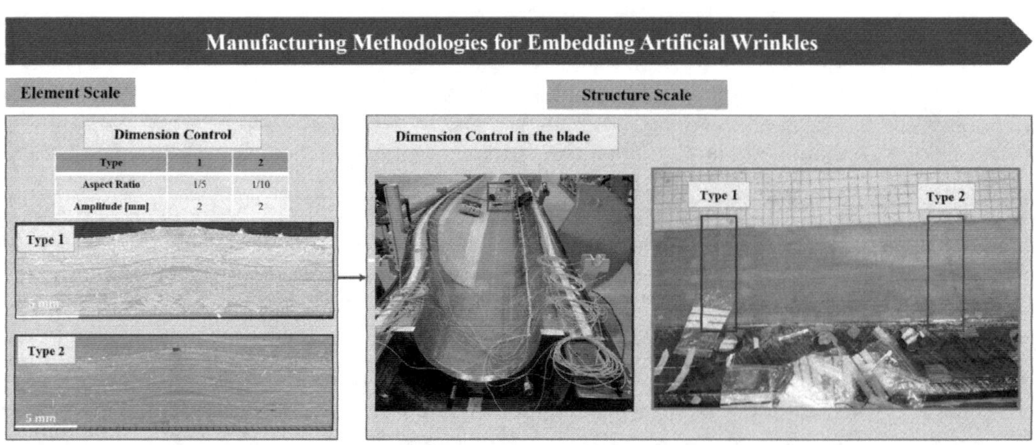

Figure 8: Defects classifications type 1 and type 2 embedded in a 12.6 m blade.

3. Conclusions

This work presents a protocol to select a manufacturing method for evaluating wrinkles impact on the fatigue life of a wind turbine blade. Different out-of-plane wrinkle defects are derived from the manufacturing methods presented. Of the three methods presented, one method fulfilled the criteria concerning the feasibility of manufacturing up-scaling and subsequently expected fatigue delamination crack growth during high-cycle fatigue loading. After evaluating the fatigue damage mechanisms, delamination is found to be the driving mechanism of fatigue failure. The proposed methods were implemented and tested, demonstrating the concept for future studies to investigate how different wrinkles severity classifications can impact the structural performance of wind turbine blades.

3.1. Acknowledgements

The authors acknowledge the support from the Danish Energy Agency through the EUDP grant 64018-0068 in the ReliaBlade project. They also thank the team at Large Scale Facility at DTU Wind and Energy Systems, particularly Knud Stenbæk Nielsen and Christen Malte Markussen, for their valuable technical assistance and specialized knowledge of defects in composites structures, which was essential to the success of this research.

4. References

[1] Katnam, K.B., Comer, A.J., Roy, D., Da Silva, L.F.M. and Young, T.M., 2015. Composite repair in wind turbine blades: an overview. The Journal of Adhesion, 91(1-2), pp.113-139.

[2] Hassan, M.H., Othman, A.R. and Kamaruddin, S., 2017. A review on the manufacturing defects of complex-shaped laminate in aircraft composite structures. The International Journal of Advanced Manufacturing Technology, 91, pp.4081-4094.

[3] Galappaththi, U.I.K., De Silva, A.M., Draskovic, M. and Macdonald, M., 2013, March. Strategic quality control measures to reduce defects in composite wind turbine blades. In International Conference on Renewable Energies and Power Quality (ICREPQ'13), Bilbao (Spain) (pp. 20-22).

[4] Thor, M., Sause, M.G. and Hinterhölzl, R.M., 2020. Mechanisms of origin and classification of out-of-plane fiber waviness in composite materials—A review. Journal of Composites Science, 4(3), p.130.

[5] Cimini Jr, C.A. and Tsai, S.W., 1999. Ply waviness on in-plane stiffness of composite laminates. Proceedings of the 12th ICCM, Paris, France, pp.7-9.

[6] Kulkarni, P., Mali, K.D. and Singh, S., 2020. An overview of the formation of fibre waviness and its effect on the mechanical performance of fibre reinforced polymer composites. Composites Part A: Applied Science and Manufacturing, 137, p.106013.

[7] Hayman, B., Berggreen, C. and Pettersson, R., 2007. The effect of face sheet wrinkle defects on the strength of FRP sandwich structures. Journal of Sandwich Structures and Materials, 9(4), pp.377-404.

[8] Nartey, M., Zhang, T., Gong, B., Wang, J., Peng, S., Wang, H. and Peng, H.X., 2020. Understanding the impact of fibre wrinkle architectures on composite laminates through tailored gaps and overlaps. Composites Part B: Engineering, 196, p.108097.

[9] Bender, J.J., Hallett, S.R. and Lindgaard, E., 2019. Investigation of the effect of wrinkle features on wind turbine blade sub-structure strength. Composite Structures, 218, pp.39-49.

[10] Leong, M., Hvejsel, C.F., Thomsen, O.T., Lund, E. and Daniel, I.M., 2012. Fatigue failure of sandwich beams with face sheet wrinkle defects. Composites science and technology, 72(13), pp.1539-1547.

[11] Mukhopadhyay, S., Nixon-Pearson, O.J. and Hallett, S.R., 2018. An experimental and numerical study on fatigue damage development in laminates containing embedded wrinkle defects. International Journal of Fatigue, 107, pp.1-12.

[12] Chen, X., Semenov, S., McGugan, M., Madsen, S.H., Yeniceli, S.C., Berring, P. and Branner, K., 2021. Fatigue testing of a 14.3 m composite blade embedded with artificial defects–damage growth and structural health monitoring. Composites Part A: Applied Science and Manufacturing, 140, p.106189.

[13] Haselbach, P.U., Zahle, F., Berring, P., Semenov, S., Voltá, L., Roqueta, I. and Verelst, D.R., 2020, September. Blade research and demonstration platform. In Journal of Physics: Conference Series (Vol. 1618, No. 5, p. 052073). IOP Publishing.

[14] Haselbach, P.U., Semenov, S. and Berring, P., 2020, October. DTU's blade research and demonstration platform. In IOP Conference Series: Materials Science and Engineering (Vol. 942, No. 1, p. 012043). IOP Publishing.

Mass determination and model prediction of retired blades from wind turbine repowering or dismantling using a GIS database

A G Silverman[1*], Y Henao[1], L C Bank[1], and T R Gentry[1]
[1] Georgia Institute of Technology, Atlanta, GA 30332, USA
[*] (corresponding author)
E-mail: aevasilverman@gmail.com

Abstract. Existing estimations of waste from wind energy infrastructure that is headed for, flowing through, or having reached the terminus of various post-processing pathways have primarily relied on reported capacity to extrapolate the material weight of turbine components. This data can be used to project future streams of composite blade material coming from wind farm repowering and decommissioning and inform policies to optimize or improve certain blade End of Life (EoL) options. However, rated capacity alone is insufficient to quantify or characterize the dynamics of US wind fleet retirement, since turbines are often repowered with new blades but their capacity remains the same. This research demonstrates an alternative method, comparing various mass estimation techniques and identifying blade models that have been retired or are soon to enter waste pathways due to turbine repowering by spatiotemporal comparison of periodic versions of the United States Geological Survey (USGS) Wind Turbine Database (USWTDB). These analyses are used to compile a list of turbine and blade models that will be at the forefront of national repowering and decommissioning movements in the near future. Mass of future waste flows are totalled and can help inform protocols and frameworks for blade material EoL processes.

1. Background

Significant investment in wind energy since the early 2000s indicates that there will soon be a large volume of wind energy related waste heading for processing [1, 2]. In addition to landfill disposal, there are a variety of ways in which blade waste may be discarded or repurposed for material value: incineration, mechanical recycling, thermal recycling, chemical recycling, or cement co-processing among others [3]. By far, landfill disposal of wind blades is cheapest and most simple, offering an attractive option to economic stakeholders [4]. However, landfill disposal of wind energy material poses environmental harm and degrades the value of wind energy as a clean, renewable source [5]. Several groups have expressed interest in implementing a circular economy solution to the prevailing issue of wind energy waste which would enhance the engagement of the wind energy sector with environmental goals [6]. In 2023, Vestas, the largest global wind turbine manufacturer, reported research that could establish a process which would allow the company to break down the epoxy resin of old blades, generating new raw materials [7]. Previous research by the Re-Wind Network and others indicates there may be additional reuse options for EoL blades, such as girders for pedestrian bridges or as poles for electrical transmission lines [8, 9] amongst others. These options utilize the existing structural features of the blade but verify their capacities for appropriate alternative loading conditions.

Blades are generally considered to be in the EoL stage once they can no longer perform their original functionality of generating the greatest possible profit for wind farms [10]. A variety of factors can lead to the decrease in generated value - over their useable life, blades may experience loading fatigue or material erosion, which reduces the performance of turbines. Wind farm owners can make the choice to 'repower' by removing underperforming blades or turbines and fitting new machinery onto existing towers. Generally, farms are permitted to use wind blades until their Design End of Life (DEoL) which is about 20 years. After this time has elapsed, DEoL can be extended by re-permitting or retrofitting. The variety of criteria for blade EoL makes it difficult to predict the future decisions of wind farm repowering or decommissioning.

Previous research has investigated circular economy systems and used GIS to develop EoL removal and disposal routes. A global wind inventory for future decommissioning (GoWInD) has been proposed using QGIS to create a framework where the decommissioning of individual wind infrastructure components is handled with spatiotemporal considerations, aiding the logistics of transportation, and recycling or disposal [11]. Other research has used USWTDB data to project the total mass of future blade waste based on existing national capacity and conversion from power rating to blade mass [3]. Prospective material flows as well as total waste inventory can be deduced from existing data on wind infrastructure and numerical models relating blade size and rated power [12].

The total capacity of US wind energy is reported by a number of organizations. The USWTDB web viewer reports the total GW to be 138 GW [13]. The International Energy Agency (IEA) reports 143.9 GW of onshore wind energy capacity in 2022 while American Clean Power (ACP) reports a capacity of 140 GW [14, 15]. The U.S. Energy Information Administration (EIA) reports 137.6 GW in June of 2022, while Wood Mackenzie, a global consultant and research group, reports 148 GW in 2022 [16]. There is clearly a discrepancy in these reports indicating that total national capacity can be (and is) quantified in a variety of ways.

Data used in this research was obtained from the USWTDB repository. Versions of the data from 2014 and 2022 were used. Both database versions include information on rated power, rotor diameter, location, and a unique identification number for cross referencing among other fields. This research employs Quantum GIS (QGIS), a free and opensource mapping software system, and MATLAB, to investigate the characteristics of blades approaching their EoL stage.

2. Identification of dismantled turbines and potential EoL blades by temporal comparison

The USGS maintains current and historic repositories of geographic locations of installed wind turbines with fields for information on the features of each turbine. Instead of describing whole farms, which are often constructed entirely of the same model of turbine, the database compiles reports of individual turbine locations and verifies existence with satellite imagery [13]. Due to this method of documentation, several locations are missing complete data entry, as characterization of each turbine in the US is tedious, but at a minimum each existing turbine is listed with its longitude and latitude. With these records it is possible to compare the change in on- and offshore existing turbine locations. New installations and past removals can be identified, and changes in the information about turbine type, blade length, and capacity can be used to determine which locations have been repowered.

In this study, the July 2022 USWTDB version (the most recent version at the time this project was started) was compared to the March 2014 version. The 2014 data was selected because it contained relatively thorough data compared to prior versions. Additionally, the 2022 data listed a field specifically for cross-referencing to the 2014 version, which meant that the matching of geographic locations of turbines could be further verified. The comparison of these two years was used to group turbines into 4 categories: 'new' turbines installed after 2014, turbines that were dismantled before 2022, turbines that have been repowered between 2014 and 2022, and turbines that have not been repowered but existed in 2014 and 2022. Figure 1 uses a Venn diagram to represent the categorization of all turbines in these two lists.

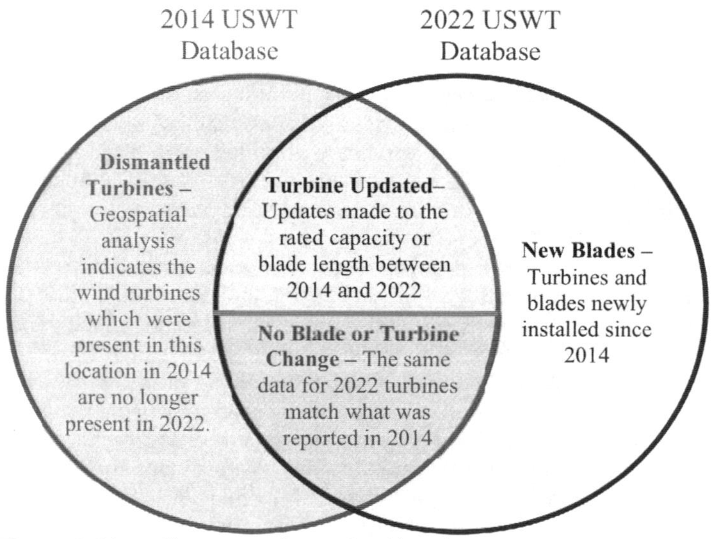

Figure 1: Venn diagram analogy of turbines between 2014 and 2022

Locations were matched by longitude and latitude with a resolution of 0.0001 degrees, which is roughly equal to 11.1 meters [17]. This resolution was necessary because the recorded location points were not identical, likely due to discrepancy from areal imagery across the different years. These location-based matches were verified with comparison of the 5-digit cross identification field provided by the USGS for the 2014 repository.

Figure 1 depicts the sectioning of turbines across the two data versions. The full circle on the left is all 2014 turbines while the right circle is all 2022 turbines. The far-left crescent represents locations that were dismantled between 2014 and 2022 since they are listed in the first but are not present in the second. Blades from this section by implication have been sent to EoL processing. In the far-right crescent are new turbines that were installed after 2014. In the bottom of the overlap between 2014 and 2022 are turbines which have no recorded change in their rated capacity or rotor diameter, they are considered to have not yet been repowered. By contrast the locations that reported an increase of 20kW or greater or whose rotor diameter increased by more than 2 meters between 2014 and 2022 are considered to have been repowered. Figure 2 uses another Venn diagram to represent the criteria for determining which turbines have been repowered.

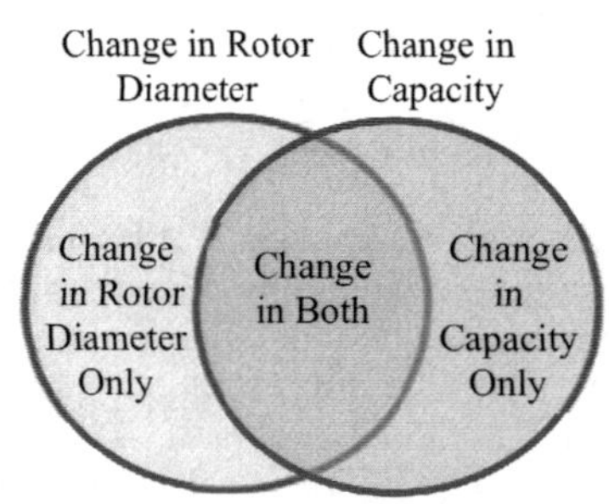

Figure 2: Venn diagram analogy for all turbines considered to be repowered

In Figure 2, the left circle represents all turbines whose rotor diameter was recorded to increase by 2 meters or more between 2014 and 2022 while the right circle is all turbines whose capacity increased by 20kW or more. In the left crescent are turbines who *only* record changes in rotor diameter while the right is turbines with *only* a recorded increase in power rating, the intersection contains turbines who saw an increase in rotor diameter *and* capacity change. Thus, the overall criteria for a turbine to be considered repowered is to record a reasonable increase in its rotor diameter *or* capacity between the two years. An increase in rated capacity is a direct indication of repowering while an increase in blade length would enhance the performance of turbines without being reflected in the capacity value. All turbines whose information did not change, or whose change was less than the criteria values, are still contenders for future repowering projects as their performance had not substantially changed since installation. The not-yet-repowered turbines and their attached blades are those due for updates in the near future and are the source of the next generation of blade waste material.

2.1. Mass Estimation

In this research, the weight of blade waste is compared using two methods. The first approach uses a high estimate of 15 metric tonnes per MW of power produced and a low estimate uses 10 tonnes per MW [2, 18]. The second technique uses a polynomial relationship to determine the weight of blades from their length according to the Equation 1 (where W is blade weight in tonnes and L is length in meters) [19]:

$$W = 0.0036L^2 + 0.0258L \tag{1}$$

The attempt to estimate the mass of various groups introduces an issue as the USWTDB has many entries missing data for the capacity rating and rotor diameter fields among others. In fact, in 2014 only 87% of turbines listed their rated capacity and 85% listed rotor diameter. Due to the missing information, the total weight was found by treating the ratio of weight determined to the percent of information present as a proportion of the total weight, described in Equation 2

$$\frac{Sum\ of\ Reported\ Capacity}{Actual\ Percent\ Reported} = \frac{Total\ Capacity}{100\%\ Reported} \tag{2}$$

Using this method, the total weight of blade material from 2014 and 2022 in each of the four categories (new, dismantled, repowered, not repowered) was estimated using the available information on each turbine's capacity rating and blade length and extrapolated to develop the total mass in each group in Table 1. The weight estimates for dismantled turbines used information taken from the 2014 lists while all other categories took information from the 2022 data.

Table 1: Mass estimates of each year and categorical group using rated capacity and blade length

Group	Number of Turbines	Low Tonnage Estimate from Power Rating	High Tonnage Estimate from Power Rating	Tonnage Estimate Based on Blade Lengths
2022 Turbines	72,130	1,449,000	2,173,500	2,132,800
2014 Turbines	48,976	658,900	988,350	883,480
New	34,035	829,300	1,243,950	1,344,100
Dismantled	10,881	39,710	59,565	49,844
Not Repowered- No Change	30,830	491,300	736,950	605,830
Repowered - Change in Blades OR Capacity	7,265	127,900	191,850	192,960

Between 2014 and 2022 the number of online wind turbines increased by nearly 50%, but the capacity more than doubled. This reflects the engineering advancement in turbine efficiency and airfoil design of blades. Blade weights calculated from length for all the 2022 turbines, all 2014 turbines, dismantled turbines, and not repowered turbines (first, second, fourth, and fifth rows respectively) fell within the high and low tonnage estimates resolved from capacity rating. By contrast, new and repowered turbines (third and sixth rows respectively) saw that the weight estimate based on length exceeded the high estimate based on rated capacity. Newly installed turbines and those that have been repowered are implied to be fitted with the most up-to-date technology. The length-based estimate for all 2022 turbines heavily favored the high tonnage capacity-based estimate, that is, the length estimate was only 1.8% less than the high tonnage capacity estimate. This signals that using an estimate of 10-15 tonnes of blade material per MW of power may be minimizing the actual weight of potential blade waste. At the very least there is an observable mismatch between weight estimates derived from blade length and turbine power rating for the most recent wind energy technologies that is not observable for older turbines.

In Table 1, the weight of repowered turbines was calculated with 2022 data so as to present the most recently available information at these locations, but it is also pertinent to look at the change in weight between 2014 and 2022, and even within the different attributes that signal repowering has occurred. Table 2 analyzes the change in weight based on blade-length calculations while Table 3 considers the change in capacity rating.

Table 2: Change in blade weight of repowered turbines from 2014 and 2022 estimated from length

Group	Weight in Tonnes in 2014	Weight in Tonnes in 2022	% Change in Weight
ALL Change in Capacity	78,384	98,456	25.61
ALL Change in Blades	130,830	169,910	29.87
Change in Capacity ONLY	22,034	22,036	~0
Change in Blades ONLY	67,154	84,703	26.13
Change in Both Blades AND capacity	55,165	75,409	36.70
Repowered – Change in Blades OR Capacity	154,050	192,960	25.26

Table 3: Low and high estimates of blade weights estimated from information on rated capacity

Group	2014 Low Tonnage Estimate	2014 High Tonnage Estimate	2022 Low Tonnage Estimate	2022 High Tonnage Estimate	% Change in Weight
ALL Change in Capacity	58,820	88,230	64,390	96,585	9.469
ALL Change in Blades	108,400	162,600	112,300	168,450	3.598
Change in Capacity ONLY	14,050	21,075	14,680	22,020	4.484
Change in Blades ONLY	57,270	85,905	57,270	85,905	0
Change in Both Blades AND Capacity	43,960	65,940	48,830	73,245	11.08
Repowered-Change in Blades OR Capacity	123,300	184,950	127,900	191,850	3.731

The consistent observation that blade weights increased for turbines identified as repowered affirms that there have been updates made in these locations between 2014 and 2022. However, the overall increase in weight of repowered turbines determined from the change in blade lengths was 25% while the increase in weight developed from rated capacity was only 3.7%. The length-based estimation for 2014 blades (that were removed) consistently falls within the high and low estimates derived from capacity rating. For the 2022 repowered turbines, or the new turbine and blade models that were fitted to existing towers, the length-based estimate is greater than the high tonnage estimate for every category except for the case where *only* blades were updated on an existing turbine. This indicates that for modern turbines and blades, capacity-based estimation doesn't completely account for the mass of newly installed blades. In both Table 2 & 3, the category with the greatest increase in blade weight was for turbines whose blade length and capacity reported changes between the two years, which could be explained by the fact that updating turbine *and* blades onto a pre-existing tower reflects an extreme form of update so increase in blade material was substantial.

2.2. Turbine and blade models due for update

It is also of interest to understand what turbine models are most frequently updated and what they are updated to. Using histograms, Figure 3(a) displays the top 10 original models that were updated after 2014 while 3(b) reports the most common turbines that are the result of repowering.

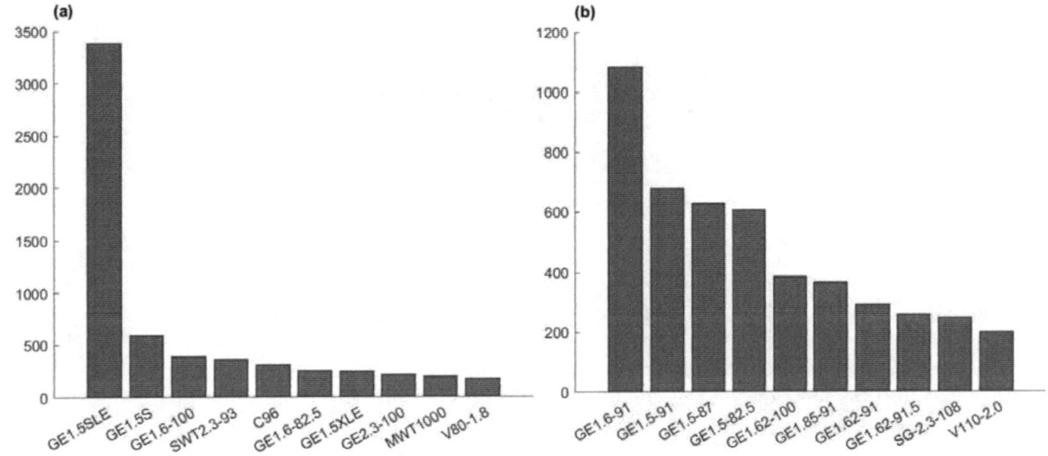

Figure 3: (a) the 2014 turbine models that were updated and (b) the 2022 turbine models that are the result of repowering

Figure 3(a) states the most frequently observed models that have already been repowered. Although the specific turbines identified by this process have been repowered recently, implying they will not be repowered again soon, the models found on this list can be understood to signal the present trend in what is being updated by most repowering projects. Figure 3(b) indicates which models farms are most often updating to, once old technology has been removed. The most commonly updated turbine model between 2014 and 2022 was the GE1.5SLE, this turbine is commonly fitted with 77.0-meter rotors (also known as the GE37 blade [20]). The most common result of the updates were various GE turbines in the 1.5-1.8 series fitted with rotors at least 82.5 meters in diameter. Around 3,500 GE1.5SLE, 500 GE1.5S, and 250 GE1.5XLE turbines were repowered between 2014 and 2022 and the 8 models representing the most likely outcome of repowering on 2022 were GE turbines and blades, so it seems as though these updates match up. Other turbines models that were updated include the Siemens Gamesa SWT2.3-93, the Clipper C96, and the Vestas V80-1.8.

The next cohort to examine are the turbines who persisted between 2014 and 2022 but were not updated – these turbines are due for update and will be at the forefront of repowering efforts in the coming decade. Figure 4 shows the top 10 turbines in the not-yet-repowered groups in a histogram. Comparing Figures 3 & 4, it can be observed that the most frequently repowered turbine and the most popular model due for future repowering is the GE1.5 turbine fitted with GE37 blades. These are locations with short blades, underproducing power compared to industry standards, that will have to modernize in the coming years. Although the GE37 is most often fitted to turbines rated for 1.5MW, there are a total of 6,612 instances of 77.0-meter rotors fitted to turbines in the 1.5-1.7 series. Since the Re-Wind network is a blade-centric research project rather than a turbine-centric one, it is of more interest to consider the blade model that will be entering post-processing streams than the turbine model they are coming from.

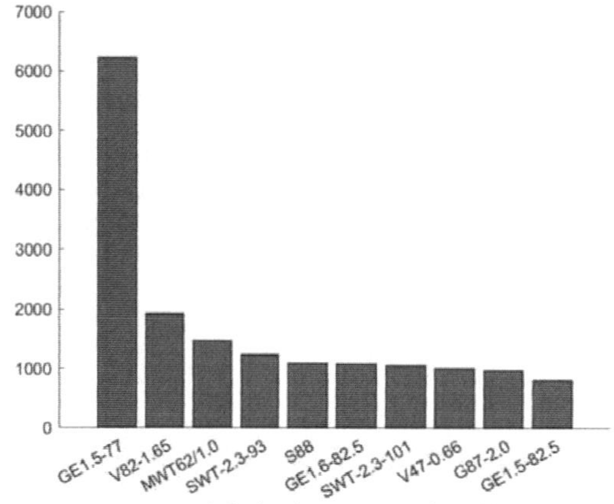

Figure 4: Top 10 most common models in the group of turbines due for future repowering

Notably on the list of to-be-repowered turbines are Vestas V82 and V47 turbines, Siemens Gamesa SWT2.3-93 and SWT2.3-101 turbines which use B45 and B49 blades, and GE 82.5-meter rotor diameters fitted to turbines of various ratings. Other assorted models come from smaller wind turbine manufacturers and the problems associated with repowering these models is unlikely to proliferate. It is pertinent to consider the spatial distribution of these models in the US so as to understand which regions should be expecting large inflows of blade waste, Figure 5 displays the geographic locations of each of the top models from Figure 4.

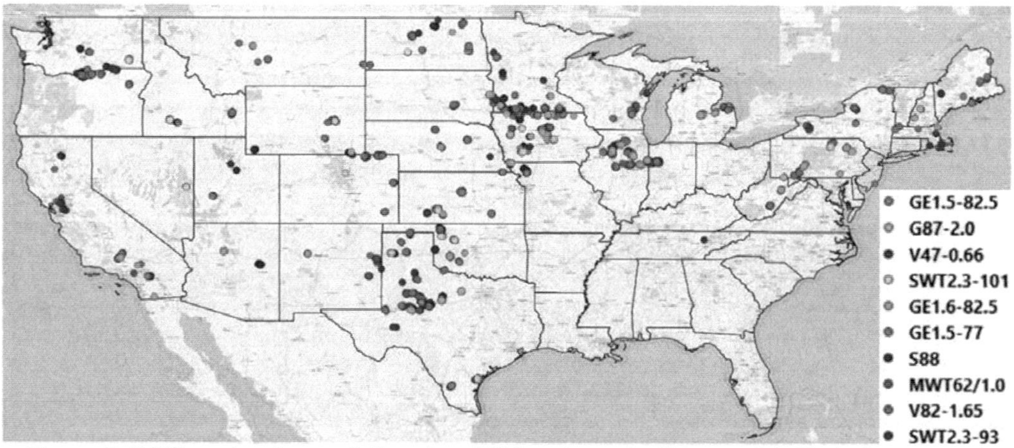

Figure 5: QGIS rendered map of all onshore US locations of turbine models likely for repowering

3. Discussion

In the interest of establishing economic circularity of wind blade materials, it is crucial to understand the features and constituents of future repowered and decommissioned blade groups. Making predictions on the type and total mass of blades coming out of service helps stakeholders plan for uses and processing procedures for blades, tailored to the specifics of the structure and material composites. In the above analysis, blade weight estimates from US turbines that have been completely dismantled, repowered, or installed after 2014 were determined, as well as a cohort of turbines with blades that are expected to be repowered in the near future. It is estimated that there are roughly 163,010 to 244,515 tonnes of blade waste currently in landfills or having been sent to EoL processing. This number comes from combining the weight of blades from dismantled turbines and the weight of blades identified as having already been repowered (where the weight is taken from 2014 data). In the next decade or so an additional 491,300 to 736,950 tonnes of blade material can be expected to enter waste streams from the cohort of turbines that are older and are due for repowering.

Significant turbine and blade models ideal for repowering in the near future have been identified. Most notably are GE1.5 MW turbines associated with the GE37 blade model, which represent 9.23% of current online blades and roughly 99,350-149,025 tonnes of material. In addition to the GE37 are Vestas V47 and V82 blades, Siemens Gamesa B45 and B49s, Mitsubishi MWT62/1.0 turbines, and other GE blades corresponding to 82.5-meter rotors (among other turbines and blades). These models are recognized as key groups to be removed and replaced from operating turbines due to their collective features of underperformance, mechanical issues, or outdatedness. Future work should investigate the rates at which certain blade models are replaced over time and should analyse additional years of data to make these predictions. For more details see [21].

Acknowledgments
Support for this research was provided by the U.S. National Science Foundation (NSF) under Grant 2016409.

References

[1] Green Energy Choices: The Benefits, Risks and Trade-Offs of Low-Carbon Technologies for Electricity Production, 2023, *UN Environmental Programme*

[2] Jensen J P and Skelton K, 2018, Wind turbine blade recycling: Experiences, challenges and possibilities in a circular economy, *Renew. and Sust. Energy Rev., Elsevier,* **97**, pp. 165-176, https://doi.org/10.1016/j.rser.2018.08.041

[3] Cooperman A, Eberle A, and Lantz E, 2021, Wind turbine blade material in the United States: Quantities, costs, and end-of-life options, *Resources, Conservation, and Recycling,* **168**, https://doi.org/10.1016/j.resconrec.2021.105439

[4] Lui P, Meng F, and Barlow C Y, 2022, Wind turbine blade end-of-life options: An economic comparison, *Resources, Conservation, and Recycling*, **180**, https://doi.org/10.1016/j.resconrec.2022.106202

[5] Meng F, Olivetti E A, Zhao Y, Chang J C, and Pickering S J, 2018, Comparing Life Cycle Energy and Global Warming Potential of Carbon Fiber Composite Recycling Technologies and Waste Management Options, *ACS Sust. Chemistry & Eng.,* **8**, pp. 9854–9865, https://doi.org/10.1021/acssuschemeng.8b01026

[6] Ghosh T, Hanes R, Key A, Walzberg J, and Eberle A, 2022, The Circular Economy Life Cycle Assessment and Visualization Framework: A Multistate Case Study of Wind Blade Circularity in United States, *Resources, Conservation, and Recycling*, **185**, https://doi.org/10.1016/j.resconrec.2022.106531

[7] Vestas Unveils Circularity Solution to End Landfill for Turbine Blades, 2023, *Vestas Global Leader in Sustainable Energy,*

[8] Ruane K, Soutsos M, Huynh A, Zhang Z, Nagle A, McDonald K, Gentry T R, Leahy P, Bank L C, 2023, Construction and Cost Analysis of BladeBridges Made from Decommissioned FRP Wind Turbine Blades. *Sustainability* **15**, https://doi.org/10.3390/su15043366.

[9] Alshannaq A A, Bank L C, Scott D W, and Gentry T R, 2021, Structural Analysis of a Wind Turbine Blade Repurposed as an Electrical Transmission Pole *J. Compos. Constr.*, **25**, https://doi.org/10.1061/(ASCE)CC.1943-5614.0001136

[10] Bank L C, Delaney E, McKinley J, Gentry R, and Leahy P, 2021, Defining the Landscape for Wind Blades at the End of Service Life, *CompositesWorld*, Accessed 16 May 2023, https://www.compositesworld.com/articles/defining-the-landscape-for-wind-blades-at-the-end-of-service-life

[11] Haces-Fernandes F, 2020, GoWInD: Wind Energy Spatiotemporal Assessment and

Characterization of End-of-Life Activities, *Energies,* 2020, 13(22), 6015, https://doi.org/10.3390/en13226015
[12] Lui P and Barlow C Y, 2017, Wind turbine blade waste in 2050, *Waste Management*, **62**, pp. 229-240, https://doi.org/10.1016/j.wasman.2017.02.007
[13] USWTDB Viewer, *Eerscmap.usgs.gov*, Accessed 15 April 2023, https://eerscmap.usgs.gov/uswtdb/viewer/#3/37.25/-96.25
[14] Renewables Data Explorer – Data Tools, *IEA*, Accessed 15 April 2023 https://www.iea.org/data-and-statistics/data-tools/renewables-data-explorer.
[15] Wind Power Facts and Statistics. *ACP*, Accessed 15 April 2023, https://cleanpower.org/facts/wind-power/#:~:text=Wind%20power%20capacity%20totals%20nearly,of%2043%20million%20American%20homes.
[16] In the First Half of 2022, 24% of U.S. Electricity Generation Came from Renewable Sources. *Homepage - U.S. Energy Information Administration (EIA)*, Accessed 15 April 2023 https://www.eia.gov/todayinenergy/detail.php?id=53779#:~:text=In%20June%202022%2C%20the%20United,June%202021%20and%20June%202022.
[17] Approximate Metric Equivalents for Degrees, 2019, Accessed 15 April 2023 https://www.usna.edu/Users/oceano/pguth/md_help/html/approx_equivalents.htm
[18] Albers H, Greiner S, Seifert H, and Kuehne U, 2009, Recycling of wind turbine rotor blades. Fact or fiction?, *DEWI-Magazin,* **34**
[19] Delaney E, 2022, Developing a GIS-based decision-making framework for the repurposing of decommissioned wind turbine blades, *School of Natural and Built Environment: Queen's University Belfast,* PhD Thesis
[20] Bauer L, 2023, *Wind Turbines Models*, Accessed 15 April 2023, https://en.wind-turbine-models.com/models
[21] Silverman A, Henao Y, Bank L C, and Gentry R, GIS analysis of wind blade decomissioning and repowering, 2023, http://dx.doi.org/10.13140/RG.2.2.17051.54562

IOP Conf. Series: Materials Science and Engineering 1293 (2023) 012031 doi:10.1088/1757-899X/1293/1/012031

Material aspects of wooden towers for offshore wind turbines

E. Kristofer Gamstedt[1], Rhodel Bengtsson[1], Sara Florisson[1], Hans Bernhoff[2]

[1] Uppsala University, Division of Applied Mechanics, Department of Materials Science and Engineering, Box 35, SE-751 03 Uppsala, Sweden

[2] Uppsala University, Division of Electricity, Department of Electrical Engineering, Box 65, 751 03 Uppsala, Sweden

E-mail: kristofer.gamstedt@angstrom.uu.se

Abstract. Possible new innovative materials for Counter Rotating Axis Floating Tilted Turbines are studied and discussed. The 40 MW version of the Counter Rotating Axis Floating Tilted Turbine (CRAFTT) will reach as far as 80 m below sea surface and up to 400 m above. The CRAFTT is an integrated design for floating offshore wind with two turbines on the same tilted shaft where the lower turbine is mounted directly on a rotating mast integrated with floater. The upper turbine will reach altitudes of 400 m. The system is designed to be a direct drive system, eliminating need for gearbox, taking advantage of the double air-gap speed of generator. With the generator placed at lower end as ballast the incentive to reduce weight for wings, tower and blades increase. Furthermore, wood is an attractive option as it enables both low CO_2 impact production and higher degree of reusability. However, fatigue properties from both mechanical and thermal cycling needs to be addressed in order to evaluate new structural materials in the context of floating wind turbines. Starting from scratch without any preconceived notions, one could consider timber as a potential option for the tower. In such a preliminary and qualitative deliberation, one can consider that the use of wood as the main load-carrying material in large structures has been proven during the last decade by the development of new high-rise wooden buildings, with even higher buildings with timber as the main structural component expected in the future, The tall wooden buildings have been made possible since wood has the advantage of having high specific mechanical properties, i.e. high strength and stiffness with respect to density in the grain direction, in addition to being renewable. Another advantage is that wood is less sensitive to fatigue than many metallic materials, since its hierarchical microstructure prevents the propagation dominant cracks when loaded in the longitudinal direction. Design against fatigue is crucial in wind turbine structures given the inevitable cyclic loading. As all materials, wood certainly has its drawbacks, the foremost being its sensitivity to moisture, which is of obvious concern in off-shore applications. Moisture has a softening effect, resulting in creep, and moisture may trigger chemical or microbial degradation. The development of barrier coatings of aluminium has shown to be very efficient in e.g. high-voltage cables and food packaging, making them impermeable to moisture and air. Such techniques should be applicable also in wood constructions. This presentation highlights the main points specific for wood as a construction material in the design of wood towers for wind turbines in offshore locations, which need to be addressed in design.

1. Introduction

Through the initiative of WorldWideWind [1], the concept of the Counter Rotating Axis Floating Tilted Turbine (CRAFTT) has been proposed. The design of generators [2] and the dynamics [3] have just recently been addressed. For a new concept, many additional engineering questions arise, for example: What would be a suitable material for the wind turbine tower? Previous long-term experiences in wood towers for vertical axis wind turbines are positive [4, 5], and shows that wood towers is a viable alternative to conventional steel towers on land. These two types are shown in Figure 1. For larger turbines offshore, the demands on long-term properties are more severe. Recent developments of large-scale wooden buildings show promise regarding upscaling to some degree. One example is the 17-floor building Mjøstårnet, where the load carrying frame is made of timber [6]. Another is the 100 m Hannover-Marienwerder wind turbine tower made of wood [7]. There are even prospects of building wooden skyscrapers in the future [8]. What makes offshore installations special compared with towers on land is that maintenance and repair is more difficult, and the inevitable presence of water and hence moisture. Therefore, a longer expected lifetime is desired to increase inspection intervals. The main concerns are then fatigue and creep over time. Secondly, moisture is a well-known challenge for all load-carrying wooden structures. Degradation in the presence of water can be addressed, as illustrated by the fact that wood is historically the go-to material to build ships. If properly cared for, microbial degradation at sea can be avoided, but the softening of the wood is an issue that should be carefully considered in design, especially for wooden offshore structures. The sorption of moisture inevitably leads to loss of stiffness and strength, increased creep deformations and decreased fatigue resistance [9, 10]. Since this effect may be significant, two alternatives present themselves. Either the moisture is kept out by an impermeable barrier, or the effects of moisture on the mechanical properties is accounted for the design process. Treating the bulk of the wood to make it hydrophobic is generally not an option, since the hydrophilic wood polymer (most notably hemicellulose) is omnipresent and abundant in the material, and any chemical impregnation would neither be practical nor environmentally justifiable.

(a)

(b)

Figure 1. (a) Conceptual illustration of CRAFTT [2], and (b) a vertical-axis wind turbine with a wooden tower [4].

In the following, we will address these concerns qualitatively, and suggest some approaches to approach them. First, fatigue is discussed, then the effect of moisture on deformation. Fatigue and failure of potentially weak links, such as joints of different geometries must be addressed specifically. Here, we focus on the material aspects of load carrying members and not on joints and geometrical features.

2. Fatigue in wood

The first design step is to prevent unwanted oscillations by dynamic FE analyses [4], preceded by validating analytic calculations if possible [3]. If eigenfrequencies coincide with the cyclic excitation, high loads and stresses will develop, which can lead to failure. Secondly, design against static failure is performed, where the maximum stresses should be well below strength values. In Eurocode for design of wood structures, the ultimate limit state concerns strength, including parameters for load variation, moisture and material variations with corresponding partial coefficients [11]. In contrast, the serviceability limit state concerns also stiffness and the resulting deformations. However, limit state design is prioritized, and design against deformations are generally based on numerical calculations and not by following validated design codes for wooden structures. Fatigue is not dealt with to the same extent in buildings, as the static loading dominates, and the wood is known to be relatively fatigue resistance, in contrast to many metallic materials, which can show individual cracks propagate until they attain a critical size leading to catastrophic failure. Wood is a fibrous aligned material, bearing resemblance to fibre composites, another fatigue resistant material, which is preferred in wind turbine rotor blades specifically because their resistance to fatigue, cf. e.g. [12]. The microstructures of aligned fibre reinforced plastics and wood alike tend to prevent or slow down damage accumulation and crack growth during cyclic loading. The wood grains (longitudinal direction) should then be aligned in the main loading direction, corresponding locally to the direction of the maximum principal stress.

Since wood is a porous anisotropic material, all material directions (longitudinal, transverse and radial) need to be considered, in addition compression or tensile loading [13]. Multiaxial fatigue analysis requires a large number of material parameters that are very costly to quantify. One can therefore start to consider uniaxial fatigue along the grain (longitudinal direction), provided the wood is adequately oriented in the structural members. An example of an S-N curve for birch in tension is shown in Figure 2(a). Fatigue damage is seldom observed in longitudinal loading. However, in off-axis loading, cracks develop at an earlier stage and can grow progressively until failure. Such a transverse fatigue crack growth is shown in Figure 2(b). S-N curves are readily implemented in most commercial FE softwares. It is advisable to identify a practical fatigue limit from S-N data, and use the corresponding stress value in a limit stress criterion.

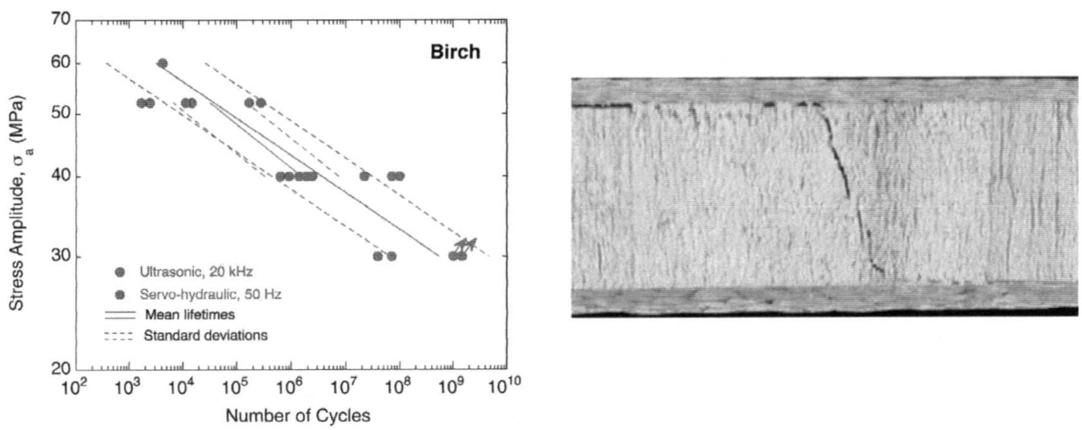

(a) (b)

Figure 2. (a) S-N curves of birch wood up to 10^9 cycles [13], and (b) fatigue damage in balsa wood predominantly along the grain direction (thickness 32 mm) [14].

As is the case for strength, stiffness and creep, also the fatigue behavior is susceptible to the moisture level, and furthermore to moisture variations (mechanosorptive effects). Thus, testing under service conditions is generally needed, since predictive schemes have not been validated. Just as for various types of composite materials, each new type of wood needs to be characterized separately. Although there are ways to predict wood stiffness and static strength as a function of density based on their porous microstructure [15], the generalization to fatigue strength has, to our knowledge, not been investigated.

Just like static strength, the presence of knots will most likely affect the fatigue strength, since the knots act as defects from which fatigue damage can initiate. The size effect in fatigue of wood is largely unexplored, and similar techniques as for size effects from strength grading could be adopted [16].

3. Effects of moisture

Although moisture is present to a very high degree in native wood, the mechanical performance of a living tree is excellent. If cracks develop due to severe loads in e.g. storms, living wood tissue can heal and grow reaction wood which makes the tree recover and continue to carry load. For dried timber, however, the presence of moisture normally reduces strength and stiffness and leads to viscoelastic deformations compared with those of fully dry wood. The effect is significant in the transverse directions (radial, tangential and in shear). This can be attributed to the breakage of hydrogen bonds in the presence of water molecules in hydrophilic biopolymers in the cell wall, most notably hemicellulose. The stress transfer between the load carrying cellulose microfibrils is then impaired.

For large offshore wood structures, the self-weight and the influence of moisture will be considerable. For long-term service, this means that, in addition to fatigue from load fluctuations, there will be significant static loads which most likely well lead to creep deformations. If these deformations become too large, the dynamic properties of the structure will change and stress redistrubtions can lead to failure. Creep is therefore an issue to consider in design, especially since creep is significantly accelerated in the presence of moisture in wood [17]. Furthermore, a particularity of wood is that a changing moisture content leads to accelerated creep, even beyond the rate for the highest moisture contents. This effect, known as mechanosorptive creep, is currently under significant investigation, since it has important engineering implications in a variable climate. To fully account for this effect in predictive models, the characterization of many material parameters is needed [18], and is yet to be practical in engineering design. For a constant climate, the approach by Bengtsson et al. [19] may be adopted. Creep data in the main material directions are then fitted to an appropriate number of parameters (see Figure 3).

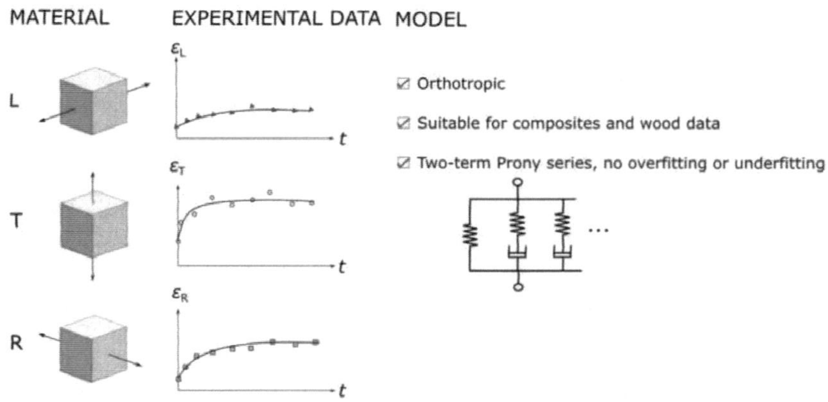

Figure 3. Schematic illustration of anisotropic creep model approach (constant moisture and temperature conditions) [19].

Although the load is predominantly unidirectional for slender structures, small misalignments and proximity to joints can lead to non-negligible off-axis stresses, which is why an orthotropic creep model is recommended, just as is the case for fibre composites. There is a striking lack of shear creep data for wood, despite its importance in overall dimensional stability of structures. A simplified method to obtain the shear creep has been proposed [20]. This would then make the orthotropic creep tensor complete, ready to be implemented in FE codes. It should also be noted that most published creep data is limited in time for practical reasons, and extrapolation for practical usage times is inevitable. It is the same situation for fatigue, although a time-temperature-moisture superposition can be envisaged to accelerate testing by increasing the surrounding temperature [21].

To mitigate creep in wood in moist environments, the obvious action is to dry out the wood and seal it with a barrier to prevent any further moisture absorption. This approach is for instance taken for wood fibre based insulators in power transformers, before sealing the system with a dielectric hydrophobic oil in a water-tight container. Another application area where large-scale sealing with a barrier coating has proven successful is aluminium coating in the extrusion process in manufacturing of high-voltage underwater cables. Such long cables which span the bottom of seas have an immense surface, and the coating rather effectively prevent the growth of water trees over time. It is therefore not far-fetched to believe that it would be possible to effectively coat dried wood with aluminium sheaths, preventing moisture ingress over prolonged times. A wooden core fully enclosed by aluminium sheaths forms a sandwich structure, not unlike those proven useful in aircraft wings with a balsa wood core. Aluminium has also high specific mechanical properties, which has made it a preferred material in components in structures subjected to fatigue, most notably in aeronautical and automotive applications. These technologies are yet to be explored to minimize moisture absorption over extended periods of time in wood members in offshore towers for wind turbines.

4. Concluding remarks

This presentation highlights some of main material aspects that needs to be addressed if wood should be used in towers for large offshore wind turbines, such as in the CRAFTT concept. A qualitative assessment of the key features of wood under long-term static and cyclic loading in a moist climate is thus presented. Since inspection and repair intervals will be longer offshore, fatigue is more of concern than for wooden wind turbine towers on land. The higher levels of moisture necessitate more effective ways to prevent moisture uptake in the wood, and models to predict creep deformations due to high self-weight of large towers. Since wood is a highly variable material, a significant amount of testing is needed to determine material parameters that can be used in design. Also, models for fatigue and creep need to be balanced, i.e. not too detailed to require an unpractical number of material parameters to be characterized, and not too simplified so that they fail to make predictions at a reasonable accuracy. It is hoped that this short review will provide some ideas of which research topics that could be prioritized in order to facilitate the use of wood in towers for offshore wind turbines. Since wood is renewable in contrast to steel, commonly used in towers, a more sustainable use of raw materials can be achieved for large future wind farms. Experiences from the ongoing rapid development of wood frames for high-rise buildings can also help pave way for the use of wood towers for wood turbines at sea.

References

[1] WordWideWind: https://worldwidewind.no/
[2] Simonsson I, Eriksson S and Bernhoff H 2023 Design and simulation of a 40 MW PM generator for the CRAFT concept *Proc. EERA DeepWind Conference* (Trondheim, Norway)
[3] Goude L and Hedlund Peters B 2023 Dynamics of a CRAFT: A simulation study on a counter rotating vertical axis foating tilting wind turbine. MSc thesis, Dept Electrical Engineering, Uppsala University
[4] Möllerström E, Ottermo F, Hylander J and Bernhoff H 2014 *Wind Energy* **28** 277
[5] Apelfröjd S, Eriksson S and Bernhoff H 2016 *Energies* **9** 570

IOP Conf. Series: Materials Science and Engineering 1293 (2023) 012031 doi:10.1088/1757-899X/1293/1/012031

[6] Abrahamsen R 2017 Mjøstårnet - Construction of an 81 m tall timber building. *Internationales Holzbau-Forum IHF.*

[7] TimberCompositeTechnology: https://ticomtec.de/referenzen/timbertower-sonderkonstruktion-deutschland/

[8] Niiler E 2022 Wooden Skyscrapers are on the rise. *Wall Street Journal* April 11

[9] Wang J, Cao X and Liu H 2021 *Eur J Wood Wood Products* **79** 245

[10] Gerhards CC 1982 *Wood Fiber Sci* **1** 4

[11] Eurocode 5: Design of timber structures 2004, EN 1995, EC 5

[12] Gamstedt EK and Andersen SI 2001 *Fatigue degradation and failure of rotating composite structures - Materials characterisation and underlying mechanisms* RISO-R-1261(EN)

[13] Clorius C, Pedersen M, Hoffmeyer P and Damkilde L 2000 *Wood Sci Technol* **34** 21

[14] Emami S 2017 PhD thesis, University of Dayton, School of Engineering.

[15] Gibson LJ, Ashby MF, Karam GN, Wegst U and Shercliff HR 1995 Proc Roy Soc London Ser A Math Phys Sci **450** 141

[16] Schneeweiß G and Felber S 2013 Am J Mater Sci **3** 41

[17] Kojima Y and Yamamoto H 2005 J Wood Sci **51** 462

[18] Dubois F, Husson JM, Sauvat N and Manfoumbi N 2012 Mech Time-Depend Mater **16** 439

[19] Bengtsson R, Afshar R and Gamstedt EK 2022 *Wood Sci Technol* **56** 1585

[20] Bengtsson R, Bergeron L, Afshar R, Mousavi M and Gamstedt EK 2023 Mech Time Depend Mater, available online: https://doi.org/10.1007/s11043-023-09604-0

[21] Peng H, Zhan T, Jiang J, Zhang Y, Cao J and Lu, J. 2022 Wood Mater Sci Eng **17** 911

Mode mixity for the fracture toughness obtained by climbing drum peel tests

K M Jespersen[1*] and H L Toftegaard[1]

[1]Department of Wind and Energy Systems, Technical University of Denmark, Frederiksborgvej 399, 4000 Roskilde, Denmark

E-mail: kmun@dtu.dk*

Abstract. Wind turbine blades include numerous interfaces, which are known to be critical regions for early failure. Thus, reliable testing of interfaces and a solid understanding of the tests used are of utmost importance. The climbing drum peel (CDP) test is a standardized test for characterization of adhesively bonded interfaces, which in literature has been referred to as a possible simple alternative to mode I double cantilever beam testing of composites. The current work uses finite element modeling of the CDP setup to show that it is actually a mixed mode test, and therefore one cannot use it to obtain the mode I fracture toughness. It can, however, be used to compare different interfaces, keeping in mind that the comparison is done for a mode mixity of approximately 40°, and that one adherend has to be thin (around 0.5mm).

1. Introduction

Wind turbine blades contain numerous interfaces and are one of the most critical locations for failure [1]. Thus, it is of great importance to be able to test and model the interface strength (fracture toughness) to be used in the blade design. The properties of an interface are often defined by the fracture toughness calculated as the energy release rate. An interface with a crack considered in two dimensions (2D) can be loaded in opening (mode I), shearing (mode II), or a mixture of the two modes (mixed mode) at the crack tip. An interface is often weakest in mode I loading (e.g. [2]). Therefore, a mixed mode loading condition at the crack tip can lead to a higher measured fracture toughness than if the same interface was tested in mode I. If the fracture toughness experimentally measured in mixed mode loading conditions is used in place of the mode I fracture toughness in an analytical or numerical model, the results might thus be non-conservative and could have catastrophic consequences.

A common test approach for experimentally measuring the mode I fracture toughness of an interface is the double cantilever beam (DCB) test (ASTM D5528 [3]). In the DCB test it is either required to measure the crack length, which is not an easy task, or do a compliance calibration, which requires additional initial test steps [3]. In addition, the crack front in a DCB test is closer to parabolically shaped than straight, even though a straight crack is assumed in the calculations of the fracture toughness [4, 5]. An alternative to the standard DCB test is the DCB test with uneven bending moments (DCB-UBM) [6]. Using bending moments instead of forces eliminates the necessity to measure the crack length, and the DCB-UBM test can be used to test for any mode mixity by changing the ratio between the bending moments applied to the two adherends. However, this test requires a special-made fixture, and as for the DCB test, the crack front is not straight. Thus, in some cases, it could be useful with a simpler test than the DCB and DCB-UBM tests to obtain the fracture toughness experimentally.

The climbing drum peel (CDP) test is a simple standard test (ASTM D1781 [7]) used in the sandwich community to measure the interface strength of an adhesive bond between the skin and core of a sandwich structure [7]. Daghia and Cluzel [4] showed some experimental indications of the CDP test being more prone to have a straight crack front than the DCB test. They argued that it was due to the kinematics of the CDP test. In the standard for the CDP test, the peel torque is used to define the interface strength. The peel torque is geometry dependent and not a real material parameter. Thus, it can only be used to compare different cases with similar geometry and not as an input to a model or similar. Literature has shown that the energy release rate (fracture toughness) can be calculated by simple energy considerations for the CDP test [4, 8, 9]. This idea was presented for sandwich structures [8, 9] and later laminates [4] as an alternative to the DCB test for the mode I fracture toughness, which is interesting due to the simplicity of the CDP test.

Generally, the fracture toughness obtained from the CDP test is presented as the mode I fracture toughness G_{Ic} [4, 8–10]. However, Thouless and Jensen [11] have also shown that the mode mixity in a regular peel test varies for different peel angles and such a peel test is thus not actually a mode I test. Although a simple peel test is slightly different from the CDP test, the load is essentially applied in a similar manner, where the peel strip is pulled at an angle to the adherend. The mode mixity is also to some extent discussed in some of the later work by Daghia et al [12] on the double drum peel, but to the author's knowledge not for the CDP test. The finite element (FE) modelling in the current study will show that the fracture toughness obtained by the CDP test is actually mixed mode. Thus, it is important to properly understand the mode mixity in the CDP test before using it as asimple alternative or supplement to the DCB test.

In the current paper, a 2D FE model including all the components (steel bands, drum, flange, and test specimen) in between the actuator and the load cell in the test machine is presented. The FE model is used to estimate the fracture toughness using the J-integral obtained by a contour integral approach. The corresponding mode mixity is obtained from the stress intensity factors extracted from the J-integral by the interaction integral method. The results are compared to analytical calculations and experiments.

2. Materials and specimens

Figure 1 shows an illustration of the specimen geometry used for the CDP experiments and FE analyses in the current study. The peel strip and the adherend were both made from the same unidirectional non-crimp fabric based glass fiber composite with a small amount of backing, where the peel strip is one ply of the adherend. The glass fiber composite was modeled as an isotropic material with a Young's Modulus of $E_p = E_a = 40$ GPa and Poisson's ratio of $\nu_p = \nu_a = 0.3$. A pre-crack was made in the specimens at the end that was attached to the drum, as also illustrated in Figure 1. The steel used to model the steel bands, flange, and drum (see Figure 2) was modeled with a Young's modulus of $E_s = 210$ GPa and a Poisson's ratio of $\nu_s = 0.3$.

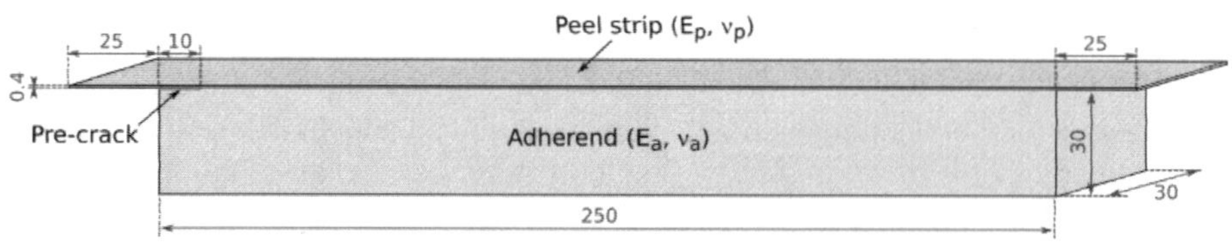

Figure 1: Specimen geometry (all numbers are in millimeters)

3. Climbing drum peel test experiments

The CDP tests were carried out according to the ASTM D1781 standard [7]. Figure 2 shows the experimental setup for the CDP test of the glass fiber test specimens considered in this study. The peel strip at the uncracked end of the specimen was attached to the upper grip of the test machine where the load cell was also placed. The peel strip at the pre-cracked end of the specimen was then inserted into the drum with flanges and steel bands. The steel bands were attached to the actuator of the test machine with a lever-type bar, as indicated in Figure 2. The tests were carried out in a universal testing machine (Instron 88R1362) with a 5kN load cell. The lower grip of the actuator holding the lever bar with the steel bands was pulled in displacement control with a speed of 25 mm/min. The load in the load cell was measured throughout the test. The diameter of the drum, flange, and the steel band thickness were 101.6 mm, 126.89 mm, and 0.11 mm, respectively. The average specimen weight was 388 g, and the drum and flange had a combined weight of 3332 g.

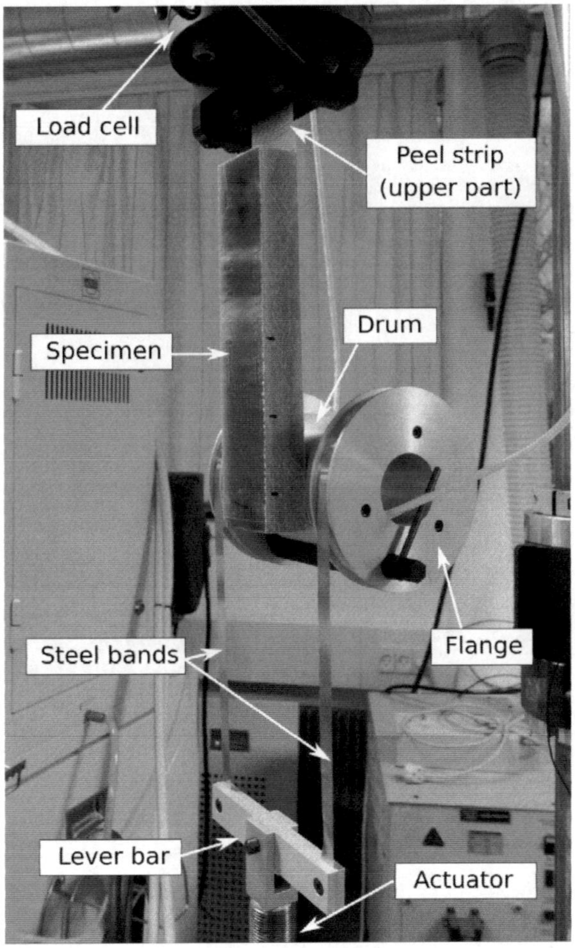

Figure 2: Experimental climbing drum peel setup

Figure 3 shows a representative curve for the load-displacement curves measured experimentally. It is seen that the curve starts from an initial value (a) which approximately corresponds to the weight of the specimen, drum, flanges, and steel bands. In the first part of the test (b) the steel bands are gradually tightened until they begin to pull and thus rotate the drum.

When the drum rotates, the peel strip is gradually bent upwards to eventually wrap around the drum. This is reflected through a plateau with a small overshoot (c) in load, observed in all the experiments, followed by a stable plateau load (d). The plateau is commonly referred to as the "winding stage" in literature, since this is where the drum is "winding" up the peel strip prior to any crack propagation. After this stage, the force increases steadily through (e) until delamination initiates (f). The second load plateau (f) is where the delamination progresses in a steady state manner [4, 9].

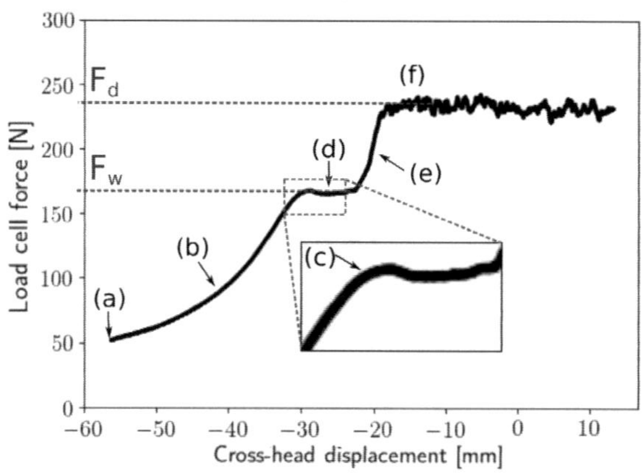

Figure 3: Representative experimental load-displacement curve

3.1. Analytical estimation of fracture toughness from experiments

The ASTM D1781 standard [7] approach was made for testing the adhesion at the bonding interface between the skin and core of a sandwich plate. In the standard, the average peel torque, T, is estimated as the final output of the test by Eq. 1.

$$T = \frac{(F_d - F_w)(r_2 - r_1)}{w} \quad (1)$$

where F_d and F_w are the forces measured by the load cell during winding (plateau (d) in Figure 3) and debonding (plateau (f) in Figure 3), respectively, and w is the specimen width. The radii r_1 and r_2 are related to the drum and flange diameters r_{d1} and r_{d2} and the thicknesses of the peel strip and steel bands h_p and h_s as follows:

$$\begin{aligned} r_1 &= r_{d1} + \frac{h_p}{2} \\ r_2 &= r_{d2} + \frac{h_s}{2} \end{aligned} \quad (2)$$

However, ashighlighted in literature [4, 9], the peel torque is not a real material parameter, and can only be used for comparing different bonding solutions. Daghia and Cluzel [4] suggested using the test for estimating the fracture toughness of monolithic laminates, which can be calculated as in Eq. 3 (see [4] for details).

$$G_c = \frac{\Delta E_{debond}}{\Delta A} = \frac{(F_d - F_w)}{w} \frac{(r_2 - r_1)}{r_1} \quad (3)$$

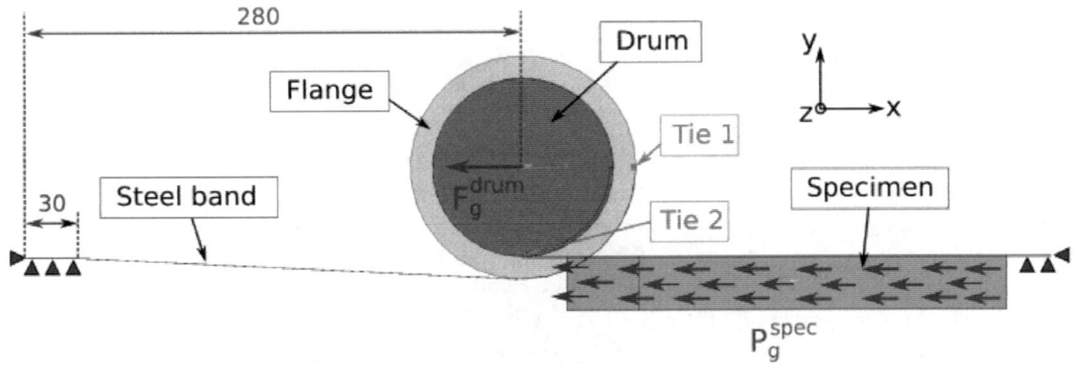

(a) Gravity step

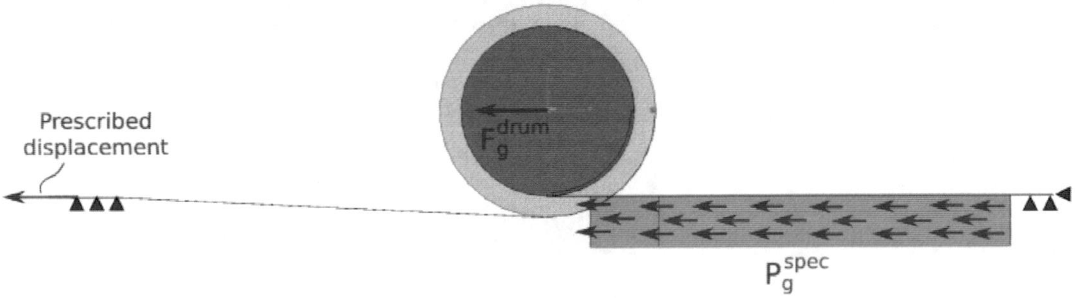

(b) Load step

Figure 4: Model load steps, boundary conditions, and dimensions

In Eq. 3, ΔE_{debond} is the change in dissipated energy during debonding and ΔA is the corresponding debonded area. Note that Daghia and Cluzel originally stated the fracture toughness as G_{Ic}, but the current work will show that it is not pure mode I, and therefore it is instead stated as the total mixed mode fracture toughness G_c. This value is calculated for the experiments and compared to the corresponding predictions of the FE model.

4. Finite element model

The current section outlines the FE model developed. The Abaqus model files can be downloaded from [13].

4.1. Model geometry

The developed 2D FE model includes all the parts in between the actuator and load cell of the test machine, i.e. test specimen, drum, flange, and steel bands as illustrated in Figure 4 (a). The steel band, flange, drum, and specimen are modeled as individual instances only connected through local tie constraints and contact definitions, as detailed below. Note that since it is modeled in 2D, the flange overlaps with the test specimen. However, it has no influence on the results since contact is not modeled between these parts. Modeling it this way makes it possible to avoid going into 3D while still realistically modeling the test setup.

4.2. Loads and constraints

The loads and constraints are applied in two analysis steps as shown in Figure 4. The gravitational loads on the drum and specimen were applied in an initial load step (Figure 4 (a)) to ensure that the full gravitational load was applied before applying the load to the steel bands since loads are gradually applied over an analysis step in Abaqus. The weight of the drum is applied to its center (F_g^{drum}) and the specimen weight is applied as an evenly distributed load across the whole specimen (P_g^{spec}). Both of the loads due to gravity are applied acting in the negative x-direction and calculated from the weight using a gravitational constant of $9.82\,\mathrm{m/s^2}$, which gives $F_g^{drum} = 32.72N$, $F_g^{spec} = 3.810N$, and $P_g^{spec} = F_g^{spec}/V_{spec} = 1.667 \cdot 10^{-5} \frac{N}{mm^3}$.

In the gravity step (Figure 4 (a)) the steel band was constrained in the x-direction to avoid excessive movement of the part, which led to the model not converging. A prescribed displacement was applied to the steel bands in a second load step as shown in Figure 4 (b). In the load step (b), a displacement is applied to the steel band instead of the constraint in the x-direction, as also seen in Figure 4. For both load steps, the specimen strip was gripped on the right-hand side and the steel band was constricted in the y-direction.

The end of the steel band was fixed to the flange using a tie constraint (tie 1), and the end of the peel strip of the specimen was fixed to the drum with a tie constraint (tie 2), as indicated in Figure 4 (a). Two hard contacts were modeled between the contacting regions of the steel band and flange and the drum and specimen. The contact between the steel band and flange allowed for separation when the load is applied and the drum is rotated. The rotations of the drum and flange were locked together using a kinematic constraint applied to a reference point in their center.

4.3. Mesh

The mesh used in the FE model in the regions of interest is shown in Figure 5. The mesh was refined in the contact regions to obtain stable convergence of the solution and avoid overlapping of the interfaces. The crack tip mesh was refined in a circular region around the crack tip with the mesh becoming increasingly finer towards the crack tip. The model had a total of 85658 linear plane stress elements (type CPS4R, CPS3) and it took roughly 5 hours to run the analysis on 16 cores of 1 CPU on a computational cluster. The analysis time could be shortened by increasing the maximum time step increment, which was set to 0.001s to obtain high resolution in the output results. However, for the purpose of this work, high resolution in the output data was desired and the analysis time was not an issue. Therefore, only limited effort was put into reducing the analysis time of the model.

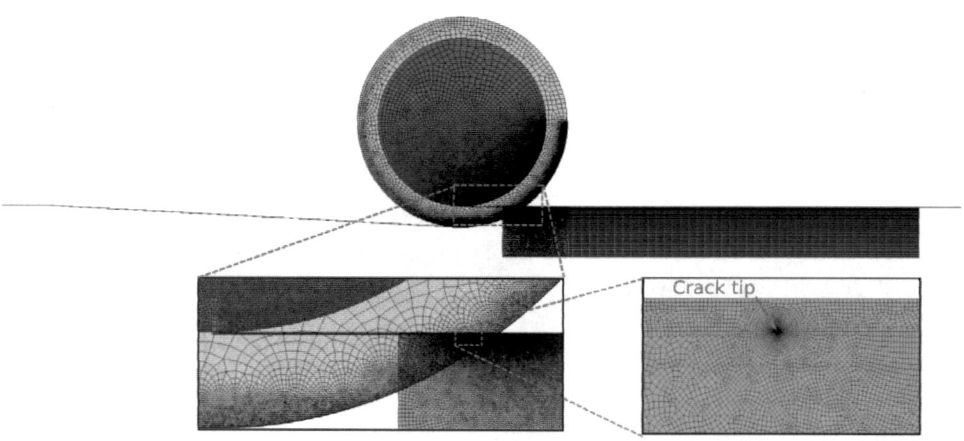

Figure 5: Mesh used in the finite element model

IOP Conf. Series: Materials Science and Engineering 1293 (2023) 012032 doi:10.1088/1757-899X/1293/1/012032

4.4. Crack

The J-integral was calculated in Abaqus using a contour integral. The J-integral is evaluated along a chosen number of contours around the crack tip, following the local circular mesh. The value of the J-integral extracted from each contour converges a distance away from the crack tip. Therefore, the values at the 16th (outhermost) contour, where a stable value was achieved, was used. The interaction integral approach was used to obtain the mode I and II stress intensity factors (K_I and K_{II}) to estimate the mode mixity. The mode mixity is in the current study calculated by Eq. 4. Note that it is required to consider a length scale for the mode mixity if the two interfaces are not the same material (bi-material) [14]. However, in the current study the material of the peel strip and adherend is similar and thus the expression in Eq. 4 is sufficient.

$$\phi = \tan^{-1}\left(\frac{K_{II}}{K_I}\right) \tag{4}$$

In the experimental setup the upper specimen end is pinned (allowing for rotations), which means that the specimen rotates a few degrees during the test. Apparently, this rotation affects the values of K_I and K_{II} predicted by the FE model due to the implementation in Abaqus. When calculating the mode mixity for the CDP setup with a pinned upper end based on K_I and K_{II} from Abaqus, the value differs significantly from the corresponding mode mixity calculated analytically based on Suo & Hutchinson [14]. However, if the upper specimen end was gripped, allowing for no rotation, the mode mixity based on K_I and K_{II} from Abaqus matched the analytical solution well. As we do not see a reason for why the FE results should not match the analytical solution, the above indicated an issue with the values obtained from Abaqus for the pinned specimen end boundary condition. To investigate this interpretation, an alternative stress based measure for the mode mixity calculated from the stresses σ_{12} and σ_{22} [15] in front of the crack tip was plotted as in Figure 6. The plot is not meant as an alternative way to obtain the mode mixity, but simply a means to look into the above issue. Based on Figure 6, it is seen that for the stress based mode mixity there is no difference for the two different boundary conditions. Therefore, a difference is not expected for the mode mixity based on K_I and K_{II}. Thus, the rotation of the specimen in the model apparently causes an issue in the way K_I and K_{II} are calculated in Abaqus. Since the issue is caused by the rotation, and the two cases should be equivalent regarding mode mixity, the results for the clamped specimen end are used. It should be noted that the predicted fracture toughness (J-integral) is unaffected by the change in boundary conditions.

As an additional check, the bending moment and normal forces were calculated from σ_{11} extracted in the peel strip a distance of 0.5 mm away from the crack tip in the FE model, and the mode mixity was estimated based on Suo and Hutchinson [14] for verification. The stresses in the peel strip were extracted in a coordinate system in the tangent direction of the deformed strip. At the considered location σ_{22} and σ_{12} from the FE model were close to zero and also assumed to be zero in the analytical mode mixity calculations.

4.5. Parameter dependency

The main geometric parameter that can vary from test to test is the length of the steel bands. To check for dependency on the results, analyses were run for an initial steel band length of 155, 280, and 530mm. This had no significant effect on the results. Another parameter that changes during the test is the crack length. The initial crack length was the same for all the test specimens but the crack length will change during the test as delamination progresses, and should still give a similar value for the fracture toughness. To look further into this, the initial crack lengths of 10, 15, and 20mm were modeled and compared.

The current study focuses on a unidirectional glass fiber composite. However, for completeness, the effect of material stiffness on the mode mixity was checked by running the

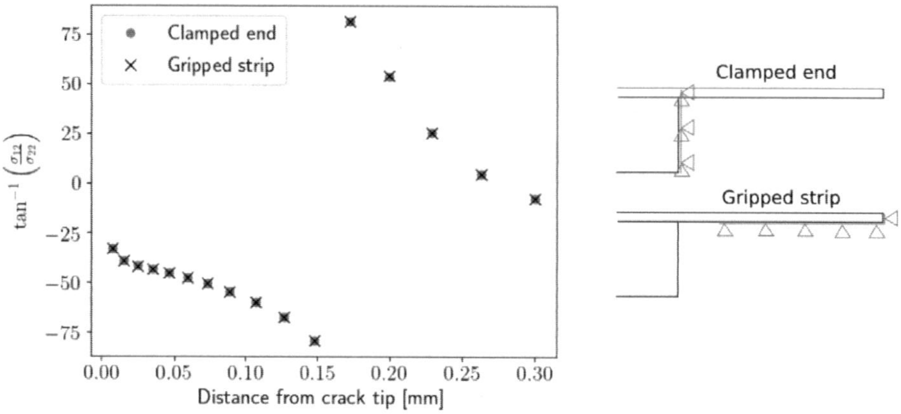

Figure 6: Ratio of σ_{12} over σ_{22} compared for fixed and pinned boundary conditions in the specimen end gripped by the test machine

analysis for a unidirectional carbon fiber composite material with a stiffness of 153GPa [16].

5. Results

In the following, the FE predictions of the displacement of the overall test setup, the fracture toughness, and the mode mixity are presented. The force-displacement curve predicted by the FE model is compared to experiments. In addition, the fracture toughness obtained by the J-integral in the FE model is compared to the fracture toughness predicted analytically using the experimental data. Finally, the effect of the initial crack length is discussed.

5.1. Displacement of CDP test setup

The force measured by the load cell (load cell force) is plotted as a function of the cross-head displacement in Figure 7 (a) comparing one representative experiment (solid line) and the FE model (dashed line). The load cell force for the FE model is the x-direction reaction force in the right-hand side grip of the specimen in Figure 4. Furthermore, the displacement obtained from the FE model has been offset to fit the experimental cross-head displacement. They are initially offset since it is necessary to tighten the steel bands in the experimental setup, whereas the steel bands are already tightened from the start in the FE model. This also explains why the curve for the FE model starts at the winding stage. It is seen that delamination experimentally occurs around a load cell force of $F_{lc} = 230$ N. Figure 7 (b) shows a close-up of the experiment during delamination where a small gap is visible near the crack tip during propagation.

Figure 8 shows the deformation of the test specimen near the crack tip at four different load levels. The small insert plot indicates where on the FE force-displacement curve the image is from. As seen in Figure 8 (a) the specimen is slightly tilted (less than 1°) as a result of the applied gravitational loads before any load is applied to the steel bands. Applying further load the peel strip starts to wind around the drum and at the end of the winding stage is nicely wrapped around the drum as seen in Figure 8 (b). Loading up to the experimental load of delamination to initiate, the peel strip separates partly from the drum close to the crack tip as seen in Figure 8 (c). This is in agreement with observations from the literature [17] and was also observed in the experiments of the current study (Figure 7 (b)). However, it is difficult to measure the exact curvature in the experiments due to the presence of the flange. At the stage in (c) the tilt of the specimen is roughly 2°, which is also in line with experimental observations where the measured angle was somewhere in between 2° and 2.5° as observed by the authors for

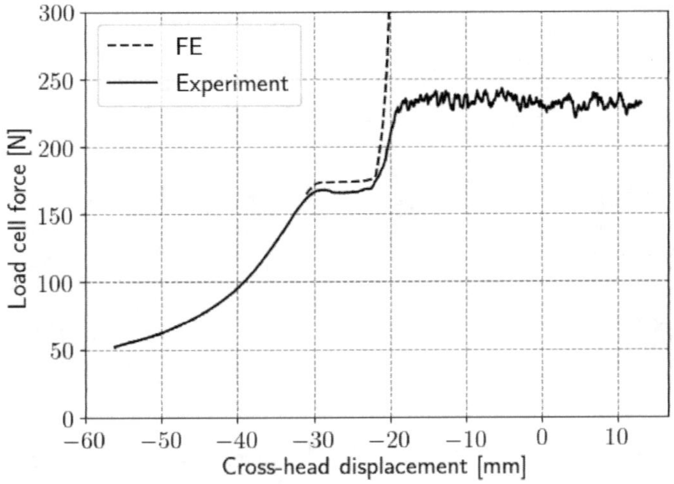

(a) Experimental and FE load-displacement curves

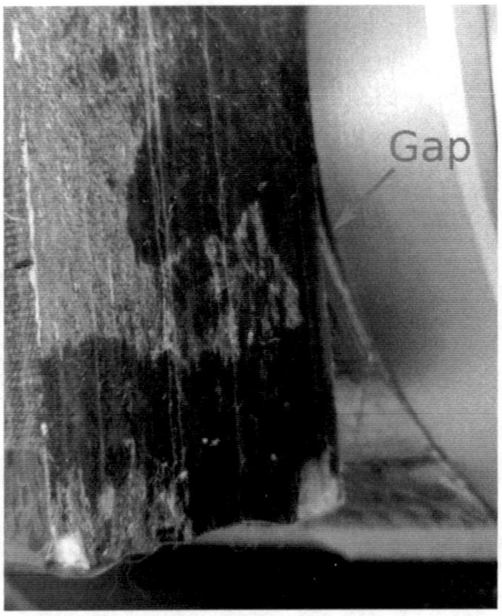

(b) Experiment during delamination

Figure 7: (a) Experimental and FE load-displacement curves compared and (b) close-up photo of the experiment during delamination

the setup in Figure 2. Figure 8 (d) shows how increasing the load further increases the space without contact between the drum and the peel strip. Thus, it is expected that the extent of non-conformance of the peel strip to the drum shape increases with increasing interface strength. However, as will be shown in the following results, the fracture toughness predicted by the FE and analytical models match almost perfectly even though the peel strip does not conform perfectly to the drum shape during propagation.

5.2. Fracture toughness and mode mixity comparisons

Figure 9 shows the fracture toughness as a function of the load cell force predicted by the FE model using the J-integral and the analytical approach estimating the critical energy release rate G_c from Eq. 3. The curve with circle markers in Figure 9 shows the analytically predicted fracture toughness using the winding force obtained from the FE model (F_w^{FE} = 173.4N). A very good agreement is obtained between the fracture toughness predicted by the J-integral in the FE model and the corresponding analytical prediction. This confirms the applicability of the simple analytical approach to estimate the fracture toughness in the CDP test. A couple of papers [4, 17] suggest that if the peel strip shape does not conform to the shape of the drum, the analytical expression for the fracture toughness may not apply. However, our analysis shows that for the case considered here, one can still use the analytical expression even if the peel strip does not conform to the drum.

The winding force is slightly higher for the FE prediction than the experimentally obtained value (see Figure 7 (a)), which affects the predicted fracture toughness. The curve with triangle markers in Figure 9 shows the analytical prediction using the experimentally measured winding force (F_w^{exp} = 165.3N). It is seen that a difference in the winding force of 8.1N between the FE model and experiment leads to a constant difference of 66.2 $\frac{J}{m^2}$ in the fracture toughness. At a load cell force of 230N (where experimental delamination occurs), this gives a difference of 14.4% between the FE-predicted (463.2 $\frac{J}{m^2}$) and the experimentally measured (533.1 $\frac{J}{m^2}$) fracture toughness obtained from Eq. 3. The difference in winding force could indicate that some additional effects like permanent deformation, cracking, and similar effects occur in the

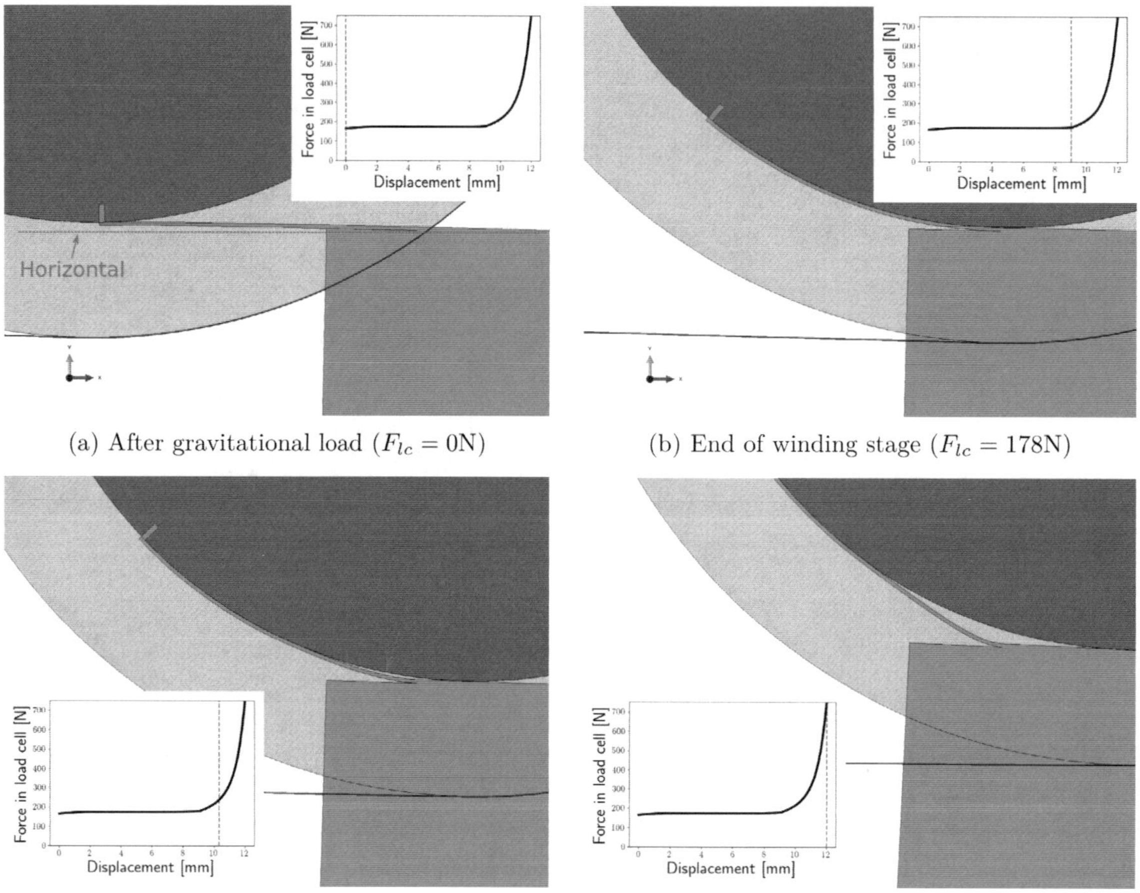

(a) After gravitational load ($F_{lc} = 0$N)

(b) End of winding stage ($F_{lc} = 178$N)

(c) At experimental delamination ($F_{lc} = 230$N)

(d) Increasing load further ($F_{lc} = 750$N)

Figure 8: Deformation of peel strip in FE model at (a) no load but after gravitation step, (b) just before the end of the winding stage, (c) at the approximate load required for experimental delamination to initiate, and (d) at a higher applied load for demonstration. Videos of the deformation can be downloaded from [13]

experiment, but are not taken into account in the FE model. This is not a problem for the analytical approach when estimating the fracture toughness based on the experiments, since these effects are "included" in the experimentally measured winding force and subtracted when the fracture toughness is estimated by Eq. 3. Nevertheless, the FE model is expected to give applicable results for the mode mixity in the CDP test.

The J-integral and mode mixity predicted by the FE model are shown in Figure 10 as a function of the displacement (a) and load cell force (b). The dashed grey line shows the point at which experimental delamination occurred corresponding to $F_{ls} = 230$N. It is seen from Figure 10 that the J-integral increases linearly with the force whereas it looks more parabolically increasing with the displacement. Another thing that can be seen from Figure 10 (a) is that the mode mixity initially (during the winding stage) experiences a few jumps whereafter it slowly stabilizes. Looking at the same mode mixity and J-integral plotted as a function of the load cell force in Figure 10 (b), it is clear that the mode mixity quickly stablizes after the J-integral begins to increase. At the time of experimental delamination, the mode mixity is $-38.1°$. In the load range of 200-300N the mode mixity varies from $-39.7°$ and $-36.9°$. Analytical calculations

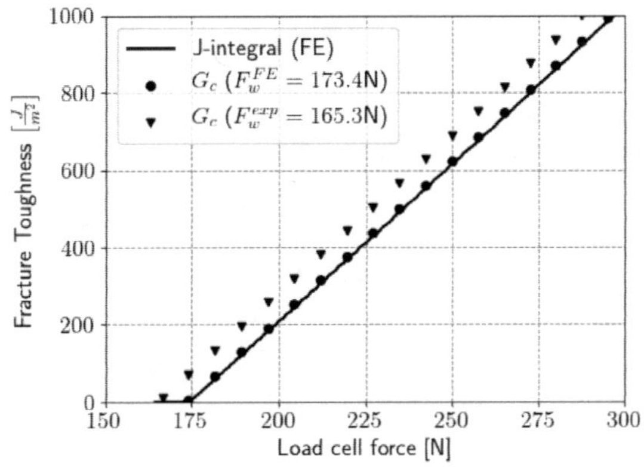

Figure 9: Comparison between the fracture energy calculated by the FE model and analytically

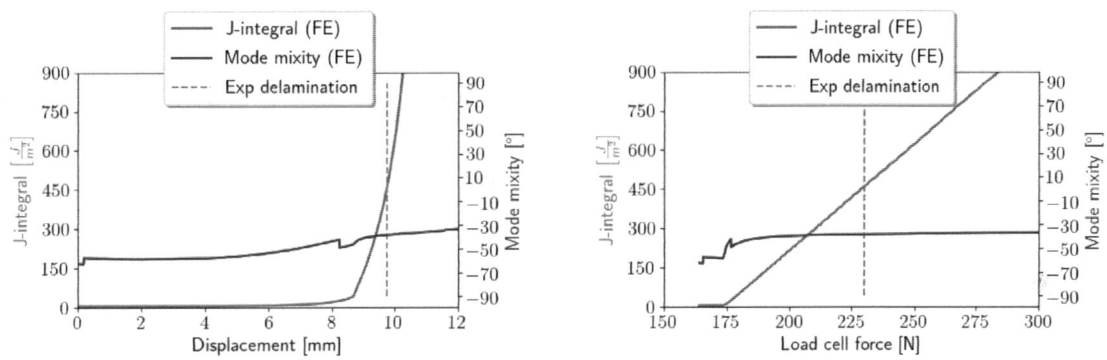

(a) Displacement of steel bands

(b) Force in load cell

Figure 10: J-integral and mode mixity as function of the displacement (a) and force (b), as predicted by the FE model

using the approach of Suo and Hutchinson [14] with the normal force (N=279.6N) and bending moment (M=-390.1N·mm) in the peel strip extracted from the FE model 0.5mm away from the crack tip, and the remaining calculated from force and moment balance gives a mode mixity of -39°. These results are also in a similar range as those obtained by Daghia et al [12] for the double drum peel test and Thouless and Jensen [11] for the regular peel test.

If changing the material to a unidirectional carbon fiber composite with a significantly higher stiffness than the glass fiber composite, a mode mixity varying between $-40.8°$ and $-37.6°$ is obtained for a load cell force of 200N and 300N. Thus, the effect of material stiffness on the mode mixity (for a similar peel strip and adherend material) is limited when considering the same geometry.

5.3. *Effect of crack length*

Figure 11 shows (a) the J-integral and (b) the mode mixity as a function of the load cell force predicted by the FE model. They are seen to be equivalent. Thus, the initial crack length has no effect on the results of neither the fracture energy nor the mode mixty, as also assumed in the

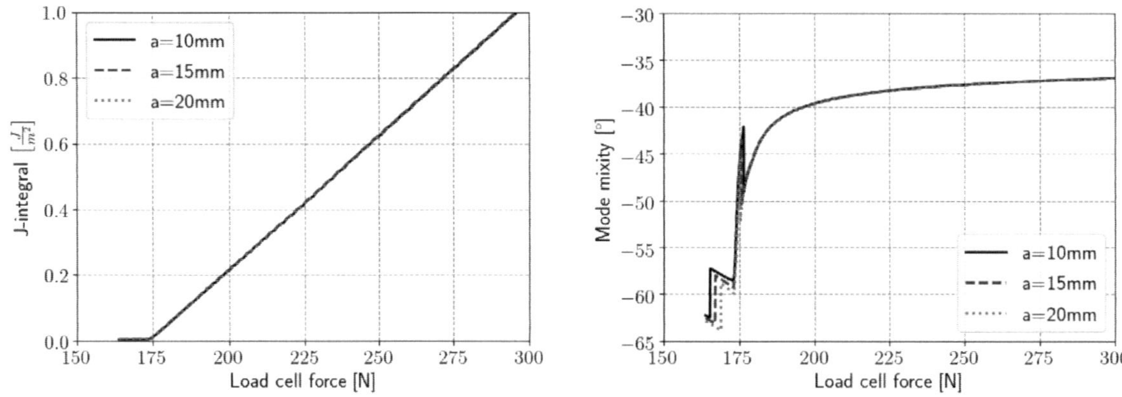

(a) J-integral (b) Mode mixity

Figure 11: J-integral (a) and mode mixity (b) as function of the force, as predicted by the FE model for different crack lengths

experimental method. However, as seen from the predicted force-displacement curves in Figure 12, increasing the crack length extends the winding stage. This is logical, as a longer peel strip has to be winded up before delamination will occur.

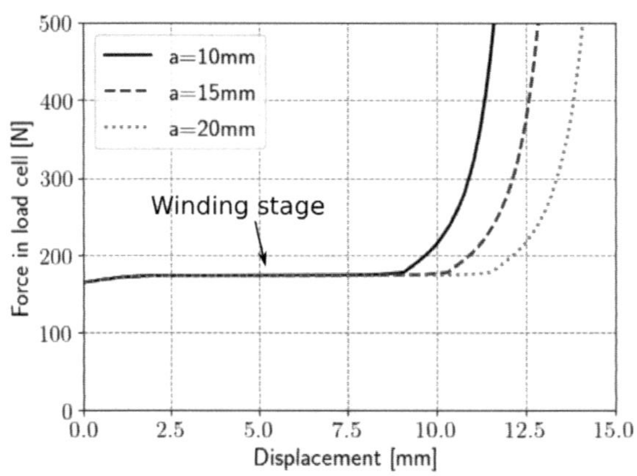

Figure 12: Effect of crack length on the force-displacement curves predicted by the FE model

6. Using the CDP test as fracture toughness test

The results presented throughout Section 5 show several important points to take into account when considering using the CDP test as an alternative to a DCB test, some of which are listed below.

- The CDP test is not pure mode I, rather the mode mixity during delamination was -38° for the material combination considered in the current study. This might vary slightly for other geometry and material combinations but is expected to reach a nearly constant mode mixity during delamination
- The simple analytical energy considerations for calculating the fracture toughness match well with the FE model and are thus sufficient to use for estimating the interfacial strength

from experiments
- Even though a gap is observed between the peel strip and the drum in the FE model, the fracture toughness matches well with the analytical calculations. Thus the conformity to the drum circle shape does not seem to affect the results for the case considered in the current study
- The fracture toughness and mode mixity as a function of the load cell force is not dependent on the crack length
- Estimation of the mode mixity for a specific test setup would be necessary, and to be feasible as part of a standard test it would likely require an analytical approach for doing so instead of FE. Inspiration could perhaps be obtained from the work by Daghia et al [12] on the double drum peel test.

Considering the above points the CDP test can be an alternative to the DCB test for comparing interfaces as long as one remembers that it will provide the mixed mode fracture toughness rather than mode I. However, in the CDP there are limitations to how thick the peel strip can be, which is not an issue in a DCB test. Nevertheless, the simplicity of the CDP test and that it is not necessary to measure the crack length could definitely make it a useful alternative or supplement to obtain the interface strength of laminates for some cases - despite not being a mode I test. To ensure that the measured fracture toughness from the CDP test is in agreement with DCB tests, mixed mode DCB testing could be carried out for the same mode mixity as observed in he CDP test. This is, however, ongoing work.

7. Conclusion
The current paper presented a finite element (FE) model to predict the fracture toughness and mode mixity of the climbing drum peel (CDP) test. The predicted load-displacement curve was shown to agree well with the experiments. The fracture toughness calculated by the J-integral in the FE model matched the analytical calculations. It was shown that the CDP test for the geometry and material used in the current study lead to a mode mixity of -38°. Thus, the CDP is a mixed mode test rather than mode I as suggested in some literature and therefore one cannot use it to obtain the mode I fracture toughness. It can, however, be used to compare different interfaces. However, one has to be aware that one will test at a given mode mixity, likely in the range of -37° to -40° depending on the material.

Acknowledgments
The work was supported by the MADEBLADES project funded by the Energy Technology Development and Demonstration Program, Grant No. 64019–0514. This support is gratefully acknowledged.

References
[1] Mishnaevsky L 2022 *Materials* **15** ISSN 19961944

[2] Liechti K M and Chai Y S 1992 *Journal of Applied Mechanics, Transactions ASME* **59** 295–304 ISSN 15289036

[3] ASTM 2021 Standard Test Method for Mode I Interlaminar Fracture Toughness of Unidirectional Fiber-Reinforced Polymer Matrix Composites ASTM D5528-13 ASTM International West Conshohocken, PA

[4] Daghia F and Cluzel C 2015 *Composites Part A: Applied Science and Manufacturing* **78** 70–83 ISSN 1359835X URL http://dx.doi.org/10.1016/j.compositesa.2015.07.010

[5] Møberg A, Budzik M K and Jensen H M 2017 *Engineering Fracture Mechanics* **175** 219–234 ISSN 00137944 URL http://dx.doi.org/10.1016/j.engfracmech.2017.01.030

[6] Sørensen B F, Jørgensen K, Jacobsen T K and Østergaard R C 2006 *International Journal of Fracture* **141** 163–176 ISSN 03769429

[7] ASTM 2021 Standard Test Method for Climbing Drum Peel for Adhesives ASTM D1781-98 ASTM International West Conshohocken, PA

[8] Okada R and Kortschot M T 2002 *Composites Science and Technology* **62** 1811–1819 ISSN 02663538

[9] Nettles A, Gregory E and Jackson J 2007 *Journal of Composite Materials* **41** 2863–2876 ISSN 0021-9983 URL `http://journals.sagepub.com/doi/10.1177/0021998307079974`

[10] Long N J, Mataira R C, Talantsev E and Badcock R A 2018 *IEEE Transactions on Applied Superconductivity* **28** ISSN 10518223

[11] Thouless M D and Jensen H M 1992 *Journal of Adhesion* **38** 185–197 ISSN 15455823

[12] Daghia F, Cluzel C, Hébrard L, Churlaud F and Courtemanche B 2018 *Composites Part A: Applied Science and Manufacturing* **113** 83–94 ISSN 1359835X URL `https://doi.org/10.1016/j.compositesa.2018.07.020`

[13] Jespersen K M 2023 2D Abaqus finite element model of the climbing drum peel test for fracture toughness and mode mixity determination [Data set] `https://10.5281/zenodo.7944569`

[14] Suo Z and Hutchinson J W 1990 *International Journal of Fracture* **43** 1–18 ISSN 03769429

[15] Agrawal A and Karlsson A M 2007 *Journal of Engineering Materials and Technology* **129** 580–587 ISSN 00944289

[16] Kumar R, Mikkelsen L P, Lilholt H and Madsen B 2021 *Materials* **14**

[17] Kawashita L F, Moore D R and Williams J G 2005 *Journal of Adhesion* **81** 561–586

IOP Conf. Series: Materials Science and Engineering 1293 (2023) 012033 doi:10.1088/1757-899X/1293/1/012033

Numerical modeling of fiber orientation in additively manufactured composites

Berin Šeta[1], Michael Sandberg[1,2], Marco Brander[1], Md. Tusher Mollah[1], Deepak Kumar Pokkalla[3], Vipin Kumar[3], Jon Spangenberg[1]

[1]Department of Civil and Mechanical Engineering, Technical University of Denmark, Lyngby, Denmark

[2]Department of Mechanical and Production Engineering, Aarhus University, Aarhus, Denmark

[3]Manufacturing Science Division, Oak Ridge National Laboratory, TN 37932, USA

E-mail: berse@dtu.dk

Abstract. Additive manufacturing has undergone a significant transformation, evolving from a mere prototyping technique to a reliable and proven manufacturing technology that can produce products of varying sizes and materials. The incorporation of fibers in additive manufacturing processes has the potential to improve a range of material properties, including mechanical, thermal, and electrical properties. However, this improvement is largely dependent on the orientation of the fibers within the material, with the properties being enhanced primarily in the direction of fiber orientation. As a result, accurately predicting and controlling the fiber orientation during the extrusion or deposition process is critical. Various methods are available to control fiber orientation, such as manipulating the nozzle shape, extrusion and nozzle speed, the gap between the nozzle and substrate, as well as fiber features like aspect ratio and volume fraction. At the same time, the presence and orientation of fibers can significantly impact the flow pattern and extrusion pressure conditions, ultimately affecting the formation of printed strands in a manner distinct from those without fibers. For that reason, our study utilizes computational fluid dynamics to anticipate and comprehend the printing conditions that would result in favorable fiber orientations and strand shapes, incl. corner printing. Our findings may be utilized to determine optimal toolpaths for 3D printing composites, as well as printing conditions that will facilitate the achievement of the desired fiber orientation within individual strands.

1. Introduction

Fused-deposition modeling (FDM), also known as Material Extrusion Additive Manufacturing (MEX-AM), has gained significant traction as a highly efficient technology for producing micro-components and large-scale structures. Its adaptability has led to widespread adoption in diverse sectors such as energy, construction, medicine, aerospace, bio-composites, and consumer products (1). Furthermore, significant attention is presently dedicated to exploring the potential of additive manufacturing in the design of wind turbine blades for large rotors, with a specific emphasis on blade tips. Additionally, recycling fibers extracted from wind turbine blades holds promise as innovative filaments suitable for additive manufacturing (2). As FDM has recently transitioned from a prototyping technique to industrial-scale manufacturing, it must meet stringent quality requirements for various applications. Consequently, in certain instances, relying solely on a single component material, such as polymer, may prove insufficient, necessitating the addition of fibers to reinforce the material and enhance its thermal, electrical, and mechanical properties (3). Given the considerable freedom offered by 3D printing in manufacturing parts, the specific printing method employed significantly influences the orientation of the fibers within the material. As fiber orientation plays a vital role in improving mechanical and other thermo-physical properties, it becomes a crucial consideration in the discussion (4).

Numerical simulations using computational fluid dynamics have been commonly employed to predict the flow of molten polymer and understand the formation of 3D printed strands (5; 6; 7). However, there has been a limited amount of research focusing on describing the behavior of fibers during 3D printing, particularly regarding their orientation. Previous 2D models have identified the existence of core/skin regions, but they were unable to specify the spatial distribution of these zones (8; 9). Only recently, the first 3D numerical simulation of the additive manufacturing process known as compression molding was published, demonstrating the presence of skin/core regions and their spatial distribution under certain printing conditions (10; 11; 12). These simulations were further complemented by parametric studies conducted by different authors, who explored the influence of various printing conditions on both strand morphology and fiber orientation (13; 14; 15). Additionally, understanding fiber orientation during multi-layer 3D printing has been crucial, as previously printed strands, especially in large-area additive manufacturing, tend to experience deformation that can cause fiber reorientation (16). Experimental studies have also contributed to this body of knowledge, particularly highlighting the significance of strand width as a critical factor correlated with fiber orientation. It has been observed that wider strands exhibit lower fiber orientation in the printing direction (17). Furthermore, Yan et al. have explored fiber orientation in common 3D printed shapes, such as corners, T-junctions, and sinusoidal waves (18).

The present study aims to utilize numerical simulations to predict fiber orientation in both single-layer and multi-layer printing scenarios, with a specific focus on corner prints with a 90° angle. The manuscript begins by describing the numerical model implemented in OpenFOAM. Subsequently, the results are discussed in relation to single-layer printing, multi-layer printing, and corner prints. The manuscript concludes with a summary of our findings.

2. Numerical model

The simulation model employed in this study involved two-phase fluid simulations, with one phase representing the extrusion/deposition material (such as a thermoplastic or thermoset polymer, concrete, etc.) and the other phase representing air. The open-source software OpenFOAM was utilized, and specifically, the two-phase solver called overInterDyMFoam, based on the volume-of-fluid method with overset methodology enabled, was chosen as the initial solver. Additionally, since the material was reinforced with fibers, a macroscopic tensor approach was employed to simulate their orientation. The governing equations for mass and momentum, along with the transport equations for the phase-scalar α and the orientation tensor, are provided below:

$$\nabla \cdot \mathbf{u} = 0 \tag{1}$$

$$\rho\left(\frac{\partial \mathbf{u}}{\partial t} + \mathbf{u} \cdot \nabla \mathbf{u}\right) = \rho \mathbf{g} + \nabla \cdot \sigma + \mathbf{f}_{\sigma_\mathrm{i}} \tag{2}$$

where ρ is the density of material matrix, $\mathbf{u}$ is the velocity vector, $\mathbf{g}$ is the gravity vector, σ is the stress , $\mathbf{f}_{\sigma_\mathrm{i}}$ is the surface tension modeled as continuum surface force and ∇ is the gradient operator.

The stress tensor, for the matrix material reinforced by fibers, can be written as (13; 16):

$$\sigma = -p\mathbf{I} + 2\mu_{app}(\mathbf{D} + N_p \mathbf{A_{(4)}} \mathbf{D}) \tag{3}$$

where p is the pressure, μ_{app} is the apparent dynamic viscosity, $\mathbf{D} = ((\nabla \mathbf{u}) + (\nabla \mathbf{u})^T)/2$ is deformation rate tensor and $A_{(4)}$ is fourth-rank fiber orientation tensor. As the problem is two-way coupled, the influence of fiber on flow dynamics through stress term in the momentum equation is given through the term $N_p \mathbf{A}_{(4)} \mathbf{D}$, where N_p is the level of anisotropy dependent on fiber volume fraction and aspect ratio.

The volume-of-fluid approach considers that a phase-scalar, α, should be transported to delineate between two phases. The transport equation of α can be written as:

$$\frac{\partial \alpha}{\partial t} + \nabla \cdot (\alpha \mathbf{u}) + \nabla \cdot [\alpha \mathbf{u_c}(\mathbf{1} - \alpha)] = \mathbf{0} \quad (4)$$

where α is the phase fraction of reinforced matrix material, $1 - \alpha$ is the fraction of surrounding air and $\mathbf{u_C}$ is the interface compression velocity which controls the diffusive interface (19).

In this study, the material matrix was considered to exhibit non-Newtonian behavior, and specifically, the Bingham model was utilized to effectively control the deformation of the bottom strand. The Bingham model is commonly employed in large-scale additive manufacturing applications, particularly for modeling the flow of materials like concrete (20; 14; 21; 22), thermoset (23) or other substances that experience changes in yield stress over time. Hence, while this study and methodology have the potential for broad applicability across different matrix materials, they are specifically designed for thermoset materials. The mathematical representation of the Bingham model is as follows:

$$\mu_{app} = k + \frac{\tau}{\dot{\gamma}} \quad \text{for} \quad \dot{\gamma} \geq \dot{\gamma_c} \quad (5)$$

where k is the plastic viscosity, τ is the yield stress, $\dot{\gamma}$ is the shear rate and $\dot{\gamma_c}$ is the critical shear rate. In this study, $\dot{\gamma_c}$ was set to 0.01, similarly as in (20). It has to be highlighted here that fiber presence and orientation did not influence yield stress value as in the novel model presented in (24), and the yield stress value was assumed to be constant. The presence of fibers did not directly impact the viscosity term μ_{app} as in (25). Instead, their influence was incorporated as an additional term in the stress formulation, as shown in Equation 3.

The equation used to determine fiber orientation is expressed as the following tensor transport equation:

$$\frac{\partial \mathbf{A}}{\partial t} + (\mathbf{u} \cdot \nabla)\mathbf{A} = -(\mathbf{W} \cdot \mathbf{A} - \mathbf{A} \cdot \mathbf{W}) + \lambda_F(\mathbf{D} \cdot \mathbf{A} + \mathbf{A} \cdot \mathbf{D} - 2\mathbf{D} : \mathbf{A}_{(4)}) + 2D_r(\mathbf{I} - 3\mathbf{A}) \quad (6)$$

where $\mathbf{W} = ((\nabla \mathbf{u}) - (\nabla \mathbf{u})^T)/2$ is the vortricity tensor, D_r is the rotary diffusion coefficient accounting for fiber-fiber interaction, $\mathbf{I}$ is the unit tensor and λ_F is the shape factor of fibers. Further information regarding the fiber modeling approach can be found in (11; 13; 16). These references provide comprehensive details on the numerical simulations, including the specific initial and boundary conditions utilized in the study.

3. Results and discussion

In this section, it will be investigated how fiber orientates during different situations encountered during 3D printing. For example, by changing different processing parameters, the shape of the strand cross-section will change, and this will influence fiber orientation. On the other hand, the majority of printed structures are composed of multiple layers, where layers can be built before the first deposited layer is completely solidified. This will result in slight deformation of previously deposited layers, and hence fiber reorientation. Moreover, it will be discussed how fiber orientation during corner printing can be used in the optimization of the toolpath wrt. to the fiber orientation. The effect of viscosity and fiber geometry are studies elsewhere (13; 16).

3.1. *Effect of printing conditions on fiber orientation*

As explained in a previous study (13), the fiber orientation in 3D printing is highly influenced by various printing conditions, particularly the ratio of extrusion/nozzle speed. When the nozzle moves slowly or the material is extruded rapidly, the resulting strand shape tends to become more rectangular, which promotes lateral flow of the material. This lateral flow occurs as a squeezing effect between the substrate and the nozzle exit, leading to deformation gradients along the height of the strand. These gradients cause the reorientation of fibers and enhance their alignment in the lateral direction.

It is important to note that all printing conditions, including gap height, nozzle/extrusion speed, diameters, matrix rheology, fiber aspect ratio, and volume fraction, impact the resulting fiber orientation and overall material strength. A numerical model is used to quantify the extent of these effects and identify the most influential parameters in this regard.

To illustrate the impact of printing conditions on fiber orientation, we considered an example of different gap heights. Although previous research (13) suggests that gap height affects the distribution of fiber orientation, with fibers generally being similarly oriented in the cross-section on average, it serves as a useful example to demonstrate the noticeable differences in strand appearance. In our study, we employed the Bingham model, which is applicable to a range of materials exhibiting viscoplastic behavior with a variable yield stress value threshold that may change over time (e.g., during curing, hardening, solidification, etc.).

Initially, it is important to recognize that fibers primarily align in the extrusion direction (z-direction) within the nozzle Fig. 1. Therefore, if no deformation occurs during the placement of the strand onto the substrate, these z-oriented fibers would be aligned in the x-direction. However, due to the presence of deformation to some extent, whether intentional or unintentional, the fiber orientation observed within the nozzle is altered, resulting in a new distribution of fiber orientations in the strand. This emphasizes the significant role played by printing conditions, as the deviation from a perfectly circular strand is influenced by these conditions as well as material and gravity forces.

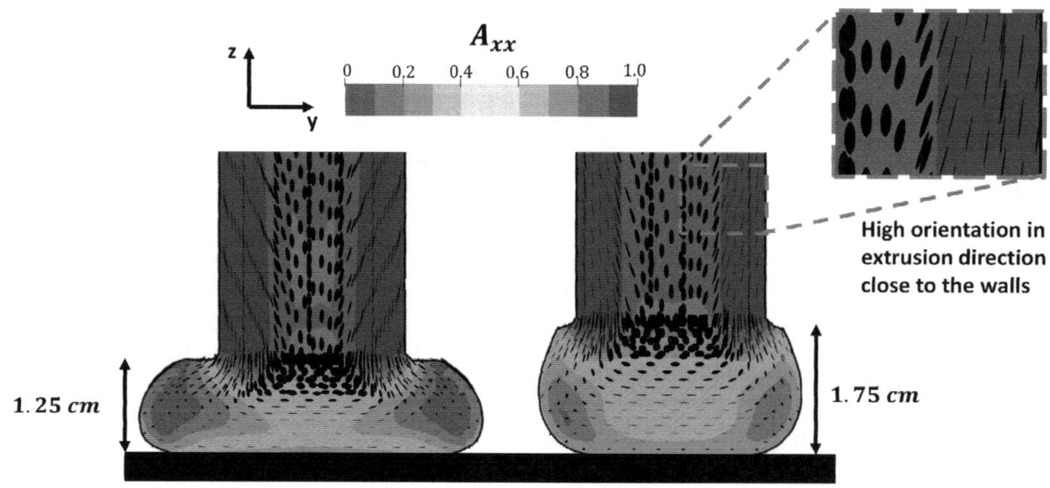

Figure 1: Fiber orientation at the end of the nozzle and during strand deposition.

In Fig. 1, it is evident that the lateral flow in the case where the gap height (h) is 1.25 cm (resulting in a wider strand) is more prominent. However, the edges of the strand, in this case, are more aligned with the printing direction compared to the case where h is 1.75 cm.

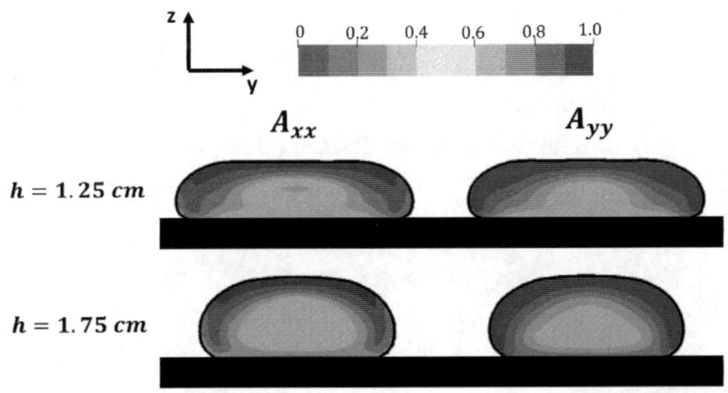

Figure 2: Fiber orientation at the end of the nozzle and during strand deposition.

This observation may seem unexpected, but it can be rationalized by considering the strand formation process. At the top of the strand, where the shearing effect from the nozzle is not yet present, there is a greater orientation of fibers in the printing direction. Furthermore, the vertical fibers (component A_{zz}) penetrate deeper into the material during the formation of the strand with h of 1.75 cm, resulting in a higher proportion of vertically oriented fibers and fewer fibers oriented in the printing direction. This unexpected behavior leads to a less oriented fiber orientation in the printing direction for strands with a larger gap height, which is intriguing considering that the lateral flow is less intense. The A_{xx} values for the case with h of 1.25 cm were 0.69, while for the case with h of 1.75 cm, they were 0.67. Although the difference is not significant, it was anticipated that with a more rounded shape, the fibers would align more in the printing direction, which was not observed.

In Fig. 2, it is evident that the fiber core region is larger in the case with a higher gap height, and the zones with highly oriented fibers are slightly smaller. This is primarily due to the fact that deformation mainly occurs in the central core region, with less impact on the sides of the strands. As the edges are already highly oriented in the printing direction, they occupy a larger spatial area in wider strands, resulting in slightly higher average fiber orientation in the printing direction for strands with h of 1.25 cm. Additionally, as shown in Fig. 3, the tall strands retain a greater presence of fibers oriented in the vertical direction, particularly in the central top region and at the very edges of the strands.

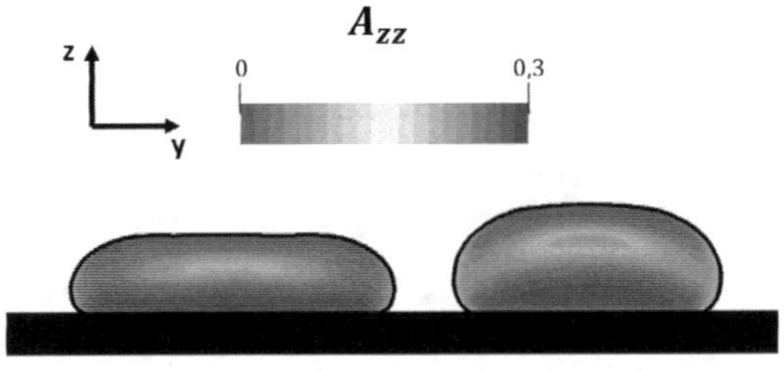

Figure 3: Fiber orientation at the end of the nozzle and during strand deposition.

3.2. Effect of multi-layer printing

When printing on a semi-solid surface, it is anticipated that the layers will experience deformation, particularly those situated at the bottom. This deformation occurs due to both the hydrostatic pressure generated by the weight of the freshly printed strands and the extrusion pressure in the nozzle during material extrusion. Having this in mind, it is crucial to comprehend the impact of this on fiber orientation and the degree to which deformation is connected to fiber reorientation. For this reason, we performed the simulations of two same cases from above, with different h values, up to two printed strands. The printed strands can be seen in Fig. 4.

Based on the simulations conducted, as shown in Table 1, the reorientation of fibers from a single-layer (SL) to a double-layer (DL) configuration was only observed in the case with a smaller h value. This can be attributed to the higher extrusion pressure exerted during printing with smaller gaps, leading to significant deformation of the first layer (see Fig. 4). The value of A_{xx} decreased from 0.69 to 0.66, primarily due to the changes in the bottom layer. Conversely, in the case where $h = 1.75$ cm, fiber orientation remained unchanged. While the bottom strand exhibited a slightly larger core region in the center, the newly deposited top strand displayed even greater alignment in the printing direction compared to a single-layer structure. When these two effects are taken into account in the calculation of average fiber orientation, it remains the same.

Table 1: Averaged fiber orientation in the cross.section

	$SL, h = 12.5$cm	$SL, h = 17.5$cm	$DL, h = 12.5$cm	$DL, h = 17.5$cm
A_{xx}	0.69	0.67	0.66	0.67
A_{yy}	0.26	0.25	0.29	0.25
A_{zz}	0.05	0.08	0.05	0.08

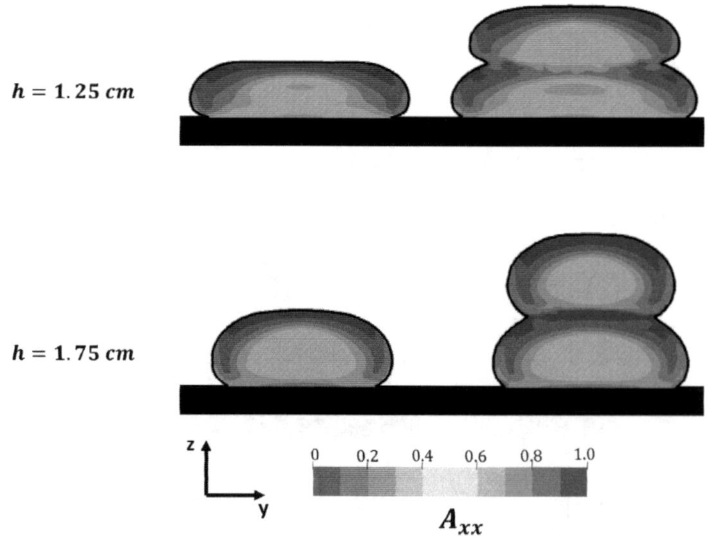

Figure 4: Fiber orientation at the end of the nozzle and during strand deposition.

3.3. Corner printing

Predicting fiber orientation during corner printing is a crucial task, considering that the optimal toolpath may be irregular since most complex shapes do not follow a straight line (26; 27). With this in mind, we are presenting numerical results for the first time, specifically focusing on corner printing at a 90 ° angle. As depicted in Fig. 5, the fibers at the cross-section of the corner exhibit an asymmetric orientation distribution. The fibers located at the outer corner are predominantly aligned in the direction from which the nozzle came, while those at the inner corner are aligned towards the direction in which the nozzle moves away.

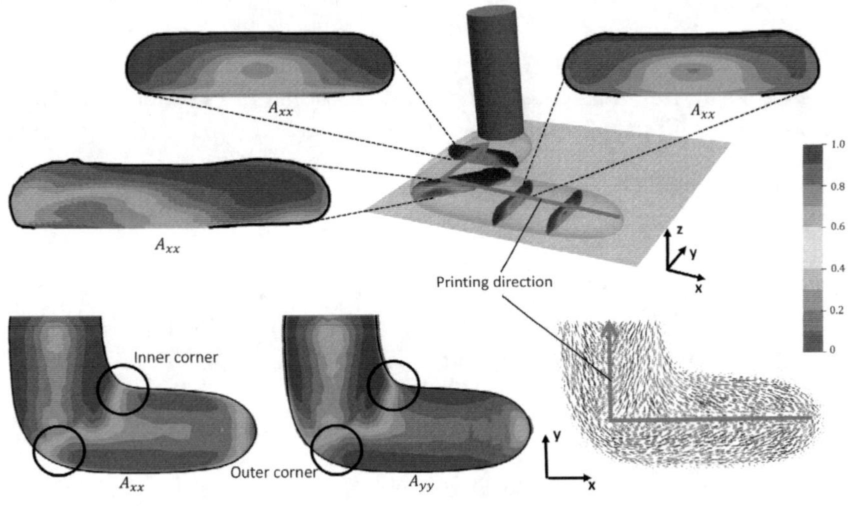

Figure 5: Fiber orientation during corner printing.

The observed effect is clearly visible in the cross-section displayed in Figure 6, which is captured at the central point of the corner. At this position, the cross-section appears significantly wider, indicating the occurrence of double deposition in the corner (28). While most of the material is compressed towards the inner corner, the fibers in that region tend to align themselves in the direction of the nozzle's movement after the corner print. This behavior can be attributed to the double shearing of the fibers by the nozzle exit, followed by the final shearing caused by the nozzle's movement in the y-direction. Consequently, the fibers orient themselves in the same direction as the nozzle's movement. Due to the over-extrusion at the inner corner, the fiber orientation at the corner becomes dependent on the printing direction. This leads to an interesting situation where, despite printing the same shape, the fiber orientation differs depending on the starting position. Understanding this concept makes it clear that strategically planning deceleration and acceleration at the corners to achieve more uniform deposition will also have a significant impact on fiber orientation at those corners.

4. Conclusions

This study has focused on investigating the orientation of fibers during various situations encountered in 3D printing. By analyzing different processing parameters and their influence on strand cross-section shape, the study has highlighted the correlation between printing conditions and fiber orientation. Additionally, the study has explored the impact of multiple layers on strand deformation and subsequent fiber reorientation. Notably, the study has examined the role of fiber orientation during corner printing and its potential for optimizing toolpath design.

The results have demonstrated that printing conditions play a significant role in determining fiber orientation. The ratio of extrusion/nozzle speed, in particular, has been shown to influence the shape of the strand and the lateral flow of the material, thereby affecting fiber orientation. Moreover, the study has emphasized that printed structures composed of multiple layers undergo deformation, leading to fiber reorientation. This observation underscores the importance of considering the effects of layer deposition on fiber orientation in 3D printing.

Furthermore, the study has presented numerical results for the first time regarding corner printing at a 90° angle. It has been observed that the fiber orientation in the cross-section of the corner is asymmetric, with fibers aligning differently at the outer and inner corners. This behavior is attributed to double shearing of the fibers by the nozzle exit and subsequent nozzle movement.

The findings of this study contribute to a better understanding of fiber orientation during 3D printing and highlight the importance of printing conditions and corner printing strategies in controlling and optimizing fiber orientation. Future research in this area could further explore the optimization of toolpaths to achieve more uniform deposition and investigate the mechanical properties of 3D printed structures considering fiber orientation as a key factor.

Acknowledgements

The authors would like to acknowledge the support of the Innovation Fund Denmark (Grant no. 0223-00084B). This manuscript has been authored by UT-Battelle, LLC under Contract No. DE-AC05-00OR22725 with the U.S. Department of Energy. The United States Government retains and the publisher, by accepting the article for publication, acknowledges that the United States Government retains a non-exclusive, paid-up, irrevocable, world-wide license to publish or reproduce the published form of this manuscript, or allow others to do so, for United States Government purposes. The Department of Energy will provide public access to these results of federally sponsored research in accordance with the DOE Public Access Plan (http://energy.gov/downloads/doe-public-access-plan).

References

[1] M. Bayat, O. Zinovieva, F. Ferrari, C. Ayas, M. Langelaar, J. Spangenberg, R. Salajeghe, K. Poulios, S. Mohanty, O. Sigmund, et al., Holistic computational design within additive manufacturing through topology optimization combined with multiphysics multi-scale materials and process modelling, Progress in Materials Science (2023) 101129.

[2] N. Su, R. S. Pierce, C. Rudd, X. Liu, Comprehensive investigation of reclaimed carbon fibre reinforced polyamide (rcf/pa) filaments and fdm printed composites, Composites Part B: Engineering 233 (2022) 109646.

[3] M. Volk, O. Yuksel, I. Baran, J. H. Hattel, J. Spangenberg, M. Sandberg, Cost-efficient, automated, and sustainable composite profile manufacture: A review of the state of the art, innovations, and future of pultrusion technologies, Composites Part B: Engineering (2022) 110135.

[4] H. L. Tekinalp, V. Kunc, G. M. Velez-Garcia, C. E. Duty, L. J. Love, A. K. Naskar, C. A. Blue, S. Ozcan, Highly oriented carbon fiber–polymer composites via additive manufacturing, Composites Science and Technology 105 (2014) 144–150.

[5] R. Comminal, M. P. Serdeczny, D. B. Pedersen, J. Spangenberg, Numerical modeling of the strand deposition flow in extrusion-based additive manufacturing, Additive Manufacturing 20 (2018) 68–76.

[6] M. T. Mollah, R. Comminal, M. P. Serdeczny, D. B. Pedersen, J. Spangenberg, Stability and deformations of deposited layers in material extrusion additive manufacturing, Additive Manufacturing 46 (2021) 102193.

[7] M. T. Mollah, R. Comminal, M. P. Serdeczny, B. Šeta, J. Spangenberg, Computational analysis of yield stress buildup and stability of deposited layers in material extrusion additive manufacturing, Additive Manufacturing 71 (2023) 103605.

[8] Z. Wang, D. E. Smith, A fully coupled simulation of planar deposition flow and fiber orientation in polymer composites additive manufacturing, Materials 14 (10) (2021) 2596.

[9] E. Bertevas, J. Férec, B. C. Khoo, G. Ausias, N. Phan-Thien, Smoothed particle hydrodynamics (sph) modeling of fiber orientation in a 3d printing process, Physics of Fluids 30 (10) (2018) 103103.

[10] V. Kumar, S. P. Alwekar, V. Kunc, E. Cakmak, V. Kishore, T. Smith, J. Lindahl, U. Vaidya, C. Blue, M. Theodore, et al., High-performance molded composites using additively manufactured preforms with controlled fiber and pore morphology, Additive Manufacturing 37 (2021) 101733.

[11] B. Šeta, M. Sandberg, M. Brander, M. T. Mollah, D. Pokalla, V. Kumar, J. Spangenberg, Modelling fiber orientation during additive manufacturing-compression molding processes, SFF symposium 2022 (2022).

[12] B. Šeta, M. Brander, M. T. Mollah, V. NLN, D. Pokkalla, P. Kim, J. Spangenberg, Modeling of additive manufacturing-compression molding process using computational fluid dynamics, in: Conference: 2022 Summer Topical Meeting: Advancing Precision in Additive Manufacturing - Knoxville, Tennessee, United States of America, 2022.

[13] B. Šeta, M. Sandberg, M. Brander, M. T. Mollah, D. Pokalla, V. Kumar, J. Spangenberg, Modeling fiber orientation and strand shape morphology in three-dimensional material extrusion additive manufacturing (Submitted).

[14] J. Reinold, V. Gudžulić, G. Meschke, Computational modeling of fiber orientation during 3d-concrete-printing, Computational Mechanics (2023) 1–21.

[15] P. Pibulchinda, E. Barocio, A. J. Favaloro, R. B. Pipes, Influence of printing conditions on the extrudate shape and fiber orientation in extrusion deposition additive manufacturing, Composites Part B: Engineering (2023) 110793.

[16] B. Šeta, M. Sandberg, M. Brander, M. T. Mollah, D. Pokalla, V. Kumar, J. Spangenberg, Numerical modeling of fiber orientation in multi-layer, material-extrusion big area additive manufacturing (Submitted).

[17] J. Yan, E. Demirci, A. Ganesan, A. Gleadall, Extrusion width critically affects fibre orientation in short fibre reinforced material extrusion additive manufacturing, Additive Manufacturing 49 (2022) 102496.

[18] J. Yan, E. Demirci, A. Gleadall, 3d short fibre orientation for universal structures and geometries in material extrusion additive manufacturing, Additive Manufacturing 69 (2023) 103535.

[19] E. Berberovic, Investigation of free-surface flow associated with drop impact: numerical simulations and theoretical modeling, Ph.D. thesis, Technische Universität (2010).

[20] R. Comminal, W. R. L. da Silva, T. J. Andersen, H. Stang, J. Spangenberg, Modelling of 3d concrete printing based on computational fluid dynamics, Cement and Concrete Research 138 (2020) 106256.

[21] N. Ranjbar, M. Mehrali, C. Kuenzel, C. Gundlach, D. B. Pedersen, A. Dolatshahi-Pirouz, J. Spangenberg, Rheological characterization of 3d printable geopolymers, Cement and Concrete Research 147 (2021) 106498.

[22] M. T. Mollah, R. Comminal, W. R. Leal da Silva, B. Šeta, J. Spangenberg, Computational fluid dynamics modelling and experimental analysis of reinforcement bar integration in 3d concrete printing (Accepted in Cement and Concrete Research, 2023).

[23] J. T. Pedersen-Bjergaard, M. M. Christensen, M. Brander, B. Seta, D. B. Pedersen, J. Spangenberg, Interlayer mechanical properties of thermoset components produced by material extrusion additive manufacturing, SFF symposium 2022 (2022).

[24] J. Férec, E. Bertevas, B. C. Khoo, G. Ausias, N. Phan-Thien, A rheological constitutive model for semiconcentrated rod suspensions in bingham fluids, Physics of Fluids 29 (7) (2017) 073103.

[25] M. Jabbari, J. Spangenberg, J. H. Hattel, Particle migration using local variation of the viscosity (lvov) model in flow of a non-newtonian fluid for ceramic tape casting, Chemical Engineering Research and Design 109 (2016) 226–233.

[26] F. Fernandez, W. S. Compel, J. P. Lewicki, D. A. Tortorelli, Optimal design of fiber reinforced composite structures and their direct ink write fabrication, Computer Methods in Applied Mechanics and Engineering 353 (2019) 277–307.

[27] F. Fernandez, J. P. Lewicki, D. A. Tortorelli, Optimal toolpath design of additive manufactured composite cylindrical structures, Computer Methods in Applied Mechanics and Engineering 376 (2021) 113673.

[28] M. T. Mollah, A. Moetazedian, A. Gleadall, J. Yan, W. E. Alphonso, R. Comminal, B. Šeta, T. Lock, J. Spangenberg, Investigation on corner precision at different corner angles in material extrusion additive manufacturing: An experimental and computational fluid dynamics analysis, in: 2022 International Solid Freeform Fabrication Symposium, 2022.

IOP Conf. Series: Materials Science and Engineering 1293 (2023) 012034 doi:10.1088/1757-899X/1293/1/012034

Online Tg estimation for blade manufacturing

Nikos Pantelelis

Synthesites, Frans Lyceumlaan 5B, 1180, Uccle, Belgium
be@synthesites.com

Abstract. Cost reduction is one of the main drivers in the production of wind turbine blades in the last few years. In this direction, the use of intelligent process monitoring to optimize the infusion, curing and bonding by measuring online the actual process can reduce energy and increase productivity by 20% while ensuring product quality. In this framework, durable non-intrusive and disposable sensors have been developed for the monitoring of resin arrival, the resin's viscosity during infusion and the on-going Tg during curing by measuring the resin's electrical resistivity and temperature.

1. Introduction

Nowadays, the need for efficient manufacturing in the wind blade industry is more evident than ever. State of the art in production involves only temperature measurements while DSC measurements are performed at the best, a few days after demoulding. Therefore, wind-blade production needs to improve its productivity, reduce production costs including repairs, ensuring product quality.

The introduction of online intelligent process monitoring systems optimised for the wind blade production can help significantly in all of these aspects. Firstly, by monitoring the resin flow and the increase of viscosity during infusion, potential flow control actions can be activated, if necessary. Then during curing, the online estimation of the Glass Transition temperature can lead to significant decrease of the cycle times by ensuring the minimum cure quality.

Towards that direction, Pantelelis and Bistekos [1] presented already in 2010, the DC-based dielectric cure monitoring technology with a clear focus on industrial applications. The Optimold system can measure directly the resin's electrical resistance and temperature using specialised sensors. Measuring electrical resistance from 10^5 Ohm up to 10^{14} Ohm is capable to follow the polymerisation from very low viscosity at high temperatures to fully cured resins at room temperature. Comparison between the present DC and the conventional AC cure monitoring technologies has demonstrated the superiority of the DC technology. Furthermore, it is cheaper, very easy to use and very reliable and robust which has been proven in various industrial installations in wind energy, automotive and aerospace production lines [2, 3, 4].

2. Intelligent process monitoring: System Components

For more than ten years, Synthesites has been developing the components for a complete suite for monitoring the complete manufacturing process in wind turbine blades: from a dedicated system to provide online resin arrival and temperature, to online viscosity and Tg after infusion and during bonding including the corresponding intelligent software to calculate and communicate with controllers and PLC. The first attempt in using these cure sensors in wind blade production took place already in 2011 [5] but it was only in 2017 that this technology attracted serious interest from blade manufacturers including leading OEMs. Nowadays, these systems are being used in everyday wind blade production.

IOP Conf. Series: Materials Science and Engineering 1293 (2023) 012034 doi:10.1088/1757-899X/1293/1/012034

The main components of this intelligent system are summarised below:

2.1. The data acquisition units

Both the Optimold cure monitoring (Figure 1a) and the Optiflow Resin arrival monitoring unit (Figure 1b) measure directly the electrical resistance of the resin by applying a DC voltage to the electrodes of the sensors. The resistance together with the resin temperature are necessary for the resin arrival as well as the online estimation of the resin's viscosity and glass transition temperature (Tg) which are the most important properties in composites processing.

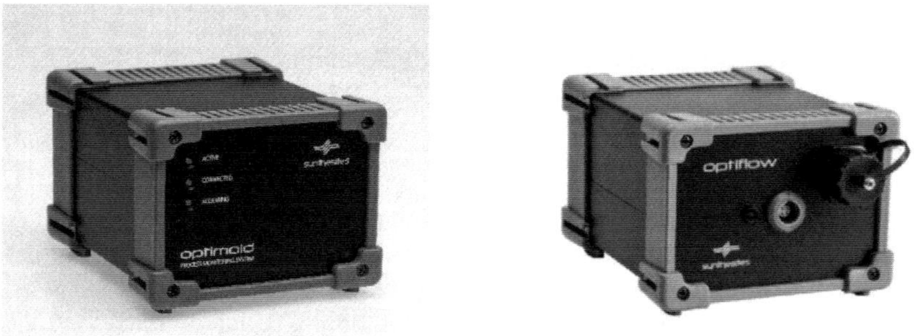

Figure 1: (a) The Optimold cure monitoring unit and (b) the Optiflow resin arrival unit.

2.2. The durable cavity sensors

A range of durable cure and resin arrival sensors have been developed to be used in direct contact with resin in the mould cavity or through the vacuum bag. Various sensor housings help to secure the sensors at the right location depending on the monitoring needs. In Figure 2, two of the available durable sensors are shown, on the left the standard in-mould sensor which should be inserted flush-mounted in the tool. The Vacuum Bag (VB) sensor, Figure 2b, can be placed easily on the vacuum bag without mould modifications on top of the laminate, where the lagging curing area is found in a typical wind blade infusion set-up. Although the cure sensor provides the resin arrival instant, a dedicated resin arrival sensor in combination with the Optiflow unit can provide the resin arrival information at lower cost allowing the use of numerous resin arrival sensors which could be particularly useful in the large moulds for wind turbine blade production.

Figure 2: (a) The durable sensors: the In-mould sensor, (b) the Vacuum-Bag (VB) sensor and (c) the inline resin arrival sensor for the feeding and the vacuum lines.

2.3. The durable inline sensor

The inline resin arrival sensor (Figure 2c) is connected to the Optiflow unit (resin arrival and temperature) and can provide a timestamp when the resin flows through the sensor, entering or exiting the mould cavity. The inline sensor can be easily integrated at the feeding or the vacuum lines without any modification in the mould. Following the resin arrival signal, Optiflow is capable for example, to trigger a valve for actively controlling the resin flow [4].

2.4. The Optiview data-acquisition and visualisation software

The Optiview data acquisition software is capable for capturing, storing and visualising all the measurements from all the connected units and sensors. Optiview provides a real-time visualisation of the on-going process not only for the raw measurements but also for their by-products such as resin arrival, viscosity, degree of cure and Glass Transition temperature. For example, in Figure 3, the resin arrival timeline for sixteen resin arrival sensors of a typical RTM application is shown where the green bar indicates that a sensor is connected but dry which becomes orange when resin arrives at the sensor.

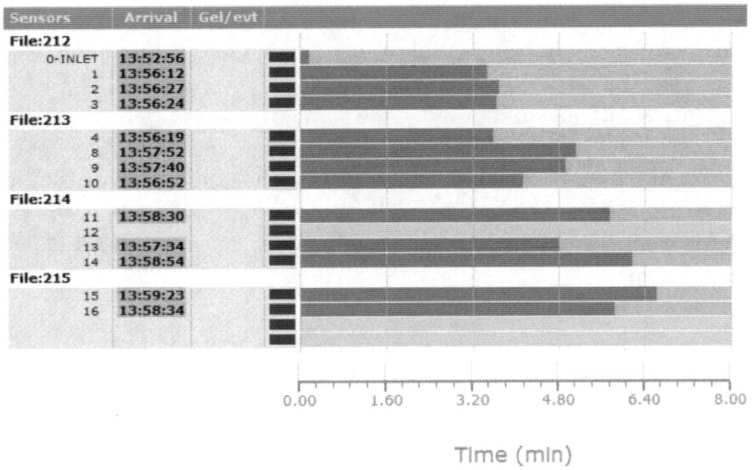

Figure 3: The resin arrival Optiview screen.

The Optiview software also provides a Human-Machine-Interface (HMI) for the automated operation at production environment as can be seen in Figure 4, allowing to eliminate any interaction with the operator, while it handles also the communication with PLC and temperature controllers.

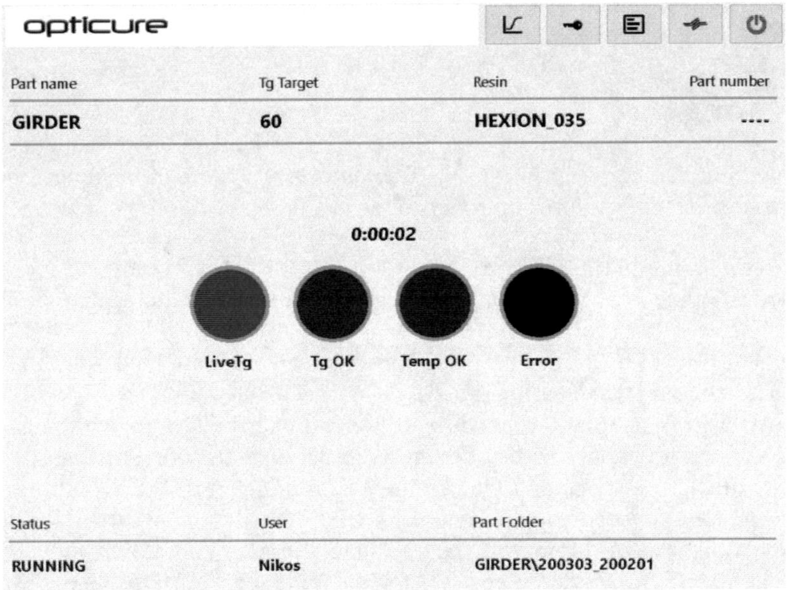

Figure 4: The Human-Machine-Interface (HMI) screen.

2.5. The Cure Simulator
The first version of the Cure Simulator was developed within an EU funded project [7] to provide a non-intrusive solution for monitoring the curing in autoclaves and other applications where it is difficult to introduce a cure sensor. In the wind-turbine blade production, the Cure Simulator has been employed for monitoring the curing at areas that they are either not easily accessible, such as the bondlines or other sensitive areas to sensor's intrusiveness such as the spar caps and the shear webs. The Cure Simulator comprises a small heated resin cell equipped with a cure sensor. The principle, depicted in Figure 5, is that the resin sample placed in the resin cell of the cure simulator, will experience the same curing as the composite at the location of the input thermocouple. The temperature of the resin cell follows accurately and instantly the input of that temperature sensor using an advanced temperature controller.
The capability of the Cure Simulator to reproduce the same curing as the laminate has been verified in several applications in aerospace and wind energy applications. A very important detail in this procedure lies on the right location of the input temperature sensor which should provide as much as possible the real temperature of the resin under curing. Besides the cure sensor elimination, the Cure Simulator can help to eliminate the operator's interaction with certain cure sensors, allowing its use in everyday production.

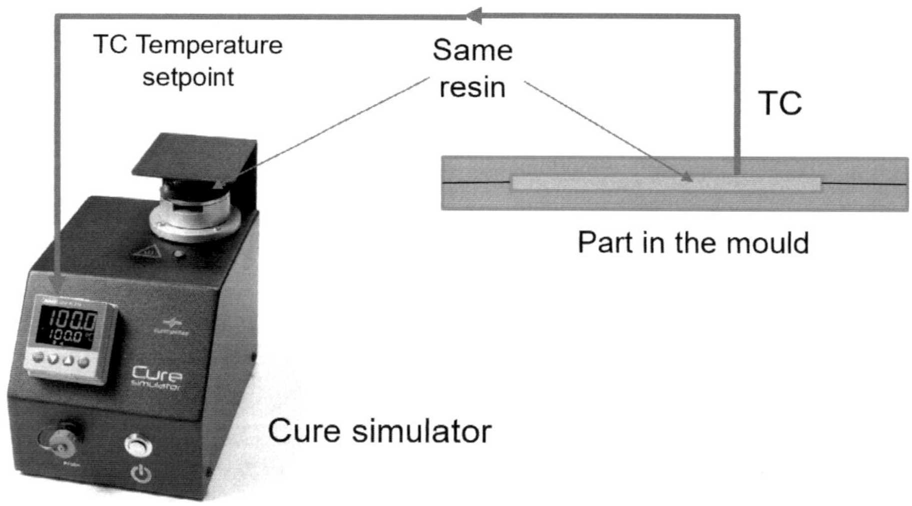

Figure 5: The Cure Simulator concept for RTM: The temperature measurement becomes continuously the setpoint for the temperature controller of the heated resin cell of the Cure Simulator.

2.6. The Online Resin State (ORS) software
Last but not least, the Online Resin State software translates in real-time the Optimold measurements of resistance and temperature to useful resin properties such as the resin's viscosity, glass transition temperature (Tg) and degree of cure. The ORS software can be combined with the Optiview HMI software to automate effectively the curing process.
The ORS software is using a calibration function for a specific resin. The calibration campaign consists of manufacturing a series of coupons made of the resin under investigation cured at specific conditions. The generated data of the resistance and temperature are correlated to the evolution of viscosity and glass transition temperature measured experimentally to produce the calibration function, which then can be used to obtain the same properties in real-time. Similar calibration procedures have been developed to check the resin mixing ratio and the storage aging, onsite. In the last 7 years more than 70 resins, all the kinds of epoxies as well as polyesters, vinylesters, reactive thermoplastics, phenolics, PFA, benzoxaxines, cyanate esters and others have been studied successfully.

3. Results

Since 2011, several industrial projects in wind blade manufacturing have been realized, verifying the accuracy of the Online Tg estimation at industrial conditions. In this framework, numerous epoxy resins and adhesives from several resin manufacturers have been studied, firstly in the lab and then at industrial conditions. In Figure 6, the resistance curves during several isothermal cure cycles of a typical epoxy adhesive used in turbine blades are shown. As expected, when increasing the curing temperature, the reaction speed increases resulting in the resistance curve reaching faster its final value. On the other hand, as temperature increases, the resistance decreases.

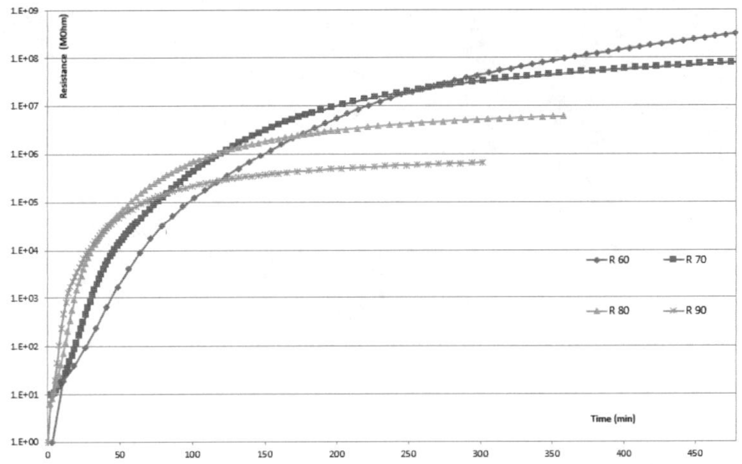

Figure 6: Resistance curves of an epoxy adhesive during isothermal curing at 60, 70, 80 and 90°C.

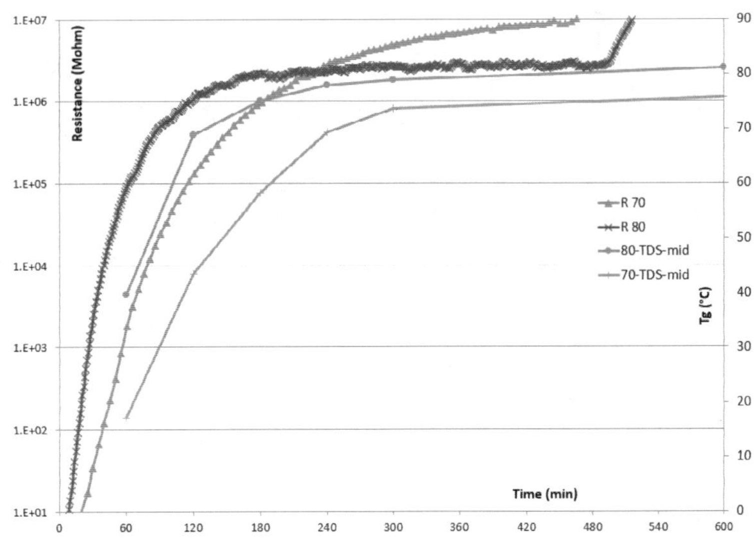

Figure 7: Correlation between Resistance and Tg from TDS of a popular epoxy infusion resin during isothermal cure at 70 and 80°C.

Similar behaviour for the resistance can be observed in the curing of a popular epoxy infusion resin (Hexion RIM 035-037) as shown in Figure 7 for 70 and 80°C together with the corresponding Tg evolution.

IOP Conf. Series: Materials Science and Engineering 1293 (2023) 012034 doi:10.1088/1757-899X/1293/1/012034

In order to test the new durable sensor and the Online Tg module at realistic conditions, numerous laboratory scale trials were performed [3] in simulated production conditions i.e. representative cure cycles and laminates from a variety of wind-blade components have been tested at the laboratory. In Figure 8, a typical vacuum infusion set-up of a thick laminate together with cure and temperature sensors can be seen.

Figure 8: A typical vacuum infusion set-up of a laminate together with cure and temperature sensors on top of the vacuum bag.

Following the successful lab-scale trials, the system was tested in the production with very good results. In Figure 9Figure 10, the first such trial in 2017 is shown. A durable sensor was placed during the infusion of a half-blade on top of the vacuum bag at the root section.

Figure 9: A Vacuum-bag sensor was used for the first time to monitor the curing on top of the vacuum bag in real production at Carbon Rotec [3].

In Figure 10, two representative lab-scale trials are shown: one 'almost' isothermal at 80°C and one highly non-isothermal from curing a rather thick laminate by heating only from the bottom of the laminate. In both cases the online T_g estimation is also depicted calculated in real-time during curing based on the real-time values of electrical resistance and temperature measured by Optimold. Especially in the non-isothermal case (Figure 10a) it is obvious that without the temperature, the electrical resistance alone cannot provide the necessary information about the curing. So with the appropriate resin calibration, the Online Tg can provide an accurate and robust estimation of the on-going T_g/degree of cure.

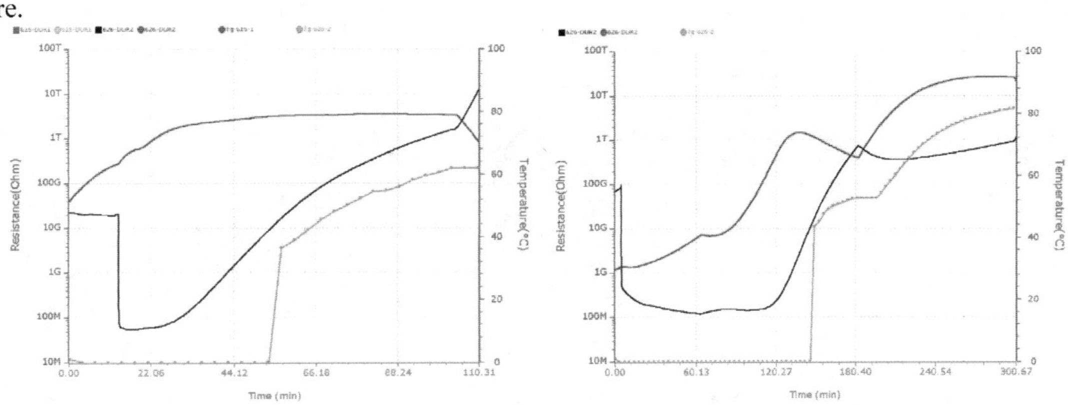

Figure 10: Typical ORS graphs showing the resistance together with temperature and T_g (a) Isothermal case, (b) Strongly non-isothermal real case.

Table 1: Overview of the various test cases and the difference between the Tg estimated online and measured right after demoulding by DSC [3].

	Trial	Duration [h]	T_g-ORS (°C)	T_g- DSC (°C)	Difference (°C)
Isothermal	80DV1	3	73.17	73.34	-0.17
	80DV3	2.5	70.30	70.91	-0.61
	80DV4	2.5	73.45	72.49	0.96
	80-120'	1.92	66.96	66.02	0.94
	80-90'-1	1.50	62.04	61.80	0.24
	80-90'-2	1.50	65.52	65.21	0.31
	80-D2-2	1.50	61.88	60.59	1.29
	60-260'	4.33	55.02	56.51	-1.49
	70-190'	3.17	64.92	65.39	-0.47
Isothermal cases, mean difference					1.61
Isothermal cases, standard deviation					2.42
Non-isothermal	TEB1-1		61.37	59.54	1.83
	TEB1-2		69.36	70.93	-1.58
	TEB2-1		60.00	58.64	1.36
	TEB2-2		70.02	70.30	-0.28
	LESW1-1		76.97	74.35	2.62
	TESW1		71.34	69.18	2.16
	Shell1-1		80.36	78.92	1.44
	Shell1-2		75.72	77.83	-2.12
	Shell2-1		79.60	77.70	1.89
Non-isothermal cases, mean difference					2.15
Non-isothermal cases, standard deviation					1.26

In Table 1, a comparison between the real-time estimated T_g and the T_g measured by DSC after demoulding are shown for various representative isothermal and non-isothermal cases. The isothermal trials include several trials with resin cured at 60, 70, 80 and 90°C such as the trial shown in Figure 10a with different duration. The trials at real blade components (non-isothermal cases) include shear web

and half-shell trials such as the one depicted in Figure 10b. As can be seen in the last column, the difference between these two values for the isothermal and the non-isothermal cases are within the DSC accuracy.
In order to facilitate the use of this system in wind blade manufacturing, Synthesites developed a new durable sensor specifically for vacuum bag applications. As can be seen in Figure 11a, the durable sensor includes a wide disc to allow its attachment on the vacuum bag, secure and easy. In order to eliminate the effect of the sensor's mass on the curing of the laminate underneath, a heated version of the Vacuum-Bag sensor was also developed as shown in Figure 11b. In both cases, the sensors were used consistently during production of blade components for acquiring the online Tg [6].

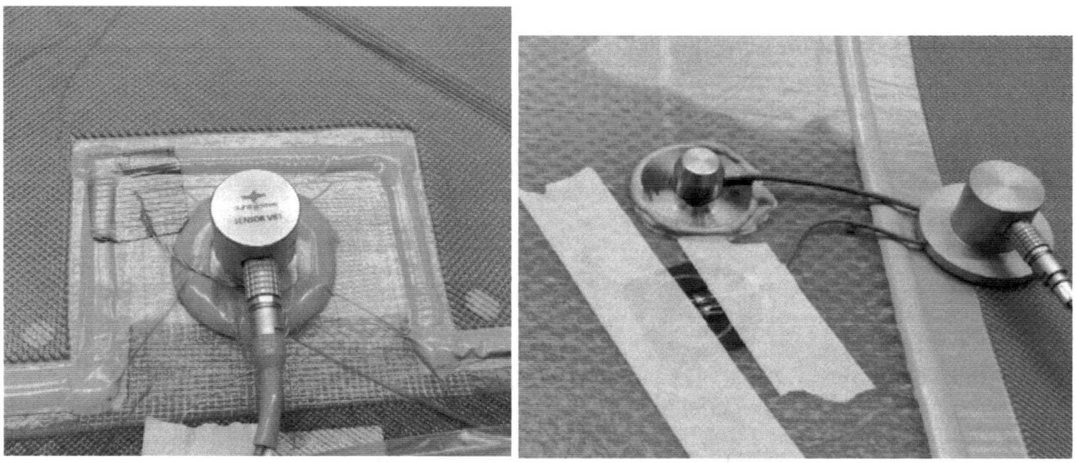

Figure 11: (a) The new durable vacuum bag sensor; (b) Installed at production

In the bonding of a blade, for the monitoring of the Tg, the Cure Simulator has been employed as can be seen in Figure 12a while a disposable Pt temperature sensor was inserted at a specific depth in the bond-line (Figure 12b). The Cure Simulator can provide a unique way to simulate and measure the curing in the middle of a bond-line which on one hand is the latest area to cure since it is quite difficult to heat from the mould or the hot air and on the other hand it cannot be accessed easily to acquire a DSC sample before it is demoulded i.e. much later than the end of heating.

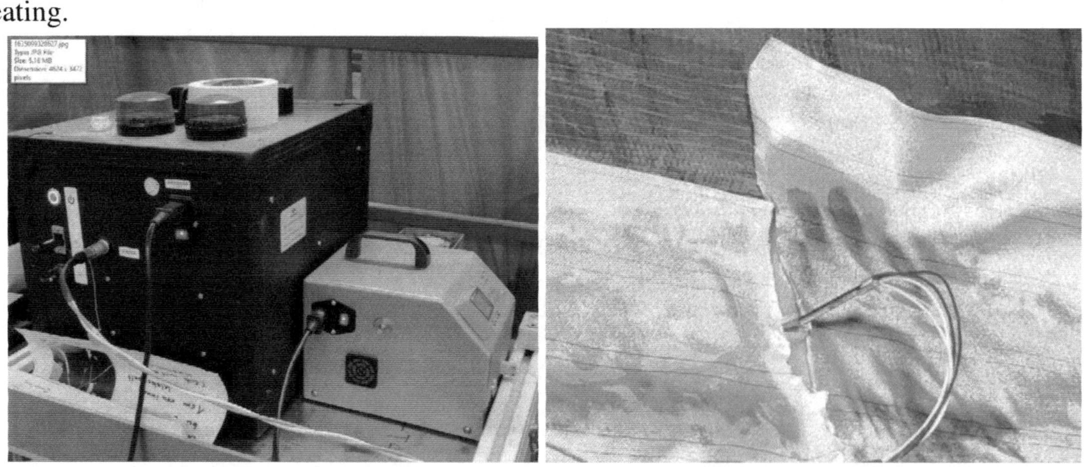

Figure 12: (a) An Opticure system with the first version of the Cure Simulator at production, (b) A special temperature sensor placed in the bond-line in blade bonding [6].

A typical bonding cycle depicted in Figure 13, shows that although there is a continuous heating of the adhesive by hot air to 95°C that temperature was hardly reached at the end of the cycle. However based on the Tg that was calculated online, after 180 min the maximum Tg was reached and cooling could have started saving significant time and energy. The accuracy of the online estimated Tg has been confirmed by DSC [6] as it is also shown in Figure 13.

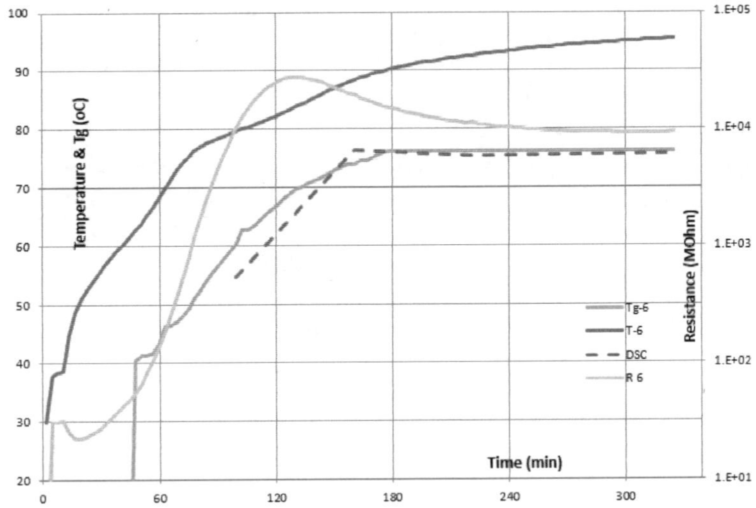

Figure 13: Typical cure cycle for the adhesive at the centre of the blade's bond-line, the resistance (right vertical axis) together with temperature and T_g (left vertical axis) and the Tg evolution that was measured by DSC (dashed line).

4. Conclusions- Next steps

Windblade production is one of the most mature composites production even though automation has not been advanced considerably due to the size of the parts. At the present work, the complete suite for the real-time T_g estimation and cure control have been analysed. The new technology was proved robust and accurate enough even at the toughest processing conditions in the manufacturing of several blade components. The use of the intelligent system in everyday production is under way by a large OEM and is expected that besides the improved quality assurance and traceability, one hour bonding time can be saved.

The next step in the intelligent platform will emerge through the TURBO R&D project which is led by DTU and involves SGRE in Denmark and it is partially funded by the European Commission. In this project, a more complete intelligent platform targets in fully automating windblade production and further advancing productivity by eliminating the needs for repairs in post-production.

5. References

[1] Pantelelis N, Bistekos E 2010 Process monitoring and control for the production of CFRP components SAMPE USA

[2] Meier R, Pantelelis N, Hauber M, Wolf C and Drechsler K 2015 Process monitoring and control for an aerospace application *International Conference in Advanced Manufacturing of Composites*

[3] Lauter C, Jaquemotte K-P, Pantelelis N, Improvement of productivity and quality in the wind energy industry through the use of an advanced sensor system, Sampe Journal **53/6** 6-10

[4] Pantelelis N, Etchells M, Lira C, Bistekos E 2019 Online Viscosity and Tg Measurement of CFRP manufactured using High-Pressure RTM *SAMPE Europe*
[5] Pantelelis N 2011 Process monitoring and control in the manufacturing of large composite structures *Windblade Manufacture Conference*
[6] Pantelelis N, Haspel B, Barros A, Thiel S and Jürges S 2022 Cost reduction through online intelligent process control *Windblade Manufacture Conference*
[7] Pantelelis N, Bistekos E, Breen D, Wilson S, Gerrits W, Wilkens S 2021 Non-intrusive intelligent cure monitoring for enhancing the manufacturing of high-temp composite structures *SAMPE Europe*

Acknowledgments

TURBO project has received funding from Horizon Europe, the European Union's Framework Programme for Research and Innovation, under Grant Agreement No. 101058054 (TURBO). Views and opinions expressed are however those of the authors only and do not necessarily reflect those of the European Union. The European Union cannot be held responsible for them.

IOP Conf. Series: Materials Science and Engineering 1293 (2023) 012035 doi:10.1088/1757-899X/1293/1/012035

Parametric and numerical Finite Element simulation of wind turbine blades subjected to thermal residual stresses

Peyman Ayoubi[1], Kaveh Rashvand[1,2,*], Pedram Ayyobi[3], Moloud Mohammadi[1]

[1] GuilAzma, Inc., Anzali Free Zone, Industrial Park No. 2, Guilan, Iran

[2] Department of Wind and Energy Systems, Technical University of Denmark, Risø Campus, Frederiksborgvej 399, 4000 Roskilde, Denmark

[3] Department of Civil Engineering, University of Minho, Campus de Azurém, 4800-058 Guimarães, Portugal

*kavra@dtu.dk

Abstract. This study aims to contribute to the ongoing efforts to enhance the reliability and durability of wind turbine blades, a critical component in wind energy generation. Specifically, this research addresses the issue of tunneling cracking and severe damage that can occur in wind turbine blades due to cohesive failure of the trailing edge. To achieve this objective, the study employs a rigorous approach, utilizing a full three-dimensional (3D) modeling strategy with finite element analysis (FEA) to simulate the behavior of wind turbine blades. The effect of cohesive materials and layered simulation methods on the thermal residual stress and crack propagation is thoroughly investigated. In particular, the study assesses the influence of carbon fiber-reinforced polymer (CFRP) and glass fiber-reinforced polymer (GFRP) materials on the phenomenon under consideration. In addition, the study undertakes a comprehensive parametric analysis to identify the independent effects of material properties and numerical simulation on thermal residual stress. Moreover, the research explores the behavior of the cohesive zone model in terms of thermal residual stress and crack propagation. The findings of this study have significant implications for researchers and practitioners in the wind energy industry. The study's outcomes can aid in the development of improved materials and simulation techniques to mitigate thermal residual stress and prevent the occurrence of tunneling cracking and other types of damage in wind turbine blades. As such, this research contributes to the broader efforts to advance the reliability, efficiency, and sustainability of wind energy generation.

1. Introduction

Fiber-reinforced polymers (FRPs) are attractive due to the weight to stiffness ratio, economical efficiency, chemical and humidity resistance, toughness, thermal, and electrical insulation. The advantages of FRP materials lead to extensive utilization in aerospace, mechanical, and civil engineering industries. In addition, adhesive bonding is required for the edges and connections of FRPs. Adhesively bonded joints offer advantages over traditional connection methods, including increased joint stiffness and higher fatigue resistance. Despite these benefits, there are also some drawbacks in terms of construction and service conditions. One of the most critical steps in the construction process is surface cleaning, which is highly sensitive. Furthermore, joint design typically results in greater shear, compressive, and tensile resistance, but may be vulnerable to peeling. Cohesive materials have a higher thermal constant than metals. Additionally, the relationship between cohesive material and temperature is nonlinear. Consequently, even minor deviations in thermal control and design conditions may result in significant changes in joint characteristics. This scenario is likely to occur under service thermal loading [1].

The term "tunneling crack" refers to damage initiation of the bond line caused by cohesive failure [2]. Carins and Peterson [3] and Struzziero and Teuwen [4] have investigated thick adhesive joints in terms of crack initiation and propagation. Samborsky et al. [5] studied mixed-mode fracture, crack growth characteristics, and fatigue growth resistance in thick adhesive joints between fiberglass laminates. The fracture behavior of the adhesive joint of the trailing edge has been evaluated by Eder and Bitsche [6]. A new approach has been proposed by Jørgensen et al. [7] to determine the residual stress. The study shows that the difference in temperature in curing and testing leads to transverse crack formation in the adhesive layer. As a result, increasing the laminate thickness has a slight effect. Rosmeier et al. [8] considered the impact of thermal residual stress on fatigue and found significant effects on the initiation and growth of tunneling cracks. Although the tunneling crack has no direct effect on the structural integrity of turbine blades, its propagation into adjacent laminates can increase the probability of severe damage.

Figure 1 represents the general turbine blade shapes and tunneling cracking formation.

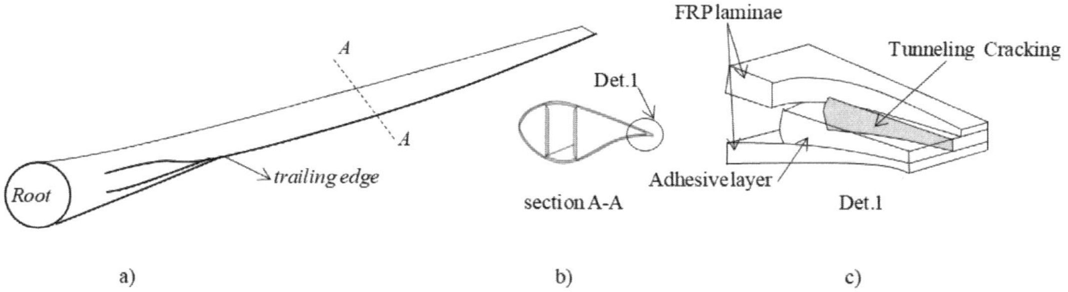

Figure 1. Turbine blade details, a) general blade, b) blade section, c) adhesive joint and tunneling cracking location

The thermal residual stress can be calculated using finite element (FE) simulation. Nielsen [9] predicts the shape distortion of thick laminates by implementing cure hardening instantaneous elastic (CHILE) and a path-dependent approach. Rosemeier et al. [10] determine the accuracy of the FE numerical model for a thick bond line. In their study, classical laminated plate theory (CLT) is implemented, along with FE plate models of different fidelities. Residual thermal stress also affects repaired wind turbine blades, leading to microstructural defects. Paul et al. [11] assess the occurrence of residual stress due to thermal cycles, by implementing a coupled temperature-displacement analysis in FE software. Hwang et al. [12] investigate the strength of the adhesive in a large turbine blade through real-scale simulation. The results indicate that by utilizing a smart curing process, the final curing temperature and hence residual thermal stress and thermal shear stress are reduced. Jørgensen et al. [13] evaluate the novel idea of embedding a buffer layer between the adhesive and the glass fiber laminate using a 2D tri-material FE model. The results show an improvement in joint design against tunneling cracks.

Despite the extensive experimental and numerical investigation, in many cases, the simplified plate model has been used as a delegacy of the full-scale turbine model. Furthermore, this simplified plate has simply supported boundary conditions which leads to contradiction with real blade loading conditions [14]. Also, Carbon fiber-reinforced polymers (CFRP) and Glass fiber-reinforced polymers (GFRP) has different thermal constant. Therefore, connection with an adhesive layer in the trailing edge can result in different distortion and displacement values. Since the cohesive zone is considered employing different damage evolution models in FE simulation, this approaches lead to different crack propagation pattern. To overcome the abovementioned issues, this study proposed an extensive parametric study implementing a full-scale FE model to achieve a comprehensive study in terms of residual thermal stress induced by curing and operational temperature.

2. Numerical model and methodology

A three-dimensional finite element model in ABAQUS 2017 software is implemented to predict the behavior of the SSP blade [14] subjected to residual thermal stress. The model includes nonlinear and thermal characteristics of the materials for CFRP/GFRP and adhesive. The model is capable of capturing all six mechanical stress components along the trailing edge. The CFRP/GFRP laminae are modeled using shell element type SC8RT and the adhesive is modeled by means of COH3D8 elements. The general model, mesh pattern, and boundary conditions are demonstrated in Figures 2 and 3. The boundary condition is considered encast (fixed in all 6 degrees of freedom) in the blade root location.

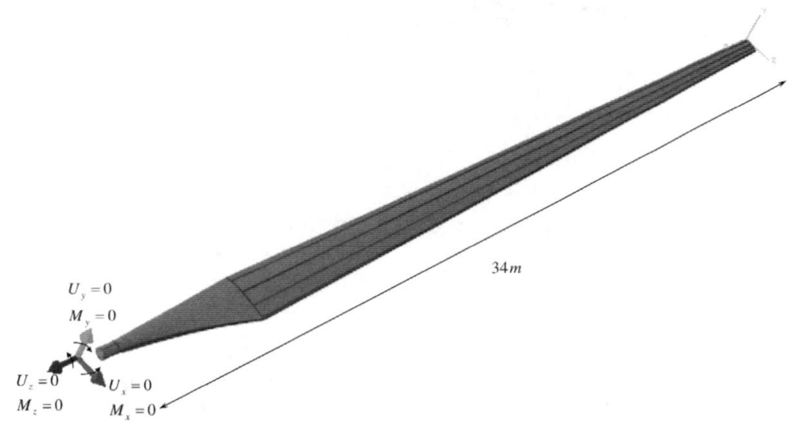

Figure 2. Model demonstration and boundary condition

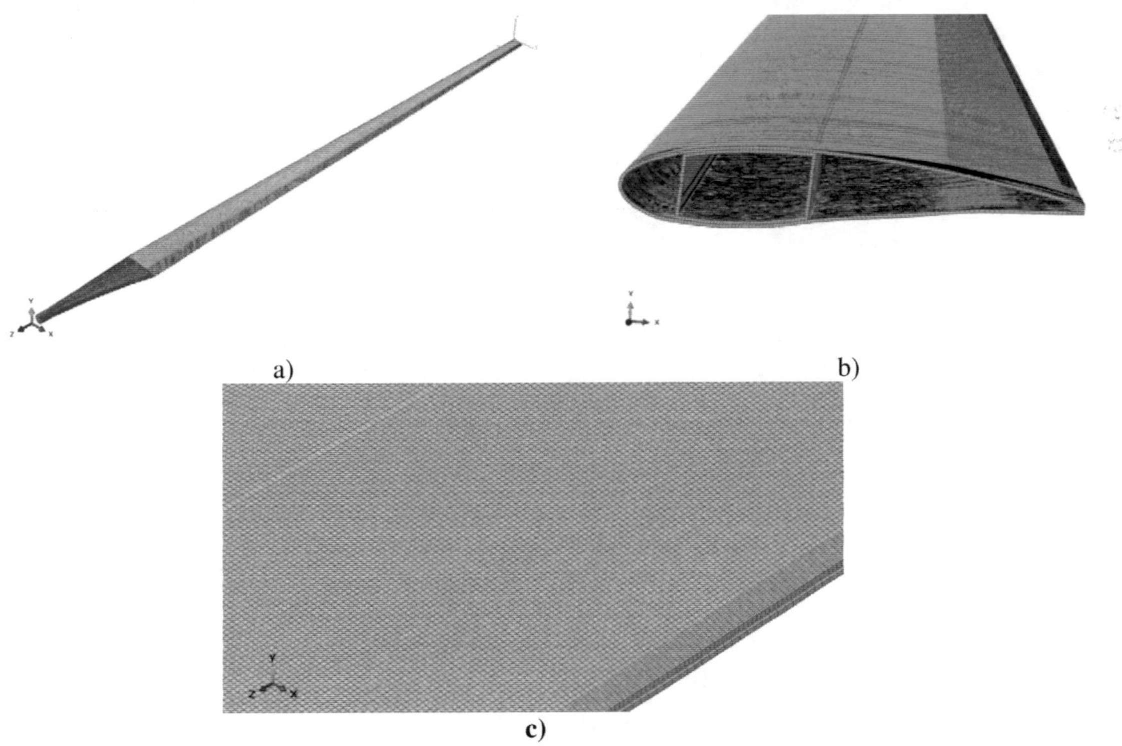

Figure 3. The mesh pattern of the blade, a) total blade, b) section of the blade, and c) mesh refinement in blade

Figure 4 shows the cohesive elements' mesh pattern and geometry. The cohesive elements are located in the trailing edge which connects the CFRP/GFRP elements.

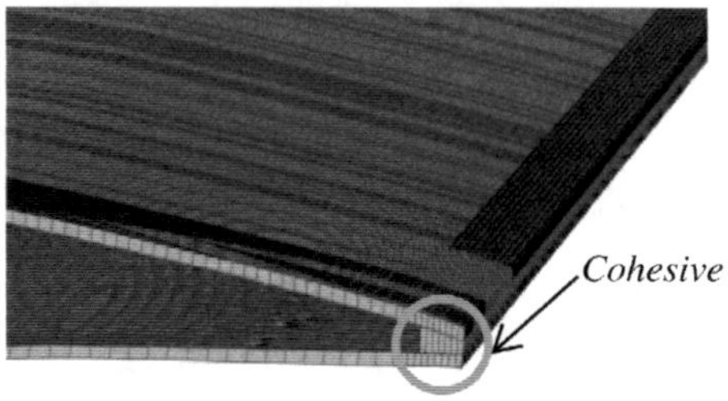

Figure 4. The Cohesive elements

In this section, we present the results related to the thermal coefficients and material properties of the materials under investigation. Table 1 provides an overview of the thermal coefficients of the different materials, namely GFRP, CFRP, and cohesive materials. These coefficients represent the rate of change of material dimensions in response to temperature variations. The values are considered equal in all directions. Table 2 tabulates the material properties of GFRP, CFRP, and cohesive materials. These properties play a crucial role in determining the mechanical behavior and performance of the materials. For GFRP and CFRP, the table includes parameters such as Young's modulus, shear modulus, Poisson's ratio, and fracture energy in crack opening as well as in crack sliding. It is worth mentioning that CFRP and GFRP matrerials are considered elastic lamina for biaxial fiber configuration. These properties provide insights into the stiffness, strength, and fracture response of the composite materials.

The detailed analysis of these material properties will shed light on the characteristics and behavior of the materials under consideration, enabling a comprehensive understanding of their performance in the subsequent sections.

Table 1. Thermal coefficient of materials

Material	α_T
GFRP [8]	5.04
CFRP [15]	2.50
Cohesive [16]	84.14

Table 2. Material properties

material	$E_1(GPa)$	$E_2(GPa)$	$G_{12}(GPa)$	υ_{12}	$G_{fI}(N.mm)$	$G_{fII}(N.mm)$
GFRP [8]	25	25	6.3	0.38	-	-
CFRP [15]	140	140	5.0	0.34	-	-
Cohesive [16]	3.42	3.42	-	0.32	1.00	0.60

By considering linear elastic behavior for CFRP/GFRP materials the relationship between stress and strain can be calculated as follows [10]

$$\begin{Bmatrix} \sigma_z^R \\ \sigma_x^R \end{Bmatrix} = \frac{E}{1-\upsilon^2} \begin{bmatrix} 1 & \upsilon \\ \upsilon & 1 \end{bmatrix} \begin{Bmatrix} \varepsilon_z^R \\ \varepsilon_x^R \end{Bmatrix} \tag{1}$$

where σ_i^R are residual stress and ε_i^R are residual strain components in the xz-plane of the adhesive layer. Moreover, E and υ are module of elasticity, and Poisson's ratio, respectively.

In the adhesive layer, the thermal residual strain is calculated as

$$\begin{Bmatrix} \varepsilon_z^R \\ \varepsilon_x^R \end{Bmatrix} = \begin{Bmatrix} \alpha_{T_1} - \alpha_{T_{2z}} \\ \alpha_{T_1} - \alpha_{T_{2x}} \end{Bmatrix} \Delta T \tag{2}$$

where α_{T1} and α_{T2i} stand for adhesive layer and bond line laminate in x and z direction, thermal coefficient and ΔT is the difference of temperature between curing T_1 and operation T_2 determined from

$$\Delta T = T_1 - T_2 \tag{3}$$

2.1. Thermal analysis of full-scale blade

The models underwent thermal loading caused by a temperature difference ΔT, as described in Equation (3). The adhesive's curing temperature was T_2 = 70 °C [10]. Since the material model used is not temperature-dependent, the operating temperature was chosen as T_1 = 23 °C. This temperature corresponds to the standard class 1 room temperature [17], at which the material properties were originally determined. It is worth mentioning that static loading is considered according to Equations (4) and (5) by Mikkelsen [18]

$$q_z^{flap} = \frac{8}{27}\rho_{air} u_R^2 \pi z \tag{4}$$

$$q_z^{edge} = \rho_{blade} A_{blade}\, g \tag{5}$$

where q_z^{flap} and q_z^{edge} are distributed flap and edge wind loads along the blade axis, respectively. ρ_{air}, ρ_{blade}, u_R, A_{blade} and g are air density, blade density, remote wind speed, area of the cross-section of blade materials, and gravitational constant, respectively. z indicates the axial blade coordinate measured from the root of the blade.

2.2. Equivalent stress

To account for the isotropic nature of the adhesive layer, the expression proposed by Betti [19] was utilized to calculate the residual equivalent stress. The following expression for calculating equivalent residual stress , σ_e^R, (Equation 6) provided an accurate means of determining the stress distribution within the adhesive layer as

$$\sigma_e^R = \left\{ \left(\sigma_z^R\right)^2 + \left(\sigma_x^R\right)^2 + \left(\sigma_y^R\right)^2 - 2\upsilon\left(\sigma_z^R\sigma_x^R + \sigma_x^R\sigma_y^R + \sigma_y^R\sigma_z^R\right) + 2(1+\upsilon)\left[\left(\tau_{zx}^R\right)^2 + \left(\tau_{xy}^R\right)^2 + \left(\tau_{zy}^R\right)^2\right]\right\}^{0.5} \tag{6}$$

where τ_{ij}^R are shear stress components in corresponding directions, respectively.

3. Results

The results obtained from the comprehensive analysis of the turbine blade under thermal loading conditions provide valuable insights into the behavior and performance of the adhesive and CFRP/GFRP materials. In this section, we present and discuss the key findings and observations from the analysis, focusing on the effects of thermal loading on the stiffness, stresses, and displacement of the blade.

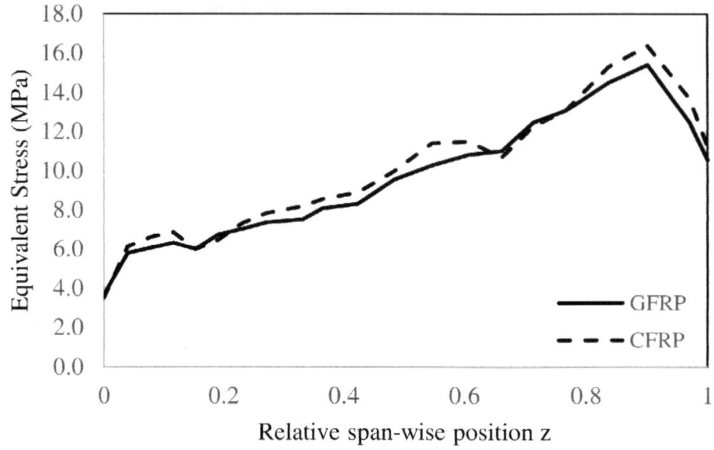

Figure 5. Stress versus relative span-wise position

The analysis of Figures 5 and 6 reveals interesting trends as we move from the root of the blade towards the tip, particularly in the region close to 80%-90% of the blade length. In the blade composed of GFRP material, it can be observed that this part of the blade experiences the highest levels of equivalent stress and displacement. Specifically, at this location, the equivalent stress reaches a magnitude of 15 MPa. Additionally, the thermal loading induces a displacement of -0.13 mm in the same region. It is noteworthy that the thermal effects diminish towards the tip of the blade due to the reduction in material length. Consequently, the impact of thermal stresses decreases accordingly.

Nevertheless, despite the lower thermal coefficient and higher stiffness of CFRP material compared to GFRP, unexpected results are observed. The CFRP blade exhibits approximately 4% higher displacement, 7% higher equivalent stress in comparison to the GFRP blade. This discrepancy can be attributed to the utilization of the same adhesive material at the trailing edge. Therefore, the choice of material and adhesive plays a crucial role in determining the stress and displacement characteristics of the blade under thermal loading conditions.

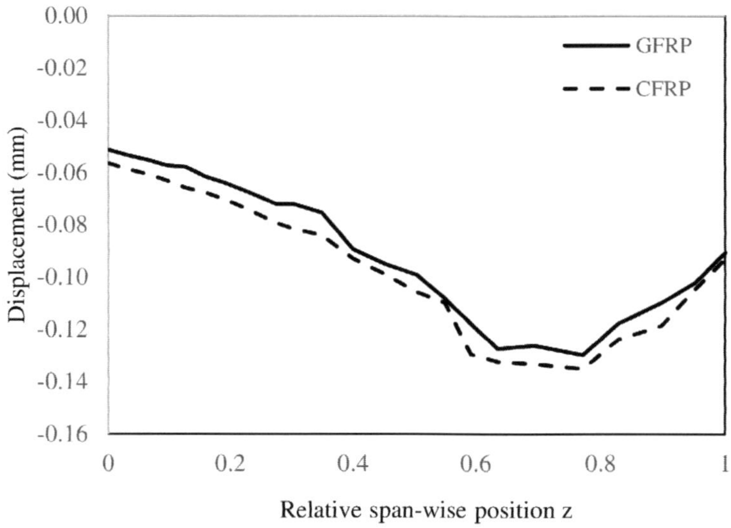

Figure 6. Relative displacement versus relative span-wise position

Figure 7 provides insights into the behavior of the blade material under thermal loading conditions. It clearly demonstrates that the application of thermal loading leads to the generation of stress within the blade material. Moreover, it is noteworthy that this stress attains its maximum magnitude in the region extending from 80% of the blade length towards the blade tip in both GFRP and CFRP blade.

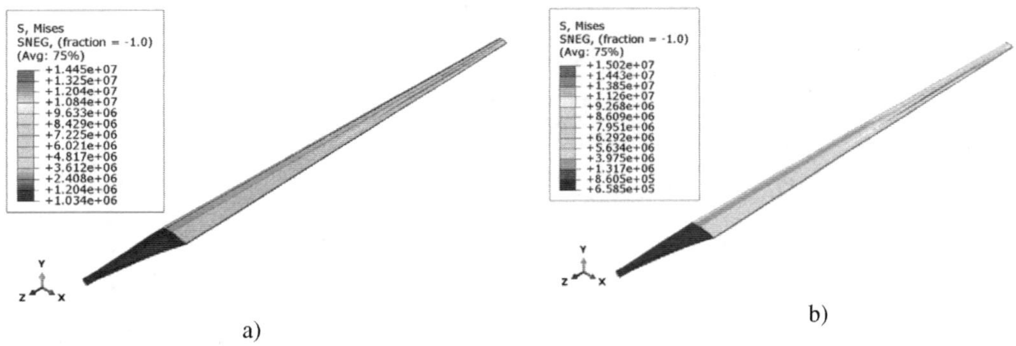

Figure 7. von Mises stress distribution in blade, a) GFRP blade, b) CFRP blade

Figure 8 presents a comprehensive visualization of the equivalent stress distribution within the adhesive material. It vividly illustrates how thermal loading influences the adhesive along its entire length, leading to the concentration of stress middle of the blade. This stress concentration phenomenon in the adhesive material in the middle of the blade is a direct consequence of the thermal loading applied to the structure. The application of thermal loading significantly affects the stiffness of the adhesive (in terms of SDEG, scalar damage in cohesive material), causing it to lose approximately 55% of its original stiffness (Figure 11a , SDEG=0.55). This reduction in stiffness makes the adhesive layer more susceptible to the applied loading conditions. Moreover, when considering the thermal loading, both the adhesive material and the GFRP and CFRP materials experience higher levels of stresses and strains compared to what was initially calculated during the design phase. These discrepancies in the loading conditions, coupled with the effects of thermal loading, can lead to unexpected failures in turbine blades. The results highlight a particular vulnerability near the blade tip, where a greater displacement is observed during the loading conditions. This increased displacement further contributes to the potential failure of the turbine blade. It is crucial to take into account these factors and accurately assess the impact of thermal loading to ensure the structural integrity and reliability of turbine blades under operating conditions.

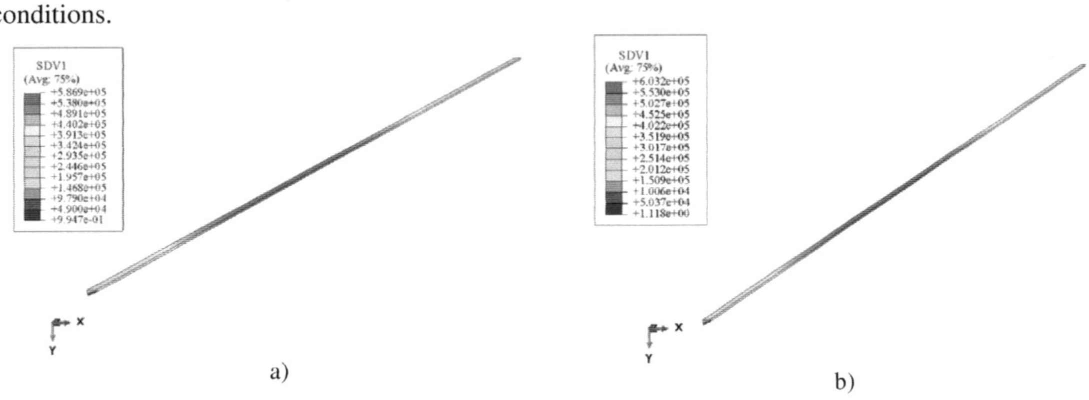

Figure 8. Equivalent stress distribution in cohesive elements a) GFRP blade, b) CFRP blade

Figures 9 and 10 provide valuable insights into the displacement behavior of the blade under different loading conditions. Specifically, Figure 9 illustrates the displacement distribution of the blade under equivalent static wind loading according to Equations (4) and (5), while Figure 10 depicts the displacement distribution resulting from a combination of thermal and wind loading, taking into account the presence of thermal residual stress. It is evident from these figures that the thermal residual stress significantly impacts the displacement distribution throughout the blade. Consequently, the blade experiences a higher level of displacement than originally anticipated. This alteration in the displacement distribution can have implications for the overall performance and structural integrity of the blade.

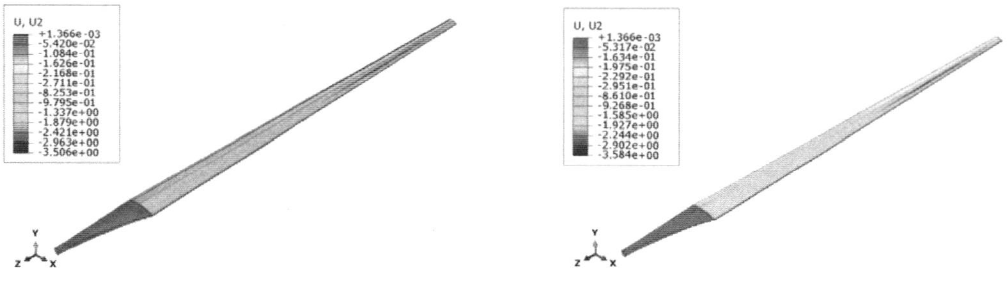

a) b)

Figure 9. Displacement of the blade subjected to wind equivalent static loading, a) GFRP blade, b) CFRP blade, (unit in m)

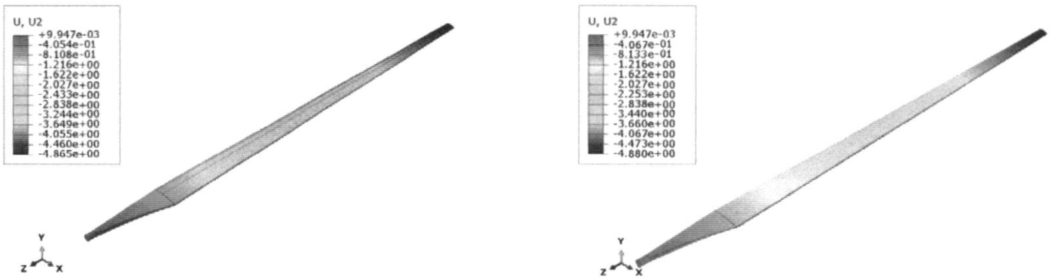

a)

b)

Figure 10. Displacement of the blade subjected to wind equivalent static and thermal loading, a) GFRP blade, b) CFRP blade, (unit in m)

By conducting a parametric study on the adhesive material (Figure 11), it was observed that reducing the shear fracture energy from 0.6 N.mm to 0.4 N.mm in the softening region of the adhesive's behavior has significant effects. Specifically, this reduction in shear fracture energy results in higher stresses being exerted on the adhesive material located at the trailing edge. Consequently, the stiffness of the adhesive decreases.

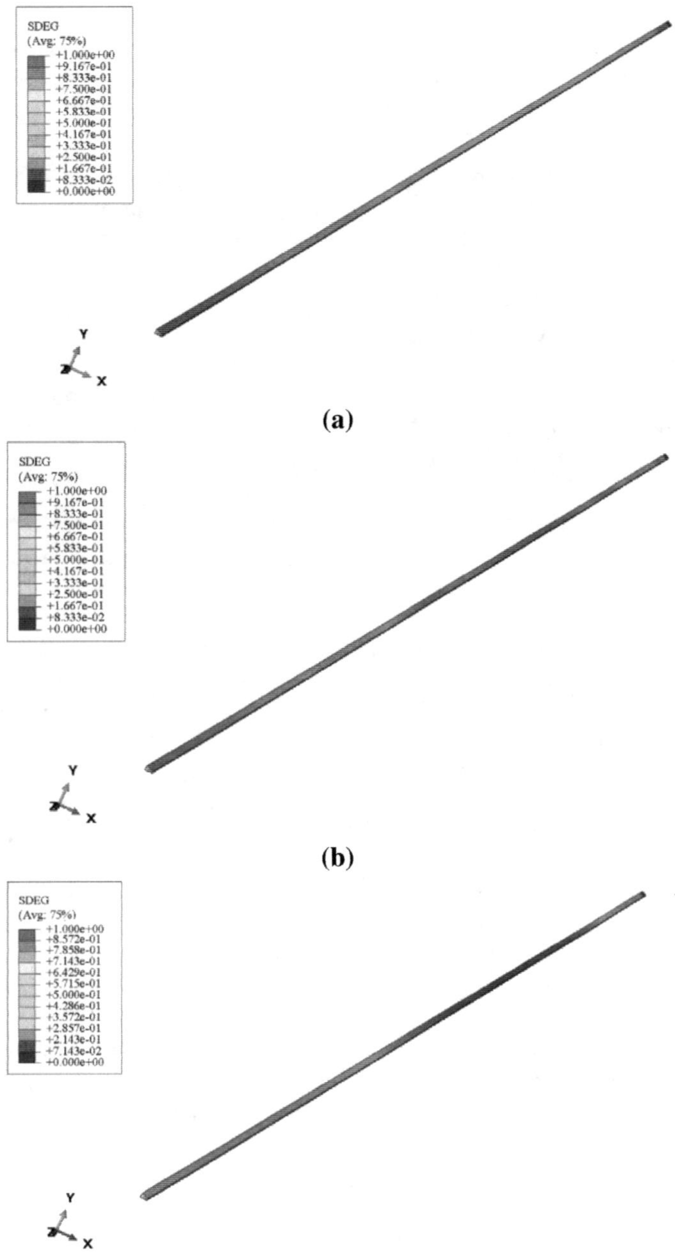

Figure 11. Scalar damage distribution in cohesive element with different fracture energy,a) 0.6 N.mm, b) 0.5 N.mm ,and c) 0.4 N.mm

Notably, in the relative spanwise position of 80%, the cohesive element with a fracture energy value of 0.4 N.mm experiences complete damage, indicated by an SDEG value of 1.0. This complete damage leads to the detachment of the connected CFRP or GFRP sheets, highlighting the critical role of the adhesive's mechanical properties in maintaining the structural integrity of the composite structure.

4. Conclusion

In conclusion, the results obtained from the analysis of the turbine blade under thermal loading conditions have provided insights into the behavior and performance of the adhesive, GFRP and CFRP materials. The findings can be summarized as follows:

- Moving from the root to the tip of the blade, the region close to 80%-90% of the blade length experienced the highest levels of equivalent stress and displacement.
- Thermal loading induced stress within the blade material, with the maximum stress magnitude occurring from 60%-80% of the blade length towards the tip.
- The thermal loading effects caused the adhesive to lose approximately 50% of its stiffness, making it more susceptible to applied loading conditions.
- Reducing the shear fracture energy of the adhesive in the softening region had significant effects, resulting in higher stresses on the adhesive material at the trailing edge and a decrease in stiffness. Complete damage was observed in the cohesive element with lower fracture energy, leading to the detachment of the connected CFRP/GFRP sheets.

These findings underscore the importance of considering thermal loading effects, material properties, and adhesive behavior in the design and analysis of turbine blades. Accurate assessment and understanding of these factors are essential for ensuring the structural reliability and performance of the blades under operating conditions. Despite efforts that have been faced with this research, to have an accurate model it is essential to develop an extensive experimental program. Moreover, material-level experiments need to model validation and verification of the behavior of the adhesive. Therefore, this parametric study is useful just to know the comparison range of the displacement and stress for CFRP/GFRP blade shape structures.

Acknowledgments

Support for this research was provided by GuilAzma, Inc. under a project for designing a stand for scanner and camera in controlling gate.

References

[1] Baldan A 2004 Adhesively-bonded joints and repairs in metallic alloys, polymers and composite materials: Adhesives, adhesion theories and surface pretreatment *J Mater Sci* **39** 1–49
[2] Antoniou A, Rosemeier M, Tazefidan K, Krimmer A and Wolken-Möhlmann G 2020 Impact of site-specific thermal residual stress on the fatigue of wind-turbine blades *AIAA Journal* **58** 4781–93
[3] Cairns D S and Peterson W M 2016 Fracture of thick adhesive joints in wind turbine blade structures *34th Wind Energy Symposium* pp 1–9
[4] Struzziero G and Teuwen J J E 2020 Residual Stresses Generation in Ultra-Thick Components for Wind Turbine Blades *Procedia CIRP* **85** 7–11
[5] Samborsky D D, Sears A T, Agastra P and Mandell J F 2011 Mixed mode static and fatigue crack growth in wind blade paste adhesives *Collection of Technical Papers - AIAA/ASME/ASCE/AHS/ASC Structures, Structural Dynamics and Materials Conference* 1–19
[6] Eder M A and Bitsche R D 2015 Fracture analysis of adhesive joints in wind turbine blades *Wind Energy* **18** 1007–22
[7] Jørgensen J B, Sørensen B F and Kildegaard C 2019 The effect of residual stresses on the formation of transverse cracks in adhesive joints for wind turbine blades *Int J Solids Struct* **163** 139–56
[8] Rosemeier M, Krimmer A, Bardenhagen A and Antoniou A 2019 Tunneling crack initiation in trailing-edge bond lines of wind-turbine blades *AIAA Journal* **57** 5462–74
[9] Nielsen M W 2012 *Predictions of process induced shape distortions and residual stresses in large bre reinforced composite laminates* (Technical University of Denmark)

[10] Rosemeier M, Krimmer A and Antoniou A 2020 Development of thermal residual stresses during manufacture of wind turbine blades *J Phys Conf Ser* **1452**
[11] Paul D, Varshney A, Mahajan P and Mishnaevsky L 2023 Post-repair residual stresses and microstructural defects in wind turbine blades: Computational modelling *Int J Adhes Adhes* **123** 103356
[12] Hwang M Y, Jang H L, Kim S, Lim J W and Kang L H 2021 Adhesive strength analysis and real-scale simulation for smart curing in a large turbine blade with carbon fiber-reinforced plastic spar cap *Renew Energy* **163** 1–14
[13] Jørgensen J B, Sørensen B F and Kildegaard C 2018 The effect of buffer-layer on the steady-state energy release rate of a tunneling crack in a wind turbine blade joint *Compos Struct* **188** 64–71
[14] Rosemeier M, Melcher D, Krimmer A, Wroblewski W and Antoniou A 2022 Validation of crack initiation model by means of cyclic full-scale blade test *J Phys Conf Ser* **2265**
[15] Aklilu G, Adali S and Bright G 2020 Tensile behaviour of hybrid and non-hybrid polymer composite specimens at elevated temperatures *Engineering Science and Technology, an International Journal* **23** 732–43
[16] Campilho R D S G 2017 *Strength prediction of adhesively-bonded joints* (CRC Press)
[17] DIN EN ISO 2008 291 2008 *Plastics–Standard Atmospheres for Conditioning and Testing (ISO 291: 2008)*
[18] Mikkelsen L P 2016 A simplified model predicting the weight of the load carrying beam in a wind turbine blade *IOP Conf Ser Mater Sci Eng* **139**
[19] Betti E 1885 *Il Nuovo Cimento, ISBN:0243470290*

IOP Conf. Series: Materials Science and Engineering 1293 (2023) 012036 doi:10.1088/1757-899X/1293/1/012036

Polymer composites with short recycled glass fibres manufactured by vacuum infusion

J Beauson[1,2], **K Munk Jespersen**[1], **M Rojas**[1], **A Canalis**[1], **R Kumar**[1], **T Løgstrup Andersen**[1], **A B Abrahamsen**[1]

[1]DTU Wind and Energy Systems, Technical University of Denmark, Frederiksborgvej 399, 4000 Roskilde, Denmark
[2]E-mail: jube@dtu.dk

Abstract. Recycling glass fibre reinforced thermoset composites remains a challenge despite years of research in the field. Among the barriers for a successful recycling, the properties of the recycled materials and their application in new products represent the main ones. With a pyrolysis recycling process, clean glass fibres can be extracted and separated from the polymer matrix. In this paper, recycled glass fibres are incorporated in a new polymer composite structure using both virgin and recycled glass fibres. Recycled glass fibres have a Young's modulus comparable to virgin glass fibres. The objective is to identify a composite lay-up using virgin and recycled glass fibres, and where the Young's modulus of recycled fibres is beneficial for the resulting composites mechanical properties. To do so, a sandwich structure composite with virgin fibres on the outer layer and recycled fibres in the inside is designed and manufactured. In order to produce flat panels despite the short and random fibres, a manufacturing procedure ensuring a good dispersion of the fibres and controlling the thickness of the panel is developed. Panels are produced incorporating two different weight fraction of recycled glass fibres. The quality of the panels are characterized using microstructure observations, fibre porosity analysis and mechanical testing. Results shows that the sandwich composites have a low porosity content and achieves higher bending stiffness than a composite only made of unidirectional virgin glass fibres.

1. Introduction

The recycling of wind turbine blades and of glass fibre reinforced thermoset polymer composites has been studied for many years, but remains challenging. A wind turbine blade is a one-piece component, which cannot easily be separated. The hollow structures made of two aerodynamic shells and a beam are composed of different composite materials. The type of material and the diversity of materials used in blades is a challenge from a recycling point of view. The aerodynamic shells are made of thin glass fibre reinforced laminate separated by a core materials made of balsa wood, PET or PVC foam. The long and slender beam is a combination of unidirectional fibre reinforced polymer composites and multiaxial fibre sandwich composite named shear webs. The fibres used are glass fibres, carbon fibres or a combination of both. The polymer resin may be epoxy, polyester or vinylester. Some blades may be manufactured as one single piece, while other blades will be assembled using adhesive joints, which cannot be easily released [1].

At end-of-life, due to the cross-linked nature of thermoset polymers, glass fibre reinforced polymer composite, as used in wind turbine blades, cannot be soften, reshaped or easily separated into fibres and matrix material. Nevertheless, recycling processes have been developed and wind turbine blades can be transformed into different types of recycled materials [2,3]. These processes can involve cutting, shredding, heat treatment and/or chemical treatment. The resulting recycled materials can be sections of blades, shredded composites or fibres; with dimension varying from a few meters down to few millimetres or microns. The applications for these recycled materials are diverse. For larger blade sections, footbridges or outdoor furniture may be considered [4–6]. With a pyrolysis recycling process, clean glass fibres can be extracted and separated from the polymer matrix [7–9]. An obvious application for recycled glass fibres is the replacement of virgin glass fibres in existing products. This option is challenged by the properties of recycled glass fibres, which are shorter, randomly oriented and with lower strength properties compared to virgin glass fibres. For thermal recycling temperatures reaching 450°C or more, the loss in fibre strength is of at least 40% [10]. To overcome this challenge, part of the research has explored restoring the strength properties of recycled glass fibres [11]. Another option is to investigate applications for recycled glass fibres, where the recycled fibres do not replace virgin fibres, but are used for their own characteristics and properties. The advantage of recycled glass fibres is that compared to other recycled materials such as shredded composite obtained with mechanical recycling, they have a clean surface. It was shown in previous publication that the adhesion between shredded composite and new polymer composite was poor [12]. Glass fibres obtained from pyrolysis are likely to have better fibre/matrix bonding.

In this paper, recycled glass fibres obtained from pyrolysis are incorporated in a new polymer composite structure using both virgin and recycled glass fibres. The idea is to take advantage of the high stiffness properties of recycled glass fibres and combine them to virgin glass fibre to provide high strength properties to the final composite. To do so, a sandwich structure composite with virgin fibres on the outer layer and recycled fibres in the inside is proposed and investigated. The objective of this paper is twofold. First, this paper aims at establishing a manufacturing procedure, which results in high quality composite materials. Second, this paper studies the impact of using recycled glass fibres on the mechanical properties of the composites produced. Some of the challenges to overcome when manufacturing composites with short randomly oriented recycled glass fibres and a vacuum infusion process are to achieve: (i) a good dispersion and distribution of the fibres in the composite, (ii) an even thickness and a flat surface and (iii) a high fibre content. To study the impact of incorporating recycled glass fibres, the glass fibres virgin and recycled are thoroughly characterized and composites are produced incorporating two different weight fraction of recycled glass fibres. The quality of the composites produced are characterized using microstructure observations, fibre porosity analysis and mechanical testing.

2. Materials and methods

2.1 Fibres and matrix properties

In this paper, the virgin fibres used are E-glass fibres in a unidirectional fibre fabric from Saertex (Textile structure 7004118) with an areal weight of 640 g/m^2. The epoxy used is the PRO SET INF 114 epoxy with the PRO SET INF 211 epoxy hardener. The recycled glass fibres are obtained from a pyrolysis process performed by the company Makeen Energy [13]. The origin of the composite processed with pyrolysis is unknown.

2.1.1 Density. To measure the density of the glass fibres, a gas pycnometer Quantachrome Ultrapyc 1200e is used, with nitrogen as the displacement medium. For the virgin fibres, the fibres are winded onto a spool and placed in the pycnometer. For the recycled glass fibres, the fibres are ground with a mortar and placed in a container. Before the measurements, the fibre samples (7.2 g of virgin fibres and 13.5 g of recycled fibres) are dried for a minimum of 12 hours. For both types of fibres, 3 repeated density measurements are performed. To measure the density of the epoxy resin, samples are collected from the inlet tube during the manufacturing of the composites. The dimension of the specimen are 2 cm long and the diameter of the tube. The samples are weighed in air, and in water and the density of the material is calculated using Archimedes law.

2.1.2 Mechanical properties. Single fibre tensile tests were performed using a semi-automated single fibre testing machine, "Favimat+", combined with Robot 2. Initially, a vibroscope test is performed to determine the linear density of the fibre. Based on the linear density (T [kg/m]) and the measured fibre density (ρ [kg/m^3]) values, the cross-sectional area ($A = T/\rho$) of the fibre is calculated. The fibres were tested at a gauge length of 30 mm. For both types of fibres, at least 100 single fibres are tested. The fibre Young's modulus is calculated according to the composite materials standards, between 0.05% and 0.25% strain. The mechanical properties of the epoxy resin are obtained from the literature.

2.2 *Manufacturing of composites*

The composites are manufactured using a vacuum infusion process. To enable a good dispersion and distribution of the recycled fibres, achieve an even thickness, a flat surface and reach high fibre content, the manufacturing procedure is made of four steps: fibre dispersion, fibre distribution, pressing and vacuum infusion. Figure 1 shows a schematic representation of the steps. In the following sections, the steps are described in more details.

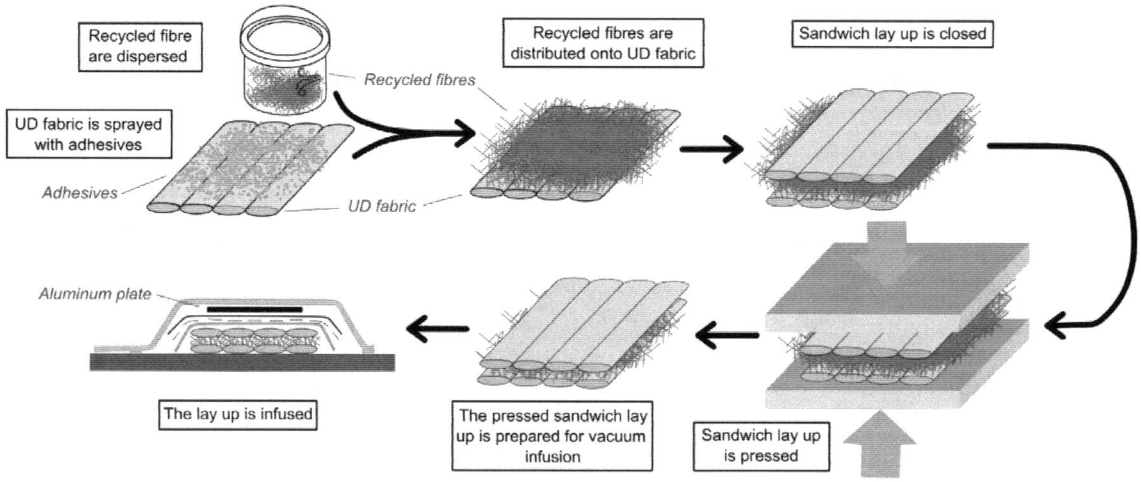

Figure 1. Schematic representation of the manufacturing steps

2.2.1 Fibre dispersion. To improve the distribution and the resin impregnation of the short recycled fibres, the fibres are dispersed using compressed air in a closed bucket. The procedure lasts for approximately 1 min. The objective of the procedure is to separate the fibre bundle and create a woollen structure of individual fibres.

2.2.2 Fibre distribution and lay-up. The laminate dimensions are 300 x 300mm. The lay-up consists of three layers in total, two outer layers of unidirectional virgin fibres and an inner layer of recycled fibres. Two laminates are prepared where the mass of recycled glass fibres is varied. These composites are named Composite #1 and Composite #2. The mass of recycled glass fibre is set to be half (Composite #1) or equal (Composite #2) to the weight of virgin fibre as illustrated in figure 2. In any case, the portion of recycled fibres is divided in two and distributed manually on the two outer layers of virgin fibre fabrics. In order to close the sandwich structure and avoid the fibres to move around, an adhesive spray is used (INFUTAC Green from Diatex). The adhesive is sprayed on the fabric and the recycled fibres are pressed against the fabric. Then, the two outer layers of unidirectional virgin fibres with the recycled glass fibres attached are assembled to close the structure into a sandwich structure containing the recycled fibres as an inner layer.

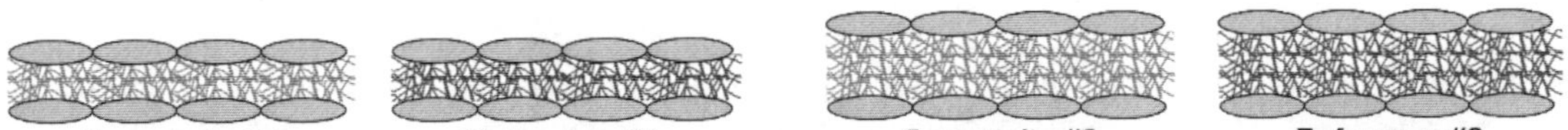

Figure 2. Schematic representation of the composites manufactured and modelled

2.2.3 Pressing. Once the sandwich lay-up is closed, it is placed into a press facility in order to compress it. The objective of this step is to reduce the thickness of the dry lay-up before the vacuum infusion. The applied force improves the compaction of the dry fibres. The pressing procedure is as follow. The fibre lay-up is pressed for 1 min at 70 kN. Considering the dimension of the lay-up, the equivalent pressure is 0.8 MPa. The procedure is repeated until the height of the lay-up could not be further decreased. After pressing, the lay-up is packed for vacuum infusion. An aluminium plate with dimension 240mm x 240mm x 3.5 mm is introduced in the vacuum bag. It is placed on top of the laminate above the release film and peel ply and below the vacuum bag. The vacuum bag encloses the whole system. The aluminium plate plays the role of a second-sided mould during the vacuum infusion. The plate helps to avoid a wavy surface for the composite.

2.2.4 Vacuum level and curing profile. The composite is infused with a vacuum level of -999 mbar. The curing cycle selected is 24 hours at 25°C, followed by 8 hours at 82°C.

2.2.5 Additional composite materials. In order to estimate the volumetric composition of the individual layers of the sandwich composites manufactured, a unidirectional fibre composite was prepared with vacuum infusion and using the same fabric and epoxy resin as in the sandwich composites. This composite is named Reference UD. To compare the mechanical properties of Composites #1 and #2, two additional reference composites are modelled, but not manufactured. These two reference composites are sandwich composites with the same fibre volume fraction and thickness of individual layers, as Composites #1 and #2, however only assuming virgin glass fibre properties. These composites are named Reference #1 and Reference #2. Reference #1 uses the characteristics of Composite #1 and Reference #2 uses the characteristics of Composite #2.

2.3 Characterization of composites

2.3.1 Microstructure characterization. To characterize the microstructure, samples with dimensions 20 mm x thickness of the plate are cut from each composite. The samples are casted in epoxy, and the sample cross-section in the plane normal to the unidirectional fibre direction is grinded and polished. Observations are made using an optical microscope (Leica, IM5000).

2.3.2 Volumetric composition. To determine the fibre, matrix and porosity content, five rectangular samples of dimensions 15 x 20 mm x thickness are cut from each composite. The samples are roughly polished, dried overnight, and weighed. The samples are then sealed with paraffin wax to include the correct porosity volume in the determinations. The sealed samples are weighed in air, and in water in order to determine the composite density. Then, the epoxy matrix in the samples is burnt off overnight at 565°C. The weight of the remaining glass fibres is recorded in order to determine the fibre weight content. Using standard equations for composite materials and based on the measured values of fibre weight content, composite density, fibre density, and matrix density, the volume content of fibres, matrix and porosity in the composites is calculated [14].

2.3.3 Static tensile test and Young's modulus. A static tensile test is performed to determine the Young's modulus of the composite. The test follows the ISO 527-5:2009(E) standard. The length of the tensile specimens is however adjusted to the dimensions of the composite manufactured. 4 tensile

specimens are tested for Composite #1 and Composite #2. The cross-sectional area of the specimens is measured at 3 different locations, and the averaged cross-sectional area is used for the calculation of stress. Tensile tests are performed with an Instron tensile test machine (Instron, Buckinghamshire, United Kingdom,) having a load cell of 500 kN and a cross-head speed of 2 mm/min. The Young's modulus is determined in the strain range from 0.05 to 0.25%.
The Young's modulus of the Reference UD plate is calculated using a rule of mixture, see equation (1), where E_{Ref} is the Young's modulus of the Reference UD composite, V_f is the fibre volume fraction, E_f and E_m the Young's modulus of the fibre and matrix respectively.

$$E_{Ref} = E_{f\ Ref} \cdot V_{f\ Ref} + E_m(1 - V_f) \tag{1}$$

2.3.4 Flexural properties. To determine the impact of the sandwich structure on the flexural properties of the composite, 3-point bending tests are performed according to the ISO 14125:1999 standard. The width of the specimens is 15 mm and the length equals the thickness multiplied with 20. 10 specimens are tested for Composite #1 and Composite #2.

The bending stiffness *[EI]* of the composites is calculated using equation (2), where E_{core} is the Young's modulus of the inner section of the composite made of recycled fibres. E_{core} is calculated using equation (4). b is equal to 15 mm for all composites and corresponds to the width of the 3-point bending test specimens, h_{core} is the thickness of the inner section of the composite. E_{Ref} is the Young's modulus of the Reference UD composite, h_{skin} is the thickness of one of the skin layers made of UD virgin glass fibres. h_{core} and h_{skin} are roughly measured from micrographs. d is the distance to the neutral axis of the composite beam and is calculated using equation (3). Figure 3 shows a schematic representation of the parameters used in equation (2) and (3).

$$[EI] = E_{core}\frac{bh_{core}^3}{12} + 2E_{Ref}\left(\frac{bh_{skin}^3}{12} + bh_{skin}d^2\right) \tag{2}$$

$$d = \frac{h_{core}}{2} + \frac{h_{skin}}{2} \tag{3}$$

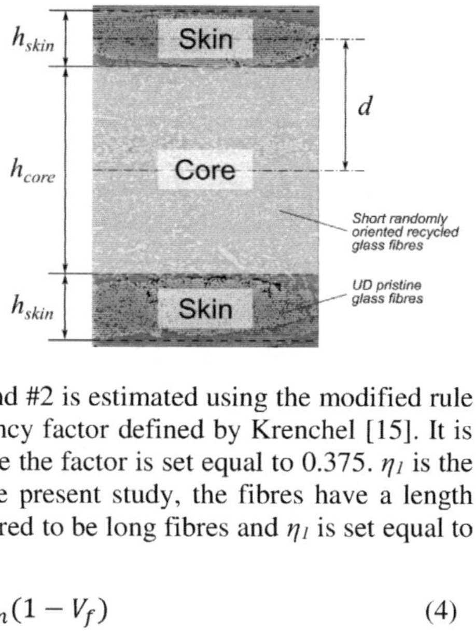

Figure 3. Schematic representation of the thickness parameters used in equations

The Young's modulus of the core section of Composite #1 and #2 is estimated using the modified rule of mixture in equation (4), where η_o is an orientation efficiency factor defined by Krenchel [15]. It is assumed that the fibres are randomly oriented in 2D, therefore the factor is set equal to 0.375. η_l is the fibre length efficiency factor as defined by Cox [16]. In the present study, the fibres have a length ranging from few mm to few cm. There are therefore considered to be long fibres and η_l is set equal to 1.

$$E_{core} = \eta_o \cdot \eta_l \cdot E_{fcore} \cdot V_{fcore} + E_m(1 - V_f) \tag{4}$$

The deflection of the specimens during the 3-point bending test is calculated using the Timoshenko-beam theory in equation (5), where P is the load applied, L is the length of the 3-point bending test specimen, *[EI]* is the bending stiffness, h_{core} is the thickness of the inner section of the composite and G_{core} is the shear properties of the core section, which is calculated using equation (6).

$$\delta = \frac{PL^3}{48[EI]} + \frac{PLh_{core}}{16G_{core}bd^2} \tag{5}$$

The shear properties of the core section of the sandwich structured composite, G_{core} is calculated using equation (6), where E_{core} is the Young's modulus of the core section, and υ is the Poisson's ratio, which is set equal to 0.4.

$$G_{core} = \frac{E_{core}}{2(1+\upsilon)} \tag{6}$$

3. Results and discussion

3.1 Fibres and matrix properties

Figure 4 shows the recycled glass fibres. The fibres have different length, ranging from a few cm up to 10 cm and are randomly oriented. The fibres are generally white, however some fibre bundles are found to have a light yellow colour and few fibre bundles are grey. The variation in length is due to the shredding process taking place before the pyrolysis process. The properties of the fibres recycled and virgin are presented in table 1. In addition, the stress strain curves of single fibre test are presented in figure 5.

Figure 4. Recycled glass fibres obtained from pyrolysis

Table 1. Fibres and matrix properties

	Young's modulus [GPa]	Strength [MPa]	Density [g.cm^{-3}]
Virgin glass fibres	81 ± 2	1977 ± 532	2.641 ± 0.002
Recycled glass fibres	86 ± 3	458 ± 170	2.618 ± 0.006
Epoxy	3.35 ± 0.08*	85.8 ± 0.4*	1.158 ± 0.000

* From literature [17]

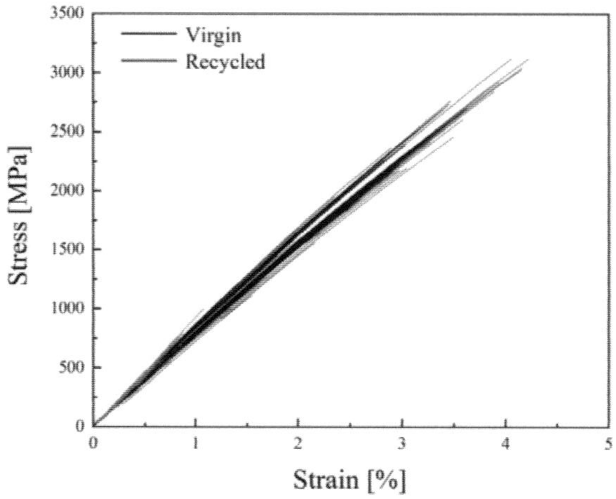

Figure 5. Stress-strain curves obtained from single fibre testing of virgin and recycled glass fibres.

3.2 Manufacturing of composites

3.2.1 Fibre dispersion. The fibre dispersion procedure is applied for approximately 1 min. The procedure is illustrated in figure 6. In this figure, the same bucket is used in all 3 images. The fibres before and after the dispersion are randomly oriented, however, before dispersion, they show a bundle morphology, whereas the fibres after dispersion show a woolly or a candy floss like structure, where fibres are separated from each other.

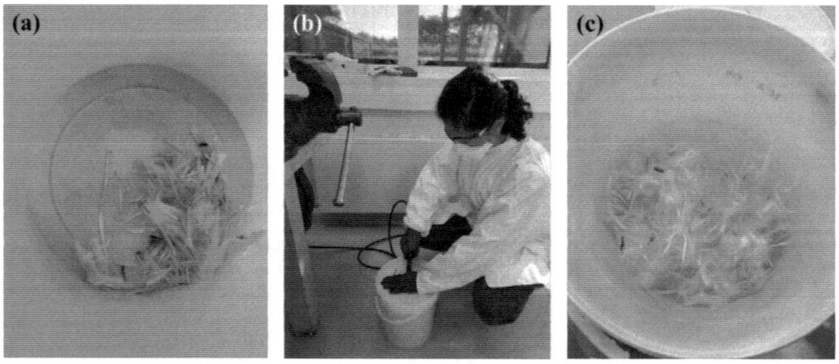

Figure 6. Illustration of the dispersion procedure (a) Recycled glass fibres before dispersion, (b) Compressed air is introduced in the bucket using a hole in the lid and (c) recycled glass fibres after dispersion

3.2.2 Fibre distribution and lay-up. Figure 7 shows some of the main steps in the distribution of the fibres and the fibre lay-up. Figure 7 (a) shows how the adhesive is centred in the middle of the fabric. Figure 7 (b) shows the resulting recycled glass fibres distributed on a virgin glass fibre fabric. Figure 7 (c) shows how the adhesive is effective in keeping the fibres in place. The measured mass of recycled glass fibres used for Composite #1 and Composite #2 is detailed in table 2. The weight fractions of recycled fibres, with respect to the total weight of fibres is calculated and presented in table 2.

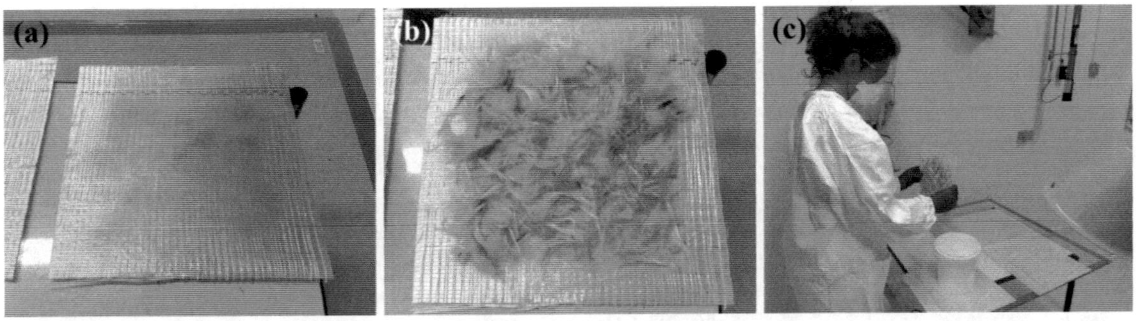

Figure 7. Illustration of the fibre distribution process (a) The adhesive is distributed on the virgin glass fibre fabric; (b) The recycled glass fibres are pressed onto the fabric with the adhesive; (c) An example of how the adhesive keeps the recycled fibres on the fabric layer

Table 2. Mass of virgin and recycled glass fibres used in the composites

Composite name	Mass of virgin glass fibres [g]	Mass of recycled glass fibres [g]	Weight fractions of recycled fibres [%]
Composite #1	120	60	33
Composite #2	117	120	51

3.2.3 Pressing. Figure 6 shows the impact of the pressing procedure. The thickness of the fibre lay-up is measured. Given the uneven surface of the lay-up, the thickness is approximately measured using a ruler placed next to the fibre lay-up. The thickness decreased from above 20 mm in figure 6 (a) down to 10 mm after two rounds of pressing in figure 6 (b). After two rounds of pressing, the thickness of the lay-up could not be further decreased. During pressing, fibres cracking could be heard.

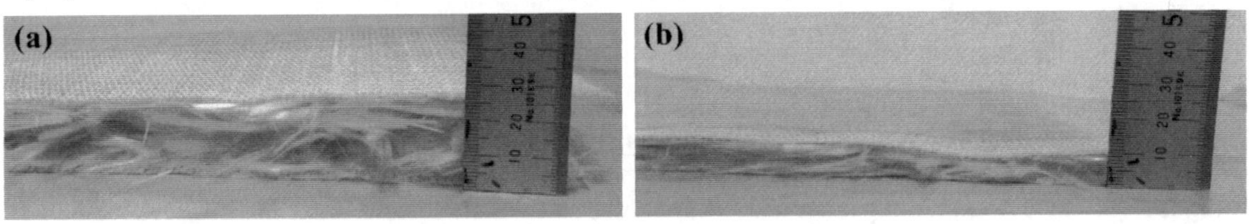

Figure 8. Illustration of the impact of pressing on the fibre lay-up for Composite #2 (a) Fibre lay-up before pressing, (b) After two pressing.

3.2.4 Vacuum infusion and curing. The composites produced are shown in figure 9, where images (a) and (b) show Composite #1 and Composite #2 respectively. As the recycled glass fibres are placed in the centre of the unidirectional fibre fabric, the edges of the composites are thinner than in the centre of the composite. To exclude this part from the characterization, 25 mm of the edges of the panel are cut out, as illustrated in figure 9 (b). Figure 9 also shows that the colour of the plate is not uniform. The black and brown areas indicate areas where recycled fibres are concentrated. Finally, figure 9 show the samples and specimens cut out for the characterization. The thickness of the composite is measured in the gauge section of the tensile test specimens and the bending test specimens. The measurements are reported in table 3 and show that Composite #1 is nearly 2 mm thinner than Composite #2. The variation in thickness is rather small for the Reference UD and Composite #1. For Composite #2 the variation in thickness is of about 10%, which is significant. This could be due to the larger amount of recycled glass fibres which needed to be distributed manually. Nevertheless, these measurements tend to indicate that the manufacturing process is quite efficient in producing uniform and flat panels. The remaining of the properties presented in table 3 are discussed in section 3.3.

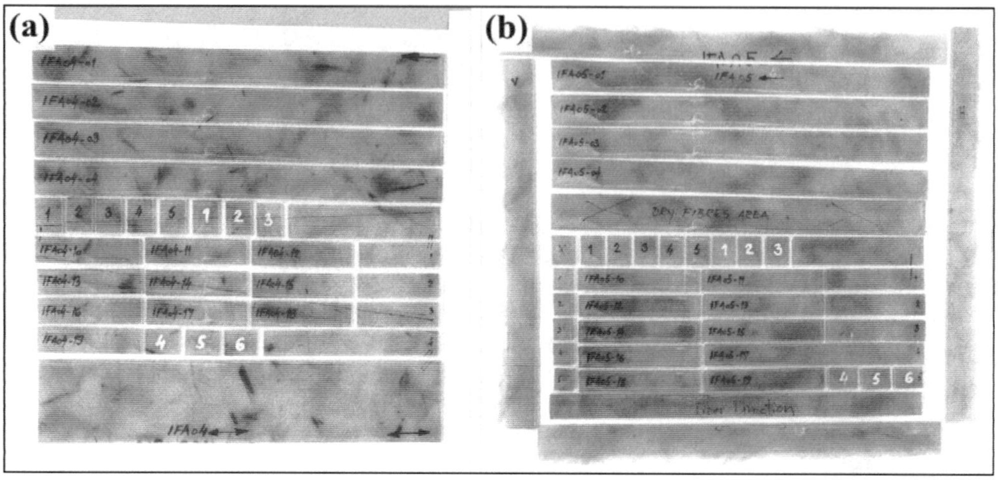

Figure 9. Manufactured composites (a) Composite #1, (b) Composite #2.

Table 3. Composites properties measured

Composite name	Thickness [mm]	FVF [%]	Porosity content [%]	Young's modulus [GPa]	3-point bending test Max. Load [kN]	3-point bending test Deflection [mm]
Reference UD	1.22 ± 0.08	47.5 ± 0.2	0.4 ± 0.1	Not measured	Not measured	Not measured
Composite #1	2.90 ± 0.07	26.8 ± 1.4	0.15 ± 0.05	17.7 ± 0.1	1.18 ± 0.05	3.7 ± 0.2
Composite #2	4.72 ± 0.44	24.3 ± 2.2	0.26 ± 0.04	13.8 ± 0.3	1.29 ± 0.11	4.1 ± 0.5

3.3 Composite characterization

3.3.1 Microstructure. Figure 10 and 11 show the microstructure of Composite #1 and #2, respectively. The overview shows the sandwich construction of the composites, where the outer layers are made of densely packed unidirectional fibre bundles, see figure 10 (c), and the inner section of the composite is made of randomly oriented fibres with lower fibre volume fractions, see figure 11 (c). There are no large porosities visible and composite #1 and #2 are found to be of good quality. The approximate thickness of the different layers measured on the micrographs is reported in table 4. The outer layer made of one layer of unidirectional fibre fabric is referred to as skin.

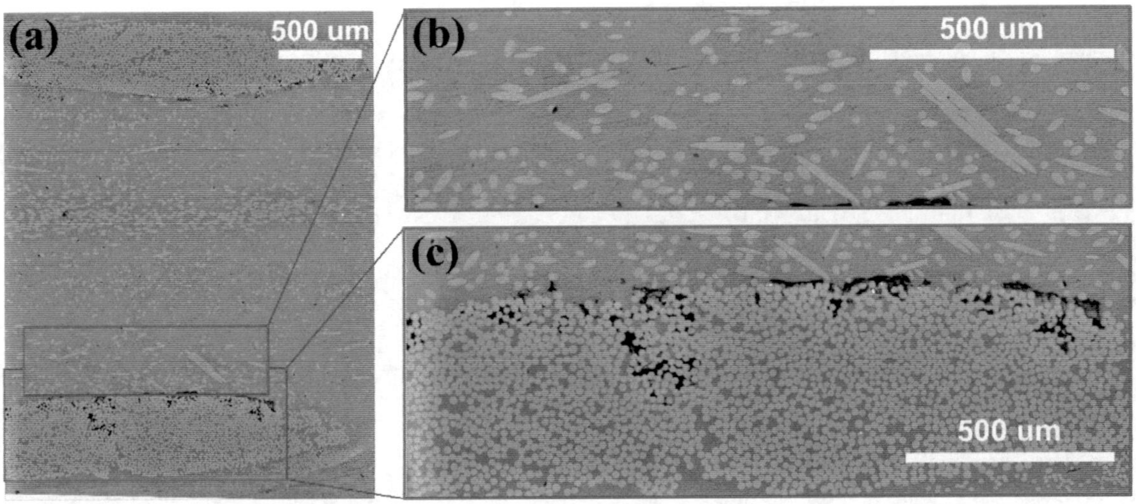

Figure 10. Microstructure of Composite #1 with (a) Overview (b) Details of the recycled fibres, and (c) Details of the region between UD virgin glass fibres and recycled fibres

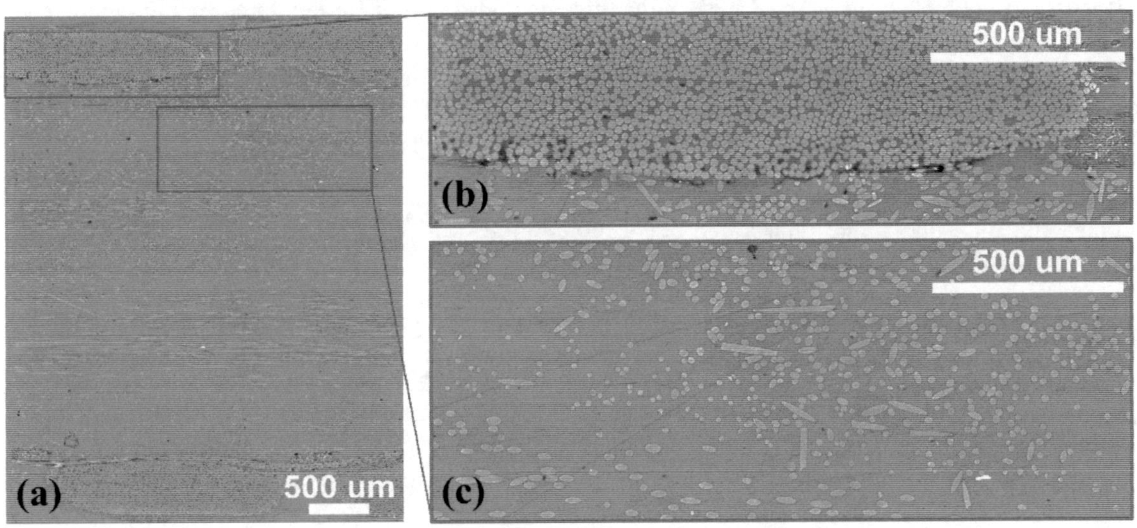

Figure 11. Microstructure of Composite #2 with (a) Overview (b) Details of the region between UD virgin glass fibres and recycled fibres, and (c) Details of the recycled fibres

Table 4. Properties of individual layers in sandwich composite

Name of layer	Thickness [mm]	FVF [%]	Young's modulus [GPa]
Skin	0.5	47.5	40.2
Core Composite #1	1.9	15.9	7.9
Core Reference #1	1.9	15.9	7.6
Core Composite #2	3.7	18.2	8.6
Core Reference #2	3.7	18.2	8.3

Figure 10 and 11 also show details of the microstructure of the two composites. In both composites, small porosities located in between the unidirectional glass fibres are observed, see figure 10 (c) and figure 11 (b). The size and the location of these porosities suggest dry and not wetted fibres or a poor impregnation of resin. These small porosities are consistently found at the interface between the unidirectional fibre bundles and the randomly oriented recycled fibres. This could indicate that the adhesive used in order to prepare the fibre mats also prevents resin to properly impregnate the fibres. The use of adhesive is an advantage for the distribution of the recycled fibres onto the virgin fibre fabric. However, in order to determine the impact of the adhesive on the composite performance, additional experiments should be conducted without the adhesive. Another observation from the detailed microstructure images are broken fibres with sharp edges, highlighting the very brittle nature of the recycled glass fibres, see figure 10 (b) and figure 11 (c). Finally, when adding the thickness measured on the microstructure images for the skin and the core, the total thickness is approximately matching the thickness measured on the composites. In the next section, the fibre volume fraction of the individual layer is calculated.

3.3.2 Volumetric composition. The fibre volume fraction and the porosity content of all composites is reported in table 3. The porosity content for all composites is low and this result is in good agreement with the microstructure observations. The fibre volume fraction is higher for the Reference UD composite than for the composite including recycled glass fibres. This is expected as the Reference UD composite only contains unidirectional fibres which can be packed to a higher degree than 2D randomly oriented fibres. The volume fraction of recycled fibres in the core section of the composite, is estimated with a rule of mixture using the thickness of the different layers as measured on the micrographs and the total fibre volume fraction measured on the three composites. The fibre volume fraction in the core section of the composite is estimated to 15.9 % for Composite #1 and 18.2% for Composite #2. The results are reported in Table 4. Composite #2 achieved a slightly higher fibre volume fraction in the core section of the composite, this indicates that the fibres may have packed to a higher degree than in Composite #1. However, given the standard deviation on the fibre volume fraction measurement and the heterogeneity of the material, these results may be considered similar. To increase the fibre volume fraction in the core section, future work may include to manufacture panels with different fibre length. Longer fibres may be easier to pack in a 2D fashion. Future work may also consider how to distribute short recycled fibres to facilitate their packing.

3.3.3 Mechanical properties. Table 3 also summarizes the experimental results obtained for the mechanical testing of the composites. The Young's modulus for the composite made with recycled fibres is obtained experimentally while the Young's modulus of the Reference UD composite is estimated theoretically and presented in table 4. The Young's modulus on the Reference UD plate is calculated using the material properties reported in table 1 and equation (1). The Young's modulus on the Reference UD plate is higher than the ones measured on Composite #1 and #2. This result is expected since Composite #1 and #2 are thicker, with a core section made of a lower fibre volume fraction and fibres randomly oriented. A theoretical estimation of the Young's modulus of the skin and the core section of Composite #1 and #2 is calculated using the fibre volume fraction and the fibre properties reported in table 1 and 4. With these Young's modulus, the bending stiffness and the deflection at max load and at 1000N are calculated. The results are presented in table 5.

Table 5. Calculated flexural properties

Composite name	Bending stiffness *[EI]* [GPa.mm^2]	Deflection at max. load [mm]	Deflection at 1000N [mm]
Composite #1	950	2.8	2.3
Reference #1	947	2.7	2.3
Composite #2	3219	2.1	1.6
Reference #2	3198	2.1	1.7

A first observation is that the calculated deflections at maximum load are smaller than the ones measured experimentally. An example of load displacement curve for a sample tested in Composite #1 is presented in figure 12. From this curve, one can see that around the maximum load the curve shows serrations probably indicating that local damages are initiating before the maximum load is reached. The theoretical estimations of deflection do not take these aspects into account. In future work, fitting the experimental curve with a second order polynomial fitting as described in [18] could provide the deflection at maximum load assuming no serration. Another observation is that the deflection of Composite #2 at maximum load is larger than the one of Composite #1. The theoretical deflection calculated indicate that it should be opposite. At maximum load, in theory, the thicker sandwich composite should have a smaller deflection. To further analyse the reasons for this observation, the use of strain gauges on the specimens would be needed to obtain more precise displacement measurements.

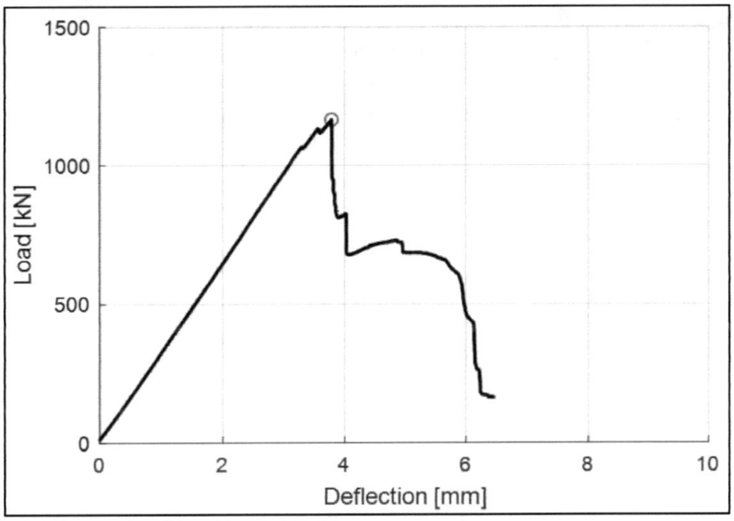

Figure 12. Example of load displacement curve obtained with the 3-point bending test

Finally, when comparing Composite #1 to Reference #1 and Composite #2 to Reference #2, the use of recycled fibres in the core section does not impact the resulting bending stiffness and deflections. The results between the manufactured and the modelled reference composites are similar. This results could be interesting for composite applications, where bending stiffness is required in order to avoid large deflection, as for example in nacelle covers. However, future work is needed to check the requirements for nacelle covers and validate Composite #1 and #2 as potential candidates for this application.

4. Conclusions

Glass fibres recovered from thermoset polymer composites with a thermal recycling process are known to be brittle. Despite the brittleness of these fibres, the objective of the present study is to identify composite structures that can take advantage of their high stiffness properties. Recycled glass fibres have Young's modulus comparable to virgin glass fibres. Sandwich structured composite were identified as a potential candidate. Sandwich composite with virgin glass fibres on the outer layer and short randomly oriented recycled fibres in the inner section were manufactured. The experimental work focused on the production of flat panels despite the short length and the random orientation of the recycled fibres. A method was developed to ensure a good dispersion of the fibres and a uniform thickness of the panels and composite panels were manufactured with vacuum infusion. Results show that the composites are of good quality, with a low porosity content. Future work, should include the manufacturing of panels without any fibre preparation in order to investigate the impact of these preparation steps on the resulting composites. The mechanical testing results show that the sandwich composites manufactured with recycled glass fibres and composites modelled with virgin fibres achieve the same flexural properties. This result is expected since the flexural properties are mainly governed by geometrical parameters. Nevertheless, this demonstration is interesting from an application point as many products are designed according to the stiffness properties of the material. Since there is no impact on the flexural properties of the composites, using recycled glass fibres instead of virgin glass fibres becomes an attractive option from a circularity perspective. Future work is needed in several aspects. From a manufacturing point of view, the adhesive used in the production of the panels resulted in regions with improper wetted fibres, which may lead to a reduction in composite performance. Experimental work performed without the use of adhesive needs to be realized in order to determine the effect of it. Future work also need to look into how to achieve higher fibre volume fraction in the inner section. This could be done with the use of fibres with a different length or with the placement of fibres in specific direction. With a higher fibre volume content in the inner section higher stiffness properties may be achieved. From a characterization point of view, the inner section made of recycled fibres requires additional characterization work to determine more precisely the fibre orientation and fibre volume fraction. In addition, the characterization of the impact of varying fibre volume fraction could also be interesting. Finally in terms of application, future work should consider the upscaling feasibility of the manufacturing process developed in this study, as well as the cost and environmental impacts of such material.

5. Acknowledgments

This research is performed during the DecomBlades project, which received funding from the Innovation Fund Denmark under Grant Agreement No. 0177-00006B.

6. References

[1] Mishnaevsky L, Branner K, Petersen HN, Beauson J, McGugan M, Sørensen BF. Materials for wind turbine blades: An overview. Materials (Basel). 2017;10(11).

[2] Jensen JP, Skelton K. Wind turbine blade recycling: Experiences, challenges and possibilities in a circular economy. Renew Sustain Energy Rev. 2018 Dec 1;97:165–76.

[3] Deeney P, Nagle AJ, Gough F, Lemmertz H, Delaney EL, McKinley JM, et al. End-of-Life alternatives for wind turbine blades: Sustainability Indices based on the UN sustainable development goals. Resour Conserv Recycl [Internet]. 2021;171(May):105642. Available from: https://doi.org/10.1016/j.resconrec.2021.105642

[4] Tasistro-Hart B, Al-Haddad T, Bank LC, Gentry R. Reconstruction of Wind Turbine Blade Geometry and Internal Structure from Point Cloud Data. In: Computing in Civil Engineering 2019: Data, Sensing, and Analytics - Selected Papers from the ASCE International Conference on Computing in Civil Engineering 2019 [Internet]. American Society of Civil Engineers (ASCE); 2019 [accessed 2021 Feb 5]. p. 130–7. Available from: https://ascelibrary.org/doi/10.1061/9780784482438.017

[5] Bank L. Re-Wind Design Catalog Fall 2021 [Internet]. 2021 [accessed 2021 Dec 16]. Available from: www.re-wind.info

[6] Bank L, Arias F, Yazdanbakhsh A, Gentry T, Al-Haddad T, Chen J-F, et al. Concepts for Reusing Composite Materials from Decommissioned Wind Turbine Blades in Affordable Housing. Recycling [Internet]. 2018 Jan 17 [accessed 2021 Feb 5];3(1):3. Available from: http://www.mdpi.com/2313-4321/3/1/3

[7] Cunliffe A, Jones N, Williams P. Recycling of fibre-reinforced polymeric waste by pyrolysis: thermo gravimetric and bench-scale investigations. J Anal Appl Pyrolysis. 2003;70:315–38.

[8] Åkesson D, Foltynowicz Z, Christeen J, Skrifvars M. Microwave pyrolysis as a method of recycling glass fibre from used blades of wind turbines. J Reinf Plast Compos [Internet]. 2012 Aug 21 [accessed 2012 Oct 31];31(17):1136–42. Available from: http://jrp.sagepub.com/cgi/doi/10.1177/0731684412453512

[9] Blazsó M. Pyrolysis for recycling waste composites. In: Goodship V, editor. Management,

recycling and reuse of waste composites. Cambridge: Woodhead Publishing in Materials; 2010. p. 102–21.
[10] Feih S, Boiocchi E, Mathys G, Mathys Z, Gibson a. G, Mouritz a. P. Mechanical properties of thermally-treated and recycled glass fibres. Compos Part B Eng [Internet]. 2011 Apr [accessed 2012 Oct 17];42(3):350–8. Available from: http://linkinghub.elsevier.com/retrieve/pii/S1359836811000217
[11] Thomason JL, Nagel U, Yang L, Saez E. Regenerating the strength of thermally recycled glass fibres using hot sodium hydroxide.pdf. Compos Part A Appl Sci Manuf. 2016;87:220–7.
[12] Beauson J, Madsen B, Toncelli C, Brøndsted P, Ilsted Bech J. Recycling of shredded composites from wind turbine blades in new thermoset polymer composites. Compos Part A Appl Sci Manuf. 2016 Nov 1;90:390–9.
[13] Makeen Energy [Internet]. 2023 [accessed 2023 Jun 9]. Available from: https://www.makeenenergy.com/business-segments/envirotech
[14] Madsen B, Løgstrup Andersen T, Lilholt H. Volumetric composition: Fiber content and porosity. In: Oksman K, Mathew AP, Bismarck A, Rojas O, Mohini S, editors. Handbook of Green Materials Vol 4 Biobased composite materials, their processing, properties, and industrial applications. World Scientific Publishing Co Pte Ltd; 2014. p. 137–53.
[15] Krenchel H. Fibre reinforcement. Theoretical and practical investigations of the elasticity and strength of fibre-reinforced materials. 1964. 159 p.
[16] Cox HL. The elasticity and strength of paper and other fibrous materials. Br J Appl Phys. 1952;3(3):72–9.
[17] Cha J, Kim J, Ryu S, Hong SH. Comparison to mechanical properties of epoxy nanocomposites reinforced by functionalized carbon nanotubes and graphene nanoplatelets. Compos Part B Eng. 2019 Apr 1;162:283–8.
[18] Kumar R, Mikkelsen LP, Lilholt H, Madsen B. Experimental Method for Tensile Testing of Unidirectional Carbon Fibre Composites Using Improved Specimen Type and Data Analysis. Materials (Basel). 2021;

IOP Conf. Series: Materials Science and Engineering 1293 (2023) 012037 doi:10.1088/1757-899X/1293/1/012037

Prediction of crack growth at trailing edge bondlines of a wind turbine rotor blade for the assessment of remaining service life

Ashish K Bangaru, Xing-Yuan Miao, Bent F Sørensen and Xiao Chen

Department of Wind Energy and Energy Systems, Technical University of Denmark, Risø Campus, 4000 Roskilde, Denmark

E-mail: akba@dtu.dk

Abstract. Debonding along trailing-edge joint of a wind turbine rotor blade is one of the often reported failure modes during the rotor blade maintenance inspections. This study focuses on buckling-induced debonding along the trailing edge under in-plane compressive loads due to the gravity forces on the rotor blade. With a priori known debond crack length, an analytical model is used for predicting debond growth rate and assessing the time that takes the crack to reach a critical length. For the growth rate prediction, required input parameters are size of the initial debonding, laminate and adhesive layer thicknesses, elastic and interfacial fracture mechanics properties, gravitational load distribution and position of the initial debonding. The damage model is implemented in excel spreadsheet tool. The predictions of the analytical model were compared with the results from 3D finite element simulations and the difference between the results (energy release rate) was 8%. This study demonstrates how damage specific tools can be developed for predicting remaining useful life of rotor blades.

1. Introduction

Modern wind turbine rotor blades typically use load-carrying composite laminates in spar caps, leading edge and trailing edge regions of their cross-section. The laminates located at the trailing edge mainly take up the bending moments associated with the gravitational loads i.e., the mass of the blade [1]. Consider point A on the trailing edge of a rotor blade, as shown in Fig. 1a. During the half rotation, due to gravity the point A is loaded in tension and for the remaining rotation, the loading type gets reversed resulting in a tension-compression cyclic loading history. With the increase in size of the rotor blades, weight significantly increases and consequently, gravitational loads become one of the key design factors. Moreover, wind turbines are designed for a service life of 20-25 years. To ensure a safe operational life of the blades, it is of great importance to be able to predict the fatigue behaviour of composite/sandwich material interfaces, and adhesive bondlines.

Adhesive joints in the rotor blade are highly susceptible to damage in the form of interfacial cracks, particularly, a joint failure in the trailing edge (see Fig. 1b) can lead to final collapse of the blade structure [2]. The trailing edge can develop at least two types of bondline damage: tunnelling cracks (cracking in the adhesive along its width direction) and interfacial debonding, i.e., cracking along the interface between laminate and adhesive layer (in the blade length direction). While tunnelling cracks alone do not cause failure [3], in some cases can lead to the formation of interfacial cracks which can eventually fail by buckling of the blade [4].

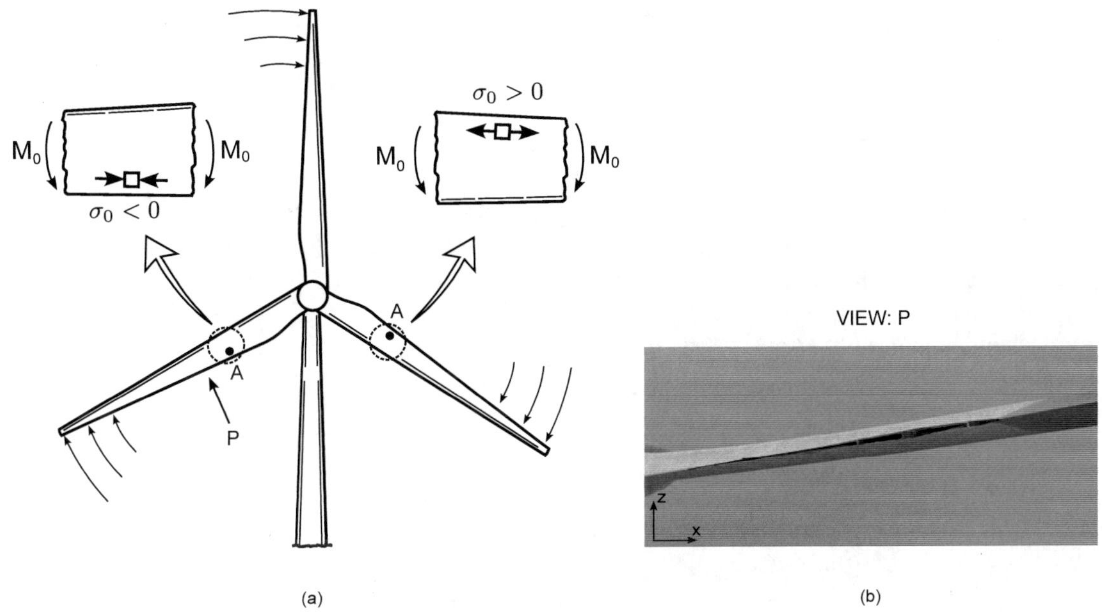

Figure 1: A wind turbine rotor blade a) with point A on the trailing edge shown to indicate how gravity induced stresses (edgewise loading) changes during its complete revolution and b) trailing edge damage seen in field service, taken from [2].

1.1. Background

Ataya and Ahmed [2] visually inspected numerous rotor blades of two different lengths 9.5 m and 14.2 m for damages at the trailing edge. The major damage type observed was the debond cracking along the laminate/adhesive interface starting from the rotor blade root section towards the tip. The damage formed was due to a combination of several factors such as extreme loading conditions, complex geometries and imperfections coming from manufacturing processes. More importantly, the length of the debond cracks varied from a few centimetres up to approx. 1.5 m. Haselbach et al. [5] tested a 34 m long blade by four-point loads and found that the local buckling of the trailing edge led to the debonding of the joint. It was concluded that the edgewise loads are the primary loads driving the failure of the trailing edge.

Full-scale blade test performed by Eder and Bitsche [6] on a 34 m blade revealed that the initiation of a debond crack occurred in the vicinity of local buckled region on the trailing edge and then propagated all along the joint. It was reported that the most critical failure mode is the normal opening of the trailing edge and a method based on virtual crack closure technique was proposed to estimate the energy release rate of the debond crack-tip [7]. However, the method did not considered the edgewise loading that could lead to local buckling of the trailing edge laminate which can lead to debonding.

Subcomponent testing on cut-section of a 3 m long trailing edge panel was performed by Chen et al. [8]. The cut panel was loaded in compression and the sequence of damage modes were documented. With the increase in the applied load, the trailing edge panel buckled followed by adhesive debonding. Post the adhesive joint failure, damage in composite laminate was observed in the form of delamination and fibre-dominant mode. A non-linear finite element model was developed in the study to account for buckling, contact and matrix/fibre damages to predict the failure evolution. However, boundary conditions considered in the numerical model were based on subcomponent testing and not the full-scale blade.

In order to predict the trailing edge damage growth, detailed yet simple models are needed that takes into account the i) correct stresses/strains that act on the blade, ii) boundary conditions, iii) layers thickness, stiffness and iv) the current damage state captured by an initial debond length parameter.

1.2. Motivation

The purpose of this study is to demonstrate a methodology for predicting damage growth at trailing edge joint of the wind turbine rotor blade. A simplified one-dimensional analytical model is presented that requires input parameters such as laminate and adhesive layers thickness, size of the initial debonding, weight distribution along the blade and elastic, fracture mechanics properties of the laminate adhesive interface. Note that the model presented here is significantly simplified with the intention to be usable it in engineering practice. By identifying and measuring critical damage parameters, these simplified yet effective predictive tools enables the end-users to estimate the approximate time needed for next damage inspection or blade-repair actions. The accuracy of the analytical model is assessed by comparing it's result with results from an accurate 3D FE model.

The paper is organized as follows: the mechanics behind the trailing edge debonding, the damage geometry and the expression for energy release rate calculation is presented in section 2. Details of the finite element modelling and the boundary conditions are presented in section 3. In section 4, results from the analytical model and finite element simulations are discussed. The predictions of debond crack-tip energy release rate as a function of the size of the initial debond were provided for the stresses acting on DTU's 86 m blade. The limitations of the damage model are discussed in section 5 and the conclusions are summarized in section 6.

2. Analytical modelling

2.1. Stresses due to gravitational load

The wind turbine rotor blade acts like a cantilever beam with its root fixed and tip left free. Thus, the rotor blade is subjected to gravity load due its own mass resulting in edgewise stresses. Let x denote the coordinate that gives the position at any point along the blade, i.e., the distance measured from the root section. The gravity load distribution along the blade length, $q(x)$ (N/m), can be obtained from the density of the blade material, ρ_{mat}, gravitational constant, g $(= 9.81\ m/s^2)$ and the cross-sectional area of the rotor blade, $A_{blade}(x)$, given as

$$q(x) = \rho_{mat} g A_{blade}(x) \tag{1}$$

subsequently, the edgewise moment per unit width can be obtained by integrating the gravity load from an arbitrary section x until the blade tip $x = L$, given as

$$M_0(x) = \int_x^L \rho_{mat} g A_{blade}(x')(x' - x)dx' \tag{2}$$

Following Mikkelsen [9], we assume a simplified cross-sectional area of the blade model wherein, the analysis of only load-carrying laminates located at spar caps (pressure side and suction side), leading and trailing edges, are taken into account. Since these laminates contribute to the overall stiffness of the blade, the simplified cross-sectional area can be written as $A_{blade}(x) = 2(A_{flap}(x) + A_{edge}(x))$ [9]. Further, we assume that the laminates at spar caps take up the bending moments in the flap-wise direction while those at leading and trailing edges take up edgewise moments.

From the expression of gravity load distribution, the shear force, $T(x)$, and the edgewise moment, $M_0(x)$ over the blade cross section can be found by solving the two differential equations,

$$\frac{dT(x)}{dx} = q(x), \ \frac{dM_0(x)}{dx} = T(x) \tag{3}$$

with the boundary conditions $T(L) = 0$ and $M_0(L) = 0$ that represent a free blade tip. The two differential equations are obtained from a force and a moment equilibrium of an infinitesimal beam element, respectively. Using the trapezoidal rule [10], the following numerical scheme is implemented with a blade split up into n elements with the nodal points: $i \in [0, 1, ...n]$,

$$T_{i+1} - T_i = \frac{1}{2}(q_{i+1} + q_i)(x_{i+1} - x_i) \text{ for } i = (n-1, n-2, ..., 0) \text{ with } T_n = 0 \tag{4}$$

$$M_{0(i+1)} - M_{0i} = \frac{1}{2}(T_{i+1} + T_i)(x_{i+1} - x_i) \text{ for } i = (n-1, n-2, ..., 0) \text{ with } M_{0n} = 0 \tag{5}$$

Let $\sigma_0(x)$ denote the in-plane edgewise stress acting on the trailing edge, at a given distance x from the root and with the simplified area moment of inertia [9],

$$I_{edge}(x) = \frac{1}{2}h_{edge}^2(x)A_{edge}(x) \tag{6}$$

where $h_{edge}(x)$ is chord length, the edgewise stresses can be written as [9]

$$\sigma_0(x) = \frac{M_0(x)}{h_{edge}(x)A_{edge}(x)} \tag{7}$$

2.2. Buckling-induced debonding

Consider a segment of length, l, along the trailing edge of a wind turbine rotor blade where there exists a local buckling driven debonding as shown in Fig. 2b. In the present study, the trailing edge consists of top and bottom laminates each with equal thickness, h_1, and a central adhesive layer with thickness h_2. We assume that both the laminates are made up of same material with a Young's modulus, E_1, and Poisson's ratio, ν_1, while the adhesive properties are denoted by E_2 and ν_2. We define two non-dimensional parameters $\Sigma = E_1/E_2$ and $\eta = h_1/h_2$.

Fig. 3 shows the equilibrium of the half symmetric sandwich panel before and after buckling. If the right end of the laminate is subjected to a uniform in-plane compressive edgewise stress, $\bar{\sigma}_0(x)$, and a moment per unit width, M_0. By assuming same axial strain in all the layers and from the force balance in the x_1-direction, the compressive stresses in the top laminate is given by

$$\sigma_1 = \sigma_0(x)\frac{(2\eta + 1)}{(2\Sigma\eta + 1)} \tag{8}$$

If this stress exceeds the critical buckling stress, $\sigma_{1,cr}$, then the laminate buckles locally and results in a strip of debond length $2a$, see figure Fig. 2b. We assume $h_1 << a$ so that the thin ply can be treated as a long, clamped Euler column of length $2a$. Further, it is assumed that outside the debonded area the adhesive and the bottom laminate are perfectly bonded and provide a stiff support to the top laminate h_1. This assumption allows us to use the bilayer solution developed by Hutchinson and Suo [11]. The critical buckling stress based on clamped-clamped Euler column of length $2a$ is given by

$$\sigma_{1,cr} = \pi^2 D/(a^2 h_1) \tag{9}$$

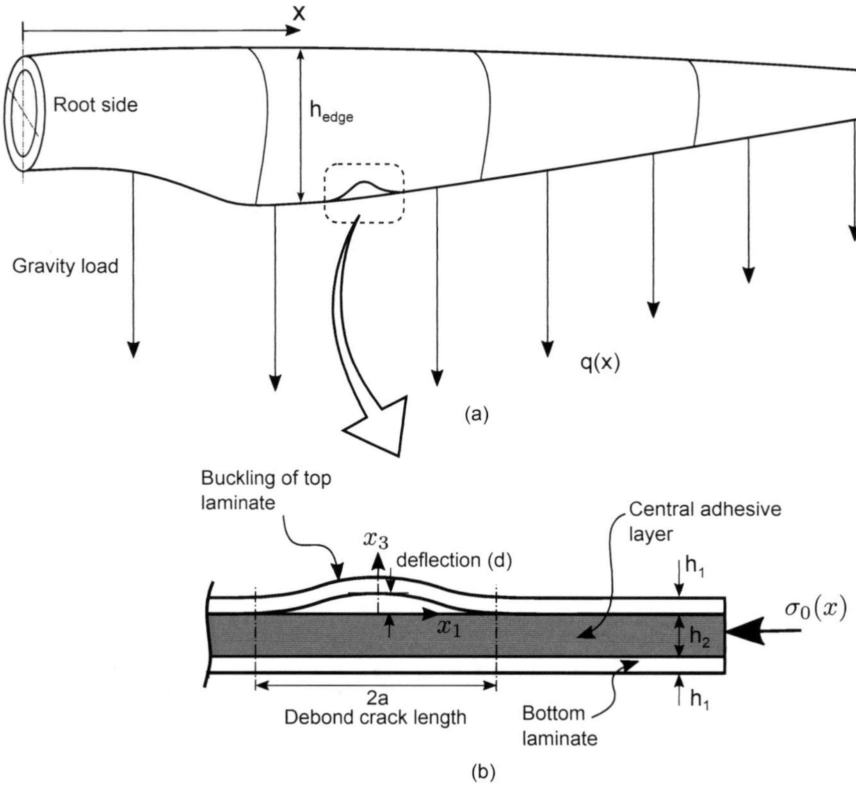

Figure 2: Schematic illustration of the damage geometry a) wind turbine rotor blade with trailing edge debonding caused due to local buckling b) one-dimensional representation of the damage geometry with layers thickness and size of the initial debonding.

where D is the flexural rigidity of the column per unit width, given by $D = E_1h_1^3/(12(1-\nu_1^2))$. In case of buckling, a moment per unit width, M_1, acts on the laminate and a change in the axial compression force ΔP per unit width near the crack-tip interfaces, see figure Fig. 3b. Here, the laminate is under axial compression and positive M_1, meaning that the debond is open and the laminate deforms away from the adhesive layer. Using the von Karman non-linear plate theory, the maximum deflection, the buckling load and the moment are obtained as

$$\frac{d_{max}}{h_1} = \sqrt{\frac{4}{3}\left(\frac{\sigma_1}{\sigma_{1,cr}} - 1\right)}\ ; \quad \Delta P = \frac{3\pi^2}{4a^2}Dd_{max}^2\ ; \quad M_1 = \frac{1}{2}\sigma_{1,cr}d_{max}h_1^2 \tag{10}$$

From the equilibrium condition of post buckling (see Fig. 3b), the relation between the moments can be written as

$$M_2 = M_1 - \frac{\Delta P}{2}(2h_1 + h_2) \tag{11}$$

In order to calculate the energy release rate per unit width at the crack-tip, we use the expression from Karihaloo and Stang [12], given as

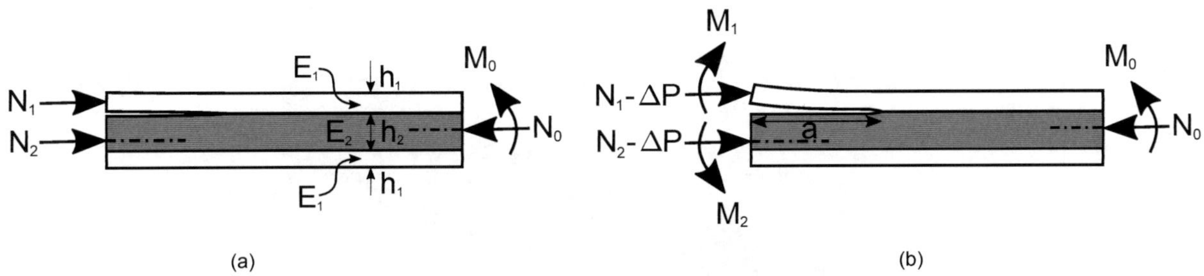

Figure 3: Equilibrium of the half symmetric trailing edge damage geometry a) before buckling of the top laminate and b) after buckling.

$$G = \frac{1-\nu^2}{2E_1}\sigma_{1,cr}^2\left(\frac{\sigma_0}{\sigma_{1,cr}}-1\right)h_1\left[\frac{\sigma_0}{\sigma_{1,cr}}(1+4\zeta+6\zeta^2+3\zeta^3)-\left\{48\left(\frac{\sigma_0}{\sigma_{1,cr}}-1\right)\right\}^{1/2}\zeta^2(1+\zeta)+(3-4\zeta-6\zeta^2+\zeta^3)\right] + \frac{2(1-\nu^2)}{E_1}M_0\frac{\zeta}{h_1}\left[\frac{M_0\zeta}{h_1^2}\frac{(1-\zeta+\zeta^2)}{(1+\zeta)^3}-\sigma_{1,cr}\left(\frac{\sigma_0}{\sigma_{1,cr}}-1\right)\right] \quad (12)$$

where $\zeta = h_1/(h_2+h_1)$. The above equation is valid when $\sigma_0 \geq \sigma_{1,cr}$.

3. Numerical modelling

A 5 m prismatic beam model from a 17 m long Vindeby blade is modelled through the finite element software Abaqus 2021. The geometrical model is reconstructed according to the cross-section geometries measured at approx. 7 m from the root. The detailed dimensions are shown in Fig. 4a. A straight through-width pre-crack is introduced in the adhesive trailing edge joint. The pre-crack is located at the middle section of the trailing edge with a length of 0.5 m, see Fig. 4b. Solid 8-node linear brick elements (Abaqus type C3D8R) were used in the model. The typical mesh sizes are $25 \times 22 \times 6$ mm for spar cap regions, $25 \times 11 \times 14$ mm for sandwich panels, and $25 \times 4 \times 0.8$ mm for trailing edge bondlines. The mesh size is chosen based on mesh sensitivity analysis performed in [8, 13, 14]. The mesh sizes result in a total element number of 162,600. The energy release rate at the crack tips were computed using the virtual crack closure technique (VCCT) in Abaqus.

A rigid body reference node is used to represent the node set at the end of the blade section model. The reference node is set at the elastic centre of the cross-section. The model is loaded in displacement control. A compression load is created by prescribing a displacement u_x to the left end reference node (in the x-direction). The right end of the model is fixed in all directions. Frictionless contact conditions are assigned to the two surfaces of the crack plane.

4. Results

4.1. Analytical and FE results

An old Vindeby blade model with layers thickness $h_1 = 3.25$ mm, $h_2 = 1$ mm and properties $E_1 = 32$ GPa, $\nu_1 = 0.29$, $E_2 = 3$ GPa and $\nu_2 = 0.3$ was used for the comparison purpose. The analytical energy release rate was compared with the VCCT energy release rate value obtained from FEM for the applied compressive stress, $\sigma_0 = 64$ MPa ($\sigma_0 > \sigma_{1,cr} = 49$ MPa). The results were $G_{anl.} = 1341.3$ J/m^2 and $G_{FE} = 1453.2$ J/m^2, i.e., a difference of approximately 8%.

4.2. Gravitational loads, edgewise stresses and the energy release rate

To demonstrate the methodology, the example turbine considered is the 10 MW DTU reference turbine [16]. Table 1 shows the blade details. The blade material analysed is a glass fibre

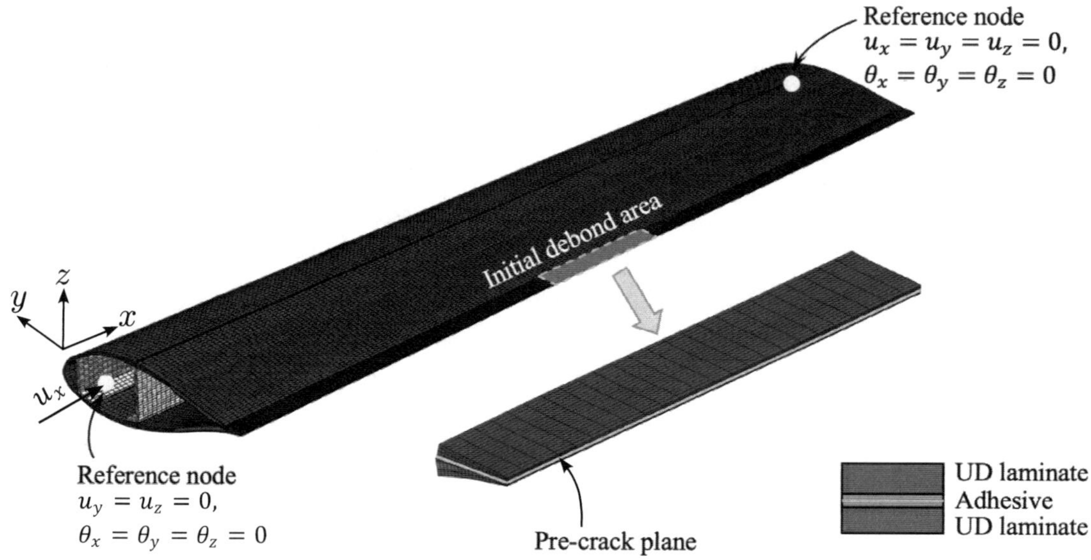

Figure 4: A 5 m prismatic beam finite element model. A straight through-width crack is introduced in the middle section of the adhesive trailing edge joint (orange area marked in the blade section model). The pre-crack is 0.5 m long and the load and the boundary conditions are applied on the two reference nodes (shown as yellow coloured dots). The model is built from one of the measured dimensions of the Vindeby blade cut-out cross-section. The data can be found in [15].

composite made up of unidirectional E-glass and epoxy resin.

Table 1: Data of DTU 10 MW reference turbine

Property/Dimension	Value
Generator size, P_{max}	10 MW
Rotor diameter, D	178 m
Blade length, L	86 m
Stiffness of GFRP, E_{11}	40 GPa
Density of GFRP, ρ_{mat}	2000 kg/m^3
Thickness of top/bottom laminate, h_1	0.5 mm
Thickness of adhesive layer, h_2	1.3 mm
Young's modulus of the top laminate, E_1	21.8 GPa
Young's modulus of the adhesive layer, E_2	3.35 GPa
Poison's ratio, ν_1, ν_2	0.478, 0.3

The internal shear forces T due to gravity load variation and the bending moment per unit width values at a given position x from the root section can be obtained numerically by solving the governing differential equations of the Bernoulli-Euler beam theory. Fig. 5 shows the variation of shear force on the 10 MW DTU reference blade [16]. Here, the total mass includes the mass of the load-carrying GFRP laminates and also the mass of the aerodynamic shell and the shear web.

The edgewise stresses along the blade can be found from the moment $M_{edge}(x)$ working on a cross-section using the eq. (4). Fig. 6 shows the variation of edgewise stresses calculated for

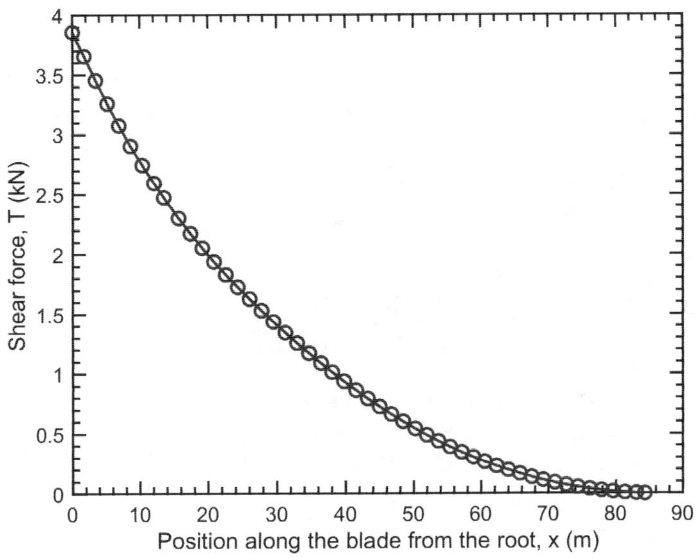

Figure 5: The variation in the shear force due to gravity load acting on the 10 MW DTU wind turbine rotor blade [16].

the 10 MW DTU reference blade [16]. The maximum value is at the root section and zero at the tip.

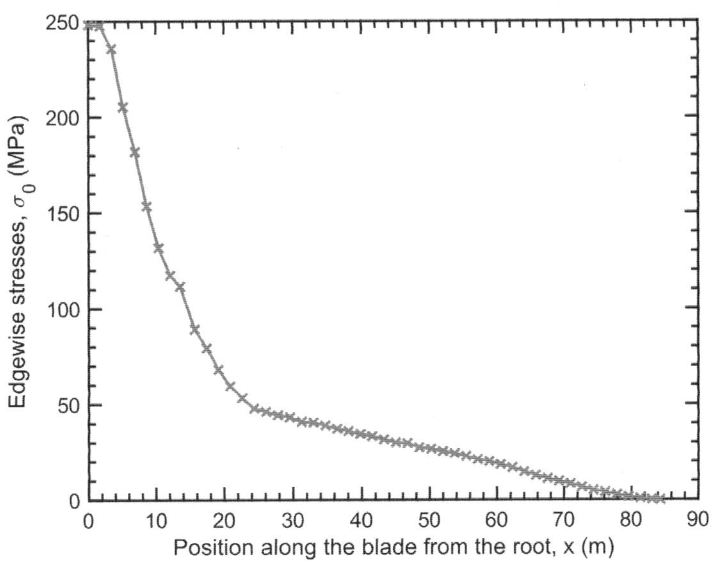

Figure 6: The distribution of edgewise stresses acting on the 10 MW DTU wind turbine [16].

4.3. Damage growth prediction model

Depending on the location and the initial size of the debonding, the post-buckling response of the trailing edge laminate depends on the stress acting at the point, layers thicknesses and their Young's moduli. Fig. 7 illustrates the required input information classified into A) Structural blade model B) Damage parameters and C) Materials test data in order to predict the debond growth. The damage model we refer to is an excel spreadsheet tool that gives the output information as debond crack-tip energy release rate, for a fixed known initial size and number of cycles required to reach a predefined critical crack length.

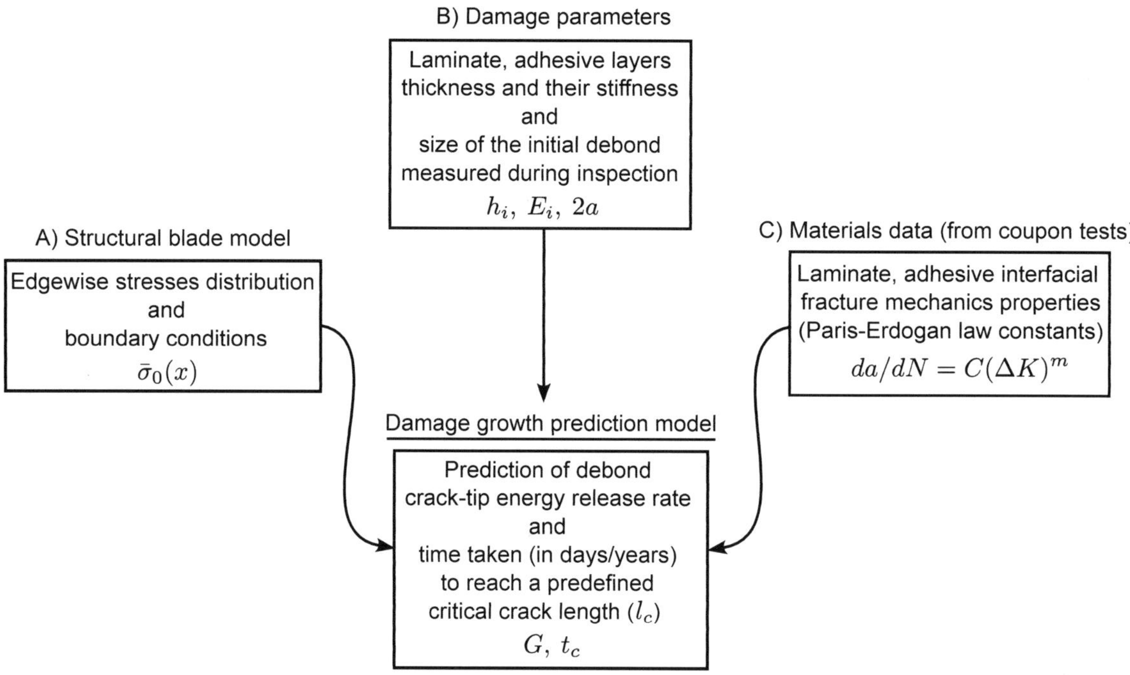

Figure 7: A block diagram showing various fields of input information that goes into the trailing edge damage growth prediction model.

The energy release rate for several initial debond lengths in the range of 0.1 m - 1 m and for various stress levels in the 10 MW DTU reference blade is shown in Fig. 8. The results for G are only valid as long as the debond buckles ($\sigma_0 \geq \sigma_{1,cr}$). In case of no buckling, i.e., $\sigma_0 < \sigma_{1,cr}$, $G = 0$. From Fig. 8, it can be noticed that for low compressive stresses, $40 - 70$ MPa, the minimum size of the debond length $2a$ must be greater than approximately 0.2 m in order to get buckling of the top laminate. While in case of high stresses $80 - 130$ MPa, $2a$ should be greater than approximately 0.14 m to get buckling. It is seen that for fixed σ_0, G increases with increasing a.

After buckling of the top laminate and consequently, with the initiation of debond crack along the interface of the top laminate and the adhesive layer, it is assumed that the debond crack-tip propagates further. Additionally, assuming that the crack-tip grows in normal crack opening under the tension-compression cyclic loading, then from the interfacial properties of the bondline of E-glass fibre composites and epoxy resin, approximate crack growth time for the debond crack-tip to reach a critical length can be estimated. For illustration, a critical length of $l_c = 5$ m is considered in the present study and the turbine rotational speed is assumed as 15 rpm. Firstly, from Goutianos and Sørensen [17], we have the relation between ΔK and G_{max}

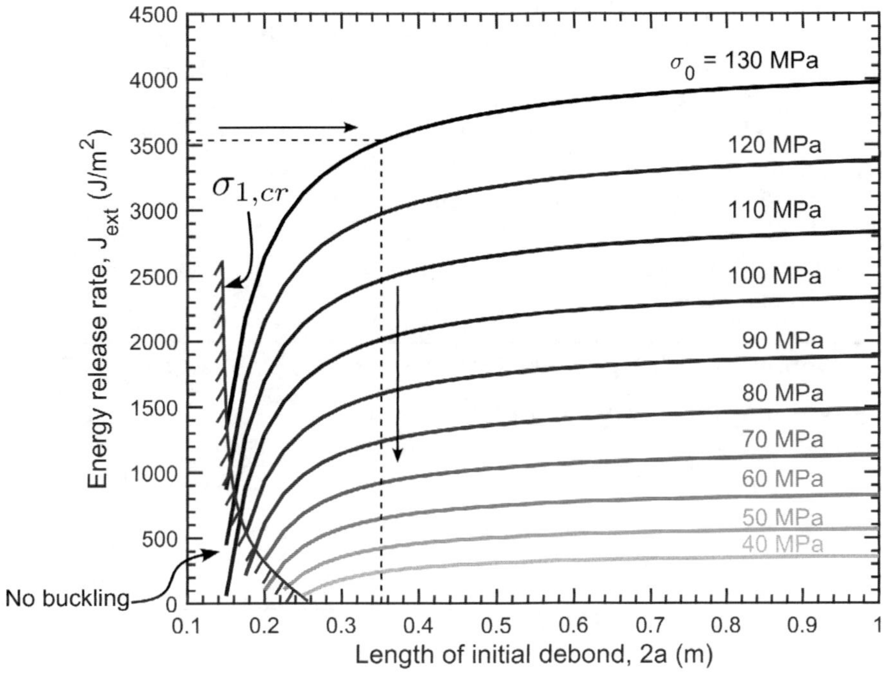

Figure 8: The energy release rate, G, shown as a function of the initial debond length for various stress levels that are dependent on the position, x, along the blade length. The steps for determining critical length l_c for the case of $\sigma_0 = 130$ MPa are denoted by dashed lines.

as $\Delta K = \sqrt{G_{max}E}(1 - R)$, where E is the Young's modulus and R is the load ratio. For the present problem, the debond only buckles when $\sigma_1 \geq \sigma_{1,cr}$. When $\sigma_1 < \sigma_{1,cr}$, the debond crack remains closed. Then $G = 0$. This holds true also when the laminate is loaded in tension. Thus, effectively, $R = 0$. Using the Paris-Erdogan law, the crack growth rate is written as $da/dN = C(\Delta K)^m$, here C and m are Paris-Erdogan law constants obtained from the materials interface characterization tests. We use the below equations to calculate the time in days/years

$$\begin{aligned} 2a_{i+1} &= 2a_i + \Delta a \\ (da/dN)_{i+1} &= C(\sqrt{G_{max}(2a_{i+1})E_1})^m \\ N_{i+1} &= \frac{\Delta a}{(da/dN)_{i+1}} \\ t_{i+1} &= \frac{N_{i+1}}{f} \end{aligned} \tag{13}$$

where f is the loading frequency. Firstly, the step size Δa is chosen with a fixed initial value and then $2a$ is calculated by adding Δa iteratively. Since G (see eq.(12)) is dependent on $\sigma_{1,cr}$ which in turn is dependent on $2a$, therefore, G is calculated in an incremental approach. Thus, for a given $2a_i$, the $G_{max}(2a_i)$ is obtained and subsequently, $(da/dN)_i$ is estimated. Thereafter, from the current debond length, the number of cycles, N_i and the time taken t_i to grow the step size is calculated. For a given critical length l_c, the time taken to reach this critical length, t_c, is estimated. The procedure is repeated with several different Δa values until a convergence in t_c is obtained.

Table 2 provides the time estimation approximately in days and also in years for an initially detected debond crack-tip to propagate up to 1 m length.As an example to demonstrate debond growth rate predictions, two stress levels $\sigma_0 = 40$ MPa and 60 MPa are considered in Table 2. For each stress level, two different initially detected debond lengths $2a$ were chosen to show the time taken for the detected crack to reach 1 m length. It can be seen that for 60 MPa stress level, when $2a$ varies from 0.2 m to 0.225 , the time considerably reduces from 3167 days to 2 days.Thus, it can be deduced that the higher the stress level, the earlier the detection of debond crack allows sufficient time to take preventive measures or repair actions.

Table 2: Illustration of trailing edge damage growth from the current state of a given initial debond length and a stress level

Debond crack-tip growth time to reach 1 m length			
Initial debond length (m) $2a$	**Applied stress (MPa)** $\sigma_0(x)$	**Time In Days**	**Years**
0.25	40	59822	164
0.3	40	116	0.3
0.2	60	3167	8.7
0.225	60	2	0

5. Discussion

An important finding from Fig. 8 is the determination of critical energy release rate, G_c, where an initially detected debond crack can grow in unstable manner. Thus, one can estimate the critical crack length l_c for that specific stress level in the blade beyond which the detected debond crack-tip propagates . In Fig. 8, dashed lines are indicated for the case $\sigma_0 = 130$ MPa to denote $G_c = 3520$ J/m^2 and $l_c = 0.35$ m. Therefore, for any initial debond crack detected along the blade during the inspection and for a known stress level at that position, the limiting crack length and the approximate time to reach that limit crack length can be estimated.

5.1. Limitations of the model

i) The damage framework presented is a 1D problem which is based on Euler's theory of column buckling. The analysis takes into account only the thickness of the layers and does not include width and its variations in each layer along the cross-section of the rotor blade. Therefore, an extended framework of this model would be a 2D plate buckling analysis problem.
iii) A constant thickness in layers (laminate and adhesive) is assumed all along the blade length.
ii) Constraining effects from the sandwich/core panel are not included.
iii) The analysis is based on only in-plane loads coming from the self-weight/gravity of the blade, the other external loads such as out-of-plane loads are not taken into account.
iv) Since G increases with increasing debond length a, static crack growth will be unstable. When J_{ext} reaches G_c, this could be used as a criterion for determining critical length a_c.

6. Concluding remarks

Buckling driven debonding in the trailing edge of a wind turbine rotor blade is assessed using an analytical model that takes input from the current damage state. The parameters required are a) size of the initial debonding, b) laminate and adhesive layer thicknesses, c) elastic and interfacial fracture mechanics properties and d) weight distribution of the material along the rotor blade. The proposed approach has been demonstrated on 10 MW DTU reference turbine by extracting gravitational loads and edgewise stresses. The energy release rate predictions have been provided for several initial debond lengths and for various stress levels acting along the rotor blade. An approximate time estimation is provided for the debond crack-tip to reach a (predefined) critical length.

The methodology demonstrated in the current study can be used in a damage specific predictive tools that helps in the prognosis of the service life of the wind turbine rotor blades.

Acknowledgements
This study is carried out as a part of the research work in the 'Methodology for assessment of remaining life time of wind turbine blades based on damage state (ReLife)' (EUDP case no. 64019-0543). The author would like to acknowledge the master student Joshua Joel Fernandes for his contribution on the geometrical modelling of the blade segment model. Further, the research assistant Mohammad Hedayatzadeh is acknowledged for his assistance in measuring the dimensions of the Vindeby blade cut-out cross-section.

References

[1] Leon Mishnaevsky Jr et al. "Materials for wind turbine blades: An overview". *Materials* 10.11 (2017), p. 1285.

[2] Sabbah Ataya and Mohamed MZ Ahmed. "Damages of wind turbine blade trailing edge: Forms, location, and root causes". *Engineering Failure Analysis* 35 (2013), pp. 480–488.

[3] Jeppe B Jørgensen, Bent F Sørensen, and Casper Kildegaard. "Tunneling cracks in full scale wind turbine blade joints". *Engineering Fracture Mechanics* 189 (2018), pp. 361–376.

[4] Philipp Ulrich Haselbach and Kim Branner. "Initiation of trailing edge failure in full-scale wind turbine blade test". *Engineering Fracture Mechanics* 162 (2016), pp. 136–154.

[5] Philipp Ulrich Haselbach, Martin Alexander Eder, and Federico Belloni. "A comprehensive investigation of trailing edge damage in a wind turbine rotor blade". *Wind Energy* 19.10 (2016), pp. 1871–1888.

[6] Martin Alexander Eder and RD Bitsche. "Fracture analysis of adhesive joints in wind turbine blades". *Wind Energy* 18.6 (2015), pp. 1007–1022.

[7] Martin Alexander Eder et al. "A practical approach to fracture analysis at the trailing edge of wind turbine rotor blades". *Wind Energy* 17.3 (2014), pp. 483–497.

[8] Xiao Chen et al. "Understanding progressive failure mechanisms of a wind turbine blade trailing edge section through subcomponent tests and nonlinear FE analysis". *Composite Structures* 214 (2019), pp. 422–438.

[9] Lars P Mikkelsen. "A simplified model predicting the weight of the load carrying beam in a wind turbine blade". *IOP Conference Series: Materials Science and Engineering*. Vol. 139. 1. IOP Publishing. 2016, p. 012038.

[10] Arieh Iserles. *A first course in the numerical analysis of differential equations*. 44. Cambridge university press, 2009.

[11] John W Hutchinson and Zhigang Suo. "Mixed mode cracking in layered materials". *Advances in applied mechanics* 29 (1991), pp. 63–191.

[12] Bhushan Lal Karihaloo and Henrik Stang. "Buckling-driven delamination growth in composite laminates: Guidelines for assessing the threat posed by interlaminar matrix delamination". *Composites Part B: Engineering* 39.2 (2008), pp. 386–395.

[13] Xiao Chen, Xiaolu Zhao, and Jianzhong Xu. "Revisiting the structural collapse of a 52.3 m composite wind turbine blade in a full-scale bending test". *Wind Energy* 20.6 (2017), pp. 1111–1127.

[14] Jing Tang, Xiao Chen, and Ke Yang. "Evaluating structural failure of load-carrying composite box beams with different geometries and load conditions". *Applied Composite Materials* 26 (2019), pp. 1151–1161.

[15] Joshua Joel Fernandes et al. *Data sets of measured cross-sectional area coordinates and material properties of an old Vindeby blade.* Zenodo, Oct. 2023. DOI: 10.5281/zenodo.8428164. URL: https://doi.org/10.5281/zenodo.8428164.

[16] Christian Bak et al. "The DTU 10-MW reference wind turbine". *Danish wind power research 2013.* 2013.

[17] Stergios Goutianos and Bent F Sørensen. "Fatigue crack growth rate at material and geometry transitions in glass-epoxy composites". *Composite Structures* 275 (2021), p. 114445.

IOP Conf. Series: Materials Science and Engineering 1293 (2023) 012038 doi:10.1088/1757-899X/1293/1/012038

Residual stresses in thick composites considering through-thickness temperature gradients during manufacturing

Yixing Wang, Claudio Balzani

Leibniz University Hannover, Institute for Wind Energy Systems,
Appelstr. 9A, 30167 Hannover, Germany

E-mail: research@iwes.uni-hannover.de

Abstract. In the composite materials of wind turbine rotor blades, residual stresses form during manufacturing due to the different thermo-chemo-mechanical behaviors of the fibers and the matrix. These residual stresses decrease the ultimate and fatigue strength of the blade because the failure stress and cycles to failure are reached at lower levels of external loading. It is thus of utmost importance to consider residual stresses in the design stage to obtain reliable blade designs.

This work deals with the prediction of residual stresses in a thick composite representative for a spar cap in a wind turbine rotor blade. A model is formulated for the thermo-chemo-mechanically coupled problem. An emphasis is given on the non-uniform and time-variable temperature distribution across the thickness of the composite. The different sources of residual stresses (e.g., chemical shrinkage, non-uniform distribution of different thermal expansion coefficients of fibers and matrix, exotherm, etc.) are analyzed, quantified, and compared. The impact of the time-dependent temperature distribution on the formation of residual stresses is highlighted by means of a representative numerical example.

1. Introduction

In the composite materials of wind turbine rotor blades, residual stresses are a phenomenon that is relevant to consider. Residual stresses not only decrease the ultimate and fatigue strength of the rotor blade [1–3], but can also lead or at least contribute to the formation of unwanted fiber waviness or undulations [4–6]. However, the calculation of residual stresses is subject to intensive research. Some authors used a viscoelastic model with a group of Maxwell elements [7–10]. Others assumed a linear relation between the stiffness of the matrix and the curing degree, and calculated the residual stress from the curing process [11–13].

Residual stresses are generated during the manufacturing process, and if it is to be analyzed, the manufacturing process must be considered and a numerical model needs to be built in order to simulate the composite behavior during the curing process [14]. In this paper, the following mechanisms are considered: First, there is chemical shrinkage in the matrix due to the network formation on curing. Then, there is a non-uniform distribution of thermal expansion coefficients due to different thermo-mechanical characteristics of the fibers and the matrix. There is also a non-uniform and time-dependent temperature distribution, especially in thick laminates. This is caused by the exothermal curing reaction in the matrix and the imbalance of heat exchange

IOP Conf. Series: Materials Science and Engineering 1293 (2023) 012038 doi:10.1088/1757-899X/1293/1/012038

across the surfaces, i.e., there is heat transfer into the laminate on the tool side and heat transfer out of the composite on the opposite side. A thermo-chemo-mechanical strong coupling approach is employed to calculate the residual stresses in a thick UD glass/epoxy laminate that is representative for a spar cap of a wind turbine rotor blade.

2. Curing kinetics model

In the wind energy industry, resin transfer molding technology is commonly used for the manufacturing of rotor blades [15]. Therein, dry fiber mats are placed in the mold and packed into a vacuum bag. Then a vacuum is drawn so that the resin, which was previously mixed with the hardener, flows in and wets the fiber mats. To decrease the viscosity and therefore increase the flowability of the resin and to initiate the curing process, the mold is heated, and the heat is transferred to the resin and fiber system. The temperature is slowly increased to accelerate curing and the temperature is fixed at the final curing temperature for a specified time period until the matrix is fully cured. After curing is finished, the heating system is switched off and the composite cools down to room temperature. To simulate this process, the chemical, thermal and mechanical are modeled in the following.

2.1. Chemical model

The chemical analysis covers the polymerization reaction in the matrix, i.e., the formation of molecular chain networks. These networks result in an increase of stiffness, i.e., the change from the liquid state to the solid state. This so-called curing process is measured by means of the curing degree denoted by β. The curing degree is a function of time defined by the expression

$$\beta(t) = \frac{Q(t)}{Q_{ex}} \, . \tag{1}$$

Herein, $Q(t)$ is the specific heat generation due to the exothermic reaction up to the current instance of time t and Q_{ex} is the specific total exothermic heat generation. Hence, the curing degree is described by the ratio between the current amount of heat generation and the total heat generation during the exothermic curing reaction. It holds that $0 \leq \beta \leq 1$. A value of $\beta = 0$ means that the curing reaction did not start yet while a value of $\beta = 1$ means that the curing reaction is already finished. The rate equation of the curing degree is defined by

$$\frac{d\beta(t)}{dt} = (\kappa_1 + \kappa_2 \beta^m)(1-\beta)^n \, , \tag{2}$$

which uses a Kamal-Sourour reaction type, see [16, 17]. The coefficients κ_1, κ_2, m, and n can be calculated using the relationships with $i = 1, 2$

$$\begin{aligned} \kappa_i &= A_i \exp\left(-\frac{E_i}{RT}\right) \, , \\ m &= C_1 \exp\left(-C_2 T\right) \, , \\ n &= C_3 \exp\left(-C_4 T\right) \, . \end{aligned} \tag{3}$$

Herein, A_1, A_2, E_1, E_2, C_1, C_2, C_3, and C_4 are material-dependent parameters, R is the gas constant, and T is the absolute temperature. The parameter set used in this study are from [16] and can be seen in Table 1 in the appendix.

2.2. Thermal model

Heat is introduced by the tool heating system and the exothermic reaction. The heat conduction through the material requires a transient thermal analysis. Here, Fourier heat conduction analysis is used [18]. In addition to the heating from the outside, the heat generated from the inside by the curing exotherm also needs to be considered. The governing equation for the transient thermal analysis is given by

$$\rho c_p \frac{\partial T}{\partial t} = \frac{\partial}{\partial x}\left(k_{ij}\frac{\partial T}{\partial x_j}\right) + \rho Q_{ex}\frac{\mathrm{d}\beta}{\mathrm{d}t}\,, \tag{4}$$

where $i, j = 1, 2, 3$ are the directions of the Cartesian coordinate system, x_j is the position vector element of a point of interest in Cartesian coordinates, ρ is the mass density of the composite material, c_p is the specific heat capacity, and k_{ij} is the tensor element of thermal conductivity.

2.3. Mechanical constitutive equations

As the curing reaction proceeds, the stiffness of the epoxy resin increases. At the beginning of the curing process, i.e., $\beta = 0$, the stiffness of the matrix is denoted by E_{m0}. When the curing is completed, i.e., $\beta = 1$, the stiffness of the matrix is denoted by E_{m1}. The initial stiffness of the matrix is supposed to be zero (liquid resin), but as an exact zero is numerically critical due to a singularity issue, it is assumed to be sufficiently small, i.e., $E_{m0} = \varepsilon E_{m1}$, with $\varepsilon = 10^{-3}$.

A linear relationship between the elastic moduli and the curing degree is further assumed in this paper. Hence, the stiffness of the matrix at an arbitrary instance of time t denoted by $E_m(t)$ is proportional to the curing degree β (which is a function of time) and it holds

$$E_m(t) = (\varepsilon + (1-\varepsilon)\,\beta)\,E_{m1}\,. \tag{5}$$

3. Multi-scale multi-physics model for residual stress calculation

For the multi-physical problem that we need to solve, the following calculation steps are employed. First, the transient thermal analysis is executed to obtain the through-thickness temperature distribution as a function of time. Then, the temperature distribution in each time step is fed into the chemical analysis to calculate the curing degree. The curing degree is then put into the thermo-mechanical and chemo-mechanical couplings to calculate the stresses in the fibers and the matrix.

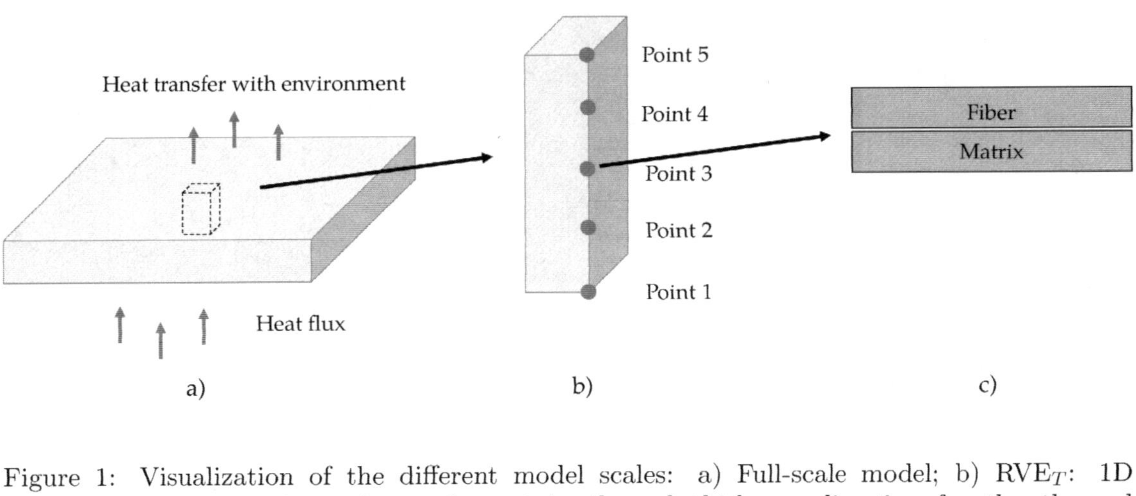

Figure 1: Visualization of the different model scales: a) Full-scale model; b) RVE$_T$: 1D mesoscopic representative volume element in through-thickness direction for the thermal analysis; c) RVE$_\sigma$: 1D microscopic representative volume element in fiber direction for the mechanical analysis.

IOP Conf. Series: Materials Science and Engineering 1293 (2023) 012038 doi:10.1088/1757-899X/1293/1/012038

To reduce the computational cost, the concept of a Representative Volume Element (RVE) is followed here, see e.g. [19–22]. It is assumed that the major processes related to heat exchange take place in through-thickness direction, and the heat flux in the laminate plane is considered negligible. Hence, we reduce the full-scale model of Figure 1a) to 1D RVE model in through-thickness direction for the thermal analysis, which is named RVE_T, see Figure 1b). It is a mesoscopic model because the laminate is treated in a smeared manner, i.e., the fibers are not resolved. The RVE_T model is evaluated in a discrete number of evaluation points (points 1-5 in Figure 1) that are distributed in an equidistant manner across the laminate thickness. After the through-thickness temperature distribution in each time step is used to calculate the curing degree and with it the matrix stiffness, a 1D RVE model in fiber direction is used to calculate the stresses in the fibers and the matrix, respectively, see Figure 1c). This microscopic model that explicitly resolves the fibers is named RVE_σ.

3.1. Chemical residual stress

During the curing process, the epoxy resin undergoes a chemical shrinkage of approximately 5% [18]. If a rigid connection between the fibers and the matrix was assumed, see Figure 2a), the resultant strain in longitudinal direction in the 1D model would constant. Consequently, the fibers would be heavily compressed. However, the fibers cannot withstand such large compression without micro-buckling. Hence, there must be a strong shear deformation in the close vicinity of the fibers. This area is called interphase layer in the following and has a very small thickness, see Figure 2b). The initial stiffness of the interphase layer is very small and increases as the curing process progresses. It is assumed that the stiffness of the interphase layer denoted by $E_i(t)$ is proportional to the curing degree, but smaller than the matrix stiffness, because the 3D network formation in the interphase layer is disturbed by the fiber surface. It is assumed that $E_i(t) = 0.1 \cdot E_m(t)$.

3.2. Thermal residual stress

For quick manufacturing processes in composite manufacturing, the curing temperature needs to be relatively high. After curing, the temperature must be cooled down to room temperature. There is a big difference in coefficient of thermal expansion between the fibers and the matrix. Due to higher thermal shrinkage in the matrix during cool-down, compression strains and stresses form in the fibers while tensile strains and stresses form in the matrix. For the thermo-mechanical cool-down analysis, one needs to know the temperature distribution at the end of the curing process. At this instance of time, the resin will be cooled down to room temperature. The

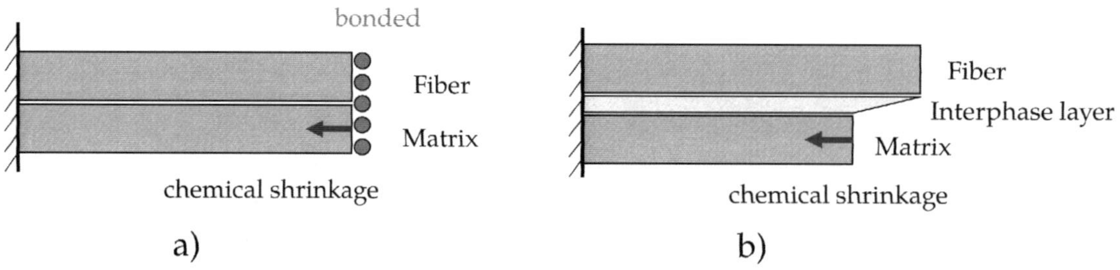

Figure 2: Two methods to calculate the chemical residual stress in the microscopic RVE_σ model: a) Constant strain in longitudinal direction; b) Strain difference between fibers and matrix with interphase (shear) layer connection.

respective temperature distribution is an outcome of the transient thermo-chemo-mechanical curing analysis and is an initial value condition for the calculation of the thermal residual stress. It will be seen in the simulation results that the temperature distribution at the beginning of the cool-down process is nonlinear in the through-thickness direction, resulting in nonlinear distributions of thermo-mechanically triggered residual stresses from cool-down.

3.3. Numerical example
The numerical example is a 5 cm thick uni-directional composite laminate, which is representative for a spar-cap section of a wind turbine blade. The material parameters used for the analysis are given in the appendix, see Table 1, Table 2, and Table 3. There is a heating source from the bottom (considered the tool surface), which gives a heat flux of 400 $J/m^2/s$. After heating for 10 hours, the heat source will be removed (tool heating system switched off). On the top of the laminate there is a heat exchange with the surrounding air and the heat transfer coefficient is 8 $J/m^2/K$.

4. Results and discussion
The multi-field coupled simulation described above was implemented in Abaqus by means of a user subroutine. The simulations are carried out without considering the exothermal peak as a first step. The exothermal peak is then included to allow for a thorough evaluation of its relevance for residual stress calculation.

4.1. Curing kinetics
The first simulation was executed without consideration of exotherm. The results are shown in Figure 3 (temperature on the left, curing degree on the right). The temperature curves are very smooth in all points observed in thickness direction and have a parabolic shape. After heating for 10 hours and removing the heat source, the temperatures decrease. The curing degree plotted against time has a sigmoidal shape. This means that the curing reaction is slow at the beginning, then quickly accelerates (high slope after a short time) and slows down again shortly before the curing process is finished. It can also be observed that the higher the temperature, the quicker the curing process takes place, which is in line with practical experience in composite manufacturing.

As a next step, the exotherm is included in the simulation. The simulation results are shown in Figure 4. As the curing process proceeds, the temperature of the material increases gradually. There is an additional increase after about 3 h associated with the exothermic peak. Because of the higher temperature, the curing process accelerates and is finished after approximately 9 h in point 5 (at the top surface). In this point, the curing process was not finished after more than 12 h when the exotherm was not considered, cf. Figure 3. This highlights the importance to include the exotherm, as the temperature distribution and curing degree are more accurate in this case, and with it the residual stress prediction as well.

It can be seen in Figure 4 that the curing process starts after approximately 2 h, although the temperature is already quite high earlier. This is because of the chosen material properties that were adopted from [16]. In wind energy, a quicker curing process would be advantageous to decrease the tool occupation time and increase the blade output frequency.

4.2. Residual stresses
As the chemical shrinkage progresses, the epoxy resin compresses the fiber through the interphase layer. The resulting fiber stress is shown on the left of Figure 5. As the laminate is internally in equilibrium, the norm of the matrix stress is the same, but has the opposite sign. The increase of the fiber stress with time is consistent with the increase of the curing degree, which is shown on the right of Figure 4.

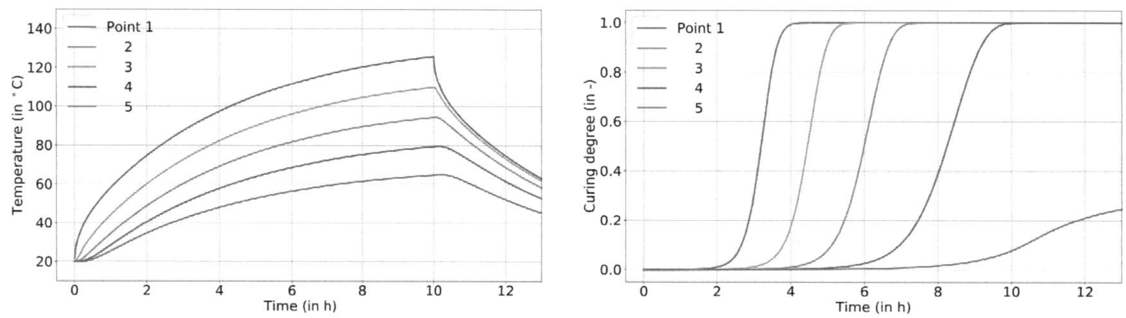

Figure 3: Simulation without consideration of the exothermic reaction: Temperature (left) and curing degree (right) plotted against time.

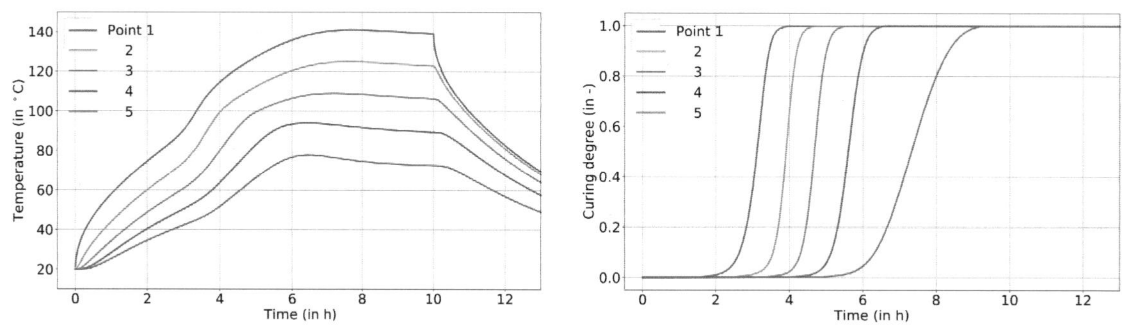

Figure 4: Simulation with consideration of the exothermic reaction: Temperature (left) and curing degree (right) plotted against time.

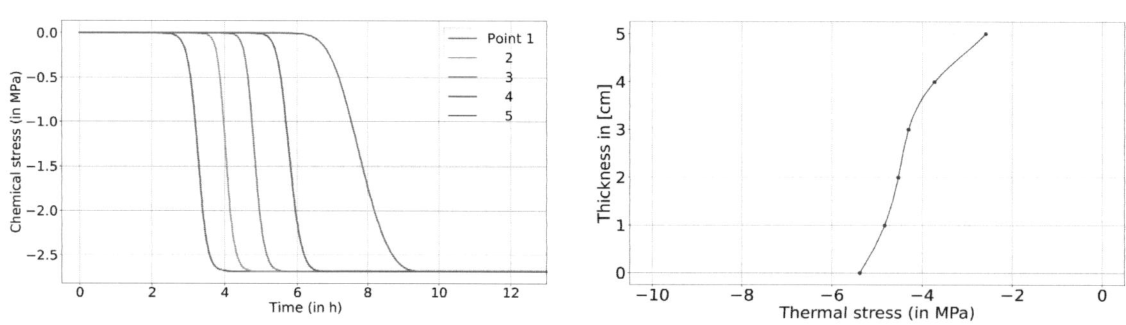

Figure 5: Residual stresses in the fibers. Fiber stresses due to chemical shrinkage (left) and after cool-down (right).

Due to the uneven heat exchange with the tool at the bottom and the surrounding air at the top, the temperature distribution is non-linear throughout the manufacturing process, not only during curing, but also during the cool-down procedure. Due to the lower temperature at the top, the time needed for curing increases from the bottom to the top, respectively. After curing is finished, the tool heating is switched off and the composite cools down. Due to different coefficients of thermal expansion in the fibers and the cured matrix (the matrix will experience larger thermal shrinkage than the fibers), the fibers experience significant compressive residual stresses on cool-down, see on the right of Figure 5.

The (compressive) residual stresses in the fibers are non-uniform across the thickness. This is a consequence of non-uniform through-thickness temperature distributions during manufacturing and different time periods for the completion of curing processes along the thickness. The through-thickness gradient in residual stress results in shear stresses between different layers (which are not considered yet), and, consequently, may trigger undulations and fiber waviness – or may at least amplify originally existing undulations and fiber waviness. A detailed evaluation of the coupling between residual stresses and formation of undulations is work in progress and will be investigated in future publications.

5. Conclusions

In this paper, a multi-physics and multi-scale approach was proposed for the calculation of residual stresses in thick composite laminates that form during the manufacturing processes. One essential part of the model is the inclusion of a transient thermal analysis revealing the required boundary conditions for the evaluation of the curing degree inside the composite structure, which in turn enters the mechanical constitutive equations. A fully chemo-thermo-mechanically coupled procedure was proposed comprising exothermal peak, chemical shrinkage, thermal shrinkage, and curing state-dependent constitutive equations for the matrix.

The material properties used in this paper were partially extracted from literature, and partially based on assumptions, so that physically meaningful results were obtained. Especially the thermal behavior is difficult to measure but will be calibrated against measurements recorded during the manufacturing of a spar-cap of a real wind turbine blade. This is work in progress and will be available soon.

Due to an imbalance in heating (heat entry on the bottom tool side and heat exchange with the surrounding air on the upper surface) and the presence of an exothermic reaction, the temperature profile in thickness direction is a function of time. The temperature profile is non-linear and shows clear temperature gradients in thickness direction that in turn changes with time. Consequently, the curing process is non-uniform across the thickness and takes place quicker at locations with higher temperatures. Due to the non-linear temperature profile, the chemical shrinkage on curing and the thermal shrinkage on cool-down is non-uniform and non-linear across the thickness as well, resulting in a nonlinear distribution of microscopic residual stresses. These are compressive in nature in the fibers and tensile in the matrix. Due to compressive residual stresses in the fibers and a vertical gradient of residual stresses, residual shear stresses are introduced that may contribute to the formation or amplification of undulations. This effect will be investigated in upcoming studies.

6. Acknowledgments

This research was supported by the German Federal Ministry for Economic Affairs and Climate Action in the frame of the research project *ReliaBlade* (grant no. 0324335B). The authors acknowledge the financial support.

7. References

[1] Agius S L, Joosten M, Trippit B, Wang C H and Hilditch T 2016 *Composites Part A: Applied Science and Manufacturing* **90** 125–36
[2] Hallett S R, Jiang W-G, Khan B and Wisnom M R 2008 *Composites Science and Technology* **68** 80–9
[3] Arteiro A, Catalanotti G, Melro A R, Linde P and Camanho P P 2015 *Composites Part A: Applied Science and Manufacturing* **79** 127–37
[4] Mizukami K, Mizutani Y, Todoroki A and Suzuki Y 2016 *Composites Part B: Engineering* **86** 84–94
[5] Kugler D and Moon T J 2002 *Journal of Composite Materials* **36** 1451–79

[6] Thor M, Sause M G R and Hinterhölzl R M 2020 *J. Compos. Sci.* **4** 130
[7] Simo J C and Hughes T J R 2000 *Computational Inelasticity* New York Springer
[8] Saseendran S, Berglund D and Varna J 2020 *Advanced Manufacturing: Polymer* & Composites Science **6** 1–16
[9] Zobeiry N, Vaziri R and Poursartip A 2006 *Journal of Engineering Materials and Technology* **128** 90–5
[10] Pettermann H E and DeSimone A 2018 *Mechanics of time-dependent materials* **22** 421–33
[11] Hui X, Xu Y and Zhang W 2020 Front. *Mech. Eng.* **15** 475–83
[12] Johnston A, Vaziri R and Poursartip A 2001 *Journal of Composite Materials* **35** 1435–69
[13] Ding A, Li S, Sun J, Wang J and Zu L 2016 *Journal of Reinforced Plastics and Composites* **35** 807–23
[14] Sayer F, Antoniou A, Goutianos S, Gebauer I, Branner K and Balzani C 2020 *IOP Conf. Ser.: Mater. Sci. Eng.* **942** 12006
[15] Brøndsted P, Lilholt H and Lystrup A 2005 *Annu. Rev. Mater. Res.* **35** 505–38
[16] Bang K G, Kwon J W, Lee D G and Lee J W 2001 *Journal of Materials Processing Technology* **113** 209–14
[17] Boey F and Qiang W 2000 *Polymer* **41** 2081–94
[18] Lienhard J H 2020 *A heat transfer textbook* New York Dover Publications
[19] Schröder J, Balzani D and Brands D 2011 *Arch Appl Mech* **81** 975–97
[20] Auenhammer R M, Mikkelsen L P, Asp L E and Blinzler B J 2021 *Composite Structures* **256** 113136
[21] Conti S and Hackl K 2015 *Analysis and Computation of Microstructure in Finite Plasticity* vol 78 (Cham: Springer International Publishing)
[22] Zhang Y, Xia Z and Ellyin F 2004 *Composites Science and Technology* **64** 1613–21

8. Appendix

Table 1: Curing kinetics-related material parameters used in this paper (adopted from [16]).

Symbol	Value	Unit
A_1	$2.154 \cdot 10^9$	min^{-1}
A_2	$2.336 \cdot 10^8$	min^{-1}
E_1	$8.180 \cdot 10^4$	J/mol
E_2	$6.682 \cdot 10^4$	J/mol
C_1	18.76	(-)
C_2	0.0089	(-)
C_3	$6.9023 \cdot 10^{-3}$	(-)
C_4	-0.0132	(-)
R	8.314	J/mol/K

Table 2: Material parameters used for the analysis of thermal residual stresses.

Description	Symbol	Value	Unit
exotherm for composites	Q_{ex}	$36 \cdot 10^3$	J/kg
conductivity in fiber direction	k_x	0.6935	J/(m K)
conductivity in other directions	k_y, k_z	0.3088	J/(m K)
specific heat capacity	c_p	893.23	J/(kg K)
density	ρ	1850	kg/m^3

IOP Conf. Series: Materials Science and Engineering 1293 (2023) 012038 doi:10.1088/1757-899X/1293/1/012038

Table 3: Material parameters used for the analysis of the mechanical behavior of the laminate.

Description	Symbol	Value	Unit
Thermal expansion of fiber	α_f	$5.1 \cdot 10^{-6}$	K^{-1}
Thermal expansion of matrix	α_m	$50 \cdot 10^{-6}$	K^{-1}
Young modulus of fiber	E_f	$7.2 \cdot 10^{10}$	Pa
Young modulus of the matrix at beginning of the curing process	E_{m0}	$2.5 \cdot 10^{6}$	Pa
Young modulus of the matrix at end of the curing process	E_{m1}	$2.5 \cdot 10^{9}$	Pa

Review existing strategies to improve circularity, sustainability and resilience of wind turbine blades – A comparison of research and industrial initiatives in Europe

Kathrin Julia Kramer[1] and Justine Beauson[2]

[1] Institute of Production Technology and Systems, Leuphana University of Lüneburg, Universitätsallee 1, 21335 Lüneburg, Germany, [2] DTU Wind and Energy Systems, Technical University of Denmark, Frederiksborgvej 399, 4000 Roskilde, Denmark

Kathrin.kramer@leuphana.de

Abstract. To become the first climate-neutral continent by 2050, the EU is committing to ambitious targets such as reducing greenhouse gas emissions by at least 55% by 2030 (compared to 1990) and increasing the share of renewable energies in the EU energy mix to 45% by 2030. To meet these targets, the wind industry has to grow significantly while ensuring enough resources across the supply chains to scale the wind energy market and by striving to reduce its already shallow environmental impact. As material extraction and the production of wind turbines cause the most emissions, costs and risks on the scarcity of resources, the implementation of a circular economy could support the wind industry in meeting the EU target and improving its sustainability and resilience. However, the circular economy is a complex concept, and it can be challenging to translate it into precise action points, objectives and measurements. To clarify how a circular economy can support the wind industry, this paper takes the example of wind turbine blades, and it establishes a structured overview of research results and industrial initiatives aiming at implementing circularity for improving sustainability and resilience. The overview is used to investigate if objectives and clear actions are stated and how those differ between research and industry. By identifying gaps, future research and industry initiatives can be directed towards closing the bigger picture. The results show that many initiatives are ongoing, but only some circular strategies are comprehensively investigated, and clear objectives and measurements often remain to be included. The industry and research progressed the furthest on recycling. Future research and industry activities should further follow the path of closing the loop but need to also concentrate on reducing material use, extending the lifetime of blades and enabling a second lifecycle of blades.

1. Introduction

The wind industry is crucial for achieving climate neutrality globally [1]. As such, an immense expansion of wind capacity and, with this, an increase of resource consumption (e.g. material, human, capital) across the wind supply chain is required [3,2]. Simultaneously, the industry has to strive to reduce its environmental impact further, even though being already one of the energy sources with the lowest greenhouse gas emissions today [4]. A circular economy could positively influence the sustainability and resilience of the industry's supply chains by decoupling from virgin material consumption [5–7]. It should be noted that several definitions of a circular economy exist that differ in their scope and objectives [8]. Within this paper, a circular economy is defined as a system-wide concept that reflects the macro, meso and micro levels and aims at using as few materials for as long as possible at their highest possible value to contribute to sustainability and resilience positively [5,8]. It considers designing out waste and pollution through a circular design of products, business models and supply chains and following a waste hierarchy – reduce, reuse, recycle, recover [6,9,8]. Material extraction and production for a wind turbine cause most emissions along the lifecycle [10]. As such, a circular economy that narrows, slows and closes resource flows could lead to reducing emissions. In addition, high supply dependencies exist for virgin materials (e.g. rare earth materials) that could be reduced through a transition to a circular economy [11]. A circular economy requires the stakeholders throughout the value chain to agree on common goals as the success of circularity relies on the involvement of all stakeholders in the product lifetime. Identifying dependencies among the stakeholders and agreeing on common goals is necessary. A structured analysis of the progress of research and industrial development towards the implementation of a circular economy in the wind energy industry is missing.

This paper focuses on the efforts of research and industries dedicated to making onshore and offshore wind turbine blades more sustainable, resilient, and circular. The selection of wind turbine blades as a component to be studied is intended to reduce the complexity of the investigation. In addition, wind turbine blades have attracted a lot attention due to the challenges in recycling them. Therefore, a lot of scientific and industrial literature is available in comparison to other components made of metals. Components made of metals such as the tower are generally assumed to be recycled and the scientific literature is nearly inexistent on this topic [10]. Despite the limitation of the scope to wind turbine blades, interrelations across the system are made. The paper is structured into three additional sections, with the next section outlining the research methodology. Section 3 draws an overview, identifies the gaps and analyses the efforts from research and industries contributing to a sustainable and resilient circular economy. It investigates if objectives and clear actions are stated and how those differ between research and industry. By identifying gaps, future research and industry initiatives can be outlined that direct towards closing the bigger picture. Section 4 provides a conclusion and outlook.

2. Research methodology

The scope of this research is the installed onshore and offshore wind turbines in Europe. The research will focus on wind turbine blades to enable precise identification of objectives and actions for implementing a circular economy. Concentrating on a key component reduces the complexity, overcoming an often-discussed issue that a circular economy is too complex and concrete actions are difficult to grasp [12]. Wind turbine blades were chosen as the object of research as they are of interest from a circular economy perspective: Blades are manufactured using the scarce material balsa wood [13] and emissions can be reduced by limiting the extraction of virgin materials [14]. Moreover, suitable recycling processes [15] and further strategies to narrow, slow and close the resource flow still need to be developed or to become more mature.

The research methodology, shown in figure 1, aims to identify gaps within and across research and industry initiatives for improving the circularity, sustainability and resilience of wind turbine blades. It focuses on identifying and analysing objectives and actions for a circular economy.

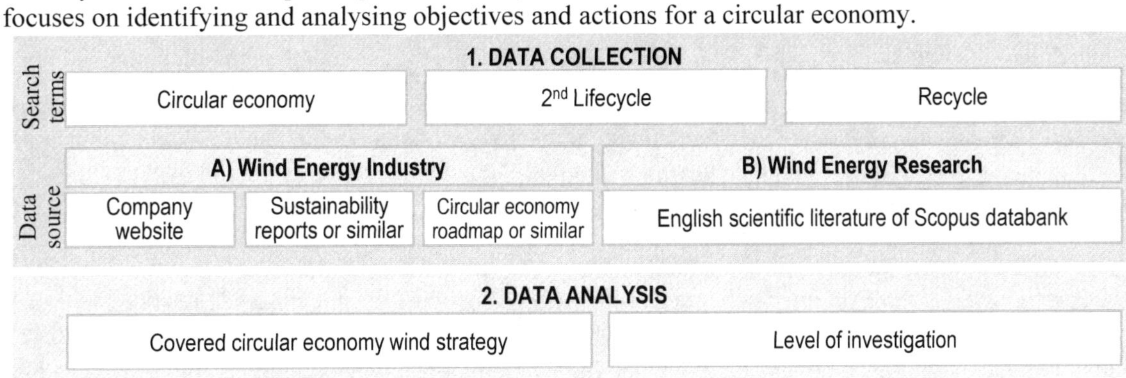

Figure 1. The research methodology of this paper

The first part of the research methodology consists of the data collection on wind energy research and industry initiatives on objectives and actions for a circular economy (Section 2.1). The second part is the data analysis of the identified data (Section 2.2).

2.1. Data collection

Data related to objectives and actions for a circular economy and circular wind strategies are collected in industry and research. Two themes are prioritized, circular economy strategies dedicated to a 2nd lifecycle of wind turbines and circular economy strategies dedicated to recycling. Strategies related to a 2nd lifecycle have received little attention so far, but the need for further investigations is mentioned by literature review papers [15–17]. As recycling gained attention by research and industry, it is interesting to investigate the link between industry and academia. For industry, the focus of investigation is on the communicated intentions and targets towards a circular economy and for academia, scientific publications and research results are reviewed.

To investigate the industry initiatives, suppliers for blade manufacturers, blade manufacturers and operators are considered as they represent the main actors in the wind turbine blade value chain. The suppliers Owens Corning, 3B Fibreglass, and Olin are selected. In addition, the manufacturers Vestas, Siemens Gamesa Renewable Energy (SGRE) and LM Wind Power and the operators Ørsted, RWE and Vattenfall are chosen for further investigation as those companies have major market shares in the European market [19,18]. Data collected are publicly available information from the companies' websites, sustainability reports, circular economy reports or similar. The data were gathered during 23.02.2023-23.05.2023. Using secondary data collection instead of primary data collection (e.g. through workshops or interviews) is preferable at this stage as the officially communicated strategy with objectives and actions can be identified.

For all nine companies selected, the latest sustainability report was downloaded from the companies' website. When available, additional objectives and actions were collected from the websites and circularity roadmaps of the companies. The yearly sustainability reports are non-technical reports, they present the companies' strategies, objectives and achievements related to sustainable development goals, including circular economy objectives. These reports are intended for communication purposes and the title of the report is generally customized with a statement promoting the vision of the company, such as "Leading the energy transition" [20]. There are no standard table of content for such report and each report is structured differently. Data from the reports are collected when circular economy or one of the circular wind strategies according to Velenturf [21] are mentioned. To make the collection of data on objectives as systematic as possible, each objective was identified as quantitative or qualitative one. An example of a quantitative objective is "We aim to send no wind turbine blades to landfill (committed since 2021)" [22]. An example of a qualitative objective is "we aim to limit to the minimum the use of balsa wood and have a substantial volume of the PET from a recycled source" [23]. While the quantitative objective mentions a quantity and a timeline and may act as a target, the qualitative objective expresses an intention. In additions, the actions planned to support the objective were also collected. An example of an action is "to clarify contracts on responsibility of blade waste and training the companies site personnel and contractors on how to comply with our landfill ban" [22].

To investigate the scientific research, a systematic literature review on the established database Scopus was conducted between 02.02.2023-07.03.2023 [24]. A search for English scientific articles and conference papers on Scopus with the search string ("circular economy" AND "wind" AND "blade*") has resulted in 20 relevant papers. Additionally, four literature review papers about a comprehensive range of circular strategies were found, providing mostly details on recycling [15,25,16,17]. Next to that eight literature review papers purely focusing on recycling were identified. It shows that the concept of a circular economy is not yet fully embedded in the research on wind turbine blades. Therefore, a further search is conducted without the term "circular economy" and instead by using the strategies according to Velenturf [21]. In addition, English scientific papers on the 2nd lifecycle of wind turbine blades and recycling are collected. For the 2nd lifecycle, a search on Scopus was carried out with ("blade*" AND "wind") AND either reuse ("reuse" OR "re-use" OR "reus*"), upgrade ("upgrade"), refurbish ("refurbish*" OR "re-furbish*" OR "overhaul" OR "recondition"), remanufacture ("remanufactur*" OR "re-manufactur*") or repurpose ("repurpose" OR "re-purpose" OR "repurpos*"). 130 papers were identified. For recycling a search on Scopus with ("blade*" AND "wind") AND either recycle ("recycl*" OR "scrap"), recycle composites ("recycl*" AND "composite*"), recycle balsa wood ("recycl*" AND "balsa wood"), recycle resin ("recycl*" AND "resin*"), recycle fibre ("recycl*" AND "fibre*") and recycle PET ("recycl*" AND "PET") was conducted and led to 183 scientific publications.

2.2. Data analysis

The second part of the research methodology is the data analysis of the identified data, i.e identified scientific articles and companies' sustainability reports. In section 3.1, the content of the data is analysed and visualized on two different perspective, which are the circular wind strategies as defined by Velenturf [21] and the level of investigation according to Kramer and Schmidt [26].

Velenturf [21] formulates 18 circular wind strategies as a result of reviewing literature, workshops and including feedback from different stakeholders of the offshore wind industry. The strategies are assigned to either narrowing, slowing, closing resource flows or the integration into the environment. Narrowing resource flows includes the strategies design for circularity, dematerialise, prevent waste, and modularisation. Maintain & repair, reuse & repurpose, refurbish & remanufacture, disassemble, and lifetime extension is allocated to slowing resource flows. Repowering is a strategy that belongs to slowing and closing of resource flows. The closing of resource flows is represented by the strategies recertify, decommission, recycle. Re-mine and recover energy are at the interface of closing and integrating. Finally, the strategy restore site and landfill are considered to be strategies belonging to the integration into the environment. Her framework aims for a comprehensive overview of the process steps along the lifetime. Some strategies are equal to the circular economy strategies used in research on conceptualizing a circular economy [6,27]. For example, repair, reuse, repurpose, refurbish, remanufacturing, recycle and recovery are common stated circular economy strategies, also known as R-principles [27]. Design for circularity is highlighted as the overall aim to design out waste and pollution by enabling the R-principles across different levels of the system [6]. Next to those strategies, Velenturf states strategies that could be more seen as facilitating strategies. For example, modularisation is seen as a strategy to ease assembly, disassembly and transport and thus could promote the establishment of the R-principles that slow and close the resource flows [28]. Definitions of each strategy are given in [21] and by the stated references. Strategies that are predominately used in this paper are introduced in the following [21]:

- Design for circularity: Considers the entire lifecycle and foresees more radical, not just incremental, changes in design to enable the comprehensive range of circular strategies, e.g. design for repair, design for reuse, design for upgradeability or design for recycling.
- Lifetime extension: Keep the wind turbine "in use beyond designed service life" at the site.
- Reuse & repurpose: Reuse is the structural reuse of components, e.g. blades, for the same function and repurposing is a structural reuse with a different function.
- Refurbish & remanufacture: Refurbishment is the replacement or repairment of some components that leads to an overall "upgrade" of the multi-component product. Remanufacturing aims at bringing the product's function at least to the originally manufactured quality by following a fully documented industrial process.
- Recycle: Recycling considers the material level and aims at "turning waste into a new substance or product" [29]. Recycled materials are not directly sourced from the natural environment as it is the case for primary resources, but instead from waste.

The levels of investigation according to Kramer and Schmidt [26] differ between organisational, product and processes level. The organisational level reflects tasks related to organisational management, information systems, and (digital) technologies. For instance, stakeholder involvement, choice of technologies and data management or choice of a business model. The product level, in the case of this paper, considers the design of the blade structure and its materials. The processes level includes actions related to the supply chain processes. Hence, tasks on the design, planning, managing, and execution of material, information, and financial flows within a supply chain [30]. This includes processes on plan, source, make, deliver, use, return, recover, and enable as outlined by Vegter et al. [31]. In this paper, processes related to blades are considered. For instance, the sourcing of materials, manufacturing, delivering, installation, decommissioning, refurbishing, repurposing, recycling and reverse logistics of blades.

For clustering the collected data from industry and academia the following procedure was taken: For industry, the authors of this paper have analysed the data according to the covered circular wind strategies and level of investigation independently and discussed when findings deviated. For research, the circular wind strategies used in the title, abstract or key words formed the basis. For the literature on a 2nd lifecycle, the authors have checked the papers according to the above set definitions of the circular wind strategies. For classifying the identified papers according to the covered level(s) of investigation, the authors followed the above-mentioned definitions. The collected data from industry and academia can cover multiple strategies and levels of investigation.

For the research output, to visualize the results, a graphical overview is obtained by counting and categorizing all research papers identified. This way, the most researched or most targeted circular economy strategies can be easily seen. The development over time is also investigated. To evaluate evolving trends, an overview of the number of research papers per year of publication considering all circular economy strategies and the level of investigation is prepared. Furthermore, a more detailed analysis of the addressed objectives and actions for a 2nd lifecycle of blades (Section 3.2), and recycling of blades (Section 3.3) is provided. Hence, the actions in industry that are linked to objectives can be compared to the existing research. Finally, a broader discussion takes place in Section 3.4.

3. Results and discussion

3.1. Circular wind strategies

The analysis of the nine companies selected is visualized in figure 2. The figure displays the circular economy strategies and indicates the number of companies contributing to them based on the content of their respective public available information, in particular from the latest sustainability report and website. The content, objectives and actions used for figure 2 is gathered and presented in a table in Appendix 1. Some companies mention a wide scope of a circular economy [32,23], but objectives and actions are predominantly set on recycling. From figure 2, it appears that all companies are mentioning and setting qualitative and/or quantitative objectives on recycling for wind turbine blades. The second most important topic addressed falls in the category of reduced or no landfill. In third and fourth position, two strategies come forward: dematerialise & prevent waste as well as design for circularity.

Analysing the level of investigation, it appears that most of the content belongs to the blade-related processes, material, and organisational level. This seems reasonable as recycling gains most attention and this strategy tackles the material design and the development of new recycling technologies. The blade structure level was not considered as often.

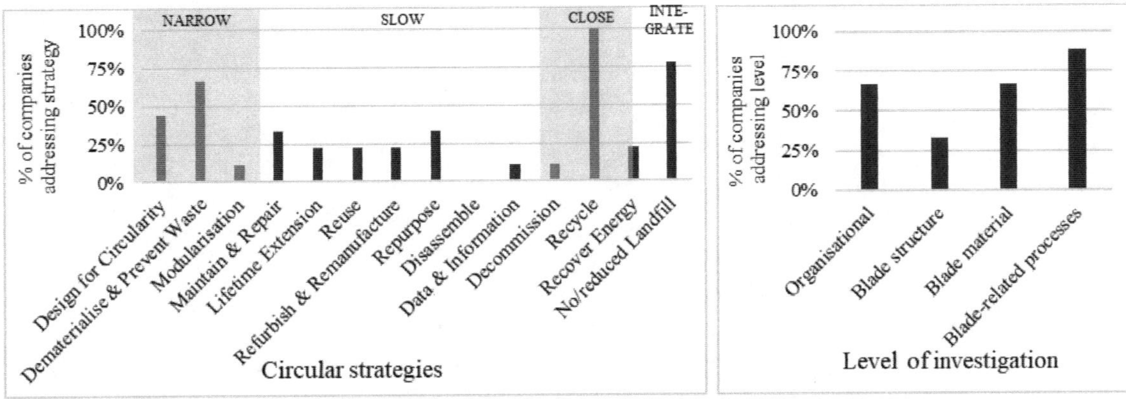

Figure 2. Industries' documentation on circular economy and circular wind strategies for wind turbine blades

In wind turbine blade research, researchers of 20 journal or conference papers use the term "circular economy" either in the title, abstract or keywords. Most of those papers address recycling (10 papers), followed by repurposing (5 papers) and end-of life (EoL) waste volume predictions (5 papers), as displayed in figure 3. The number of publications has slightly increased in recent years, nevertheless the number of papers using the term circular economy is a niche in comparison to 22,710 scientific English papers on wind turbine blades found on Scopus (on 25.05.2023). The papers have addressed different levels, mainly material (8 papers), blade-related processes (7 papers) and the blade structure (6 papers).

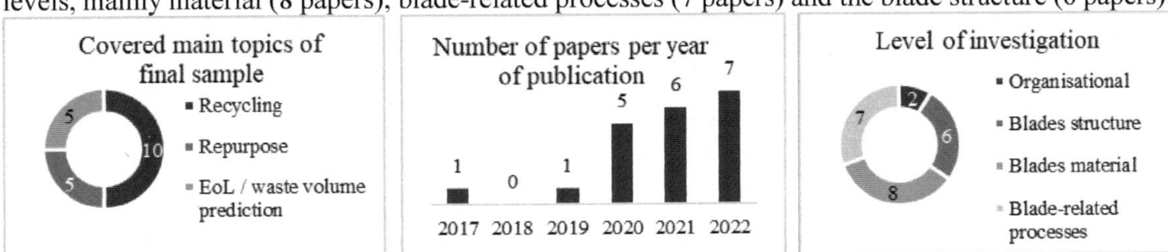

Figure 3. Research on circular economy for wind turbine blades

Figure 4 shows the results on scientific publications on wind turbine blades using the term of at least one circular wind strategy in their title, abstract or keywords. It shows a focus on the strategies "maintain and repair" (1,483 papers), "lifetime extension" (457 papers) and "decommissioning" (601 papers). Research on maintenance and repair as well as lifetime extension eventually slows down the material use within the 1st use phase. The 2nd lifecycle and the strategies reuse, refurbish, remanufacturing, and repurposing have not gained the same attention. Furthermore, the narrowing strategies have only being rarely investigated. For the strategies "repower" and "restore site" it is reasonable that those haven't been researched for blades as those strategies are more likely to be researched from a turbine perspective.

IOP Conf. Series: Materials Science and Engineering 1293 (2023) 012039 doi:10.1088/1757-899X/1293/1/012039

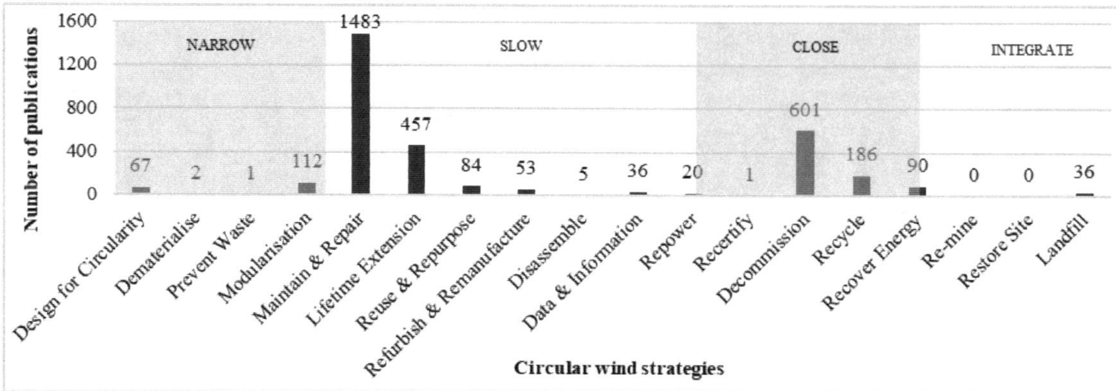

Figure 4. Research on circular wind strategies for wind turbine blades

Comparing between industry and research indicates that the industry communicates mostly on recycling and research focuses on maintain and repair. However, the 20 research publications that used the term "circular economy" were mostly on recycling. Narrowing and 2nd lifecycle strategies have not gained a lot attention by industry and academia. The 2nd lifecycle is closer analysed in the next section (3.2). In addition, as recycling has gained a lot attention by industry and research, a more detailed analysis of the link between research and industry is conducted (section 3.3). Hence, the next sections zoom in at the objectives and actions for enabling a 2nd lifecycle of blades and recycling of blades' materials.

3.2. 2nd lifecycle

After reaching the end of a wind turbine's first lifecycle – hence after the original designed lifetime or lifetime extension – a reuse of the turbine or its components (e.g. blades) should be prioritized according the EU Waste Hierarchy [33]. Therefore, enabling a 2nd (3rd, and so on) lifecycle of the turbine. If the entire turbine cannot be reused, then some components could be reused at the same site in the course of a repowering project or at a different site. The 2nd lifecycle comprises of strategies that foresee a reuse with the same function (direct reuse, refurbishment, remanufacture) or a different function (repurpose).

Objectives by the industry collected on the topic of a 2nd lifecycle of wind turbine blades are presented in table 1. The only quantitative objective formulated among the industries is the one by Vestas. SGRE formulates a qualitative objective on lifetime extension, which is not the same as 2nd lifecycle, however it was still collected as it is part of the higher ranked circular economy strategies. The structural reuse of components could be part of achieving the objective of lifetime extension. However, no information is provided on this. The actions by Vestas to support its objective on reuse of component and a 2nd lifecycle is to create new repair loops for minor components to regionalize their repair and refurbishment infrastructure where possible.

Table 1. Industries' objectives on a 2nd lifecycle for blades

Objective	**Company**
To achieve a 55% total refurbished component utilization by 2030 and 75% by 2040	Vestas
Ensuring an additional 10 years of income: safety and availability of the upgraded turbine until year 30	SGRE

Looking at the research publications, in total, 130 papers use key words linked to a 2nd lifecycle of turbines by applying reuse, refurbish, remanufacturing, and/or repurposing. After analysing the content, only 27 papers were identified that address the 2nd lifecycle of wind turbine blades. Figure 5 shows the identified papers according to the addressed circular wind strategy and their level of investigation. Most papers are linked by the authors to reuse (20 papers), followed by repurpose (10 papers), and remanufacturing (4 papers). No paper was found on refurbishment. When checking the content of the papers, none of those deal with remanufacturing or refurbishment and 23 papers focus on repurposing. Two papers research the modularisation of blades. Two papers provide waste volume predictions that are aimed for planning recycling capacities. However, this information could also be beneficial for the planning of capacities for a 2nd lifecycle. When looking at the level of investigations, 5 papers have addressed the organisational, 11 papers the blade structure, 6 papers the material, and 7 publications the blade-related processes. The development over time shows a slight increase of publications since 2018, howsoever, the total number of publications is still very limited. As such no detailed trends in regards to the circular wind strategies or the level of investigation can be derived from the sample.

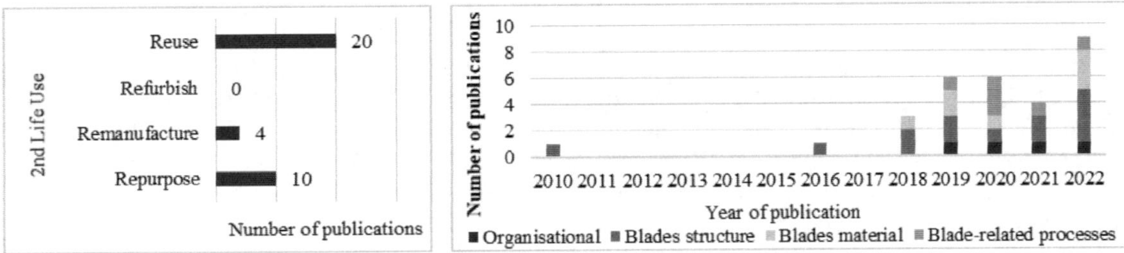

Figure 5. Papers on a 2nd lifecycle of blades by addressed circular wind strategies and level of investigation

The inconsistent use of the terms reuse, refurbish, remanufacturing, and repurposing shows a lack in common definitions of the used terms. This is underpinned by looking at the example of the term "reuse": Mishnaevsky [16] uses the terms reuse and refurbishment and assigns it as primary recycling to recycling of blades. He foresees reuse for the same use, either through lifetime extension, refurbishment or reusing in different structures (e.g. bus shelters). Woo and Whale [17] see reuse to take place after decommissioning of a turbine and providing a 2nd lifecycle for the entire turbine or parts. The authors mention remanufacturing processes in relation to recycling and refurbishment in relation to reuse and lifetime extension. Beauson et al. [15] frame reuse as continued use of a blade as blade after decommissioning. Also, Martinez-Marquez et al. [25] mention remanufacturing, refurbishment, and repurposing, but do not distinguish consistently between remanufacturing and refurbishment throughout their conducted review. The term "reuse" is not defined by them.

Similar to the industry activities, concrete actions in research are rare. On reusing with the same function (i.e. direct reuse, refurbishment or remanufacturing) only a paper by Kaczmarek et al. [34] present structural modularization for providing the possibility to reuse and recombine modules. Further actions could be derived when looking at repair, as indicated by Mishnaevsky [16]. On repurposing, industry hasn't set any quantitative objectives or actions. Owens Corning mention that they aim to "repurpose or recycle the remaining waste, including recycling waste back into our own processes whenever possible" [35]. However, as a fibre manufacturer they eventually relate this objective to the fibre and not the repurposing of wind turbine blades. Research has investigated different actions recently. For example, research on repurposing blades as playground, bridge or bus shelter was carried out [36].

It becomes clear, while analysing the collected data, that uncertainties concerning the future development and technical feasibility for a 2nd lifecycle exist. Mishnaevsky [16] refers to a technical lifetime of blades of up to 50 years. Beauson et al. [15] expect that manufacturers are designing blades for suiting the designed lifetime and as such reducing residual life. Researchers highlight the importance of reuse for achieving a circular economy [15,16]. However, for instance, Kwon et al. [36] skip the pathway of reuse as blade entirely by stating safety concerns, without providing explanations. Joustra et al. [37] write that blades of a turbine cannot easily be installed at a different turbine and as such they conclude that material recycling remains the only suitable option. Research by Martinez-Marquez et al. [25] state that structural repair procedures are not well developed, but Mishnaevsky outlines a technical readiness level (TRL) of 9 for repair and reuse. Furthermore, it is noted by Martinez-Marquez et al. that with an increase in size of blades a structural reuse becomes more difficult [25]. Ortegon [38] mentions that remanufacturing would be possible. This is supported by Martinez-Marquez et al., who refer to a TRL of 10 for refurbishment/remanufacturing [25]. Due to the different use of terms across the studies as well as few provided details, an assessment of the potential for a 2nd lifecycle is difficult to conduct.

Investigations on the technical feasibility and the historical and expected development of the residual lifetime across turbine and blade types could provide clarity. Also, an assessment of the secondary market and predictions of supply and demand for a 2nd lifecycle of blades are currently missing in research. The gathered knowledge can be used to assess if more research in this field is required and if so, objectives and actions can be formulated.

3.3. *Recycling*

When a blade cannot be reused anymore, recycling of the materials is considered as the next step of the EU Waste Hierarchy [33]. Table 2 shows, the quantitative objectives by the industry collected on the topic of wind turbine blade recycling. The wording of those objectives (e.g. "100% recyclable turbine" or "zero waste") as well as the targeted achievement date (e.g. 2030 or 2040) differs slightly. Some additional qualitative objectives were mentioned by the remaining companies, these are not listed in table 1, but shown in the Appendix 1.

Figure 7 shows a schematic representation of the lifecycles of a wind turbine. The stakeholders analysed in this paper: the wind turbine blade material supplier, the wind turbine blade manufacturer and the operators are represented in the figure together with the objectives stated in their respective sustainability reports. From this figure, it is clear that all stakeholders involved in the wind turbine's lifetime are committed to recycle wind turbine blades. It seems that the industries from the wind energy sector are joining forces to solve the challenges of recycling wind turbine blades. As mentioned in the introduction, this aspect is key in realizing and implementing circular economy strategies.

The actions from industries are to support research and be part of research projects to develop recycling solutions. An additional action from Ørsted is to clarify contracts on responsibility of blade waste and training Ørsted site personnel and contractors on how to comply with a landfill ban for blades.

Table 2. Objectives on recycling for blades

Objectives	**Company**
Zero Glass to Landfill (objective pursued since 2017)	3B Fibreglass
By 2030, we will send zero waste to landfill	Owens Corning
To create a rotor that can be 100% recycled by 2030	Vestas
To ensure a 100% recyclable turbine marketed by 2040 (currently 94%)	SGRE
To manufacture zero waste blades by 2030	LM Wind Power
Achieve a recycling rate of wind turbine blades of 50% by 2025, and 100% by 2030	Vattenfall
We aim to have 100 % recyclable blades installed for new assets by 2030	RWE
We aim to send no wind turbine blades to landfill (committed since 2021)	Ørsted

Results from the identified research are presented in figure 6. In total, 183 papers used keywords related to recycling of composites, balsa wood, resin, fibre and/or PET. Out of those, 112 papers are appropriate for analysing as they address wind turbine blade recycling. The right part of figure 6 outlines that the overall research interest has increased since 2017. A peak is shown in 2022, representing 27.6% of papers from the sample. A reason for this peak could be the announcement of the industry to call for a ban on old turbine blades in landfills that was announced in June 2021 [39].

Looking at the terms used by authors in the title, abstract and keywords, recycling of composites (86 papers) is the most researched topic within recycling of wind turbine blades. For 35 papers, researchers have used the term "recycle resin" and in 33 papers the term "recycle fibre". Research on recycling PET (2 papers) and balsa wood (3 papers) is rarely shown. Detailed reviews on the state of the art of recycling are given for example in [15], [16], [40] and [41]. Researchers are investigating the replacement of currently used materials for the manufacturing of new blades. For the fibres and resin of composites, research paths are biocomposites, thermoplastic matrices, and modified epoxies [15]. For the core, mostly balsa wood is used. Referring to above, research on alternative materials for the core is rare. Further, the manufacturing, recycling and reuse of materials from existing and new blades is being researched. In particular for materials of existing blades, the development of suitable recycling solutions is required as decommissioning volumes are expected to increase. Recycling methods can be clustered into mechanical (e.g. shredding), thermal (e.g. pyrolysis), or chemical recycling (e.g. solvolysis) [15,16]. However, recovering energy of composites to use in the cement kiln production is the only existing solution which runs at an industrial scale and indicates a positive environmental impact over landfill [42]. The use of recycled materials from blades in the manufacturing of new blades has not been achieved in scale [15]. An interesting comparison can be made on the different recycling solutions needed for existing blade reaching end-of-life today and the ones built with a recyclable resin and that will be recycled in a few decades from now. Recycling processes for existing blades, such as mechanical recycling for the construction of sound insulation panels close the loop as long as the sound insulation are not sent to landfill at the end-of-life. These solutions rely on the proper recycling of the following application made of the materials. Recycling processes for future blades pushes for the separation and the recovery of the individual material constituents with the objective, in some cases, of reusing these materials in the production of new blades. In this case, the loop is closed and a new loop starts.

When distinguishing between the covered level of investigation, it shows that 12 papers address the organisational, 15 papers the blade structure, 65 papers the blade material and 47 papers the blade-related processes. When looking at the development over time, research on blades materials and processes starts in 2005. The blade structure and organisational level was first mentioned in 2010 for the structure and 2012 on the organisational level. Especially, since 2018 the number of publications on materials has increased. In all levels of investigation, a peak appears in 2022. The topic recycling of composites has a steady research output since 2010 with first publications dated back to 2005. First publications on recycling resins also date back to 2005, but more regular annual output appears since 2016. The term "recycle fibre" is first used by a publication from 2010, however, 57.6% of the publications using this term were published in 2021 and 2022. The three publications on recycling of balsa wood are from 2020, 2022, and 2023. The two publications on recycling of PET are from 2005 and 2023.

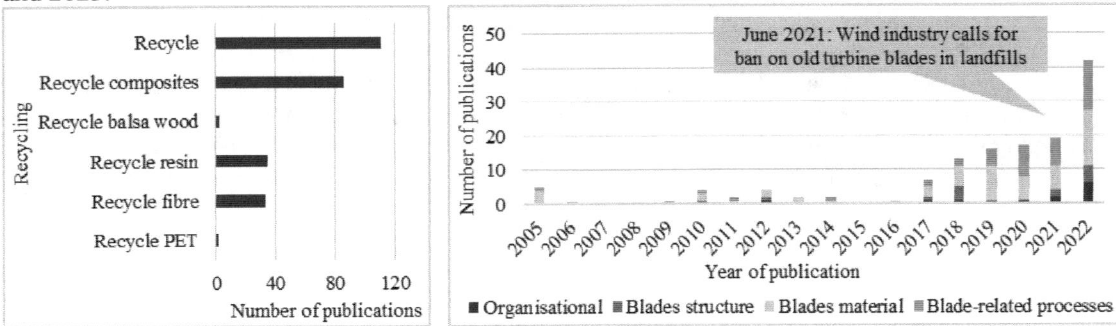

Figure 6. Papers on recycling of blades by addressed circular wind strategies and level of investigation

A comparison of the industry and research activities is, due to the different nature of data, difficult to conduct. It seems that research is more focusing on material development and the related manufacturing or recycling processes for recycling composites and with this the fibre and resin. In the collected data, companies do not communicate a lot on these engineering activities, but rather on organisational actions. For example, choosing to collaborate in research projects or training personnel. The first mentioned task implies to work with research on the above-mentioned topics. The recycling of the core material (e.g. balsa wood) has not been addressed prominently by research or industry. Eventually the interest will raise as, for example, the Netherlands have proposed the recycling of balsa wood as one criterion within the circular economy-related auction criteria in their upcoming auctions for offshore wind [44,43]. PET, a fossil-based material, has also not received a lot attention so far. With the required phase-out of fossil fuels, alternative materials and the reuse of existing PET from different industries could be a field for further research.

3.4. Broader discussion

For a broader discussion of the findings, figure 7 summarises some of the key objectives for a circular economy from industry and puts those into context of the different strategies of a circular economy. Qualitative objectives are marked with a yellow box and quantitative ones with a blue box. The figure also sketches how the different lifecycle processes and circular strategies on narrowing, slowing and closing resource flows are interlinked. It should be noted that a circular economy does not foresee a complete closed system, but rather cooperation with other companies within and across industrial sectors [9]. Moreover, in practice there is not one ideal circular economy and instead it depends on the local environment, e.g. available reverse supply chain, suitable refurbishment and recycling facilities, country's energy mix or potential synergies with other industries.

First of all, waste and pollution are to be designed out of the material extraction and processing, manufacturing, remanufacturing, refurbishment, installation, operation, decommissioning, reusing, repurposing, and recycling processes.

The material extraction and processing of the fibre, resin and core material should be reduced to an absolute minimum and be replaced by recyclable or regenerative materials. Within the production of the materials and the blade itself, manufacturing waste should be eliminated. A blade can be produced either through manufacturing, refurbishment or remanufacturing. When installing a wind turbine, manufactured, refurbished, remanufactured or directly reused blades can be used. The three last mentioned strategies are preferred from a circular economy perspective as they eventually have a lower environmental impact than manufacturing a new blade. However, currently it is not possible to disassembly a blade after a use phase for inspecting and returning it to an as-new condition as it is for example being done for the engines from Rolls-Royce [45]. Hence, remanufacturing for blades is, as of today, technically not possible. Furthermore, due to the technical progress and launch of larger and larger turbines and with this, larger blades, a reuse of older and smaller blades might be less attractive. For instance, an onshore wind turbine is designed with an operational lifetime of 20 years according to the IEC standard [46]. An aim can be to extend the lifetime, as such operating the turbine at the same site for longer than originally designed. During the operational phase, maintenance and repair work are carried out. After the end of the first lifecycle, the turbine is decommissioned. The decommissioned blades preferable enter, in a cascading order, the direct reuse, refurbishment and remanufacturing loops. If the blades cannot be reused as a blade, they could be repurposed, i.e. reused with a different function. When finally reaching the end of the structure's life, recycling of the material is foreseen. The materials of the blades are aimed to be recycled with the same quality of material properties so they could be reused for the production of blades.

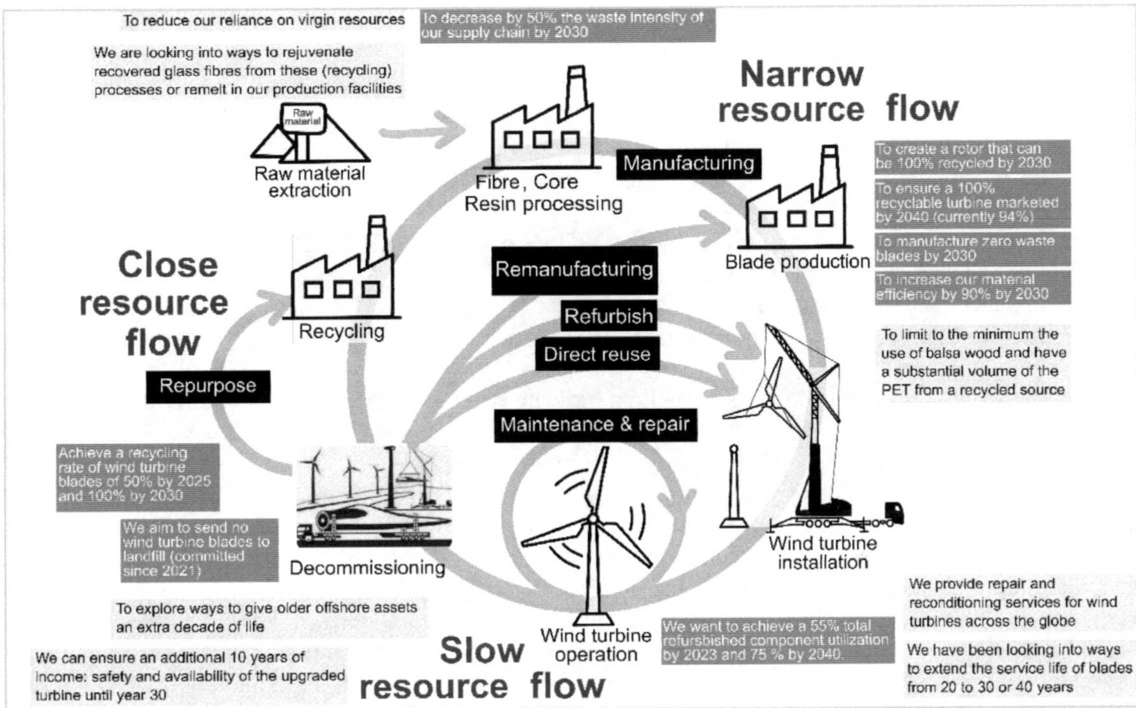

Figure 7. A schematic circular economy for wind turbine blades with narrowing, slowing and closing resource flows

The presented objectives, that have their origin from different companies, address a wide range of processes, e.g. reduction of virgin materials, design fully recycle turbines or enable a lifetime extension. However, not all of these objectives have a quantitative nature and the level of ambition differs across the companies. Finding a solution for the recycling of composites is an aligned objective the entire industry has and research is investigating different solutions for the already installed blades and designs for the new to be produced blades. This is one of the few topics addressed in the reports of the companies where a lot of data is available and a general consensus on the target exists. Interesting to note is that objectives are set by companies that lead to benefits for their suppliers and customers. For instance, wind turbine blades manufacturers can develop recyclable blades design and may participate in developing recycling solutions (however they cannot develop the technology). Glass fibre manufacturer may develop the technology to remelt recycled fibres and may also support the development of recycling technology by providing their needs. The establishment of feedback loops on lessons learned, e.g. from the operation, disassemble or recycling processes can in return support the material and blade manufacturer on new designs. For the lifetime extension objectives by the manufacturers, who also offer services along the operational phase, and by operators are set. Most research publications were found on repair and maintenance as well as lifetime extension. A few objectives on further slowing (e.g. establish refurbishment of components) and narrowing (e.g. decrease waste intensity) were identified. The research coverage on those objectives was limited.

4. Conclusion and outlook

The paper provides a comprehensive review of objectives and actions from industry and research for transforming to a circular economy. The example of wind turbine blades is used. The review shows that different circular wind strategies and levels of a system are addressed by different stakeholders. The findings were put into perspective by sketching out the lifecycle processes in a circular economy for wind turbine blades and outlining the interdependencies between the stakeholders (refer to figure 7). The main results and observations from this review are the following:

1) Despite the different nature of the content, we see good agreement between different supply chain industry players and research on recycling.
2) Second lifecycle of blades lacks attention from research and suffers from lack of definition and characterization.
3) The interdependencies across stakeholders, product lifecycle phases, and system levels are to be considered for transforming to a circular economy for wind turbine blades.

This review highlights three major challenges standing on the way:
1) Lack of standard and definitions.
2) Lack of traceability and transparency.
3) Second lifecycle shows potential but lacks attention and policies.

The review's findings, in particular on recycling, indicate two potential enablers for a circular economy:
1) Promoting a transition of industry through policy change.
2) Progress on circular economy strategies through alignment of objectives across stakeholders.

In the following the major challenges and enablers identified herein are described and with this a research agenda is derived: The provided review illustrates that there is no standard in used terms for articulating circular economy objectives and actions, neither in industry, nor in research. Industry and research often do not define their understanding of the used terms. For instance, the term "refurbishment" is sometimes used interchangeably with "remanufacturing", even though remanufacturing for blades, when referring to the above set definition, does not make sense as of today. The review of scientific publications on wind turbine blades has shown that only a few researchers use the term "circular economy" that could indicate that the concept is not yet embedded into the research field. The lack of definition will also have major impact on the calculation of recyclability rate. Recyclability rate are not presented in this paper, however, they are being increasingly used in regulations to promote circularity and the recycling of product before the products are produced and used. If definitions are unclear, recyclability calculation will be highly unreliable.

There is also a lack of traceability and transparency on the objectives and actions. As mentioned the non-standardised companies' sustainability reports were used as main source for analysing objectives and actions for a circular economy. Due to the lack of standardisation of the reports, but also of the used terms, aspects could be understood differently if analysed by someone else. Also, it cannot be guaranteed that all relevant information on the website were identified. To provide transparency, the underlying data is provided in Appendix 1. The reviewed companies do not provide a comprehensive report on the progress of the objectives' fulfilment. There is also no standard yet on how to measure progress. The data quality might improve in the future through the Corporate Sustainability Reporting Directive that aims to standardise reporting on sustainability for companies being active in the EU [47]. Further standardisation could be achieved through agreeing on circular design criteria as targeted by the Global Alliance for Sustainable Energy [48]. In future research methods to collect primary data (e.g. expert interviews) could be used. Additionally, further companies, along the entire supply chain should be investigated. In particular, to include companies working on the refurbishment, resell, decommissioning and recycling of blades could add further insights. For the identified data on academic research, further findings could appear if other databanks (e.g. Web of Science) are used. In addition, only a detailed look into recycling and a 2nd lifecycle took place, but the other circular wind strategies could also be reviewed in more depth. Further insights on current activities and intentions of research could be assessed through reviewing ongoing research projects. In those, more broader objectives on sustainability, resilience and circularity might be outlined.

The 2nd lifecycle shows potential but lacks attention from researchers and policies. As stated above terms are not well defined and the potential of reuse has not been assed. There is also a lack in traceability and transparency. For instance, North Europe, in particular Denmark and North Germany have an aging fleet of onshore turbines. Predictions of waste volumes have not materialised so far that mostly assumed a decommissioning after 20 years [49]. Thus, annual expected quantities are highly uncertain, which complicates the planning of strategies at the end of a lifecycle. When and where which turbines will be decommissioned is of importance for the capacity planning of reverse logistics, decommissioning, direct reuse, refurbishment, repurposing, and recycling in this region. An assessment if blades were reused could provide insights on the expected quantities of subsequent strategies such as recycling. Additionally, an assessment if the reuse of blades is a strategy for this region that should be promoted from a circular economy perspective was not conducted yet. Further insights could help to set or redefine objectives by the industry, define research needs, and inform policy for eventually required changes.

It looks as if a transition of industry could be promoted through policy change. It seems that the call for a landfill ban in June 2021 has triggered circular economy strategies, in particular recycling, across the industry. Further investigations on the role of policy and public perception for transforming to a circular economy, could bring valuable insights for further progress. The achieved progress and ongoing activities on recycling of blades indicate that progress on circular economy strategies is achieved through an alignment of objectives across stakeholders. Research on the transition process towards more sustainability, resilience and circularity could help to understand the mechanism and thus provide insights on how to further accelerate the transition in the wind industry. For the herein conducted analysis, a new dimension of analysis could be added by investigating the development of the companies' strategies over time. For instance, when was circular economy first communicated by the companies and how did it evolve over the years.

It remains the question if all necessary objectives have already been defined and set objectives are progressive enough to stick within planetary boundaries and reduce dependencies on materials to actually achieve the build-out of the wind industry. There is no consensus on what the circularity objectives should be across the wind industry, except of promoting recycling. As there is not one ideal circular economy, empirical research on the narrowing, slowing, and closing of resource flows for different regions and with different stakeholders across the onshore and offshore wind industry as well as other industries could provide more insights. Hence, taking into consideration different interdependencies along the lifecycle and stakeholders with a system-wide perspective. The formulation of aligned objectives with stakeholders across supply chains, research, and policy would be the first step towards more progress on transforming to a circular economy and thus decoupling from virgin material consumption.

References

[1] International Renewable Energy Agency, 2019. Future of Wind – A Global Energy Transformation paper, Abu Dhabi, 88 pp.
[2] Hutchinson, M., Zhao, F., 2023. Global Wind Report 2023, Brussels, 120 pp.
[3] European Commission, 2023. European Green Deal: EU agrees stronger legislation to accelerate the rollout of renewable energy, Brussels, 2 pp.
[4] United Nations Economic Commission for Europe, 2021. Life Cycle Assessment of Electricity Generation Options, Genf, 107 pp.
[5] Chaouni Benabdellah, A., Zekhnini, K., Cherrafi, A., 2021. Sustainable and Resilience Improvement Through the Design for Circular Digital Supply Chain, in: Dolgui, A., Bernard, A., Lemoine, D., Cieminski, G. von, Romero, D. (Eds.), Advances in Production Management Systems. Artificial Intelligence for Sustainable and Resilient Production Systems, vol. 633. Springer International Publishing, Cham, pp. 550–559.
[6] 2019. Completing the Picture: How the Circular Economy Tackles Climate Change. www.ellenmacarthurfoundation.org/publications. Accessed 21 February 2022.
[7] Negri, M., Cagno, E., Colicchia, C., Sarkis, J., 2021. Integrating sustainability and resilience in the supply chain: A systematic literature review and a research agenda. Bus Strat Env 30 (7), 2858–2886.
[8] Kirchherr, J., Reike, D., Hekkert, M., 2017. Conceptualizing the circular economy: An analysis of 114 definitions. Resources, Conservation and Recycling 127, 221–232.

[9] Farooque, M., Zhang, A., Thürer, M., Qu, T., Huisingh, D., 2019. Circular supply chain management: A definition and structured literature review. Journal of Cleaner Production 228, 882–900.
[10] Bonou, A., Laurent, A., Olsen, S.I., 2016. Life cycle assessment of onshore and offshore wind energy-from theory to application. Applied Energy 180, 327–337.
[11] Alves Dias, P., Bobba, S., Carrara, S., Plazzotta, B., 2020. The role of rare earth elements in wind energy and electric mobility: An analysis of future supply/demand balances. Publications Office of the European Union, Luxembourg, 1 online resource.
[12] Corvellec, H., Stowell, A.F., Johansson, N., 2021. Critiques of the circular economy. Journal of Industrial Ecology, 1–12.
[13] The Economist Newspaper Limited, 2021. The wind-power boom set off a scramble for balsa wood in Ecuador. https://www.economist.com/the-americas/2021/01/30/the-wind-power-boom-set-off-a-scramble-for-balsa-wood-in-ecuador. Accessed 15 June 2023.
[14] Chiesura, G., Stecher, H., Pagh Jensen, J., 2020. Blade materials selection influence on sustainability: a case study through LCA. IOP Conf. Ser.: Mater. Sci. Eng. 942 (1), 12011.
[15] Beauson, J., Laurent, A., Rudolph, D.P., Pagh Jensen, J., 2022. The complex end-of-life of wind turbine blades: A review of the European context. Renewable and Sustainable Energy Reviews 155 (July), 111847.
[16] Mishnaevsky, L., 2021. Sustainable End-of-Life Management of Wind Turbine Blades: Overview of Current and Coming Solutions. Materials (Basel, Switzerland) 14 (5).
[17] Woo, S.M., Whale, J., 2022. A mini-review of end-of-life management of wind turbines: Current practices and closing the circular economy gap. Waste management & research: the journal of the International Solid Wastes and Public Cleansing Association, ISWA 40 (12), 1730–1744.
[18] Rystad Energy, 2023. The State of the European Wind Energy Supply Chain: A «what-would-it-take» analysis of the European supply chain's ability to support ambitious capacity targets towards 2030, 56 pp.
[19] EurObserv'ER, 2022. WIND ENERGY BAROMETER, 7 pp.
[20] Vestas Wind Systems A/S, 2022. Sustainability Report 2021: Leading the energy transition, 95 pp.
[21] Velenturf, A.P.M., 2021. A Framework and Baseline for the Integration of a Sustainable Circular Economy in Offshore Wind. Energies 14 (17), 5540.
[22] Ørsted A/S, 2023. Green energy to power lasting positive impact: Sustainability report 2022, 45 pp.
[23] LMWind, 2022. Sustainability Performance 2021, 30 pp.
[24] Martín-Martín, A., Orduna-Malea, E., Thelwall, M., López-Cózar, E. D., 2018. Google Scholar, Web of Science, and Scopus. A systematic comparison of citations in 252 subject categories. Journal of Informetrics 12 (4), 1160–1177.
[25] Martinez-Marquez, D., Florin, N., Hall, W., Majewski, P., Wang, H., Stewart, R.A., 2022. State-of-the-art review of product stewardship strategies for large composite wind turbine blades. Resources, Conservation & Recycling Advances 15 (17), 200109.
[26] Kramer, K.J., Schmidt, M., 2022. Circular supply chain management for the wind energy industry – Conceptional ideas towards more circularity, in: Plapper, P. (Ed.), Digitization of the work environment for sustainable production. GITO Verlag, pp. 61–80.
[27] Potting, J., Hekkert, M., Worrell, E., Hanemaaijer, A., 2016. Circular Economy: Measuring innovation in product chains, The Hague.
[28] Mignacca, B., Locatelli, G., Velenturf, A., 2020. Modularisation as enabler of circular economy in energy infrastructure. Energy Policy 139, 111371.
[29] Defra - Department for Environment, Food & Rural Affairs, 2011. Guidance on Applying the Waste Hierarchy 2011. https://assets.publishing.service.gov.uk/government/uploads/system/uploads/attachment_data/file/69403/pb13530-waste-hierarchy-guidance.pdf. Accessed 15 June 2023.
[30] The Supply Chain Council, 2012. SCOR supply chain operations reference model, 11th ed., 2 volumes.

[31] Vegter, D., van Hillegersberg, J., Olthaar, M., 2020. Supply chains in circular business models: processes and performance objectives. Resources, Conservation and Recycling 162, 105046.
[32] 3B fibreglass, 2022. Sustainability Report 2021, 57 pp.
[33] European Union, 2008. DIRECTIVE 2008/98/EC OF THE EUROPEAN PARLIAMENT AND OF THE COUNCIL of 19 November 2008 on waste and repealing certain Directives.
[34] Kaczmarek, S., Hogreve, S., Greten, A., Schröder, M., Tracht, K., 2016. Improving Design Efforts and Assembly Efficiency of Rotor Blade Carriers through Modularisation. Procedia CIRP 50, 76–81.
[35] Owens Corning, 2022. BUILDUNG A SUSTAINABLE FUTURE: 2021 SUSTAINABILITY REPORT, 376 pp.
[36] Kwon, E., Pehlken, A., Thoben, K.-D., Bazylak, A., Shu, L.H., 2019. Visual Similarity to Aid Alternative-Use Concept Generation for Retired Wind-Turbine Blades. Journal of Mechanical Design 141 (3), 505.
[37] Joustra, J., Flipsen, B., Balkenende, R., 2021. Structural reuse of wind turbine blades through segmentation. Composites Part C: Open Access 5, 100137.
[38] Ortegon, K., Nies, L.F., Sutherland, J.W., 2012. Remanufacturing: An Alternative for End of Use of Wind Turbines, in: Dornfeld, D.A., Linke, B.S. (Eds.), Leveraging Technology for a Sustainable World. Springer Berlin Heidelberg, Berlin, Heidelberg, pp. 155–160.
[39] Mathis, W., 2021. Wind Industry Calls for Ban on Old Turbine Blades in Landfills. https://www.bloomberg.com/news/articles/2021-06-16/wind-industry-calls-for-ban-on-old-turbine-blades-in-landfills?leadSource=uverify%20wall#xj4y7vzkg. Accessed 15 June 2023.
[40] Rathore, N., Panwar, N.L., 2023. Environmental impact and waste recycling technologies for modern wind turbines: An overview. Waste management & research: the journal of the International Solid Wastes and Public Cleansing Association, ISWA 41 (4), 744–759.
[41] Khalid, M.Y., Arif, Z.U., Hossain, M., Umer, R., 2023. Recycling of wind turbine blades through modern recycling technologies: A road to zero waste. Renewable Energy Focus 44, 373–389.
[42] Nagle, A.J., Delaney, E.L., Bank, L.C., Leahy, P.G., 2020. A Comparative Life Cycle Assessment between landfilling and Co-Processing of waste from decommissioned Irish wind turbine blades. Journal of Cleaner Production 277, 123321.
[43] Ministry of Economic Affairs and Climate Policy - Minister for Climate and Energy, 2023. Ministerial Order for granting the permit for IJmuiden Ver wind Wind Farm Site Alpha: Draft, 29 pp.
[44] Ministry of Economic Affairs and Climate Policy - Minister for Climate and Energy, 2023. Ministerial Order for granting the permit for IJmuiden Ver Wind Farm Site Beta: Draft.
[45] Rolls-Royce, 2021. Rolls-Royce to expand mtu Aiken Plant to add remanufacturing and overhaul capabilities.
[46] International Electrotecnical Commission (IEC), 2019. IEC 61400-1:2019 RVL, Wind energy generation systems - Part 1: Design requirements. https://webstore.iec.ch/publication/64648. Accessed 5 July 2023.
[47] European Union, 2022. DIRECTIVE (EU) 2022/2464 OF THE EUROPEAN PARLIAMENT AND OF THE COUNCIL of 14 December 2022 amending Regulation (EU) No 537/2014, Directive 2004/109/EC, Directive 2006/43/EC and Directive 2013/34/EU, as regards corporate sustainability reporting.
[48] 2022. Circular Design criteria Working Group Position Paper: Reducing Raw materials' Environmental and Social impacts in Electrical Energy Technologies. https://www.sustainable-energy.eco/resources/documents/Circular_Design_WG_position_paper_11112022.pdf. Accessed 19 June 2023.
[49] Lichtenegger, G., Rentizelas, A.A., Trivyza, N., Siegl, S., 2020. Offshore and onshore wind turbine blade waste material forecast at a regional level in Europe until 2050. Waste management (New York, N.Y.) 106, 120–131.
[50] Olin, 2023. Resilience - Sustainability report 2022, 13 pp.

[51] SGRE, 2021. Consolidated Non-Financial Statement 2020 (formerly Sustainability Report), 182 pp.
[52] SGRE, Delivering sustainability through actions - Sustainability vision 2040, 11 pp.
[53] Vattenfall, 2023. LEADING THE WAY TO FOSSIL FREEDOM – Annual and Sustainability report 2022, 196 pp.
[54] RWE, 2022. Focus on tomorrow - Sustainability report 2021, 142 pp.
[55] RWE, 2023. Together we turn passion into sustainability - Sustainability Strategy report 2022, 46 pp.

Appendix 1.

Company	Objectives	Actions	Level of investigation	Circular wind strategy
3B Fibreglass	**Zero Glass to Landfill (objective that they pursue since 2017).**	1. Virgin raw material reduction: long term vision = use pre post consumer mineral waste and external glass waste to replace virgin raw materials. 2. Internal Recycling of glass waste: ongoing investigation 3. Involved in several research project	materials, processes	no landfill, reduce, recycle
Olin	No objectives.	By utilizing unique technologies, together with our partners, we are ready to recover materials from spent blades and convert them into new epoxies that can be reused in new wind turbine blades. We are excited to bring our expertise and unique asset footprint to this new endeavor, and realize breakthrough sustainable material solutions for existing wind blades and those of the future.	material, processes	recover, recycle
Owens Corning	By 2030, we will establish viable circular economy business models involving our materials and how they are used.	1. This overall objective is achieved by the further stated objectives below. 2. Collaborating up and down the supply chain, with customers, suppliers, communities, academics, policy makers, government entities, and other organizations. 3. The circular economy team, established in 2020, defines goals and prioritizes projects that accelerate our circular economy ambitions. 4. Source materials and serve customers in ways that minimize transportation and its impact.	organisational, processes	reduce, design for circularity
Owens Corning	**By 2030, we will send zero waste to landfill.**	Using the following plan: Reduce waste intensity by 50% by improving efficiency and process design. Repurpose or recycle the remaining waste, including recycling waste back into our own processes whenever possible.	processes	reduce, recycle, no landfill, repurpose
Owens Corning	By 2030, expand the use of recycled materials in our manufacturing operations and our products and decrease virgin materials used in our products.	1. Engage in the ZEBRA (Zero WastE Blade ReseArch) project to develop the first 100% recyclable wind turbine blade. 2. We have begun a pilot program to build capability for recycling within our Composites business with the aim of enabling circularity 3. We are also investigating the potential in processes such as chemical recycling, which could prove valuable as we try to perform tasks such as separating fibers from resins and chemicals.	organisational, processes, material	recycle, reduce

Owens Corning	By 2030, we are seeking innovative technologies and business models for our product materials and packaging to be reused and repurposed indefinitely.	1. Internal partnerships among R&D, commercial, and corporate development to shape the vision and execution in this area. 2. Engage with external partners to develop end-of-life solutions for our products, as well as the products where our materials are used.	organisa-tional, processes, material	reuse, recycle, repurpose
LM	No objectives.	Optimizing the use of natural resources and improve existing blade designs to facilitate more recycled materials.	blade structure, processes	design for circularity , recycle
LM	**To manufacture zero waste blades by 2030 (committed since 2021).**	1. We aim to send no excess manufacturing materials and packaging to landfill and incineration without energy recovery by 2030. 2. Support research in finding application for recycled materials (Decomblades, Zebra) 3. R&D activities: Manufacturing waste is reduced through further optimization of our manufacturing systems.	processes	no landfill, recycle, reduce
LM	To advance towards the end goal of fully recyclable blades.	Supporting research: Finally, the Zero wastE Blade ReseArch (ZEBRA) project sees us advance towards the end goal of fully recyclable blades.)	organi-sational, structure, material, processes	recycle
LM	To limit to the minimum the use of balsa wood and have a substantial volume of the PET from a recycled source.	We have progressively transitioned from balsa wood to PET. Currently, around 95% of our blades use PET.	blade structure, material, processes	reduce, recycle
LM	No objectives.	To deliver waste back to suppliers, for recycling into new materials that will be supplied to the wind industry or other sectors.	processes	recycle
SGRE	Ensuring an additional 10 years of income: safety and availability of the upgraded turbine until year 30 (service available on request for customers).	1. Providing repair and reconditioning services for wind turbines across the globe. 2. Efforts are being made to improve our component upgrades and lifetime extension service offerings, spare parts and parts refurbishment offerings for service and maintenance operations on our customers' turbines.	processes	lifetime extension, repair, refurbish
SGRE	**To ensure a 100% recyclable turbine marketed by 2040 (currently 94%).**	1. Support research on recycling processes and in finding application for recycled materials (e.g. Decomblades, FiberEUse) 2. Ensuring that all materials in the turbine can be recycled (e.g. RecyclableBlade) 3. Modular wind turbine design is also an environmental benefit since it eases dismantling and optimizes waste treatment methods.	materials, processes	recycle, design for circularity , modular., decom-missionin g

SGRE	To make sure that the blades from the wind projects that are being decommissioned do not end in landfill.	Finding ways to ensure these blades are either recycled or repurposed	not stated	recycle, repurpose, no landfill
SGRE	**Reduce waste generation to achieve 6.0 t/MW waste generation rate by 2023.**	1. Monitoring environmental data and visualizing these for better analysis. 2. Creating transparency & opportunities for best practice sharing.	organisa-tional	reduce, data & info.
Vestas	**To achieve a 55% total refurbished component utilization by 2030 and 75% by 2040.**	1. Creating new repair loops for minor components 2. Regionalizing our repair and refurbish-ment infrastructure where possible	processes	repair, refurbish
Vestas	**To create a rotor that can be 100% recycled by 2030. Achieve zero-waste wind turbines by 2040.**	1. Support research in the development of recycling processes for blade materials (CETEC and DecomBlades) 2. Support research in finding application for recycled materials (Decomblades, SUSWind)	organisa-tional, processes	recycle, no landfill
Vestas	**To increase our material efficiency by 90% by 2030.**	1. Raising awareness on waste within our factories 2. Optimising our blade design and production to create less waste, 3. Sourcing more efficient manufacturing kits, 4. Including circularity metrics in our key performance indicators	organisa-tional, blade structure, processes	reduce
Vestas	**To decrease by 50% the waste intensity of our supply chain by 2030.**	1. To ask the Vestas strategic suppliers to report on their waste generation. 2. To ask the Vestas strategic suppliers to set waste reduction targets for their own operations, and ask their own suppliers to set waste reduction targets.	organisa-tional	reduce
Vestas	**1. To reduce the amount of manufacturing waste ending up in landfill to less than 1% (from 7% today) by 2030** **2. To reduce the amount of waste incinerated to less than 1% (from 38% today) by 2030** **3. To reduce the amount waste incinerated with energy recovery to less than 5% (from 11% today) by 2030**	Mapping out waste streams in manufacturing, construction, and service	organisa-tional	reduce landfill, reduce incinerat., reduce recovery
Ørsted	To reduce our reliance on virgin resources.	By optimising how we design offshore wind farms, and by increasing our use of components made from recycled materials	structure, material, processes	reduce, design for circularity
Ørsted	To explore ways to give older offshore assets an extra decade of life.	By improving the way we repair, reuse, and repurpose assets and components so we can extend their lifetime.	processes	lifetime extension, reuse, repair, repurpose

Ørsted	We aim to divert as much as possible of the waste we produce from disposal.	Training of personal "published a waste management guide and launched online and in-person training courses at all our operational sites to improve waste management."	organisational	recycle, no landfill
Ørsted	To close the material loop, and create a circular value chain.	By finding new ways to recycle our renewable energy assets so we can feed in materials from old assets into the production of new assets	materials, processes	recycle
Ørsted	**We aim to send no wind turbine blades to landfill (committed since 2021).**	1. Support research: DecomBlades, a three-year innovation project that seeks to commercialise sustainable recycling routes for wind turbine blades, carried out by ten project partners from the wind and recycling industry as well as academia 2. Clarifying contracts on responsibility of blade waste and training Ørsted site personnel and contractors on how to comply with our ban.	organisa-tional, processes	no landfill, recycle
RWE	**We aim to have 100 % recyclable blades installed for new assets by 2030.**	Cooperation with SGRE (RecyclableBlade) and Veolia North, Pilot Project: Kaskasi offshore wind farm	organisa-tional	recycle
RWE	**Aim to increase recovery rate to over 90% by 2030 (share of our total waste that could be reused). Aim to be fully circular by 2050.**	1. Our operating units will now work on specific actions to implement our Group goal. 2. Measure circularity progress. We plan to report on our recycling rate in the future.	organisa-tional	recycle
Vatten-fall	**Achieve a recycling rate of wind turbine blades of 50% by 2025, and 100% by 2030.**	1. Installing recyclable blades (Hollandse Kust Zuid project) 2. Supporting research into material recycling of composite waste and collaborating with partners and industry actors to find new solutions for recycling and reuse	processes	recycle
Vatten-fall	**Immediate landfill ban on decommissioned blades from Vattenfall's wind farms.**			no landfill

Sources: Owens Corning: Sustainability Report 2021 [35], circular economy information on company's website (last accessed on 22.05.2023); Olin: Sustainability Report 2022 [50]; 3B Fibreglass: Sustainability Report 2021 [32]; Vestas: Sustainability Report 2021 [20]; Circularity Roadmap and Zero Waste on company's website (last accessed on 22.05.2023); LM Wind Power: Sustainability Report 2021 [23], Zero Waste Blades on company's website (last accessed on 22.05.2023); SGRE: Consolidated Non-Financial Statement 2020 [51], Sustainability Vision 2040 (2018) [52], Sustainability page on company's website (last accessed on 22.05.2023); Ørsted: Sustainability Report 2022 [22], information on sustainability and circular resource use on company's website (last accessed on 22.05.2023); Vattenfall: Sustainability Report 2022 [53], sustainable use of resources on company's website (last accessed on 22.05.2023); RWE: Sustainability Report 2021 [54], Sustainability Strategy Report 2022 [55].

Stretch broken carbon fiber for primary wind turbine structure

Cairns, D.S, Booth, E. Ridgard, C.

Mechanical and Industrial Engineering Department, Montana State University, Bozeman, MT

dcairns@montana.edu

Abstract. Stretch broken carbon fiber (SBCF) composites offer significant benefits over conventional continuous fiber composites. These include the greatly enhanced formability of unidirectional tape form products prior to cure without any compromise on the mechanical performance after curing. The recent work performed under the US Army funded Stretch Broken Carbon Fiber for Primary Aircraft Structure program by Montana State University (MSU) is reviewed, along with the results achieved by previous efforts to develop this technology. The achievement of shorter stretch break lengths to improve formability is a principal objective of the work and a successful achievement of the recent efforts. In addition, alternate means of manufacturing SBCF prepregs are discussed, including conventional spooled fiber presentation suitable for running on current prepregging equipment, as well as a newly developed direct to prepreg approach, where the stretch breaking of the fibers is done in-line with the prepregging process. While aerospace primary structure has been the focus of current research efforts, applications in other industries such as wind turbine blades are also obvious areas of applications for SBCF technology.

1. Introduction

Carbon fiber composite materials are used extensively in aerospace and other structures because of the combination of strength, stiffness and low density which they offer. For conventional carbon fiber composites, made with continuous fibers, the materials are however limited in terms of formability to produce complex shapes. This has led to many structures being designed as 'hybrids' with composite skins of relatively simple shapes combined with interior structures of more complex shape such as frames, stringers and spars being made from metallic materials such as aluminum alloys. Aside from the weight penalty which is incurred because of the use of metals, the combination of carbon and aluminum has led to significant corrosion issues where the aluminum becomes the anode in contact with the carbon fiber cathode. Attempts to mitigate this by the use of e.g. glass fiber isolation plies, have been only partially successful and fleets of aircraft such as the early F18 models are now experiencing problems sufficient to require them being taken out of service to be repaired at costs measured into the $US billions. There is thus significant impetus to produce carbon fiber composite materials which are more conformable, which would allow the metallic substructure to be replaced by composites in a cost effective manner. Stretch broken carbon fiber (SBCF) is a technology which offers this potential and is the subject of this paper.

1.1. SBCF basic principles

Stretch breaking of fibers using sets of nip rollers driven at differential speeds is not a new technology and has been employed with various textiles for many years. For carbon fibers, the same technology is used to generate random breaks at natural flaws already present such that when combined with a resin matrix, the material can be formed into complex shapes in a similar manner to the traditional stretch forming of aluminum sheet. Breaking of carbon fibers is governed by three flaw types, surface flaws, internal microstructure flaws and the basic Carbon-Carbon theoretical cohesive strength as shown in Figure 1.

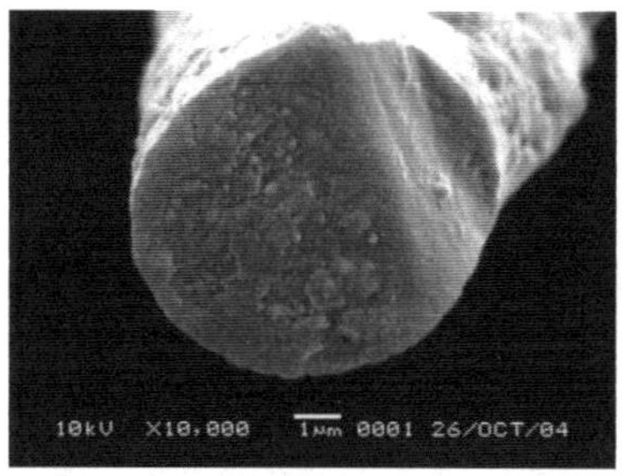

a. "Nick", flaw on surface (most deleterious to filament strength)

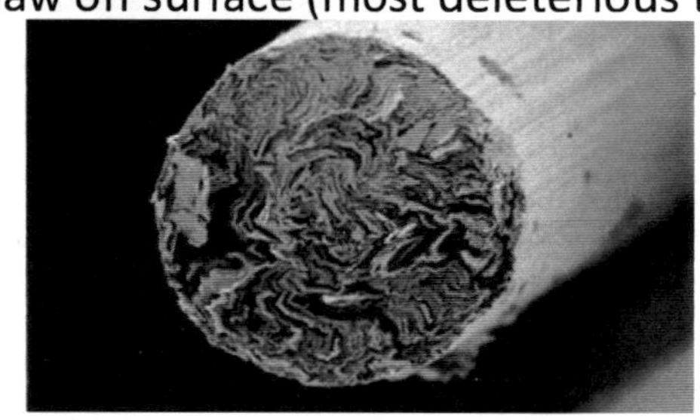

b. Internal microstructural flaw (highly oriented carbon structure in thefiber)

c. Carbon-Carbon Cohesive Strength (Very High)

Fiber flaws a., b., and c. are governed by individual probability distributions with respect to flaw size and frequency.

Figure 1. Carbon fiber flaw types affecting filament strength.

It is noted that the primary difference between high strength aerospace grade carbon fibers and lower industrial strength carbon fibers are the flaw distributions of Figure 1. Strength is a flaw driven property, Stiffness (Young's modulus) is a bulk property governed by the fiber microstructure. The random breaking at "natural flaws" is an important concept which warrants some discussion compared to other aligned, discontinuous fiber material forms. The tensile strength of carbon fiber composites along the axial fiber direction is governed primarily by fiber flaw size and distribution. During loading, individual fibers break, and the load is distributed to the surrounding filaments. Once a critical number of breaks occur, ultimate tensile strength is achieved. This was identified and analyzed in the classic paper by Rosen as shown in Figure 2. [1].

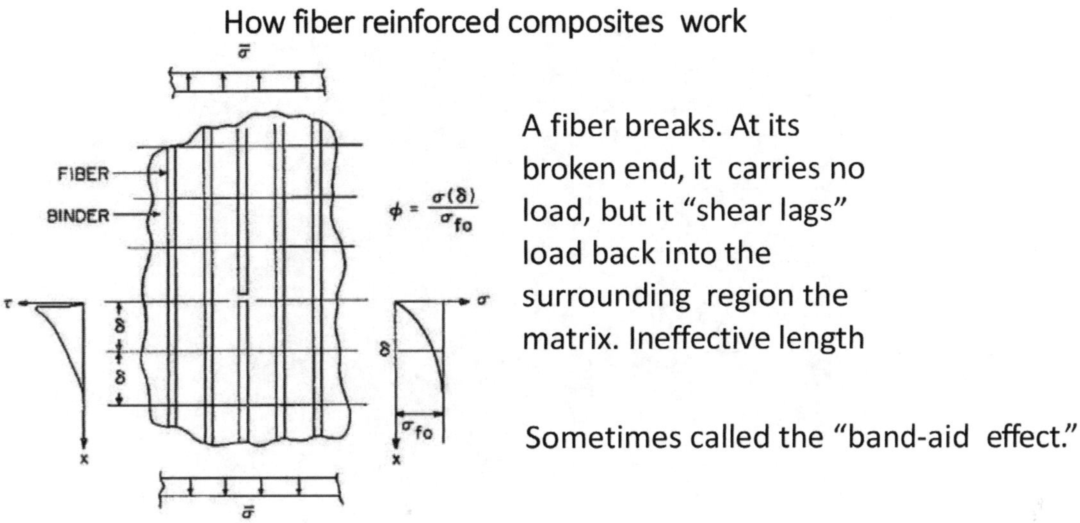

Figure 2. Local load distribution in a fiber reinforced composite [1].

Once cured however, the strength and stiffness properties of the material are the same as those of the equivalent continuous fiber product as will be discussed in 1.2 below.

Figures 1 and 2 illustrate the principle and an actual end result respectively.

Figure 3. Schematic of a stretch broken carbon fiber material.

Figure 4. SEM of a stretch broken carbon fiber material showing break locations.

1.2. Previous SBCF Development Work

Previous programs have developed methodologies for the production and evaluation of SBCF materials, references [2] [3] [4] [5] with some degree of success. Figure 5 shows some of the parts produced under these prior programs.

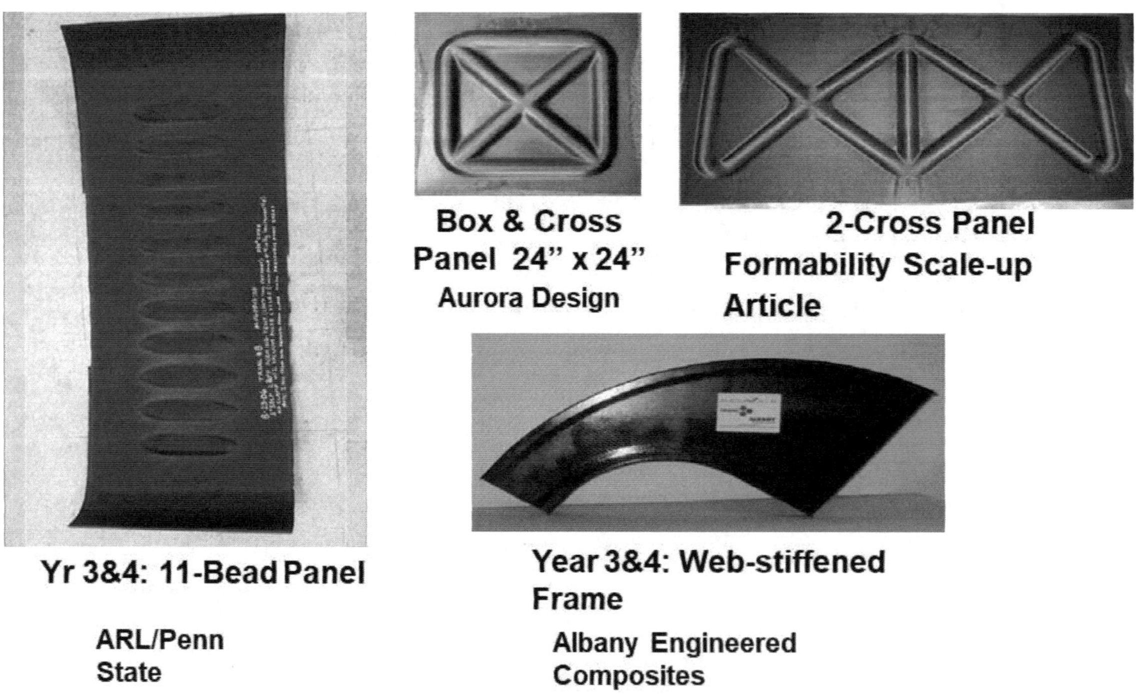

Figure 5. SBCF parts from prior programs.

The reasons that the materials have yet to be widely adopted include the fact that the stretch break length of the legacy materials was insufficiently short to allow the formability of more complex geometries and the fact that the economics and cost of production of the materials was prohibitive in lieu of investment in scale-up. As described in section 2, progress has been made in addressing these issues. Nevertheless, the previous work has demonstrated the potential for SBCF plus also illustrated the lack of impact on strength properties once the materials are cured, as shown in Figure 6. The retention of the strength of the continuous fiber material is an expected result given that the ineffective shear lag length is of the order of ten filament diameters, which for IM7 fiber are of the order of 5 microns, such that unless multiple break points align in a local region, shear transfer into adjacent filaments ensures continuity of the load path. There is also a theory that by pre-breaking at such natural flaws before applying and curing a matrix, there could be property enhancements since less damage will occur compared to when such defects fail surrounded by a cured resin matrix.

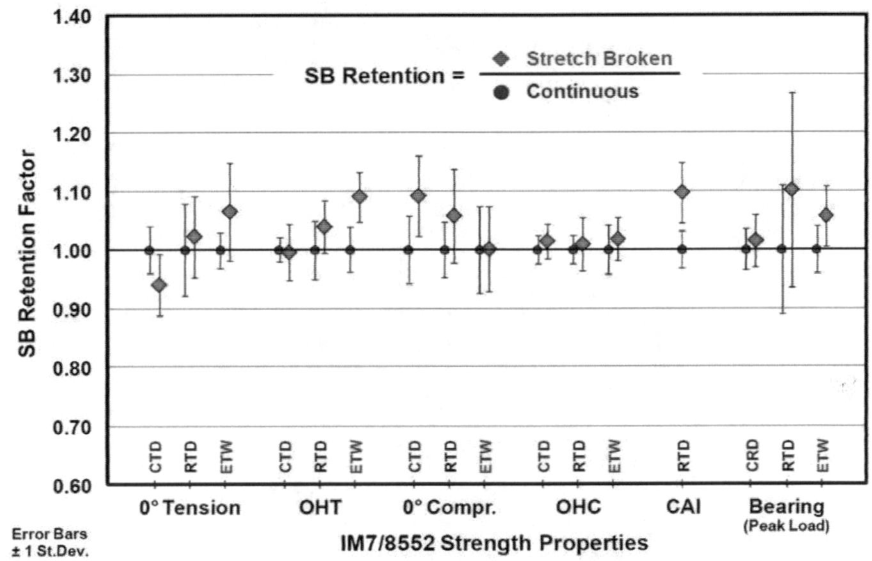

Figure 6. Mechanical property comparison of SBCF and continuous fiber materials.

It is noteworthy to state that most discontinuous fiber composite materials forms where the fibers have been cut or chopped to achieve the formable shorter fibers have a 20% or more reduction in tensile strength properties compared to continuous because the natural flaws are still present, combined with the flaws from the chopping or cutting.

2. SBCF recent developments

The recent and ongoing work on SBCF at MSU has a number of objectives including:

- Reducing the stretch break length from those achieved in previous programs
- Developing an alternative manufacturing technique called direct to prepreg (DTP) as well as the more conventional spooled fiber presentation used on previous programs
- Scaling up both manufacturing techniques to achieve a cost effective product
- Demonstrating that either technique can produce more complex parts than achieved previously both by lab scale forming tests and full sized demonstration parts

- Generating mechanical property data to confirm that the SBCF materials are equivalent to the baseline continuous fiber products

2.1. Materials
The current work has utilized IM7 12K fiber from Hexcel, both sized and unsized, 8552 resin from Hexcel and Cycom 977-3 resin from Solvay. The combination of the fiber and the two resins covers many applications in military aerospace primary structure and extensive databases exist. The SBCF is made on two stretch breaking machines at MSU, known respectively as Bobcat 2 and SB2, and converted to unidirectional prepreg on a pilot line to produce material for testing and evaluation. In terms of the ultimate product forms, the DTP process produces unidirectional tape with the prepregging achieved at the same time as stretch breaking occurs. The more traditional spooled single tow presentation can also produce unidirectional tape using a more conventional prepregging process but can also be used for 2D or 3D weaving, braiding or stitched multiaxial reinforcements. Figure 7 shows the 'reel' 'ribbon' or 'pancake' presentation of the DTP material. A much larger machine called Bobcat 3 is under construction which will allow much wider DTP prepreg to be made, plus also be capable of simultaneously stretch breaking significantly more tows for the traditional spooled presentation. It should be emphasized that the stretch breaking technology under development is not limited to the aforementioned fiber and resins, nor is it intended that it be limited to aerospace primary structure applications only. Other markets such as automotive are clearly potential users but will require different materials such as rapid cure resin systems or thermoplastic matrices.

Figure 7. Direct to prepreg SBCF reel presentation.

2.2. Stretch break length improvement
Figure 8 shows the fiber length distribution of MSU SBCF materials compared with those of legacy SBCF materials from previous programs. It can be seen that not only is the mean stretch break length of the MSU SBCF significantly less than those of the legacy materials, but the distribution is tighter and the upper tail representing a number of long fibers has been eliminated. It is believed that the presence

of this small percentage of long fibers has been detrimental to the formability of the legacy materials as much as the higher mean lengths. The improvement in stretch break length has been achieved by the development of techniques different from those of previous programs and are the subject of US provisional patent application number 63/070,151, entitled ‘Techniques, Method and Device for the Manufacture of Stretch Broken Carbon Fiber’.

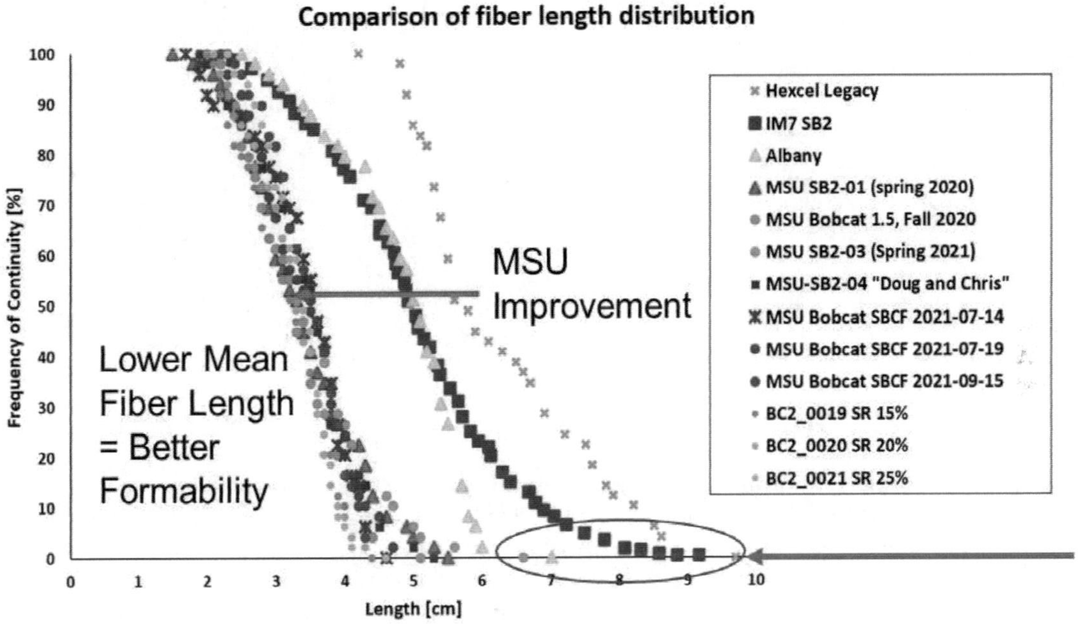

Figure 8. SBCF fiber length distributions.

2.3. Tow tenacity issues with short break length SBCF

Experience from previous programs has shown that SBCF tow is initially fragile and requires special techniques to apply sizing in order to produce a handleable product. The shorter stretch break length of the MSU SBCF, while improving formability, has presented additional challenges which have however been overcome using procedures different from those employed on past programs. Figure 9 shows the tensile tenacity of MSU single tow SBCF as a function of temperature. The tow has tenacity up to around 60°C, allowing it to be woven or prepregged, but at higher temperatures loses tenacity in order for stretch forming to take place during part manufacture.

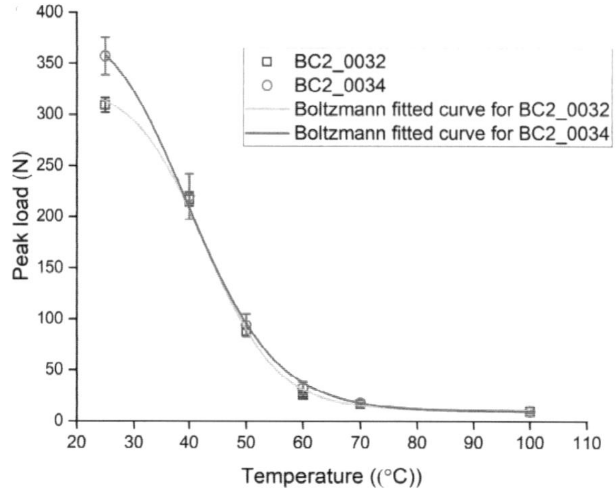

Figure 9. Tensile tenacity of single tow MSU SBCF as a function of temperature.

3. Laboratory scale evaluation testing for formability and other characteristics

A variety of test methods have been developed by MSU as there are no standard test methods for the formability of composites. These include a formability 'plunger' test, ply on ply and ply on tool friction test, a dome formability test based on the metals Ericksen test plus creep and mechanical true stress strain curve measurement. This work is described in detail in a series of papers to be presented at CAMX 2022, references [5], [6] [7] [8] [9] and will therefore not be covered explicitly in this paper.

4. Processes for part forming and manufacturing

The emphasis of the current program in terms of part processing is autoclave curing with process cycles similar to those used for the production of parts containing continuous carbon fiber prepregs. It is however intended to explore other processes such as press molding later in the program, but this is not the focus of the current work. Figure 10 shows an outline of the intended geometries to be used in the short to medium term. All of these features and sub-structures could be utilized in wind turbine blades structures. The tapered sine wave spar example of Complex structure is particularly interesting to optimize single spar wind turbine blade design. With shorter fiber lengths in the present SBCF materials, it is also expected that it will prove possible to attempt forming into more aggressive geometries than demonstrated in the previous work, including beaded panels with narrower bead widths, representative of what would be envisaged to prevent buckling in the shear webs of spars, ribs and frames, without the need to use separate mechanically or adhesively attached stiffeners. Designed experiments are in the process of being conducted to confirm optimum cure cycles for different resin systems, which will likely include intermediate temperature 'forming' holds below the point at which the thermoset resins begin to advance, before completing the cure in a conventional fashion. There are three basic mechanisms which permit forming for uncured prepreg – overall layup slippage against the tool (drawing in metals terminology), interlayer slippage and membrane stretching. Only SBCF permits the latter. It is intended to utilize all three mechanisms and employ only 'natural clamping' as provided by the autoclave pressure. The end results will be evaluated in terms of parameters such as edge wrinkling and thinning of the complex features such as beads. The metrics and information gained will be ultimately employed in producing complex full scale parts. The results of this ongoing work will be the subject of a future paper.

Geometry type	Example	Purpose	Metric
Dome Forming	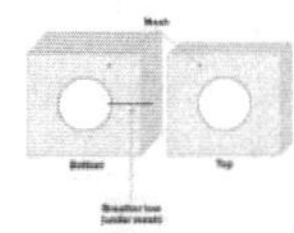	Stretch forming evaluation; cure cycle optimization	Forming depth; forming limit with aid of in-situ LVDT measurement
Simple Beads		Combined stretch & slip forming (simple)	Material thinning
Beaded Panel		Combined stretch & slip forming (complex)	Material thinning AND forming limit
Complex structure	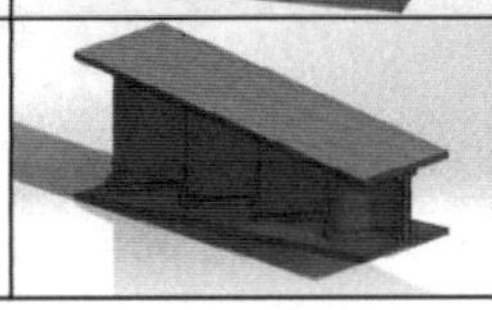	Demonstration of SBCF potential for challenging geometries	Geometric tolerance, part quality, defects, etc.

Figure 10. Outline of proposed basic formability and manufacturing study geometries.

In Figure 11 is shown the cross section of a modern wind turbine blade. It consists of multiple built up structures and complex shapes which can benefit from SBCF.

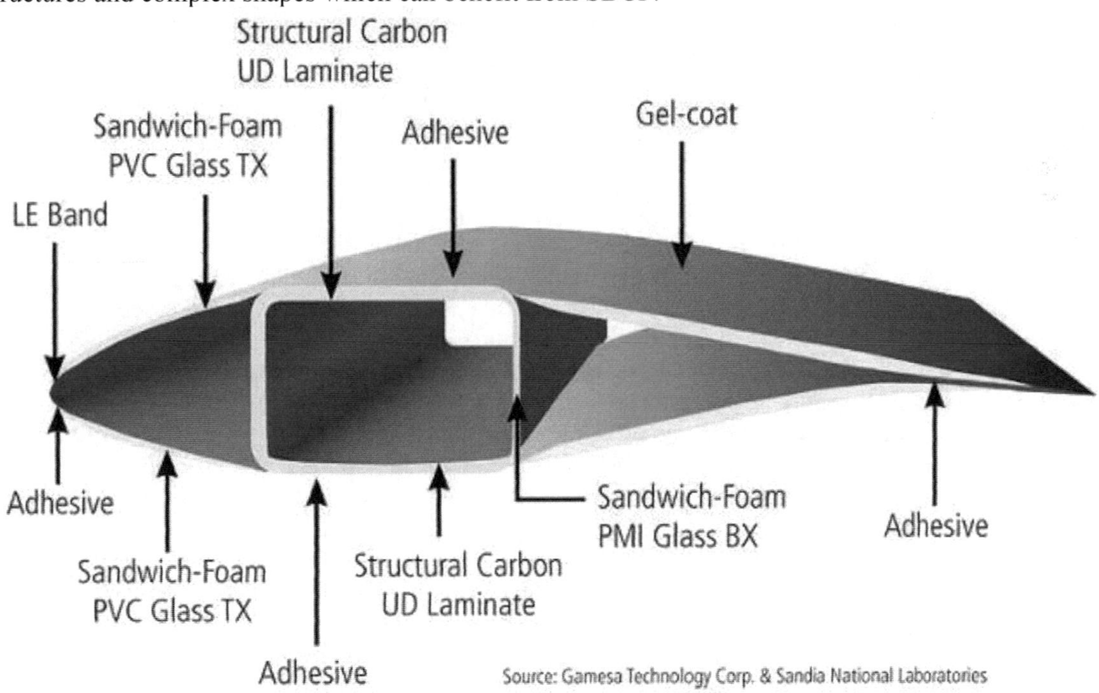

Figure 11. External and internal complex structure of modern wind turbine blade (cross section).

Figure 12 shows variations in complex geometry along the length of a modern wind turbine blade with complex geometry along the length.

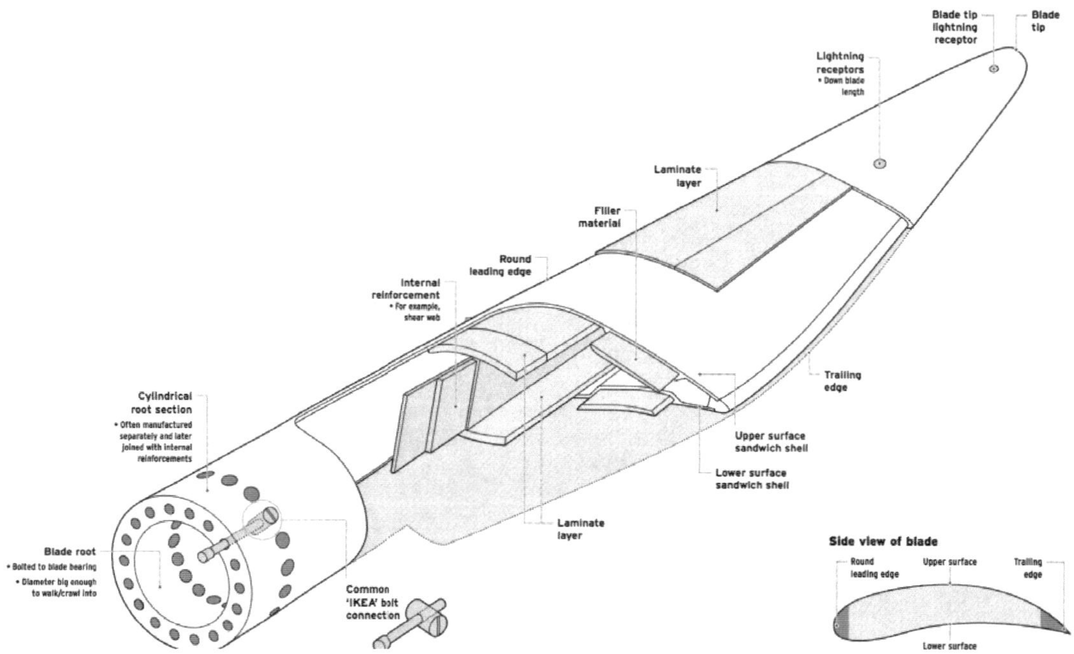

Figure 12. External and internal complex structure of modern wind turbine blade [11] (along the length).

The shapes and substructures shown in Figure 10-12 are applicable to modern composite wind turbine blades to reduce the cost of manufacturing, durability and damage tolerance, and operation costs compared to continuous fiber composites.

5. Conclusions and future work

Advances have been made in the development of a more conformable and potentially more cost effective stretch broken carbon fiber which can be used for unidirectional prepregs, woven prepregs plus other reinforcement forms. The advances include the achievement of a shorter break length with a tighter distribution than previous efforts, together with the development of processes suitable for the production of either a conventional spooled single tow presentation or a direct to prepreg 'ribbon' format. A great benefit for SBCF could be realized if, by stretch breaking industry fibers, the natural flaws shown in Figure 1 could be mitigated via the stretch breaking process, such that the properties of industrial or commercial grade carbon fiber products could approach aerospace carbon fibers. Additional equipment is in the process of being manufactured which will further scale up production capability and improve cost effectiveness. The production of large demonstration articles representative of aerospace primary structure will be the subject of a future paper, and these result should be directly applicable to wind composite wind turbine blade structures. In addition to the production of lighter, more cost effective structures, these developments address a serious corrosion issue arising from the use of composite skins over aluminum substructure which has now become apparent for older aircraft but will also be manifest for both later and future composite primary structures.

References

[1] B. Walter Rosen, "Tensile Failure of Fibrous Composites", AIAA Journal, Vol. 2, No. 4, NOVEMBER 1964.

[2] Abdallah, M. at. al, “Development of Stretch Broken Carbon Fiber (SBCF) Materials”, 2005 International SAMPE Symposium, Long Beach CA.
[3] Dillon, G.P. & Waldzyk, D., “Comparison of Continuous Fiber and Stretch Broken Carbon Fiber (SBCF) Materials in Forming Processes”, 2005 International SAMPE Symposium, Long Beach CA.
[4] Dillon, G.P. & Barsoum, M., “Characterization of Stretch Broken Carbon Fiber Materials for Automated Forming Processes”, 2005 International SAMPE Symposium, Long Beach CA.
[5] Ng, S.J. et. al, “Characterization of Stretched Broken Carbon Fiber (SBCF) Material Forms”, 2005 International SAMPE Symposium, Long Beach CA.
[6] Tasnia Javin Nur: Novel experimental unit to evaluate inter-ply and tool-ply friction in carbon fiber reinforced polymer composites prepregs, CAMX Anaheim CA, October 2022.
[7] Yoni Shchemelinin: Bulge and dome testing to investigate the formability of continuous and stretch broken carbon fiber prepreg laminates, CAMX Anaheim CA, October 2022.
[8] Dalton Nold: Tensile Properties of Stretch Broken Carbon Fiber Prepreg, CAMX Anaheim CA, October 2022
[9] Matthew Egloff: A Predictive Forming Model for Stretch-Broken Carbon Fiber Composites as a Function of State of Stress, Forming Rate, and Resin Viscosity, CAMX Anaheim CA, October 2022
[10] Riad Morshed Rezaul: Effect of process variables on the uncured handleability and formability of stretch broken carbon fiber, CAMX Anaheim CA, October 2022
[11] de Vries, Eize, How to service and maintain a wind turbine blade, 1 July 2012.

IOP Conf. Series: Materials Science and Engineering 1293 (2023) 012041 doi:10.1088/1757-899X/1293/1/012041

Study on recycling wind turbine blades into reinforcement for filaments used in 3D printing

J James Ezhilan[1*], Ajith Damodaran[1], and K G Ashok[1]

[1]Department of Mechanical Engineering, Easwari Engineering College, Chennai, India

*jamesezhilan02@gmail.com

Abstract. Heavy applications like power production through wind energy requires light weight but strong material like composites with customizable properties. When the lifetime of the wind turbine blades ends, the parts are dumped in landfills and results in environmental pollution. Natural fibers are great option for improving the biodegradability of the conventional plastic which also ends up in the landfills. Fused Filament Fabrication (FFF) is chosen to combine the benefits of both materials, as the technique is highly customizable and sustainable. Wind turbine blade wastes are recycled using mechanical grinding. Recycled Fiber glass (FG) material is tested for contamination with Fourier Transform Infrared Spectroscopy (FTIR). Wood fiber (WF) is also added in order to improve biodegradability of the materials. Filaments are produced using a single screw extruder with various combinations of 9 wt% fiber content and recycled pellets. Tensile test shows comparable performance of reinforced filaments with recycled Polypropylene (PP) filaments. 6% WF + 3% FG sample withstood up to 380 MPa Young's modulus.

1. Introduction

In the energy crisis of the 21st century, renewable energy sources are the only option to produce more and cleaner energy. Among the various renewable energy production methods, wind energy is the cheapest and easiest to harvest electricity [1]. Over the time the wind energy extraction is optimized into wind turbines. The concept of the wind turbines is to produce electricity using the wind to turn the turbine blades which operates dynamo to produce alternating current. Wind turbines are well suited for developing countries that have more windy areas because of its cost of running and maintenance. But the capital cost for a wind turbine is high. In order to increase the productivity of the wind turbine, it has to be installed in groups called wind farms. Wind turbines nowadays produce 2-3 MW energy on average [2]. Wind turbines are made from various materials from metal to composites. The stem of the wind turbines is made from steel and concrete, since the stem won't be moved over the time and the conventional materials keeps the cost low. But the blades are made from the composite materials, in order to produce tailored properties high strength to weight ratio, low drag coefficient. The wind turbine blades are primarily made from Fiberglass (FG) and Epoxy resin [3–5]. The Wind turbine blades (WTB) are extremely tough but they have a maximum lifespan of 25 years. In reality the WTB will be in operating conditions for about 15-20 years, due to various reasons for failure [6]. After their life time the blades cannot be used for any material reuse purposes, since the shape and composition of the blades prevent these use cases. So, the blades get dumped in landfills for the lack of recycling methods [7].

There are some innovative methods that are proposed for recycling the blades but no method is implemented in large scale. Some studies propose non-conventional designs reusing the WTB for structural purposes after their end of life. These studies show promising applications like furniture and roofing, but require remodeling the blades. Remodeling consumes same amount of time and work as new construction, so material recycling is required to implement WTB recycling in large-scale [8]. Methods like Pyrolysis will increase the use case of the waste blades but the environmental impact of the methods is restricting the extensive implementation of the method. The optimizations in pyrolysis improve the emission control and the efficiency of energy recovery [9,10]. Another method of using the recycled materials as alternatives for construction materials is explored and it is found that the composition of the composites increases the process complexity. So, the segmentation of the blade is proposed as a technique to use different parts of the blade to test its properties for using as structural elements [11]. Mechanical grinding is a successful technique to recycle the FG from waste blades, since it reduces the wastes into powder form to be used as reinforcement in other thermoplastics in order to improve the performance of the materials. The study where the FG is used to reinforce the polypropylene material indicates the improvements in mechanical strength and the participation of FG in the improvement [12–14]. Also using the FG reinforcements in Additive manufacturing is explored extensively by many researchers. Since, the additive manufacturing improves the efficiency and ease of the recycling process. A study performed using Fused deposition modelling (FDM) to reuse the recycled FG from WTB wastes as reinforcements for conventional materials. The study shows the performance of the reinforced filaments for FDM in terms of improvements in tensile strength [15]. It also shows the possibilities of mechanical grinding in recycling of the WTB. The production of FG reinforced filaments opens wide range of applications for WTB recycling. The filament production using recycled material as reinforcements is explored well by researchers [16–20]. The natural fibers such as wood fibers, starch fibers etc., as reinforcements helps to increase the biodegradability of the composite materials. Natural fiber reinforced filaments prove the possibilities but it has one major problem that is strength reduction [21,22]. Since, most natural fibers are hydrophilic and all the thermoplastics are hydrophobic. So, the bonding is not complete which leaves much space inside the material that causes strength failures [23–26]. By using FG as reinforcement for the natural fiber additions to the thermoplastics and fabricating the material in the 3D printable filament form, the recycling efficiency and applications of the WTB wastes increases. This study explores the methods of recycling the FG from WTB wastes and using it as reinforcement for the recycled plastic-natural fiber composites.

2. Materials and Methods

The study explores the mechanical grinding method to recycle the FG from the WTB, since the mechanical grinding is the cost effective and simple method among other methods. The process flow of the study methodology is represented in a flow chart in figure 1. The damaged/scrap WTB is either dumped in nearby landfill or shipped back to scrapyards where they stay for while then ends up in the landfills. The recovered FG is used as reinforcement for the conventional plastic pellets. In order to improve the biodegradability of the materials, natural fiber (wood fiber-WF) is blended with FG and

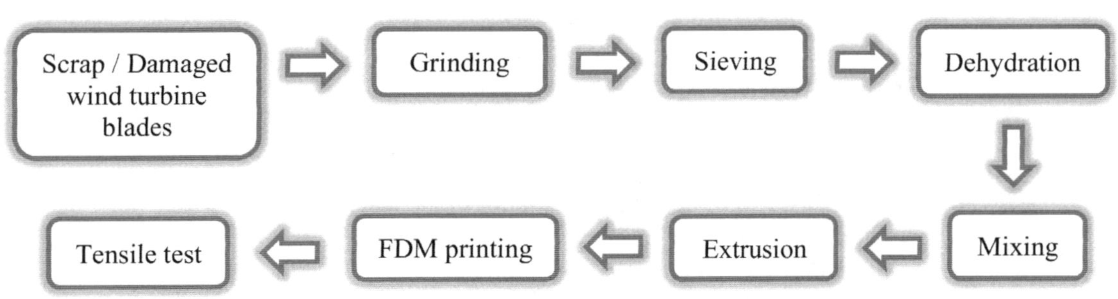

Figure 1. Process flow chart

pellets. The extrusion method creates the 3D printable filaments, which can be used to print any desired products easily in a FDM printer.

2.1. Materials

2.1.1. Fiber glass. The FG material is obtained from a mid-size wind turbine whose blades are burnt in a lightning strike in a wind farm located in Tirunelveli, Tamilnadu. The damaged WTB from landfills are collected and the mechanical grinding is done using a grinding wheel. The waste WTB contains various impurities like dirt, plant matter, paint, etc. In order to avoid the segregation and cleaning processes along with chemical treatments, mechanical grinding method is used to collect the FG from the WTB parts directly. The ground powder is collected and sieved to remove irregular particles. Composite with various strength and compositions used throughout the blade's structure. So, samples from different parts of the blade like Beam, Sheet and Resin are obtained and tested. The samples from Beam and Sheet are denoted as FG-B and FG-S respectively. The sample collected is shown in figure 2. Fiber glass powder is dried at 60 ^{0}C for 2 hours before mixed with pellets for extrusion. FG additives are chosen in weight percentage (wt %) 0, 3, 6, 9. The 0 % FG indicates the 9 % presence of WF, 3 % FG with 6 % WF, 6 % FG with 3 % WF and 9 % with 0 % WF. Thus, the wt % of additives always stays 9 % in total. Two different mixtures of additives and pellets sample is shown in figure 3. The 9 % threshold is set for the extrusion machine, since the two different material mixing might clog the extrusion and the single step melting of the pellets will be easier with less percentage of additives.

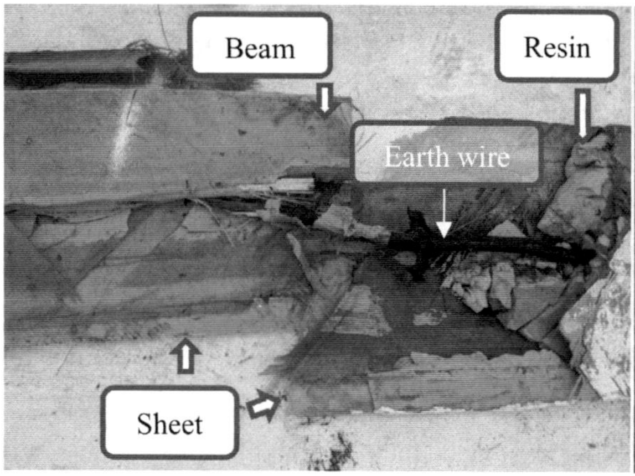

Figure 2. Damaged edge of the wind turbine blade

Figure 3. 6% FG and 3% WF (Left) & 6% WF and 3% FG (Right)

2.1.2. Wood fiber. The natural fiber to be added with plastic pellets is selected as wood fiber from Burma teak (Tectona grandis). The teak timber is one of the most preferred for its great strength and longevity. It is commonly used by many local carpenters for manufacturing good quality furniture like cot, chair, table, etc. Since the timber can only be processed by conventional manufacturing techniques, the wastage created from manufacturing is excessive. Also, the high density if the timber allows intricate wood works to be carved on it, which creates high amount of fine saw dust all over the workshop. The wood wastes are later collected and thrown in landfills. Some cases might use the saw dust for fertilizers and other uses them creating engineered wood. But the fine dust is usually left in landfills since the retrieval of the improper disposal· is nearly impossible. So, it will be a suitable material for the reinforcement in the composite material. Studies show the compatibility of the wood fiber as reinforcement using several wood fibers [27,28]. The existence of commercial wood/PLA filament for 3D printing indicates the significance of the wood fiber in composites. The saw dust from the timber is collected and the powder is screened and cleaned with filters. The saw dust is in fine powder from, so filters are enough to control the particle size. The saw dust is dried at 100 ^{0}C for 2 hours before extrusion.

2.1.3. Recycled PP. Polypropylene (PP) is a non-toxic thermoplastic which is used in food packaging and medical industry. The bio compatibility of the PP helps in the disposal of the material after its life cycle. However, the recycling ability of the material is well studied by researchers [12,23,29]. So, the recycled PP is very effective in the composite field as matrix material for various fiber reinforcements and it is a very good choice for this study. Recycled PP pellets are obtained from a recycling plant (Ambika Plastics near Chennai, Tamilnadu). The recycled pellets were dyed in black colour from the recycling plant. It is a common occurrence in the recycling industries, because the pure colour cannot be obtained after recycling due to contaminations. The diameter of the pellets' cross section varies from 0.8 mm to 3 mm. Length of the pellets varies from 3 mm to 6 mm. The pellets are dried at 60 ^{0}C for 2 hours to remove any moisture content. Since the plastic is recycled, the operating temperature of the extruder should be decreased to 200 ^{0}C due to the thermal degradation of the plastic.

2.2. Methods

2.2.1. FTIR. The fiber glass samples are tested for any contamination or moisture absorption to predict the material behaviour during extrusion and printing. The FTIR results are compared with the virgin fiber glass without any additives from literature, in order to find the type of contamination [30]. FTIR helps to find out any organic substance in the sample by calculating their absorbance of the IR waves. The different wavelength absorptions show different organic bonds in the test material.

2.2.2. Extrusion. The mixed materials are extruded into filaments using an in-house developed single screw filament extruder machine. The design of the machine is based on the 'Recyclebot' but all the parts are redesigned for improved functionality and metric standards. The machine is built using local available parts and 3D printed parts, which makes the machine highly scalable as shown in figure 4. The additives are mixed with plastic pellets thoroughly and preheated before fed into the extruder, while the heaters are set at 200 ^{0}C along with preheating time of 10 mins for the nozzle to reach proper temperature. No thermal degradation is observed from the extrusion due to the additives. The filaments are extruded properly at the rate of 2 m per minute.

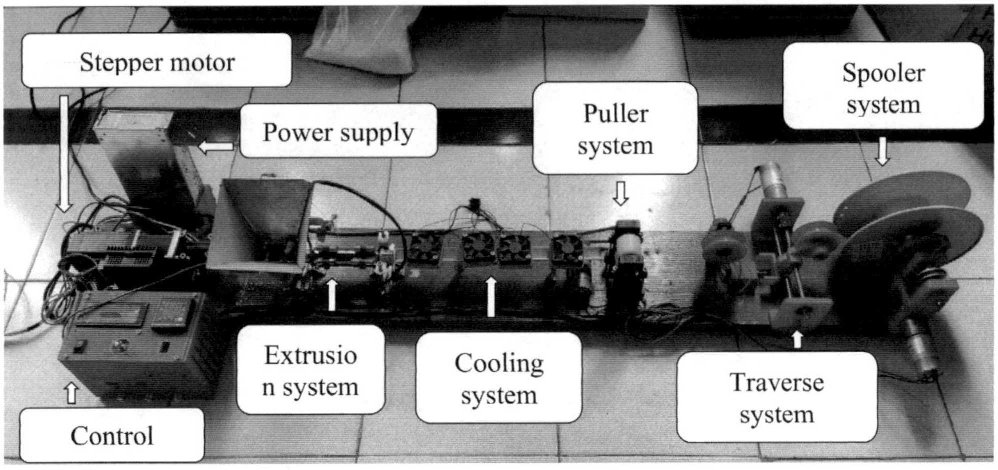

Figure 4. Single screw extruder with parts marked

2.2.3. 3D Printing. The reinforced filament then used in a FDM machine (Creality Ender 3 pro) with the CAD model designed for the prints. The print parameters are set in the slicing software according to the base material PP and test specimens. Due to inconsistency in the filament diameter printing process was often interrupted. The filament diameter variation causes the print head stepper motor to slip, few layers were printed but the flow of the filament is cut in the nozzle and resulting in clogging. The print speed slowed down to check the filament flow but the print didn't get affected. The low surface finish of the filament is the main cause for the improper printing.

2.2.4. Tensile Testing. The fabricated filament is tested directly in the Tensile testing machine (Initial gauge length – 100 mm, feed rate – 1 mm/s, max load – 20 kN) with the help of specially prepared clamps for mounting the filaments without breaking them. This method of testing the filaments is carried out as pilot study for composites before they can be printed on the FDM machines [15,31]. A clamp setup was designed for mounting the filaments in the vice. The filaments are tested with longitudinal load along the gauge length for its tensile strength and modulus of elasticity. The test parameters and test results are used to plot stress vs strain graphs and modulus of elasticity. There were 5 samples for each composition tested along with a recycled PP filament that is extruded from the same extruder machine as the samples and a commercial grade PLA filament.

3. Results and discussion

3.1. FTIR results

The primary component of the fiber glass is sand and it is superheated to create glass and fiberglass. So, the figure 5 shows prominent peak in samples' graphs as Si-O-Si bond. The peak shows up on the samples from Beam, Sheet and also in the Resin at 1050 cm^{-1}. Since the epoxy resin is mixed with some fiber glass to create the strength needed in the packing between sheets and beam. The use of epoxy resin in the samples from Beam and Sheets are evident from the peak at 2800 – 3050 cm^{-1}, which indicates the CH_3 and CH_2 bonds in the sample. The final peak at 3700 cm^{-1} indicates the O-H bonding, which confirms the presence of the plastic reinforcements in the composite.

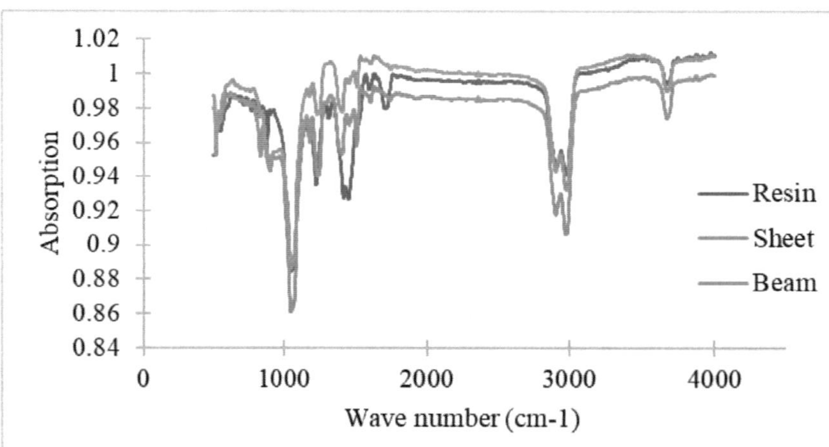

Figure 5. Average results of FTIR test from various parts of WTB

3.2. Visual inspection

The images from the optical microscopes are taken at 10x zoom level are shown in figure 6. The images show the individual fiber glass particles clearly. The fiber length is averaging in 1.2 µm. And also, from the images the resin particles can be seen clumping together. It is happening because of the thermal degradation of the resin particles, since the resin is a thermosetting polymer and the grinding process heated the particles while shaving them out. The visual inspection shows that without chemical treatment, the resin particles cannot be completely removed from the FG samples. The resin particles may affect the bonding of the FG with the PP matrix, resulting in the poor surface finish and low strength of the filaments.

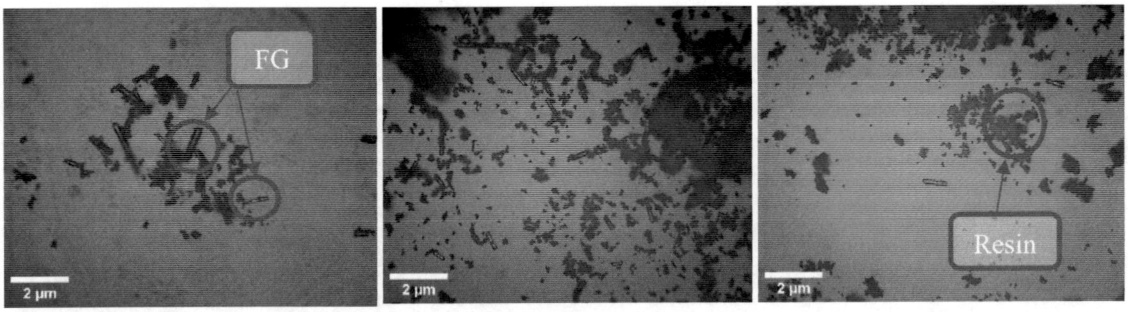

(a) (b) (c)

Figure 6. Optical microscopic images of FG samples

3.3. Filament quality

The filament quality is directly affected by two main factors, namely Extrusion pressure and Puller speed. The extrusion pressure is controlled by the extruder motor speed. But the extruder motor speed doesn't take effect immediately like puller motor speed. The extruder motor speed change creates extrusion pressure difference near the nozzle after some time delay. That's why it affects indirectly. The extrusion pressure pushes the filament through the nozzle and the puller motor pulls the filament while it is cooled by cooling fans. The push and pull balance is very important in determining the diameter of the filament. Also, the filament has to be cooled below its glass transition temperature before reaching the puller, otherwise the puller rollers will deform the filament like figure 7(a). While increasing the filament diameter by reducing the puller motor speed or increasing the extrusion pressure, the filament quality started to drop dramatically. The filament became filled with void and brittle after the alteration. When the conditions are reversed and the perfect balance between puller speed and extruder speed is achieved, the diameter of the filament is back to around 1.6 mm. The problem is recreated multiple times to find out the cause and every time the diameter is start to increase the improper cooling and brittle filament is obtained. But the decrease in diameter is happening smoothly at every speed combination.

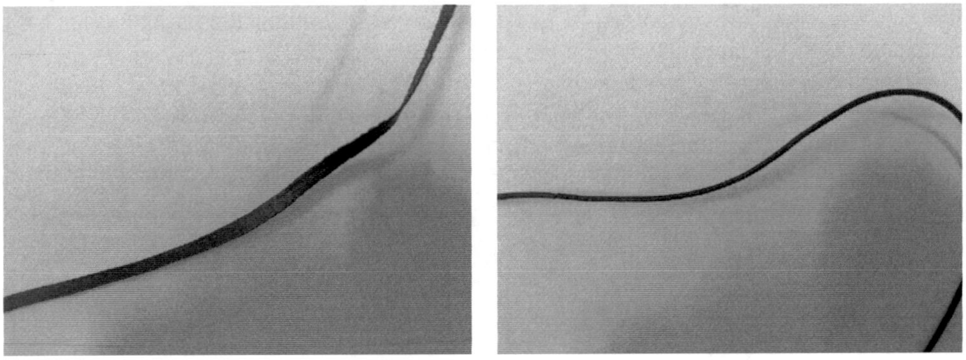

(a) (b)

Figure 7. (a) Bad extrusion and (b) Good extrusion

So, it is found that the diameter inconsistency is due to the improper bonding between the fiber and matrix. Figure 8(b) shows the diameter measurement using Vernier calliper. Lower fiber content might improve the situation, since the pure PP filaments can be extruded at any diameter without losing the filament quality. The diameter measurements were made multiple times in a 10 m filament. The diameter distribution of the filaments throughout its length is shown in figure in a form of a histogram chart 8(a). The surface quality of the filament varies throughout the extrusion 3 times distinctively in all the ratios of fiber reinforcements. The initial, mid and final stages of the given mixture vary by significant margin.

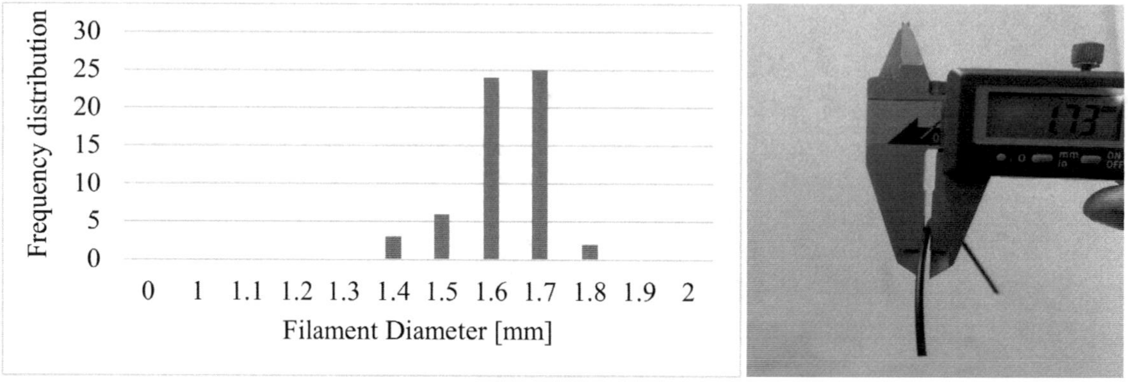

Figure 8. (a) Filament diameter frequency distribution, (b) Diameter measurement using Vernier calliper

The fine powders of FG and WF didn't stick with the PP pellets thoroughly and the particles didn't go through the extrusion chamber in the same speed as the pellets. As a result, the initial extrusion brought low concentration of reinforcement with them. So, the filament quality resembled the pure PP filaments. After the extrusion built up perfectly, the mixture is obtained perfectly and the filament quality is exact representation of the composite. After the extrusion pressure drops at the end of the extrusion, the remaining reinforcements carried by the remaining pellets. So, the filament is very rough, filled with void and brittle at this point.

3.4. Tensile test results

Figure 9 shows the tensile test results of the reinforcement samples and pure samples. The decrease in strength and the plastic deformation shows the improper bonding between the fiber and matrix in the material. But the behaviour of the recycled PP is also very different from the commercial PLA filament despite having no fiber in the material. The main difference between those materials came down to the thickness of the filaments. The commercial filament is thicker (1.75 mm) than the recycled PP filament which meant that the filament might slip on the clamp without providing good tension before the test starts. After the test initiated the filament began to grip. The material started to plastically deform at 0.05%. But the recycled PP filament and reinforced filaments are thinner (around 1.6 mm) so they provided enough tension from the beginning. But the reinforced samples failed after 0.1% strain, the recycled PP filament actually didn't fail even at 215% strain. The test is stopped after 255% elongation. The graph is in 'serrated' pattern which indicates that some particles prevent failure. Which is not

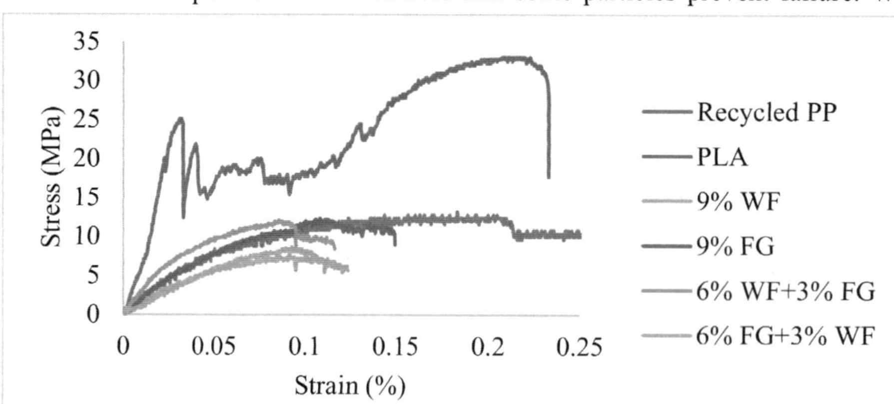

Figure 9. Tensile test results

desired in the working products. Unwanted elongation might cause strangling or stinging effects in material failure. So, the samples might be weaker in strength performance but the material behaviour is predictable and desirable. Among the samples, the 6% WF+3% FG samples show the best results in terms of strength and the Young's modulus of the material is also higher than the rest of the samples. Figure 10(b) shows that it has modulus of elasticity as 378 MPa which is the closest one to the recycled PP (386 MPa). It is due to the minimal amount of FG present in the material. The WF is reinforced by the FG particles, while the WF particles bond with the matrix it also takes the FG with it. Because the 9% WF samples itself couldn't create good bonding with the matrix. And same can be said for the 9% FG samples. The higher concentration of the FG also didn't help the material. However further studies in this regard is needed. It is found that in reinforcement the natural fiber has to be more than the FG content.

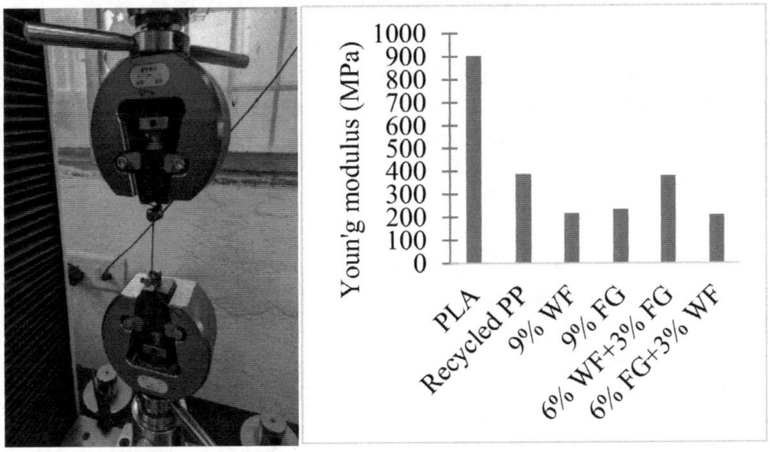

(a) (b)

Figure 10. (a) Tensile test setup, (b) Young's modulus comparison

4. Conclusion

Wind turbine blade waste was effectively used as a reinforcement in making filaments for FDM process Since the thermosetting nature of the epoxy resin affects the bonding of the fiber and matrix materials, it is challenging to determine the true potential of the recycling method. So, further in-depth studies are required to improve the efficiency of the process. 6% WF+3% FG sample shows promising tensile strength good surface quality. It also shows that the natural fiber needs to be in higher concentration than the FG reinforcement, which is a critical finding in this study.

4.1. Future work and Applications

Double melt method and NaOH treatments for FG and WF will be investigated. Further commercial products will be fabricated as a case study. The applications of these reinforced and recycled filaments in 3D printing are endless, yet the ease of access of the FDM machines makes these filaments suitable for small-scale companies and personal use. Studies shows promising results while making small machine parts and household items [20]. However, these recycling studies help to accelerate conventional processes to adapt recycled materials. This study helps to ensure the feasibility of simple mechanical recycling of WTB and the conversion into FDM filaments is possible and useful in personal and industrial level.

References

[1] Das A, Jani H K, Nagababu G and Kachhwaha S S 2021 Wind and solar power deployment in India: Economic aspects and policy implications *African Journal of Science, Technology,*

Innovation and Development **13** 357–375
[2] Li X, Gao Q, Cao Y, Yang Y, Liu S, Wang Z L and Cheng T 2022 Optimization strategy of wind energy harvesting via triboelectric-electromagnetic flexible cooperation *Appl Energy* **307** 118311
[3] Schubel P J and Crossley R J 2012 Wind turbine blade design *Energies (Basel)* **5** 3425–49
[4] Mishnaevsky L, Branner K, Petersen H N, Beauson J, McGugan M and Sørensen B F 2017 Materials for wind turbine blades: An overview *Materials* **10**
[5] Jani H K, Singh Kachhwaha S, Nagababu G and Das A 2022 A brief review on recycling and reuse of wind turbine blade materials *Mater Today Proc* **62** 7124–7130
[6] Boopathi K, Mishnaevsky L, Sumantraa B, Premkumar S A, Thamodharan K and Balaraman K 2022 Failure mechanisms of wind turbine blades in India: Climatic, regional, and seasonal variability *Wind Energy* **25** 968–979
[7] Heng H, Meng F and McKechnie J 2021 Wind turbine blade wastes and the environmental impacts in Canada *Waste Management* **133** 59–70
[8] Joustra J, Flipsen B and Balkenende R 2021 Structural reuse of high end composite products: A design case study on wind turbine blades *Resour Conserv Recycl* **167**
[9] Du C, Jin G, Zhang L, Tong B, Wang B, Zhang G and Cheng Y 2022 Zero–Waste Recycling of Fiber/Epoxy from Scrap Wind Turbine Blades for Effective Resource Utilization *Polymers (Basel)* **14**
[10] Rani M, Choudhary P, Krishnan V and Zafar S 2021 A review on recycling and reuse methods for carbon fiber/glass fiber composites waste from wind turbine blades *Compos B Eng* **215**
[11] Joustra J, Flipsen B and Balkenende R 2021 Structural reuse of wind turbine blades through segmentation *Composites Part C: Open Access* **5**
[12] Colucci G, Simon H, Roncato D, Martorana B and Badini C 2017 Effect of recycling on polypropylene composites reinforced with glass fibres *Journal of Thermoplastic Composite Materials* **30** 707–723
[13] Cousins D S, Suzuki Y, Murray R E, Samaniuk J R and Stebner A P 2019 Recycling glass fiber thermoplastic composites from wind turbine blades *J Clean Prod* **209** 1252–63
[14] Pender K and Yang L 2020 Regenerating performance of glass fibre recycled from wind turbine blade *Compos B Eng* **198**
[15] Rahimizadeh A, Kalman J, Fayazbakhsh K and Lessard L 2019 Recycling of fiberglass wind turbine blades into reinforced filaments for use in Additive Manufacturing *Compos B Eng* **175**
[16] Gaikwad V, Ghose A, Cholake S, Rawal A, Iwato M and Sahajwalla V 2018 Transformation of E-Waste Plastics into Sustainable Filaments for 3D Printing *ACS Sustain Chem Eng* **6** 14432–14440
[17] Osman M A and Atia M R A 2018 Investigation of ABS-rice straw composite feedstock filament for FDM *Rapid Prototyp J* **24** 1067–1075
[18] Baechler C, Devuono M and Pearce J M 2013 Distributed recycling of waste polymer into RepRap feedstock *Rapid Prototyp J* **19** 118–125
[19] Mamanpush S H, Li H, Englund K and Tabatabaei A T 2018 Recycled wind turbine blades as a feedstock for second generation composites *Waste Management* **76** 708–14
[20] Zhong S, Pearce J, Tightening J P and Pearce J M 2018 *Resources, Conservation and Recycling* **128** 48–58
[21] Rakesh P, Diwakar V, Venkatesh K, Savannananavar R N and Engineer S 2020 A Concise Report on properties of Hybrid Composites manufactured from glass and natural fibers *Mater Today Proc* **22**
[22] Nagarjun J, Kanchana J, RajeshKumar G, Manimaran S and Krishnaprakash M 2022 *Journal of Natural Fibers* **19** 4662–4674
[23] Najafi S K, Mostafazadeh-Marznaki M and Chaharmahali M 2010 Effect of Thermo-Mechanical Degradation of Polypropylene on Hygroscopic Characteristics of Wood Flour-Polypropylene Composites *J Polym Environ* **18** 720–726

[24] Yousefian H, Ben Azouz K and Rodrigue D 2016 *J Polym Environ* **24** 48–55
[25] Mohd Amin A M, Mohd Sauid S and Ku Hamid K H 2015 Polymer-Starch Blend Biodegradable Plastics: An Overview *Adv Mat Res* **1113** 93–98
[26] Darwish L R, Farag M, El-Wakad M T and Emara M 2013 The use of starch matrix-banana fiber composites for biodegradable maxillofacial bone plates *International Conference on Biology, Medical Physics, Medical Chemistry, Biochemistry and Biomedical Engineering* 70–76
[27] Bhayana M, Singh J, Sharma A and Gupta M 2023 A review on optimized FDM 3D printed Wood/PLA bio composite material characteristics *Mater Today Proc*
[28] Furtos G, Molnar L, Silaghi-Dumitrescu L, Pascuta P and Korniejenko K 2022 *Journal of Natural Fibers* **19** 6676–6691
[29] Raj M M, Patel H V, Raj L M, Patel N K and Patel R 2013 Studies on mechanical properties of recycled polypropylene blended with virgin polypropylene *Journal of Applied Polymer Science* **87** 747-751
[30] Bonon A J, Weck M, Bonfante E A and Coelho P G 2016 Physicochemical characterization of three fiber-reinforced epoxide-based composites for dental applications *Materials Science and Engineering C* **69** 905–13
[31] Olesik P, Kozioł M, Konik D and Jała J 2019 The use of shredded car windscreen waste as reinforcement of thermoplastic composites for 3D (FDM) printing *Composites theory and practice* **19** 30–33

IOP Conf. Series: Materials Science and Engineering 1293 (2023) 012042 doi:10.1088/1757-899X/1293/1/012042

Quantification of process-induced effects on fatigue life of short-glass-fiber-filled adhesive used in wind turbine rotor blades

M Rosemeier[1], T Holst[1] and A Antoniou[2]

[1]Department of Rotor Blades, Fraunhofer Institute for Wind Energy Systems IWES, Am Seedeich 45, 27572 Bremerhaven, Germany

[2]Department of Mechanics, National Technical University of Athens NTUA, 9, Iroon Polytexneiou str, 15780, Zografou, Greece

E-mail: malo.rosemeier@iwes.fraunhofer.de

Abstract. Cracks in the adhesive joints of wind turbine blades can often cause severe structural blade damage. Effects induced by the manufacturing process can have a significant impact on the cycle number toward crack initiation in the bonding material. This research compared two experimental series of neat adhesive, dog-bone-shaped specimens with different degrees of cure, fiber orientation, and porosity content, which were subjected to static and cyclic tension-tension loading. To enhance the comparability of the two coupon series, the experimental data was normalized. Normalization factors were introduced to compensate for the different properties of the two series. These factors were derived from correction models. After normalization, the S-N values of the two data series were in good agreement in the high-cycle fatigue regime. In the very-low-cycle fatigue regime, however, there was still a lack of agreement in some cases, including the static strength. Moreover, the failure mechanisms were investigated by analyzing images from the fracture surfaces and dedicated samples using X-ray microscopy. The correction models presented can be applied to estimate the design strength of adhesives. This knowledge can then be used to make rotor blade structures more resistant to crack initiation.

1. Introduction

Cracks in adhesive joints of wind turbine rotor blades can often cause structural damage [1]. During operation, tunneling cracks can be initiated in the adhesive layer, and propagate into the surrounding laminate, making structural repairs necessary [2]. Such tunneling cracks can be initiated early in the operating life and even earlier still, during manufacture [3]. Beside the impact of thermal residual stresses that develop in the adhesive layer during the manufacture [4], other process-induced effects of the manufacture can have a significant impact on the stress-life of a bond line.

The mechanical properties of the adhesive depend on several parameters. Some polymeric materials exhibit anisotropic properties due to the addition of short glass-fibers as reinforcement [5]. Holst et al. [6] experimentally investigated the relation between short-glass-fiber orientation and the resulting coefficient of thermal expansion (CTE) and static tensile strength. Moreover, the manufacturing process itself induces effects that affect the performance, i.e., the curing cycle and the degree of cure produced [7]; the mixing of the adhesive resin and hardener and the resulting porosity and specimen geometry; and the adhesive mold-injection process which results in voids of different sizes; cf. Wentingmann et al. [8], Fernandez et al. [9], and Zarouchas and Nijssen [10]. Kuhn et al. [11] aimed to clarify the impact of these effects on the stress-strain relationship and stress-life (S-N) curves. Taking the assumptions made by Fiedler [12] as their basis, Rosemeier and Antoniou [13] normalized the experimental data set of neat epoxy to a normalized ratio between test temperature T and degree of cure in terms of the glass transition temperature T_g. Recently, Drummer et al. [14] confirmed the linear relation between static strength R_t and T/T_g for several experimentally investigated polymers.

To enhance understanding of the failure behavior of adhesives, this research investigates how manufacturing effects such as porosity [8], short-fiber orientation [6], and degree of cure [13] affect the stress-life of a short-glass-fiber-filled epoxy adhesive. To this end, two series of specimens with different degrees of cure, indicated by a measured T_g, fiber orientation, and porosity were manufactured and compared in respect of their S-N behavior at a load ratio of $R = 0.1$. Based on the models mentioned above, the differences between the two series was reduced in the HCF regime.

2. Methods

2.1. Material, specimen manufacture, and testing

The fatigue life of Hexion's EPIKOTE^{TM} resin MGS^{TM} BPR 135G3 and EPIKURE^{TM} curing agent MGS^{TM} BPH 137G [15] was investigated. This type of adhesive is based on epoxy resin and filled with short glass-fibers.

Two sets of specimens were considered, which were manufactured by different methods: (i) machine-assisted (MA) specimens and (ii) manually manufactured (MN) specimens. The MA specimen plates were manufactured by means of a mixing and dosing machine. A bead of adhesive was applied and pressed to a thickness of 8 mm between two aluminum plates, which were pretreated with a release agent. For the MN specimens, the resin and hardener were mixed by speed mixer and the adhesive paste was applied manually to a first glass fiber-reinforced polymer (GFRP) plate, which was pretreated with a release agent, in a meander pattern aligned in the transverse, or s-direction, of the specimen (Fig. 1). Then, another pretreated GFRP plate was placed on top to adjust the adhesive layer to a thickness of 8 mm, cf. [16]. The MA plates were cured at 75°C for 4 hours without exothermicity. The MN plates were cured at 80°C for 4 hours with an exothermic peak of up to 120°C. All specimen series were then cut with a water-jet. Beside their standard thickness, the specimens had a standard dog-bone geometry according to [17] (Fig. 1), and were used for both the static characterization with a stress ratio of $R = 1.0$, and the tension-tension fatigue characterization with a stress ratio of $R = 0.1$. The surface and edges of the MA specimens were smoothed with 180-grit and 1000-grit sandpaper. The specimen dimensions were measured with a dial gauge with a tolerance of $\pm 1.6\,\mu$m.

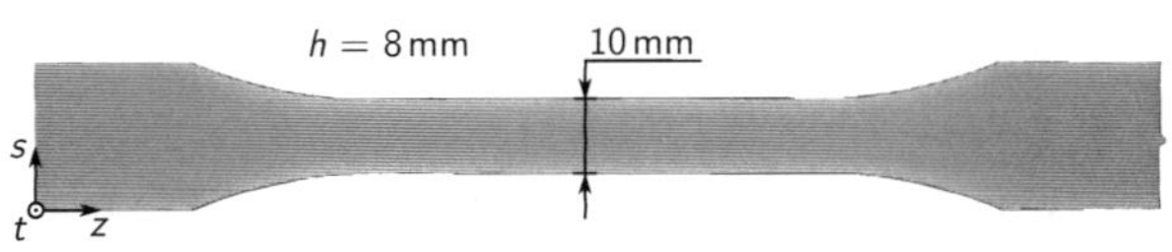

Figure 1. Specimen dimensions.

Static tensile and cyclic tests were conducted on calibrated 25 kN and 50 kN uni-axial servo-hydraulic tension-compression test machines, respectively, each with a class 0.5 load cell according to [18] with a measurement uncertainty of 0.18% and 0.16%, respectively. The frequency for the cyclic tests was kept between 1 Hz to 2 Hz to prevent a temperature increase caused by hysteretic energy dissipation. The target ambient temperature during the cyclic tests and a series of static tests was the standard class 1 room temperature of 23°C [19]. The surface temperature of each specimen was measured using a 100 Ω platinum Resistance Temperature

Detector (RTD) from YAGEO with a tolerance of ±0.1°C. Type GFLAB strain gauges from Tokyo Sokki Kenkyujo with a relatively low stiffness were installed parallel and transverse to the load axis, one on each side of a set of the static specimens. The glass transition temperature T_g of each specimen tested was measured via Differential Scanning Calorimetry (DSC) using a DSC 204 F1 from Netzsch. Statistical evaluation was used to determine a standard deviation of 0.6°C for the determination procedure. For further details, see Appendix A.

The maximum stress σ_{max}, the stress ratio R, the cycles to failure N, the average surface temperature T, and the glass transition temperature T_g of each tested specimen are summarized in Table 1.

Table 1. Summary of static and cyclic test results.

No.	σ_{max} [MPa]	σ^{n*}_{max} [MPa]	R	N	T [°C]	T_g [°C]	A_v/A_{tot} [%]	A_f/A_{tot} [%]
MA_S1	67.0	65.6	1.0	1	23.9	72.2	0.0	— —
MA_S2	66.7	65.7	1.0	1	24.0	71.7	0.0	— —
MA_S3	66.4	66.5	1.0	1	23.9	72.3	0.0	— —
MA_S4	64.6	66.0	1.0	1	24.0	73.6	0.0	— —
MA_S5	64.1	66.1	1.0	1	23.9	74.2	0.0	— —
MA_C1	49.4	51.9	0.1	1,930	22.4	73.0	0.5	4.1
MA_C2	46.1	45.8	0.1	3,710	23.9	72.3	0.0	4.7
MA_C3	44.5	44.7	0.1	5,200	22.4	71.0	0.5	6.4
MA_C4	41.7	39.6	0.1	13,700	24.1	70.9	0.1	5.0
MA_C5	39.5	39.4	0.1	16,500	24.4	73.0	0.0	5.0
MA_C6	37.9	35.8	0.1	23,400	23.5	70.3	0.0	6.2
MA_C7	36.2	37.3	0.1	64,600	23.4	72.8	0.0	4.6
MA_C8	34.6	34.6	0.1	182,000	22.5	70.9	0.0	9.3
MA_C9	33.0	32.8	0.1	976,000	23.4	71.8	0.0	13.2
MN_S1	39.5	56.9	1.0	1	24.7	78.2	7.6	— —
MN_S2	36.3	53.6	1.0	1	24.7	79.1	7.0	— —
MN_S3	35.3	53.5	1.0	1	24.7	78.0	11.6	— —
MN_C1	26.8	43.3	0.1	858	22.4	78.6	5.6	0.0
MN_C2	24.5	42.7	0.1	9,040	22.1	79.4	9.2	0.0
MN_C3	25.8	43.1	0.1	18,700	22.1	79.6	4.6	0.0
MN_C4	23.3	38.3	0.1	22,600	22.2	78.2	4.7	0.0
MN_C5	21.8	36.6	0.1	75,700	22.8	79.5	3.5	0.0
MN_C6	18.6	33.4	0.1	163,000	23.1	80.0	5.8	0.0
MN_C7	18.0	30.8	0.1	245,000	22.5	78.4	2.4	0.0

* Normalized values: MA specimens normalized for test temperature and glass transition temperature (T/T_g); MN specimens normalized for porosity (P), fiber orientation (F), and T/T_g.

2.2. Short-fiber orientation correction

The short-fiber orientation in the adhesive plates manufactured was estimated according to the method described in [6]. A cube-shaped thermo-mechanical analysis (TMA) sample was taken at four measuring points of an MN plate and four measuring points of an MA plate. According to

[6], the determination of the linear CTE along all spatial axes allows the short-fiber orientation in the different types of specimen plates to be estimated on average in the form of a smeared fiber orientation tensor. The estimation method yielded a fiber orientation tensor entry of $\alpha_{zz}^{\mathrm{MN}} = 0.08$ in the longitudinal, or z-direction, of the specimen for the MN specimens and an entry of $\alpha_{zz}^{\mathrm{MA}} = 0.75$ for the MA specimens. Taking the relationship between tensile strength and fiber orientation from [6], which investigated the same material, the fiber orientation information in the longitudinal, or z-direction, of the specimen was used to adjust the tensile strength of the MN data points by a factor of $R_\mathrm{t}(\alpha_{zz}^{\mathrm{MA}} = 0.75)/R_\mathrm{t}(\alpha_{zz}^{\mathrm{MN}} = 0.08) = 70\,\mathrm{MPa}/58\,\mathrm{MPa} = 1.2$.

2.3. Temperature and degree-of-cure correction

The experimental results for the cyclic specimens listed in Table 1 were normalized to remove effects stemming from the average test temperature T and the degree of cure, the latter being affected by process parameters such as mix ratio, exothermicity, or post-curing cycle. On the basis of the hypothesis of a linear relationship between the static strength R_t and T/T_g as shown in [12, 14], the data points were adjusted according to

$$\sigma_{\max}^{\mathrm{n}} = m_\mathrm{R}\left(T^\mathrm{n}/T_\mathrm{g}^\mathrm{n} - T/T_\mathrm{g}\right) + \sigma_{\max}, \tag{1}$$

where T^n stands for the normalization temperature, $\sigma_{\max}^{\mathrm{n}}$ denotes the maximum stress under normalized conditions, and $\sigma_{\max}$ stands for the actual maximum stress determined by the maximum force over the cross-sectional area $F_{\max}/A$. The gradient of the linear approximation is denoted as m_R. Temperature values inserted into (1) must be in Kelvin. Therefore, a gradient of $m_\mathrm{R} = 469.2\,\mathrm{MPa\,K\,K^{-1}}$ (Fig. 2) was used, which was derived from the static MA specimens listed in Table 1. For the normalization of the data points, we chose the average T/T_g value of the MA specimens: $T^\mathrm{n}/T_\mathrm{g}^\mathrm{n} = 0.859\,\mathrm{K\,K^{-1}}$ as highlighted by the dotted vertical line in Fig. 2.

2.4. Porosity correction

To take account of the stress increase within the critical fracture cross-section, the sizes of the voids within the fracture surface were measured. The machine force measured during the test was ratioed to the cross-sectional area actually remaining. The ratios between the cross-sectional areas of voids on the fracture surface and the total area $A_\mathrm{v}/A_\mathrm{tot}$ are given in Table 1. Neither the eccentric positioning nor the 3D shape of the voids was taken into consideration in the normalization of the stress data.

2.5. X-ray microscopy

For 3D X-ray microscopy (XRM), the sample MN_S3 was scanned in a 360° rotation conducted with the ZEISS Xradia 520 Versa system operated at the MAPEX Center for Materials and Processes, University of Bremen, Germany. The sample was imaged with 16.2 µm per voxel, using a setting of 80 kV, 7 W, and ZEISS filter LE1.

Correction of ring artifacts and reconstruction of the spatial information on the linear attenuation coefficient of the sample was done using the ZEISS software. The images were post-processed through thresholding using OpenCV [20].

3. Results

3.1. Test temperature and glass transition temperature dependence

We observed that the static strength R_t of the MA specimens increased with decreasing glass transition temperature T_g, see Fig. 2. Although the normalized strength values for porosity (P) and fiber orientation (F) of the MN specimens lay significantly below the trend line of the MA specimens, the gradient observed for the normalized values of the MN specimens is similar although the statistical basis of the data was small, only three data points being available.

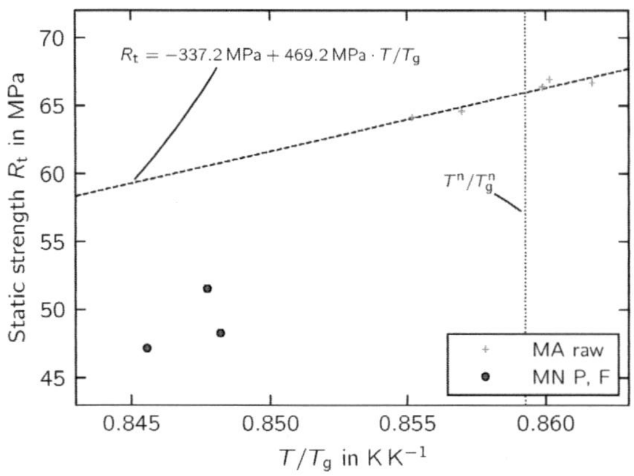

Figure 2. Static strength as a function of T/T_g; specimens manufactured by means of machine-assisted application (MA) and manual application (MN) normalized for porosity (P) and fiber orientation (F).

3.2. Stress-strain curves

The raw stresses (Fig. 3a) of the static MN and MA specimens were adjusted (Fig. 3b). In a first step, the cross-sectional area for the MN specimens was adjusted to take account of porosity. In a second step, the effect of the fiber orientation was normalized such that the orientation of the MN specimens matched that of the MA specimens. And in a third step, the effect of the test temperature and glass transition temperature was normalized for the two series to match T^n/T_g^n given in Fig. 2. In a final step, we adjusted the stress-strain curve of the MN specimens to match the stress-strain gradient ($d\sigma/d\varepsilon$) of the MA specimens between $\varepsilon = 0\,\mu\text{m}\,\text{m}^{-1}$ and $\varepsilon = 5000\,\mu\text{m}\,\text{m}^{-1}$.

Now, the two curves should be comparable, meaning that all manufacturing effects have been ruled out enhancing the direct comparison between MN and MA specimens. As a result, the normalized MN curves indicate behavior which is more brittle than that indicated by the MA curves. Local stress and strain effects due to stress concentration are not captured in these global curves. Thus we can derive an experimental stress concentration factor of $K_e = 64.8/54.2 = 1.19$ from the remaining stress/strain gap between the MN and MA specimens. Note, however, that K_e already includes the porosity correction.

3.3. Stress-life curves

After we had normalized the values of the cyclic MN specimens, their stress-life was in good agreement with the S-N trend of the normalized values for the MA specimens (Fig. 4a). The normalized maximum stress values σ_{max}^n are given in Table 1. Both data sets were approximated by probabilistic S-N models. As the normalized MN data exhibited a rather linear trend in a double-log grid, we first approximated these data according to Basquin's linear model [21], which is still state-of-the-art in the design of rotor blades, cf. [22]. As the MA data was rather sigmoidal in shape, we approximated this data with a non-linear model suggested by Stüssi [23, 24], which was slightly modified and reformulated by Rosemeier and Antoniou [13], cf. [25]. For the sake of comparison, the Stüssi model was also used to approximate the normalized MN data.

In the cycle ranges where the S-N data of the two specimen series overlapped (gray shaded areas in Fig. 4b), the differences between the Stüssi and Basquin models approximating the normalized values for the MN specimens ranged between 5% in the HCF regime and −20% in the static and VLCF regime. Furthermore, for the static case in particular, these two models differed from a Stüssi model which approximates the stress-life of the MA specimens (Fig. 4b).

The largest impact on the stress level came from the degree-of-cure and temperature correction followed by the fiber orientation correction, and the porosity correction. The porosity correction appeared to have greater impact on the static strength than on the cyclic stress level.

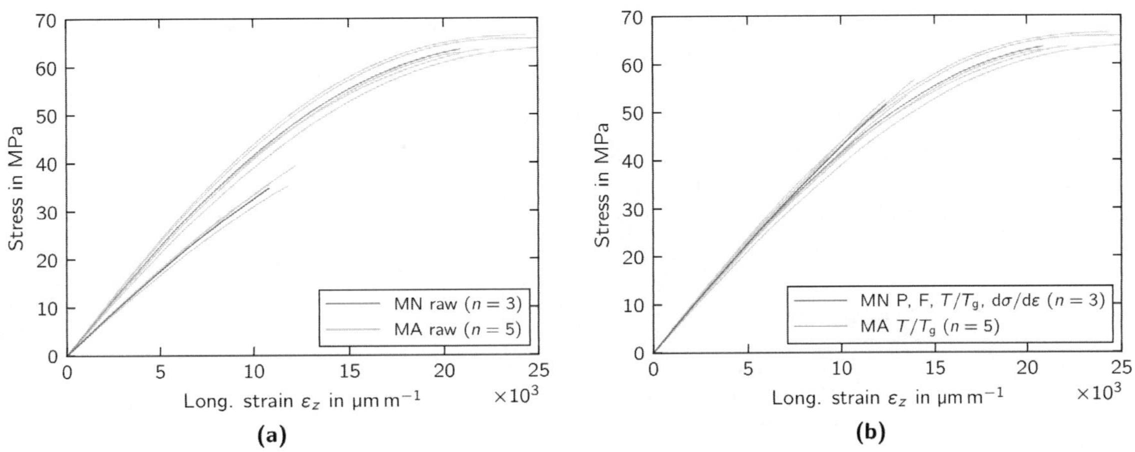

Figure 3. Static stress-strain curves of neat adhesive specimens manufactured by means of machine-assisted application (MA) and manual application (MN): (a) raw data and (b) data normalized for porosity (P), fiber orientation (F), test temperature and glass transition temperature (T/T_g), and stress-strain gradient ($d\sigma/d\varepsilon$).

The standard deviation was taken into consideration as the quality criterion for an S-N model approximating the experimental data points. Considering the normalized MA data, the probabilistic Stüssi model produced a relatively low standard deviation of $R^{\sigma}_{t,MA,S} = 0.88\,\text{MPa}$. Considering the normalized MN data, the probabilistic Stüssi model produced a lower standard deviation ($R^{\sigma}_{t,MN,S} = 1.99\,\text{MPa}$) than the probabilistic Basquin model ($R^{\sigma}_{t,MN,B} = 2.78\,\text{MPa}$).

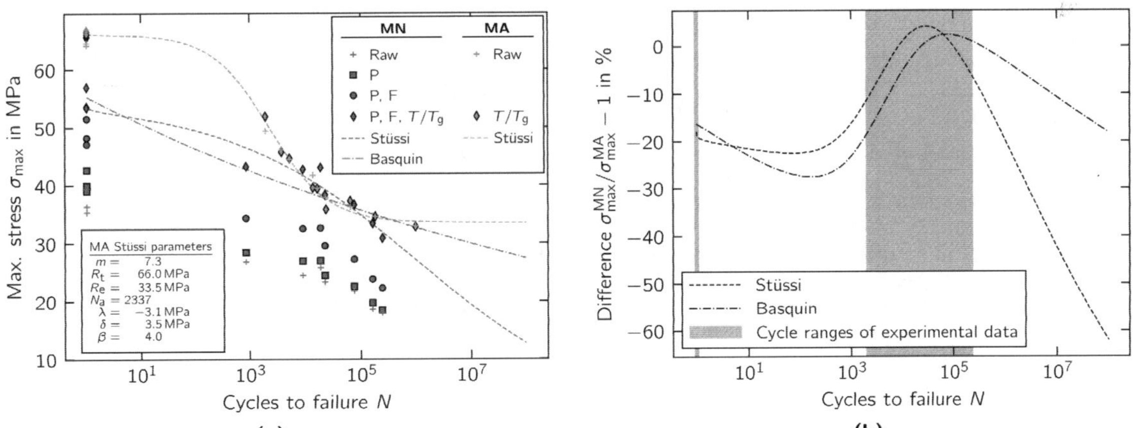

Figure 4. Effect of manufacturing process on stress-life of neat adhesive tested at $R = 1$ and $R = 0.1$; specimens manufactured by means of machine-assisted application (MA) and manual application (MN): (a) adjusted for porosity (P), fiber orientation (F), test temperature and glass transition temperature (T/T_g), and respective S-N model approximations, and (b) differences in the model approximations between the MN and MA data sets.

Next, a trial was carried out to approximate the stress-life probabilistically using both data sets. As the correction models applied in this research did not sufficiently capture the effects in the static and VLCF regimes, the static MN data points were not included in the approximation; this is discussed further in Sec. 4. Taking all adjusted cyclic data points, an approximation of the probabilistic Stüssi model then produced an overall standard deviation of $R^{\sigma}_{\mathrm{t,MN+MA,S}} = 2.14\,\mathrm{MPa}$ (Fig. 5). As the scatter of the MN data points was greater, this approximation yielded a larger standard deviation than the Stüssi approximation, which considered only the corrected MA data points. The endurance limit for the 50% failure probability of the MA+MN Stüssi model was lower ($R_{\mathrm{e}} = 26.0\,\mathrm{MPa}$) than that of the MA Stüssi model ($R_{\mathrm{e}} = 33.5\,\mathrm{MPa}$).

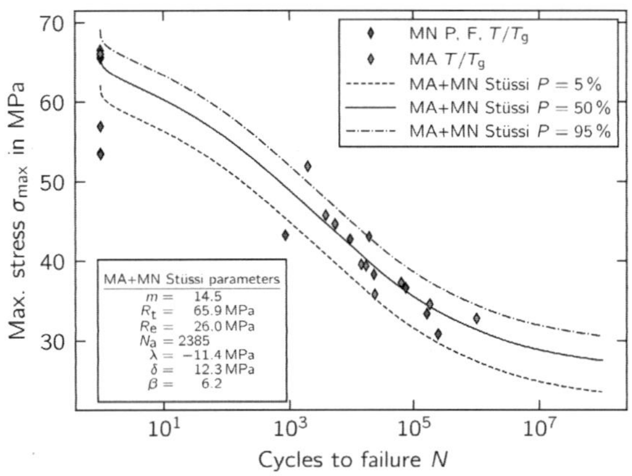

Figure 5. Probabilistic Stüssi prediction considering all normalized data points of MA specimens and normalized data points of cyclic MN specimens.

3.4. Fracture surfaces

Considering the fracture surfaces of the MA specimens (Fig. 6), we observed two types of crack initiation regions: (a) in the internal specimen structure, and (b) at the surface of the specimen.

While the three static specimens with the highest strength, i.e., MA_S1, MA_S2, and MA_S3, exhibited a largely internal crack initiation, the two specimens with lowest static strength, i.e., MA_S4 and MA_S5, had crack initiation closer to the side surface, and in particular toward a specimen edge.

When considering the cyclic specimens, three specimens, i.e., MA_C1, MA_C3, and MA_C4, had an initiation location at a void at the surface big enough to be visible to the human eye. The cross-sectional area of the void measured between $0.08\,\mathrm{mm}^2$ and $0.37\,\mathrm{mm}^2$. The other cyclic specimen also exhibited crack initiation at the surface but concentrated more at an edge. The more load cycles a specimen was subjected to, the larger the fatigue zone, also called plasticization zone (Ger. Schwingbruchfläche [26]), around the crack initiation location became. The ratios between the cross-sectional area of the fatigue zone on the fracture surface and the total area $A_{\mathrm{f}}/A_{\mathrm{tot}}$ are given in Table 1.

Considering the fracture surfaces of the MN specimens (Fig. 7), we observed no such clear areas of crack initiation as for the MA specimens, i.e., no fatigue zone.

A common feature of all fracture surfaces of the MN specimens is that they had relatively large voids. The cross-sectional area of the voids measured between $1.92\,\mathrm{mm}^2$ and $9.28\,\mathrm{mm}^2$. Many voids were adjacent or close to the specimen surface. Only the fracture surface of specimen MN_C7, which was subjected to the highest number of load cycles, had a void which seemed to be located in the internal structure of the specimen.

3.5. Fiber orientation

As methodologically shown in [6], an XRM scan of specimen MN_S3 confirmed the fiber orientation tensor entries of $\alpha_{zz} = 0.08$, $\alpha_{ss} = 0.87$, and $\alpha_{tt} = 0.05$, which suggests that

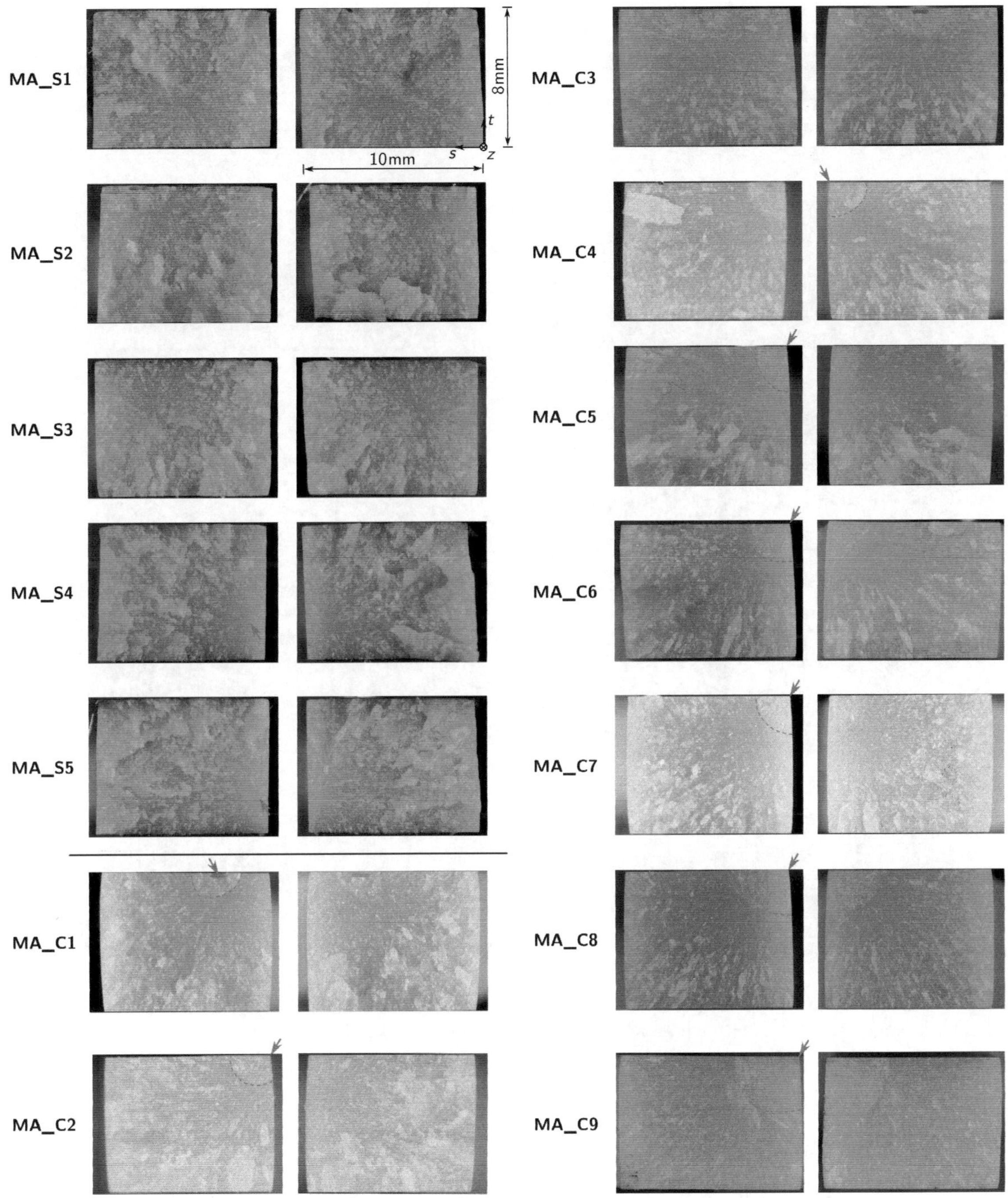

Figure 6. Fracture surfaces of static (S) and cyclic (C) machine-assisted (MA) manufactured specimens post-mortem; red arrows indicate point crack initiation, red-dashed lines indicate border of fatigue zone.

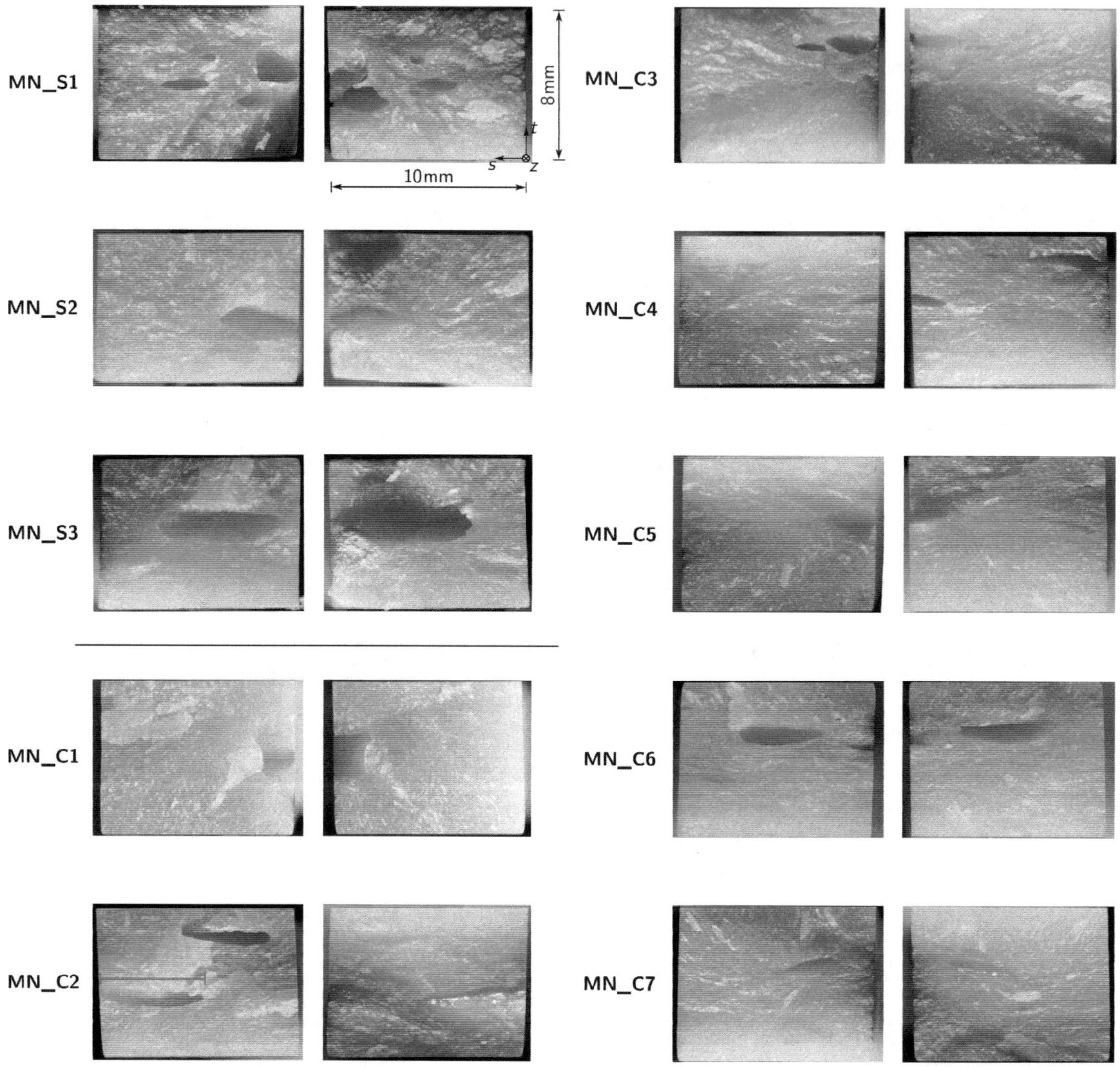

Figure 7. Fracture surfaces of static (S) and cyclic (C) manually applied (MN) manufactured specimens post-mortem; red bars indicate length of critical voids.

the short fibers were mainly oriented in the transverse, or s-direction, of the specimen (Fig. 8b, c). We observed that the fiber orientation varied in the vicinity of larger voids, i.e., the fiber orientation changed from mainly s-direction (Fig. 8b) to an orientation in the specimen's z-direction (Fig. 8d). Moreover, we observed sharp edges, e.g., lobes or notches, at the large void close to the fracture surface (Fig. 8e). Smaller voids with a cylindrical morphology along the main fiber direction (Fig. 8b) and elliptic cross-section (Fig. 8c) were also present in the volume scanned.

4. Discussion

4.1. Effect of voids

The correction steps presented in this research appear to be applicable in the HCF regime. Despite all the correction factors applied, there still seem to be differences in the VLCF regime,

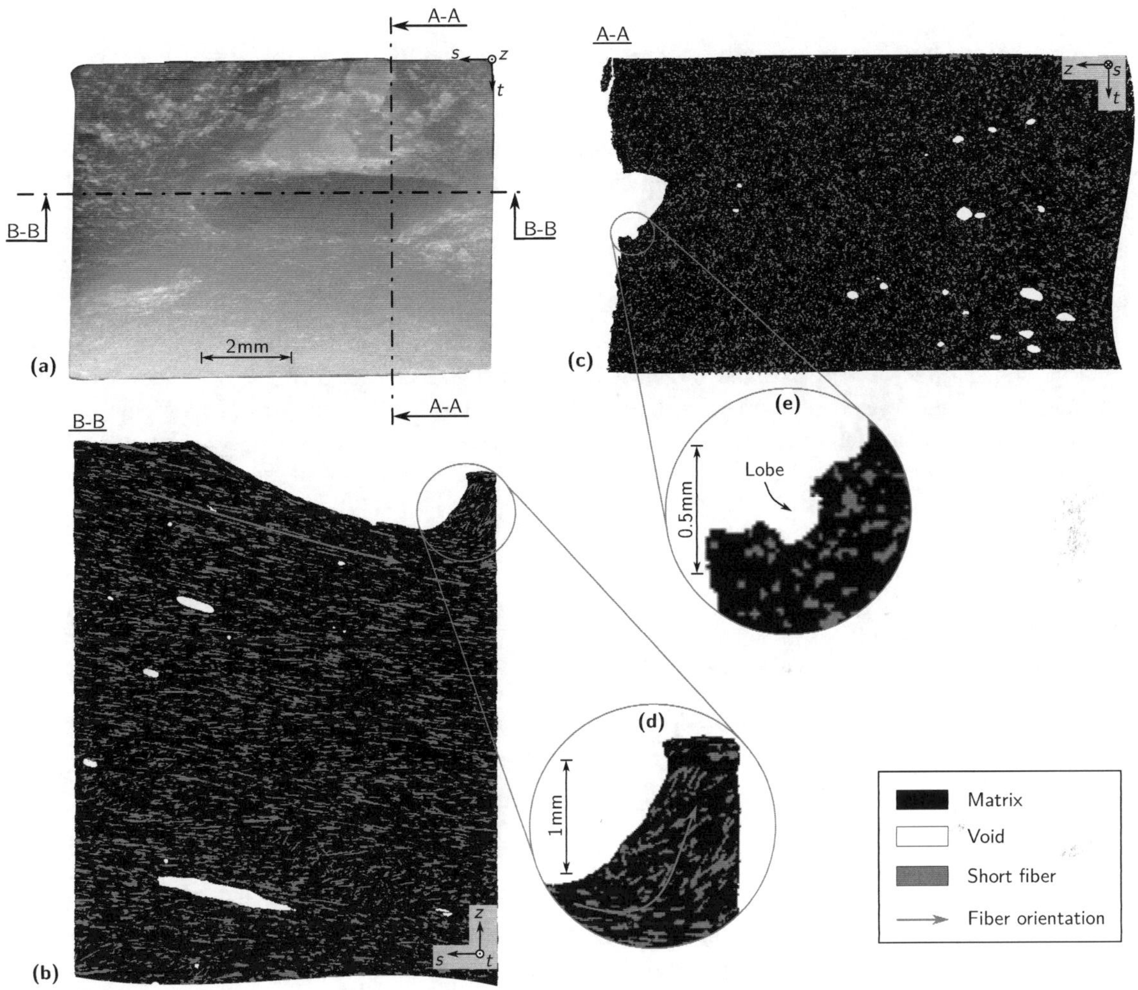

Figure 8. Views of specimen MN_S3: (a) fracture surface, (b) thresholded XRM image through B-B plane, (c) thresholded XRM image through A-A plane, and (d,e) zoom into void edges.

including in the static strengths of the MA and MN specimens (Fig. 4). This discrepancy can be partially explained.

Considering the relative sizes of the cross-sectional areas of the voids at the fracture surface $A_{\mathrm{v}}/A_{\mathrm{tot}}$ (Table 1) of the static specimens MN_S1, MN_S2, and MN_S3, they tend to be larger than those of all cyclic specimens MN_CX with the exception of specimen MN_C2. The stress concentration increases non-linearly with void size, cf. [27, 28]. This tendency is in agreement with our observations.

Considering the eccentric position of the void within the *st*-plane, i.e., whether the void is close to one specimen surface: The specimens MN_C2 and MN_C3 were subjected to the same stress level after correction. In the fracture surface of MN_C2 one void which extends approx. 50% in *s*-direction and is very close to the surface. Similarly, in the fracture surface MN_C3 there is also one void which extends approx. 25% in *s*-direction, also very close to the surface. Assuming voids are equidistant from the surface and circular in shape, a specimen with a void of a larger diameter is generally subjected to a higher stress concentration than a specimen with a void of a smaller diameter, cf. [28, 29]. This tendency is also in agreement with our observations.

This research has not investigated nor taken into account for correction several effects which affect the stress concentration at a void edge: the eccentric position of the void within the *st*-plane and the three-dimensional (3D) shape and size of the void, as well as the interaction of multiple voids, cf. [28]. To quantify these effects on the stress concentration by means of models, XRM scans of all specimens would be required, preferably in the pre-mortem state, in order to capture the full 3D shape.

In general, fatigue experiments lead us to assume that a void subjected to a cyclic loading produces an effect which concentrates stress to a lesser extent than is predicted by theoretical elastic analysis [30].

The fiber orientation can change in the vicinity of large voids, and also locally, thereby affecting the stiffness and resistance to crack initiation locally (Fig. 8d).

Considering the effect of void size and shape on crack initiation in MN specimens: A correlation cannot be derived from images of the fracture surface. XRM scan images in the sz-plane or zt-plane, preferably in the pre-mortem state, could reveal more insights into the original void shape and possible sharp edges.

4.2. Effect of glass transition temperature
In contrast to what is evident in pure resin materials [7, 14], the static strength of short-fiber-glass-filled adhesives increased with decreasing degree of cure in terms of the glass transition temperature T_{g} (Fig. 2). Although it is not clear why this happens, we assume that residual stresses develop at the resin-fiber interface. This could result in premature crack initiation at higher resin stiffness, which increases with degree of cure. Also, the increase in cross-linking with increasing degree of cure can have an effect on the load transfer between and around the short fibers.

4.3. Effect of fiber orientation
Considering the effect of fiber orientation on the fracture surface: The MN specimens had a transverse fiber orientation in the specimen's s-direction. Here, no fatigue zone was observed. In contrast, the MA specimens had a longitudinal fiber orientation in the specimen's z-direction and although smaller voids were observed in the fracture surface, a fatigue zone was clearly visible. One explanation could be that the crack propagation rate is faster when a crack propagates parallel to the main short-fiber direction, which is the case for the MN specimens, cf. [31].

The XRM images in Fig. 8 show in detail that the orientation of the short fibers is locally affected by the formation of voids as the adhesive is being applied manually during the manufacturing process. The short fibers rearrange in the direction of the void contour. This local fiber orientation differs from the estimated global smeared short-fiber orientation on which the normalization of the experimental data is based. This can be considered as another uncertainty in the analysis. For a more detailed view, the local short-fiber orientation in these areas can be determined using additional XRM scans in combination with software-assisted fiber segmentation. However, this increases the experimental and analytical effort.

4.4. Location of crack initiation under static and cyclic loading
Cyclic MA specimens without voids exhibited crack initiation at one specimen edge, while the best-performing static MA specimens exhibited initiation in the inside. This observation indicates that the mechanisms that lead to crack initiation after one cycle and n cycles must be different. Several effects can play a role, e.g., stiffness degradation, or relaxation during cyclic loading. These effects can "equalize" the stress level within the structure and prevent a crack being initiated at a local micro-defect, as is the case in the static MA specimens. Consequently,

the specimen edge, being the most vulnerable part, becomes the location most susceptible to crack initiation.

The stress-strain curves (Fig. 3b) represent a global behavior as they are derived from machine force and strain gauge measurements. That is, they do not capture the local strain behavior at the strain concentration. If such strain concentrations are located at the specimen surface, they could be analyzed using Digital Image Correlation (DIC) techniques during cyclic loading, c.f. [32].

4.5. Cycle number to crack initiation

With the MA specimens, we observed that the fatigue zone grew larger, the lower the stress level and the higher the cycle number toward crack initiation. To obtain the actual cycle number toward crack initiation, the crack propagation cycle number from initiation until final rupture would be of interest. This correlation leads to an overestimation of the stress-life, which might be even greater at a lower stress level or in the VHCF regime. A method to quantify the crack propagation cycle number is required. This would allow the actual cycle number toward crack initiation to be determined by backward extrapolation.

5. Conclusions

This research is the first to be published in openly accessible English-language literature about how manufacturing process parameters such as curing cycle and temperature, process-induced effects such as fiber orientation, and effects of defects such as voids, can be quantified simultaneously and taken into account by models for a short-glass-fiber-filled adhesive material widely used in the wind industry. These effects were taken into account by correction models which describe their contribution to the mechanical performance of an adhesive material, under both static and cyclic loading.

The different S-N behavior in the HCF regime of two series of specimens with different degrees of cure, fiber orientation, and porosity was explained by correction models. A method whereby the manufacturing effects can be removed from the data on material strength, both static and fatigue, through the normalization of two different sets of experimental data was presented by way of example. The normalization of the degree of cure and fiber orientation had the largest impact on the stress-life.

All correction models applied were derived on the basis of simple standardized tests which can be performed quickly and with a limited budget. The correction models presented can be applied to estimate the design strength of adhesives. This knowledge can then be used to make rotor blade structures more resistant to crack initiation.

Future work should enhance the porosity correction method to model the stress concentration in more detail on the basis of the void eccentricity and shape, in order to explain the remaining post-normalization differences, including the static strength R_t, between the two series in respect of the VLCF regime.

Supplemental material

The data used for the figures are available at `https://doi.org/10.5281/zenodo.8214266` [33].

Acknowledgments

We acknowledge the support provided by the German Federal Ministry for Economic Affairs and Climate Action (BMWK) within the ReliaBlade project (0324335A) and the OptiMUm project (03EE2017). Furthermore, we thank our colleagues Holger Bannies, Henning Schnellen, Nils Englisch, Lisa Schudack, Stefan Krause, Martina Karalus, Konstantin Kinsvater, and Robert Flindt, who prepared the specimens and conducted the experiments. Also, we thank Michael

Kuhn from Leibniz University Hannover for providing the test samples used to determine the DSC standard deviation. Moreover, we acknowledge the use of the ZEISS Xradia 520 system at MAPEX, University of Bremen, Germany (DFG 434618658) and we thank Wolf-Achim Kahl from MAPEX, who conducted the XRM scan. Finally, we thank our colleague Florian Sayer, who obtained the funding for this research.

Appendix A. Derivation of DSC standard deviation

The average standard deviation for the glass transition temperature of $T_g^\sigma = 0.6$°C was derived statistically. To derive the accuracy of a DSC measurement and determine the T_g, three DSC samples per specimen with a mass of between 5 and 20 mg were extracted from adjacent sites on the fracture surface. Three specimens from the measurement campaign presented in [8, 34] were analyzed. Table A1 lists the results.

Table A1. Summary of DSC measurement statistics.

No.	σ_{max} [MPa]	T [°C]	T_{g1} [°C]	T_{g2} [°C]	T_{g3} [°C]	T_g^σ [°C]
073	74.9	21.5	72.1	72.1	71.2	0.5
067	76.0	21.5	75.5	75.1	76.4	0.7
041	77.4	22.0	76.5	76.6	75.5	0.6

References

[1] Mishnaevsky Jr L, Frost-Jensen Johansen N, Fraisse A, Fæster S, Jensen T and Bendixen B 2022 *Energies* **15** 1767
[2] Rosemeier M, Krimmer A, Bardenhagen A and Antoniou A 2019 *AIAA Journal* **57** 5462–5474
[3] Rosemeier M, Gebauer T and Antoniou A 2022 *20th European Conference on Composite Materials (ECCM20)-Composites Meet Sustainability* vol 5 ed Vassilopoulos A and Michaud V
[4] Rosemeier M, Melcher D, Krimmer A, Wroblewski W and Antoniou A 2022 *Journal of Physics: Conference Series–The Science of Making Torque from Wind (TORQUE 2022)* vol 2265 (IOP Publishing) p 032045
[5] Voss H and Walter R 1985 *Journal of materials science letters* **4** 1174–1177
[6] Holst T, Antoniou A, Englisch N, Manousides N and Balzani C 2023 *AIAA SCITECH 2023 Forum*
[7] Antoniou A, Rosemeier M, Tazefidan K, Krimmer A and Wolken-Möhlmann G 2020 *AIAA Journal* **58** 4781–4793
[8] Wentingmann M, Manousides N, Antoniou A and Balzani C 2021 *Materials & Design* **212** 110213
[9] Fernandez G, Vandepitte D, Usabiaga H and Debruyne S 2018 *International Journal of Fatigue* **117** 340–351
[10] Zarouchas D and Nijssen R 2016 *Journal of Adhesion Science and Technology* **30** 1413–1429
[11] Kuhn M, Manousides N, Antoniou A and Balzani C 2023 *Fatigue & Fracture of Engineering Materials & Structures* **46** 1121–1139
[12] Fiedler B 2009 *Hochleistungs-Faserverbundwerkstoffe mit Duroplastmatrix, Werkstoffe Struktur Eigenschaften Modellierung* (Hamburg, Germany: TuTech Innovation GmbH) ISBN 978-3-930400-26-3
[13] Rosemeier M and Antoniou A 2022 *AIAA Journal* **60** 951–961
[14] Drummer J, Gibhardt D, Körbelin J and Fiedler B 2022 *Composites Science and Technology* 109486
[15] Hexion Inc 2016 Technical Datasheet: EPIKOTE™ Resin MGS© BPR 135G-Series and EPIKURE™ Curing Agent MGS BPH 134G-137GF
[16] Holst T, Antoniou A and Englisch N 2022 *ECCM20 - The 20th European Conference on Composite Materials* vol 5 - Applications and structures pp 137–144
[17] DIN EN ISO 2012 *527-2, Plastics - Determination of tensile properties - Part 2: Test conditions for moulding and extrusion plastics* (Deutsches Institut für Normung)
[18] ISO 2018 *ISO 7500-1, Metallic materials - Verification of static uniaxial testing machines - Part 1: Tension/compression testing machines - Verification and calibration of the force-measuring system* (Geneva, Switzerland: International Organization for Standardization)

IOP Conf. Series: Materials Science and Engineering 1293 (2023) 012042 doi:10.1088/1757-899X/1293/1/012042

[19] DIN EN ISO 2008 *291, Plastics–Standard Atmospheres for Conditioning and Testing (ISO 291: 2008), German Version EN ISO* (Deutsches Institut für Normung)
[20] Bradski G 2000 *Dr. Dobb's Journal of Software Tools*
[21] Basquin O H 1910 *Proceedings of the thirteenth annual meeting* vol 10 (Atlantic City, New Jersey, USA: American Society for Testing Materials) pp 625–630 URL `https://pdfcoffee.com/basquin-the-exponential-law-of-endurance-testspdf-pdf-free.html`
[22] DNV GL A/S 2015 DNVGL-ST-0376 - Rotor blades for wind turbines URL `https://rules.dnvgl.com/docs/pdf/DNVGL/ST/2015-12/DNVGL-ST-0376.pdf`
[23] Stüssi F 1955 *Mitteilungen der Technischen Kommission des Verbandes Schweizerischer Brückenbau- und Stahlhochbau-Unternehmungen (TKVSB)* vol 13 (Zurich, Switzerland: VSB)
[24] Stüssi F 1955 *Tragwerke aus Aluminium* (Berlin/ Göttingen/ Heidelberg, Germany: Springer-Verlag)
[25] Kohout J 2022 *Journal of the Brazilian Society of Mechanical Sciences and Engineering* **44** 1–7
[26] Haibach E 2006 *Betriebsfestigkeit* 3rd ed (Berlin, Heidelberg, Germany: Springer)
[27] Isida M 1955 *Trans. Jpn. Soc. Mech. Eng.* **21** 507
[28] Pilkey W D, Pilkey D F and Bi Z 2020 *Peterson's stress concentration factors* 4th ed (John Wiley & Sons)
[29] Sjöström S 1950 On the stresses at the edge of an eccentrically located circular hole in a strip under tension (report no. 36) Tech. rep. Aeronautical Research Institute
[30] Suresh S 2003 *Fatigue of materials* 2nd ed (Cambridge, UK: Cambridge University Press)
[31] Tanaka K, Kitano T and Egami N 2014 *Engineering Fracture Mechanics* **123** 44–58
[32] Toasa Caiza P D and Uematsu Y 2023 *Applications in Engineering Science* **15** 100131
[33] Rosemeier M, Holst T and Antoniou A 2023 Supplemental Material to Article "Quantification of process-induced effects on fatigue life of short-glass-fiber-filled adhesive used in wind turbine rotor blades" URL `https://doi.org/10.5281/zenodo.8214266`
[34] Wentingmann M, Manousides N, Antoniou A and Balzani C 2022 *Polymer Testing* 107648

IOP Conf. Series: Materials Science and Engineering 1293 (2023) 012043 doi:10.1088/1757-899X/1293/1/012043

Viability of recycled fibres extracted from EoL composites

P. Laurikainen and E. Sarlin

Materials Science and Environmental Engineering, Tampere University
P.O. Box 589, 33014 Tampere University, Finland

essi.sarlin@tuni.fi

Abstract. Recycling thermoset-based composites is a technological and economic challenge. The predominating recycling methods (mechanical recycling and pyrolysis) downgrade the fibre properties (length, strength) and are energy-intensive. To avoid downcycling and to ensure economic feasibility, technological solutions need to be developed to enable recycling of long reinforcing fibres that can also be used in demanding applications. Thermochemical recycling, where reinforcing fibres are extracted from the depolymerized matrix, can be considered as an advanced option for recovery of higher quality fibres from EoL composites. Additional steps are required to use these fibres in new composite structures. Thermochemical recycling removes the sizing from the fibre surface making it difficult to handle and resulting in poor fibre-matrix compatibility especially in the case of glass fibres. In this study, we discuss the re-sizing of recycled fibres. We have focused on fibres extracted from glass fibre reinforced composites (GFRPs) from EoL wind turbine blades and carbon fibre reinforced composites (CFRPs) collected from aeronautics industry. We demonstrate the steps for a batch re-sizing process and study the effects of the re-sizing on the properties of composites made of recycled fibres. The interfacial properties are analysed with micro-mechanical testing using the microbond method – both for thermoset and thermoplastic composites – and with tensile tests for thermoplastic composites. Re-sizing significantly eases the handling of the fibres, but the interfacial compatibility is improved only in the case of GFRPs as thermochemically recycled carbon fibres (rCF) exhibit very good interfacial performance even without sizing. This study highlights that with proper recovery and resizing processes, recycled fibres can be viable feedstocks for various applications and – with future improvements on the fibre recovery methods – even approach the applicability of virgin fibres.

1. Introduction

The volume of European composite production is the largest in short (< 2 mm) fibre reinforced composites, which have lower requirements for (mechanical) performance than long (2-5 mm) and endless (> 50 mm) fibre reinforced composites. With proper recycling, recycled composites can represent a valuable feedstock for short and long fibre reinforced composites with high part and production volumes and less restrictive performance requirements. Although this inherently includes some degree of down cycling, and cannot be therefore considered fully circular, much of the technical and economic added value of the material can be preserved.

One of the first hindrances of the circular economy trend is the poor consumer acceptance for the use of recycled composite feedstock. Mitigation strategies could include standardized verification processes and open availability of data, among others. Naturally, it also requires sufficient capacity of reliable and reproducible recycling technologies and reprocessing processes to ensure a steady supply of the material.

This ties into a lack of systemic value chains, actors within these value chains and circular economy business models, which slows down recycling in composite industry. The integration of different stakeholders into circular economy is poor partly due to their fragmented economic and geographical nature. This leads to poor communication between the stakeholders and to a less than unified front, making it more difficult to affect policymakers' decisions and to achieve rational changes in regulatory and legal frameworks. The circular economy business models are traditionally based on 'pushing' the recycled materials towards the demand side in the value chain. The FiberEUse project [1] promoted the implementation of a model where the requirements and specifications of the materials to be recycled come from the demand side generating a 'pull' for the secondary raw materials. This model would better ensure the characteristics and functionalities of the recycled material are suitable for high added value products. This kind of demand-driven circular economy requires strong communication between the actors within the value chain and possible new innovations to support it, e.g., through novel ICT solutions. If a strong demand from the end-users (customers) for sustainable composites is realized, we believe that the probability of a demand-driven circular economy scenario might become reality as is gradually happening in the plastics industry. Here, the technological and commercial viability of fibres extracted from EoL composite structures is discussed based on both previous and on-going research efforts on the topic.

2. Composite waste streams and waste management

European composite industry is mostly producing glass fibre reinforced composites (GFRP, > 95%) the rest being mainly carbon and natural fibres reinforced composites (CFRP and NFRP, respectively). The volumes are divided roughly between short fibre reinforced thermoplastics (51%) and other thermoplastic and thermoset composites [2]. Thermoplastic composites are mostly used in the transport and electrical application sectors of industry. Thermoset composites, which are typically reinforced with long or endless fibres, cover a share of approx. 45% of the European market excluding the CFRPs [2] and are prominently used in construction, transportation and electrical sector. The CFRP composites production volume has increased from 36.000 tonnes in 2017 to 52.000 tonnes in 2022 [2] but still remains less than 2% of the total European composite production.

Compared to other waste streams, the composite waste stream is relatively small. The composite waste stream is generally also quite unstable in composition and properties, unpredictable in volumes and geographically dispersed. Due to increasingly strict and varied regulations on waste collection and management, international composite waste collection does not seem practical. Instead, local cooperation of waste management companies could be a more realistic scenario to ensure higher collection volumes. The local collaboration and application of a demand driven circular economy approach could help in this attempt. A revision of the EU's End-of-Waste criteria to enable the easier use of secondary raw materials would also promote the implementation of robust composite recycling systems.

Current composite waste management practices vary considerably in Europe. Landfilling and incineration remain as the most typical routes while mechanical recycling is also used in smaller extent. Using the classification of European Waste Framework (EWF) directive, the use of composites in cement kilns is located somewhere between incineration and recycling. This 'recycling' method is compliant with EU's legislation and requires relatively small investments to incorporate into existing systems, thus offering a tempting choice for the composite waste management. For example, in Finland where the wind power infrastructure is fairly new and still increasing and where the local composite industry is small in production volume, the use of composite waste in cement industry was recently identified as the only feasible route for composite recycling [3]. In this process, the resin combusts providing energy for the cement manufacturing process while the reinforcing fibres are left to the cement as fillers. This can lead to some reductions in the CO_2 emissions of the final cement product. However, the use of composite waste in cement process fuel is technically restricted and composites containing E-glass are shown to have a negative effect on the cement properties. All in all, the technical added value of composites is fully lost in this scheme of recycling.

The composite waste can be divided roughly in three categories: small components that typically end up in municipal waste streams (e.g., electrical components such as switches or housings, sport equipment, etc.), large infrastructures – e.g., wind blades or boats that are dismantled in a controlled process – and industrial composite production waste. These waste streams have very different properties and potential for recycling. While the first group is a very heterogeneous fraction of the composite waste and has the least potential to be recycled without significant improvements to sorting and collection, it is also the smallest fraction. The large structures have better – if not always fully – known composition simplifying the recycling process. Considering the production volumes and the most typical composite applications, the volume of this fraction is large. It has estimated that the annual amount of End-of-Life (EoL) wind blades increases to 50.000 tons in Europe by 2030 [4] and the amount of disposed leisure boats is approx. 10.000 tons [5]. However, these composites have undergone ageing during their use and the residual material properties are poorly known. A composite components' lifetime is very long, typically 10 to 30 years, and they are typically used in environments where they are exposed to varying external stresses. Furthermore, the waste stream volume and geographic locations are somewhat unpredictable. The composition of industrial process waste is well-known, the material can be considered pristine, and its volume and location are the least problematic to predict. Further, the amount of the process waste is far from negligible, annually up to 50.000 tons [6]. Naturally, the reuse of in-house collected production waste should be promoted to reduce the amount of externally managed process waste to minimum. In cases where in-house use is not possible and established procedural framework should be established to avoid recycling gaps.

3. Composite recycling and reprocessing

The composite recycling technologies can be classified in three categories: mechanical, thermal and chemical recycling. Mechanical recycling (cutting, shredding, sieving) results in recyclates that have a size distribution dependent on the process parameters ranging from shredding powder into larger particles. The composition of individual particles resembles the original composition of the composite, although variation has been found to exist between different size fractions. The use of the particles and dust as fillers or reinforcement is somewhat limited due to low cost of virgin feedstock, processing issues (e.g., the increase in resin viscosity when dust is added) and decrease in the mechanical properties of the final composite parts (poor compatibility between the recyclates and the virgin matrix). However, for thermoplastic composites mechanical recycling is the most straightforward process as the recyclates can be used as such, together with virgin raw materials. The application of recycled feedstock is mostly in short fibre reinforced composites but especially the thermoplastic recyclates can be also used in long fibre reinforced ones. The smallest fractions are typically incinerated. Electrofragmentation can be classified as a mechanical recycling method and results in recyclates with higher fibre fraction and longer fibre length [7]. Electrofragmentation is actively studied in the RECREATE project [8].

Thermal recycling, i.e. pyrolysis, fluidized-bed pyrolysis or microwave-assisted pyrolysis, can be positioned in the EWF directive somewhere between recycling and recovery, as the resin part is lost from the material circulation when it decomposes to volatile molecules during the process. However, the fibres are made available for further use. The process leaves char on the fibre surfaces that might negatively affect the final properties of a product made with the secondary feedstock. Thus, the char is typically removed in a subsequent process, which has the unwanted side effect of degrading the fibre properties. Typical reduction in fibre strength due to thermal recycling is 50-90% which is a significant feasibility challenge. With process optimization, CFRP pyrolysis can result in remarkably lower (only 5%) strength reduction in rCF compared to virgin [9, 10]. Thermal recycling is more energy intensive than mechanical recycling, further reducing its feasibility for glass fibres (GF). The volatilized resin can, however, be used as energy input to the system somewhat reducing the costs. Thermal recycling is equally applicable to thermoset and thermoplastic composites and the process is quite insensitive to material composition. The purified fibres can be potentially used in long or endless fibre reinforced composites with higher added value, as also demonstrated in the FibreEUse project. However, the process is more effective if the density of the recycled material is low, i.e., the composite has been shredded to small particles, which is even a technical requirement in microwave-assisted pyrolysis.

Chemical recycling is based on solvolysis, i.e., degrading the resin into monomers using a solvent, most typically water, a catalyst and increased temperature lower than in thermal methods. No char is formed or left on the fibres. The reactors are expensive, and the process is more sensitive to the composition of the recycled material than in other methods. Thus, chemical recycling is yet to become a larger scale possibility for composite recycling. However, if it is combined with a thermal recycling process, additional value can be gained. Successful pilots for these thermochemical processes are already patented (EP 3114191, 2018 by Korec srl) and implemented in industry. In Korec's process, a controlled CO_2 atmosphere is used to enable the collection of both clean fibres and the resin as a liquid fraction applicable to further use as a thermoset feedstock, for which the method has been originally developed. This pilot system was also applied in the FiberEUse project.

After shredding the composite into smaller particles or extracting the fibres for further use, the secondary raw materials must be reprocessed. When it comes to mechanically recycled materials, the reprocessing might not differ significantly from the use of virgin feedstock, i.e., compounding thermoplastic granulates or mixing fillers and thermoset resins. However, the relatively clean fibres from thermal or chemical recycling are not applicable as such. The recycling process removes the original surface treatment, sizing, that is applied by the fibre manufacturer to ensure the processability and compatibility of the fibres with the matrix. This part of the value chain is absent from the value chain of traditional (virgin) reinforcements and a natural actor for the part is missing. The resizing (i.e., sizing the recycled fibres) falls between the actor specialized in the recycling technology and the actor compounding the recycled fibres with a polymer matrix. Neither of these actors generally has the infrastructure or expertise to apply the sizing of the fibres, both of which exist practically exclusively with fibre manufacturers.

4. Resizing recycled fibres

In many composite recycling studies, where extracted fibres are discussed, the surface treatments of the fibres are mentioned being highly important for the properties of the final composite, but the topic is not addressed further [11]. In the FiberEUse project, the aspects of the resizing of recycled fibres were studied for different kind of waste streams (GFRPs from EoL wind turbine blades and CFRPs collected from aeronautical industry) and output compositions (short fibre reinforced thermoplastic composites and long fibre reinforced thermoset composites, for the latter the results available in [12]). In this chapter, the resizing process, and the reprocessing of resized fibres for the FiberEUse demo cases are discussed and the results for the use case validation are reported. This is intended as a practical example of the multitude of scientific and process related aspects of resizing, and, perhaps more importantly, of functional products from the recovered fibres.

4.1. Materials and methods

The composites were shredded, recycled by the thermochemical process patented by Korec, and resized in a batch process with a chemistry compatible with the matrix material. The resulting fibres were clean, and their mechanical properties are reported in [10].

In general, sizing is a water solution comprising mainly film former and a coupling agent, the latter of which is not used for carbon fibres. These represent the main components for improving interfacial properties. Other possible sizing components are lubricants, antistatic agents, wetting agents and cross-linking agents, which are often included in a commercial sizing to improve, for example, storage properties, handling, and resin flow [13]. A batch process for the resizing where fibres are immersed in the sizing solution was considered the most realistic for the resizing at laboratory scale. The case is very different from continuous virgin fibres treated in very fast roll-to-roll processes. Here, the preceding thermal recycling processes as well as the following compounding the resized fibres with a thermoplastic matrix are also batch processes. However, in other cases, such as reinforcement non-woven preparation, or a fibre spray-up process, the resizing could be incorporated into the process together with a possible binder or as the binder. This would enable more industrial scale options for the use of recovered fibres in thermoset applications and such options are currently explored in the RECREATE project.

In FiberEUse the focus was primarily on improving short fibre reinforced thermoplastic compounds with a simple sizing formulation. Detailed formulations are given in [14] together with the process parameters. The following aspects were studied when optimizing the sizing composition and process: the total solid content of the sizing solution and rinsing after the immersion in the sizing solution. In general, a thin sizing layer and lack of large fibre clusters are preferable for mechanical performance, whereas a higher packing density and fibre agglomeration are advantageous for the compounding process. The resizing process was optimized against these aspects. The resized fibres were analysed with Fourier transformation infrared spectroscopy (FTIR), thermogravimetric analysis (TGA) and scanning electron microscopy (SEM). The resized fibres were compounded with copo-polypropylene (copo-PP) and polyamide 6 (PA, only rCF) thermoplastic matrices into tensile test specimens and tested (ISO 1172 (method A) standard). The fibre fraction of the compounds was 20 wt% and verified by the burning-off method (ISO 1172 (method A)). Reference compounds were prepared with recycled fibres without resizing.

After the selection of the most optimal resizing process, larger batched of thermoplastic composites for the use case demonstrators of the FiberEUse project were produced. The same copo-PP and PA matrices were used but in addition, a homo-PP was used with rCFs. The compounding process was carried out with a twin-screw extruder. For more details about the compounding, see [12]. The final fibre fractions of the compounds were verified to be within ±2 wt% from the intended value (20 wt% or 30 wt%) via TGA. The mechanical properties of the compounds were characterized by tensile tests to compare the influence of sizing (fibres with and without sizing) and the influence of the origin of the fibre (recycling process). The comparison was done to material specifications used for component design in the automotive sector. After tensile testing, composite fracture surfaces were studied with SEM. Finally, the use case demonstrators were manufactured, and the components were tested with use case specific validation procedures.

4.2. Optimizing the sizing concentration

Based on the FTIR analysis of the resized rGF, the surfaces contained both the film former and the coupling agent. FTIR and TGA results both indicated that higher total solid content (concentration) in the solution also increased the amount of sizing deposited on the fibres. Too high silane concentrations led to self-polymerization of the silane – instead of interaction with the fibre surface – and the silane was mostly removed by the rinsing. The SEM analysis revealed a slightly uneven sizing layer but differences between different concentrations were not observed. In the case of recycled carbon fibres (rCF), the sizing layer was more homogeneous, smoothing out also the original surface grooves of the fibres. TGA revealed that the average sizing layer thickness followed the resizing solution concentration if the fibres were rinsed after the immersion to remove excess sizing. Without rinsing the amount of sizing in the fibres was higher and random. More detailed characterization of the resized fibres can be found from [12, 14]. Table 1 presents collected key results from microscale interfacial testing (microbond) performed using these sizing formulations. A detailed description of the experimental setup and analysis of the results can be found in [15].

The mechanical properties of the rGF reinforced PP composites were improved after resizing and SEM analysis revealed better fibre-matrix contact for the resized rGF. As expected, no further improvement was observed with increasing solution concentration and the maximum values were reached at a range of 0.5 wt% (tensile modulus) to 2 wt% (tensile strength). For the rCF reinforced PP and PA composites, no significant difference was observed in the tensile properties as a function of sizing concentration indicating that after thermochemical recycling the surface of recycled carbon fibres is also by itself compatible with thermoplastic matrices. However, the fractography revealed that the distribution of the fibres was improved with resizing. Further, the binding and packing of the resized fibres are highly important for the handleability and industrially feasible processing of the recycled fibres. The optimal resizing for rGF was concluded to be 1 wt% total solid content without rinsing and for rCF 5 wt% total solid content without rinsing. In the FiberEUse project, this aspect was more critical than the exact mechanical properties, and therefore a relatively high sizing concentration was chosen as 'optimal'.

Table 1. Influence of the sizing formulations on the interfacial strength of various fibre-matrix combinations. Data partially from optimisation trials and partially collected from [16,17].

Fibre (+sizing)	Matrix	IFSS [MPa]	stdev [MPa]
rGF/GF	Epoxy	35.2	8.9
rGF (1 w%, m-PP FF†)		31.7	2.1
GF (1 w%, EP FF‡)		58.2	3.2
rGF (5 w%, m-PP FF)		28.9	2.0
rCF		43.6	2.5
rCF (1 w%, m-PP FF)		25.0	2.2
CF	PP	8.4	4.0
rCF		20.1	9.9
rCF (1 w%, m-PP FF)		17.3	6.1
rCF (5 w%, m-PP FF)		17.3	5.0
rCF	PA6	66.5	6.0

† maleated polypropylene film former
‡ epoxy film former

4.3. *Compounding secondary raw materials for use cases*

For the compounding experiments and the FiberEUse project use case demonstrators, larger batches of resized fibres were prepared. The fibre fraction of the compounds was also different from the resizing process optimization step. Thus, the results of the resizing process optimization study and the compounding study are not directly comparable.

The modulus of the rGF-PP composite was improved by 38% and the tensile strength by 58% via resizing and better fibre-matrix adhesion was evident from fracture surfaces. However, the properties were lower than the corresponding automotive reference specifications and the main issue was assumed to be the cohesive properties of the recycled glass fibres [18].

The resizing increased the modulus of the rCF-PA compounds by 100% and the tensile strength by 45%. An even distribution and good fibre-matrix contact was seen in SEM images. The resized composites achieved almost the required level for the mechanical properties, the strength being only 8% below the automotive reference material.

In homo-PP matrix, the resizing of rCF increased the modulus by 37% and the tensile strength by 40% and, similarly to rCF-PA compounds, resizing improved the compatibility of the fibres in the matrix. However, the dispersion of the fibres was good even without the sizing. Compared with automotive reference, the material was stiffer (23% higher modulus) but less strong (13% lower tensile strength). The results were very similar with copo-PP matrix. Based on the results and the experience gathered during the compounding trials, it was clear that the resizing improves especially the processability of the rCF compounds having simultaneously positive effect on the mechanical properties.

Based on the compounding trials, small changes were made to the formulation of the materials, namely the fibre fraction, and the design of the use case demonstrators. However, the required changes were relatively small and no new design concepts or other significant changes were needed.

4.4. Verification of the demonstrator properties

Three different automotive use case demonstrators were produced from the recycled fibres. Details of these demonstrators can be found from [19]. **A cowl top** is a stiff and strong component that should break after absorbing a certain amount of energy at dynamic conditions and is produced by injection moulding. The current design is produced with short GF (30 wt%) reinforced PP and the new design with rGF was based on similar composition. In the validation phase, the demonstrator must withstand specific impact loads without breaks and cracks, weathering tests without deformations and peeling off and prove sufficient thermal stability. All criteria were passed.

The second demonstrator was **a pedal bracket**, which is a security part with high structural requirements both in stiffness and strength. The reference structure is manufactured by injection moulding with short GF reinforced PA (fibre fraction 40 wt%) and the secondary raw material design was based on PA and 20 wt% rCF. In the validation test, the part was subjected to varying loads applied to different areas of the brake and gas pedals. To pass the test, the applied load must induce deformations less than a vehicle manufacturer defined requirement. Also, the component must withstand loads up to 3000 N without any break at extreme temperatures (80 °C and −35 °C). The new bracket manufactured with recycled fibres fulfilled the requirements.

A front-end carrier, which was the third demonstrator, must support load and prevent torsions. The reference structure is manufactured by injection moulding with long GF reinforced PP (fibre fraction 40 wt%) and the secondary raw material design was based on PP and 20 wt% rCF. The prototype must fulfil static load cases with combined compression and a traction load up to 2300 N loading without any breaks and achieving a specific stiffness. The front-end carrier produced with recycled feedstock met the requirements and was lighter than the original version.

5. Conclusions

To ensure that circular economy will be viable scheme for composite materials, the secondary composite feedstock produced from recycled materials must be able to compete with virgin raw materials in environmental, economic, and technical aspects. Further, each step of the recycling and reprocessing value chain must be covered by an industrial actor with suitable expertise. Currently, this is not yet the case for thermal recycling of composites, where the reinforcing fibres are extracted for further use. Especially, resizing (sizing of recycled fibres) is shown to be a necessary step of the reprocessing route to achieve recycled fibres that can be processed into high performance composites, but currently the industry is lacking actors that could combine it either with fibre extraction, compounding (thermoplastic composites) or thermoset composite manufacturing. Possible options for fibre recovery incorporated resizing include reinforcement mats, non-wovens, recycled fibre prepregs or other comparable preforms. At least for thermochemically recovered carbon fibres with high mechanical properties, a polymer binder can replace majority of the functions of the sizing enabling straightforward, cost-effective, and safe approaches with few process steps from composite waste to new feedstock. Similar approaches could be available for extracted glass fibres with the added step of resizing. These aspects must be studied further, as is done in RECREATE [8] and DeremCo [20] projects.

Acknowledgements

This work is done as a part of the following EU-funded projects: H2020-IND-CE funded (#730323) project *Large scale demonstration of new circular economy value-chains based on the reuse of end-of-life fiber reinforced composites*, FiberEUse; HORIZON-CL4-2021-RESILIENCE funded (#101058756) project *Recycling technologies for circular reuse and remanufacturing of fiber-reinforced composite materials*, RECREATE; and I3-2021-INV1-MANU funded (#101084037) project *De&Remanufacturing for Circular Economy Investments in the Composite Industry*, DeremCo. The content of this publication does not represent the official position of the European Commission and is entirely the responsibility of the authors. The information presented here has been

Funded by the
European Union

thoroughly researched and evaluated and is believed to be accurate and correct. However, the authors cannot be held legally responsible for any errors. There are no warranties, expressed or implied, made with respect to the information provided. The authors will not be liable for any direct, indirect, special, incidental, or consequential damages arising out of the use or inability to use the content of this publication.

References

[1] H2020-IND-CE funded (#730323) project *Large scale demonstration of new circular economy value-chains based on the reuse of end-of-life fiber reinforced composites, FiberEUse* (2017-2021). www.fibereuse.eu/

[2] E. Witten, V. Mathes: Der europäische Markt für Faserverstärkte Kunststoffe / Composites 2022. AVK – Industrievereinigung Verstärkte Kunststoffe e.V. https://www.avk-tv.de/

[3] Finnish Ministry of the Environment funded pilot project *Collection and utilisation of composite plastic waste, Kimura* (2021–2022). www.plastics.fi/projektit/kimura_-projekti/

[4] Innovation Fund Denmark's Grand Solutions program co-funded project *DecomBlades*. https://decomblades.dk/

[5] FP7-TRANSPORT funded (#218609) project *Recycling Thermoset Composites of the SST, EURECOMP* (2009-2012). https://eurecomp.eu/

[6] Woidasky, J.: Weiterentwicklung des Recyclings von faserverstärkten Verbunden. Proceedings of Berliner Recycling und Rohstoffkonferenz, 241–260 (2013).

[7] Rouholamin, D., Shyng, T., Savage, L., Ghita, O.: A comparative study into mechanical performance of glass fibres recovered through mechanical grinding and high voltage pulse power fragmentation. In: 16th European conference on composite materials, ECCM 2014 (2014).

[8] HORIZON-CL4-2021-RESILIENCE funded (#101058756) project *Recycling technologies for circular reuse and remanufacturing of fiber-reinforced composite materials, RECREATE* (2022-2026). https://www.recreatecomposites.eu/

[9] Meyer, L.O., Schulte, K., Grove-Nielsen, E.: CFRP-recycling following a pyrolysis route: process optimization and potentials. J. Compos. Mater. 43(9), 1121–1132 (2009).

[10] García-Arrieta, S., Sarlin, E., De La Calle, A., Dimiccoli, A., Saviano, L., Elizetxea, C.: Thermal Demanufacturing Processes for Long Fibers Recovery. In: Colledani, M. and Turri, S. (Eds): Systemic Circular Economy Solutions for Fiber Reinforced Composites, Springer, Cham (2023).

[11] Rani, M., Choudhary, P., Krishnan, V., Zafar, S.: A review on recycling and reuse methods for carbon fiber/glass fiber composites waste from wind turbine blades, Composites Part B: Engineering, Vol 215, 108768 (2021).

[12] Laurikainen, P., Palola, S., De La Calle, A., Elizetxea, C., García-Arrieta, S., Sarlin, E.: Fiber Resizing, Compounding and Validation. In: Colledani, M. and Turri, S. (Eds): Systemic Circular Economy Solutions for Fiber Reinforced Composites, Springer, Cham (2023).

[13] Kim, J., Mai, Y.: Engineered Interfaces in Fiber Reinforced Composites, 1st edn. Elsevier, Amsterdam (1998).

[14] Matrenichev, V., Lessa Belone, M.C., Palola, S., Laurikainen, P., Sarlin, E.: Resizing approach to increase the viability of recycled fiber-reinforced composites. Materials 13(24), 5773 (2020).

[15] Laurikainen, P., Kakkonen, M., von Essen, M., Tanhuanpää, O., Kallio, P., Sarlin, E.: Identification and compensation of error sources in the microbond test utilising a reliable high-throughput device. Composites Part A. 137, 105988. 2020.

[16] Laurikainen, P., Dsouza, R., Kakkonen, M., Kanerva, M., Sarlin, E.: Exploring the role of fibre sizing to the fatigue of glass fibre composites using a novel, reliable micro-fatigue test. Composites Part A 167, 107425, (2023).

[17] Palola, S.: Exploring mechanical adhesion in fiber reinforced composites with aramid and

recycled carbon fibers. Tampere University Dissertations 678, ISSN 2490-0028, (2022). http://urn.fi/URN:ISBN:978-952-03-2586-2
[18] Ginder, R.S., Ozcan, S.: Recycling of commercial e-glass reinforced thermoset composites via two temperature step pyrolysis to improve recovered fiber tensile strength and failure strain. Recycling 4(2), 24 (2019).
[19] García-Arrieta, S., López Benito, I., García, M., Bonaiti, G., Ollo Escudero, O., Elizetxea, C.: Use Case 2: Thermal Recycling of Long Fibers. In: Colledani, M. and Turri, S. (Eds): Systemic Circular Economy Solutions for Fiber Reinforced Composites, Springer, Cham (2023).
[20] I3-2021-INV1-MANU funded (#101084037) project *De&Remanufacturing for Circular Economy Investments in the Composite Industry, DeremCo* (2022-2025). https://deremco.afil.it/

AUTHOR INDEX

Institute of Physics
Dirac House, Temple Back
Bristol BS1 6BE UK

ISSN: 1757-8981
ISBN 978-1-7138-8504-7

CURRAN ASSOCIATES INC.
proceedings.com